# COMMUTATIVE ALGEBRA

VOLUME II

Oscar Zariski and Pierre Samuel

T0176085

DOVER PUBLICATIONS
Garden City, New York

# COMMUTATIVE ALGEBRA

*Bibliographical Note*

This Dover edition, first published in 2019, is an unabridged republication of the work originally published in the University Series in Higher Mathematics by D. Van Nostrand Company, Inc., New York, in 1960.

*Library of Congress Cataloging-in-Publication Data*

Names: Zariski, Oscar, 1899–1986, author. | Samuel, Pierre, 1921– author. | Cohen, Irvin Sol, 1917– author.
Title: Commutative algebra / Oscar Zariski and Pierre Samuel ; with the cooperation of I.S. Cohen.
Description: Dover edition [2019 edition]. | Mineola, New York : Dover Publications, Inc., 2019. | Series: Dover books on mathematics | Originally published: Princeton, N.J. : Van Nostrand Company, Inc., 1958–1960. Reprinted: New York : Springer-Verlag, 1975–1976. | Includes indexes. Contents: Introductory concepts—Elements of field theory—Ideals and modules—Noetherian rings—Valuation theory—Polynomial and power series rings—Local algebra.
Identifiers: LCCN 2019021023 | ISBN 9780486836614 (v. 1) | ISBN 9780486838601 (v. 2) | ISBN 0486836614 (v. 1) | ISBN 0486838609 (v. 2)
Subjects: LCSH: Commutative algebra.
Classification: LCC QA251.3 .Z37 2019 | DDC 512/.44—dc23
LC record available at https://lccn.loc.gov/2019021023

Manufactured in the United States of America
83860903
www.doverpublications.com

# PREFACE

This second volume of our treatise on commutative algebra deals largely with three basic topics, which go beyond the more or less classical material of volume I and are on the whole of a more advanced nature and a more recent vintage. These topics are: (a) valuation theory; (b) theory of polynomial and power series rings (including generalizations to graded rings and modules); (c) local algebra. Because most of these topics have either their source or their best motivation in algebraic geometry, the algebro-geometric connections and applications of the purely algebraic material are constantly stressed and abundantly scattered throughout the exposition. Thus, this volume can be used in part as an introduction to some basic concepts and the arithmetic foundations of algebraic geometry. The reader who is not immediately concerned with geometric applications may omit the algebro-geometric material in a first reading (see "Instructions to the reader," page vii), but it is only fair to say that many a reader will find it more instructive to find out immediately what is the geometric motivation behind the purely algebraic material of this volume.

The first 8 sections of Chapter VI (including § 5bis) deal directly with properties of places, rather than with those of the valuation associated with a place. These, therefore, are properties of valuations in which the value group of the valuation is not involved. The very concept of a valuation is only introduced for the first time in § 8, and, from that point on, the more subtle properties of valuations which are related to the value group come to the fore. These are illustrated by numerous examples, taken largely from the theory of algebraic function fields (§§ 14, 15). The last two sections of the chapter contain a general treatment, within the framework of arbitrary commutative integral domains, of two concepts which are of considerable importance in algebraic geometry (the Riemann surface of a field and the notions of normal and derived normal models).

The greater part of Chapter VII is devoted to classical properties of polynomial and power series rings (e.g., dimension theory) and their applications to algebraic geometry. This chapter also includes a treatment of graded rings and modules and such topics as characteristic (Hilbert) functions and chains of syzygies. In the past, these last two topics represented some final words of the algebraic theory, to be followed only by deeper geometric applications. With the modern development of homological methods in commutative algebra, these topics became starting points of extensive, purely algebraic theories, having a much wider range of applications. We could not include, without completely disrupting the balance of this volume, the results which require the use of truly homological methods (e.g., torsion and extension functors, complexes, spectral sequences). However, we have tried to include the results which may be proved by methods which, although inspired by homological algebra, are nevertheless classical in nature. The reader will find these results in Chapter VII, §§ 12 and 13, and in Appendices 6 and 7. No previous knowledge of homological algebra is needed for reading these parts of the volume. The reader who wants to see how truly homological methods may be applied to commutative algebra is referred to the original papers of M. Auslander, D. Buchsbaum, A. Grothendieck, D. Rees, J.-P. Serre, etc., to a forthcoming book of D. C. Northcott, as well, of course, as to the basic treatise of Cartan-Eilenberg.

Chapter VIII deals with the theory of local rings. This theory provides the algebraic basis for the local study of algebraic and analytical varieties. The first six sections are rather elementary and deal with more general rings than local rings. Deeper results are presented in the rest of the chapter, but we have not attempted to give an encyclopedic account of the subject.

While much of the material appears here for the first time in book form, there is also a good deal of material which is new and represents current or unpublished research. The appendices treat special topics of current interest (the first 5 were written by the senior author; the last two by the junior author), except that Appendix 6 gives a smooth treatment of two important theorems proved in the text. Appendices 4 and 5 are of particular interest from an algebro-geometric point of view.

We have not attempted to trace the origin of the various proofs in this volume. Some of these proofs, especially in the appendices, are new. Others are transcriptions or arrangements of proofs taken from original papers.

We wish to acknowledge the assistance which we have received from M. Hironaka, T. Knapp, S. Shatz, and M. Schlesinger in the work of checking parts of the manuscript and of reading the galley proofs. Many improvements have resulted from their assistance.

The work on Appendix 5 was supported by a Research project at Harvard University sponsored by the Air Force Office of Scientific Research.

*Cambridge, Massachusetts* OSCAR ZARISKI
*Clermont-Ferrand, France* PIERRE SAMUEL

# INSTRUCTIONS TO THE READER

As this volume contains a number of topics which either are of somewhat specialized nature (but still belong to pure algebra) or belong to algebraic geometry, the reader who wishes first to acquaint himself with the basic algebraic topics before turning his attention to deeper and more specialized results or to geometric applications, may very well skip some parts of this volume during a first reading. The material which may thus be postponed to a second reading is the following:

### CHAPTER VI

All of § 3, except for the proof of the first two assertions of Theorem 3 and the definition of the rank of a place; § 5: Theorem 10, the lemma and its corollary; § 5bis (if not immediately interested in geometric applications); § 11: Lemma 4 and pages 57-67 (beginning with part (b) of Theorem 19); § 12; § 14: The last part of the section, beginning with Theorem 34′; § 15 (if not interested in examples); §§ 16, 17, and 18.

### CHAPTER VII

§§ 3, 4, 4bis, 5 and 6 (if not immediately interested in geometric applications); all of § 8, except for the statement of Macaulay's theorem and (if it sounds interesting) the proof (another proof, based on local algebra, may be found in Appendix 6); § 9: Theorem 29 and the proof of Theorem 30 (this theorem is contained in Theorem 25); § 11 (the contents of this section are particularly useful in geometric applications).

### CHAPTER VIII

All of § 5, except for Theorem 13 and its Corollary 2; § 10; § 11: Everything concerning multiplicities; all of § 12, except for Theorem 27 (second proof recommended) and the statement of the theorem of Cohen-Macaulay; § 13.

All appendices may be omitted in a first reading.

# TABLE OF CONTENTS

# COMMUTATIVE ALGEBRA

# VI. VALUATION THEORY

§ 1. **Introductory remarks.** Homomorphic mappings of rings into fields are very common in commutative algebra and in its applications. We may cite the following examples:

EXAMPLE 1. *The reduction of integers* mod $p$. More precisely, let $p$ be a prime number; then the canonical mapping of the ring $J$ of integers onto the residue class ring $J/Jp$ maps $J$ onto a field with $p$ elements. More generally, we may consider a ring $D$ of algebraic integers (Vol. I, Ch. V, § 4, p. 265), a prime ideal $\mathfrak{p}$ in $D$, and the mapping of $D$ onto $D/\mathfrak{p}$. These examples are of importance in number theory.

EXAMPLE 2. We now give examples pertaining to algebraic geometry. Let $k$ be a field and $K$ an extension of $k$. Let $(x_1, \cdots, x_n)$ be a point in the affine $n$-space $A_n{}^K$ over $K$. With every polynomial $F(X_1, \cdots, X_n)$ with coefficients in $k$ we associate its *value* $F(x_1, \cdots, x_n)$ at the given point. This defines a homomorphic mapping of the polynomial ring $k[X_1, \cdots, X_n]$ into $K$. Now let us say that a point $(x'_1, \cdots, x'_n)$ of $A_n{}^K$ is a *specialization* of $(x_1, \cdots, x_n)$ over $k$ if every polynomial $F \in k[X_1, \cdots, X_n]$ which vanishes at $(x_1, \cdots, x_n)$ vanishes also at $(x'_1, \cdots, x'_n)$. Then (by taking differences) two polynomials $G$, $H$ with coefficients in $k$ which take the same value at $(x_1, \cdots, x_n)$ take also the same value at $(x'_1, \cdots, x'_n)$. This defines a mapping of $k[x_1, \cdots, x_n]$ onto $k[x'_1, \cdots, x'_n]$ ($\subset K$), which maps $x_i$ on $x'_i$ for $1 \leq i \leq n$. Such a mapping, and more generally *any homomorphic mapping* $\varphi$ *of a ring* $R$ *into a field*, such that $\varphi(x) \neq 0$ for some $x \in R$, is called a *specialization* (of $k[x_1, \cdots, x_n]$ into $K$ in our case). Note that this definition implies that $\varphi(1) = 1$ if $1 \in R$. If, as in the above example, the specialization is the identity on some subfield $k$ of the ring, then we shall say that the specialization is *over* $k$.

EXAMPLE 3. From function theory comes the following example: with any power series in $n$ variables with complex coefficients we associate its constant term, i.e., its value at the origin.

Since any integral domain may be imbedded in its quotient field, a homomorphic mapping of a ring $A$ into a field is the same thing as a homomorphic mapping of $A$ *onto* an integral domain. Thus, by Vol. I, Ch. III, § 8, Theorem 10 a necessary and sufficient condition that a homomorphism $f$ of a ring $A$ map $A$ into a field is that *the kernel of $f$ be a prime ideal.*

From now on we suppose that we are dealing with a ring $A$ which is an *integral domain.* Let $K$ be a field containing $A$ (not necessarily its quotient field), and let $f$ be a specialization of $A$. An important problem is to investigate whether $f$ may be extended to a specialization defined on as big as possible a subring of $K$. An answer to this question will be given in § 4. We may notice already that this problem is not at all trivial.

EXAMPLE 4. Consider, in fact, a polynomial ring $k[X, Y]$ in two variables over a field $k$, and the specialization $f$ of $k[X, Y]$ onto $k$ defined by $f(a)=a$ for $a$ in $k$, $f(X)=f(Y)=0$ ("the value at the origin"). The value to be given to the rational function $X/Y$ at the origin is not determined by $f$ (since it appears as $0/0$). We have $k[X/Y, Y]\supset k[X, Y]$, and any maximal ideal $\mathfrak{P}$ in $k[X/Y, Y]$ which contains $Y$ contains also $X$ and thus contracts to the maximal ideal $(X, Y)$ in $k[X, Y]$. Since there are infinitely many such maximal ideals $\mathfrak{P}$ (they are the ideals generated by $h(X/Y)$ and $Y$, where $h(t)$ is any irreducible polynomial in $k[t]$) it follows that $f$ admits infinitely many extensions to the ring $k[X, Y, X/Y]$.

However, there are elements of $K$ to which the given specialization $f$ of $A$ may be extended without further ado and in a unique fashion. Consider, in fact, the elements of $K$ which may be written in the form $a/b$ with $a$ in $A$, $b$ in $A$, and $f(b)\neq 0$. These elements constitute the *quotient ring* $A_{\mathfrak{p}}$ where $\mathfrak{p}$ is the kernel of $f$ and is a prime ideal. For such an element $a/b$ let us write $g(a/b)=f(a)/f(b)$. It is readily verified that $g$ is actually a mapping: if $a/b=a'/b'$ with $f(b)\neq 0$ and $f(b')\neq 0$, then $f(a)/f(b)=f(a')/f(b')$ since $ab'=ba'$ and since $f$ is a homomorphism. One sees also in a similar way that $g$ is a homomorphism of $A_{\mathfrak{p}}$ extending $f$ (see Vol. I, Ch. IV, § 9, Theorem 14). Since $g$ takes values in the same field as $f$ does, $g$ is a specialization of $A_{\mathfrak{p}}$. The ring $A_{\mathfrak{p}}$ is sometimes called the *specialization ring* of $f$; it is a local ring if $A$ is noetherian (Vol. I, Ch. IV, § 11, p. 228).

In Example 1 this local ring is the set of all fractions $m/n$ whose denominator $n$ is not a multiple of $p$. In Example 2 it is the set of all rational functions in $X_1, \cdots, X_n$ which are "finite" at the point $(x_1, \cdots, x_n)$ (i.e., whose denominator does not vanish at this point). In Example 3 it is the power series ring itself, as a power series with non-zero constant term is invertible.

On the other hand there are (when the specialization $f$ is not an isomorphic mapping) elements of $K$ to which $f$ cannot be extended by any means. These elements are those which can be written under the form $a/b$, with $a$ and $b$ in $A$, with $f(a) \neq 0$ and $f(b)=0$, for the value $g(a/b)$ of $a/b$ in an extension $g$ of $f$ must satisfy the relation $g(a/b)\cdot f(b)=f(a)$ (since $(a/b)\cdot b=a$), but this is impossible. The elements $a/b$ of the above form are the inverses of the non-zero elements in the maximal ideal of the specialization ring of $f$.

We are thus led to studying the extreme case in which all elements of $K$ which are not in $A$ are of this latter type. In this case $A$ is identical with the specialization ring of $f$, and every element of $K$ which is not in $A$ must be of the form $1/x$, where $x$ is an element of $A$ such that $f(x)=0$.

## § 2. Places

DEFINITION 1. *Let $K$ be an arbitrary field. A place of $K$ is a homomorphic mapping $\mathcal{P}$ of a subring $K_{\mathcal{P}}$ of $K$ into a field $\Delta$, such that the following conditions are satisfied:*

(1) *if $x \in K$ and $x \notin K_{\mathcal{P}}$, then $1/x \in K_{\mathcal{P}}$ and $(1/x)\mathcal{P}=0$;*

(2) $x\mathcal{P} \neq 0$ *for some $x$ in $K_{\mathcal{P}}$.*

In many applications of ideal theory (and expecially in algebraic geometry) a certain basic field $k$ is given in advance, called *the ground field*, and the above arbitrary field $K$ is restricted to be an extension of $k$: $k \subset K$. In that case, one may be particularly interested in places $\mathcal{P}$ of $K$ which reduce to the identity on $k$, i.e., places $\mathcal{P}$ which satisfy the following additional condition:

(3) $c\mathcal{P} = c$ for all $c$ in $k$ (whence $k$ is a subfield of $\Delta$).

Any place $\mathcal{P}$ of $K$ which satisfies (3) is said to be a place of $K$ *over* $k$, or a place of $K/k$.

EXAMPLES OF PLACES:

EXAMPLE 1. Let $A$ be a *UFD*, and $a$ an irreducible element in $A$. The ideal $Aa$ is a prime ideal, whence $A/Aa$ is an integral domain. Denote by $\Delta$ its quotient field. The canonical homomorphism of $A$ onto $A/Aa$ is a specialization $f$ of $A$ into $\Delta$. The specialization ring $B$ of $f$ is the set of all fractions $x/y$, with $x \in A$, $y \in A$, $y \notin Aa$ (i.e., $y$ prime to $a$). We denote by $g$ the extension of $f$ to $B$. The homomorphic mapping $g$ is a *place*: in fact, by the unique factorization, any element $z$ of the quotient field $K$ of $A$ which does not belong to $B$ can be written in the form $y/x$, with $y \in A$, $x \in A$, $y \notin Aa$, $x \in Aa$; then its inverse $1/z=x/y$ belongs to $B$ and satisfies the relation $g(1/z)=0$.

We call the place $g$ which is thus determined by an irreducible element $a$ of $A$ an $a$-adic place (of the quotient field of $A$).

EXAMPLE 2. A similar example may be given if one takes for $A$ a Dedekind domain and if one considers the homomorphic mapping $f$ of $A$ into the quotient field of $A/\mathfrak{p}$ ($\mathfrak{p}$ denoting a prime ideal of $A$). The extension $g$ of $f$ to the local ring $A_\mathfrak{p}$ of $f$ is again a place [notice that $A_\mathfrak{p}$ is a *PID* (Vol. I, Ch. V, § 7, Theorem 16), to which the preceding example may be applied]. This place is called the *$\mathfrak{p}$-adic place of $A$.*

We shall show at once the following property of places: if $\mathscr{P}$ is a place of $K$, then $\mathscr{P}$ has no proper extensions in $K$. Or more precisely: if $\varphi$ is a homomorphic mapping of a subring $L$ of $K$ (into some field), such that $L \supset K_\mathscr{P}$ and $\varphi = \mathscr{P}$ on $K_\mathscr{P}$, then $L = K_\mathscr{P}$. We note first that, by condition (1), the element 1 of $K$ belongs to $K_\mathscr{P}$. It follows then from condition (2) that $1\mathscr{P}$ must be the element 1 of $\Delta$. Now, let $x$ be any element of $L$. We cannot have simultaneously $1/x \in K_\mathscr{P}$ and $(1/x)\mathscr{P} = 0$, for then we would have $1 = 1\varphi = (x \cdot 1/x)\varphi = x\varphi \cdot (1/x)\varphi = x\varphi \cdot 0 = 0$, a contradiction. It follows therefore, by condition (1), that $x \in K_\mathscr{P}$. Hence $L = K_\mathscr{P}$, as asserted.

It will be proved later (§ 4, Theorem 5′, Corollary 4) that the above is a *characteristic property* of places.

We introduce the symbol $\infty$ and we agree to write $x\mathscr{P} = \infty$ if $x \notin K_\mathscr{P}$. The following assertions are immediate consequences of conditions (1) and (2) above:

(a) if $x\mathscr{P} = \infty$ and $y\mathscr{P} \neq \infty$, then $(x \pm y)\mathscr{P} = \infty$;
(b) if $x\mathscr{P} = \infty$ and $y\mathscr{P} \neq 0$, then $(xy)\mathscr{P} = \infty$;
(c) if $x \neq 0$, then $x\mathscr{P} = 0$ if and only if $(1/x)\mathscr{P} = \infty$.

If $x \in K_\mathscr{P}$ we shall call $x\mathscr{P}$ the *$\mathscr{P}$-value of $x$*, or *the value of $x$ at the place $\mathscr{P}$*, and we shall say that $x$ is finite at $\mathscr{P}$ or has finite $\mathscr{P}$-value if $x\mathscr{P} \neq \infty$, i.e., if $x \in K_\mathscr{P}$. The ring $K_\mathscr{P}$ shall be referred to as *the valuation ring of the place $\mathscr{P}$.*

It is clear that the elements $x\mathscr{P}$, $x \in K_\mathscr{P}$, form a subring of $\Delta$. It is easily seen that this subring is actually a field, for if $\alpha = x\mathscr{P} \neq 0$, then, by condition (1), also $1/x \in K_\mathscr{P}$, and hence $1/\alpha = (1/x)\mathscr{P}$. We call this field the *residue field of $\mathscr{P}$*. The elements of $\Delta$ which are not $\mathscr{P}$-values of elements of $K$ do not interest us. Hence we shall assume that the residue field of $\mathscr{P}$ is the field $\Delta$ itself.

If $K$ is an extension of a ground field $k$, if $\mathscr{P}$ is a place of $K/k$ and if $s$ is the transcendence degree of $\Delta$ over $k$ ($s$ may be an infinite cardinal), we call *$s$ the dimension of the place $\mathscr{P}$*, over $k$, or in symbols: $s = \dim \mathscr{P}/k$. If $K$ has transcendence degree $r$ over $k$, then $0 \leqq s \leqq r$. The place $\mathscr{P}$ of

$K/k$ is *algebraic* (over $k$) if $s=0$; *rational* if $\Delta=k$. On the other extreme we have the case $s=r$. In this case and under the additional assumption that $r$ is finite, $\mathcal{P}$ is an isomorphism (Vol. I, Ch. II, § 12, Theorem 29), and furthermore it follows at once from condition (1) that $K_{\mathcal{P}}=K$, whence $\mathcal{P}$ is merely a $k$-isomorphism of $K$. Places which are isomorphisms of $K$ will be called *trivial* places of $K$ (or trivial places of $K/k$, if they are $k$-isomorphisms of $K$).

It is obvious that the trivial places $\mathcal{P}$ of $K$ are characterized by the condition $K_{\mathcal{P}}=K$. On the other hand, if $\mathcal{P}$ is a place of $K$ and $K_1$ is a subfield of $K$, then the restriction $\mathcal{P}_1$ of $\mathcal{P}$ to $K_1$ is obviously a place of $K_1$. Therefore, if $K_1 \subset K_{\mathcal{P}}$ then $\mathcal{P}_1$ is a trivial place of $K_1$. In particular, if $K$ has characteristic $p \neq 0$, then any place $\mathcal{P}$ of $K$ is trivial on the prime subfield of $K$ (for $1 \in K_{\mathcal{P}}$).

From condition (1) of Definition 1 it follows that if an element $x$ of $K_{\mathcal{P}}$ is such that $x\mathcal{P} \neq 0$, then $1/x$ belongs to $K_{\mathcal{P}}$ and hence $x$ is a unit in $K_{\mathcal{P}}$. Hence the kernel of $\mathcal{P}$ consists of all non-units of the ring $K_{\mathcal{P}}$. The kernel of $\mathcal{P}$ is therefore a maximal ideal in $K_{\mathcal{P}}$; in fact it is the only maximal ideal in $K_{\mathcal{P}}$. (However, the valuation ring $K_{\mathcal{P}}$ of a place $\mathcal{P}$ is not necessarily a local ring, since according to our definition, a local ring is noetherian (Vol. I, Ch. IV, § 11, p. 228), while, as we shall see later (§ 10, Theorem 16), a valuation ring need not be noetherian.) The maximal ideal in $K_{\mathcal{P}}$ will be denoted by $\mathfrak{M}_{\mathcal{P}}$ and will be referred to as the prime ideal of the place $\mathcal{P}$. The field $K_{\mathcal{P}}/\mathfrak{M}_{\mathcal{P}}$ and the residue field $\Delta$ of $\mathcal{P}$ are isomorphic.

Let $L$ be a subring of $K$. Our definition of places of $K$ implies that if $L$ is the valuation ring of a place $\mathcal{P}$ of $K$, then $L$ contains the reciprocal of any element of $K$ which does not belong to $L$; and, furthermore, $L$ must contain $k$ if $L$ is the valuation ring of a place of $K/k$. We now prove that also the converse is true:

THEOREM 1. *Let $L$ be a subring of $K$. If $L$ contains the reciprocal of any element of $K$ which does not belong to $L$, then there exists a place $\mathcal{P}$ of $K$ such that $L$ is the valuation ring of $\mathcal{P}$. If, furthermore, $K$ contains a ground field $k$ and $L$ contains $k$, then there also exists a place $\mathcal{P}$ of $K/k$ such that $L$ is the valuation ring of $\mathcal{P}$.*

PROOF. Assume that $L$ contains the reciprocal of any element of $K$ which does not belong to $L$. Then it follows in the first place that $1 \in L$. We next show that the non-units of $L$ form an ideal. For this it is only necessary to show that if $x$ and $y$ are non-units of $L$, then also $x+y$ is a non-unit, and in the proof we may assume that both $x$ and $y$ are different from zero. By assumption, either $y/x$ or $x/y$ belongs to $L$. Let, say, $y/x \in L$. Then $x+y=x(1+y/x)$, and since $1+y/x \in L$ and $x$

is a non-unit in $L$, we conclude that $x+y$ is a non-unit in $L$, as asserted. Let, then, $\mathfrak{M}$ be the ideal of non-units of $L$, and let $\mathscr{P}$ be the canonical homomorphism of $L$ onto the field $L/\mathfrak{M}$. Then condition (1) of Definition 1 is satisfied, with $K_{\mathscr{P}}=L$ (while $\Delta$ is now the field $L/\mathfrak{M}$), for if $x \in K$ and $x \notin L$, then $1/x \in L$, whence $1/x \in \mathfrak{M}$ and therefore $(1/x)\mathscr{P}=0$. It is obvious that also condition (2) is satisfied, since $L/\mathfrak{M}$ is a field and since $\mathscr{P}$ maps $L$ onto $L/\mathfrak{M}$.

Assume now that the additional condition $k \subset L$ is also satisfied. Then the field $L/\mathfrak{M}$ contains the *isomorphic* image $k\mathscr{P}$ of $k$. We may therefore identify each element $c$ of $k$ with its image $c\mathscr{P}$, and then also condition (3) is satisfied. Q.E.D.

An important property of the valuation ring $K_{\mathscr{P}}$ of a place $\mathscr{P}$ is that *it is integrally closed in $K$.* For let $x$ be any element of $K$ which is integrally dependent on $K_{\mathscr{P}}$: $x^n+a_1x^{n-1}+\cdots+a_n=0$, $a_i \in K_{\mathscr{P}}$. Dividing by $x^n$ we find $1=-a_1(1/x)-a_2(1/x)^2-\cdots-a_n(1/x)^n$. If $x \notin K_{\mathscr{P}}$, then $1/x \in K_{\mathscr{P}}$, $(1/x)\mathscr{P}=0$, and hence equating the $\mathscr{P}$-values of both sides of the above relation we get $1=0$, a contradiction. Hence $x \in K_{\mathscr{P}}$, and $K_{\mathscr{P}}$ is integrally closed in $K$, as asserted.

DEFINITION 2. *If $\mathscr{P}$ and $\mathscr{P}'$ are places of $K$ (or of $K/k$), with residue fields $\Delta$ and $\Delta'$ respectively, then $\mathscr{P}$ and $\mathscr{P}'$ are said to be isomorphic places (or $k$-isomorphic places) if there exists an isomorphism $\psi$ (or a $k$-isomorphism $\psi$) of $\Delta$ onto $\Delta'$ such that $\mathscr{P}'=\mathscr{P}\psi$.*

A necessary and sufficient condition that two places $\mathscr{P}$ and $\mathscr{P}'$ of $K$ (or of $K/k$) be isomorphic (or $k$-isomorphic) is that their valuation rings $K_{\mathscr{P}}$ and $K_{\mathscr{P}'}$ coincide. It is obvious that the condition is necessary. Assume now that the condition is satisfied, and let $\varphi$ be the canonical homomorphism of $K_{\mathscr{P}}$ onto $K_{\mathscr{P}}/\mathfrak{M}_{\mathscr{P}}$. Then $\mathscr{P}^{-1}\varphi$ is an isomorphism of $\Delta$ onto $K_{\mathscr{P}}/\mathfrak{M}_{\mathscr{P}}$, and similarly $\mathscr{P}'^{-1}\varphi$ is an isomorphism of $\Delta'$ onto $K_{\mathscr{P}}/\mathfrak{M}_{\mathscr{P}}$. Hence $\mathscr{P}^{-1}\mathscr{P}'(=\mathscr{P}^{-1}\varphi\cdot\varphi^{-1}\mathscr{P}')$ is an isomorphism $\psi$ of $\Delta$ onto $\Delta'$, showing that $\mathscr{P}$ and $\mathscr{P}'$ are isomorphic places. If, moreover, $\mathscr{P}$ and $\mathscr{P}'$ are places of $K/k$, then $\psi$ is a $k$-isomorphism of $\Delta$ onto $\Delta'$, whence $\mathscr{P}$ and $\mathscr{P}'$ are $k$-isomorphic places.

It is clear that $k$-isomorphic places of $K/k$ have the same dimension over $k$.

Isomorphic *algebraic* places of $K/k$ will be referred to as *conjugate places* (over $k$) if their residue fields are subfields of one and the same algebraic closure $\bar{k}$ of $k$. In that case, these residue fields are conjugate subfields of $\bar{k}/k$.

If $\mathscr{P}$ is a place of $K/k$, where $k$ is a ground field, then $K$ and the residue field $\Delta$ of $\mathscr{P}$ have the same characteristic (since $k \subset \Delta$). Conversely, assume that $\mathscr{P}$ is a place of $K$ such that $K$ and $\Delta$ have the same characteristic $p$. (Note that this assumption is satisfied for any place $\mathscr{P}$ of $K$ if $K$ has characteristic $\neq 0$, for in that case the restriction of $\mathscr{P}$ to the prime subfield of $K$ is an isomorphism.) Let $\Gamma$ denote the prime subfield of $K$. We know that if $p \neq 0$ then the restriction of $\mathscr{P}$ to $\Gamma$ is an isomorphism. If $p=0$ and if $J$ denotes the ring of integers in $\Gamma$, then $J \subset K_{\mathscr{P}}$ (since $1 \in K_{\mathscr{P}}$) and the restriction of $\mathscr{P}$ to $J$ must be an isomorphism (for otherwise $\Delta$ would be of characteristic $\neq 0$). Hence again the restriction of $\mathscr{P}$ to $\Gamma$ is an isomorphism (and we have $\Gamma \subset K_{\mathscr{P}}$). *It follows at once* (as in the proof of the last part of Theorem 1) *that $\mathscr{P}$ is isomorphic to a place of $K/\Gamma$.* We thus see that the theory of places over ground fields is essentially as general as the theory of arbitrary places $\mathscr{P}$ *in the equal characteristic case* (i.e., in the case in which $K$ and $\Delta$ have the same characteristic).

§ 3. **Specialization of places.** Let $\mathscr{P}$ and $\mathscr{P}'$ be places of $K$. We say that *$\mathscr{P}'$ is a specialization of $\mathscr{P}$* and we write $\mathscr{P} \rightarrow \mathscr{P}'$, if the valuation ring $K_{\mathscr{P}'}$ of $\mathscr{P}'$ is contained in the valuation ring $K_{\mathscr{P}}$ of $\mathscr{P}$, and we say that *$\mathscr{P}'$ is a proper specialization of $\mathscr{P}$* if $K_{\mathscr{P}'}$ is a proper subring of $K_{\mathscr{P}}$. If both $\mathscr{P}$ and $\mathscr{P}'$ are places of $K/k$ and $\mathscr{P}'$ is a specialization of $\mathscr{P}$, then we shall write $\mathscr{P} \xrightarrow{k} \mathscr{P}'$.

It is clear that $\mathscr{P} \rightarrow \mathscr{P}'$ if and only if either one of the following conditions is satisfied: (a) $x\mathscr{P}' \neq \infty$ implies $x\mathscr{P} \neq \infty$; (b) $x\mathscr{P}=0$ implies $x\mathscr{P}'=0$ (for, $x\mathscr{P}=0$ implies $(1/x)\mathscr{P}=\infty$, whence $(1/x)\mathscr{P}'=\infty$, or $x\mathscr{P}'=0$). Hence we have, in view of (b):

$$\mathscr{P} \rightarrow \mathscr{P}' \Leftrightarrow K_{\mathscr{P}} \supset K_{\mathscr{P}'} \quad \textit{and} \quad \mathfrak{M}_{\mathscr{P}} \subset \mathfrak{M}_{\mathscr{P}'}. \tag{1}$$

In particular, if both $\mathscr{P}$ and $\mathscr{P}'$ are places of $K/k$ and $\mathscr{P} \xrightarrow{k} \mathscr{P}'$, then we conclude at once with the following result: *If $x_1, x_2, \cdots, x_n$ are any elements of $K$ which are finite at $\mathscr{P}'$ (and therefore also at $\mathscr{P}$), then any algebraic relation, over $k$, between the $\mathscr{P}$-values of the $x_i$ is also satisfied by the $\mathscr{P}'$-values of the $x_i$.* Thus, our definition of specialization of places is a natural extension of the notion of specialization used in algebraic geometry.

Every place of $K$ is a specialization of any trivial place of $K$. Furthermore, isomorphic places are specializations of each other. Conversely, if two places $\mathscr{P}$ and $\mathscr{P}'$ are such that each is a specialization of the other, then they are isomorphic places. As a generalization of the last statement, we have the following theorem:

THEOREM 2. *Let $\mathscr{P}$ and $\mathscr{P}'$ be places of $K$, with residue fields $\Delta$ and $\Delta'$ respectively. Then $\mathscr{P} \rightarrow \mathscr{P}'$ if and only if there exists a place $\mathscr{Q}$ of $\Delta$ such that $\mathscr{P}' = \mathscr{P}\mathscr{Q}$ on $K_{\mathscr{P}'}$.*

PROOF. Assume that $\mathscr{P} \to \mathscr{P}'$. We set $\Delta_{\mathscr{Q}} = K_{\mathscr{P}'}\mathscr{P}$ and we observe that since $K_{\mathscr{P}'} \subset K_{\mathscr{P}}$, $\Delta_{\mathscr{Q}}$ is a subring of $\Delta$. On the other hand, we have, by (1), that $\mathfrak{M}_{\mathscr{P}}$ is a prime ideal in $K_{\mathscr{P}'}$. Let now $\varphi$ and $\varphi'$ denote the canonical homomorphisms of $K_{\mathscr{P}'}$ onto $K_{\mathscr{P}'}/\mathfrak{M}_{\mathscr{P}}$ and $K_{\mathscr{P}'}/\mathfrak{M}_{\mathscr{P}'}$ respectively, and let $\mathscr{P}_1$ be the restriction of $\mathscr{P}$ to $K_{\mathscr{P}'}$. Since $\mathfrak{M}_{\mathscr{P}}$ is the kernel of $\mathscr{P}_1$, the product $\mathscr{P}_1^{-1}\varphi$ is an isomorphism of $\Delta_{\mathscr{Q}}$ onto $K_{\mathscr{P}'}/\mathfrak{M}_{\mathscr{P}}$. Similarly $\varphi'^{-1}\mathscr{P}'$ is an isomorphism of $K_{\mathscr{P}'}/\mathfrak{M}_{\mathscr{P}'}$ onto $\Delta'$. Since $\mathfrak{M}_{\mathscr{P}} \subset \mathfrak{M}_{\mathscr{P}'}$, $\varphi^{-1}\varphi'$ is a homomorphism of $K_{\mathscr{P}'}/\mathfrak{M}_{\mathscr{P}}$ onto $K_{\mathscr{P}'}/\mathfrak{M}_{\mathscr{P}'}$. We set $\mathscr{Q} = \mathscr{P}_1^{-1}\varphi \cdot \varphi^{-1}\varphi' \cdot \varphi'^{-1}\mathscr{P}' = \mathscr{P}_1^{-1}\mathscr{P}'$. Then $\mathscr{Q}$ is a homomorphism of $\Delta_{\mathscr{Q}}$ onto $\Delta'$. If $\xi$ is an element of $\Delta$ which is not in $\Delta_{\mathscr{Q}}$ and $x$ is some fixed element of $K_{\mathscr{P}}$ such that $x\mathscr{P} = \xi$, then $x \notin K_{\mathscr{P}'}$, $(1/x)\mathscr{P}' = 0$, and hence $(1/\xi)\mathscr{Q} = 0$. We have thus proved that $\mathscr{Q}$ is a place of $\Delta$, with residue field $\Delta'$, and that $\mathscr{P}_1\mathscr{Q} = \mathscr{P}'$. Hence $\mathscr{P}'$ and $\mathscr{P}\mathscr{Q}$ coincide on $K_{\mathscr{P}'}$.

Conversely, if we have $\mathscr{P}' = \mathscr{P}\mathscr{Q}$ on $K_{\mathscr{P}'}$, where $\mathscr{Q}$ is a place of $\Delta$, then it is clear that $x\mathscr{P}' \neq \infty$ implies $x\mathscr{P} \neq \infty$, whence $K_{\mathscr{P}'} \subset K_{\mathscr{P}}$, and $\mathscr{P}'$ is a specialization of $\mathscr{P}$. This completes the proof.

We note that $\mathscr{P}'$ and $\mathscr{P}\mathscr{Q}$ coincide not only on $K_{\mathscr{P}'}$ but also on $K_{\mathscr{P}}$, in the following sense: *if $x \in K_{\mathscr{P}}$ and $x \notin K_{\mathscr{P}'}$ (whence $x\mathscr{P} \in \Delta$ and $x\mathscr{P}' = \infty$), then $(x\mathscr{P})\mathscr{Q} = \infty$.* For, if $x \notin K_{\mathscr{P}'}$, then $(1/x)\mathscr{P}' = 0$, and hence $(1/x)\mathscr{P}\mathscr{Q} = 0$ (since $\mathscr{P}' = \mathscr{P}\mathscr{Q}$ on $K_{\mathscr{P}'}$), i.e., $(1/x\mathscr{P})\mathscr{Q} = 0$ and $(x\mathscr{P})\mathscr{Q} = \infty$, as asserted.

We note also that in the special case of isomorphic places $\mathscr{P}$, $\mathscr{P}'$, $\mathscr{Q}$ is an isomorphism of $\Delta$, i.e., $\mathscr{Q}$ is a trivial place of $\Delta$.

It is clear that the place $\mathscr{Q}$ whose existence is asserted in Theorem 2 is uniquely determined by $\mathscr{P}$ and $\mathscr{P}'$ and that if both $\mathscr{P}$ and $\mathscr{P}'$ are places *over* $k$, then also $\mathscr{Q}$ is a place over $k$ (i.e., a place of $\Delta/k$).

COROLLARY. *If $\mathscr{P}$ and $\mathscr{P}'$ are places of $K/k$ and $\mathscr{P} \xrightarrow{k} \mathscr{P}'$, then* $\dim \mathscr{P}'/k \leqq \dim \mathscr{P}/k$. *Furthermore, if the residue field $\Delta$ of $\mathscr{P}$ has finite transcendence degree over $k$ and $\mathscr{P}'$ is a specialization of $\mathscr{P}$ over $k$, then* $\dim \mathscr{P}'/k = \dim \mathscr{P}/k$ *if and only if $\mathscr{P}$ and $\mathscr{P}'$ are $k$-isomorphic places.*

We shall now investigate the following question: given a place $\mathscr{P}$ of $K$, find all the places of $K$ of which $\mathscr{P}$ is a specialization. From Theorem 1 (§ 2) it follows at once that any ring (in $K$) which contains the valuation ring of a place of $K$ is itself a valuation ring of a place of $K$. Hence our question is equivalent to the following: find all the subrings of $K$ which contain $K_{\mathscr{P}}$. The answer to this equation is given by the following theorem:

THEOREM 3. *Any subring of $K$ which contains $K_{\mathscr{P}}$ is necessarily the quotient ring of $K_{\mathscr{P}}$ with respect to some prime ideal of $K_{\mathscr{P}}$. If $\mathfrak{M}_1$ and $\mathfrak{M}_2$ are ideals in $K_{\mathscr{P}}$, then either $\mathfrak{M}_1$ contains $\mathfrak{M}_2$ or $\mathfrak{M}_2$ contains $\mathfrak{M}_1$ (and hence* *the set of rings between $K_{\mathscr{P}}$ and $K$ is totally ordered by set-theoretic inclusion $\subset$). If $\mathscr{P}$ is a place of $K/k$ and if tr.d. $K/k=r\neq\infty$, then $K_{\mathscr{P}}$ has only a finite number of prime ideals, and the number of prime ideals of $K_{\mathscr{P}}$ (other than $K_{\mathscr{P}}$ itself) is at most equal to $r-s$, where $s=\dim \mathscr{P}/k$.*

PROOF. Let $L$ be a ring between $K_{\mathscr{P}}$ and $K: K_{\mathscr{P}}<L<K$. Then $L$ is the valuation ring $K_{\mathscr{Q}}$ of a place $\mathscr{Q}$ of which $\mathscr{P}$ is a specialization and hence the prime ideal $\mathfrak{M}_{\mathscr{Q}}$ of $\mathscr{Q}$ is also a prime ideal in $K_{\mathscr{P}}$. Any element of $K_{\mathscr{P}}$ which is not in $\mathfrak{M}_{\mathscr{Q}}$ is a unit in $K_{\mathscr{Q}}$ (since $\mathfrak{M}_{\mathscr{Q}}$ is the ideal of non-units of $K_{\mathscr{Q}}$ and since $K_{\mathscr{P}}\subset K_{\mathscr{Q}}$). Hence the quotient ring of $K_{\mathscr{P}}$ with respect to the prime ideal $\mathfrak{M}_{\mathscr{Q}}$ (i.e., the set of all quotients $a/b$, where $a, b \in K_{\mathscr{P}}$ and $b \notin \mathfrak{M}_{\mathscr{Q}}$) is contained in $K_{\mathscr{Q}}$. On the other hand, we now show that any element $x$ of $K_{\mathscr{Q}}$ belongs to the above quotient ring. This is obvious if $x \in K_{\mathscr{P}}$. Assume that $x \notin K_{\mathscr{P}}$. If we set $y=1/x$, then $y \in K_{\mathscr{P}}$ (since $K_{\mathscr{P}}$ is a valuation ring). Furthermore, $x \notin \mathfrak{M}_{\mathscr{Q}}$ (since $\mathfrak{M}_{\mathscr{Q}}\subset K_{\mathscr{P}}$), and hence $x$ is a unit in $K_{\mathscr{Q}}$. Therefore also $y$ is a unit in $K_{\mathscr{Q}}$, and so $y \notin \mathfrak{M}_{\mathscr{Q}}$. It follows that $x(=1/y)$ belongs to the quotient ring of $K_{\mathscr{P}}$ with respect to $\mathfrak{M}_{\mathscr{Q}}$. This proves the first part of the theorem.

Let $\mathfrak{M}_1$ and $\mathfrak{M}_2$ be any two proper ideals in $K_{\mathscr{P}}$ (not necessarily prime ideals) and assume that $\mathfrak{M}_1 \not\subset \mathfrak{M}_2$. Let $x$ be an element of $\mathfrak{M}_1$, not in $\mathfrak{M}_2$, and let $y$ be any element of $\mathfrak{M}_2$, $y\neq 0$. Then $x/y \notin K_{\mathscr{P}}$, and hence $y/x \in K_{\mathscr{P}}$, $y \in \mathfrak{M}_1$ (since $\mathfrak{M}_1$ is an ideal and $x \in \mathfrak{M}_1$). Hence $\mathfrak{M}_2\subset\mathfrak{M}_1$.

Assume now that $\mathscr{P}$ is a place of $K/k$ and that tr.d. $K/k=r\neq\infty$. Let $\mathfrak{M}_1$ and $\mathfrak{M}_2$ be two prime ideals in $K_{\mathscr{P}}$ and let us assume that, say, $\mathfrak{M}_1>\mathfrak{M}_2$. Let $L_i$, $i=1, 2$, be the quotient ring of $K_{\mathscr{P}}$ with respect to $\mathfrak{M}_i$, and let $\mathscr{P}_i$ be a place of $K$ whose valuation ring is $L_i$. We have $L_2>L_1$, and hence $\mathscr{P}_1$ is a *proper* specialization of $\mathscr{P}_2$. On the other hand, $\mathscr{P}$ is a specialization of $\mathscr{P}_1$. It follows by Theorem 2, Corollary, that $\dim \mathscr{P}/k \leqq \dim \mathscr{P}_1/k < \dim \mathscr{P}_2/k \leqq r$. This shows that the number of prime ideals of $K_{\mathscr{P}}$ is finite and that the number of prime ideals in $K_{\mathscr{P}}$, other than $K_{\mathscr{P}}$ itself, is at most $r-s$. This completes the proof of the theorem.

DEFINITION 1. *The ordinal type*† *of the totally ordered set of proper prime ideals $\mathfrak{q}$ of $K_{\mathscr{P}}$ ($\mathfrak{q}\neq(0)$, $\mathfrak{q}\neq K_{\mathscr{P}}$; $\mathfrak{q}_1$ precedes $\mathfrak{q}_2$ if $\mathfrak{q}_1>\mathfrak{q}_2$) is called the rank of the place $\mathscr{P}$.*

† In most axiomatic systems of set theory it is possible to attach to every totally ordered set $E$ a well-defined object $o(E)$ in such a way that we have $o(E)=o(F)$ if and only if $E$ and $F$ are isomorphic ordered sets (i.e., if there exists a one-to-one mapping $f$ of $E$ onto $F$ such that the relations $x\leqq y$ and $f(x)\leqq f(y)$ are equivalent). The object $o(E)$ is called the *ordinal type* of $E$. Furthermore, if $E$ is isomorphic to the set $\{1, 2, \cdots, n\}$ (i.e., if $E$ is a finite, totally ordered set with $n$ elements), we shall identify its ordinal type with its cardinal number $n$.

COROLLARY 1. *If $K$ has finite transcendence degree $r$ over $k$, then any place $\mathscr{P}$ of $K/k$ has rank $\leqq r-s$, where $s=\dim \mathscr{P}/k$.*

The rank of a place $\mathscr{P}$ of $K$ is zero if and only if $\mathscr{P}$ is a trivial place of $K$.

The rank of $\mathscr{P}$ is 1 if and only if $\mathscr{P}$ is not a trivial place of $K$ and is not a proper specialization of any non-trivial place of $K$. A necessary and sufficient condition that a place $\mathscr{P}$ be of rank one is that its valuation ring be a maximal (proper) subring of $K$. We shall see later (§ 4, Theorem 4, Corollary 3) that any maximal (proper) subring of $K$ is in fact the valuation ring of a place of $K$, provided the subring is a *proper ring*, i.e., not a field.

We shall have occasion to use in § 6 the following corollary:

COROLLARY 2. *If $a_1, a_2, \cdots, a_m$ are elements of $K$, not all zero, then for at least one integer $j$, $1 \leqq j \leqq m$, it is true that $(a_i/a_j)\mathscr{P} \neq \infty$, $i=1, 2, \cdots, m$, $a_j \neq 0$.*

Since $K$ is the quotient field of $K_{\mathscr{P}}$, it is sufficient to consider the case in which all the $a_i$ are in $K_{\mathscr{P}}$. In that case we take for $a_j$ the element which generates the greatest ideal in the set of principal ideals $(a_i)$.

If $\mathscr{P}$ is of finite rank $m$, there are exactly $m-1$ rings $L_i$ between $K_{\mathscr{P}}$ and $K$, and we have $K_{\mathscr{P}} < L_1 < L_2 < \cdots < L_{m-1} < K$. If $\mathscr{P}_i$ is a place of $K$ whose valuation ring is $L_i$, then $\mathscr{P}_i$ is of rank $m-i$, $\mathscr{P}_i$ is a specialization of $\mathscr{P}_j$ if $i<j$ $(i=0, 1, \cdots, m-1;\ \mathscr{P}_0=\mathscr{P})$. We have thus a *specialization chain for* $\mathscr{P}$:

$$\mathscr{P}_{m-1} \rightarrow \mathscr{P}_{m-2} \rightarrow \cdots \rightarrow \mathscr{P}_1 \rightarrow \mathscr{P}, \tag{2}$$

which joins a place $\mathscr{P}_{m-1}$ of rank 1 to the given place $\mathscr{P}$ of rank $m$. This chain is *maximal* in the sense that it cannot be refined by insertion of other places which are not isomorphic to any of the $m$ places $\mathscr{P}_i$. We shall call the chain (2) a *composition chain for* $\mathscr{P}$. Any place $\mathscr{P}^\star$ of which $\mathscr{P}$ is a specialization is isomorphic to one of the places $\mathscr{P}_i$ (assuming of course that $\mathscr{P}^\star$ is not a trivial place of $K$), and if

$$\mathscr{P}'_{m-1} \rightarrow \mathscr{P}'_{m-2} \rightarrow \cdots \rightarrow \mathscr{P}'_1 \rightarrow \mathscr{P}$$

is any other composition chain for $\mathscr{P}$, then $\mathscr{P}_i$ and $\mathscr{P}'_i$ are isomorphic places $(i=0, 1, \cdots, m-1)$.

If $r=\text{tr.d. } K/k \neq \infty$, then of particular importance are the places which are of dimension $r-1$. It is clear that the rank of such a place is 1 (Corollary 1). The $(r-1)$-dimensional places of fields of algebraic functions of $r$ independent variables are of particular importance in the theory of algebraic varieties. A discussion of these places will be found in § 14.

§ 4. **Existence of places.** We shall prove the following existence theorem:

THEOREM 4. *Let* $\mathfrak{o}$ *be a subring of* $K$ *containing* 1, *and let* $\mathfrak{A}$ *be an ideal in* $\mathfrak{o}$, *different from* $\mathfrak{o}$. *Then there exists a place* $\mathscr{P}$ *of* $K$ *such that* $K_{\mathscr{P}} \supset \mathfrak{o}$ *and* $\mathfrak{M}_{\mathscr{P}} \supset \mathfrak{A}$.

PROOF. Let $M$ denote the set of all subrings $R_i$ of $K$ such that $\mathfrak{o} \subset R_i$ and $R_i \mathfrak{A} \neq R_i$. The set $M$ is non-empty, since $\mathfrak{o} \in M$. We partially order the rings $R_i$ by set-theoretic inclusion. Let $\{R_\alpha\}$ be a totally ordered subset $N$ of $M$, and let $R$ be the join of the rings $R_\alpha$. We cannot have a relation of the form $1 = a_1 \xi_1 + a_2 \xi_2 + \cdots + a_m \xi_m$, $a_i \in \mathfrak{A}$, $\xi_i \in R$, for the $\xi$'s would then belong to some $R_\beta$, $R_\beta \in N$ (since $N$ is linearly ordered), and we would have $R_\beta \mathfrak{A} = R_\beta$, a contradiction (since $R_\beta \in M$). It follows that $R \mathfrak{A} \neq R$, and hence $R \in M$. We have therefore proved that every totally ordered subset $N$ of $M$ has an upper bound $R$ in $M$. By Zorn's lemma, $M$ contains, then, maximal elements. We shall prove that every maximal element of $M$ is the valuation ring of a place $\mathscr{P}$ of $K$, satisfying the required conditions.

Let $L$ be a maximal element of $M$. The ring $L$ satisfies, then, the following conditions (1) $\mathfrak{o} \subset L$, $L\mathfrak{A} \neq L$; (2) if $L'$ is any subring of $K$ such that $L < L'$, then $L' \mathfrak{A} = L'$. The remainder of the proof will be based on the following lemma:

LEMMA. *Let* $R$ *be a subring of a field* $K$, *containing* 1, *and let* $\mathfrak{I}$ *be a proper ideal in* $R$. *Then for any element* $x$ *of* $K$ *at least one of the extended ideals* $R[x]\mathfrak{I}$, $R[1/x]\mathfrak{I}$ *is a proper ideal of* $R[x]$, $R[1/x]$ *respectively.*

PROOF OF LEMMA. Assume the contrary: $R[x]\mathfrak{I} = R[x]$, $R[1/x]\mathfrak{I} = R[1/x]$. That means that we have two representations of the element 1 of $R$:

$$(1) \qquad 1 = \sum_{i=0}^{n} a_i x^i, \quad a_i \in \mathfrak{I}, \quad 0 \leqq i \leqq n;$$

$$(1') \qquad 1 = \sum_{j=0}^{m} b_j / x^j, \quad b_j \in \mathfrak{I}, \quad 0 \leqq j \leqq m.$$

We shall suppose that the relations (1) and (1') are of the smallest possible degrees $n$ and $m$. Let, say, $m \leqq n$. We multiply (1) by $1 - b_0$ and (1') by $a_n x^n$:

$$1 - b_0 = (1 - b_0)a_0 + \cdots + \cdot(1 - b_0)a_n x^n,$$

$$(1 - b_0)a_n x^n = a_n b_1 x^{n-1} + \cdots + a_n b_m x^{n-m}.$$

Thus,

$$1 - b_0 = (1 - b_0)a_0 + \cdots + (1 - b_0)a_{n-1}x^{n-1} + a_n b_1 x^{n-1} + \cdots + a_n b_m x^{n-m},$$

or

$$1 = \sum_{\nu=0}^{n-1} c_\nu x^\nu, \quad c_\nu \in \mathfrak{I},$$

and this is a relation of the same form as (1) and of degree less than $n$, contrary to our assumption that (1) is of lowest possible degree.

We now apply the lemma to the case $R = L$, $\mathfrak{I} = L\mathfrak{A}$. If $x$ is any element of $K$, and if we set $L' = L[x]$, $L'' = L[1/x]$, then the lemma tells us that at least one of the following two relations must hold: $L'\mathfrak{I} \neq L'$, $L''\mathfrak{I} \neq L''$. This implies by the maximality property of $L$, that either $L = L'$ or $L = L''$, i.e., either $x \in L$ or $1/x \in L$. Hence $L$ is a valuation ring of a place $\mathscr{P}$ of $K$ (§ 2, Theorem 1).

The prime ideal $\mathfrak{M}_{\mathscr{P}}$ of $\mathscr{P}$ is the ideal of non-units of $L$, whence $\mathfrak{M}_{\mathscr{P}} \supset L\mathfrak{A} \supset \mathfrak{A}$, and since $L \supset \mathfrak{o}$ the proof of the theorem is now complete.

We note that if $\mathscr{P}$ is a trivial place of $K$ then $\mathfrak{M}_{\mathscr{P}} = (0)$. Hence if the ideal $\mathfrak{A}$ is not the zero ideal, any place $\mathscr{P}$ satisfying the conditions of the theorem is necessarily non-trivial.

COROLLARY 1. *If $\mathfrak{o}$ is an integral domain, not a field, and if $K$ is a field containing $\mathfrak{o}$ as subring, then there exist non-trivial places $\mathscr{P}$ of $K$ such that $K_{\mathscr{P}} \supset \mathfrak{o}$.*

For $\mathfrak{o}$ contains ideals different from (0) and $\mathfrak{o}$.

COROLLARY 2. *A field $K$ possesses only trivial places if and only if $K$ is an absolutely algebraic field, of characteristic $p \neq 0$ (i.e., if and only if $K$ is an algebraic extension of the prime field of characteristic $p \neq 0$).*

For, the absolutely algebraic fields, of characteristic $p \neq 0$, are the only fields with the property that all their subrings are fields, whereas the valuation ring of a non-trivial place is not a field.

COROLLARY 3. *If $\mathfrak{o}$ is a proper ring and a maximal subring of a field $K$, then $\mathfrak{o}$ is the valuation ring of a place $\mathscr{P}$ of $K$.*

This follows at once from Corollary 1. Note that $\mathscr{P}$ is then necessarily of rank 1 (see § 3, Definition 1).

Of great importance for applications to algebraic geometry is the following consequence of our existence theorem:

THEOREM 5. *If $\mathfrak{o}$ is an integral domain contained in a field $K$ and if $\mathfrak{m}$ is a prime ideal in $\mathfrak{o}$, $\mathfrak{m} \neq \mathfrak{o}$, then there exists a place $\mathscr{P}$ of $K$ such that $K_{\mathscr{P}} \supset \mathfrak{o}$ and $\mathfrak{M}_{\mathscr{P}} \cap \mathfrak{o} = \mathfrak{m}$.*

PROOF. Let $\mathfrak{o}'$ denote the quotient ring of $\mathfrak{o}$ with respect to $\mathfrak{m}$ and let $\mathfrak{m}' = \mathfrak{o}'\mathfrak{m} = $ ideal of non-units in $\mathfrak{o}'$. From our assumptions on $\mathfrak{m}$ it follows that $\mathfrak{m}' \neq \mathfrak{o}'$. Hence there exists a place $\mathscr{P}$ of $K$ such that

$K_{\mathscr{P}} \supset \mathfrak{o}'$, $\mathfrak{M}_{\mathscr{P}} \cap \mathfrak{o}' \supset \mathfrak{m}'$. Since $\mathfrak{m}'$ is a maximal ideal in $\mathfrak{o}'$ and since $1 \notin \mathfrak{M}_{\mathscr{P}}$, it follows that $\mathfrak{M}_{\mathscr{P}} \cap \mathfrak{o}' = \mathfrak{m}'$. Hence $\mathfrak{M}_{\mathscr{P}} \cap \mathfrak{o} = \mathfrak{m}$, since $\mathfrak{m}' \cap \mathfrak{o} = \mathfrak{m}$.

The following is essentially an equivalent formulation of Theorem 5:

THEOREM 5′. (*The extension theorem*). *If* $\mathfrak{o}$ *is an integral domain and* $K$ *is a field containing* $\mathfrak{o}$, *then any specialization* $\varphi$ *of* $\mathfrak{o}$ *can be extended to a place* $\mathscr{P}$ *of* $K$. *In particular, if* $k$ *is a subfield of* $K$ *then any place of* $k$ *can be extended to a place of* $K$.

For if $\mathfrak{m}$ denotes the kernel of $\varphi$ then $\mathfrak{m} \neq \mathfrak{o}$ (by definition of specializations), and there exists a place $\mathscr{P}$ of $K$ such that $K_{\mathscr{P}} \supset \mathfrak{o}$ and $\mathfrak{M}_{\mathscr{P}} \cap \mathfrak{o} = \mathfrak{m}$. If $\psi$ denotes the restriction of $\mathscr{P}$ to $\mathfrak{o}$, $\psi^{-1}\varphi$ is an isomorphism of $\mathfrak{o}\psi$ onto $\mathfrak{o}\varphi$ (since $\mathfrak{m}$ is the kernel of both $\varphi$ and $\psi$). This isomorphism can be extended to an isomorphism of the residue field $\Delta$ of $\mathscr{P}$ into some field containing $\mathfrak{o}\varphi$. If $\mathscr{Q}$ is such an extension, then the place $\mathscr{P}\mathscr{Q}$ of $K$ is an extension of $\varphi$.

We now give a number of important consequences of Theorems 5 and 5′.

For applications to algebraic function fields, or, more generally, to fields $K$ in which a subfield $k$ has been specified as ground field, it is important to analyze Theorem 5′ in the special case $\varphi = 1$ (whence $\mathfrak{m} = (0)$), with reference to the following question: does there exist in this case a *non-trivial* place which is an extention of $\varphi$? If $\mathscr{P}$ is such a place then $K_{\mathscr{P}}$ contains the quotient field of $\mathfrak{o}$ in $K$, and the restriction of $\mathscr{P}$ to that quotient field is also the identity. Therefore, we may as well assume that $\mathfrak{o}$ is a field, say $\mathfrak{o} = k$, and the non-trivial places $\mathscr{P}$ which we are seeking are the places of $K/k$. If $K$ is an algebraic extension of $k$, then $K_{\mathscr{P}} \supset k$ implies $K_{\mathscr{P}} = K$, since $K_{\mathscr{P}}$ must be integrally closed in $K$ and since every element of $K$ is integrally dependent on $k$. Hence *if* $K$ *is an algebraic extension of* $k$, *then* $K/k$ *possesses only trivial places.* On the other hand, assume that $K$ has positive transcendence degree over $k$. Then if $x$ is any transcendental element of $K$ over $k$, the polynomial ring $k[x]$ is a proper ring (i.e., not a field) and admits at least one specialization $\varphi$ over $k$ which is not an isomorphism (in fact, there are infinitely many such specializations of $k[x]$, for each irreducible polynomial in $k[x]$ can be used to define a $\varphi$). We have therefore the following.

COROLLARY 1. *If* $K$ *is a field extension of a ground field* $k$, *then* $K/k$ *has non-trivial places if and only if* $K$ *has positive transcendence degree over* $k$.

To this corollary we can now add the following very useful additional result:

COROLLARY 2. *If a field* $K$ *has positive transcendence degree over a subfield* $k$, *then there exist algebraic places of* $K/k$.

For consider the set $M$ of all valuation rings in $K$ which belong to places of $K/k$ (i.e., valuation rings which contain $k$). By Corollary 1, $M$ is non-empty. By Theorem 1, § 2, the intersection of any descending chain of valuation rings in $K$ is again a valuation ring. Hence, by Zorn's lemma, $M$ contains minimal elements (it is understood that $M$ is partially ordered by set-theoretic inclusion). Let $R$ be a minimal element of $M$ and let $\mathscr{P}$ be a place of $K/k$ such that $K_{\mathscr{P}}=R$. We assert that $\mathscr{P}$ is algebraic over $k$. For, assuming the contrary, i.e., assuming that the residue field $\Delta$ of $\mathscr{P}$ has positive transcendence degree over $k$, then it would follow from Corollary 1 that there exists a non-trivial place $\mathscr{Q}$ of $\Delta/k$. Then the composite place $\mathscr{P}'=\mathscr{P}\mathscr{Q}$ is a place of $K/k$ whose valuation ring is a proper subset of $R$, a contradiction.

COROLLARY 3. *If $\varphi$ is a specialization of an integral domain $\mathfrak{o}$, and if $K$ is a field containing $\mathfrak{o}$, then there exists a place of $K$ which is an extension of $\varphi$ and whose residue field is algebraic over the quotient field of $\mathfrak{o}\varphi$.*

Let $k$ be the quotient field of the $\varphi$-transform $\mathfrak{o}\varphi$ of $\mathfrak{o}$. We fix a place $\mathscr{P}$ of $K$ which is an extension of $\varphi$ and whose residue field $\Delta$ therefore contains $k$. If $\Delta$ is algebraic over $k$ then $\mathscr{P}$ is the desired place. If $\Delta$ is not algebraic over $k$, then we fix, by Corollary 2, an algebraic place $\mathscr{Q}$ of $\Delta/k$. The composite place $\mathscr{P}'=\mathscr{P}\mathscr{Q}$ of $K$ is an extension of $\varphi$ (since $\mathscr{Q}$ is the identity of $\mathfrak{o}\varphi$) and its residue field is algebraic over $k$ (since $\mathscr{Q}$ is an algebraic place of $\Delta/k$).

COROLLARY 4. *Let $\mathfrak{o}$ be an integral domain and let $K$ be a field containing $\mathfrak{o}$ as subring. If a specialization $\varphi$ of $\mathfrak{o}$ is such that it has no proper extensions within $K$, then $\varphi$ is a place of $K$* (this is the converse of a result proved in the beginning of § 2).

This is a direct consequence of Theorem 5′.

The two corollaries that follow have already been proved in the preceding chapter in the more general case of arbitrary commutative rings with identity. However, as in the case of domains they are very simple consequences of Theorem 5, we give here a second proof of these results.

COROLLARY 5. *Let $\mathfrak{O}$ and $\mathfrak{o}$ be integral domains such that $\mathfrak{o}$ is a subring of $\mathfrak{O}$ and such that every element of $\mathfrak{O}$ is integrally dependent on $\mathfrak{o}$. Then for every prime ideal $\mathfrak{m}$ in $\mathfrak{o}$ there exists a prime ideal $\mathfrak{M}$ in $\mathfrak{O}$ such that $\mathfrak{M}\cap\mathfrak{o}=\mathfrak{m}$.*

The assertion being trivial if $\mathfrak{m}=\mathfrak{o}$, we assume $\mathfrak{m}\neq\mathfrak{o}$. If $K$ is the quotient field of $\mathfrak{O}$, there exists a place $\mathscr{P}$ of $K$ such that $K_{\mathscr{P}}\supset\mathfrak{o}$ and $\mathfrak{M}_{\mathscr{P}}\cap\mathfrak{o}=\mathfrak{m}$ (Theorem 5). Since $K_{\mathscr{P}}$ is integrally closed in $K$ and $\mathfrak{O}$ is integral over $\mathfrak{o}$, it follows from $K_{\mathscr{P}}\supset\mathfrak{o}$ that $K_{\mathscr{P}}\supset\mathfrak{O}$. Hence $\mathfrak{M}_{\mathscr{P}}\cap\mathfrak{O}$ is a prime ideal $\mathfrak{M}$ in $\mathfrak{O}$, and we have $\mathfrak{M}\cap\mathfrak{o}=\mathfrak{m}$.

COROLLARY 6. *The rings $\mathfrak{O}$ and $\mathfrak{o}$ being as in the preceding corollary, let $\mathfrak{a}$ be an ideal in $\mathfrak{o}$. Then if $\mathfrak{a} \neq \mathfrak{o}$, we have $\mathfrak{O}\mathfrak{a} \neq \mathfrak{O}$.*

Since $\mathfrak{o}$ contains an identity, there exists a prime ideal $\mathfrak{m}$ in $\mathfrak{o}$ such that $\mathfrak{a} \subset \mathfrak{m} \neq \mathfrak{o}$ (for instance, there exist maximal ideals containing $\mathfrak{a}$). By Corollary 5, let $\mathfrak{M}$ be a prime ideal in $\mathfrak{O}$ such that $\mathfrak{M} \cap \mathfrak{o} = \mathfrak{m}$. Then clearly $\mathfrak{M} \neq \mathfrak{O}$, and since $\mathfrak{O}\mathfrak{a} \subset \mathfrak{O}\mathfrak{m} \subset \mathfrak{M}$, it follows that $\mathfrak{O}\mathfrak{a} \neq \mathfrak{O}$.

Place-theoretic properties of integrally closed domains are of particular importance in the arithmetic theory of algebraic varieties. Many of these properties are based on the following theorem:

THEOREM 6. *If $\mathfrak{o}$ is an integral domain and $K$ is a field containing $\mathfrak{o}$, the intersection of all the valuation rings $K_{\mathscr{P}}$ of places $\mathscr{P}$ of $K$ such that $K_{\mathscr{P}} \supset \mathfrak{o}$ is the integral closure of $\mathfrak{o}$ in $K$.*

PROOF. Since every $K_{\mathscr{P}}$ is integrally closed, every $K_{\mathscr{P}}$ containing $\mathfrak{o}$ contains the integral closure $\bar{\mathfrak{o}}$ of $\mathfrak{o}$. So we have only to show that if $x$ is an element of $K$ which does not belong to $\bar{\mathfrak{o}}$, then there exists a place $\mathscr{P}$ of $K_{\mathscr{P}}$ such that $K_{\mathscr{P}} \supset \mathfrak{o}$ and $x \notin K_{\mathscr{P}}$. To show this, we consider the ring $\mathfrak{o}' = \mathfrak{o}[y]$, where $y = 1/x$. Our basic remark is to the effect that *$y$ is a non-unit in $\mathfrak{o}'$.* For, if $y$ were a unit in $\mathfrak{o}'$, then we would have a relation of the form: $1/y = x = a_0x^{-n} + a_1x^{-n+1} + \cdots + a_n$, $a_i \in \mathfrak{o}$, or $x^{n+1} - a_nx^n - \cdots - a_0 = 0$, and hence $x$ would be integrally dependent on $\mathfrak{o}$, contrary to assumption. Since $y$ is a non-unit in $\mathfrak{o}'$, the ideal $\mathfrak{o}'y$ is different from $\mathfrak{o}'$. By Theorem 4, there exists, then, a place $\mathscr{P}$ of $K$ such that $K_{\mathscr{P}} \supset \mathfrak{o}'$, $\mathfrak{M}_{\mathscr{P}} \supset \mathfrak{o}'y$. Hence $y$ is also a non-unit in $K_{\mathscr{P}}$, and consequently $x \notin K_{\mathscr{P}}$.

COROLLARY. *Let $\mathfrak{o}$ be an integral domain and let $K$ be a field containing $\mathfrak{o}$. If $\mathfrak{o}$ is integrally closed in $K$, then $\mathfrak{o}$ is the intersection of all the valuation rings $K_{\mathscr{P}}$ of places $\mathscr{P}$ of $K$ such that $K_{\mathscr{P}} \supset \mathfrak{o}$.*

REMARK. If $K$ is a field of algebraic functions over a ground field $k$, then all the results established in this section continue to hold if by a "place of $K$" we always mean a "place of $K/k$," *provided that $k \subset \mathfrak{o}$.* For, every place $\mathscr{P}$ such that $K_{\mathscr{P}} \supset \mathfrak{o}$ is $k$-isomorphic to a place of $K/k$.

## § 5. The center of a place in a subring.

Let $\mathfrak{o}$ be an integral domain, let $K$ be a field containing $\mathfrak{o}$ and let $\mathscr{P}$ be a place of $K$. We say that *$\mathscr{P}$ is finite on $\mathfrak{o}$* if $\mathscr{P}$ has finite value at each element of $\mathfrak{o}$, or—equivalently—if $\mathfrak{o} \subset K_{\mathscr{P}}$. If $\mathscr{P}$ is finite on $\mathfrak{o}$ then the restriction of $\mathscr{P}$ to $\mathfrak{o}$ is a specialization of $\mathfrak{o}$. If this specialization is the identical mapping of $\mathfrak{o}$ onto itself, then we shall say that $\mathscr{P}$ is a place of $K$ over $\mathfrak{o}$.

Let $\mathscr{P}$ be a place of $K$ which is finite on $\mathfrak{o}$. The set $\mathfrak{p} = \mathfrak{M}_{\mathscr{P}} \cap \mathfrak{o}$ of those elements of $\mathfrak{o}$ at which $\mathscr{P}$ has value zero is clearly a prime ideal in

$\mathfrak{o}$. This prime ideal is called *the center of* $\mathscr{P}$ *in* $\mathfrak{o}$. The center $\mathfrak{p}$ is always different from $\mathfrak{o}$ since $1 \notin \mathfrak{M}_{\mathscr{P}}$; it is the zero ideal if and only if the restriction of $\mathscr{P}$ to $\mathfrak{o}$ is an isomorphism (in particular, $\mathfrak{p}=(0)$ if $\mathscr{P}$ is a place of $K$ over $\mathfrak{o}$). It is clear that the residue class ring $\mathfrak{o}/\mathfrak{p}$ is isomorphic to the subring $\mathfrak{o}\mathscr{P}$ of the residue field $\Delta$ of $\mathscr{P}$.

Since any element of $\mathfrak{o}$ which is not in the center $\mathfrak{p}$ of $\mathscr{P}$ in $\mathfrak{o}$ is a unit in the valuation ring $K_{\mathscr{P}}$, it follows that $\mathscr{P}$ is also finite on the local ring $\mathfrak{o}_{\mathfrak{p}}$ of the specialization induced by $\mathscr{P}$ in $\mathfrak{o}$, and it is clear that the center of $\mathscr{P}$ in $\mathfrak{o}_{\mathfrak{p}}$ is the maximal ideal $\mathfrak{p}\mathfrak{o}_{\mathfrak{p}}$ in $\mathfrak{o}_{\mathfrak{p}}$. Conversely, if $\mathfrak{p}$ is a prime ideal in $\mathfrak{o}$, different from $\mathfrak{o}$, and if $\mathscr{P}$ is a place of $K$ such that (1) $\mathscr{P}$ is finite on $\mathfrak{o}_{\mathfrak{p}}$ and (2) the center of $\mathscr{P}$ in $\mathfrak{o}_{\mathfrak{p}}$ is the maximal ideal $\mathfrak{m}$ in $\mathfrak{o}_{\mathfrak{p}}$, then $\mathscr{P}$ is also finite on $\mathfrak{o}$ and has center $\mathfrak{p}$ in $\mathfrak{o}$ (since $\mathfrak{m} \cap \mathfrak{o} = \mathfrak{p}$). Note that condition (1) by itself is only equivalent to the following condition: $\mathscr{P}$ *is finite on* $\mathfrak{o}$ *and its center in* $\mathfrak{o}$ *is contained in* $\mathfrak{p}$.

Isomorphic places have the same center in any ring $\mathfrak{o}$ on which they are finite. On the other hand, if we have two places $\mathscr{P}$ and $\mathscr{Q}$ such that $\mathscr{Q}$ is a specialization of $\mathscr{P}$, then if $\mathscr{Q}$ is finite on $\mathfrak{o}$ also $\mathscr{P}$ is finite on $\mathfrak{o}$ (since $K_{\mathscr{P}} \supset K_{\mathscr{Q}}$) and the center of $\mathscr{P}$ in $\mathfrak{o}$ is contained in the center of $\mathscr{Q}$ in $\mathfrak{o}$ (for $\mathfrak{M}_{\mathscr{P}} \subset \mathfrak{M}_{\mathscr{Q}}$).

Theorem 5 (§ 4) said that any prime ideal (different from (1)) in a subring $\mathfrak{o}$ of a field $K$ is the center in $\mathfrak{o}$ of a place of $K$. A more precise result can be proved:

THEOREM 7. *Let* $\mathfrak{o}$ *be a subring of a field* $K$, $\mathfrak{p}$ *and* $\mathfrak{q}$ *two prime ideals in* $\mathfrak{o}$ *such that* $\mathfrak{p} \subset \mathfrak{q}$. *Suppose that* $\mathscr{P}$ *is a place of* $K$ *with center* $\mathfrak{p}$ *in* $\mathfrak{o}$. *Then there exists a place* $\mathscr{Q}$ *of* $K$ *which is a specialization of* $\mathscr{P}$ *and which admits* $\mathfrak{q}$ *as a center in* $\mathfrak{o}$.

PROOF. Without loss of generality we may assume that $K_{\mathscr{P}}/\mathfrak{M}_{\mathscr{P}}$ is the residue field of $\mathscr{P}$. Consider now the subring $\mathfrak{o}/\mathfrak{p}$ of the residue field $K_{\mathscr{P}}/\mathfrak{M}_{\mathscr{P}}$ of $\mathscr{P}$, the prime ideal $\mathfrak{q}/\mathfrak{p}$ of $\mathfrak{o}/\mathfrak{p}$, and the canonical homomorphism of $\mathfrak{o}/\mathfrak{p}$ onto $(\mathfrak{o}/\mathfrak{p})/(\mathfrak{q}/\mathfrak{p})$. By Theorem 5′ (§ 4), this homomorphism can be extended to a place $\mathscr{R}$ of the field $K_{\mathscr{P}}/\mathfrak{M}_{\mathscr{P}}$. The product $\mathscr{Q}=\mathscr{P}\mathscr{R}$ is then a place of $K$. Its valuation ring contains $\mathfrak{o}$, and its center on $\mathfrak{o}$ is obviously $\mathfrak{q}$.

COROLLARY. *Let* $\mathfrak{O}$ *be an integral domain,* $\mathfrak{o}$ *a subring of* $\mathfrak{O}$ *over which* $\mathfrak{O}$ *is integral,* $\mathfrak{P}$ *a prime ideal in* $\mathfrak{O}$, $\mathfrak{p}$ *the prime ideal* $\mathfrak{P} \cap \mathfrak{o}$, *and* $\mathfrak{q}$ *a prime ideal in* $\mathfrak{o}$ *containing* $\mathfrak{p}$. *Then there exists a prime ideal* $\mathfrak{Q}$ *in* $\mathfrak{O}$ *containing* $\mathfrak{P}$ *and such that* $\mathfrak{Q} \cap \mathfrak{o} = \mathfrak{q}$.

For, let $K$ be a field containing $\mathfrak{O}$. There exists a place $\mathscr{P}$ of $K$ with center $\mathfrak{P}$ in $\mathfrak{O}$. Then the center of $\mathscr{P}$ in $\mathfrak{o}$ is $\mathfrak{p}=\mathfrak{o} \cap \mathfrak{P}$. Theorem 7 shows the existence of a specialization $\mathscr{Q}$ of $\mathscr{P}$ with center $\mathfrak{q}$ in $\mathfrak{o}$. Since $\mathfrak{O}$ is integral over $\mathfrak{o}$, the valuation ring of $\mathscr{Q}$ contains $\mathfrak{O}$. Thus $\mathscr{Q}$ admits a center $\mathfrak{Q}$ in $\mathfrak{D}$, and this center is a prime ideal containing $\mathfrak{P}$. Furthermore, we have $\mathfrak{Q} \cap \mathfrak{o} = \mathfrak{q}$, since $\mathfrak{q}$ is the center of $\mathcal{Q}$ in $\mathfrak{o}$.

REMARK. This corollary has already been proved in Vol. I, Ch. V, p. 259, without any assumption on zero divisors.

The places $\mathcal{P}$ of a field $K$ which have given center $\mathfrak{p}$ in a given subring $\mathfrak{o}$ of $K$ are among the places of $K$ whose valuation ring contains the quotient ring $\mathfrak{o}_\mathfrak{p}$, but they are those which satisfy the additional condition $\mathfrak{M}_\mathcal{P} \cap \mathfrak{o}_\mathfrak{p} = \mathfrak{p}\mathfrak{o}_\mathfrak{p}$. By Theorem 6, § 4, we know that the integral closure of $\mathfrak{o}_\mathfrak{p}$ in $K$ is the intersection of all the valuation rings $K_\mathcal{P}$ which contain $\mathfrak{o}_\mathfrak{p}$. We shall now prove the following stronger result.

THEOREM 8. *Let $\mathfrak{o}$ be an arbitrary subring of a field $K$ and let $\mathfrak{p}$ be a given prime ideal in $\mathfrak{o}$, different from $\mathfrak{o}$. Let $\mathfrak{D}$ be the quotient ring of $\mathfrak{o}$ with respect to $\mathfrak{p}$. If $N$ denotes the set of all valuation rings $R$ in $K$ which belong to places $\mathcal{P}$ of $K$ having center $\mathfrak{p}$ in $\mathfrak{o}$, then*

$$\bigcap_{R \in N} R = \text{integral closure of } \mathfrak{D} \text{ in } K.$$

PROOF. It will be sufficient to show that every valuation ring $S$ in $K$ which contains $\mathfrak{D}$ contains as subset some member of $N$. Let $\mathcal{Q}$ be a place of $K$ such that $S = K_\mathcal{Q}$ and let $\mathfrak{M}_\mathcal{Q} \cap \mathfrak{o} = \mathfrak{q}$, where $\mathfrak{q}$ is a prime ideal in $\mathfrak{o}$. Since $S \supset \mathfrak{D}$, $\mathfrak{q}$ is the contraction of some prime ideal in $\mathfrak{D}$ (namely of $\mathfrak{M}_\mathcal{Q} \cap \mathfrak{D}$), and hence $\mathfrak{q} \subset \mathfrak{p}$. By Theorem 7 (where $\mathfrak{q}$ and $\mathfrak{p}$ have now to be interchanged) there exists a place $\mathcal{P}$ of $K$ which is a specialization of $\mathcal{Q}$ and admits $\mathfrak{p}$ as center in $\mathfrak{o}$. Then $K_\mathcal{P} \subset S$, and since $K_\mathcal{P} \in N$, the proof is complete.

COROLLARY. *If $\mathfrak{o}$ is integrally closed in $K$, then* $\bigcap_{R \in N} R = \mathfrak{o}_\mathfrak{p}$.

For in that case also $\mathfrak{o}_\mathfrak{p}$ is integrally closed in $K$.

As an application of the notion of the center of a place we shall now give a complete answer to the following question: given a Dedekind domain $R$, find all the places of the quotient field of $R$ which are finite on $R$.

THEOREM 9. *Let $R$ be a Dedekind domain, $K$ its quotient field. The non-trivial places of $K$ which are finite on $R$ are the $\mathfrak{p}$-adic places of $R$ (see* § 2, Example 2) *and these places are all of rank* 1.

PROOF. Let $\mathcal{P}$ be a non-trivial place of $K$ which is finite on $R$. Since $\mathcal{P}$ is non-trivial, and since $K$ is the quotient field of $R$, the center of $\mathcal{P}$ in $R$ is a proper prime ideal $\mathfrak{p}$. The valuation ring of $\mathcal{P}$ contains the quotient ring $R_\mathfrak{p}$. In order to show that these two rings are equal, we need only prove that $R_\mathfrak{p}$ is a *maximal subring* of $K$, and this will prove Theorem 9.

It has been proved (Vol. I, Ch. V, § 6, Theorem 15) that there exists an element $m$ of $R_{\mathfrak{p}}$ such that every element of $R_{\mathfrak{p}}$ may be written as $um^q$ where $u$ is a unit in $R_{\mathfrak{p}}$ and $q$ a non-negative integer. It follows, upon division, that every element of $K$ may also be written under the form $vm^s$, where $v$ is a unit in $R_{\mathfrak{p}}$ and $s$ an integer. Let $S$ be a subring of $K$ properly containing $R_{\mathfrak{p}}$. Then $S$ contains some element $vm^s$, with $s<0$. Thus, since $S$ contains $R_{\mathfrak{p}}$, it contains $m^{-1}=(m^{-s-1}v^{-1})(vm^s)$; hence $S$ contains $m^{-n}$ for every integer $n$, and therefore also every element $um^q$ ($u$ a unit in $R_{\mathfrak{p}}$, $q$—any integer). It follows that $S=K$. Q.E.D.

COROLLARY 1. *The only non-trivial places of the field of rational numbers are the p-adic ones ($p$, a prime number).*

In fact, the valuation ring of such a place must contain the ring $J$ of ordinary integers.

COROLLARY 2. *Let $k$ be a field, and $K=k(X)$ the field of rational functions in one indeterminate $X$ over $k$. The non-trivial places of $K/k$ are:*

(a) *The $p(X)$-adic places ($p(X)$, an irreducible polynomial in $k[X]$).*
(b) *The place $\mathscr{R}$ whose valuation ring consists of all fractions $a(X)/b(X)$ ($a$, $b$: polynomials) such that $\partial a \leqq \partial b$.*

(Equivalent places may be obtained by replacing in the rational functions $f(X)$ either

(a) $X$ by a root of the irreducible polynomial $p(X)$ or
(b) $1/X$ by 0.)

Let $\mathscr{R}$ be a non-trivial place of $K/k$. If its valuation ring $K_{\mathscr{R}}$ contains $X$, it contains $k[X]$, and we are in case (a). Otherwise $1/X$ is in $K_{\mathscr{R}}$, and is a non-unit in this ring. Thus $K_{\mathscr{R}}$ contains the polynomial ring $k[1/X]$, and the center of $\mathscr{R}$ in this ring must be a prime ideal containing $1/X$, i.e., it must be the principal ideal $(1/X)$. Then the valuation ring of $\mathscr{R}$ consists of all fractions $a'(1/X)/b'(1/X)$ ($a'$, $b'$: polynomials over $k$) such that $b'(0)\neq 0$. The verification of the fact that this is the valuation ring described in (b) may be left to the reader.

REMARK. The last corollary expresses the fact that the non-trivial places of $k(X)/k$ correspond to the elements of the algebraic closure $\bar{k}$ of $k$ (more precisely to the classes of conjugate elements of $\bar{k}$) and to the symbol $\infty$: the value of the rational function $f(X)$ at the place $\mathscr{R}$ corresponding to $x$ in $\bar{k}$ (to $\infty$) being $f(x)$ ($f(\infty)$). Notice that all these places have dimension 0 and rank 1, and that their valuation rings are quotient rings of polynomial rings. The places of $K/k$, where $K$ is a field of rational functions in several variables over $k$, are of more complicated types (see § 15).

COROLLARY 3. *An integrally closed local domain $R$ in which the ideal of non-units is the only proper prime ideal is the valuation ring of a place of rank* 1.

For, $R$ is a Dedekind domain (Vol. I, Ch. V, § 6, Theorem 13), and if $\mathfrak{p}$ is the ideal of non-units in $R$ then $R=R_{\mathfrak{p}}$. Note that $R$ is a *discrete* valuation ring of rank 1 (in the sense of Vol. I, Ch. V, end of § 6, p. 278; see also § 10 of this chapter, Theorem 16, Corollary 1).

We shall conclude this section with the derivation of another criterion for a domain to be a valuation ring. Let $\mathfrak{o}$ be an integral domain, $\mathfrak{q}$ a prime ideal in $\mathfrak{o}$, and let $\mathcal{P}$ be a place of the quotient field $K$ of $\mathfrak{o}$ which is finite on $\mathfrak{o}$ and has center $\mathfrak{q}$. Since $\mathfrak{M}_{\mathcal{P}} \cap \mathfrak{o} = \mathfrak{q}$, the integral domain $\mathfrak{o}/\mathfrak{q}$ can be canonically identified with a subring of the residue field $\Delta$ of $\mathcal{P}$. Thus $\Delta$ is an extension of the quotient field $\Delta_0$ of $\mathfrak{o}/\mathfrak{q}$. We shall say that the place $\mathcal{P}$ is of the *first* or *of the second kind*, with respect to $\mathfrak{o}$, according to whether the transcendence degree of $\Delta$ over $\Delta_0$ is zero or positive.

THEOREM 10. *Given an integrally closed integral domain $\mathfrak{o}$ and a prime ideal $\mathfrak{q}$ in $\mathfrak{o}$, $\mathfrak{q} \neq \mathfrak{o}$, a necessary and sufficient condition for the quotient ring $\mathfrak{o}_{\mathfrak{q}}$ to be a valuation ring is that there should not exist a place $\mathcal{P}$ of the quotient field of $\mathfrak{o}$ such that $\mathcal{P}$ has center $\mathfrak{q}$ and is of the second kind with respect to $\mathfrak{o}$.*

For the proof of Theorem 10 we shall first prove a general lemma:

LEMMA. *Let $\mathfrak{o}$ be an integrally closed integral domain, let $K$ be the quotient field of $\mathfrak{o}$ and let $\mathfrak{q}$ be a prime ideal in $\mathfrak{o}$. If an element $t$ of $K$ is a root of a polynomial $f(X)=a_0X^n+a_1X^{n-1}+\cdots+a_n$, where the coefficients $a_i$ are in $\mathfrak{o}$ but not all in $\mathfrak{q}$, then either $t$ or $1/t$ belongs to the quotient ring $\mathfrak{o}_{\mathfrak{q}}$.*

PROOF. The element $1/t$ is a root of the polynomial $a_0+a_1X+\cdots+a_nX^n$. Our assumptions are therefore symmetric in $t$ and $1/t$. There exists a place $\mathcal{P}$ having center $\mathfrak{q}$. We shall show that $t \in \mathfrak{o}_{\mathfrak{q}}$ or $1/t \in \mathfrak{o}_{\mathfrak{q}}$ according as $t\mathcal{P} \neq \infty$ or $(1/t)\mathcal{P} \neq \infty$. Let, say, $t\mathcal{P} \neq \infty$. Let us assume that $a_0, a_1, \cdots, a_{j-1} \in \mathfrak{q}$, $a_j \notin \mathfrak{q}$; here $j$ is some integer such that $0 \leqq j \leqq n$. If $j=0$, then the equation $f(t)=0$, upon division by $a_0$, implies that $t$ is integrally dependent on $\mathfrak{o}_{\mathfrak{q}}$, and hence $t \in \mathfrak{o}_{\mathfrak{q}}$ since $\mathfrak{o}_{\mathfrak{q}}$ is integrally closed (Vol. I, Ch. V, § 3, p. 261). We cannot have $j=n$, for in the contrary case the existence of a place $\mathcal{P}$ having center $\mathfrak{q}$ and such that $t\mathcal{P} \neq \infty$ would imply that $a_n\mathcal{P}=0$, $a_n \in \mathfrak{q}$, a contradiction. We shall therefore assume that $0<j<n$.

Let

$$\xi = a_0t^j+a_1t^{j-1}+\cdots+a_{j-1}t+a_j$$
$$\eta = a_{j+1}+a_{j+2}/t+\cdots+a_n/t^{n-j-1}.$$

Let $\mathscr{P}$ be any place which is finite on $\mathfrak{o}$. If $t\mathscr{P} \neq \infty$, then also $\xi\mathscr{P} \neq \infty$, and also $\eta\mathscr{P} \neq \infty$ since $\xi t + \eta = 0$. If $t\mathscr{P} = \infty$, then $\eta\mathscr{P} \neq \infty$, and since $\xi + \eta/t = 0$, it follows that $\xi\mathscr{P} = 0$. Hence, in all cases we have $\xi\mathscr{P} \neq \infty$ and $\eta\mathscr{P} \neq \infty$. Since this holds for all places which are finite on $\mathfrak{o}$, it follows that *the elements* $\xi$ *and* $\eta$ *both belong to* $\mathfrak{o}$. Now, by assumption, there exists a place $\mathscr{P}$ having center $\mathfrak{q}$ and such that $t\mathscr{P} \neq \infty$. For such a place $\mathscr{P}$ we will have $\xi\mathscr{P} \neq 0$ since $a_i\mathscr{P} = 0$, $i = 0, 1, \cdots, j=1$, and $a_j\mathscr{P} \neq 0$ (in view of the assumption made on the coefficients $a_0, a_1, \ldots, a_j$). Therefore the element $\xi$ of $\mathfrak{o}$ does not belong to $\mathfrak{q}$, and consequently $t = -\eta/\xi \in \mathfrak{o}_{\mathfrak{q}}$. This completes the proof of the lemma.

We note the following consequence of the lemma:

COROLLARY. *Let* $\mathfrak{o}$ *be an integrally closed integral domain, let* $K$ *be the quotient field of* $\mathfrak{o}$ *and let* $\mathfrak{q}$ *be a prime ideal in* $\mathfrak{o}$. *If an element* $t$ *of* $K$ *is such that neither* $t$ *nor* $1/t$ *belongs to the quotient ring* $\mathfrak{o}_{\mathfrak{q}}$ *and if* $\tilde{\mathfrak{o}}$ *denotes the ring* $\mathfrak{o}[t]$, *then the extended ideal* $\tilde{\mathfrak{q}} = \tilde{\mathfrak{o}}\mathfrak{q}$ *is prime, the contracted ideal* $\tilde{\mathfrak{q}} \cap \mathfrak{o}$ *coincides with* $\mathfrak{q}$, *and the* $\tilde{\mathfrak{q}}$*-residue of* $t$ *is transcendental over* $\mathfrak{o}/\mathfrak{q}$.

For, $\tilde{\mathfrak{o}}\mathfrak{q}$ consists of all elements of the form $\pi_0 t^n + \pi_1 t^{n-1} + \cdots + \pi_n$, $\pi_i \in \mathfrak{q}$, $n$ an arbitrary integer $\geqq 0$. If $\pi_0 t^n + \pi_1 t^{n-1} + \cdots + \pi_n = a \in \mathfrak{o}$, then it follows from the lemma that $a \in \mathfrak{q}$, showing that $\tilde{\mathfrak{o}}\mathfrak{q} \cap \mathfrak{o} = \mathfrak{q}$. Hence the integral domain $\mathfrak{o}/\mathfrak{q}$ can be regarded as a subring of $\tilde{\mathfrak{o}}/\tilde{\mathfrak{q}}$. If we have a relation of the form $\xi_0 \bar{t}^n + \xi_1 \bar{t}^{n-1} + \cdots + \xi_n = 0$, where $\xi_i \in \mathfrak{o}/\mathfrak{q}$ and $\bar{t}$ is the $\tilde{\mathfrak{q}}$-residue of $t$, and if we fix an element $a_i$ in $\mathfrak{o}$ such that $\xi_i$ is the $\mathfrak{q}$-residue of $a_i$, then $a_0 t^n + a_1 t^{n-1} + \cdots + a_n \in \tilde{\mathfrak{q}}$, i.e., there must exist elements $\pi'_1, \pi'_2, \cdots, \pi'_h, \pi_0, \pi_1, \cdots, \pi_n$ in $\mathfrak{q}$ such that $\sum_{i=1}^{h} \pi'_i t^{n+i} + (a_0 - \pi_0)t^n + (a_1 - \pi_1)t^{n-1} + \cdots + (a_n - \pi_n) = 0$. Therefore, by the lemma, we must have $a_i - \pi_i \in \mathfrak{q}$, $\bar{a}_i = \xi_i = 0$, showing that $\bar{t}$ is transcendental over $\mathfrak{o}/\mathfrak{q}$. Hence $\mathfrak{o}/\mathfrak{q}[\bar{t}]$ is an integral domain, and since this ring is the residue class ring $\tilde{\mathfrak{o}}/\tilde{\mathfrak{q}}$, it follows that $\tilde{\mathfrak{q}}$ is a prime ideal.

[In terms of dimension theory: $\dim \tilde{\mathfrak{q}} = 1 + \dim \mathfrak{q}$.]

The proof of Theorem 10 is now immediate. The necessity of the condition is obvious, for if $\mathfrak{o}_{\mathfrak{q}}$ is a valuation ring, any place $\mathscr{P}$ which is finite on $\mathfrak{o}$ and has center $\mathfrak{q}$ necessarily has $\mathfrak{o}_{\mathfrak{q}}$ as valuation ring, and thus the residue field of $\mathscr{P}$ coincides (to within an isomorphism) with the quotient field of $\mathfrak{o}/\mathfrak{q}$. To prove the sufficiency of the condition, we assume that $\mathfrak{o}_{\mathfrak{q}}$ is not a valuation ring and we show that there exists a place $\mathscr{P}$ of $K$ which has center $\mathfrak{q}$ and is of second kind with respect to $\mathfrak{o}$. For this purpose, we consider an element $t$ of $K$ such that neither $t$ nor $1/t$ belongs to $\mathfrak{o}_{\mathfrak{q}}$ (such an element exists since $\mathfrak{o}_{\mathfrak{q}}$ is not a valuation ring) and we pass to the ring $\tilde{\mathfrak{o}} = \mathfrak{o}[t]$ and to the ideal $\tilde{\mathfrak{q}} = \tilde{\mathfrak{o}}\mathfrak{q}$. By the above corollary, $\tilde{\mathfrak{q}}$ is a prime ideal, different from $\tilde{\mathfrak{o}}$. Let $\mathscr{P}$ be a place of $K$

which is finite on $\tilde{\mathfrak{o}}$ and has center $\tilde{\mathfrak{q}}$ in $\tilde{\mathfrak{o}}$. Then it follows from the corollary that the center of $\mathcal{P}$ in $\mathfrak{o}$ is $\mathfrak{q}$ and that $\mathcal{P}$ is of the second kind with respect to $\mathfrak{o}$ (since the residue field of $\mathcal{P}$ contains $\tilde{\mathfrak{o}}/\tilde{\mathfrak{q}}$).

The following consequence of Theorem 10 has been useful in the geometric applications of valuation theory:

COROLLARY OF THEOREM 10. *Let* $\{\mathfrak{o}_\alpha\}$, $\alpha \in A$, *be a collection of subrings of a field K, integrally closed in K and indexed by a set A, and let for each* $\mathfrak{o}_\alpha$ *a proper prime ideal* $\mathfrak{q}_\alpha$ *in* $\mathfrak{o}_\alpha$ *be given. Assume that the following conditions are satisfied:* (a) *if* $\mathfrak{o}_\alpha \subset \mathfrak{o}_\beta$ *then* $\mathfrak{q}_\beta \cap \mathfrak{o}_\alpha = \mathfrak{q}_\alpha$; (b) *for any two rings* $\mathfrak{o}_\alpha$, $\mathfrak{o}_\beta$ $(\alpha, \beta \in A)$ *there exists a third ring* $\mathfrak{o}_\gamma$ *in the collection such that* $\mathfrak{o}_\alpha \subset \mathfrak{o}_\gamma$ *and* $\mathfrak{o}_\beta \subset \mathfrak{o}_\gamma$. *Let* $\mathfrak{O} = \bigcup_{\alpha \in A} \mathfrak{o}_\alpha$, $\mathfrak{Q} = \bigcup_{\alpha \in A} \mathfrak{q}_\alpha$. *Then* $\mathfrak{O}_\mathfrak{Q}$ *is a valuation ring if and only if there does not exist a place* $\mathcal{P}$ *of K which satisfies, for each* $\alpha$, *the following conditions:* $\mathcal{P}$ *has center* $\mathfrak{q}_\alpha$ *in* $\mathfrak{o}_\alpha$ *and is of the second kind with respect to* $\mathfrak{o}_\alpha$.

From condition (b) it follows that $\mathfrak{O}$ is a ring, integrally closed in $K$, and (a) implies that the set $\mathfrak{Q}$ is a proper prime ideal in $\mathfrak{O}$. Any place $\mathcal{P}$ of $K$ which has center $\mathfrak{Q}$ in $\mathfrak{O}$ has center $\mathfrak{q}_\alpha$ in $\mathfrak{o}_\alpha$ for each $\alpha \in A$; and conversely. The residue class ring $\mathfrak{O}/\mathfrak{Q}$ can be regarded, canonically, as the union of the rings $\mathfrak{o}_\alpha/\mathfrak{q}_\alpha$. It follows that a place $\mathcal{P}$ of $K$ which has center $\mathfrak{Q}$ in $\mathfrak{O}$ is of the second kind with respect to $\mathfrak{O}$ if and only if $\mathcal{P}$ is of the second kind with respect to each of the rings $\mathfrak{o}_\alpha$, and the corollary now follows from Theorem 10.

**§ 5bis. The notion of the center of a place in algebraic geometry.** The concept of center of a place has been first introduced in algebraic geometry, and in fact the theorems given in the preceding section are merely generalizations of similar theorems concerning algebraic varieties. We shall briefly review here the algebro-geometric background of the material presented in the preceding section. For further details, see Chapter VII, § 3.

If $K$ is a field, *the n-dimensional affine space* $A_n^K$ *over* $K$ is the set of all points $(z_1, z_2, \cdots, z_n)$ (i.e., ordered $n$-tuples) whose (non-homogeneous) coördinates $z_1, z_2, \cdots, z_n$ are elements of $K$. We now assume that $K$ is an algebraically closed field and that it contains a ground field $k$. If $\mathfrak{A}$ is an ideal in the polynomial ring $k[X_1, X_2, \cdots, X_n]$ $(=k[X])$ in $n$ indeterminates, with coefficients in the ground field $k$, *the variety of* $\mathfrak{A}$ is the set of all points $(z)(=(z_1, z_2, \cdots, z_n))$ in $A_n^K$ such that $f(z)=0$ for every polynomial $f(X)$ in $\mathfrak{A}$. An *algebraic affine variety in* $A_n^K$ *(defined over k)* is any subset of $A_n^K$ which is the variety of some ideal in $k[X]$. If $V$ is a variety in $A_n^K$, defined over $k$, the polynomials in $k[X]$ which vanish at *all* points of $V$ obviously form an ideal. This ideal, called *the ideal of the variety* $V$, is the greatest ideal in $k[X]$ whose variety is $V$. It is clear that the ideal of a variety $V$ coincides with its own radical and is therefore (see Vol. I, Ch. IV, § 4, Theorem 5) an intersection of prime ideals. If the ideal of $V$ is itself a prime ideal, then $V$ is said to be *irreducible* (over $k$) (cf. Ch. VII, § 3).

Let $V$ be an affine variety in $A_n^K$, defined and irreducible over the ground field $k$, and let $\mathfrak{p}$ be the prime ideal of $V$ in $k[X]$. The residue class ring $k[X]/\mathfrak{p}$ is called *the coördinate ring of* $V$. We shall denote this ring by $k[V]$. If $x_i$ denotes the $\mathfrak{p}$-residue of $X_i$, then $k[V]= k[x_1, x_2, \cdots, x_n]\ (=k[x])$. The point $(x_1, x_2, \cdots, x_n)$ is called a *general point* of $V$ over $k$. The quotient field $k(x)$ of $k[x]$ is called *the function field* of $V$, over $k$, and will be denoted by $k(V)$. The *dimension* $r$ *of* $V$ is the transcendence degree of $k(V)$ over $k$. We have of course $0 \leqq r \leqq n$.

Since the $\mathfrak{p}$-residues $x_i$ of the $X_i$ are not generally elements of $K$, the general point $(x)$ is not always actually a point of the space $A_n^K$ However, if $K$ has transcendence degree $\geqq r$ over $k$, there always exist $k$-isomorphisms of $k(V)$ into $K$ (since $K$ is algebraically closed). If $\tau$ is one such isomorphism, and if $x_i\tau = z_i$, then also the point $(z_1, z_2, \cdots, z_n)$ of $A_n^K$ is called a general point of $V$ over $k$. It is now a standard procedure in algebraic geometry to assume once and for all an algebraically closed field $K$ *which has infinite transcendence degree over* $k$ (a so-called *universal domain* $K$). This guarantees that *any* irreducible variety $V$, over $k$, in $A_n^K$ (*$n$ arbitrary*) carries general points (which are actually points of the affine space $A_n^K$).

Let $\mathscr{P}$ be a place of $k(V)/k$ *such that the residue field of* $\mathscr{P}$ *is contained in* $K$ (which is not a serious restriction on $\mathscr{P}$, at least if $K$ is a universal domain, for in that case every place of $k(V)/k$ is isomorphic to a place $\mathscr{P}$ satisfying the above condition). *If* $\mathscr{P}$ *is finite on the coördinate ring* and if, say, $x_i\mathscr{P} = z_i$ $(z_i \in K)$, then the point $(z)$ is called *the center of the place* $\mathscr{P}$ *on* $V$. (It is obvious that $(z)$ is indeed a point of $V$, for if a polynomial $f(X)$ belongs to the ideal of $V$ then $f(x)=0$ and hence $f(z)=f(x)\mathscr{P}=0$.) The elements $g(x)$ of $k[V]$ which vanish at the point $(z)$ form a prime ideal $\mathfrak{p}$, *the prime ideal of* $(z)$ *in* $k[V]$. We have $g(x) \in \mathfrak{p}$ if and only if $g(x)\mathscr{P}=0$, i.e., if and only if $g(x) \in \mathfrak{M}_{\mathscr{P}}$. Hence *the prime ideal of the center of* $\mathscr{P}$ *on the variety* $V$ *is merely the center of* $\mathscr{P}$ *in the coördinate ring* $k[V]$ of $V$.

By *the dimension of a point* $P=(z_1, z_2, \cdots, z_n)$, over $k$ (in symbols: dim $P/k$, or dim $(z)/k$) we mean the transcendence degree of $k(z)$ over $k$. Two points $(z)$ and $(z')$ in $A_n^K$ are said to be *$k$-isomorphic* if there exists a $k$-isomorphism $\tau$ of the field $k(z)$ onto the field $k(z')$ such that $z_i\tau = z'_i$, $1 \leqq i \leqq n$. For instance, any two general points of our irreducible variety $V$, over $k$, are $k$-isomorphic, and any general point of $V$, over $k$, has dimension $r$ over $k$, where $r = \dim V$. We now list some of the properties of the center of a place on $V$. (We remind the reader that a place $\mathscr{P}$ of $k(V)$ admits a center on an affine variety if and only if $\mathscr{P}$ is finite on $k[V]$.)

PROPERTY 1. A place $\mathscr{P}$ of $k(V)/k$ is trivial if and only if its center on $V$ is a general point of $V$ over $k$.

The proof is straightforward and may be left to the reader.

PROPERTY 2. If $Q$ is the center on $V$ of a place $\mathscr{P}$ of $k(V)/k$ then $\dim Q/k \leqq \dim \mathscr{P}/k \leqq \dim V$, and $\mathscr{P}$ is trivial if and only if $\dim \mathscr{P}/k = \dim V$.

Obvious.

Given two points $Q = (z_1, z_2, \cdots, z_n)$ and $Q' = (z_1', z_2', \cdots, z_n')$ in $A_n^K$, $Q'$ is said to be *a specialization of $Q$ over $k$* if there exists a specialization $\varphi$ of the ring $k[z]$ onto the ring $k[z']$ such that $\varphi$ is the identity on $k$ and $z_i\varphi = z'_i$. Notation: $Q \xrightarrow{k} Q'$. If $Q \xrightarrow{k} Q'$ then $\dim Q'/k \leqq \dim Q/k$. If we have both $Q \xrightarrow{k} Q'$ and $Q' \xrightarrow{k} Q$, then $Q$ and $Q'$ are $k$-isomorphic points, and conversely. If $Q \xrightarrow{k} Q'$ and $\dim Q'/k = \dim Q/k$, then again $Q$ and $Q'$ are $k$-isomorphic points, for any proper $k$-homomorphism of the integral domain $k[z]$ lowers the transcendence degree of the domain. (See Vol. I, Ch. II, § 12, Theorem 29).

PROPERTY 3. *Let $\mathscr{P}$ and $\mathscr{Q}$ be places of $k(V)/k$ and let $P$ and $Q$ be their respective centers on $V$. If $\mathscr{P} \xrightarrow{k} \mathscr{Q}$ then also $P \xrightarrow{k} Q$.*

Obvious.

PROPERTY 4. *Let $P$ and $Q$ be points of $V$ such that $P \xrightarrow{k} Q$. Suppose that $\mathscr{P}$ is a place of $k(V)/k$ which admits $P$ as center on $V$. Then there exists a place $\mathscr{Q}$ of $k(V)/k$ which is a specialization of $\mathscr{P}$ over $k$ and has center $Q$ on $V$.*

This is the analogue of Theorem 7, § 5, and the proof is the same.

If $Q$ is a point of V and $\mathfrak{p}$ is the prime ideal of $Q$ in the coördinate ring $k[V]$, then the quotient ring of $k[V]$ with respect to $\mathfrak{p}$ is called *the local ring of $V$ at $Q$* (or also briefly: *the local ring of $Q$* (on $V$)). This ring shall be denoted by $\mathfrak{o}(Q; V)$, and the maximal ideal in that ring shall be denoted by $\mathfrak{m}(Q; V)$.

PROPERTY 5. *If $Q$ is the center on $V$ of a place $\mathscr{P}$ of $k(V)/k$ then $\mathfrak{o}(Q; V) \subset k(V)_{\mathscr{P}}$ and $\mathfrak{m}(Q; V) = \mathfrak{M}_{\mathscr{P}} \cap \mathfrak{o}(Q; V)$. Conversely, if these two conditions are satisfied for a given point $Q$ on $V$ and a given place $\mathscr{P}$ of $k(V)$, then the center of $\mathscr{P}$ on $V$ is a point $k$-isomorphic to $Q$. If only* *condition* $\mathfrak{o}(Q; V) \subset k(V)_{\mathscr{P}}$ *is satisfied, then* $Q$ *is a specialization, over* $k$, *of the center of* $\mathscr{P}$ *on* $V$.

Obvious.

It follows that every point $Q$ of $V$ is the center of some place of $k(V)/k$.

PROPERTY 6. *If* $Q$ *is a point of* $V$ *then the integral closure of* $\mathfrak{o}(Q; V)$ *is the intersection of all the valuation rings which belong to places* $\mathscr{P}$ *of* $k(V)/k$ *having center* $Q$ *on* $V$.

This is a particular case of Theorem 8, § 5.

To be able to speak of the center of a place $\mathscr{P}$ of $k(V)/k$ also in the case in which $\mathscr{P}$ is not finite on $k[V]$, it is only necessary to adjoin to $V$ its points at infinity and to consider thus the enlarged *projective* variety $V^\star$. We shall discuss this question later in the next chapter (see Ch. VII, § $4^{\text{bis}}$). At this stage it will suffice to say that if $V$ is regarded as a variety in the projective $n$-space, then every place of $k(V)$ has a well-defined center on $V$. This is important, since it allows one to introduce the concept of a birational correspondence in a purely valuation-theoretic fashion. Two irreducible varieties $V$ and $V'$, over $k$, are birationally equivalent if their function fields $k(V)$ and $k(V')$ are $k$-isomorphic. In that case, after fixing a definite $k$-isomorphism between $k(V)$ and $k(V')$, we may identify these two fields. Assuming therefore that $k(V)=k(V')$, we can set up a correspondence $T$ between the points of $V$ and $V'$ in the following fashion: a point $Q$ of $V$ and a point $Q'$ of $V'$ are corresponding points if there exists a place $\mathscr{P}$ of $k(V)(=k(V'))$ whose center on $V$ is $Q$ and whose center on $V'$ is $Q'$. Such a correspondence $T$ is called a *birational correspondence*. The fact that every point of $V$ is the center of at least one place guarantees that in a birational correspondence between two birationally equivalent varieties to every point of one variety corresponds *at least* one point of the other variety.

§ **6. Places and field extensions.** Let $K$ be a field and $K^\star$ an overfield of $K$. It follows easily from our definition of a place that if $\mathscr{P}^\star$ is a place of $K^\star$ then the restriction of $\mathscr{P}^\star$ to $K$ is a place of $K$. If $\mathscr{P}$ and $\mathscr{P}^\star$ are places of $K$ and $K^\star$ respectively, we say that $\mathscr{P}^\star$ is an *extension* of $\mathscr{P}$ if $\mathscr{P}$ is the restriction of $\mathscr{P}^\star$ to $K$. Our object in this section is to study the extensions in $K^\star$ of a given place $\mathscr{P}$ of $K$.

LEMMA 1. *If* $\mathscr{P}^\star$ *is an extension of* $\mathscr{P}$, *then* $K^\star_{\mathscr{P}^\star} \cap K = K_{\mathscr{P}}$. *Conversely, if this last relation holds for given places* $\mathscr{P}$ *and* $\mathscr{P}^\star$ *of* $K$ *and* $K^\star$ *respectively, then there exists an extension* $\mathscr{P}_1^\star$ *of* $\mathscr{P}$ *which is isomorphic to* $\mathscr{P}^\star$. *The relation* $K^\star_{\mathscr{P}^\star} \cap K = K_{\mathscr{P}}$ *implies* $\mathfrak{M}_{\mathscr{P}^\star} \cap K = \mathfrak{M}_{\mathscr{P}}$ *and is equivalent to "*$K^\star_{\mathscr{P}^\star} \supset K_{\mathscr{P}}$ *and* $\mathfrak{M}_{\mathscr{P}^\star} \supset \mathfrak{M}_{\mathscr{P}}$."

PROOF. The first part of the lemma is self-evident. Assume now that $K^{\star}_{\mathscr{P}^{\star}} \cap K = K_{\mathscr{P}}$, and let $\mathscr{P}_1$ be the restriction of $\mathscr{P}^{\star}$ to $K$. Then $K_{\mathscr{P}_1} = K_{\mathscr{P}}$, and hence $\mathscr{P}$ and $\mathscr{P}_1$ are isomorphic places of $K$. Hence $\mathscr{P} = \mathscr{P}_1 f$, where $f$ is an isomorphism of the residue field $\Delta_1$ of $\mathscr{P}_1$ onto the residue field $\Delta$ of $\mathscr{P}$. Extend $f$ to an isomorphism $f^{\star}$ of the residue field of $\mathscr{P}^{\star}$ and set $\mathscr{P}_1{}^{\star} = \mathscr{P}^{\star} f^{\star}$. Then $\mathscr{P}^{\star}$ and $\mathscr{P}_1{}^{\star}$ are isomorphic places, and $\mathscr{P}_1{}^{\star}$ is an extension of $\mathscr{P}$, which proves the second part of the lemma. Furthermore, it is clear that $\mathfrak{M}_{\mathscr{P}^{\star}} \cap K = \mathfrak{M}_{\mathscr{P}}$, and this proves one half of the last part of the lemma. Assume now that we have $K^{\star}_{\mathscr{P}^{\star}} \supset K_{\mathscr{P}}$ and $\mathfrak{M}_{\mathscr{P}^{\star}} \supset \mathfrak{M}_{\mathscr{P}}$ for two given places $\mathscr{P}$ and $\mathscr{P}^{\star}$ of $K$ and $K^{\star}$ respectively. If $x$ is any element of $K$, not in $K_{\mathscr{P}}$, then $1/x$ belongs to $\mathfrak{M}_{\mathscr{P}}$, hence $1/x \in \mathfrak{M}_{\mathscr{P}^{\star}}$, and therefore $x \notin K^{\star}_{\mathscr{P}^{\star}}$. This completes the proof of the lemma.

Note in particular the case in which $\mathscr{P}$ is a trivial place of $K$ ($\mathscr{P}$ = an isomorphism of $K$). If $\mathscr{P}$ is the identity automorphism of $K$, then the extensions of $\mathscr{P}$ to $K^{\star}$ are the places of $K^{\star}/K$. It follows from Lemma 1 that if $\mathscr{P}$ is an arbitrary trivial place of $K$, then any extension of $\mathscr{P}$ to $K^{\star}$ is isomorphic with a place of $K^{\star}/K$.

The *existence* of extensions to $K^{\star}$ of any given place $\mathscr{P}$ of $K$ is assured by the extension theorem (Theorem 5′, § 4), where $\mathfrak{o}$, $K$ and $\varphi$ are now to be identified with $K_{\mathscr{P}}$, $K^{\star}$ and $\mathscr{P}$ respectively.

We shall generally denote by $\Delta$ (or by $\Delta^{\star}$) the residue field of a place $\mathscr{P}$ of $K$ (or of a place $\mathscr{P}^{\star}$ of $K^{\star}$). If $\mathscr{P}$ is the restriction of $\mathscr{P}^{\star}$ in $K$, then $\Delta \subset \Delta^{\star}$, and the transcendence degree of $\Delta^{\star}$ over $\Delta$ shall be called the *relative dimension of* $\mathscr{P}^{\star}$ and shall be denoted by $\dim_K \mathscr{P}^{\star}$. In the special case in which $\mathscr{P}^{\star}$ is a place of $K^{\star}/K$, we have $\Delta = K$, and our definition is in agreement with our earlier definition of the dimension of $\mathscr{P}^{\star}/K$.

LEMMA 2. *Let $\mathscr{P}^{\star}$ be a place of $K^{\star}$ and let $\mathscr{P}$ be the restriction of $\mathscr{P}^{\star}$ to $K$. Let $x_1, x_2, \cdots, x_m$ be elements of $K^{\star}_{\mathscr{P}^{\star}}$ and let $\xi_1, \xi_2, \cdots, \xi_m$ be their $\mathscr{P}^{\star}$-values (in $\Delta^{\star}$). If the $x_i$ are linearly dependent over $K$, then the $\xi_i$ are linearly dependent over $\Delta$.*

PROOF. We have, by assumption, a relation of the form $a_1x_1 + a_2x_2 + \cdots + a_mx_m = 0$, where the $a_i$ belong to $K$ and are not all zero. We select a coefficient $a_j$ which satisfies the following conditions: $a_j \neq 0$ and $(a_i/a_j)\mathscr{P} \neq \infty$ for $i = 1, 2, \cdots, m$ (see Theorem 3, Corollary 2, § 3). Dividing the above linear relation by $a_j$ and passing to the $\mathscr{P}^{\star}$-values, we find $u_1\xi_1 + u_2\xi_2 + \cdots + u_m\xi_m = 0$, where $u_i = (a_i/a_j)\mathscr{P} \in \Delta$. Since the $u_i$ are not all zero ($u_j$, for instance, is 1), the lemma is proved.

COROLLARY 1. *The relative dimension of $\mathscr{P}^{\star}$ is not greater than the transcendence degree of $K^{\star}/K$.*

For let $\{\xi_i\}$ be a transcendence basis of $\Delta^\star/\Delta$ and let $x_i$ be an element of $K$ such that $x_i\mathscr{P}^\star = \xi_1$. By assumption, any finite set of monomials in the $\xi_i$ consists of elements which are linearly independent over $\Delta$. Hence, by the above lemma, the corresponding monomials in the $x_i$ are also linearly independent over $K$, i.e., the $x_i$ are algebraically independent over $K$.

COROLLARY 2. *If $K^\star$ is a finite algebraic extension of $K$, of degree $n$, then also $\Delta^\star$ is a finite algebraic extension of $\Delta$, and we have $[\Delta^\star:\Delta] \leqq [K^\star:K]$.*

*The integer $[\Delta^\star:\Delta]$ is called the relative degree of $\mathscr{P}^\star$ with respect to $\mathscr{P}$ (or with respect to $K$).*

THEOREM 11. *For any place $\mathscr{P}$ of $K$ there exist extensions $\mathscr{P}^\star$ in $K^\star$ such that* $\dim_K \mathscr{P}^\star$ *is any preassigned cardinal number* $\geqq 0$ *and* $\leqq$ *transcendence degree of $K^\star/K$.*

PROOF. Let $\{y_j\}$ be a transcendence basis of $K^\star/K$ and let $\{u_j\}$ be a set of indeterminates over $\Delta$, in (1, 1) correspondence with the set $\{y_j\}$. Let $f$ be the (uniquely determined) homomorphism of the polynomial ring $K_\mathscr{P}[\{y_j\}]$ onto the polynomial ring $\Delta[\{u_j\}]$ such that $y_jf = u_j$ and $f = \mathscr{P}$ on $K_\mathscr{P}$. By Theorem 5′, § 4, $f$ can be extended to a place $\mathscr{P}^\star$ of $K^\star$. Then $\mathscr{P}^\star$ is an extension of $\mathscr{P}$, and since the residue field of $\mathscr{P}^\star$ contains the elements $u_j$ it follows that $\dim_K \mathscr{P}^\star$ is greater than or equal to the transcendence degree of $K^\star/K$. It follows by Corollary 1 of the preceding lemma that $\dim_K \mathscr{P}^\star$ *is exactly equal to the transcendence degree of $K^\star/K$.*

We now observe that there also exist extensions $\mathscr{P}^\star$ of $\mathscr{P}$ having relative dimension zero. This follows directly from Theorem 5′, Corollary 3 (§ 4).

To complete the proof of the theorem, let $\alpha$ be any cardinal number between 0 and the transcendence degree of $K^\star/K$. We fix a subset $L = \{x_i\}$ of $K^\star$ which has cardinal number $\alpha$ and which consists of elements which are algebraically independent over $K$. Let $K'$ be the subfield of $K^\star$ which is generated over $K$ by the elements $x_i$ of $L$. Since $K'/K$ has transcendence degree $\alpha$, it follows by the preceding proof that there exists an extension $\mathscr{P}'$ of $\mathscr{P}$ in $K'$ such that the relative dimension of $\mathscr{P}'$ (over $K$) is equal to $\alpha$. Again by the preceding proof, there exists an extension $\mathscr{P}^\star$ of $\mathscr{P}'$ in $K^\star$ whose relative dimension (*over* $K'$) is zero. Then it is clear that $\mathscr{P}^\star$ is an extension of $\mathscr{P}$ and that the relative dimension of $\mathscr{P}^\star$ (over $K$) is equal to $\alpha$. This completes the proof of the theorem.

COROLLARY. *If $K$ is a field of algebraic functions of $r$ independent variables, over a ground field $k$, there exist places of $K/k$ of any dimension $s$, $0 \leqq s \leqq r$.*

This follows from the preceding theorem if we replace $K^\star$ and $K$ by $K$ and $k$ respectively and take for $\mathscr{P}$ the identity automorphism of $k$.

**§ 7. The case of an algebraic field extension.** We shall now study the case in which $K^\star$ is an algebraic extension of $K$. Let $\mathscr{P}$ be a place of $K$ and let $\mathscr{P}^\star$ be an extension of $\mathscr{P}$ to $K^\star$. We denote by $K^\star_{\mathscr{P}}$ the integral closure of $K_{\mathscr{P}}$ in $K^\star$. If we denote by $\mathfrak{P}^\star$ the ideal $\mathfrak{M}_{\mathscr{P}^\star} \cap K^\star_{\mathscr{P}}$, then the contraction of $\mathfrak{P}^\star$ to $K_{\mathscr{P}}$ is a maximal ideal in $K_{\mathscr{P}}$, namely the ideal $\mathfrak{M}_{\mathscr{P}}$ of non-units of $K_{\mathscr{P}}$. It follows from Vol. I, Ch. V, § 2, Complement (2) to Theorem 3, that $\mathfrak{P}^\star$ is a maximal ideal in $K^\star_{\mathscr{P}}$.

THEOREM 12. *Let $K^\star$ be an algebraic extension of $K$, let $\mathscr{P}^\star$ be an extension of a place $\mathscr{P}$ of $K$ and let $K^\star_{\mathscr{P}}$ be the integral closure of $K_{\mathscr{P}}$ in $K^\star$. If $\mathfrak{P}^\star = K^\star_{\mathscr{P}} \cap \mathfrak{M}_{\mathscr{P}^\star}$, then $K^\star_{\mathscr{P}^\star}$ is the quotient ring of $K^\star_{\mathscr{P}}$ with respect to $\mathfrak{P}^\star$.*

PROOF. It is clear that the quotient ring in question is contained in $K^\star_{\mathscr{P}^\star}$. Now, let $\alpha \neq 0$ be any element of $K^\star_{\mathscr{P}^\star}$ and let $a_0\alpha^n + a_1\alpha^{n-1} + \cdots + a_n = 0$, $a_i \in K$, $a_0 \neq 0$, be the minimal equation for $\alpha$ over $K$. Let $j$ be the smallest of the integers $0, 1, \cdots, n$, such that $(a_i/a_j)\mathscr{P} \neq \infty$, $i = 0, 1, \cdots, n$. Then it is clear that $(a_i/a_j)\mathscr{P} = 0$, if $i < j$. If we set $b_i = a_i/a_j$, then we have $b_0\alpha^n + b_1\alpha^{n-1} + \cdots + b_n = 0$, and the $b_i$ are elements of $K^\star_{\mathscr{P}}$, not all in $\mathfrak{P}^\star$ (since $b_j = 1$). Since $K^\star_{\mathscr{P}}$ is integrally closed, it follows from the lemma in § 5 that either $\alpha$ or $1/\alpha$ belongs to the quotient ring of $K^\star_{\mathscr{P}}$ with respect to $\mathfrak{P}^\star$. Were $\alpha$ not in this quotient ring, $1/\alpha$ would be a non-unit in that ring, whence we would have $(1/\alpha)\mathscr{P}^\star = 0$, $\alpha\mathscr{P}^\star = \infty$, which is impossible. This completes the proof.

COROLLARY 1. *If $\mathscr{P}_1{}^\star$ and $\mathscr{P}_2{}^\star$ are two non-isomorphic extensions of $\mathscr{P}$, then $\mathfrak{M}_{\mathscr{P}_1{}^\star} \cap K^\star_{\mathscr{P}} \neq \mathfrak{M}_{\mathscr{P}_2{}^\star} \cap K^\star_{\mathscr{P}}$.*

Obvious.

COROLLARY 2. *If $\mathfrak{P}^\star$ is any maximal ideal in $K^\star_{\mathscr{P}}$, then the quotient ring of $K^\star_{\mathscr{P}}$ with respect to $\mathfrak{P}^\star$ is the valuation ring of a place $\mathscr{P}^\star$ of $K^\star$ which is an extension of $\mathscr{P}$.*

For, by Theorem 4, § 4, there exists a place $\mathscr{P}^\star$ of $K^\star$ such that $K^\star_{\mathscr{P}^\star} \supset K^\star_{\mathscr{P}}$ and $\mathfrak{M}_{\mathscr{P}^\star} \supset \mathfrak{P}^\star$. Since $K^\star_{\mathscr{P}}$ is integrally dependent on $K_{\mathscr{P}}$ and since $\mathfrak{M}_{\mathscr{P}}$ is the only maximal ideal in $K_{\mathscr{P}}$, it follows that $\mathfrak{P}^\star \cap K_{\mathscr{P}} = \mathfrak{M}_{\mathscr{P}}$. Therefore $K^\star_{\mathscr{P}^\star} \supset K_{\mathscr{P}}$ and $\mathfrak{M}_{\mathscr{P}^\star} \supset \mathfrak{M}_{\mathscr{P}}$. This shows that $\mathscr{P}^\star$ is, to within an isomorphism, an extension of $\mathscr{P}$ (§ 6, Lemma 1). Since $\mathfrak{P}^\star = K^\star_{\mathscr{P}} \cap \mathfrak{M}_{\mathscr{P}^\star}$, the corollary follows from the theorem just proved above.

Before stating the next corollary we give the following definition:

DEFINITION. *If $K^\star$ is a normal extension of a field K, then two places $\mathscr{P}_1^\star$, $\mathscr{P}_2^\star$ of $K^\star$ are said to be conjugate over K if there exists a K-automorphism s of $K^\star$ such that $\mathscr{P}_2^\star = s\mathscr{P}_1^\star$.*†

COROLLARY 3. *Let $K^\star$ be a finite normal extension of K and let $\mathscr{P}$ be a place of K. If $\mathscr{P}^\star$ and $\mathscr{P}'^\star$ are extensions of $\mathscr{P}$ in $K^\star$, then $\mathscr{P}'^\star$ is isomorphic to a conjugate of $\mathscr{P}^\star$.*

Let $\mathfrak{P}^\star$ and $\mathfrak{P}'^\star$ be the centers of $\mathscr{P}^\star$ and $\mathscr{P}'^\star$ in the ring $K^\star_{\mathscr{P}}$. Since $K^\star_{\mathscr{P}}$ is integral over $K_{\mathscr{P}}$ and since $\mathfrak{P}^\star$ and $\mathfrak{P}'^\star$ both lie over the ideal $\mathfrak{M}_{\mathscr{P}}$ in $K_{\mathscr{P}}$, it follows by V, § 9, Theorem 22, that $\mathfrak{P}^\star$ and $\mathfrak{P}'^\star$ are conjugate prime ideals over $K$. Consequently some conjugate $\mathscr{P}_1^\star$ of the place $\mathscr{P}^\star$ will have center $\mathfrak{P}'^\star$ in $K^\star_{\mathscr{P}}$, and hence $\mathscr{P}_1^\star$ and $\mathscr{P}'^\star$ are isomorphic since, by Theorem 12, these two places have the same valuation ring.

The above corollary can be extended to infinite normal extensions $K^\star$ of $K$. The proof is as follows:

Given the two extensions $\mathscr{P}^\star$ and $\mathscr{P}'^\star$ of $\mathscr{P}$ to $K^\star$, let $M$ denote the set of all pairs $(F, s)$ such that: (1) $F$ is a field between $K$ and $K^\star$ and is a normal extension of $K$; (2) $s$ is a $K$-automorphism of $F$; (3) if $\mathscr{P}_F$ and $\mathscr{P}'_F$ are the restrictions of $\mathscr{P}^\star$ and $\mathscr{P}'^\star$ to $F$ then $\mathscr{P}'_F = s\mathscr{P}_F$. If $(F, s)$ and $(G, t)$ are two such pairs, we write $(F, s) < (G, t)$ if $F < G$ and $s$ is the restriction of $t$ to $F$. Then $M$ becomes a partially ordered set. It is clear that $M$ is an inductive set and hence, by Zorn's lemma, $M$ contains maximal elements. Let $(F_0, s_0)$ be a maximal element of $M$. To prove the corollary we have only to show that $F_0 = K^\star$. Assuming the contrary, we take an element $x$ in $K^\star$, not in $F_0$, and we adjoin to $F_0$ the element $x$ and all its conjugates over $K$. We thus obtain a field $F_1$

† In § 2 (p. 6) we have defined conjugate algebraic places of a field $K$ over a ground field $k$. In the present definition we have introduced the concept of conjugate places, with respect to a field $K$, of a normal extension of $K$. The two definitions agree whenever they are both applicable, namely when $K$ is a normal algebraic extension of $k$ and when we are dealing with places of $K$ over $k$. In fact, let $\mathscr{P}_1$ and $\mathscr{P}_2$ be two places, over $k$, of a normal algebraic extension $K$ of $k$. If these places are conjugate in the sense of the present definition, then it is obvious that they have the same residue field and are isomorphisms of $K^\star$ onto that common residue field; they are therefore conjugate over $k$ also in the sense of the definition of § 2. (Observe that both places must be trivial, in view of § 4, Theorem 5′, Corollary 1.) Conversely, assume that $\mathscr{P}_1$ and $\mathscr{P}_2$ are places of $K/k$ (necessarily algebraic) which are $k$-conjugate in the sense of the definition given in § 2, and let $\Delta_1$ and $\Delta_2$ be their residue fields. Since both $\mathscr{P}_1$ and $\mathscr{P}_2$ must be trivial places, $\Delta_1$ and $\Delta_2$ are $k$-isomorphic normal extensions of $k$. Since they are subfields of one and the same algebraic closure $\bar{k}$ of $k$, they must coincide. Therefore if we set $s = \mathscr{P}_2\mathscr{P}_1^{-1}$, then $s$ is an automorphism of $K/k$ and have $\mathscr{P}_2 = s\mathscr{P}_1$, i.e., $\mathscr{P}_1$ and $\mathscr{P}_2$ are also conjugate in the sense of the present definition.

which is a normal extension of $K$ and such that $F_0 < F_1 \subset K^\star$. Let the restrictions of $\mathscr{P}^\star$ to $F_0$ and $F_1$ be respectively $\mathscr{P}_0$ and $\mathscr{P}_1$; similarly, let $\mathscr{P}'_0$ and $\mathscr{P}'_1$ be the restrictions of $\mathscr{P}'^\star$ to $F_0$ and $F_1$ respectively. We fix an automorphism $s_1$ of $F_1$ such that $s_1$ is an extension of $s_0$, and we set $\mathscr{P}''_1 = s_1^{-1}\mathscr{P}'_1$. Since $\mathscr{P}_0 = s_0^{-1}\mathscr{P}'_0$, it follows that $\mathscr{P}_1$ and $\mathscr{P}''_1$ are both extensions of $\mathscr{P}_0$. By the finite case of the corollary we have therefore that $\mathscr{P}''_1 = \tau\mathscr{P}_1$, where $\tau$ is a suitable $F_0$-automorphism of $F_1$. Then $\mathscr{P}'_1 = s_1\tau\mathscr{P}_1$, showing that $(F_1, s_1\tau) \in M$. This is a contradiction with the maximality of $(F_0, s_0)$, since $F_0 < F_1$ and $s_0$ is the restriction of $s_1\tau$ to $F_0$.

A similar argument could be used to prove that also Theorem 22 of Vol. I, Ch. V, § 9, holds for infinite normal algebraic extensions. On the other hand, the above proof of the corollary already establishes Theorem 22 in the infinite case, for every prime ideal is the center of some place.

COROLLARY 4. *If $K^\star$ is a finite algebraic extension of $K$ and $\mathscr{P}$ is a place of $K$, then the number of non-isomorphic extensions of $\mathscr{P}$ in $K^\star$ is not greater than the degree of separability* $[K^\star : K]_s$.

This is an immediate consequence of Theorem 12, Corollary 3 if $K^\star$ is a normal extension of $K$. In the general case, it is sufficient to pass to the least normal extension $K_1^\star$ of $K$ which contains $K^\star$ and to observe that: (a) every extension $\mathscr{P}^\star$ of $\mathscr{P}$ in $K^\star$ is the restriction of an extension of $\mathscr{P}$ in $K_1^\star$ (for $\mathscr{P}^\star$ has an extension in $K_1^\star$); (b) two extensions of $\mathscr{P}$ in $K_1^\star$ which differ by a $K^\star$-automorphism of $K_1^\star$ have the same restriction in $K^\star$; (c) if $G$ and $H$ are the Galois groups of $K_1^\star/K$ and $K_1^\star/K^\star$ respectively, then the index of the subgroup $H$ of $G$ is equal to the degree of separability $[K^\star : K]_s$.

In view of the intrinsic importance of the above corollary, we shall give below another proof which makes no use of the theorems developed in this section. The proof will be based on the following lemma which expresses the *independence* of any finite set of places such that none is a specialization of any other place in the set.

LEMMA 1. *If $\mathscr{P}_1, \mathscr{P}_2, \cdots, \mathscr{P}_s$ are places of a field $K$ such that $K_{\mathscr{P}_i} \not\supset K_{\mathscr{P}_j}$ if $i \neq j$, then there exist $s$ elements $\xi_1, \xi_2, \cdots, \xi_s$* in $K$ *such that $\xi_i\mathscr{P}_i \neq 0, \infty$ and $\xi_i\mathscr{P}_j = 0$ if $i \neq j$ $(i, j = 1, 2, \cdots, s)$.*

PROOF. We first consider the case $s = 2$. Since $K_{\mathscr{P}_1} \not\subset K_{\mathscr{P}_2}$, there exists an element $x$ in $K$ such that $x\mathscr{P}_1 \neq \infty$, $x\mathscr{P}_2 = \infty$. If $x\mathscr{P}_1 \neq 0$, we set $\xi_1 = 1/x$. If $x\mathscr{P}_1 = 0$, we set $\xi_1 = 1/(x+1)$. In a similar fashion we can find $\xi_2$.

We assume now that $s > 2$ and we use induction with respect to $s$. By our induction hypothesis, there exists an element $x$ such that $x\mathscr{P}_1 \neq 0, \infty$,

$x\mathcal{P}_i=0$; $i=2, 3, \cdots, s-1$. We show that there exists an element $y_s$ such that $y_s\mathcal{P}_1\neq 0, \infty$, $y_s\mathcal{P}_i=0$, $i=2, 3, \cdots, s-1$, and $y_s\mathcal{P}_s\neq\infty$. If $x\mathcal{P}_s\neq\infty$, there is nothing to prove. If $x\mathcal{P}_s=\infty$, we set $y_s=x/(x-1)$ if $x\mathcal{P}_1\neq 1$, and $y_s=x/(x+1)$ if $x\mathcal{P}_1=1$ and the characteristic of the residue field of $\mathcal{P}_1$ is $\neq 2$. If the characteristic is 2 and $x\mathcal{P}_1=1$, we set $y_s=(x^3+x^2+x)/(x^3+x+1)$.

In a similar fashion we find, for each $i=2, 3, \cdots, s$, an element $y_i$ such that $y_i\mathcal{P}_1\neq 0, \infty$, $y_i\mathcal{P}_i\neq\infty$ and $y_i\mathcal{P}_j=0$, if $j\neq 1, i (i=2, 3, \cdots, s)$. If we then set $\xi_1=y_2y_3\cdots y_s$, we have $\xi_1\mathcal{P}_1\neq 0, \infty$; $\xi_1\mathcal{P}_i=0$, $i=2, 3, \cdots, s$. The existence of $\xi_2, \xi_3, \cdots, \xi_s$ is proved in a similar manner.

The above Corollary of Theorem 12 can now be proved as follows: Let $\mathcal{P}_1^\star, \mathcal{P}_2^\star, \cdots, \mathcal{P}_s^\star$ be non-isomorphic extensions of $\mathcal{P}$ in $K^\star$. *Since each $\mathcal{P}_i^\star$ has relative dimension zero, no $\mathcal{P}_i^\star$ is a specialization of any $\mathcal{P}_j^\star$ if $i\neq j$.* There exist then elements $\xi_1, \xi_2, \cdots, \xi_s$ in $K^\star$ satisfying the conditions of the above lemma (with $\mathcal{P}_i$ replaced by $\mathcal{P}_i^\star$). *We assert that for any integer $e\geqq 0$ the elements $\xi_i^{p^e}$ are linearly independent over $K$* (here $p$ is the characteristic of $K$; if $p=0$, we set $p^e=1$). For assume that we have a linear relation of the form $a_1\xi_1^{p^e}+a_2\xi_2^{p^e}+\cdots+a_s\xi_s^{p^e}=0$, where the $a_i$ are in $K$ and are not all zero. Upon dividing by one of the coefficients we may assume that one of the coefficients, say $a_j$, is equal to 1, while the remaining coefficients have finite $\mathcal{P}$-values. But then, passing to the $\mathcal{P}_j^\star$-values, we find the absurd relation $1=0$.

Since for a suitable integer $e$ the elements $\xi_i^{p^e}$ are all separable over $K$, it follows that $s\leqq[K^\star:K]_s$, establishing the corollary.

We shall need later on the following *approximation theorem* which expresses the independence of places in a much stronger form than does Lemma 1.

LEMMA 2. *If $\mathcal{P}_1, \mathcal{P}_2, \cdots, \mathcal{P}_s$ are places of a field $K$, such that $K_{\mathcal{P}_i}\not\supset K_{\mathcal{P}_j}$ if $i\neq j$, then given $s$ arbitrary elements $a_1, a_2, \cdots, a_s$ belonging to the residue fields of $\mathcal{P}_1, \mathcal{P}_2, \cdots, \mathcal{P}_s$ respectively, there exists an element $u$ in $K$ such that $u\mathcal{P}_i=a_i$, $i=1, 2, \cdots, s$.*

PROOF. Using the elements $\xi_1, \xi_2, \cdots, \xi_s$ of Lemma 1 we set $\zeta_i=\xi_i/(\xi_1+\xi_2+\cdots+\xi_s)$. The $s$ elements $\zeta_i$ have then the following properties: $\zeta_i\mathcal{P}_i=1$, $\zeta_i\mathcal{P}_j=0$ if $i\neq j$. We shall make use of the $\zeta$'s in the present proof, in the following fashion: instead of proving the existence of an element $u$ satisfying the conditions of the lemma, we shall prove that for each $i=1, 2, \cdots, s$ there exists an element $u_i$ such that $u_i\mathcal{P}_i=a_i$, $u_i\mathcal{P}_j\neq\infty$ if $i\neq j$. For, once this is proved, the element $u=u_1\zeta_1+u_2\zeta_2+\cdots+u_s\zeta_s$ will satisfy all our requirements.

Let us prove, for instance, that there exists an element $u_1$ such that $u_1\mathscr{P}_1 = a_1$, $u_1\mathscr{P}_j \neq \infty$ if $j \neq 1$. We begin with the case $s=2$. Let $z_1$ be an arbitrary element of $K$ such that $z_1\mathscr{P}_1 = a_1$. If $z_1\mathscr{P}_2 \neq \infty$, we set $u_1 = z_1$. If $z_1\mathscr{P}_2 = \infty$, then we may set $u_1 = z_1/(1+z_1\zeta_2)$.

We now assume that $s>2$ and we use induction with respect to $s$. There exists then an element $z_1$ in $K$ such that $z_1\mathscr{P}_1 = a_1$, $z_1\mathscr{P}_j \neq \infty$, $j=2, 3, \cdots, s-1$. If also $z_1\mathscr{P}_s \neq \infty$, we set $u_1 = z_1$. If $z_1\mathscr{P}_s = \infty$, we may set $u_1 = z_1/(1+z_1\zeta_s)$.

This completes the proof of the lemma.

We shall conclude our study of extensions of places in algebraic field extensions by a theorem which is of importance for applications, since it covers a situation which occurs whenever two integral domains are given, one of which is integrally dependent on the other.

THEOREM 13. *Let $\mathfrak{O}$ be an integrally closed integral domain, and let $\mathfrak{O}^\star$ be an integral domain which is integrally dependent on $\mathfrak{O}$. Let $\mathfrak{q}$ be a prime ideal in $\mathfrak{O}$ and let $\mathfrak{q}^\star$ be a prime ideal in $\mathfrak{O}^\star$ which lies over $\mathfrak{q}$. If $\mathscr{P}$ is a place of the quotient field $K$ of $\mathfrak{O}$ which has center $\mathfrak{q}$ in $\mathfrak{O}$, then at least one of the extensions of $\mathscr{P}$ to the quotient field $K^\star$ of $\mathfrak{O}^\star$ has center $\mathfrak{q}^\star$ in $\mathfrak{O}^\star$.*

PROOF. Since $\mathfrak{O}^\star$ is integrally dependent on $\mathfrak{O}$, $K^\star$ is an algebraic extension of $K$. We also observe that we may replace $\mathfrak{O}^\star$ by its integral closure $\bar{\mathfrak{O}}^\star$ in $K^\star$, since there is at least one prime ideal in $\bar{\mathfrak{O}}^\star$ which lies over $\mathfrak{q}^\star$ (Vol. I, Ch. V, § 2, Theorem 3). Hence we may assume that $\mathfrak{O}^\star$ is integrally closed.

We first consider the case in which $K^\star$ is a finite normal extension of $K$. We fix an extension $\mathscr{P}'^\star$ of $\mathscr{P}$ in $K^\star$ and we denote by $\mathfrak{q}'^\star$ the center of $\mathscr{P}'^\star$ in $\mathfrak{O}^\star$. Since both $\mathfrak{O}$ and $\mathfrak{O}^\star$ are integrally closed and since both $\mathfrak{q}'^\star$ and $\mathfrak{q}^\star$ lie over $\mathfrak{q}$, the prime ideals $\mathfrak{q}'^\star$ and $\mathfrak{q}^\star$ are conjugate over $K$ (Vol. I, Ch. V, § 9, Theorem 22). If, say, $\mathfrak{q}'^\star = \tau(\mathfrak{q}^\star)$, where $\tau$ is a $K$-automorphism of $K^\star$, then the place $\mathscr{P}^\star = \tau\mathscr{P}'^\star$ is an extension of $\mathscr{P}$ and has center $\mathfrak{q}^\star$.

If $K^\star$ is a finite extension of $K$, not necessarily normal, we consider the least normal extension $K'$ of $K$ which contains $K^\star$ and we denote by $\mathfrak{O}'$ the integral closure of $\mathfrak{O}^\star$ in $K'$. There exists a prime ideal $\mathfrak{q}'$ in $\mathfrak{O}'$ such that $\mathfrak{q}' \cap \mathfrak{O}^\star = \mathfrak{q}^\star$, and by the preceding case, there exists an extension $\mathscr{P}'$ of $\mathscr{P}$ in $K'$ such that $\mathfrak{M}_{\mathscr{P}'} \cap \mathfrak{O}' = \mathfrak{q}'$. Then if $\mathscr{P}^\star$ is the restriction of $\mathscr{P}'$ to $K^\star$, the place $\mathscr{P}^\star$ will be an extension of $\mathscr{P}$ with center $\mathfrak{q}^\star$.

Now, let $K^\star$ be an arbitrary algebraic extension of $K$. Our theorem is equivalent with the assertion that $K^\star_{\mathscr{P}}\mathfrak{q}^\star \neq (1)$, where $K^\star_{\mathscr{P}}$ is the integral closure in $K^\star$ of the valuation ring $K_{\mathscr{P}}$. For, if there exists an extension $\mathscr{P}^\star$ of $\mathscr{P}$ which has center $\mathfrak{q}^\star$, then $K^\star_{\mathscr{P}}\mathfrak{q}^\star \subset \mathfrak{M}_{\mathscr{P}^\star}$ and therefore $1 \notin K^\star_{\mathscr{P}}\mathfrak{q}^\star$. Conversely, if $K^\star_{\mathscr{P}}\mathfrak{q}^\star \neq (1)$, then the ideal $K^\star_{\mathscr{P}}\mathfrak{q}^\star$ in

$K^\star_{\mathscr{P}}$ is contained in a maximal ideal $\mathfrak{P}^\star$ of $K^\star_{\mathscr{P}}$. By Theorem 12, Corollary 2, the quotient ring of $K^\star_{\mathscr{P}}$ with respect to $\mathfrak{P}^\star$ is the valuation ring of an extension $\mathscr{P}^\star$ of $\mathscr{P}$. The prime ideal $\mathfrak{M}_{\mathscr{P}^\star}$ of $\mathscr{P}^\star$ contracts in $\mathfrak{O}^\star$ to a prime ideal which contains $\mathfrak{q}^\star$ (since $\mathfrak{P}^\star \supset \mathfrak{q}^\star$) and contracts to the ideal $\mathfrak{q}$ in $\mathfrak{O}$. Hence $\mathfrak{q}^\star = \mathfrak{M}_{\mathscr{P}^\star} \cap \mathfrak{O}^\star$ (see Vol. I, Ch. V, p. 259, complement 1 to Theorem 3), and thus $\mathfrak{q}^\star$ is the center of $\mathscr{P}^\star$.

Now, the proof that $K^\star_{\mathscr{P}}\mathfrak{q}^\star \neq (1)$ is achieved by observing that if $K^\star_{\mathscr{P}}\mathfrak{q}^\star = (1)$, then $1 = \sum_{i=1}^{h} a_i^\star q_i^\star$, $a_i^\star \in K^\star_{\mathscr{P}}$, $q_i \in \mathfrak{q}^\star$, and from this relation one concludes easily that there exists an intermediate ring $\mathfrak{O}'$ between $\mathfrak{O}$ and $\mathfrak{O}^\star$ with the following properties: the quotient field $K'$ of $\mathfrak{O}'$ is a finite algebraic extension of $K$, and if $\mathfrak{q}' = \mathfrak{q}^\star \cap \mathfrak{O}'$ then $K'_{\mathscr{P}}\mathfrak{q}' = (1)$, where $K'_{\mathscr{P}}$ is the integral closure of $K_{\mathscr{P}}$ in $K'$. The relation $K'_{\mathscr{P}}\mathfrak{q}' = (1)$ is, however, in contradiction with the fact that our theorem holds true for the finite algebraic extension $K'$ of $K$. This completes the proof of the theorem.

COROLLARY. *The assumption and notations being the same as in* Vol. I, Ch. V, § 13, *Theorem* 34 (*the theorem of Kummer*), *given any place* $\mathscr{P}$ *of* $K$ *which has center* $\mathfrak{p}$ *in* $R$ *and given any irreducible factor* $f_i(X)$ of $\bar{F}(X)$, *there exists an extension* $\mathscr{P}'$ *of* $\mathscr{P}$ *to* $K'$ *such that* $y\mathscr{P}'$ *is a root of* $f_i(X)$.

Apply the theorem to the case in which $\mathfrak{O}^\star = R'$, $\mathfrak{p}^\star = R'\mathfrak{p} + R'F_i(y)$.

## § 8. Valuations.

Let $K$ be a field and let $K'$ denote the multiplicative group of $K$, i.e., let $K'$ be the set of elements of $K$ which are different from zero. Let $\Gamma$ be an additive abelian *totally ordered* group.

DEFINITION. *A valuation of* $K$ *is a mapping* $v$ *of* $K'$ *into* $\Gamma$ *such that the following conditions are satisfied:*

(a) $$v(xy) = v(x) + v(y)$$

(b) $$v(x+y) \geqq \min \{v(x), v(y)\}$$

For any $x$ in $K'$, the corresponding element $v(x)$ of $\Gamma$ is called the *value of* $x$ in the given valuation. The set of all elements of $\Gamma$ which are values of elements of $K'$ is clearly a subgroup of $\Gamma$ and is called the *value group* of $v$. The elements of $\Gamma$ which do not belong to the value group do not interest us. We shall therefore assume that $\Gamma$ itself is the value group of $v$, i.e., that $v$ is a mapping of $K'$ onto $\Gamma$.

A valuation $v$ is *non-trivial* if $v(a) \neq 0$ for some $a$ in $K'$; in the contrary case $v$ is said to be a *trivial* valuation.

Condition (a) signifies that $v$ is a homomorphism of the multiplicative group $K'$ onto the additive group $\Gamma$. Hence $v(1)=0$; $v(-1)+v(-1)=v(1)=0$, and hence $v(-1)=0$ since $\Gamma$ is a totally ordered group. More generally, if an element $w$ of $K'$ is a root of unity, say if $w^n=1$, then $nv(w)=0$, whence $v(w)=0$ (for $\Gamma$ is totally ordered).

From $v(-1)=0$ it follows that $v(-x)=v(x)$, and hence, by (b):

(b′) $$v(x-y) \geqq \min \{v(x), v(y)\}$$

We also note the following consequences of the properties (a), (b) and (b′):

(1) $$v(y/x) = v(y)-v(x), \quad x \neq 0$$

(2) $$v(1/x) = -v(x), \quad x \neq 0$$

(3) $$v(x) < v(y) \Rightarrow v(x+y) = v(x).$$

To prove (3), we first observe that $v(x+y) \geqq v(x)$, by (b). On the other hand, if we write $x$ in the form $(x+y)-y$ and apply (b′), we find $v(x) \geqq \min \{v(x+y), v(y)\}$. Hence $v(x) \geqq v(x+y)$, since, by assumption, $v(x) < v(y)$. Combining with the preceding inequality $v(x+y) \geqq v(x)$ we find (3).

The following are easy generalizations of (b) and (3):

(4) $v\left(\sum_{i=1}^{n} x_i\right) \geqq \min \{v(x_1), v(x_2), \cdots, v(x_n)\}$ for all $x_i \in K$;

(5) $v\left(\sum_{i=1}^{n} x_i\right) = \min \{v(x_1), v(x_2), \cdots, v(x_n)\}$ if the minimum is reached by only one of the $v(x_j)$.

Relation (4) follows by a straightforward induction. To prove (5), let $i$ be the unique value of the index $j$ for which $v(x_j)$ attains its minimum. We have

$$v\left(\sum_{j \neq i} x_j\right) \geqq \min_{j \neq i} \{v(x_j)\} > v(x_i),$$

and now (5) follows from (3).

Let $v$ and $v'$ be two valuations of $K$, with value groups $\Gamma$ and $\Gamma'$ respectively. We shall say that $v$ and $v'$ are *equivalent* valuations if there exists an order preserving isomorphism $\varphi$ of $\Gamma$ onto $\Gamma'$ such that $v'(x)=[v(x)]\varphi$ for all $x$ in $K'$. *We shall make no distinction between equivalent valuations*; we agree in fact to identify any two valuations of $K$ if they are equivalent.

If a particular subfield $k$ of $K$ has been specified as ground field, then a valuation $v$ of $K$ is said to be a valuation *over* $k$, or *a valuation of* $K/k$, if $v(c)=0$ for all $c$ in $k$, $c\neq 0$, i.e., if $v$ is trivial on $k$.

The set of elements $x$ of $K$ such that $v(x)\geqq 0$ is clearly a ring. This ring will be denoted by $R_v$ and will be called the *valuation ring* of $v$.

Since, for every $x$ in $K$, we have either $v(x)\geqq 0$ or $v(x)\leqq 0$, i.e., either $v(x)\geqq 0$ or $v(1/x)\geqq 0$ (by (2)), it follows that either $x$ or $1/x$ belongs to the valuation ring. This justifies the name "valuation ring" (see Theorem 1, § 2).

The "*divisibility* relation in $K$ with respect to $R_v$," i.e., the relation $y|x$ defined by the condition that there exists an element $z$ in $R_v$ such that $x=yz$, is equivalent to the relation "$v(x)\geqq v(y)$." This follows at once from ($a$).

In order that both $x$ and $1/x$ belong to $R_v$ it is necessary and sufficient that $v(x)\geqq 0$ and $-v(x)\geqq 0$, i.e., that $v(x)=0$. In other words: the multiplicative group of *units* in $R_v$ coincides with the *kernel* of the homomorphism $v$ of $K'$ onto $\Gamma$.

The non-units in $R_v$ are therefore the elements $y$ in $K$ such that $v(y)>0$. It follows directly from (a) and (b′) that the set of non-units in $R_v$ is a prime ideal. We shall denote this prime ideal by $\mathfrak{M}_v$ and refer to it as *the prime ideal of the valuation* $v$. Notice that any element of $K$ which does not belong to $R_v$ is the reciprocal of an element of $\mathfrak{M}_v$. Since $\mathfrak{M}_v$ is the set of all non-units in $R_v$, it is a maximal ideal in $R_v$, in fact the greatest proper ideal in $R_v$.

In the case of a non-trivial valuation, $\mathfrak{M}_v$ is not the zero ideal, and $R_v$ is a proper subring of $K$. For a trivial valuation $v$ we have $R_v=K$, $\mathfrak{M}_v=(0)$.

Since $\mathfrak{M}_v$ is a maximal ideal, $R_v/\mathfrak{M}_v$ is a field. This field will be called the *residue field* of the valuation $v$ and will be denoted by $D_v$, or simply by $D$. The image of an element $x$ of $R_v$ under the canonical homomorphism $R_v\rightarrow R_v/\mathfrak{M}_v$ will be called the *v-residue of x*.

If $v$ is a valuation of $K$ over a ground field $k$, then $k\subset R_v$ and $k$ can be canonically identified with a subfield of the residue field $D$ of $v$. The transcendence degree of $D/k$ is called the *dimension* of the valuation $v$ (over $k$).

It is obvious that equivalent valuations of $K$ have the same valuation ring and the same residue field. *Conversely, if two valuations $v$ and $v'$ of $K$ have the same valuation ring, then they are equivalent.* For let $\Gamma$ and $\Gamma'$ be the value groups of $v$ and $v'$ respectively, and assume that $R=R_v=R_{v'}$. The two valuations $v$ and $v'$ are homomorphisms of $K'$ onto $\Gamma$ and $\Gamma'$ respectively. By assumption, they have the same kernel, namely the set of units in $R$. Hence $v^{-1}v'$ is an isomorphism $\varphi$ of $\Gamma$

onto $\Gamma'$. The elements of positive value are the same in both valuations, namely they are the non-units of $R$. Hence $\varphi$ transforms the set of positive elements of $\Gamma$ *onto* the set of positive elements of $\Gamma'$ and is therefore order preserving. Since $v' = v\varphi$, our assertion is proved.

§ **9. Places and valuations.** Let $v$ be a valuation of $K$, with value group $\Gamma$. It has been pointed out in the preceding section that if $x$ is an element of $K$, not in $R_v$, then $1/x$ belongs to $R_v$ ($1/x$ belongs then even to $\mathfrak{M}_v$). Now, we know from § 2 that this property of $R_v$ characterizes valuation rings of places of $K$. Hence *every valuation $v$ of $K$ determines a class of isomorphic places $\mathscr{P}$ of $K$ such that $K_{\mathscr{P}} = R_v$.* These places are non-trivial if and only if $v$ is non-trivial. If $\mathscr{P}$ is any place in the class determined by a given valuation $v$, and if $x$ is any element of $K$, then the relations

$$x\mathscr{P} = 0, \quad x\mathscr{P} = \infty, \quad x\mathscr{P} \neq 0, \infty$$

are respectively equivalent to the relations

$$x \in \mathfrak{M}_{\mathscr{P}}, \quad x \notin K_{\mathscr{P}}, \quad x \in K_{\mathscr{P}} - \mathfrak{M}_{\mathscr{P}},$$

and therefore are also respectively equivalent to the relations

$$v(x) > 0, \quad v(x) < 0, \quad v(x) = 0,$$

since $K_{\mathscr{P}} = R_v$ and $\mathfrak{M}_{\mathscr{P}} = \mathfrak{M}_v$.

We now show that, *conversely, every place $\mathscr{P}$ of $K$ is associated (in the above fashion) with a valuation of $K$.* The case of a trivial place $\mathscr{P}$ is trivial, and we shall therefore assume that $\mathscr{P}$ is non-trivial. Let $E$ denote the set of units in $K_{\mathscr{P}}$ ($E = K_{\mathscr{P}} - \mathfrak{M}_{\mathscr{P}}$). Then $E$ is a subgroup of the multiplicative group $K'$ of $K$. Let $\Gamma$ denote the quotient group $K'/E$ and let us write the group operation in $\Gamma$ additively. Let $v$ be the canonical homomorphism of $K'$ *onto* $\Gamma$. Then condition (a) of the definition of valuations is satisfied for $v$. We now introduce a relation of order in the group $\Gamma$. It will be sufficient to define the set $\Gamma_+$ of positive elements of $\Gamma$. *We define $\Gamma_+$ as the transform of $\mathfrak{M}_{\mathscr{P}}$ by $v$.* Since $\mathfrak{M}_{\mathscr{P}}$ is closed under multiplication, *$\Gamma_+$ is closed under addition.* Since $\mathfrak{M}_{\mathscr{P}}$ is an ideal in $K_{\mathscr{P}}$ and since $E$ is a subset of $K_{\mathscr{P}}$, it follows that $\mathfrak{M}_{\mathscr{P}}$ is the set-theoretic sum of a family of $E$-cosets in $K'$. Hence $\mathfrak{M}_{\mathscr{P}}$, with the zero element deleted, is the full inverse image of $\Gamma_+$ under $v^{-1}$. Or, in other words: if $y \in K'$, $y \notin \mathfrak{M}_{\mathscr{P}}$, then $v(y) \notin \Gamma_+$. Now, let $\alpha$ be any element of $\Gamma$ and let $\alpha = v(x)$, $x \in K'$. If $\alpha \in \Gamma_+$, then $x \in \mathfrak{M}_{\mathscr{P}}$. In that case, $1/x \notin \mathfrak{M}_{\mathscr{P}}$ and hence $-\alpha = v(1/x) \notin \Gamma_+$. If $\alpha \notin \Gamma_+$ and $\alpha \neq 0$, then $x \notin \mathfrak{M}_{\mathscr{P}}$ and $x \notin E$, whence $x \notin K_{\mathscr{P}}$. But then $1/x \in \mathfrak{M}_{\mathscr{P}}$ and $-\alpha =$

$v(1/x) \in \Gamma_+$. We have thus proved that $\Gamma_+$ satisfies all the conditions for the set of positive elements of an ordered group.

It remains to show that condition (b) of the definition of valuations is satisfied. We have to show that if $x, y \in K'$ and $v(x) \leqq v(y)$, then $v(x+y) \geqq v(x)$, or—what is the same—that $v(1+y/x) \geqq 0$. But that is obvious, since the assumption $v(x) \leqq v(y)$ implies that $y/x$ is an element of $K_{\mathscr{P}}$, and hence also $1+y/x$ belongs to $K_{\mathscr{P}}$.

Since by our construction of $v$ the valuation ring of $v$ is the ring $K_{\mathscr{P}}$, the proof is complete.

It is clear that if $\mathscr{P}$ is a place of $K$ and $v$ is the corresponding valuation of $K$, then the residue fields of $\mathscr{P}$ and $v$ are isomorphic. In particular, if $K$ contains a ground field $k$ and if $\mathscr{P}$ is a place of $K/k$, then the residue fields of $\mathscr{P}$ and $v$ are $k$-isomorphic, and hence $\mathscr{P}$ and $v$ have the same dimension. Note that, for a given valuation $v$ a particular place associated with $v$ is the canonical homomorphism of $R_v$ onto $D_v$ $(=R_v/\mathfrak{M}_v)$.

Although places and valuations are closely related concepts, they are nevertheless *distinct* concepts. The *value* of an element $x$ at a place $\mathscr{P}$ is, roughly speaking, the analogue of the value of a function at a point, while the value of $x$ in the corresponding valuation $v$ is the analogue of the *order* of a function at a point. We shall, in fact, adopt this function theoretic teminology when we deal with places and valuations. If, namely, $\mathscr{P}$ is a place and $v$ is the corresponding valuation, then for any $x$ in $K$ we shall refer to $v(x)$ as *the order of $x$ at $\mathscr{P}$*. If $\alpha = v(x)$ and $\alpha$ is positive (whence $x\mathscr{P} = 0$), then we say that *$x$ vanishes at $\mathscr{P}$ to the order $\alpha$*. If $\alpha$ is negative (whence $x\mathscr{P} = \infty$), then we say that *$x$ is infinite at $\mathscr{P}$* to the order $-\alpha$. The order of $x$ at $\mathscr{P}$ is zero if and only if $x\mathscr{P} \neq 0, \infty$.

It must be pointed out explicitly that the above definition of the order of the elements of $K$ at a given place $\mathscr{P}$ of $K$ presupposes that among the (infinitely many) equivalent valuations determined by $\mathscr{P}$ one has been selected and fixed in advance. Without a fixed choice of $v$, the definition of the order is ambiguous. The ambiguity may remain even if the value group $\Gamma$ is fixed, for $\Gamma$ may very well possess non-identical order preserving automorphisms.

It is well known that, with the exception of the additive group of integers, every totally-ordered abelian group does possess such automorphisms. Hence, it is only when the value group is the group of integers that the order of any element of $K$ at the given place $\mathscr{P}$ is determined without any ambiguity. There is, of course, one *canonical* valuation $v$ associated with a class of isomorphic places $\mathscr{P}$, and that is the canonical mapping of $K'$ onto $K'/E$, where $E$ is the set of units of $K_{\mathscr{P}}$. However, in practice one replaces $K'/E$ by some isomorphic ordered group of a more concrete type (for instance, by a subgroup of the additive group of real numbers, if $v$ is of rank 1; see § 10 below) and when that is done then the ambiguity referred to above reappears.

If a particular subfield $k$ of $K$ has been specified as a ground field then the valuations $v$ of $K/k$ are characterized by the condition that $k$ is contained in $R_v$. It follows that the valuations of $K/k$ are associated with the places of $K/k$.

The following theorem seems, in some respects, to be an analogue of the extension theorem for places (Theorem 5′, § 4) but is actually a much more trivial result:

THEOREM 14. *Let $\mathfrak{o}$ be an integral domain, $K$ the quotient field of $\mathfrak{o}$, and let $v_0$ be a mapping of $\mathfrak{o}$ (the zero excluded) into a totally ordered abelian group $\Gamma$ satisfying the following conditions:*

$$v_0(xy) = v_0(x)+v_0(y), \tag{1}$$

$$v_0(x+y) \geqq \min\{v_0(x), v_0(y)\}. \tag{2}$$

*Then $v_0$ can be extended to a valuation $v$ of $K$ by setting $v(x/y)=v_0(x)-v_0(y)$, and this valuation $v$ is the unique extension of $v_0$ to $K$.*

PROOF. If $y/x=y'/x'$ then $xy'=x'y$, $v_0(x)+v_0(y')=v_0(x')+v_0(y)$, i.e., $v_0(x)-v_0(y)=v_0(x')-v_0(y')$, and this shows that $v$ is well defined and is, of course, the unique valuation of $K$ which coincides with $v_0$ on $A$. Furthermore, $v$ satisfies conditions (a) and (b) of the definition of valuations. For, we have:

$$\begin{aligned} v\left(\frac{x}{y}\cdot\frac{x'}{y'}\right) &= v_0(xx')-v_0(yy') = v_0(x)+v_0(x')-[v_0(y)+v_0(y')] \\ &= [v_0(x)-v_0(y)]+[v_0(x')-v_0(y')] \\ &= v\left(\frac{x}{y}\right)+v\left(\frac{x'}{y'}\right), \end{aligned}$$

i.e., condition (a) is satisfied. We also have:

$$\begin{aligned} v\left(\frac{x}{y}+\frac{x'}{y'}\right) &= v_0(xy'+x'y)-v_0(yy') \\ &\geqq \min\{v_0(xy'), v_0(x'y)\}-v_0(yy') \\ &= \min\left\{v\left(\frac{x}{y}\right)+v_0(yy'), v\left(\frac{x'}{y'}\right)+v_0(yy')\right\}-v_0(yy') \\ &= \min\left\{v\left(\frac{x}{y}\right), v\left(\frac{x'}{y'}\right)\right\}, \end{aligned}$$

showing that condition (b) is also satisfied.

By analogy with § 5 we say that a valuation $v$ of a field $K$ is *non-negative* on a subring $A$ of $K$ if the valuation ring $R_v$ contains $A$, i.e., if each element of $A$ has non-negative order for $v$. In this case the set $A \cap \mathfrak{M}_v$ of all elements of $A$ which have positive orders for $v$ is a *prime ideal* $\mathfrak{p}$ in $A$; it is called the *center of* $v$ *in* $A$. The ideal $\mathfrak{p}$ is also the center of the (equivalent) places associated with $v$. It follows that if $A$ is a subring of a field $K$ and if $\mathfrak{p}$ is a prime ideal in $A$, then there exists a valuation $v$ of $K$ having $\mathfrak{p}$ as center in $A$.

In the algebro-geometric case, when dealing with a valuation $v$ of the function field $k(V)$ of an irreducible variety $V/k$, and assuming that $v$ is non-negative on the coördinate ring $k[V]$, we shall mean by the center of $v$ on $V$ the irreducible subvariety of $V/k$ which is defined by the prime ideal $\mathfrak{M}_v \cap k[V]$. Thus, while the center of a place $\mathcal{P}$, which is finite on $k[V]$, is a *point* $Q$ *of* $V$, the center of the corresponding valuation is the irreducible subvariety of $V$ which has $Q$ as general point over $k$.

EXAMPLES OF VALUATIONS:

EXAMPLE (1). A finite field $K$ admits only trivial valuations. In fact, all its non-zero elements are roots of unity.

EXAMPLE (2). Let $A$ be *UFD*, $K$ its quotient field. Given a non-zero element $x$ in $K$, we consider the (unique) factorization

$$x = u \prod_{p \in P} p^{v_p(x)},$$

$u$ denoting a unit in $A$, and $P$ a maximal set of mutually non-associated irreducible elements in $A$. For a given $x \neq 0$ in $K$, there is always only a finite number of elements $p$ in $P$ such that $v_p(x) \neq 0$, and the integers $v_p(x)$ are all $\geqq 0$ if and only if $x \in A$. The uniqueness of such a factorization shows immediately that $v_p(xy) = v_p(x) + v_p(y)$. Denoting by $m_p$ the integer min $(v_p(x), v_p(y))$, the fact that $x+y$ may be written in the form $a \prod p^{m_p}$ with $a$ in $A$, shows that $v_p(x+y) \geq \min (v_p(x), v_p(y))$. In other words, for each $p$ in $P$, $v_p$ is a *valuation* of $K$. Its valuation ring is obviously the quotient ring $A_{Ap}$, and its center in $A$ is the prime ideal $Ap$. This valuation is called the *p-adic valuation* of $K$. Its value group is the additive group of integers.

EXAMPLE (3). Let $R$ be a Dedekind domain, $K$ its quotient field. By Theorem 9, § 5, we know that if $v$ is a non-trivial valuation of $K$ which is non-negative on $R$, then the valuation ring $R_v$ of $v$ is the quotient ring $R_{\mathfrak{p}}$ of $R$ with respect to a proper prime ideal $\mathfrak{p}$ in $R$, and that in fact for every proper prime ideal $\mathfrak{p}$ in $R$ the quotient ring $R_{\mathfrak{p}}$ is a valuation ring. Let then $\mathfrak{p}$ be any proper prime ideal in $R$ and let $v_{\mathfrak{p}}$ denote the (unique) valuation of $K$ whose valuation ring is $R_{\mathfrak{p}}$. In the course of the proof of Theorem 9 we have seen that every non-zero element $x$ of $R_{\mathfrak{p}}$ is of the form $\varepsilon t^n$, where $\varepsilon$ is a unit in $R_{\mathfrak{p}}$ and $t$ is some fixed element of $R$ which belongs to $\mathfrak{p}$ but not to $\mathfrak{p}^2$. In other words, we have shown that $R_{\mathfrak{p}}$ is a unique factorization domain, that $t$ is an irreducible element in $R_{\mathfrak{p}}$ and that every other irreducible element of $R_{\mathfrak{p}}$ is an associate of $t$. It follows, as a special case of the preceding example, that if we set $v(\varepsilon t^n)=n$, then $v$ is a valuation of $K$ and $R_{\mathfrak{p}}$ is the valuation ring of $v$. Therefore $v=v_{\mathfrak{p}}$ (up to equivalence). The center of $v_{\mathfrak{p}}$ in $R$ is obviously the prime ideal $\mathfrak{p}$. This valuation $v$ is called the *$\mathfrak{p}$-adic valuation* of the quotient field $K$ of $R$. We have therefore shown that *every valuation $v$ of the quotient field $K$ of a Dedekind domain $R$ such that $v$ is non-negative on $R$ is (or, is equivalent to) a $\mathfrak{p}$-adic valuation of $K$, where $\mathfrak{p}$ is a suitable prime ideal in $R$, and that the value group* of *$v$ is (or is order isomorphic with) the additive group of integers.*

In particular, all the non-trivial valuations of the field of rational numbers, are equivalent to $p$-adic valuations, where $p$ is a prime number. Similarly, each non-trivial valuation of the field $k(X)/k$ of rational functions of one variable is equivalent to a valuation of the following type:

(a) a $p(X)$-adic valuation, where $p(X)$ is an irreducible polynomial in $k[X]$;

(b) the valuation defined by $v_\infty(f(X)/g(X))=\text{deg.}\, f(X)-\text{deg.}\, g(X)$. (See Theorem 9, Corollary 2, § 5).

The above analysis can be applied to fields of algebraic numbers (finite algebraic extensions of the field of rational numbers). If $K$ is such a field and $v$ is a non-trivial valuation of $K$, then the valuation ring $R_v$ contains the ring $J$ of ordinary integers and therefore $R_v$ must also contain the integral closure of $J$ in $K$, i.e., the ring $\mathfrak{o}$ of algebraic integers in $K$. Since $\mathfrak{o}$ is a Dedekind domain (Vol. I, Ch. V, § 8, p. 284), $v$ is a $\mathfrak{p}$-adic valuation of $K$, where $\mathfrak{p}$ is a prime ideal in $\mathfrak{o}$, and the value group of $v$ is the additive group of integers. The center of $v$ in $J$ is a prime ideal $Jp$, where $p$ is a prime number and $\mathfrak{p} \cap J=Jp$. Given a prime number $p$, there is only a finite number of prime ideals $\mathfrak{p}$ in $\mathfrak{o}$ such that $\mathfrak{p} \cap J=p$ (they are the prime ideals of $\mathfrak{o}p$). Hence, there is only a finite number of mutually non-equivalent valuations $v$ of $K$ in which a given prime number $p$ has positive value $v(p)$.

§ **10. The rank of a valuation.** Let $K$ be a field and let $v$ be a valuation of $K$. By the *rank of $v$* we mean the rank of any place $\mathscr{P}$ such that $K_{\mathfrak{p}} = R_v$ (see § 3, Definition 1). We proceed to interpret the rank of $v$ directly in terms of the value group $\Gamma$ of $v$.

A non-empty subset $\Delta$ of $\Gamma$ is called a *segment* if it has the following property: if an element $\alpha$ of $\Gamma$ belongs to $\Delta$, then all the elements $\beta$ of $\Gamma$ which lie between $\alpha$ and $-\alpha$ (the element $-\alpha$ included) also belong to $\Delta$. A subset $\Delta$ of $\Gamma$ is called an *isolated subgroup of* $\Gamma$ if $\Delta$ is a segment and a *proper* subgroup of $\Gamma$.

It is clear that the set of all segments of $\Gamma$ is totally ordered by the relation of set-theoretic inclusion. We shall say, namely, that $\Delta_1$ precedes $\Delta_2$ if the segment $\Delta_1$ is a proper subset of the segment $\Delta_2$. We proceed to prove that *the ordinal type of the set of all isolated subgroups of* $\Gamma$ *is equal to the rank of* $v$. This assertion is included in the theorem stated and proved below.

If $A$ is any subset of the valuation ring $R_v$, we shall denote by $Av$ the set of all elements $\alpha$ of $\Gamma$ which are of the form $v(x)$, $x \in A$, $x \neq 0$, and by $-Av$ the set of elements $-\alpha$, $\alpha \in Av$. We denote by $\Gamma_A$ the complement in $\Gamma$ of the union of the two sets $Av$ and $-Av$.

THEOREM 15. *If* $\mathfrak{A}$ *is a proper ideal in* $R_v$ *(i.e.,* $\mathfrak{A} \neq (0), R_v$*), then* $\Gamma_{\mathfrak{A}}$ *is a segment in* $\Gamma$. *The mapping* $\mathfrak{A} \rightarrow \Gamma_{\mathfrak{A}}$ *transforms in* (1, 1) *order-reversing fashion the set of all proper ideals* $\mathfrak{A}$ *in* $R_v$ *onto the set of all segments of* $\Gamma$ *which are different from* $\Gamma$. *The segment* $\Gamma_{\mathfrak{A}}$ *is an isolated subgroup of* $\Gamma$ *if and only if* $\mathfrak{A}$ *is a proper prime ideal of* $R_v$.

PROOF. If $\mathfrak{A}$ is a proper ideal in $R_v$, the set $\mathfrak{A}v$ is non-empty and contains only positive elements of $\Gamma$. Hence $\Gamma_{\mathfrak{A}}$ is non-empty (it contains the zero of $\Gamma$) and is a proper subset of $\Gamma$.

Since $\mathfrak{A}R_v \subset \mathfrak{A}$, we have $\mathfrak{A}v + \Gamma_+ \subset \mathfrak{A}v$. In other words: if $\alpha \in \mathfrak{A}v$ and $\beta > \alpha$, then $\beta \in \mathfrak{A}v$. This shows that $\Gamma_{\mathfrak{A}}$ is a segment.

Since $\mathfrak{A}$ is an ideal, we have $xE \subset \mathfrak{A}$ for all $x$ in $\mathfrak{A}$. Here $E$—the set of units in $R_v$—is the kernel of the mapping $v$ of $K'$ onto $\Gamma$. Hence $\mathfrak{A}$ consists of $E$-cosets and is therefore the full inverse image of $\mathfrak{A}v$ under $v^{-1}$. Hence the mapping $\mathfrak{A} \rightarrow \Gamma_{\mathfrak{A}}$ is univalent. It is obvious that if $\mathfrak{A}$ and $\mathfrak{B}$ are ideals in $Rv$ and $\mathfrak{A} \supset \mathfrak{B}$, then $\Gamma_{\mathfrak{A}} \subset \Gamma_{\mathfrak{B}}$. Hence the mapping $\mathfrak{A} \rightarrow \Gamma_{\mathfrak{A}}$ reverses order.

Let $\Delta$ be an arbitrary segment of $\Gamma$, different from $\Gamma$, and let $L$ be the set of all positive elements of $\Gamma$ which do not belong to $\Delta$. We set $\mathfrak{A} = Lv^{-1}$. The fact that $\Delta$ is a segment implies that $L + \Gamma_+ \subset L$. Hence $\mathfrak{A}R_v \subset \mathfrak{A}$. Furthermore, if $x, y \in \mathfrak{A}$ and if, say, $v(x) \leqq v(y)$, then $v(x-y) \geqq v(x) \in L$, and hence $v(x-y) \in L$ (since $\Delta$ is a segment) and $x - y \in \mathfrak{A}$ (since $\mathfrak{A} = Lv^{-1}$). We have proved that $\mathfrak{A}$ is an ideal. Since $L$ is non-empty and does not contain the zero of $\Gamma$, $\mathfrak{A}$ is a proper ideal. Thus everything is proved, except the last part of the theorem.

We observe that an ideal $\mathfrak{A}$ is prime if and only if its complement in $R_v$ is closed under multiplication. Hence $\mathfrak{A}$ is prime if and only if the set of non-negative elements of $\Gamma_{\mathfrak{A}}$ is closed under addition. But since $\Gamma_{\mathfrak{A}}$ is a segment, this property of the set of non-negative elements of $\Gamma_{\mathfrak{A}}$ is equivalent to the group property of $\Gamma_{\mathfrak{A}}$. Hence $\Gamma_{\mathfrak{A}}$ is a subgroup of $\Gamma$ (necessarily isolated) if and only if $\mathfrak{A}$ is a proper prime ideal of $R_v$. This completes the proof of the theorem.

In the sequel we shall also speak of the *rank* of any ordered abelian group $\Gamma$; we mean by that the ordinal type of the set of all isolated subgroups of $\Gamma$.

THEOREM 16. *The valuation ring $R_v$ is noetherian if and only if the value group $\Gamma$ of $v$ is the additive group of integers.*

PROOF. We first show that if $R_v$ is noetherian then $v$ must be of rank 1. For suppose that $v$ is of rank greater than 1. Since the null-group is an isolated subgroup of $\Gamma$, there must exist an isolated subgroup $\Delta$ different from (0). Fix a positive element $\alpha$ in $\Delta$. Then $\alpha < 2\alpha < \cdots < n\alpha < \cdots$. Since $\Delta$ is a proper subgroup of $\Gamma$ we can find in $\Gamma$ a positive element $\beta$ which does not belong to $\Delta$. Since $\Delta$ is a segment and since the elements $n\alpha$ belong to $\Delta$, it follows that $\beta > n\alpha$, $n = 1, 2, \cdots$. We thus have in $\Gamma$ a strictly descending sequence $\beta$, $\beta - \alpha$, $\beta - 2\alpha$, $\cdots$ of *positive* elements. Such a sequence determines an infinite strictly descending sequence of segments of $\Gamma$, and therefore, by Theorem 15, we have an infinite strictly ascending sequence of ideals in $R_v$. Hence $R_v$ is not noetherian.

Let now $v$ be of rank 1. If $R_v$ is noetherian, there must be a least positive element in $\Gamma$, say $\alpha$. Then if $n$ is any integer, no element of $\Gamma$ can lie between $n\alpha$ and $(n+1)\alpha$, for in the contrary case there would also be elements between 0 and $\alpha$. Hence the set of all multiples $n\alpha$ of $\alpha$ ($n = 0, \pm 1, \pm 2, \cdots$) is a segment. Since this set is also a subgroup of $\Gamma$, it follows that this set coincides with $\Gamma$, for otherwise $v$ would be of rank $> 1$. We have thus proved that if $R_v$ is noetherian, then $\Gamma$ is isomorphic with the additive group of integers. The converse is obvious, for the group of integers contains no infinite strictly descending sequence of segments.

We give another proof of Theorem 16, which does not make use of Theorem 15. We first observe that the following holds in any valuation ring $R_v$: *if an ideal $\mathfrak{A}$ in $R_v$ has a finite basis, then $\mathfrak{A}$ is a principal ideal.* For if $\{x_1, x_2, \cdots, x_n\}$ is a basis of $\mathfrak{A}$ and if, say, $x_1$ is an element of the basis having least value in $v$, then $x_i/x_1 \in R_v$, and hence $\mathfrak{A}$ is the principal ideal $(x_1)$. Let us suppose now that $R_v$ is noetherian. By the above remark, $R_v$ is then a principal ideal ring. Let $t$ be a generator of the maximal ideal $\mathfrak{M}_v$ of $R_v$. Then any element of $R_v$ which is not divisible by $t$ is a unit. A familiar and straightforward argument shows that no element of $R_v$ (different from zero) can be divisible by *all* powers of $t$ (if $x=t^n a_n$, $a_n \in R_v$, $n=1, 2, \cdots$, then the principal ideals $(a_1)$, $(a_2), \cdots, (a_n), \cdots$ would form a strictly ascending chain). It follows that every element $x$ of $R_v$, $x \neq 0$, can be put (uniquely) in the form $at^n$, where $n \geqq 0$ and $a$ is a unit. This shows that the principal ideals $(t^n)$, $n=1, 2, \cdots$, are *all* the proper ideals of $R_v$. Hence the maximal ideal $(t)$ of $R_v$ is the only proper *prime* ideal of $R_v$, whence $v$ is of rank 1. Furthermore, it is immediately seen that if $K'$ denotes, as usual, the multiplicative group of the field $K$ and $E$ is the set of units in $R_v$, then the quotient group $K'/E$, written additively, is isomorphic to the group of integers. The given valuation $v$ is necessarily equivalent to the valuation $v'$ obtained by setting $v'(at^n)=n$, if $a$ is a unit.

A valuation of rank 1 is said to be *discrete* if its value group is the additive group of integers. Thus, Theorem 16 states that a valuation ring $R_v$ is noetherian if and only if $v$ is a discrete valuation of rank 1.

COROLLARY 1. *An integrally closed local domain in which the ideal of non-units is the only proper prime ideal is a discrete valuation ring of rank* 1.

This follows from § 5, Theorem 9, Corollary 3.

COROLLARY 2. *If $R$ is an integrally closed noetherian domain and $\mathfrak{p}$ is a minimal prime ideal in $R$, then the quotient ring $R_\mathfrak{p}$ is a discrete valuation ring of rank* 1.

For, the ring $R_\mathfrak{p}$ satisfies then the assumptions of the preceding corollary (cf. Vol. I, Ch. V, § 6, Theorem 14, Corollary).

We add another important result concerning noetherian integrally closed domains $R$. Let $S$ denote the set of minimal prime ideals in $R$. If $\mathfrak{p} \in S$, we denote by $v_\mathfrak{p}$ the unique valuation of the quotient field $K$ of $R$ which is non-negative on $R$ and has center $\mathfrak{p}$. By Corollary 2, the valuation ring of $v_\mathfrak{p}$ is $R_\mathfrak{p}$, and each $v_\mathfrak{p}$ is discrete, of rank 1.

COROLLARY 3. *Let $K$ be the quotient field of an integrally closed noetherian domain $R$. If $w$ is any element of $K$, $w \neq 0$, then* (1) *there is only a finite number of prime ideals $\mathfrak{p}$ in the set $S$ such that $v_\mathfrak{p}(w) \neq 0$;* (2) *$w$ belongs to $R$ if and only $v_\mathfrak{p}(w) \geqq 0$ for all $\mathfrak{p}$ in $S$; furthermore* (3) *$w$ is a unit in $R$ if and only if $v_\mathfrak{p}(w)=0$ for all $\mathfrak{p}$ in $S$.*

If $w \in R$, then $Rw=\mathfrak{p}_1^{(n_1)} \cap \mathfrak{p}_2^{(n_2)} \cap \cdots \cap \mathfrak{p}_s^{(n_s)}$, where $s \geqq 0$, the $\mathfrak{p}_i$ are minimal prime ideals in $R$, $n_i \geqq 1$ and $s=0$ if and only if $w$ is a unit (see Vol. I, Ch. V, § 6, Theorem 14, Corollary 1). If follows at once that $v_{\mathfrak{p}_i}(w)=n_i$, $i=1, 2, \cdots, s$, and $v_\mathfrak{p}(w)=0$ if $\mathfrak{p} \in S$ and $\mathfrak{p} \neq \mathfrak{p}_1$, $\mathfrak{p}_2, \cdots \mathfrak{p}_s$. This proves (1) in the case in which $w \in R$ and therefore also in the general case. If $w \in K$, $w \neq 0$, we write $w = w_1/w_2$, $w_i \in R$. If $v_{\mathfrak{p}}(w_1) \geqq v_{\mathfrak{p}}(w_2)$ for all $\mathfrak{p}$ in $S$, then in view of the relations

$$Rw_1 = \bigcap_{\mathfrak{p} \in S} \mathfrak{p}^{(v_{\mathfrak{p}}(w_1))}$$

$$Rw_2 = \bigcap_{\mathfrak{p} \in S} \mathfrak{p}^{(v_{\mathfrak{p}}(w_2))},$$

it follows that $Rw_1 \subset Rw_2$ and hence $w_1/w_2 \in R$. This proves (2). The last part of the corollary is now obvious.

We now go back to the study of general valuations and we add first some remarks about isolated subgroups, which we shall presently make use of.

Let $\Delta$ be an isolated subgroup of $\Gamma$. It is immediately seen that the canonical homomorphism of $\Gamma$ onto $\Gamma/\Delta$ defines a total ordering in $\Gamma/\Delta$, in the following fashion: an element of $\Gamma/\Delta$ shall be, by definition, *non-negative* if it corresponds to a non-negative element of $\Gamma$. From now on, when we speak of $\Gamma/\Delta$ as a totally *ordered* group we mean that $\Gamma/\Delta$ has been ordered in the above fashion.

In the canonical homomorphism of $\Gamma$ onto $\Gamma/\Delta$, the isolated subgroups of $\Gamma$ which contain $\Delta$ correspond in (1, 1) fashion to the isolated subgroups of $\Gamma/\Delta$. Since every isolated subgroup of $\Gamma$ either contains or is contained in $\Delta$, it follows that *if $\xi$ is the rank of $\Delta$ and $\eta$ is the rank of $\Gamma/\Delta$, then the rank of $\Gamma$ is $\xi + \eta$.*

In § 3, we have defined specialization of places. The valuation-theoretic interpretation of this concept leads to the notion of *composite valuations.* Let $v$ be a valuation of $K$, of rank $> 1$. There exists then another valuation $v_1$ of $K$ such that $R_v < R_{v_1}$. Let $\mathscr{P}$ and $\mathscr{P}_1$ be the places of $K$ which are defined respectively by the canonical homomorphism of $R_v$ onto $R_v/\mathfrak{M}_v$ and of $R_{v_1}$ onto $R_{v_1}/\mathfrak{M}_{v_1}$. Then $\mathscr{P}$ is a proper specialization of $\mathscr{P}_1$ and we have $\mathscr{P} = \mathscr{P}_1\bar{\mathscr{P}}$, where $\bar{\mathscr{P}}$ is a place of $R_{v_1}/\mathfrak{M}_{v_1}$. Let $\bar{v}$ be the valuation of $R_{v_1}/\mathfrak{M}_{v_1}$ determined by $\bar{\mathscr{P}}$. We then say that $v$ is a *composite valuation*, that it is *composite with the valuations* $v_1$ and $\bar{v}$ and we write $v = v_1 \circ \bar{v}$.

Let $\mathfrak{P}$ denote the prime ideal of $v_1$. We know (§ 3) that $\mathfrak{P}$ is also a prime ideal in $R_v$. If, then, $\Gamma$ is the value group of $v$, $\mathfrak{P}$ determines an isolated subgroup $\Delta$ of $\Gamma$ (see Theorem 15). We shall now prove the following theorem:

THEOREM 17. *The value group $\Gamma_1$ of $v_1$ and the group $\Gamma/\Delta$ are isomorphic (as ordered groups). Similarly, the value group $\bar{\Gamma}$ of $\bar{v}$ and the group $\Delta$ are isomorphic.*

PROOF. Let $E$ and $E_1$ denote, respectively, the set of units in $R_v$ and

$R_{v_1}$ respectively. We first observe that $E_1$ *is the full inverse image of* $\Delta$ *under* $v^{-1}$. For if $x$ is any element of $E_1$, then $x = y/z$, where $y$ and $z$ are elements of $R_v$, *not in* $\mathfrak{P}$ (since $R_{v_1}$ is the quotient ring of $R_v$ with respect to $\mathfrak{P}$). Then $v(z)$ is a non-negative element of $\Gamma$ which does not belong to $\mathfrak{P}v$, and hence, by the definition of $\Delta$, $v(z)$ must belong to $\Delta$. Similarly for $v(y)$. Since $\Delta$ is a group, it follows that $v(x) \in \Delta$. Conversely, if $x$ is an element of $K'$ such that $v(x)$ belongs to $\Delta$, then neither $v(x)$ nor $v(1/x)$ belongs to $\mathfrak{P}v$. Since $\mathfrak{P}$ is the full inverse image of $\mathfrak{P}v$ under $v^{-1}$, it follows that neither $x$ nor $1/x$ can belong to $\mathfrak{P}$. Hence $x$ is a unit in $R_{v_1}$. This establishes our assertion that $E_1$ is the full inverse image of $\Delta$ under $v^{-1}$.

We can therefore assert that

(a) *the restriction of* $v$ *to* $E_1$ *is a homomorphism of the multiplicative group* $E_1$ *onto the additive group* $\Delta$, *and the kernel of this homomorphism is* $E$.

Now, $v$ and $v_1$ are homomorphisms of $K'$ onto $\Gamma$ and $\Gamma_1$ respectively, with kernels $E$ and $E_1$. Since $E_1 \supset E$, it follows that $v^{-1}v_1$ is a homomorphism of $\Gamma$ onto $\Gamma_1$. By (a), the kernel of this homomorphism is precisely the isolated subgroup $\Delta$. Hence $\Gamma_1$ and $\Gamma/\Delta$ are isomorphic as groups. If $\alpha$ is a non-negative element of $\Gamma$, then the set $\alpha v^{-1}$ is contained in $R_v$, hence also in $R_{v_1}$, and therefore the element $\alpha v^{-1}v_1$ is non-negative. Hence the groups $\Gamma_1$ and $\Gamma/\Delta$ are isomorphic also as ordered groups and this completes the proof of the first part of the theorem.

Now consider the product $\mathscr{P}_1\bar{v}$. This transformation into $\bar{\Gamma}$ is defined for those and only those elements $x$ of $K$ for which $x\mathscr{P}_1 \neq 0, \infty$. Hence the domain of $\mathscr{P}_1\bar{v}$ is $E_1$, and the range of $\mathscr{P}_1\bar{v}$ is the value group $\bar{\Gamma}$ of $\bar{v}$. The transformation $\mathscr{P}_1\bar{v}$ is clearly a homomorphism (of the multiplicative group $E_1$ onto the additive group $\bar{\Gamma}$). Its kernel consists of those elements $x$ for which $x\mathscr{P}_1$ has value zero in $\bar{v}$, i.e., of those elements $x$ for which $x\mathscr{P}_1\bar{\mathscr{P}} \neq 0, \infty$. Since $\mathscr{P}_1\bar{\mathscr{P}} = \mathscr{P}$, we conclude that the kernel of $\mathscr{P}_1\bar{v}$ is $E$. Comparing this result with (a), we conclude that $\bar{\Gamma}$ and $\Delta$ are isomorphic as groups. An element $x$ of $E_1$ is mapped by $v$ into a non-negative element of $\Delta$ if and only if $x$ belongs to $R_v$. On the other hand, an element $x$ of $E_1$ is mapped by $\mathscr{P}_1\bar{v}$ into a non-negative element of $\bar{\Gamma}$ if and only if $x\mathscr{P}_1\bar{\mathscr{P}} \neq \infty$, i.e., if and only if $x\mathscr{P} \neq \infty$, hence again if and only if $x \in R_v$. This shows that $\bar{\Gamma}$ and $\Delta$ are isomorphic also as ordered groups, and this completes the proof of the theorem.

COROLLARY. *Rank of* $v$ = *rank of* $\bar{v}$ + *rank of* $v_1$.

The only valuations encountered in most applications (and, in particular, in algebraic geometry) are valuations *of finite rank* (see § 3, Definition 1, Corollary 1), and we shall now derive some properties of such valuations.

An *archimedean* totally ordered (additive) group $\Gamma$ is one satisfying the following condition: if $\alpha$ and $\beta$ are any two elements of $\Gamma$ and $\alpha > 0$, then there exists an integer $n$ such that $n\alpha > \beta$. Let $\Gamma$ be archimedean and let $\Delta$ be an isolated subgroup of $\Gamma$. It follows at once from the above definition that if $\Delta$ contains a positive element $\alpha$ then $\Delta$ coincides with $\Gamma$, contrary to the fact that an isolated subgroup of $\Gamma$ is, according to our definition, a proper subgroup of $\Gamma$. Hence (0) is the only isolated subgroup of $\Gamma$, and $\Gamma$ is therefore of rank 1. Conversely, suppose that $\Gamma$ is a totally ordered group of rank 1, and let $\alpha$ be a positive element of $\Gamma$. The set of all elements $\pm\beta$, where $\beta$ is a non-negative element of $\Gamma$ such that $n\alpha > \beta$ for a suitable $n$ (depending on $\beta$), is a segment and a subgroup of $\Gamma$, and this set does not consist only of the element 0, for $\alpha$ belongs to the set. Since $\Gamma$ is of rank 1, it follows that the above set coincides with $\Gamma$, and hence $\Gamma$ is archimedean. We have thus proved that *an ordered group is archimedean if and only if it is of rank* 1.

The following well-known argument shows that *every archimedean ordered abelian group $\Gamma$ is isomorphic to a subgroup of the ordered additive group of real numbers* (and therefore valuations of rank 1 are frequently referred to as *real* valuations).

We fix a positive element $\alpha$ of $\Gamma$. If $\beta$ is any element of $\Gamma$ we divide the set of all rational numbers $m/n$ $(n > 0)$ into two classes $C_1$ and $C_2$, as follows: $m/n \in C_1$ if $m\alpha < n\beta$, and $m/n \in C_2$ if $m\alpha \geqq n\beta$. The fact that $\Gamma$ is archimedean insures that neither $C_1$ nor $C_2$ is empty. It is then seen immediately that the pair of classes $C_1$, $C_2$ defines a Dedekind cut in the set of rational numbers. If $b$ is the real number defined by this Dedekind cut, we set $\varphi(\beta) = b$. It is then easily verified that $\varphi$ is an order preserving isomorphism of $\Gamma$ into the set of real numbers. Note that $\varphi$ depends on the choice of the fixed positive element $\alpha$ of and that $\varphi(\alpha) = 1$.

We have proved earlier (§ 7, Lemma 2) an approximation theorem expressing the independence of any finite set of places, provided no place in the set is a specialization of any other place in the set. For valuations *of rank* 1 we have the following stronger approximation theorem:

THEOREM 18. *Let $v_1, v_2, \cdots, v_h$ be rank* 1 *valuations of a field $K$, with value groups $\Gamma_1, \Gamma_2, \cdots, \Gamma_h$ respectively. (We may assume that each $\Gamma_i$ consists of real numbers.) Given h arbitrary elements $u_1, u_2, \cdots, u_h$ of*

*K and h arbitrary elements* $\alpha_1, \alpha_2, \cdots, \alpha_h$ *of* $\Gamma_1, \Gamma_2, \cdots, \Gamma_h$ *respectively, there exists an element u of K such that*

$$v_i(u-u_i) = \alpha_i, \quad i = 1, 2, \cdots, h. \tag{1}$$

PROOF. It will be sufficient to prove the following: given any integer $m$, there exists an element $x$ in $K$ such that

$$v_i(x-u_i) \geqq m, \quad i = 1, 2, \cdots, h. \tag{2}$$

For, assume that this has already been proved. We then fix an integer $m$ such that $m > \alpha_i$, $i=1, 2, \cdots, h$, and for each $i$ we fix an element $x_i$ in $K$ such that $v_i(x_i)=\alpha_i$. By assumption, there exists an element $y$ in $K$ such that $v_i(y-x_i) \geqq m$, $i=1, 2, \cdots, h$. Since $y=(y-x_i)+x_i$ and $v_i(y-x_i) > v_i(x_i)$, we conclude that $v_i(y)=\alpha_i$, $i=1, 2, \cdots, h$. Now let $x$ be an element of $K$ satisfying the inequalities (2) and let $u=x+y$. We have $u-u_i=(x-u_i)+y$ and $v_i(y)=\alpha_i<m\leqq v_i(x-u_i)$. Hence $v_i(u-u_i)=v_i(y)=\alpha_i$, $i=1, 2, \cdots, h$, i.e., $u$ satisfies relations (1).

Since the valuations $v_i$ are of rank 1, Lemma 1 of § 7 is applicable. There exists therefore a set of elements $\eta_1, \eta_2, \cdots, \eta_h$ in $K$ such that $v_i(\eta_i)=0$ and $v_i(\eta_j)>0$ if $i\neq j$, for $i, j=1, 2, \cdots, h$. We replace the elements $\eta_i$ by the following elements $\zeta_i$ (compare with the proof of Lemma 2, § 7):

$$\zeta_i = \eta_i/(\eta_1+\eta_2+ \cdots +\eta_h), \quad i = 1, 2, \cdots, h.$$

Then it remains true that $v_i(\zeta_i)=0$ and $v_i(\zeta_j)>0$ if $i\neq j$, but furthermore we have that the $v_i$-residue of $\zeta_i$ is equal to the element 1 of the residue field $R_{v_i}/\mathfrak{M}_{v_i}$. Hence $v_i(\zeta_i-1)>0$, where 1 now stands for the element 1 of $K$.

We now fix a positive integer $n$ satisfying the following conditions:

$$nv_i(\zeta_i-1)+v_i(u_i) \geqq m, \quad i = 1, 2, \cdots, h; \tag{3}$$

$$nv_j(\zeta_i)+v_j(u_i) \geqq m, \quad i\neq j;\ i, j = 1, 2, \cdots, h. \tag{4}$$

(*Note.* If for some $i$ we have $u_i=0$, then the corresponding equation (3) (or (4)) imposes no condition on the integer $n$, for $v_i(0)$ is interpreted then as $+\infty$.)

Consider the following elements $\xi_i$ of $K$:

$$\xi_i = 1-(1-\zeta_i{}^n)^n, \quad i = 1, 2, \cdots, h.$$

We have: $v_i(\xi_i-1)=nv_i(1-\zeta_i{}^n) \geqq nv_i(1-\zeta_i)$, whence, by (3):

$$v_i[u_i(\xi_i-1)] \geqq m. \tag{5}$$

We also have: $\xi_i=\zeta_i{}^n f(\zeta_i)$, where $f$ is a polynomial with coefficients in the prime ring. Hence, if $i \neq j$ then $v_j(\xi_i) \geqq nv_j(\zeta_i)$, and therefore, in view of (4):

$$(6) \qquad v_j(u_i\xi_i) \geqq m.$$

If we now set $x = u_1\xi_1 + u_2\xi_2 + \cdots + u_h\xi_h$, then it follows at once from (5) and (6) that the element $x$ satisfies the inequalities (2). This completes the proof of the theorem.

The above approximation theorem holds also for valuations of arbitrary rank provided the valuations $v_1, v_2, \cdots, v_h$ are *independent* in the sense of the following definition: *the valuations $v_1, v_2, \cdots, v_h$ are said to be independent if no two of them are composite with one and the same non-trivial valuation.* We shall prove therefore the following:

THEOREM 18′. *The approximation theorem* (*Theorem* 18) *remains valid if the valuations $v_1, v_2, \cdots, v_h$ are independent* (*and not necessarily of rank* 1).

PROOF. It will be sufficient to prove the existence of an element $w$ in $K$ such that the inequalities

$$(7) \qquad v_i(w - u_i) > \alpha_i, \quad i = 1, 2, \cdots, h,$$

hold (the $\alpha_i$ and $u_i$ being arbitrary, as in Theorem 18). For assume that this has already been proved. We then fix an element $x_i$ in $K$ such that $v_i(x_i) = \alpha_i$ and an element $y$ in $K$ such that $v_i(y - x_i) > \alpha_i$, $i = 1, 2, \cdots, h$. We have then $v_i(y) = v_i(y - x_i + x_i) = \alpha_i$. We then determine an element $x$ in $K$ such that $v_i(x - u_i) > \alpha_i$ and we set $u = x + y$. Then $v_i(u - u_i) = v_i(x - u_i + y) = \alpha_i$, since $v_i(y) = \alpha_i < v_i(x - u_i)$.

To prove the existence of an element $w$ satisfying the $h$ inequalities (7) we proceed as follows:

We set $\alpha_{ij} = \alpha_i - v_i(u_j)$ if $u_j \neq 0$ and $\alpha_{ij} = 0$ if $u_j = 0$ $(i, j = 1, 2, \cdots, h)$. Let $\beta_i = \max\{\alpha_{i1}, \alpha_{i2}, \cdots, \alpha_{ih}\}$. If $\beta_i > 0$ then we denote by $\Delta_i$ the greatest isolated subgroup of $\Gamma_i$ which does not contain $\beta_i$ ($\Delta_i$ exists: it is the union of all the isolated subgroups of $\Gamma_i$ which do not contain $\beta_i$). If $\beta_i \leqq 0$ we take for $\Delta_i$ the zero of $\Gamma_i$. If $\Delta_i \neq (0)$ we denote by $v'_i$ the valuation of $K$ whose value group is the group $\Gamma'_i = \Gamma_i/\Delta_i$ and with which $v_i$ is composite. If $\Delta_i = (0)$, we set $v'_i = v_i$. Let $\beta'_i$ be the coset $\beta_i + \Delta_i$.

It is clear, by the definition of $\Delta_i$, that if $\beta'_i > 0$ then the zero of $\Gamma'_i$ is the only isolated subgroup of $\Gamma'_i$ which does not contain $\beta'_i$. Now any positive element $\gamma'$ of $\Gamma'_i$ determines a smallest isolated subgroup containing $\gamma'$: it is the subgroup of $\Gamma'_i$ consisting of all the elements $\pm\delta'$ such that $\delta' \geqq 0$ and such that $n\gamma' > \delta'$ for some integer $n$. It follows that *for any positive element $\gamma'$ of $\Gamma'_i$ there exists an integer $n$ (depending* *on* $\gamma'$) *such that* $n\gamma' > \beta'$, *and this is true for* $i = 1, 2, \cdots, h$. Going back to the value groups $\Gamma_i$ we can express this property as follows: *if* $\gamma$ *is any positive element of* $\Gamma_i$, *not in* $\Delta_i$, *then there exists an integer* $n$ *such that* $n\gamma > \beta_i$. Another fact that has to be taken into account is the following: *If* $i \neq j$ *then* $K_{v'_i} \not\supset K_{v'_j}$. For, in the contrary case, both $v_i$ and $v_j$ would be composite with the non-trivial valuation $v'_i$. From this fact follows, by Lemma 2, § 7, the existence of elements $\zeta_1, \zeta_2, \cdots, \zeta_h$ in $K$ such that $v'_i(\zeta_i - 1) > 0$ and $v'_j(\zeta_i) > 0$ if $j \neq i$ ($i, j = 1, 2, \cdots, h$). Hence, in view of the above mentioned property, we can find an integer $n$ such that

$$nv_i(\zeta_i - 1) > \beta_i, \quad nv_j(\zeta_i) > \beta_j, \quad \text{if} \quad j \neq i, \quad i, j = 1, 2, \cdots, h.$$

From the definition of the elements $\beta_i$ it follows then that we have for all $i$ such that $u_i \neq 0$:

$$nv_i(\zeta_i - 1) + v_i(u_i) > \alpha_i,$$
$$nv_j(\zeta_i) + v_j(u_i) > \alpha_j, \quad \text{if} \quad j \neq i.$$

Hence, if we consider the elements $\xi_i = 1 - (1 - \zeta_i^n)^n$ introduced in the proof of Theorem 18, we find that if $u_i \neq 0$ then $v_i(u_i\xi_i - u_i) > \alpha_i$ and $v_j(u_i\xi_i) > \alpha_j$, and that therefore the element $w = u_1\xi_1 + u_2\xi_2 + \cdots + u_h\xi_h$ satisfies the inequalities (7). This completes the proof of the Theorem.

REMARK. Concerning the notion of independent and dependent valuations we point out the following criterion: *two valuations* $v$ *and* $v'$ *of* $K$ *are dependent if and only if some proper prime ideal of* $K_v$ *coincides with a prime ideal of* $K_{v'}$. The "only if" is obvious. On the other hand, if $K_v$ and $K_{v'}$ have in common a proper prime ideal $\mathfrak{p}$, then $v$ is composite with a non-trivial valuation $v_1$ such that $\mathfrak{M}_{v_1} = \mathfrak{p}$. Similarly, $v'$ is composite with a valuation $v'_1$ such that $\mathfrak{M}_{v_1'} = \mathfrak{p}$. From $\mathfrak{M}_{v_1} = \mathfrak{M}_{v_1'}$ follows $K_{v_1} = K'_{v_1}$, $v_1 = v'_1$ and hence $v$ and $v'$ are dependent.

We add some final remarks concerning (A) *discrete* ordered groups of finite rank and (B) *the rational rank* of a valuation.

(A) Let $\Gamma$ be a totally ordered (abelian) group of finite rank $n$ and let $\Gamma_0 = (0), \Gamma_1, \cdots, \Gamma_{n-1}$ be its isolated subgroups: $\Gamma_0 < \Gamma_1 < \cdots < \Gamma_{n-1} < \Gamma$. It is clear that the quotient groups $\Gamma_{i+1}/\Gamma_i$, $i = 0, 1, \cdots, n-1$ ($\Gamma_n = \Gamma$), are groups of rank 1. If each of these quotient groups is isomorphic to the group of integers, then the ordered group $\Gamma$ is said to be a *discrete* group. A discrete ordered group of rank 1 is, then, a group isomorphic to the group of integers. A valuation is called *discrete* if its value group is discrete.

We now observe, quite generally, that given a finite set of ordered groups $G_1, G_2, \cdots, G_m$, then the direct product $G^\star = G_1 \times G_2 \times \cdots \times G_m$ can be ordered *lexicographically*, as follows: $\alpha^\star = (\alpha_1, \alpha_2, \cdots, \alpha_m) > 0 (\alpha_i \in G_i)$, if the first $\alpha_i$ which is not zero is positive. If $H$ is an isolated subgroup of $G_s (1 \leqq s \leqq m)$, then the elements $\alpha^\star$ of $G^\star$ such that $\alpha_1 = \alpha_2 = \cdots = \alpha_{s-1} = 0$, $\alpha_s \in H$, form an isolated subgroup of $G^\star$, and in this fashion all the isolated subgroups of $G^\star$ can be obtained. It follows at once that *the rank of $G^\star$ is equal to the sum of the ranks of $G_m, G_{m-1}, \cdots, G_1$* (in this order).

With this observation in mind, we now show that a *discrete totally ordered group $\Gamma$, of rank $n$, is isomorphic to the direct product $G_0 \times G_0 \times \cdots \times G_0$ ($n$ times), where $G_0$ is the group of integers.* We sketch the proof. Let $\varphi_i$ be the isomorphism of $\Gamma_{i+1}/\Gamma_i$ onto $G_0$, where $\Gamma_0, \Gamma_1, \cdots, \Gamma_{n-1}$ are the isolated subgroups of $\Gamma$ and where $\Gamma_n = \Gamma$. For each $i = 0, 1, 2, \cdots, n-1$, we fix in $\Gamma_{i+1}$ a positive element $\alpha_{n-i}$ such that the $\Gamma_i$-coset of $\alpha_{n-i}$ is mapped by $\varphi_i$ into the integer 1. Then each element $\alpha$ of $\Gamma$ can be expressed in one and only one way as a linear combination of $\alpha_1, \alpha_2, \cdots, \alpha_n$, with integral coefficients: $\alpha = m_1\alpha_1 + m_2\alpha_2 + \cdots + m_n\alpha_n$. It is then found that $\alpha > 0$ if and only if the first of the non-zero coefficients $m_i$ is positive. Hence the mapping $\varphi$: $\alpha \rightarrow (m_1, m_2, \cdots, m_n)$ is an order preserving isomorphism of $\Gamma$ onto the direct product $G_0 \times G_0 \times \cdots \times G_0$ ($n$ times).

It should be noted that the isomorphism $\varphi$ which we have just constructed depends on the choice of the $n$ elements $\alpha_{n-i}$. Suppose that $\alpha'_1, \alpha'_2, \cdots, \alpha'_n$ is another set of elements of $\Gamma$ with the property that $\alpha'_{n-i} \in \Gamma_{i+1}$ and the $\Gamma_i$-coset of $\alpha'_{n-1}$ is mapped by $\varphi_i$ into 1, and let $\varphi'$ denote the isomorphism similar to $\varphi$ and relative to this new set of elements $\alpha'_1, \alpha'_2, \cdots, \alpha'_n$. Since $\alpha'_{n-i} - \alpha_{n-i} \in \Gamma_i$ it follows that

$$\alpha'_{n-i} = \alpha_{n-i} + q_{n-i,\, n-i+1}\alpha_{n-i+1} + \cdots + q_{n-i,\, n}\alpha_n, \quad i = 0, 1, \cdots, n-1'$$

where the $q_{\nu\mu}$ are integers. If we then write $\alpha = m'_1\alpha'_1 + m'_2\alpha_2, + \cdots + m'_n\alpha'_n$, then the following are the equations of the order preserving automorphism $\varphi^{-1}\varphi'$ of $G_0 \times G_0 \times \cdots \times G_0$:

$$m_1 = m'_1; \quad m_2 = q_{12}m'_1 + m'_2, \quad m_3 = q_{13}m'_1 + q_{23}m'_2 + m'_3, \quad \text{etc.}$$

(B) In addition to the rank of a valuation $v$ we also introduce the so-called *rational rank* of $v$. If $\Gamma$ is the value group of $v$ and $\alpha_1, \alpha_2, \cdots, \alpha_m$ are elements of $\Gamma$, we say that the $\alpha$'s are *rationally dependent* if there exist integers $n_1, n_2, \cdots, n_m$, not all zero, such that $n_1\alpha_1 + n_2\alpha_2 + \cdots + n_m\alpha_m = 0$. In the contrary case, the $\alpha$'s are said to be *rationally independent.*

DEFINITION. *The maximum number of rationally independent elements of $\Gamma$ is called the rational rank of $v$* (the rational rank of $v$ may be infinite).

LEMMA. *Let $v$ be a valuation of $K/k$ and let $x_1, x_2, \ldots, x_s$ be elements of $K$, different from zero. If $x_1, x_2, \cdots, x_s$ are algebraically dependent over $k$, then $v(x_1), v(x_2), \cdots, v(x_s)$ are rationally dependent.*

PROOF. Let $f(X_1, X_2, \cdots, X_s)$ be a non-zero polynomial in $k[X]$ such that $f(x_1, x_2, \cdots, x_s)=0$. As has been pointed out in § 8, the valuation axioms imply then that there must exist a pair of distinct terms in the polynomial $f(X)$, say $aX_1^{i_1}X_2^{i_2} \cdots X_s^{i_s}$ and $bX_1^{j_1}X_2^{j_2} \cdots X_s^{j_s}$, such that $v(ax_1^{i_1}x_2^{i_2} \cdots x_s^{i_s})=v(bx_1^{j_1}x_2^{j_2} \cdots x_s^{j_s})$, where $a$, $b$ are non-zero elements of $k$. Since $v(a)=v(b)=0$, it follows that $(i_1-j_1)v(x_1)+(i_2-j_2)v(x_2)+ \cdots +(i_s-j_s)v(x_s)=0$, and this establishes the lemma, since the $s$ integers $i_\nu-j_\nu$ are not all zero.

COROLLARY. *If $K/k$ is a field of algebraic functions of $r$ independent variables, then the rational rank of any valuation of $K/k$ is not greater than $r$.*

NOTE. We observe that the rank of a valuation $v$ is never greater than the rational rank of $v$ whenever the rational rank is finite. To show this we have only to show the following: if $\Gamma_0<\Gamma_1 <\cdots <\Gamma_{h-1}$ is a finite, strictly ascending chain of isolated subgroups of $\Gamma$ and if for each $i=1, 2, \cdots, h$ we fix an element $\alpha_i$ which belongs to $\Gamma_i$ and not to $\Gamma_{i-1}$ ($\Gamma_h=\Gamma$), then $\alpha_1, \alpha_2, \cdots, \alpha_h$ are rationally independent. Assume then that we have a relation $m_1\alpha_1+m_2\alpha_2+ \cdots +m_g\alpha_g=0$, where the $m_i$ are integers, $m_g\neq 0$ and $g\leq h$. Then $m_g\alpha_g \in \Gamma_{g-1}$, and since $\Gamma_{g-1}$ is a segment and $m_g\neq 0$ it follows that $\alpha_g \in \Gamma_{g-1}$, a contradiction. In particular, a valuation of rational rank 1 is necessarily a real valuation. Its value group may be assumed to consist of rational numbers and for that reason a valuation of rational rank 1 is sometimes called a *rational* valuation.

**§ 11. Valuations and field extensions.** Let $K$ be a field and let $K^\star$ be an overfield of $K$. If $v^\star$ is a valuation of $K^\star$, the restriction $v$ of $v^\star$ to $K$ is clearly a valuation of $K$ ($v$ *may* be trivial even if $v^\star$ is non-trivial). The valuation ring of $v$ is then given by $R_{v^\star} \cap K$, and the valuation $v^\star$ is said to be *an extension* of $v$. If $v^\star$ is an extension of $v$ and if $\mathscr{P}^\star$ is any place of $K^\star$ whose valuation ring is $R_{v^\star}$, then the restriction $\mathscr{P}$ of $\mathscr{P}^\star$ to $K$ is a place of $K$ whose valuation ring is $R_v$. It follows that the results of §§ 6–7 on extensions of places, when translated into the language of valuation theory, yield corresponding results on extensions of valuations. However, in the valuation-theoretic interpretation of these results it must be observed that isomorphic places are associated with one and the same valuation, and corresponding formal changes must be made in the statements of those results. Any reference to isomorphic places should be replaced by a reference to *one* valuation, while any mention of "non-isomorphic places" should be replaced by that of "distinct valuations". In particular, we point out explicitly the following changes:

In § 6, Lemma 1: The relation $R_{v^\star} \cap K = R_v$ is not only a necessary but also a sufficient condition for $v^\star$ to be an extension of $v$.

In § 7, Theorem 12, Corollary 3: The field $K^\star$ is now a normal algebraic extension of $K$, and the result is to the effect that if $v$ is any valuation of $K$, then any two extensions $v_1^\star$ and $v_2^\star$ of $v$ in $K^\star$ are conjugate over $K$($v_1^\star$ and $v_2^\star$ are conjugate valuations of $K^\star$, over $K$, if $v_2^\star = sv_1^\star$, where $s$ is a $K$-automorphism of $K^\star$).

Our principal object in this section is to derive some partial but basic results on extensions of valuations, in which the value groups of the valuations come into play. We shall be mainly concerned with finite algebraic extensions of $K$.

Let $v$ be a valuation of a field $K$ and let $v^\star$ be an extension of $v$ in some overfield $K^\star$ of $K$. Let $\Gamma$ and $\Gamma^\star$ be the value groups of $v$ and $v^\star$ respectively. It is clear that $\Gamma$ is (or can be canonically identified with) a subgroup of $\Gamma^\star$.

LEMMA 1. *If $K^\star$ is an algebraic extension of $K$, then every element of the quotient group $\Gamma^\star/\Gamma$ has finite order (and the two groups $\Gamma$ and $\Gamma^\star$ have therefore the same rational rank).*

PROOF. Let $\alpha^\star$ be an arbitrary element of $\Gamma^\star$. We have to show that there exists an integer $s \neq 0$ such that $s\alpha^\star \in \Gamma$. We fix an element $z$ of $K^\star$ such that $v^\star(z) = \alpha^\star$. Let $z^n + a_1 z^{n-1} + \cdots + a_n = 0$ $(a_i \in K)$ be a relation of algebraic dependence for $z$ over $K$. At least two terms in this relation must have equal value in $v^\star$ (see § 8). Let, say, $v^\star(a_i z^{n-i}) = v^\star(a_j z^{n-j})$, $i \neq j$, $a_i \neq 0$, $a_j \neq 0$ $(a_0 = 1)$. Then $(j-i)v^\star(z) = v^\star(a_j/a_i) \in \Gamma$, and this proves the lemma.

LEMMA 2. *If $K^\star$ is an algebraic extension of $K$, then the valuations $v$ and $v^\star$ (or—equivalently—their value groups $\Gamma$ and $\Gamma^\star$) have the same rank.*

PROOF. We have to exhibit an order preserving (1, 1) mapping of the set of all isolated subgroups $\Delta^\star$ of $\Gamma^\star$ onto the set of all isolated subgroups $\Delta$ of $\Gamma$. We define such a mapping as follows: if $\Delta^\star$ is any isolated subgroup of $\Gamma^\star$, let $\Delta = \Delta^\star \cap \Gamma$. It is obvious that $\Delta$ is a segment and a subgroup of $\Gamma$, and to show that $\Delta$ is an isolated subgroup of $\Gamma$ we have only to show that $\Delta \neq \Gamma$. We fix an element $\alpha^\star$ in $\Gamma^\star$ such that $\alpha^\star \notin \Delta^\star$. By Lemma 1, we have $s\alpha^\star \in \Gamma$ for some integer $s$. On the other hand, $s\alpha^\star \notin \Delta^\star$ (since $\Delta^\star$ is a segment and since $\alpha^\star \notin \Delta^\star$). Hence, *a fortiori*, $s\alpha^\star \notin \Delta$, showing that $\Delta \neq \Gamma$.

We next show that our mapping $\Delta^\star \to \Delta$ is univalent. We observe that if $\alpha^\star$ is any element of $\Delta^\star$, then all integral multiples of $\alpha^\star$ belong to $\Delta^\star$, while, by Lemma 1, some multiple $s\alpha^\star$, $s \neq 0$, belongs to $\Gamma$ and hence also to $\Delta$. Conversely, if $\alpha^\star$ is an element of $\Gamma^\star$ such that $s\alpha^\star \in \Delta$ for some integer $s \neq 0$, then $s\alpha^\star \in \Delta^\star$ and therefore $\alpha^\star \in \Delta^\star$ (since $\Delta^\star$ is a segment). We have thus shown that $\Delta^\star$ is uniquely determined by $\Delta$ as the set of all $\alpha^\star$ in $\Gamma^\star$ such that $s\alpha^\star \in \Delta$ for some integer $s \neq 0$. Hence, our mapping $\Delta^\star \to \Delta$ is univalent, and it is clearly order preserving. Finally, if $\Delta$ is an arbitrary isolated subgroup of $\Gamma$, then it is immediately seen that the set $\Delta^\star$ of elements $\alpha^\star$ in $\Gamma^\star$ such that $s\alpha^\star \in \Delta$, for some integer $s \neq 0$, is an isolated subgroup of $\Gamma^\star$ and that $\Delta^\star \cap \Gamma = \Delta$. Hence our mapping is *onto* the set of isolated subgroups of $\Gamma$, and the lemma is proved.

COROLLARY. *If $K^\star$ is a finite algebraic extension of $K$ then $v^\star$ is discrete if $v$ is discrete* (we recall that it is implicit in our definition of a discrete valuation that any such valuation is of finite rank).

For, let $n$ be the relative degree $[K^\star:K]$. The proof of Lemma 1 shows that if we let $N = n!$, then $N\alpha^\star \in \Gamma$ for all $\alpha^\star$ in $\Gamma^\star$. Let $\Delta_i^\star$ and $\Delta_{i+1}^\star$ be two consecutive isolated subgroups of $\Gamma^\star (\Delta_i^\star < \Delta_{i+1}^\star)$ and let $\Delta_i$ and $\Delta_{i+1}$ be the corresponding isolated subgroups of $\Gamma$. The mapping $\alpha^\star \to N\alpha^\star$ $(\alpha^\star \in \Gamma^\star, N\alpha^\star \in \Gamma)$ transforms $\Delta_i^\star$ and $\Delta_{i+1}^\star$ into $\Delta_i$ and $\Delta_{i+1}$ respectively, and furthermore we know from the proof of Lemma 2 that $N\alpha^\star \in \Delta_i$ if and *only if* $\alpha^\star \in \Delta_i^\star$. Hence our mapping $\alpha^\star \to N\alpha^\star$ induces an order preserving *isomorphism* of $\Delta_{i+1}^\star/\Delta_i^\star$ into $\Delta_{i+1}/\Delta_i$. Since the latter quotient group is, by assumption, isomorphic to the group of integers, it follows that also $\Delta_{i+1}^\star/\Delta_i^\star$ is isomorphic to the group of integers, and hence the valuation $v^\star$ is discrete.

LEMMA 3. *Let $x_1^\star, x_2^\star, \cdots, x_m^\star$ be elements of $K^\star$ such that $m$ elements $v^\star(x_i^\star)$ of $\Gamma^\star$ belong to distinct cosets of $\Gamma$. Then the $x_i^\star$ are linearly independent over $K$.*

PROOF. Assume that there is a relation of the form $\sum_{i=1}^{m} u_i x_i^\star = 0$, where the $u_i$ are elements of $K$, not all zero. Then at least two terms in this relation must have equal (and least) value in $v^\star$. Let, say, $v^\star(u_s x_s^\star) = v^\star(u_t x_t^\star)$, where $s \neq t$ and $u_s u_t \neq 0$. Then $v^\star(x_s^\star) - v^\star(x_t^\star) = v^\star(u_t) - v^\star(u_s) \in \Gamma$, in contradiction with our assumption on the $v^\star$-values of the $x_i^\star$.

COROLLARY. *If $K^\star$ is a finite algebraic extension of $K$, of degree $n$, then the index of the subgroup $\Gamma$ of $\Gamma^\star$ is finite and is not greater than $n$.*

On the basis of this corollary we can now give the following definition:

DEFINITION. *Let $K^\star$ be a finite algebraic extension of $K$ and let $v$ and*

*$v^\star$ be valuations of $K$ and $K^\star$ respectively, such that $v^\star$ is an extension of $v$. Let $\Gamma$ and $\Gamma^\star$ be the value groups of $v$ and $v^\star$ respectively. Then the index $e$ of the subgroup $\Gamma$ of $\Gamma^\star$ is called the reduced ramification index of $v^\star$ with respect to $v$, or relative to $v$ (or with respect to $K$).*

If $K^\star$ is a finite algebraic extension of $K$, we can speak of *the relative degree* of a valuation $v^\star$ of $K^\star$, meaning by this the relative degree of any place associated with $v^\star$ (see § 6). If $v$ is the restriction of $v^\star$ to $K$, then the residue field $R_v/\mathfrak{M}_v$ of $v$ is (or can be canonically identified with) a subfield of the residue field $R_{v^\star}/\mathfrak{M}_{v^\star}$ of $v^\star$, and the relative degree of $v^\star$ is the relative degree $[R_{v^\star}/\mathfrak{M}_{v^\star}:R_v/\mathfrak{M}_v]$. We know that this relative degree is at most equal to $[K^\star:K]$ (§ 6, Lemma 2, Corollary 2).

The relative degree of $v^\star$ shall be denoted by $f$. If $K^\star$ is a *separable* extension of $K$ we also define the *ramification index of $v^\star$ relative to $v$* as the product $ep^s$, where $p^s$ is the inseparable factor of $f$.

It is easy to see that the above terminology agrees with terminology introduced for Dedekind rings in the preceding chapter. For, assume that we have the following special case: $K$ is the quotient field of a Dedekind domain $R$ and $v$ is the $\mathfrak{p}$-adic valuation of $K$ defined by a proper prime ideal $\mathfrak{p}$ in $R$. If $R'$ denotes the integral closure of $R$ in $K^\star$, then the valuation ring of $v^\star$ contains $R'$. Since $R'$ is a Dedekind domain (Vol. I, Ch. V, § 8, Theorem 19), $v^\star$ is necessarily a $\mathfrak{P}$-adic valuation of $K^\star$, where $\mathfrak{P}$ is a prime ideal in $R'$ lying over $\mathfrak{p}$. Let $e_1$ be the reduced ramification index of $\mathfrak{P}$ with respect to $\mathfrak{p}$. If $u$ is an element of $\mathfrak{p}$ not in $\mathfrak{p}^2$, then $\Gamma$ consists of all integral multiples of $v(u)$. On the other hand, since $\mathfrak{P}$ occurs to the exponent $e_1$ in the factorization of $R'\mathfrak{p}$, it follows that $u \in \mathfrak{P}^{e_1}$, $u \notin \mathfrak{P}^{e_1+1}$, showing that $\Gamma$ consists of all multiples $me_1\alpha^\star$, $\alpha^\star \in \Gamma^\star$, where $m$ is an arbitrary integer. Hence $e_1$ is the index of $\Gamma$ in $\Gamma^\star$, and thus the reduced ramification index of $\mathfrak{P}$ with respect to $\mathfrak{p}$ is also the reduced ramification index of $v^\star$ with respect to $v$. Furthermore, it is clear that the residue fields of $v$ and $v^\star$ are isomorphic respectively with the residue fields $R/\mathfrak{p}$ and $R'/\mathfrak{P}$.

We shall need a lemma on extensions of composite valuations.

LEMMA 4. *Let a valuation $v$ of $K$, with value group $\Gamma$, be composite with valuations $v_1$ and $\bar{v}$ (where $v_1$ is a valuation of $K$ and $\bar{v}$ is a valuation of the residue field of $v_1$), and let $G$ be the isolated subgroup of $\Gamma$ which corresponds to this decomposition of $v$ into $v_1$ and $\bar{v}$. Let $v^\star$ be an extension of $v$ to an overfield $K^\star$ of $K$ and let $\Gamma^\star$ be the value group of $v^\star$. There exist isolated subgroups $H^\star$ of $\Gamma^\star$ such that $H^\star \cap \Gamma = G$, and if $v^\star = v_1^\star \circ \bar{v}^\star$ is the decomposition of $v^\star$ which corresponds to such a subgroup $H^\star$ then $v_1^\star$ is an extension of $v_1$ and $\bar{v}^\star$ is an extension of $\bar{v}$. Conversely, if $v_1^\star$ is any* *extension of* $v_1$ *to* $K^\star$ *and* $\bar{v}^\star$ *is any extension of* $\bar{v}$ *to the residue field of* $v_1^\star$, *then* $v^\star = v_1^\star \circ \bar{v}^\star$ *is an extension of* $v$ *to* $K^\star$, *and if* $H^\star$ *is the isolated subgroup of the value group of* $v^\star$ *which corresponds to the decomposition* $v_1^\star \circ \bar{v}^\star$, *then* $H^\star \cap \Gamma = G$.

PROOF. We consider the smallest segment $G^\star$ in $\Gamma^\star$ such that $G^\star \supset G$ ($G^\star$ = set of all elements of $\Gamma^\star$ which are of the form $\pm \alpha^\star$, where $0 \leqq \alpha^\star \leqq \alpha$ for some $\alpha$ in $G$). Then it is immediately seen that $G^\star$ is a subgroup of $\Gamma^\star$ and that it is a proper subgroup of $\Gamma^\star$, since $G$ is a proper subgroup of $\Gamma$. *Hence* $G^\star$ *is an isolated subgroup of* $\Gamma^\star$, and it is clear from the definition of $G^\star$ that we have $G^\star \cap \Gamma = G$ and that $G^\star$ *is the smallest of all the isolated subgroups* $H^\star$ *of* $\Gamma^\star$ *such that* $H^\star \cap \Gamma = G$.

Let now $H^\star$ be any isolated subgroup of $\Gamma^\star$ such that $H^\star \cap \Gamma = G$, and let $v^\star = v_1^\star \circ \bar{v}^\star$ be the corresponding decomposition of $v^\star$, where $v_1^\star$ is then a valuation of $K^\star$, with value group $\Gamma^\star/H^\star$, and $\bar{v}^\star$ is a valuation of the residue field of $v_1^\star$, with value group $H^\star$ (see § 10, Theorem 17). We know from the proof of Theorem 17 that $v^{\star -1} v_1^\star$ is a homomorphism of $\Gamma^\star$ onto $\Gamma^\star/H^\star$, with kernel $H^\star$. The elements of $\Gamma^\star$ which are mapped by this homomorphism into non-negative elements are those and only those which belong to the set $\Gamma_+^\star \cup H^\star$. Hence $R_{v_1^\star}$ is the full inverse image of $\Gamma_+^\star \cup H^\star$ under $v^{\star -1}$. Similarly, $R_{v_1}$ is the full inverse image of $\Gamma_+ \cup G$ under $v^{-1}$. Now, since $v$ is the restriction of $v^\star$ to $K$ and since $(\Gamma_+^\star \cup H^\star) \cap \Gamma = \Gamma_+ \cup G$, we conclude that $R_{v_1} = R_{v_1^\star} \cap K$, *showing that* $v_1^\star$ *is an extension of* $v_1$.

Let $\mathscr{P}_1$ and $\mathscr{P}_1^\star$ denote the canonical homomorphisms $R_{v_1} \rightarrow D_{v_1}$ $(= R_{v_1}/\mathfrak{M}_{v_1})$ and $R_{v_1^\star} \rightarrow D_{v_1^\star}$ respectively. The ring $R_v$ is the full inverse image of $R_{\bar{v}}$ under $\mathscr{P}_1^{-1}$, and similarly $R_{v^\star}$ is the full inverse image of $R_{\bar{v}^\star}$ under $\mathscr{P}_1^{\star -1}$. Since $R_v = K \cap R_{v^\star}$ and since we have just proved that $\mathscr{P}_1$ is the restriction of $\mathscr{P}_1^\star$ to $K$, it follows at once that $R_{\bar{v}} = R_{\bar{v}^\star} \cap D_{v_1}$, *showing that* $\bar{v}^\star$ *is an extension of* $\bar{v}$.

Conversely, assume that we are given a valuation $v_1^\star$ of $K^\star$ which is an extension of $v_1$ and a valuation $\bar{v}^\star$ of the residue field of $v_1^\star$ which is an extension of $\bar{v}$. If $v^\star = v_1^\star \circ \bar{v}^\star$, then we can repeat the reasoning of the preceding paragraph. This time we are *given* that $R_{\bar{v}} = R_{\bar{v}^\star} \cap D_{v_1}$ and from this we can *conclude* that $R_v = K \cap R_{v^\star}$, *showing that* $v^\star$ *is an extension of* $v$. Furthermore, we have that $v^{\star -1} v_1^\star$ is a homomorphism of $\Gamma^\star$ onto $\Gamma^\star/H^\star$, with kernel $H^\star$, and that $v^{-1} v_1$ is a homomorphism of $\Gamma$ onto $\Gamma/G$, with kernel $G$. Since $v^{-1} v_1$ is the restriction of $v^{\star -1} v_1^\star$ to $\Gamma$, it follows that $H^\star \cap \Gamma = G$.

This completes the proof of the lemma.

COROLLARY 1. *Assume that $K^\star$ is an algebraic extension of $K$, and let $v^\star$ be an extension of a composite valuation $v=v_1 \circ \bar{v}$ of $K$. Then there is only one decomposition $v_1^\star \circ \bar{v}^\star$ of $v^\star$ such that $v_1^\star$ and $\bar{v}^\star$ are extensions of $v_1$ and $\bar{v}$ respectively.*

For, it was shown in the course of the proof of Lemma 2 that if $K^\star$ is an algebraic extension of $K$, then for any isolated subgroup $G$ of $\Gamma$ there exists one and only one isolated subgroup $H^\star$ of $\Gamma^\star$ such that $H^\star \cap \Gamma = G$.

COROLLARY 2. *The notations being the same as in the preceding corollary, assume that $K^\star$ is a finite algebraic extension of $K$. Then the reduced ramification index of $v^\star$ relative to $v$ is the product of the reduced ramification indices of $v_1^\star$ and $\bar{v}^\star$ relative to $v_1$ and $\bar{v}$ respectively.*

For, the reduced ramification indices of $v^\star$, $v_1^\star$ and $\bar{v}^\star$ are equal respectively to the orders of the following finite abelian groups: $\Gamma^\star/\Gamma$, $(\Gamma^\star/G^\star)/(\Gamma/G)$ and $G^\star/G$. Since $G^\star \cap \Gamma = G$, the group $G^\star/G$ can be canonically identified with a subgroup of $\Gamma^\star/\Gamma$. Using the well known isomorphism theorem from group theory, we find that the groups $(\Gamma^\star/\Gamma)/(G^\star/G)$ and $(\Gamma^\star/G^\star)/(\Gamma/G)$ are isomorphic (they are both isomorphic to $\Gamma^\star/(\Gamma, G^\star)$). Hence the order of $\Gamma^\star/\Gamma$ is the product of the orders of $G^\star/G$ and $(\Gamma^\star/G^\star)/(\Gamma/G)$.

We are now ready to prove two basic results (Theorems 19 and 20 below) on extensions of valuations.

THEOREM 19. *Let $K^\star$ be a finite algebraic extension of $K$, let $v$ be a valuation of $K$ of finite rank† and let $v_1^\star, v_2^\star, \cdots, v_g^\star$ be the extensions of $v$ to $K^\star$. If $n=[K^\star:K]$ and if $n_i$ and $e_i$ are respectively the relative degree and the reduced ramification index of $v_i^\star$ with respect to $v$ then*

$$e_1n_1+e_2n_2+\cdots+e_gn_g \leqq n. \tag{1}$$

PROOF. (a) We shall first consider the case in which $v$ is of rank 1. In that case, the $g$ valuations $v_i^\star$ are also of rank 1 (Lemma 2), and the theorem of independence of valuations (§ 10, Theorem 18) is applicable to the $v_i^\star$. The value groups $\Gamma$, $\Gamma_i^\star$ of $v$, $v_i^\star$ can be assumed to consist of real numbers. For each $i$, we fix an element $\alpha_{is}$ in each of the $e_i$ cosets of $\Gamma$ in $\Gamma_i^\star$ $(s=1, 2, \cdots, e_i)$. We also fix $n_i$ elements $u_{it}$ in $K^\star$ such that the $v_i^\star$-residues of the $u_{it}$ form a basis of the residue field of $v_i^\star$ over the residue field of $v$ $(t=1, 2, \cdots, n_i)$. Next, using the independence of the valuations $v_i^\star$, we find elements $x_{is}$ and $y_{it}$ in $K^\star$

† Later on, at the end of this section, we shall prove Theorem 19 also for valuations of infinite rank, using an idea which we have found in some unpublished notes of I. S. Cohen.

($i=1, 2, \cdots, g$; $s=1, 2, \cdots, e_i$; $t=1, 2, \cdots, n_i$) satisfying the following conditions:

$$(2) \qquad v_i^\star(x_{is}) = \alpha_{is};$$

$$(2') \qquad v_j^\star(x_{is}) > \max(\alpha_{11}, \alpha_{12}, \cdots, \alpha_{21}, \alpha_{22}, \cdots, \alpha_{ge_g}), \quad \text{if} \quad j \neq i;$$

$$(3) \qquad v_i^\star(y_{it}-u_{it}) > 0;$$

$$(3') \qquad v_j^\star(y_{it}) > 0, \quad j \neq i.$$

*We assert that the* $e_1n_1+e_2n_2+\cdots+e_gn_g$ *products* $x_{is_i}y_{it_i}$ ($s_i=1, 2, \cdots, e_i$; $t_i=1, 2, \cdots, n_i$) *are linearly independent over* $K$. The proof of this assertion will establish our theorem in the case of valuations of rank 1.

Assume that our assertion is false and that we have therefore a relation of the form:

$$(4) \qquad \sum_{i,s_i,t_i} a_{is_it_i}x_{is_i}y_{it_i} = 0,$$

where the $a_{is_it_i}$ are elements of $K$, not all zero. We may assume that these elements all belong to $R_v$ and that at least one of these elements is a unit in $R_v$. We may then assume, without loss of generality, that $v(a_{111})=0$. We set

$$(5) \qquad z_s = \left(\sum_{t=1}^{n_1} a_{1st}y_{1t}\right)x_{1s}, \quad s = 1, 2, \cdots, e_1.$$

*We now observe that the* $v_1^\star$ *value of any element* $y_1$ *of* $K^\star$, *of the form* $\sum_{t=1}^{n_1} b_ty_{1t}$, $b_t \in R_v$, *belongs to* $\Gamma$. For, if $b_q$ is one of the coefficients $b_t$ which has least $v$-value, we can write:

$$y_1 = b_q \sum_{t=1}^{n_1} c_ty_{1t},$$

where all $c_t$ are in $R_v$ and $c_q=1$. Now, by (3) (for $i=1$), we have that the $v_1^\star$-residues of the $n_1$ elements $y_{1t}$ are the same as the $v_1^\star$-residues of the $u_{1t}$, and hence these residues are linearly independent over the residue field of $v$. On the other hand, the $v$-residues of the $c_t$ are not all zero (since $c_q=1$). It follows that the $v_1^\star$-residue of $\sum_{t=1}^{n_1} c_ty_{1t}$ is different from zero. Hence $v_1^\star(y_1)=v_1^\star(b_q)=v(b_q) \in \Gamma$, as asserted.

In view of this observation, we find from (5) that $v_1^\star(z_s)-v_1^\star(x_{1s}) \in \Gamma$, i.e., $v_1^\star(z_s)$ belongs to the $\Gamma$-coset determined by $\alpha_{1s}$ in $\Gamma_1^\star$ [see (2)]. Since the $e_1$ elements $\alpha_{1s}$ of $\Gamma_1^\star$ belong to distinct $\Gamma$-cosets, it follows that the $v_1^\star$-values of $z_1, z_2, \cdots, z_{e_1}$ are distinct elements of $\Gamma_1^\star$ and that consequently $v_1^\star(z_1+z_2+\cdots+z_{e_1})=\min\{v_1^\star(z_1), v_1^\star(z_2), \cdots, v_1^\star(z_{e_1})\}$. Now, since $v(a_{111})=0$, the reasoning used in the proof of the above observation shows that $v_1^\star\left(\sum_{t=1}^{n_1} a_{11t}y_{1t}\right)=0$ and that consequently $v_1^\star(z_1)=\alpha_{11}$. Therefore $v_1^\star(z_1+z_2+\cdots+z_{e_1})\leqq\alpha_{11}$, i.e.,

$$v_1^\star\left(\sum_{s_1,t_1} a_{1s_1t_1}x_{1s_1}y_{1t_1}\right) \leqq \alpha_{11}. \tag{6}$$

On the other hand, we have by (2′) and (3′) (for $m=1$) that

$$v_1^\star\left(\sum_{i=2}^{g}\sum_{s_i,t_i} a_{is_it_i}x_{is_i}y_{it_i}\right) > \alpha_{11}. \tag{7}$$

*By* (6) *and* (7) *it follows that the* $v_1^\star$ *value of the left-hand side of* (4) *is* $\leqq\alpha_{11}$, in contradiction with (4). This contradiction establishes our assertion that the $e_1n_1+e_2n_2+\cdots+e_gn_g$ products $x_{is_i}y_{it_i}$ are linearly independent over $K$.

(b) We now pass to the general case of a valuation $v$ of finite rank $m>1$ and we shall use induction with respect to $m$. We assume therefore that our theorem is true for any valuation of rank $<m$. Let $v=v'\circ\bar{v}$ be a decomposition of $v$ into valuations of rank $<m$. Let $v'^\star_1, v'^\star_2, \cdots, v_h'^\star$ be the distinct extensions of $v'$ to $K^\star$ and let $\bar{v}_{s1}^\star, \bar{v}_{s2}^\star, \cdots, \bar{v}_{sq_s}^\star$ $(s=1, 2, \cdots, h)$ be the distinct extensions of $\bar{v}$ to the residue field of $v'^\star_s$. We set $v_{st_s}^\star=v'^\star_s\circ\bar{v}_{st_s}^\star$. By Lemma 4 and Corollary 1 of that lemma, the $q_1+q_2+\cdots+q_h$ valuations $v_{st_s}^\star$ of $K^\star$ are distinct and represent all the extensions of $v$ to $K^\star$, i.e., the set $\{v_{11}^\star. v_{12}^\star, \cdots, v_{hq_h}^\star\}$ coincides with the set $\{v_1^\star, v_2^\star, \cdots, v_g^\star\}$. We denote by $n_{st_s}$ and $e_{st_s}$ the relative degree and the reduced ramification index of $v_{st_s}^\star$ with respect to $v$. What we have to prove then is the following inequality:

$$\sum_{s=1}^{h}\sum_{t_s=1}^{q_s} e_{st_s}n_{st_s} \leqq n.$$

We observe that the relative degree of $\bar{v}_{st_s}^\star$ with respect to $\bar{v}$ is equal to $n_{st_s}$, since the residue fields of $\bar{v}_{st_s}^\star$ and $\bar{v}$ coincide respectively with the residue fields of $v_{st_s}^\star$ and $v$. We denote by $\bar{e}_{st_s}$ the reduced ramification index of $\bar{v}_{st_s}^\star$ with respect to $\bar{v}$. We also denote by $n'_s$ and $e'_s$ respectively the relative degree and the reduced ramification index of $v'^\star_s$

with respect to $v'$. Since $v'$ and $\bar{v}$ are valuations of rank $<m$, we have by our induction hypothesis:

$$(8) \qquad \sum_{t_s=1}^{q_s} \bar{e}_{st_s} n_{st_s} \leqq n'_s,$$

$$(8') \qquad \sum_{s=1}^{h} e'_s n'_s \leqq n.$$

Hence

$$(8'') \qquad \sum_{s=1}^{h} \sum_{t_s=1}^{q_s} e'_s \bar{e}_{t_s} n_{st_s} \leqq n,$$

and this is the desired inequality, since, by Lemma 4, Corollary 2, we have $e_{st_s} = e'_s \bar{e}_{t_s}$. This completes the proof of the theorem.

We shall see in the next section that (a) *if the residue field $D$ of $v$ is of characteristic zero then the equality sign holds in* (1) (§ 12, Theorem 24, Corollary); and (b) *if $K^\star$ is a normal extension of $K$ and the characteristic $p$ of $D$ is different from zero, then the quotient $n/(e_1n_1+e_2n_2+\cdots+e_gn_g)$ is a power $p^\delta$ of $p$, where $\delta$ is an integer* $\geqq 0$ (§ 12, Theorem 25, Corollary). The integer $\delta$ may be referred to as the *ramification deficiency* of $v$ (this integer is defined only in the case of normal extensions $K^\star$). Here we shall only show that if we assume that (a) is valid in the case of normal extensions $K^\star$ then its general validity is an immediate consequence. For, let $\bar{K}$ be the least normal extension of $K$ which contains $K^\star$ and let $\bar{v}_{i1}, \bar{v}_{i2}, \cdots$ be the extensions of $v_i^\star$ to $\bar{K}$. Let $N=[\bar{K}:K]$, $n^\star=[\bar{K}:K^\star]$. We denote by $E_{ij}$ and $e_{ij}^\star$ the reduced ramification indices of $\bar{v}_{ij}$ relative to $v$ and $v_i^\star$ respectively. Similarly, we denote by $N_{ij}$ and $n_{ij}^\star$ the two corresponding relative degrees of $\bar{v}_{ij}$. We have $E_{ij}=e_ie_{ij}^\star$, $N_{ij}=n_in_{ij}^\star$, $\sum_i \sum_j E_{ij}N_{ij} = \sum_i e_in_i \sum_j e_{ij}^\star n_{ij}^\star$. By assumption, we have $N=\sum_{i,j} E_{ij}N_{ij}$, and $n^\star = \sum_j e_{ij}^\star n_{ij}^\star$ for $i=1, 2, \cdots, g$. Hence $N=n^\star \sum e_in_i$, whence $\sum e_in_i = n$, as asserted.

We denote by $R$ the valuation ring $R_v$ of $v$ and by $\mathfrak{P}$ the maximal ideal $\mathfrak{M}_v$ of $R_v$. Let $R_i^\star$ denote the valuation ring of $v_i^\star$. We set

$$(9) \qquad R^\star = \bigcap_{j=1}^{g} R_j^\star,$$

$$(10) \qquad \mathfrak{H}_i^\star = \bigcap_{j\neq i} (R_j^\star \mathfrak{P} \cap R^\star),$$

$$(11) \qquad \mathfrak{P}^\star = \bigcap_{i=1}^{g} \mathfrak{H}^\star = \bigcap_{j=1}^{g} R_j^\star \mathfrak{P}.$$

The $g$ rings $R_j^\star$ are the only valuation rings in $K^\star$ which belong to places of $K^\star$ having center $\mathfrak{P}$ in $R$. Hence by Theorem 8, § 5, *$R^\star$ is the integral closure of $R$ in $K^\star$*. We also observe that

$$(12) \qquad \mathfrak{P}^\star = R^\star\mathfrak{P}.$$

To prove (12) we have only to show that $\mathfrak{P}^\star \subset R^\star\mathfrak{P}$, for the opposite inclusion is obvious. Let $x^\star$ be any element of $\mathfrak{P}^\star$, and let $v_j^\star(x^\star) = \alpha_j^\star$. Since $x^\star \in R_j^\star\mathfrak{P}$, it is obvious that we can find, for each $j$, positive elements in $\Gamma$ which are not greater than $\alpha_j^\star$. Therefore we can also find a positive element in $\Gamma$ which is not greater than any of the $\alpha_j^\star$. Let $\beta$ be such an element: $\alpha_j^\star \geqq \beta, j = 1, 2, \cdots, g$. We fix an element $x$ in $\mathfrak{P}$ such that $v(x) = \beta$. Then $v_j(x^\star/x) \geqq 0, j = 1, 2, \cdots, g$, whence $x^\star/x \in R^\star$ and $x^\star \in R^\star\mathfrak{P}$, as asserted.

It is clear that $\mathfrak{P}^\star \cap R = \mathfrak{P}$ Hence the ring $R^\star/\mathfrak{P}^\star$ can be regarded as a vector space over the field $R/\mathfrak{P}$. We next prove the following lemma:

LEMMA 5. *The assumptions being the same as in Theorem* 19, *except that $v$ may now have infinite rank, the dimension of the vector space $R^\star/\mathfrak{P}^\star$ (over the field $R/\mathfrak{P}$) is not greater than $e_1n_1 + \cdots + e_gn_g$.*

PROOF. The ring $R^\star$ has exactly $g$ maximal prime ideals $\mathfrak{P}_i^\star = \mathfrak{M}_{v_1}^\star \cap R^\star$, $i = 1, 2, \cdots, g$, and each valuation ring $R_i^\star$ is the quotient ring of $R^\star$ with respect to $\mathfrak{P}_i^\star$ (Theorem 12, § 7). We know that given any element $\alpha^\star$ of the value group $\Gamma_i^\star$ of $v_i^\star$ there exists an integer $s \neq 0$ such that $s\alpha^\star \in \Gamma$ (Lemma 1). Therefore, given any element $x^\star$ of $\mathfrak{P}_i^\star$, we will have some integer $s \geqq 1$ such that $v_i^\star(x^{\star s}) \in \Gamma$. Let $y$ be an element of $\mathfrak{P}$ such that $v(y) = v_i^\star(x^{\star s})$. Then $x^{\star s}/y \in R_i^\star$ and so $x^{\star s} \in R_i^\star\mathfrak{P} \cap R^\star$. Since, on the other hand, $R_i^\star\mathfrak{P} \cap R^\star \subset \mathfrak{P}_i^\star$, *we have therefore shown that $\mathfrak{P}_i^\star$ is the radical of $R_i^\star\mathfrak{P} \cap R^\star$.* It follows that for $i \neq j$ the ideals $R_i^\star\mathfrak{P} \cap R^\star$ and $R_j^\star\mathfrak{P} \cap R^\star$ are comaximal (see Vol. I, Ch. III, § 13, Theorem 31). Furthermore, from (11) and (12) it follows that $\mathfrak{P}^\star$ is the intersection of the $g$ ideals $R_i^\star\mathfrak{P} \cap R^\star$. Hence, by Theorem 32 of III, § 13, the ring $R^\star/\mathfrak{P}^\star$ is the direct sum of the $g$ rings $\mathfrak{H}_i^\star/\mathfrak{P}^\star$. Since the $\mathfrak{H}_i^\star$ are ideals in $R^\star$, we have a direct decomposition of the vector space $R^\star/\mathfrak{P}^\star$ into the $g$ subspaces $\mathfrak{H}_i^\star/\mathfrak{P}^\star$ (over the field $R/\mathfrak{P}$), and in order to prove the lemma it will be sufficient to prove that $\mathfrak{H}_i^\star/\mathfrak{P}^\star$ has dimension $\leqq e_in_i$.

Let us consider, for instance, the space $\mathfrak{H}_1^\star/\mathfrak{P}^\star$. The subspaces of $\mathfrak{H}_1^\star/\mathfrak{P}^\star$ correspond in (1, 1) fashion to the $R$-submodules of $\mathfrak{H}_1^\star$ which contain $\mathfrak{P}^\star$. We first make some straightforward observations about the two value groups $\Gamma_1^\star$ and $\Gamma$. Let $L_1$ denote the set of non-negative elements $\alpha^\star$ of $\Gamma_1^\star$ such that $\alpha^\star < \beta$ for *all* positive elements $\beta$ of $\Gamma$. If $\alpha_1^\star$ and $\alpha_2^\star$ are two distinct elements of $L_1$, and if say $\alpha_1^\star < \alpha_2^\star$, then

$0<\alpha_2^\star-\alpha_1^\star<\alpha_2^\star$, and therefore, by definition of $L_1$, $\alpha_2^\star-\alpha_1^\star \notin \Gamma$. Thus, *distinct elements of $L_1$ belong to distinct $\Gamma$-cosets, and hence $L_1$ is a finite set, consisting of at most $e_1$ elements.*

*If $x^\star$ is any element of $\mathfrak{H}_1^\star$ then $v_1^\star(x^\star) \in L_1$ if and only if $x^\star \notin \mathfrak{P}^\star$.* For, if $x^\star \in \mathfrak{P}^\star = R^\star\mathfrak{P}$, then it is clear that $v_1^\star(x^\star) \geqq v_1^\star(y)$, for some $y$ in $\mathfrak{P}$, and hence $v_1^\star(x^\star) \notin L_1$, since $v_1^\star(y) \in \Gamma_+$. Conversely, if $v_1^\star(x^\star) \notin L_1$, then $v_1^\star(x^\star) \geqq v_1^\star(y)$, for some $y$ in $\mathfrak{P}$, and hence $x^\star = (x^\star/y)y \in R_1^\star\mathfrak{P} \cap \mathfrak{H}_1^\star = R^\star\mathfrak{P}$.

If follows from these remarks that if $\mathfrak{A}^\star$ is any $R$-submodule of $\mathfrak{H}_1^\star$ which contains $\mathfrak{P}^\star$ as a proper subset then $\mathfrak{A}^\star$ contains elements of least value and that this value is an element of $L_1$. We denote this minimum by $v_1^\star(\mathfrak{A}^\star)$.

If for a given element $\alpha^\star$ of $L_1$ there exist elements $x^\star$ in $\mathfrak{H}_1^\star$ such that $v_1^\star(x^\star)=\alpha^\star$, then the set of *all* elements $y^\star$ of $\mathfrak{H}_1^\star$ such that $v_1^\star(y^\star) \geqq \alpha^\star$ is an $R$-submodule $\mathfrak{A}^\star$ of $\mathfrak{H}_1^\star$ which contains $\mathfrak{P}^\star$ as a proper subset and is such that $v_1^\star(\mathfrak{A}^\star)=\alpha^\star$. If $0=\alpha_1^\star<\alpha_2^\star< \cdots <\alpha_s^\star (s \leqq e_1)$ are those elements of $L_1$ which are $v_1^\star$-values of elements of $\mathfrak{H}_1^\star$, then we obtain in this fashion a strictly descending chain of $R$-submodules of $\mathfrak{H}_1^\star$:

$$\mathfrak{H}_1^\star = \mathfrak{A}_1^\star > \mathfrak{A}_2^\star > \cdots > \mathfrak{A}_s^\star > \mathfrak{A}_{s+1}^\star = \mathfrak{P}^\star,$$

where $\mathfrak{A}_i^\star$ is the set of all $y^\star$ in $\mathfrak{H}_1^\star$ such that $v_1^\star(y^\star) \geqq \alpha_i^\star (i=1, 2, \cdots, s)$. It is clear that for $i=2, 3, \cdots, s+1$ the module $\mathfrak{A}_i^\star$ consists of all the elements $y^\star$ in $\mathfrak{H}_1^\star$ such that $v_1^\star(y^\star) > \alpha_{i-1}^\star$.

To prove the inequality $\dim \mathfrak{H}_1^\star/\mathfrak{P}^\star \leqq e_1 n_1$, it will be sufficient to show that for $i=2, 3, \cdots, s+1$ we have $\dim \mathfrak{A}_{i-1}^\star/\mathfrak{A}_i^\star \leqq n_1$ (since $s \leqq e_1$); here $\mathfrak{A}_{i-1}^\star/\mathfrak{A}_i^\star$ $(=\mathfrak{A}_{i-1}^\star/\mathfrak{P}^\star/\mathfrak{A}_i^\star/\mathfrak{P}^\star)$ is regarded as a vector space over $R/\mathfrak{P}$. Let then $x_1^\star, x_2^\star, \cdots, x_{n_1+1}^\star$ be any $n_1+1$ elements of $\mathfrak{A}_{i-1}^\star$. We have to show that there exist elements $u_1, u_2, \cdots, u_{n_1+1}$ in $R$, *not all in* $\mathfrak{P}$, such that $u_1x_1^\star + \cdots + u_{n_1+1}x_{n_1+1} \in \mathfrak{A}_i^\star$. We fix an element $y^\star$ in $\mathfrak{A}_{i-1}^\star$ of least value: $v_1^\star(y^\star)=\alpha_{i-1}^\star=v_1^\star(\mathfrak{A}_{i-1}^\star)$, and we set $z_j^\star=x_j^\star/y^\star$. Then the $z_j^\star$ are in the valuation ring of $v_1^\star$, and since the relative degree of $v_1^\star$ is $n_1$ it follows that we can find elements $u_1, u_2, \cdots, u_{n_1+1}$ in $R$, not all in $\mathfrak{P}$, such that $v_1^\star(u_1z_1^\star+u_2z_2^\star+ \cdots + u_{n_1+1}z_{n_1+1}^\star)>0$. Then we have $v_1^\star(u_1x_1^\star+u_2x_2^\star+ \cdots +u_{n_1+1}x_{n_1+1}^\star) > v_1^\star(y^\star)=\alpha_{i-1}^\star$, and therefore $u_1x_1^\star+ \cdots +u_{n_1+1}x_{n_1+1}^\star \in \mathfrak{A}_i^\star$. This completes the proof of the lemma.

Of particular importance is the next theorem:

THEOREM 20. *The notations and assumption being the same as in Theorem 19, (in particular, it is now again being assumed that $v$ has finite* *rank), assume also that the integral closure $R^\star$ in $K^\star$ of the valuation ring $R$ of $v$ is a finite $R$-module. Then*

$$e_1n_1+e_2n_2+\cdots+e_gn_g = n, \tag{13}$$

*and*

$$\dim R^\star/R^\star\mathfrak{P} = n. \tag{14}$$

PROOF. Let $\{w_1, w_2, \cdots, w_m\}$ be an $R$-basis of $R^\star$ *which has the least number of elements.* We assert that the $w$ *are linearly independent over* $K$. For assume that we have a relation of linear dependence: $x_1w_1+x_2w_2+\cdots+x_mw_m=0$, where the $x_i$ are elements of $K$, not all zero. An argument which has been repeatedly used before shows that we may assume that the $x_i$ belong to $R$ and that one of the $x_i$ is 1. If, say, $x_m=1$, then already $\{w_1, w_2, \cdots, w_{m-1}\}$ is an $R$-basis of $R^\star$, a contradiction.

Any element $x^\star$ of $K^\star$ satisfies an algebraic equation with coefficients in $R$ (since $K$ is the quotient field of $R$). If $a_0$ is the leading coefficient of this equation then $a_0x^\star$ is integral over $R$, whence $a_0x^\star \in R^\star$. This shows that $\{w_1, w_2, \cdots, w_m\}$ is also a basis of $K^\star/K$. *Consequently* $m=n$.

If $\bar{w}_i$ denotes the $R^\star\mathfrak{P}$-residue of $w_i$, then $\bar{w}_1, \bar{w}_2, \cdots, \bar{w}_n$ span the vector space $R^\star/R^\star\mathfrak{P}$(over $R/\mathfrak{P}$). *We assert that the $n$ vectors $\bar{w}_i$ are linearly independent over $R/\mathfrak{P}$.* We have only to show that if we have a relation of the form $x_1w_1+x_2w_2+\cdots+x_nw_n \in R^\star\mathfrak{P}$, $x_i \in R$, then the $x_i$ necessarily belong to $\mathfrak{P}$. But this follows at once from the linear independence of the $w_i$ over $R$, for we have, by assumption: $x_1w_1+x_2w_2+\cdots+x_nw_n=y_1w_1+y_2w_2+\cdots+y_nw_n$, where the $y_i$ are suitable elements of $\mathfrak{P}$, and this relation implies $x_i=y_i$, $i=1, 2, \cdots, n$.

We have therefore proved that

$$n = \dim R^\star/R^\star\mathfrak{P}. \tag{14}$$

Since we have, by Theorem 19 and Lemma 5:

$$\dim R^\star/R^\star\mathfrak{P} \leqq e_1n_1+e_2n_2+\cdots e_gn_g \leqq n, \tag{15}$$

the theorem is proved.

COROLLARY. *If $v$ is a non-discrete valuation of rank 1 and if $R^\star$ is a finite $R$-module, then all the extensions of $v$ to $K^\star$ are unramified.*

For the proof, we first show that

$$R^\star\mathfrak{P} = \{x^\star \in K^\star | v_i^\star(x^\star) > 0, \quad i = 1, 2, \cdots, g\}. \tag{16}$$

In fact, let $x^\star$ be any element of $K^\star$ such that $v_i^\star(x^\star)=\beta_i>0$, $i=1, 2, \cdots, g$. Since the value groups $\Gamma$, $\Gamma_i^\star$ are now groups of real numbers and $\Gamma$ is non-discrete, there exist positive elements of $\Gamma$ in an arbitrarily small neighborhood of zero. Hence there exists an element $\alpha$ of $\Gamma$ such that $0<\alpha<\beta_i$, $i=1, 2, \cdots, g$. Let $x$ be an element of $\mathfrak{P}$ such that $v(x)=\alpha$. Then $v_i^\star(x^\star/x)>0$, $i=1, 2, \cdots, g$, whence $x^\star \in R^\star x \subset R^\star\mathfrak{P}$. This establishes (16). We now make use of the proof of Lemma 5. From (16) it follows that the set denoted by $L_1$ in the proof of Lemma 5 consists now of the element zero only, and that consequently the integer $s$ is now equal to 1. It was shown in the proof of Lemma 5 that dim $\mathfrak{H}_1^\star/R^\star\mathfrak{P} \leqq sn_1$. Hence dim $\mathfrak{H}_1^\star/R^\star\mathfrak{P} \leqq n_1$. Similarly dim $\mathfrak{H}_i^\star/R^\star\mathfrak{P} \leqq n_i$, $i=1, 2, \cdots, g$. Hence dim $R^\star/R^\star\mathfrak{P} = \sum_{i=1}^{g}$ dim $\mathfrak{H}_i^\star/R^\star\mathfrak{P} \leqq n_1+n_2+\cdots+n_g$. Therefore, by Theorem 20, we must have $e_1=e_2=\cdots=e_g=1$.

The following example, due to F. K. Schmidt, shows that the finiteness assumption made in Theorem 20 (i.e., the assumption that $R^\star$ is a finite $R$-module) is essential, and that without this assumption the strict equality (13) may fail to hold already in the case of a valuation $v$ which is discrete and of rank 1 (and whose valuation ring $R_v$ is therefore noetherian):

Let $\mathcal{J}_p$ be the prime field of characteristic $p \neq 0$ and let

$$\{\xi_0, \xi_1, \cdots, \xi_n, \cdots\}$$

be an infinite sequence of algebraically independent elements over $k$. We set $k=\mathcal{J}_p(\xi_0, \xi_1, \cdots, \xi_n, \cdots)$ and $K=k(x, y)$, where $x$ and $y$ are algebraically independent over $k$. Consider the formal power series

$$\varphi(x) = \xi_0^p+\xi_1^p x^p+\cdots+\xi_n^p x^{np}+\cdots.$$

We assert that $\varphi(x)$ *is not algebraic over the field* $k(x)$ (or, in algebro-geometric terms: the branch $y=\varphi(x)$ is not algebraic). For assume the contrary, and let, say, $f(X, Y)$ be a non-zero polynomial in $k[X, Y]$ such that $f(x, \varphi(x))=0$. We may assume that $X$ does not divide $f(X, Y)$. Then $f(0, Y)\neq 0$, while $f(0, \xi_0^p)=0$. *Hence* $\xi_0$ *is algebraic over* $k_0$, where $k_0$ is the field generated over $\mathcal{J}_p$ by the coefficients of $f$. Let $X^s$ be the highest power of $X$ which divides $f(X, X^pY+\xi_0^p)$ (whence, necessarily, $s>0$) and let $f(X, X^pY+\xi_0^p)=X^sf_1(X, Y)$. We have

$$f_1(x, \xi_1^p+\xi_2^p x^p+\cdots+\xi_n^p x^{(n-1)p}+\cdots) = 0$$

and therefore $f_1(0, \xi_1^p)=0$. On the other hand, the coefficients of $f_1(X, Y)$ belong to $k_0(\xi_0)$, and since $\xi_0$ is algebraic over $k_0$, it follows that also $\xi_1$ *is algebraic over* $k_0$. Proceeding in this fashion, we find that all the $\xi_i$ are algebraic over $k_0$, and this is impossible since $k_0$ has finite transcendence degree over $\mathcal{J}_p$.

We now define a valuation $v$ of $k(x, y)$, as follows:

If $u=f(x, y)$ is an element of $k[x, y]$, then by the preceding result the power series $f(x, \varphi(x))$ is not zero. If $x^n$ is the lowest power of $x$ which occurs in this series, we let $v(u)=n$. If $z$ is an arbitrary element of $k(x, y)$, we write $z$ in the form $u_1/u_2$, where $u_i=f_i(x, y) \in k[x, y]$, and we let $v(x)=v(u_1)-v(u_2)$. The value group of $v$ is then the group of integers, and so $v$ is discrete, of rank 1. It is immediately seen that the residue field of $v$ is the field $k$.

Now we let $K^\star=K(y^\star)$, where $y^\star=\sqrt[p]{y}$. Then $K^\star=k(x, y^\star)$, and it is immediately seen that the extension $v^\star$ of $v$ to $K^\star$ is the valuation which is defined by the "branch"

$$y^\star = \xi_0+\xi_1x+\xi_2x^2+ \cdots +\xi_nx^n+ \cdots,$$

in a fashion similar to that in which $v$ was defined by the branch $y=\varphi(x)$. (Note that since $K^\star$ is a purely inseparable extension of $K$, $v$ has a unique extension to $K^\star$.) The two valuations $v$ and $v^\star$ have the same value group and the same residue field (namely, the field $k$). Hence the relative degree and the reduced ramification index of $v^\star$ are both equal to 1, while the degree $[K^\star:K]$ is $p$. Thus (13) fails to hold in the present case. In view of Theorem 20, we can conclude *a priori* that the integral closure $R^\star$ of $R_v$ in $K^\star$ is not a finite $R_v$-module. This can also be seen directly as follows:

If $R^\star$ has a finite $R_v$-basis, then a minimal $R_v$-basis of $R^\star$ will contain precisely $p$ elements, say $w_1, w_2, \cdots, w_p$ (see the proof of Theorem 20). Let $w_i=a_{i0}+a_{i1}y^\star+ \cdots +a_{i,p-1}y^{\star p-1}$, $a_{ij} \in K$. Since the value group $\Gamma$ of $v$ is the group of integers, there exists an integer $\rho$ such that all the products $a_{ij}x^\rho$ belong to $R_v$. From this it follows that $R^\star x^\rho \subset R_v+R_vy^\star+ \cdots +R_vy^{\star p-1}$. Now, consider the element $z=[y^\star-(\xi_0+\xi_1x+ \cdots +\xi_\rho x^\rho)]/x^{\rho+1}$. It is clear that $z \in R^\star$ (since $v^\star(z) \geqq 0$). But $zx^\rho=-(\xi_0+\xi_1x+ \cdots +\xi_\rho x^\rho)/x+y^\star/x \notin R_v+R_vy^\star+ \cdots +R_vy^{\star p-1}$, a contradiction.

An important case in which the finiteness assumption of Theorem 20 is always satisfied is the following: *$v$ is a discrete valuation of rank* 1 *and $K^\star$ is a separable extension of $K$.* This follows from the following well-known result: *if $R$ is any noetherian integrally closed domain having $K$ as quotient field, and if $K^\star$ is a finite separable extension of $K$, then the integral closure of $R$ in $K^\star$ is a finite $R$-module* (Vol. I, Ch. V, § 4, Theorem 7, Corollary 1).

It may also be observed that *for discrete valuations $v$, of rank* 1, the converse of Theorem 20 is also true, i.e., *if relation* (13) *holds, then $R^\star$ is a finite $R$-module.* To see this, we go back to the case ($a$) of the proof of

Theorem 19 and we show that if $v$ is discrete, of rank 1, and if (13) holds, then *the* $n_1e_1+n_2e_2+\cdots+n_ge_g$ *products* $x_{s_i}y_{t_i}$ *form an R-basis for* $R^\star$. We know that these products are linearly independent over $K$. If (13) holds, the number of these products is equal to $n$ $(=[K^\star:K])$ and they therefore form a basis of $K^\star/K$. Now let $z^\star$ be any element of $R^\star$ and let

$$z^\star = \sum_{i,s_i,t_i} b_{is_it_i}x_{is_i}y_{it_i}, \quad b_{is_it_i} \in K.$$

We have to show that the $b_{is_it_i}$ belong to $R$: Upon factoring out a coefficient $b_{js_jt_j}$ of least value we can write $z^\star$ in the form: $z^\star=by^\star$, $b \in K$ and

$$y^\star = \sum_{i,s_i,t_i} a_{is_it_i}x_{is_i}y_{it_i},$$

where the $a_{is_it_i}$ are elements of $R$, *not all in* $\mathfrak{P}$. We now make use of the considerations developed in the course of the proof of Theorem 19, case (a) (p. 56). As group $\Gamma$ we can now take the group of integers, and as group $\Gamma_i^\star$ the additive group of integral multiples of $1/e_i$. As representatives of the $e_i$ cosets of $\Gamma$ in $\Gamma_i^\star$ we take the rational numbers $\alpha_{is}=(s-1)/e_i$, $s=1, 2, \cdots, e_i$. By assumption, at least one of the coefficients $a_{is_it_i}$ has order zero in $v$ (and all have non-negative order). If, say $v(a_{1qr})=0$ then, as was shown in the course of the proof of Theorem 19 (see the italicized statement immediately following inequality (7), p. 57), we have $v_1^\star(y^\star)\leqq\alpha_{1q}$, and hence $v_1^\star(y^\star)<1$. On the other hand, we have that $v(b)(=v_1^\star(b))$ is an integer (since $b \in K$). Since $v_1^\star(b)+v_1^\star(y^\star)=v_1^\star(z^\star)\geqq 0$, we conclude that $v(b)$ is necessarily a non-negative integer. Hence $b \in R$, and since $b_{is_it_i}=ba_{is_it_i}$ it follows that also the $b_{is_it_i}$ belong to $R$, as asserted.

Note that this result has also been proved in Vol. I, Ch. V, § 9 (Theorem 21).

NOTE. We shall end this section by extending Theorem 19 to valuations of infinite rank. We first observe that the proof of Theorem 19, in the case of valuations of rank 1, is based solely on the fact that for such valuations the approximation theorem of § 10 (Theorem 18) is valid. However, we have seen that the approximation theorem is valid more generally for independent valuations of any rank (Theorem 18′, § 10). Hence we can assert that *Theorem* 19 *is valid whenever the* $g$ *extensions* $v_1^\star, v_2^\star, \cdots, v_g^\star$ *of* $v$ *are independent.* Our second observation is that in the inductive proof of Theorem 19 for valuations of finite rank $>1$ we have actually proved the following: *Let* $v=v'\circ\bar{v}$, *let* $v'^\star_1$, $v'^\star_2, \cdots, v'^\star_h$ *be the extensions of* $v'$ *to* $K^\star$ *and let* $\bar{v}_{s1}^\star, \bar{v}_{s2}^\star, \cdots, \bar{v}_{sq_s}^\star$

*be the extensions of* $\bar{v}$ *to the residue field* $\Delta'^{\star}_s$ *of* $v'^{\star}_s$ $(s=1, 2, \cdots, h)$. *Then if Theorem* 19 *holds for* $v'$, $K$, $K^{\star}$ *and for* $\bar{v}$, $\Delta'$, $\Delta'^{\star}_s$ $(s=1, 2, \cdots, h$; $\Delta'=$ *residue field of* $v'$), *the theorem holds also for* $v$, $K$ *and* $K^{\star}$. We shall now make use of these two observations. We shall use induction with respect to the number $g$ of extensions of $v$, i.e., we shall assume that Theorem 19 holds true in all cases in which we are dealing with a valuation $v$ which has fewer than $g$ extensions. (For $g=1$ the proof of Theorem 19 is valid as given, for in that case the approximation theorem is not needed; or—more precisely—the approximation theorem is trivial in the case of single valuations.)

We first introduce some notations and prove an auxiliary lemma. If $v$ is a valuation of a field $K$ we shall denote by $L(v)$ the set of all valuations $v'$ of $K$ such that $R_v < R_{v'} < K$. In other words, $L(v)$ is the set of all non-trivial valuations $v'$ such that $v$ is composite with and is non-equivalent to $v'$. We denote by $E(v)$ the set of distinct (i.e., non-equivalent) extensions of $v$ to $K^{\star}$. We write $v' < v$ if $v' \in L(v)$ (note that this partially orders the valuations according to *increasing rank*, or—equivalently—according to *decreasing valuation ring*). If $v' < v$ and $v^{\star}$ is any element of $E(v)$, then there exists a unique element $v'^{\star}$ in $E(v')$ such that $v'^{\star} < v^{\star}$ (Lemma 4, Corollary 1). This defines a mapping $\varphi_{v'}{}^{v}$ of $E(v)$ into $E(v')$, and it follows directly from the second part of Lemma 4 that $\varphi_{v'}{}^{v}$ maps $E(v)$ onto $E(v')$. If $v'' < v' < v$ then it is immediate that

$$\varphi_{v'}{}^{v}\varphi_{v''}{}^{v'} = \varphi_{v''}{}^{v}.$$

For fixed $v$ and a fixed extension $v^{\star}$ of $v$ to $K^{\star}$, the set of valuations $\varphi_{v'}{}^{v}(v^{\star})$, $v' \in L(v)$, coincides with the set $L(v^{\star})$. In fact, if $v'^{\star} = \varphi_{v'}{}^{v}(v^{\star})$ and $v' \in L(v)$, then $v'^{\star} < v^{\star}$ by definition of $\varphi_{v'}{}^{v}$, and hence $v'^{\star} \in L(v^{\star})$; conversely, if $v'^{\star} \in L(v^{\star})$, i.e., if $v'^{\star} < v^{\star}$, then the restriction $v'$ of $v'^{\star}$ to $K$ satisfies the relation $v' < v$, and we have $v'^{\star} \in E(v')$, whence $v'^{\star} = \varphi_{v'}{}^{v}(v^{\star})$. Another way of expressing this fact is to say that for fixed $v^{\star}$ the mapping $v' \rightarrow \varphi_{v'}{}^{v}(v^{\star})$ (where $v=$ restriction of $v^{\star}$ in $K$) is a (1, 1) mapping of $L(v)$ onto $L(v^{\star})$. Each of the two sets $L(v)$ and $L(v^{\star})$ is totally ordered, and the above mapping of $L(v)$ onto $L(v^{\star})$ is order preserving, for it maps each element of $L(v^{\star})$ into its restriction in $K$.

For each valuation $v$ of $K$ we denote by $\gamma(v)$ the number of elements in the set $E(v)$, i.e., the number of distinct extensions of $v$ to $K^{\star}$. If $v' < v$ then from the existence of the mapping $\varphi_{v'}{}^{v}$ it follows that $\gamma(v') \leqq \gamma(v)$. Since $1 \leqq \gamma(v) \leqq [K^{\star}:K]$, the function $\gamma$ can assume only a finite number of values.

LEMMA 6. *Let $v$ be a valuation of $K$ such that the set $L(v)$ has no last element, and let* $m = \max_{v' \in L(v)} \{\gamma(v')\}$. *Then* $\gamma(v) = m$.

PROOF. We fix a valuation $v'_0$ in $L(v)$ such that $\gamma(v'_0) = m$. For each $v'$ in $L(v)$ such that $v'_0 \leqq v'$ the set $E(v')$ has exactly $m$ elements, and therefore $\varphi_{v'_0}{}^{v'}$ is a (1, 1) mapping of $E(v')$ onto $E(v'_0)$. Let $v^\star$ be an extension of $v$ to $K^\star$ and let $v'_0{}^\star$ be that extension of $v'_0$ with which $v^\star$ is composite; in other words, let $v'_0{}^\star = \varphi_{v'_0}{}^{v}(v^\star)$. If $v'$ is any element of $L(v)$ such that $v'_0 \leqq v'$, then the corresponding element $v'^\star$ of $L(v^\star)$, i.e., the valuation $v'^\star = \varphi_{v'}{}^{v}(v^\star)$ is uniquely determined by $v'_0{}^\star$, and by $v'$, i.e., if $v_1{}^\star$ is another extension of $v$ to $K^\star$ which is composite with $v'_0{}^\star$ then $\varphi_{v'}{}^{v}(v_1{}^\star) = \varphi_{v'}{}^{v}(v^\star)$, for we must have $v'_0{}^\star = \varphi_{v'_0}{}^{v'}(v'^\star)$, and $\varphi_{v'_0}{}^{v'}$ is (1, 1). We now observe that since $L(v)$ and $L(v^\star)$ are in (1, 1) order preserving correspondence, also $L(v^\star)$ has no last element and that therefore

$$R_{v^\star} = \bigcap_{v'_0{}^\star \leqq v'^\star < v^\star} R_{v'^\star}. \tag{17}$$

We have just seen that the set of valuations $v'^\star$ in $L(v^\star)$ such that $v'_0{}^\star \leqq v'^\star$, where $v'_0{}^\star = \varphi_{v'_0}{}^{v}(v^\star)$, is uniquely determined by $v'_0{}^\star$. Hence it follows from (17) that there exists *only one* extension $v^\star$ of $v$ to $K^\star$ which is composite with a given valuation $v'_0{}^\star$ belonging to the set $E(v'_0)$. Since $E(v'_0)$ contains $m$ valuations, $v$ has exactly $m$ extensions. Q.E.D.

We now proceed to the proof of Theorem 19 for a valuation $v$ of arbitrary rank. Let $\gamma(v) = g$. We first observe that the case in which the $g$ extensions of $v$ are independent valuations is characterized by the condition that the mapping $\varphi_{v'}{}^{v}$ be (1,1) for any $v'$ in $L(v)$, i.e., it is characterized by the condition $\gamma(v') = g$, for all $v'$ in $L(v)$. We may therefore assume that there exist valuations $v'$ in $L(v)$ such that $\gamma(v') < g$. Let $L_1(v)$ be the set of all such valuations $v'$ and let $g' = \max_{v' \in L_1(v)} \{\gamma(v')\}$. Then $g' < g$. The intersection of all the valuation rings $R_{v'}$, $v' \in L_1(v)$, is again a valuation ring of some valuation $v'_1$ of $K$. If $L_1(v)$ has a last element, then $v'_1$ is the last element of $L_1(v)$ and hence $\gamma(v'_1) = g'$. In the contrary case it is clear that $L_1(v) = L(v'_1)$, whence $L(v'_1)$ has no last element. It follows then from Lemma 6 that $\gamma(v'_1) = g'$. Thus we have $\gamma(v'_1) = g' < g$ in both cases (showing, incidentally, that $v'_1$ necessarily belongs to $L_1(v)$ and that consequently the second case is to be ruled out), and Theorem 19 is valid for $v'_1$.

Since $v'_1 \in L(v)$, we can write $v = v'_1 \circ \bar{v}$. Since $v'_1$ has exactly $g'$ extensions to $K^\star$ and since $g' < g$, it follows by our induction hypothesis that Theorem 19 holds for $v'_1$, $K$ and $K^\star$. Let $v'^\star_1$ be *any* extension of $v'_1$ to $K^\star$, let $\Delta$, $\Delta'_1$ and $\Delta'^\star_1$ be respectively the residue field of $v$, $v'_1$ amd $v'^\star_1$ (whence $\bar{v}$ is a valuation of $\Delta'_1$, with residue field $\Delta$, and $\Delta'_1 \subset \Delta'^\star_1$). We assert that the extensions of $\bar{v}$ to $\Delta'^\star_1$ are independent. This will establish the validity of Theorem 19 for $\bar{v}$, $\Delta'_1$ and $\Delta'^\star_1$, and hence, by the preceding remark, Theorem 19 will be established for $v$, $K$ and $K^\star$.

Let $\bar{v}'^\star_1$, $\bar{v}'^\star_2$ be two distinct extensions of $\bar{v}$ to $\Delta'^\star_1$ and assume that there exists a non-trivial valuation $\tilde{v}'^\star$ of $\Delta'^\star_1$ with which both valuations $\bar{v}'^\star_1$ and $\bar{v}'^\star_2$ are composite. Set $v_i^\star = v'^\star_1 \circ \bar{v}'^\star_i$, $i=1, 2$, and $\tilde{v}^\star = v'^\star_1 \circ \tilde{v}'^\star$. Then $v_1^\star$, $v_2^\star$ are extensions of $v$, i.e., belong to $E(v)$, while $\tilde{v}^\star$ is an extension of a valuation $\tilde{v}$ of $K$ such that $v > \tilde{v} > v'_1$. Hence both $E(v)$ and $E(\tilde{v})$ consists exactly of $g$ elements. On the other hand, it is obvious that both $v_1^\star$ and $v_2^\star$ are composite with $\tilde{v}^\star$, and hence $\varphi_{\tilde{v}}^v(v_1^\star) = \varphi_{\tilde{v}}^v(v_2^\star)$ $(=\tilde{v}^\star)$. Thus $\varphi_{\tilde{v}}^v$ is not (1, 1), in contradiction with the fact that $E(v)$ and $E(\tilde{v})$ have the same number of elements.

**§ 12. Ramification theory of general valuations.** In Vol. I, Ch. V, § 10 we have developed the ramification theory of prime ideals in Dedekind domains. Now, if $R$ is a Dedekind domain, with quotient field $K$, and $K^\star$ is an algebraic extension of $K$, then any proper prime ideal $\mathfrak{p}$ in $R$ defines a discrete, rank 1 valuation $v$ of $K$, whose valuation ring is the quotient ring $R_\mathfrak{p}$ (§ 2, Example 2), and the prime ideals which lie over $\mathfrak{p}$ in the integral closure $R^\star$ of $R$ in $K^\star$ correspond to the extensions of $v$ in $K^\star$. Hence the theory developed in Vol. I, Ch. V, § 10 is identical with the ramification theory of discrete, rank 1 valuations. In this section we shall generalize that theory to arbitrary valuations.

Let $K$ be a field, $K^\star$ a finite normal and *separable* extension of $K$, and let $G$ be the Galois group of $K^\star$ over $K$. We fix a valuation $v$ of $K$ and we denote by $\Delta$ and $\Gamma$ respectively the residue field and the value group of $v$. If $v^\star$ is an extension of $v$ in $K^\star$ and $s$ is an element of $G$, then the conjugate valuation $sv^\star$ (= the automorphism $s$ of $K^\star/K$, followed by the mapping $v^\star$ of the multiplicative group $K'^\star$ of $K^\star$ onto the value group $\Gamma^\star$ of $v^\star$) is again an extension of $v$ in $K^\star$ (with the same value group $\Gamma^\star$), and we know (§ 7, Theorem 12, Corollary 3) that all the extensions of $v$ in $K^\star$ are in fact, up to equivalence, conjugates $sv^\star$ $(s \in G)$ of any one of them.

We fix an extension $v^\star$ of $v$. As usual, $R_v$ and $\mathfrak{M}_v$ will denote respectively the valuation ring and the prime ideal of $v$. Similar notations $R_{v^\star}$ and $\mathfrak{M}_{v^\star}$ will be used for $v^\star$. We shall find it convenient to denote by $v^{\star s}$ the valuation $s^{-1}v^{\star}$ $(s \in G)$. With this notation, we will have $R_{v^{\star s}} = s(R_{v^{\star}})$, $\mathfrak{M}_{v^{\star s}} = s(\mathfrak{M}_{v^{\star}})$ and

$$(1) \qquad v^{\star s}(s(x)) = v^{\star}(x), \quad 0 \neq x \in K^{\star}.$$

We denote by $\Delta^{\star}$ and $\Gamma^{\star}$ respectively the residue field and the value group of $v^{\star}$. Here $\Delta^{\star}$ is a finite algebraic extension of $\Delta$, and $\Gamma$ is a subgroup of $\Gamma^{\star}$, of finite index. We set, in agreement with previous notations:

$$(2) \qquad e = (\Gamma^{\star}:\Gamma), \quad f = [\Delta^{\star}:\Delta].$$

The integers $e$ and $f$ are the same for all the extensions of $v$. We denote by $g$ the number of distinct (i.e., non-equivalent) extensions of $v$.

We now introduce two subgroups $G_Z$ and $G_T$ of $G$ called respectively the *decomposition group* and the *inertia group* of $v^{\star}$: $G_Z$ is the set of all $s$ in $G$ such that $v^{\star s}$ is equivalent to $v^{\star}$ (i.e., has the same valuation ring as $v^{\star}$), while $G_T$ is the set of all $s$ in $G$ such that $s(x) - x \in \mathfrak{M}_{v^{\star}}$ for all $x$ in $R_{v^{\star}}$. It is obvious that $G_Z$ is a subgroup of $G$. It is easy to see that $G_T$ is *a subgroup of* $G_Z$. For if $s \in G_T$, then it follows from the definition of $G_T$ that we have $s(x) \in R_{v^{\star}}$ for any $x$ in $R_{v^{\star}}$, i.e., the valuation ring of $v^{\star s}$ is contained in the valuation ring of $v^{\star}$. Therefore the valuation rings of $v^{\star}$ and $v^{\star s}$ coincide (since all extensions of $v$ have the same relative dimension zero with respect to $K$; see italicized statement on p. 30 immediately following the proof of Lemma 1, § 7), $s \in G_Z$, showing that $G_T \subset G_Z$. Furthermore, if $s \in G_T$ and $x \in R_{v^{\star}}$, then also $y = s^{-1}(x)$ is in $R_v^{\star}$ (since $s \in G_Z$), and $s^{-1}(x) - x = y - s(y) \in \mathfrak{M}_{v^{\star}}$, whence $s^{-1} \in G_T$; and if $s, t \in G_T$ then for any $x$ in $R_{v^{\star}}$ we have $(st)(x) - x = t(s(x) - x) + (t(x) - x) \in \mathfrak{M}_{v^{\star}}$, since both $s(x) - x$ and $t(x) - x$ are in $\mathfrak{M}_{v^{\star}}$ and since $t(\mathfrak{M}_{v^{\star}}) \subset \mathfrak{M}_{v^{\star}}$. This proves that $G_T$ is a group.

Moreover it is not difficult to see that $G_T$ is an invariant subgroup of $G_Z$. For if $s \in G_T$, $t \in G_Z$ and $x \in R_{v^{\star}}$ and if we set $t(x) = y$ (whence $y \in R_{v^{\star}}$) and $s(y) - y = z$ (whence $z \in \mathfrak{M}_{v^{\star}}$), then $(tst^{-1})(x) - x = (st^{-1})(y) - x = t^{-1}(y + z) - x = t^{-1}(z) \in \mathfrak{M}_{v^{\star}}$ (since $t(\mathfrak{M}_{v^{\star}}) = \mathfrak{M}_{v^{\star}}$), and hence $tst^{-1} \in G_T$.

Let $s$ be any element of $G_Z$. Then the valuation $v^{\star s}$ defined by (1), is, by definition of $G_Z$, equivalent to $v^{\star}$. However, it is not difficult to see—and that will be important for the sequel— that $v^{\star s}$ *coincides with* $v^{\star}$, that we have therefore

$$(3) \qquad v^{\star}(s(x)) = v^{\star}(x), \quad (s \in G_Z, \quad 0 \neq x \in K^{\star}).$$

For, since $v^{\star}$ and $v^{\star s}$ are equivalent valuations, with the same value group (see (1)), $v^{\star s}v^{\star -1}$ is an order preserving automorphism $\varphi_s$ of the value group $\Gamma^{\star}$. Since $s$ has finite period, also $\varphi_s$ has finite period, and it is immediate that such an order preserving automorphism of an ordered abelian group is necessarily the identity. Thus, $\varphi_s = 1$ and $v^{\star s} = v^{\star}$.

THEOREM 21. *The field $\Delta^\star$ is a normal extension of $\Delta$. The group of automorphisms of $\Delta^\star$ over $\Delta$ is canonically isomorphic to the factor group $G_Z/G_T$.*

PROOF. We first show that every automorphism $s$ in $G_Z$ defines an automorphism $\bar{s}$ of $\Delta^\star$ over $\Delta$. Given any element $\xi$ in $\Delta^\star$, there exists an element $x$ in $R_{v^\star}$ whose $v^\star$-residue is $\xi$. If $s \in G_Z$ then also $s(x) \in R_{v^\star}$. If $x'$ is another element of $R_{v^\star}$ with $v^\star$-residue $\xi$, then $x' - x \in \mathfrak{M}_{v^\star}$ and hence also $s(x') - s(x) \in \mathfrak{M}_{v^\star}$, since $s \in G_Z$. It follows that the $v^\star$-residue of $s(x)$, for given $s$ in $G_Z$, depends only on $\xi$. We denote this residue by $\bar{s}(\xi)$. It is immediate that the mapping $\xi \rightarrow \bar{s}(\xi)$ is an automorphism $\bar{s}$ of $\Delta^\star$, and that $\bar{s}$ is an automorphism *over* $\Delta$, for if $\xi \in \Delta$ then we can choose $x$ in $R_v$ and have then $s(x) = x$. It is also clear that the mapping $s \rightarrow \bar{s}$ is a homomorphism of $G_Z$ into the group $G(\Delta^\star/\Delta)$ of automorphisms of $\Delta^\star$ over $\Delta$ and that the kernel of this homomorphism is the inertia group $G_T$ of $v^\star$. We have now to show that $\Delta^\star$ is a normal extension of $\Delta$ and that the mapping $s \rightarrow \bar{s}$ sends $G_Z$ *onto* $G(\Delta^\star/\Delta)$.

Let $\xi$ again be any element of $\Delta^\star$, different from zero. Since the places defined by the $g$ distinct extensions of $v$ are such that none is a specialization of another, it follows from Lemma 2, § 7, that we can find an element $x$ in $R_{v^\star}$ having $v^\star$-residue $\xi$ and such that $v_j^\star(x) > 0$ for each of the $g-1$ extensions $v_j^\star$ of $v$ which are different from $v^\star$. Let $x_1(=x), x_2, \cdots, x_q$ be the roots of the minimal polynomial $F(X) = X^q + a_{q-1}X^{q-1} + \cdots + a_0$ of $x$ over $K$. Since $K^\star$ is normal over $K$, all the $x_j$ belong to $K^\star$. For any $x_j$ we have $x = s(x_j)$, for a suitable $s$ in $G(K^\star/K)$, and hence, by (1): $v^{\star s}(x) = v^\star(x_j)$. Since $v^{\star s}(x) \geqq 0$ for any $s$ in $G(K^\star/K)$ (by our choice of $x$), it follows that all the roots $x_j$ and all the coefficients $a_\nu$ of $F(X)$ belong to $R_{v^\star}$. We have $F(X) = \prod_{j=1}^{q} (X - x_j)$, and taking $v^\star$-residues on both sides we find that the roots of the polynomial $\bar{F}(X) = X^q + \bar{a}_{q-1}X^{q-1} + \cdots + \bar{a}_0$ ($\bar{a}_\nu = v^\star$-residue of $a_\nu$) are the $v^\star$-residues of $x_1, x_2, \cdots, x_q$ and therefore belong to $\Delta^\star$. Since $\xi$ is among these residues and since the coefficients $\bar{a}_\nu$ of $\bar{F}(X)$ belong to $\Delta$, we have shown that all the conjugates of $\xi$ over $\Delta$ belong to $\Delta^\star$. Hence $\Delta^\star$ is a normal extension of $\Delta$.

If $\xi_2$ is any conjugate of $\xi$ over $\Delta$, and if say $\xi_2 = v^\star$-residue of $x_j$, let $s$ be an automorphism of $K^\star/K$ such that $x_j = s^{-1}(x)$. Then $v^{\star s}(x) = v^\star(x_j) = 0$ (since $\xi_2 \neq 0$), and hence $v^{\star s} = v^\star$ (since $v_1^\star(x) > 0$ for each extension $v_1^\star$ of $v$ which is different from $v^\star$) and $s \in G_Z$. Furthermore

$\bar{s}^{-1}(\xi)=\xi_2$. If we take now for $\xi$ a primitive element, over $\Delta$, of the maximal separable extension of $\Delta$ in $\Delta^\star$, then our result that every conjugate of $\xi$ over $\Delta$ is of the form $\bar{s}(\xi)$, $s \in G_Z$, implies that the homomorphism $s \to \bar{s}$ maps $G_Z$ *onto* the group $G(\Delta^\star/\Delta)$. This completes the proof of the theorem.

In the sequel we shall denote by $K_Z$ and $K_T$ respectively the fixed fields of $G_Z$ and $G_T$; $K_Z$ is the *decomposition field* of $v^\star$, and $K_T$ is the *inertia field* of $v^\star$ (relative to $K$). We shall denote by $v_Z$ and $v_T$ respectively the restriction of $v^\star$ in $K_Z$ and $K_T$, by $\Delta_Z$ and $\Delta_T$ the residue fields of the valuations $v_Z$ and $v_T$, and by $\Gamma_Z$ and $\Gamma_T$ their respective value groups. Clearly $\Delta_Z$ is a subfield of $\Delta_T$, and $\Gamma_Z$ is a subgroup of $\Gamma_T$. Furthermore, $K_T$ is a normal extension of $K_Z$, with Galois group $G_Z/G_T$, since $G_T$ is a normal subgroup of $G_Z$.

These definitions have a relative character, and it is easy to see how the decomposition field or inertia field of $v^\star$ is affected if we replace $K$ by another field $L$ *between $K$ and $K^\star$.* Namely, if we denote by $L_Z$ and $L_T$ respectively the decomposition field and the inertia field of $v^\star$, relative to $L$, then *$L_Z$ is the compositum of $K_Z$ and $L$* (least subfield of $K^\star$ which contains both $K_Z$ and $L$) and similarly *$L_T$ is the compositum of $K_T$ and $L$*:

$$L_Z = (K_Z, L), \tag{4}$$

$$L_T = (K_T, L). \tag{4'}$$

The proof is straightforward and consists simply in observing that the decomposition group and inertia group of $v^\star$ *relative* to $L$ are obviously equal respectively to $G_Z \cap G(K^\star/L)$ and $G_T \cap G(K^\star/L)$.

THEOREM 22. (a) *The valuation $v^\star$ is the only extension of $v_Z$ to $K^\star$, and the decomposition field $K_Z$ is the smallest of all fields $L$ between $K$ and $K^\star$ with the property that $v^\star$ is the only extension, to $K^\star$, of the restriction of $v^\star$ to $L$.* (b) *The field $\Delta^\star$ is purely inseparable over $\Delta_T$, $\Delta_T$ is separable and normal over $\Delta_Z$, and $\Delta_Z$ coincides with $\Delta$.*

PROOF. Since all the extensions of $v$ in $K^\star$ are conjugates of $v^\star$, it follows that $v^\star$ is the only extension of $v$ if and only if $G_Z=G$, i.e., if and only if $K_Z=K$. If $L$ is an arbitrary field between $K$ and $L$, then $K^\star$ is also a normal separable extension of $L$, and therefore it follows, by the same token, that $v^\star$ is the only extension to $K^\star$ of the restriction $v'$ of $v^\star$ to $L$ if and only if $L_Z=L$, i.e., by (4), if and only if $L \supset K_Z$. This proves part (a) of the theorem.

We have $G(K^\star/K_T)=G_T$, and therefore both the decomposition group and the inertia group of $v^\star$ relative to $K_T$ are equal to $G_T(=G_Z \cap G_T=G_T \cap G_T)$. If we now replace in Theorem 21 the field

$K$ by the field $K_T$ it follows that $G(\Delta^\star/\Delta_T)=G_T/G_T=(1)$, showing that $\Delta^\star$ is purely inseparable over $\Delta_T$. On the other hand, we have already observed that $G_T$ is an invariant subgroup of $G_Z$ and that consequently $K_T$ is a normal separable extension of $K_Z$, with Galois group $G_Z/G_T$. Hence, if we replace in Theorem 21 the fields $K$ and $K^\star$ by the fields $K_Z$ and $K_T$ respectively, we find that $G(\Delta_T/\Delta_Z)$ is canonically isomorphic with $G_Z/G_T$. Since $[\Delta_T:\Delta_Z] \leqq [K_T:K_Z]=$ order of $G_Z/G_T$, it follows that $[\Delta_T:\Delta_Z] \leqq$ order of $G(\Delta_T/\Delta_Z)$, and hence $[\Delta_T:\Delta_Z]=$order of $G(\Delta_T/\Delta_Z)$, showing that $\Delta_T$ is a normal separable extension of $\Delta_Z$.

We point out that in the course of this proof we have shown incidentally that

$$[\Delta_T:\Delta_Z] = [K_T:K_Z]. \tag{5}$$

It remains to prove that $\Delta_Z=\Delta$. Let $\xi$ be any element of $\Delta_Z$. By the cited Lemma 2 of § 7 we can find an element $x$ in $K_Z$ having $v_Z$-residue $\xi$ and such that $v'(x)>0$ for every extension $v'$ of $v$ to $K_Z$, different from $v_Z$. If $x_i$ is any conjugate of $x$ (over $K$), *different from* $x$, then $x=s(x_i)$ for some $s$ in $G$, and we have necessarily $s \notin G_Z$ since $x_i \neq x$. By (1), we have $v^\star(x_i)=v^{\star s}(x)$, and, furthermore, we have $v^{\star s}(x)>0$ since $v^{\star s}\neq v^\star$ ($s$ being outside of $G_Z$) and since therefore $v^{\star s}$ induces in $K_Z$ a valuation different from $v_Z$ ($v^\star$ being the only extension of $v_Z$ to $K^\star$). We have found therefore that $v^\star(x_i)>0$ for every conjugate $x_i$ of $x$ which is different from $x$. Consequently the trace $x+\Sigma x_i$ is an element $y$ of $K$ whose $v_Z$-residue is $\xi$ ($=v_Z$-residue of $x$). Therefore, $\xi \in \Delta$ and $\Delta_Z=\Delta$. This completes the proof of the theorem.

THEOREM 23. *The value groups $\Gamma$, $\Gamma_Z$ and $\Gamma_T$ coincide.*

PROOF. If we apply the inequality $\Sigma e_i f_i \leqq n$ (§ 11, Theorem 19 and *Note* on page 64) to the two fields $K_Z$, $K_T$ and to the valuation $v_Z$ of $K_Z$, we deduce at once from (5) that $v_Z$ has only one extension to $K_T$ (a fact that we know already) and also that $(\Gamma_T:\Gamma_Z)=1$. This proves that $\Gamma_Z=\Gamma_T$.

We shall first prove the equality $\Gamma_Z=\Gamma$ under the assumption that the $g$ extensions of $v$ to $K^\star$ are independent. It will be sufficient to show that every positive element of $\Gamma_Z$ is in $\Gamma$. Let $\alpha$ be a positive element of $\Gamma_Z$. By the approximation theorem for independent valuations (§ 10, Theorem 18′) there exists an element $x$ in $K_Z$ such that $v_Z(x)=\alpha$ and $v'(x)=0$ for every extension $v'$ of $v$ to $K_Z$, different from $v_Z$ (since from our assumption that the extensions of $v$ to $K^\star$ are independent follows *a fortiori* that also the extensions of $v$ to $K_Z$ are independent). The argument developed toward the end of the proof of the preceding theorem shows that if $x_i$ is any conjugate of $x$ over $K$, different from $x$, then $v_Z(x_i)=0$. Hence the norm $x \cdot \Pi x_i$ is an element $y$ of $K$ such that $v(y)=v_Z(x)+0=\alpha$. Therefore $\alpha \in \Gamma$ and $\Gamma_Z=\Gamma$. This completes the proof of the theorem in the case in which the extensions of $v$ to $K^\star$ are independent valuations.

In the general case we shall use induction with respect to the number $g$ of distinct extensions of $v$ to $K^\star$, for if $g=1$ then $K=K_Z$ (by Theorem 22, part (a)) and the equality $\Gamma=\Gamma_Z$ is then trivial.

If $v$ has rank 1 then the $g$ extensions of $v$ to $K^\star$ are also of rank 1 and are therefore independent. We shall therefore assume that $v$ is of rank $>1$ and we may also assume that the $g$ extensions of $v$ to $K^\star$ are not independent. We shall make use of the results proved at the end of the preceding section (§ 11, *Note*). From our assumption that the $g$ extensions of $v$ to $K^\star$ are dependent valuations follows that $\gamma(v')$ is not constantly equal to $g$ as $v'$ varies in the set $L(v)$. It was shown in § 11 that in that case there exists a decomposition $v=v' \circ \bar{v}$ of $v$ satisfying the following condition: $\gamma(v')=h<g$, and if $v'_1{}^\star, v'_2{}^\star, \cdots, v'_h{}^\star$ are the extensions of $v'$ to $K^\star$ then for each $s=1, 2, \cdots, h$ the extensions of $\bar{v}$ to the residue field $\Delta'_s{}^\star$ of $v'_s{}^\star$ are independent.

To the decomposition $v=v' \circ \bar{v}$ there corresponds a decomposition $v^\star=v'^\star \circ \bar{v}^\star$, where $v'^\star$ is one of the $h$ extensions $v'_i{}^\star$ of $v'$ to $K^\star$ and $\bar{v}^\star$ is an extension of $\bar{v}$ to the residue field $\Delta'^\star$ of $v'^\star$. We denote by $G_{Z'}$ and $G_{T'}$ respectively the decomposition group and the inertia group of $v'^\star$. It is not difficult to see that we have the following inclusions:

$$(5') \qquad G_{Z'} \supset G_Z \supset G_T \supset G_{T'}.$$

The inclusion $G_{Z'} \supset G_Z$ follows from the fact that $v'^\star$ is the only extension of $v'$ such that $v^\star$ is composite with $v'^\star$ and that, therefore, if $s \in G_Z$, then we must have $v'^{\star s}=v'^\star$, since $v^\star(=v^{\star s})$ is composite with both valuations $v'^\star$ and $v'^{\star s}$. The inclusion $G_T \supset G_{T'}$ follows from the inclusions $R_{v^\star} \subset R_{v'^\star}$, $\mathfrak{M}_{v^\star} \supset \mathfrak{M}_{v'^\star}$. Namely, if $s \in G_{T'}$ and $x$ is any element of $R_{v^\star}$, then $x \in R_{v'^\star}$ (since $R_{v^\star} \subset R_{v'^\star}$), $s(x)-x \in \mathfrak{M}_{v'^\star}$ (since $s \in G_{T'}$), and $s(x)-x \in \mathfrak{M}_{v^\star}$ (since $\mathfrak{M}_{v^\star} \supset \mathfrak{M}_{v'^\star}$), showing that $G_{T'} \subset G_T$.

We denote by $K_{Z'}$ and $K_{T'}$ respectively the decomposition field and inertia field of $v'$. We have therefore, by (5'):

$$(6) \qquad K \subset K_{Z'} \subset K_Z \subset K_T \subset K_{T'}.$$

We denote by $v_{Z'}, v_Z, v_T, v_{T'}$ the restrictions of $v^\star$ in $K_{Z'}, K_Z, K_T, K_{T'}$ respectively, and by $v'_{Z'}, v'_Z, v'_T, v'_{T'}$ the corresponding restrictions of $v'^\star$. The associated value groups will be denoted by $\Gamma_{Z'}, \Gamma_Z, \cdots$ and $\Gamma'_{Z'}, \Gamma'_Z, \cdots$ respectively.

Since $h<g$, it follows from our induction hypothesis that Theorem 23 is valid for $v'$ and $v'^{\star}$, i.e., we have

$$(7) \qquad \Gamma' = \Gamma'_{Z'} = \Gamma'_{T'},$$

where $\Gamma'$ is the value group of $v'$. In view of (6), this also implies that

$$(8) \qquad \Gamma' = \Gamma'_{Z} = \Gamma'_{T}.$$

The decomposition $v^{\star}=v'^{\star} \circ \bar{v}^{\star}$ yields a corresponding decomposition of $v_{Z'}$:

$$(9) \qquad v_{Z'} = v'_{Z'} \circ \bar{v}_{Z'},$$

where $\bar{v}_{Z'}$ is the restriction of $\bar{v}^{\star}$ to the residue field $\Delta'_{Z'}$ of $v'_{Z'}$. By Theorem 22, part (b), we have that *$\Delta'_{Z'}$ coincides with the residue field $\Delta'$ of $v'$.* Since $\bar{v}_{Z'}$ is an extension of the valuation $\bar{v}$ of $\Delta'$, it follows that $\bar{v}_{Z'}=\bar{v}$. This, in conjunction with (9) and equality (7), shows that $\Gamma=\Gamma_{Z'}$. It is therefore only necessary to show that $\Gamma_{Z'}=\Gamma_{Z}$. Thus we may replace the field $K$ by the field $K_{Z'}$. We may therefore assume that $K$ is the decomposition field of $v'^{\star}$ and that therefore $v'^{\star}$ is the only extension of $v'$ to $K^{\star}$. The valuation $\bar{v}$ has then exactly $g$ extensions to $\Delta'^{\star}$, and by our choice of $v'$ *these $g$ extensions are independent valuations.*

Let $H$ be the isolated subgroup of $\Gamma$ which corresponds to the decomposition $v=v' \circ \bar{v}$ ($H$=value group of $\bar{v}$; $\Gamma'=\Gamma/H$=value group of $v'$). Let similarly $H_Z$ be the isolated subgroup of $\Gamma_Z$ which corresponds to the decomposition $v_Z=v'_Z \circ \bar{v}_Z$ (here $\bar{v}_Z$ is the restriction of $\bar{v}^{\star}$ to the residue fields of $v'_Z$). We have therefore $H=H_Z \cap \Gamma$ (see § 11, Lemma 4). We know that $\Gamma'=\Gamma'_Z$, i.e., $\Gamma/H=\Gamma_Z/H_Z$. To prove the equality $\Gamma=\Gamma_Z$ it will therefore be sufficient to show that

$$(10) \qquad H_Z = H,$$

i.e., that the value group $H$ of $\bar{v}$ coincides with the value group $H_Z$ of its extension $\bar{v}_Z$ to the residue field of $v'_Z$. Since the extensions of $\bar{v}$ to the residue field of $v'^{\star}$ are independent it follows *a fortiori* that also the extensions of $\bar{v}$ to the residue field of $v'_Z$ are independent. Hence, given a positive element $\alpha$ of $H_Z$ we can find an element $\bar{x}$ of the residue field of $v'_Z$ such that $\bar{v}_Z(\bar{x})=\alpha$ and $\bar{v}'_Z(\bar{x})=0$ for all other extensions of $\bar{v}'_Z$ of $\bar{v}$ to the residue field of $v'_Z$. If, now, $x$ is an element of $K_Z$ whose $v'_Z$-residue is $\bar{x}$ then we will have $v_Z(x)=\alpha$ and $v_1(x)=0$ for all *other* extensions of $v$ to $K_Z$. By an argument given earlier it follows that if $y=N_{K_Z,K}(x)$ then $v(y)=\alpha$. This establishes the equality (10) and completes the proof of the theorem.

It is clear that the index of $G_Z$ in $G$ is equal to the number $g$ of extensions of $v$ to $K^\star$. Hence

$$(11) \qquad [K_Z:K] = g = (G:G_Z).$$

We denote by $f_0$ the separable factor of the relative degree $f=[\Delta^\star:\Delta]$ and we set

$$(12) \qquad f = f_0\pi^s,$$

where $\pi$ is the characteristic of $\Delta$ if the characteristic is different from zero and is 1 otherwise. Theorems 21 and 22 show that

$$(13) \qquad f_0 = [\Delta_T:\Delta_Z] = [K_T:K_Z] = \text{order of } G_Z/G_T.$$

For any $s$ in $G_T$ and for any element $a$ of $K^\star$, $a \neq 0$, we denote by $(a, s)$ the $v^\star$-residue of $s(a)/a$. (By (3), this residue is different from $\infty$ and 0 if $s \in G_Z$ and hence, *a fortiori*, also if $s$ is in $G_T$.) We have the following relations

$$(14) \qquad (a, s) = 1 \quad \text{if} \quad a \in R_v,\ a \notin \mathfrak{M}_v,\ s \in G_T;$$

$$\begin{rcases} (14') \qquad (ab, s) = (a, s)(b, s), \\ (14'') \qquad (a, st) = (a, s)(a, t). \end{rcases} a, b \in K^\star;\ s, t \in G_T$$

Relation (14) is evident, since $s(a)-a=m \in \mathfrak{M}_v$, $s(a)/a=1+m/a$, and the $v$-residue of $m/a$ is zero if $a \notin \mathfrak{M}_v$. Also relation (14′) is evident since $s(ab)=s(a)s(b)$. As to (14″), we write $\dfrac{(st)(a)}{a}=\dfrac{t(s(a))}{t(a)}\cdot\dfrac{t(a)}{a}$ and we note that $\dfrac{t(s(a))}{t(a)}=t\left(\dfrac{s(a)}{a}\right)$, and since the $v^\star$-residue of $\dfrac{s(a)}{a}$ is neither $\infty$ nor 0 $\left(\text{whence } \dfrac{s(a)}{a} \in R_v, \dfrac{s(a)}{a} \notin \mathfrak{M}_v\right)$ it follows, by (14), that $\dfrac{(st)(a)}{t(a)}$ has the same $v^\star$-residue as $\dfrac{s(a)}{a}$, since $t \in G_T$. Relations (14′) and (14″) show that the function $(a, s)$ establishes a "pairing" between the group $G_T$ and the multiplicative group of $K^\star$. For fixed $s$ in $G_T$ the mapping $a \to (a, s)$ is a homomorphism of the multiplicative group of $K^\star$ into the multiplicative group of $\Delta^\star$. We denote by $K^{\star\prime}$ and $\Delta^{\star\prime}$ these multiplicative groups and we use the customary notation $\mathrm{Hom}(K^{\star\prime}, \Delta^{\star\prime})$ for the set of all homomorphisms of $K^{\star\prime}$ into $\Delta^{\star\prime}$. This set $\mathrm{Hom}(K^{\star\prime}, \Delta^{\star\prime})$ is a group in an obvious way (if $f$ and $g$ are two homomorphisms of $K^{\star\prime}$ into $\Delta^{\star\prime}$ we define $fg$ by $(fg)(a)=f(a)g(a)$, $a \in K^{\star\prime}$). Hence, for fixed $s$ in $G_T$ the mapping $a \to (a, s)$ is an element of $\mathrm{Hom}(K^{\star\prime}, \Delta^{\star\prime})$. If we denote this element by $\varphi(s)$:

$$(15) \qquad \varphi(s):\quad a \to (a, s), \quad a \in K^{\star\prime},$$

then (14″) shows that the mapping

$$(15') \qquad \varphi\colon\ G_T \to \mathrm{Hom}\,(K^{\star\prime}, \Delta^{\star\prime})$$

is a homomorphism. Similarly, for fixed $a$ in $K^{\star\prime}$, the mapping $s \to (a, s)$ is an element of Hom $(G_T, \Delta^{\star\prime})$. If we denote this element by $\psi(a)$:

$$(16) \qquad \psi(a)\colon\ s \to (a, s), \quad s \in G_T,$$

then (14′) shows that the mapping

$$(16') \qquad \psi\colon\ K^{\star\prime} \to \mathrm{Hom}\,(G_T, \Delta^{\star\prime})$$

is a homomorphism. We shall investigate the kernels of $\varphi$ and $\psi$ in order to determine to what extent the pairing $(a, s)$ is "faithful."

The elements of the kernel of $\varphi$ are those elements $s$ of $G_T$ for which it is true that $\varphi(s)$ maps every element of $K^{\star\prime}$ into the element 1 of $\Delta^{\star\prime}$, i.e., those elements $s$ for which $(a, s)=1$ for any $a$ in $K^{\star\prime}$. Now, $(a, s)=1$ is equivalent to $v^\star\left(\frac{s(a)}{a}-1\right)>0$. Hence the kernel of $\varphi$ consists of those elements $s$ of $G_T$ which satisfy the condition

$$(17) \qquad v^\star(s(x)-x) > v^\star(x), \quad \text{for all } x \text{ in } K^{\star\prime}.$$

These elements form therefore an invariant subgroup of $G_T$. This subgroup is denoted by $G_V$ and is called the *large ramification group of* $v^\star$.

In the case of Dedekind rings treated in Chapter V, § 10, the large ramification group $G_V$ is the inverse image in $G_T$ of the subgroup $G'_1$ of $G_T/G_{V_2}$ mentioned in V, § 10, Theorem 25. It is also the set, denoted in V, § 10 (p. 295) by $H_1$, of all $s$ in $G_T$ such that $s(u)-u \in \mathfrak{M}_{v^\star}{}^2$, where $u$ is a generator of $\mathfrak{M}_{v^\star}$.

We now study the kernel of $\psi$. If $a \in K_T$ then $s(a)=a$ and therefore $(a, s)=1$ for all $s$ in $G_T$. Hence the kernel of $\psi$ contains the inertia field $K_T$. The kernel of $\psi$ also contains all the units of the valuation ring $R_{v^\star}$, by (14). *It follows now that the kernel of $\psi$ contains all the elements $a$ of $K^\star$ such that $v^\star(a) \in \Gamma$*, for if $a$ is such an element and if $b$ is an element of $K$ such that $v^\star(a)=v^\star(b)$, then $a=bc$, with $c$ a unit in $R_{v^\star}$, and since both $b$ and $c$ are in the kernel of $\psi$, also $a$ is in the kernel.

The above consideration shows that $(a, s)$ depends only on the pair $(\bar{\alpha}, \bar{s})$, where $\bar{\alpha}$ is the $\Gamma$-coset of $v^\star(a)$ and $\bar{s}$ is the $G_V$-coset of $s$. Since $v^\star$ is a homomorphism of $K^{\star\prime}$ onto $\Gamma^\star$, it follows that *the pairing $(a, s)$ defines in a natural way a pairing between the (multiplicative) group $G_T/G_V$ and the (additive) group $\Gamma^\star/\Gamma$.* The homomorphism $\varphi$, given by (15) and (15′), gives rise to an *isomorphism*

$$(18) \qquad \bar{\varphi}_1\colon\ G_T/G_V \to \mathrm{Hom}\,(\Gamma^\star/\Gamma, \Delta^{\star\prime})$$

*of $G_T/G_V$ into the group of homomorphisms of $\Gamma^\star/\Gamma$* into $\Delta^{\star\prime}$, while the homomorphism $\psi$, defined by (16) and (16′), gives rise to *a homomorphism*

$$\bar{\psi}_1\colon \quad \Gamma^\star/\Gamma \to \mathrm{Hom}\,(G_T/G_V, \Delta^{\star\prime}) \tag{19}$$

*of $\Gamma^\star/\Gamma$ into the group of homomorphisms of $G_T/G_V$ into $\Delta^{\star\prime}$.*

We point out the special case in which $\Gamma^\star/\Gamma$ is a cyclic group of order $e$ [see (2)] (we have this case, for instance, if $v$ is a discrete valuation of rank 1). If we choose a generator $\alpha$ of $\Gamma^\star/\Gamma$ (for instance, $\alpha=$ the $\Gamma$-coset of the smallest positive element of $\Gamma^\star$, if $v$ is discrete of rank 1), then any homomorphism $h$ of $\Gamma^\star/\Gamma$ into $\Delta^{\star\prime}$ is uniquely determined by the value $h(\alpha)$. Hence, if we set, for any $\sigma$ in $G_T/G_V$, $i(\sigma)=(\bar{\varphi}_1(\sigma))(\alpha)$, then *$i$ is an isomorphism of $G_T/G_V$ into the multiplicative group $\Delta^{\star\prime}$* (see Vol. I, Ch. V, § 10, Theorem 25).

We denote by $\pi$ the "characteristic exponent" of the residue field $\Delta$ of $v$, i.e., $\pi$ is equal to the characteristic $p$ of $\Delta$ if $p\neq 0$ and is equal to 1 if $p=0$. The finite abelian group $\Gamma^\star/\Gamma$ is the direct sum of a $\pi$-group $\bar{\Gamma}_\pi$ (= the set of elements $\bar{\alpha}$ such that the order of $\bar{\alpha}$ is a power of $\pi$) and a group $\bar{\Gamma}_0$ whose order is prime to $\pi$ ($\bar{\Gamma}_0=$ set of elements $\bar{\alpha}$ such that order of $\bar{\alpha}$ is prime to $\pi$). If we set

$$e = e_0\pi^t, \quad e_0 \text{ prime to } \pi, \tag{20}$$

then $\pi^t$ is the order of $\bar{\Gamma}_\pi$, and $e_0$ is the order of $\bar{\Gamma}_0$. Since 1 is the only element $\xi$ of $\Delta^{\star\prime}$ such that the order of $\xi$ is a power of $\pi$, it follows that every homomorphism of $\Gamma^\star/\Gamma$ into $\Delta^{\star\prime}$ is trivial on $\bar{\Gamma}_\pi$.

We thus have a pairing between the multiplicative group $G_T/G_V$ and the additive group $\bar{\Gamma}_0$, defining an *isomorphism* of $G_T/G_V$ into $\mathrm{Hom}\,(\bar{\Gamma}_0, \Delta^{\star\prime})$:

$$\bar{\varphi}\colon \quad G_T/G_V \to \mathrm{Hom}\,(\bar{\Gamma}_0, \Delta^{\star\prime}) \tag{21}$$

and a homomorphism of $\bar{\Gamma}_0$ into $\mathrm{Hom}\,(G_T/G_V, \Delta^{\star\prime})$

$$\bar{\psi}\colon \quad \bar{\Gamma}_0 \to \mathrm{Hom}\,(G_T/G_V, \Delta^{\star\prime}). \tag{22}$$

We shall prove later on that $\bar{\varphi}$ and $\bar{\psi}$ are actually *isomorphisms onto*. At present we only note the following: since every element of $\bar{\Gamma}_0$ has order prime to $\pi$, also every homomorphism of $\bar{\Gamma}_0$ has order prime to $\pi$; hence the order of the (finite) group $\mathrm{Hom}\,(\bar{\Gamma}_0, \Delta^{\star\prime})$† is prime to $\pi$, and consequently

(23) *The order $e'_0$ of $G_T/G_V$ is prime to $\pi$.*

† Any homomorphism of the group $\bar{\Gamma}_0$ (which is of order $e_0$) into the group $\Delta^{\star\prime}$ maps $\bar{\Gamma}_0$ into the set of $e_0$th roots of unity; since the latter set is finite, the set $\mathrm{Hom}\,(\bar{\Gamma}_0, \Delta^{\star\prime})$ is also finite.

We note that in the case of characteristic zero $\bar{\varGamma}_0$ coincides with $\varGamma^\star/\varGamma$. We now study the large ramification group $G_V$.

THEOREM 24.   *$G_V$ is a $\pi$-group, i.e., a group whose order is a power of $\pi$. (In particular, $G_V = (1)$ if $\varDelta$ has characteristic zero.)*

PROOF.   We have only to show that if $s \in G_V$ and $s$ has prime order $q$, then $q = \pi$. Assume the contrary: $q \neq \pi$. Let $L$ be the fixed field of $s$. Then $K^\star$ is a cyclic extension of $L$, of degree $q$. Let $x$ be a primitive element of $K^\star/L$ and let $X^q + a_{q-1}X^{q-1} + \cdots + a_0$, $a_i \in L$ be the minimal polynomial of $x$ over $L$. We may assume that $a_{q-1} = 0$ since $q \neq \pi$ and since therefore we can replace $x$ by $x + a_{q-1}/q$. Hence we may assume that the trace of $x$ is zero. On the other hand, if we set $s_i = s^i$, $i = 0, 1, \cdots, q-1$, then the $v^\star$-residue of $x^{s_i}/x$ is 1, since $s_i \in G_V$, and hence the $v^\star$-residue of $\sum_{i=0}^{q-1} x^{s_i}/x$ is equal to $q \neq 0$, a contradiction since the trace $\sum x^{s_i}$ is zero. This completes the proof of the theorem.

At this stage we can already obtain, as a corollary of Theorem 24, the definitive result in the case $\pi = 1$ (i.e., in the case in which $\varDelta$ has characteristic zero):

COROLLARY.   *If the residue field $\varDelta$ of $v$ has characteristic zero then the groups $G_T$ and $\varGamma^\star/\varGamma$ are isomorphic. The ramification deficiency of $v$, relative to $K^\star$, is zero, i.e., we have $efg = n$ $(n = [K^\star : K])$.*

In fact, if $\varDelta$ has characteristic zero, then $G_V = (1)$ and hence $\bar{\varphi}_1$, defined by (18), is an isomorphism of $G_T$ into the group Hom $(\varGamma^\star/\varGamma, \varDelta^\star')$. This latter group is a subgroup of the group of characters† of the abelian group $\varGamma^\star/\varGamma$. Since $\varGamma^\star/\varGamma$ has order $e$ and since $\varGamma^\star/\varGamma$ and its group of characters are isomorphic groups, it follows that $G_T$ is isomorphic with a subgroup of $\varGamma^\star/\varGamma$ and hence has order $\leqq e$. Since $n = gf \cdot$order $G_T$, it follows that $n \leqq efg$, and therefore, by § 11, Theorem 19, we must have $n = efg$, which proves all the assertions of the corollary.

We now continue with the general case.

LEMMA.   *The homomorphism $\bar{\psi}$ defined in (22) is an isomorphism (into).*

PROOF.   We have only to show that if an element $x$ of $K^{\star\prime}$ is such that $\dfrac{s(x)}{x} - 1 \in \mathfrak{M}_{v^\star}$ for every $s$ in $G_T$, then there exists a power $\pi^u$ of $\pi$ such that $\pi^u v^\star(x) \in \varGamma$. Denote by $\pi^u$ the order of $G_V$ (Theorem 24) and by $K_V$ the fixed field of $G_V$. We set $y = N_{K^\star/K_V}(x)$. It is clear that $v^\star(y) = \pi^u v^\star(x)$. On the other hand, by applying the operation

---

† For properties of the group of characters of finite abelian groups see, for instance, B. L. van der Waerden, *Moderne Algebra*, vol. 2 (p. 189), or E. Hecke, *Vorlesungen über die Theorie der algebraischen Zahlen*, p. 33.

$N_{K^\star/K_V}$ to the relation $\frac{s(x)}{x}-1\in\mathfrak{M}_{v^\star}$, we easily get $\frac{s(y)}{y}-1\in\mathfrak{M}_{v^\star}$ for every $s$ in $G_T$. It follows that the conjugates $y_i$ of $y$ over $K_T$ may be written in the form $y_i=y(1+b_i)(b_i\in\mathfrak{M}_{v^\star})$. Since $[K_V:K_T]=e'_0$ (see (23)), there are $e'_0$ conjugates $y_i$, and, by summation, we get

$$T_{K_V/K_T}(y) = y(e'_0+b)$$

with $b=\sum b_i\in\mathfrak{M}_{v^\star}$. Since $e'_0$ is prime to $\pi$, it is a unit in $R_{v^\star}$. Hence $v^\star(y)=v^\star(T(y))\in\Gamma_T$, and therefore $v^\star(y)\in\Gamma$ by Theorem 23. Q.E.D.

It follows from the lemma that the pairing

$$h:\quad G_T/G_V\times\tilde{\Gamma}_0\to\Delta^{\star\prime}$$

defined by (21) and (22) is *faithful* in the sense that 1 is the only element $\sigma$ of $G_T/G_V$ such that $h(\sigma,\tilde{\alpha})=1$ for every $\tilde{\alpha}$ in $\tilde{\Gamma}_0$, and that 0 is the only element $\tilde{\alpha}$ of the additive group $\tilde{\Gamma}_0$ such that $h(\sigma,\alpha)=1$ for every $\sigma$ in $G_T/G_V$. On the other hand, $h$ takes its values in the group $U$ of $e'_0$-th roots of unity contained in $\Delta^\star$; this group $U$ is a *cyclic group of order prime to* $\pi$.

Now the *theory of characters*† for finite abelian groups shows that, given a finite abelian group $H$, the only subgroup $H'_1$ of its character group $H'$ which "separates" the elements of $H$ (i.e., such that $\chi(h)=1$ for all $\chi$ in $H'_1$ implies $h=1$) is the character group $H'$ itself. Thus, if we regard $G_T/G_V$ as a group of characters of $\tilde{\Gamma}_0$, it is the *entire character group of* $\tilde{\Gamma}_0$. Similarly $\tilde{\Gamma}_0$ is the entire character group of $G_T/G_V$. In particular†

THEOREM 25. *The groups $\tilde{\Gamma}_0$ and $G_T/G_V$ are isomorphic (whence $G_T/G_V$ is abelian). Their orders $e_0$ and $e'_0$ are equal.*

COROLLARY. *The product efg divides the degree $n=[K^\star:K]$, and $n/efg$ is a power of $\pi$.*

In fact, $n=(G:G_Z)(G_Z:G_T)(G_T:G_V)(G_V:1)=gf_0e_0\pi^u=efg\pi^{u-s-t}$ (the notations are those of formulae (11), (12), and (20)). Since $efg\leq n$ (§ 11, Theorem 19), it follows that $u-s-t$ is $\geq 0$.

Finally, two series of subgroups of $G$, generalizing the *higher ramification groups*, may be defined. For every ideal $\mathfrak{a}$ in $R_{v^\star}$ we define

(24) $G_\mathfrak{a}$ *as the set of all s in G such that* $s(x)-x\in\mathfrak{a}$ *for every x in* $R_{v^\star}$;

(25) $H_\mathfrak{a}$ *as the set of all s in G such that* $s(x)-x\in\mathfrak{a}x$ *for every x in* $K^\star$.

The following facts are easily verified (many proofs are as in Chapter V, § 10):

(a) $H_\mathfrak{a}\subset G_\mathfrak{a}$.

(b) $H_{\mathfrak{M}_{v^\star}}=G_V$, $G_{\mathfrak{M}_{v^\star}}=G_T$, $H_{R_{v^\star}}=G_{R_{v^\star}}=G_Z$.

† See op. cit. in the footnote of the preceding page.

(c) If $\mathfrak{a} \subset \mathfrak{b}$, then $G_{\mathfrak{a}} \subset G_{\mathfrak{b}}$ and $H_{\mathfrak{a}} \subset H_{\mathfrak{b}}$.

(d) $G_{\mathfrak{a}}$ and $H_{\mathfrak{a}}$ are invariant subgroups of $G_Z$.

(e) The commutator of an element of $H_{\mathfrak{a}}$ and of an element of $H_{\mathfrak{b}}$ is in $H_{\mathfrak{ab}}$.

(f) Let the value group $\Gamma^\star$ be isomorphic to a *dense* subgroup of the group of real numbers, and be identified with such a subgroup. If $\alpha$ is a positive real number, and if $\mathfrak{a}$ is the ideal in $R_{v^\star}$ defined by $v^\star(x) \geq \alpha$, then $G_{\mathfrak{a}} = H_{\mathfrak{a}}$. In fact take any $x \neq 0$ in $R_{v^\star}$, any real number $\varepsilon > 0$, and write $x = x_1 \cdots x_n$ where $0 \leq v^\star(x_i) \leq \varepsilon$ (this is possible for $n$ large enough, since $\Gamma^\star$ is a dense subgroup of the real line). The formula

$$s(x) - x = \sum_{j=1}^{n} s(x_1) \cdots s(x_{j-1})(s(x_j) - x_j)x_{j+1} \cdots x_n$$

shows that, if $s$ is in $G_Z$, we have

$$v^\star(s(x) - x) \geq \min_j (v^\star(x) - v^\star(x_j) + v^\star[s(x_j) - x_j]).$$

Taking $s$ in $G_{\mathfrak{a}}$, this gives $v^\star(s(x) - x) \geq v^\star(x) + \alpha - \varepsilon$. As this is true for every $\varepsilon > 0$, we have $v^\star(s(x) - x) \geq v^\star(x) + \alpha$, i.e., $s(x) - x \in \mathfrak{a}x$, whence $s \in H_{\mathfrak{a}}$. Our conclusion follows then from (a).

REMARK. In the case of a *discrete* valuation $v^\star$ of rank 1, the decomposition of $x$ into a product of elements of order 1 shows, in a similar (and simpler) way that $G_{\mathfrak{M}^n_{v^\star}} \subset H_{\mathfrak{M}^{n-1}_{v^\star}}$.

(g) Let $\mathfrak{a}$ be a *principal ideal* $\mathfrak{a} = R_{v^\star}a$, contained in $(\mathfrak{M}_{v^\star})^2$. For $s$ in $G_{\mathfrak{a}}$ and $x$ in $R_{v^\star}$, we denote by $B(x, s)$ the *$v^\star$-residue of* $\dfrac{s(x) - x}{a}$. For fixed $s$, the mapping $x \to B(x, s)$ is a *derivation* of $R_{v^\star}$ (see Chapter II, § 17) with values in the additive group of $\Delta^\star$:

(26) $$B(x + y, s) = B(x, s) + B(y, s)$$

(27) $$B(xy, s) = \bar{x} \cdot B(y, s) + \bar{y}B(x, s)$$

($\bar{x}$, $\bar{y}$ denoting the $v^\star$-residues of $x$, $y$). The proofs are straightforward. On the other hand, for fixed $x$ in $R_{v^\star}$, the mapping $s \to B(x, s)$ is a homomorphism of $G_{\mathfrak{a}}$ into the additive group of $\Delta^\star$:

(28) $$B(x, ts) = B(x, s) + B(x, t)$$

PROOF. We set $s(x) = x + ay_s$ and $a = a'a''$ with $a'$, $a''$ in $\mathfrak{M}_{v^\star}$ (this is possible since $a \in (\mathfrak{M}_{v^\star})^2$). Then $ay_{ts} = s(t(x)) - x = s(x + ay_t) - x = ay_s + s(a)s(y_t) = ay_s + ay_t + s(a)[s(y_t) - y_t] + [s(a) - a]y_t$. Since $v^\star(s(a)) =$

$v^\star(a)$ ($s \in G_Z$), and since $s(y_t) - y_t \in R_{v^\star}a \subset \mathfrak{M}_{v^\star}$, the term $s(a)\cdot[s(y_t)) - y_t]$ is in $\mathfrak{M}_{v^\star}a$. Similarly, since $s(a) - a = s(a')s(a'') - a'a'' = s(a')[s(a'') - a''] + a''[s(a') - a']$, the term $(s(a) - a)y_t$ belongs to $\mathfrak{M}_{v^\star}a$. Hence $ay_{ts} \equiv ay_s + ay_t \pmod{\mathfrak{M}_{v^\star}a}$, and therefore $y_{ts} \equiv y_s + y_t \pmod{\mathfrak{M}_{v^\star}}$.

In other words, we have a *pairing* $B$ between $G_\mathfrak{a}$ and the additive group of $R_{v^\star}$, with values in the additive group of $\Delta^\star$. The *kernel* of the homomorphism $\varphi$ of $G_\mathfrak{a}$ into Hom $(R_{v^\star}, \Delta^\star)$ defined by $\varphi(s)(x) = B(x, s)$ is the set of all $s$ in $G_\mathfrak{a}$ such that $\dfrac{s(x) - x}{a} \in \mathfrak{M}_{v^\star}$ for every $x$ in $R_{v^\star}$; in other words, *this kernel* is $G_{\mathfrak{a}\mathfrak{M}_{v^\star}}$. The image $\varphi(G_\mathfrak{a})$ in Hom $(R_{v^\star}, \Delta^\star)$ is therefore a subgroup of Hom $(R_{v^\star}, \Delta^\star)$, which is isomorphic to $G_\mathfrak{a}/G_{\mathfrak{a}\mathfrak{M}_{v^\star}}$ and therefore finite. If the characteristic of $\Delta^\star$ is *zero*, no subgroup of Hom $(R_{v^\star}, \Delta^\star)$ is finite, except the subgroup (0), since such a subgroup contains, with any element $\Theta \neq 0$, all its multiples $\Theta + \Theta$, $\Theta + \Theta + \Theta, \cdots$; we therefore have $G_\mathfrak{a} = G_{\mathfrak{a}\mathfrak{M}_{v^\star}}$ in this case; more particularly, if $v^\star$ is a discrete valuation of rank 1, then we get $G_{\mathfrak{M}^2_{v^\star}} = G_{\mathfrak{M}^3_{v^\star}} = \cdots = G_{\mathfrak{M}^n_{v^\star}} = \cdots$, and this implies at once that $G_{\mathfrak{M}_{v^\star}n} = \{1\}$ for all $n > 1$ (since from $s(x) - x \in \mathfrak{M}^n_{v^\star}$, all $n$ and all $x$ follows that $s(x) - x = 0$ for all $x$, whence $s = 1$). If the characteristic $p$ of $\Delta^\star$ is $\neq 0$, then every element $\neq 0$ of Hom $(R_{v^\star}, \Delta^\star)$ is of order $p$; therefore $G_\mathfrak{a}/G_{\mathfrak{a}\mathfrak{M}_{v^\star}}$ is an *abelian group of type* $(p, \cdots, p)$ (i.e., a direct sum of cyclic groups with $p$ elements).

On the other hand, the homomorphism $\psi$ of $R_{v^\star}$ into Hom $(G_\mathfrak{a}, \Delta^\star)$ defined by $\psi(x)(s) = B(x, s)$, takes the value 0 on $(\mathfrak{M}_{v^\star})^2$ by formula (27), and also on $R_{v^\star} \cap K(G_\mathfrak{a})(K(G_\mathfrak{a})$ denoting the fixed field of $G_\mathfrak{a}$), whence *a fortiori* on $R_{v^\star} \cap K_T$. We suppose that there is *no inseparability* in the residue field extension, i.e., that $\Delta^\star$ is separable over $\Delta$; then $\Delta^\star = \Delta_T$ by Theorem 22 (b), and this means that every element of $R_{v^\star}$ is congruent mod $\mathfrak{M}_{v^\star}$ to some element of $R_{v^\star} \cap K_T$. [In the case in which $\Gamma^\star$ is *dense* (i.e., has no smallest strictly positive element), we have $\mathfrak{M}_{v^\star} = (\mathfrak{M}_{v^\star})^2$, whence $\psi$ takes everywhere the value 0. From what has been seen above, it follows that $G_\mathfrak{a} = G_{\mathfrak{a}\mathfrak{M}_{v^\star}}$ for every principal ideal $\mathfrak{a}$; we may notice that, if $\mathfrak{b}$ is a non-principal ideal in $R_{v^\star}$, then $\mathfrak{b} = \mathfrak{b}\mathfrak{M}_{v^\star}$ (still under the assumption that $\Gamma^\star$ is dense).]

In the case in which $\Gamma^\star$ admits a smallest positive element, say $v^\star(u)$ ($u \in \mathfrak{M}_{v^\star}$), then the assumption that $\Delta^\star = \Delta_T$ shows that every $x$ in $R_{v^\star}$ may be written in the form $x = z' + zu + x'$, with $z, z' \in R_{v^\star} \cap K_T$ and $x'$ in $(\mathfrak{M}_{v^\star})^2$. Denoting as usual by $\bar{z}$ the $v^\star$-residue of $z$, formula (27) shows that $\psi(x) = \psi(zu) = \bar{z}\cdot\psi(u)$. Therefore the image $\psi(R_{v^\star})$ in Hom $(G_\mathfrak{a}, \Delta^\star)$ is the $\Delta^\star$-vector subspace of Hom $(G_\mathfrak{a}, \Delta^\star)$ generated by

$\psi(u)$; in particular we have $\psi(u)=0$ if and only if $G_{\mathfrak{a}}=G_{\mathfrak{a}\mathfrak{M}_{v^\star}}$. Furthermore (still under the assumptions that $\Delta^\star$ is separable over $\Delta$ and that $\Gamma^\star$ admits a smallest element $>0$), the mapping $s \to \psi(u)(s)=B(u, s)$ defines an *isomorphism of* $G_{\mathfrak{a}}/G_{\mathfrak{a}\mathfrak{M}_{v^\star}}$ *onto an additive subgroup of* $\Delta^\star$.

(h) Let still $\mathfrak{a}$ be a *principal ideal* $R_{v^\star}a$ with $a$ in $\mathfrak{M}_{v^\star}$. For $t$ in $H_{\mathfrak{a}}$ and $x \neq 0$ in $K^\star$, we denote by $C(x, t)$ the *$v^\star$-residue of* $\dfrac{t(x)-x}{ax}$.

The mapping $C$ satisfies the following relations:

$$C(xy, t) = C(x, t)+C(y, t), \tag{29}$$

$$C(x, ts) = C(x, s)+C(x, t). \tag{30}$$

PROOF. If we set $s(x)=x(1+ax_s)$, then $C(x, s)$ is the $v^\star$-residue of $x_s$. From $s(xy)=xy(1+ax_s+ay_s+a^2x_sy_s)$ and from $a^2 \in \mathfrak{M}_{v^\star}a$, we deduce formula (29). From

$$\begin{aligned} s(t(x)) = s(x)[1+s(a)s(x_t)] &= x(1+ax_s)[(1+a(1+aa_s)(1+a(x_t)_s)x_t] \\ &\equiv x(1+ax_s+ax_t)(\text{mod. } \mathfrak{M}_{v^\star}ax), \end{aligned}$$

we deduce formula (30).

We have again a *pairing*, this time between $H_{\mathfrak{a}}$ and the multiplicative group $K^{\star\prime}$ of $K^\star$, with values in the additive group of $\Delta^\star$. Since $H_{\mathfrak{a}} \subset G_V$, we have $H_{\mathfrak{a}}=(1)$ in characteristic 0 (Theorem 24), and we may restrict ourselves to the case in which the characteristic $p$ of $\Delta^\star$ is $\neq 0$. It is easily seen that the kernel of the homomorphism $\varphi$: $H_{\mathfrak{a}} \to \text{Hom}(K'^\star, \Delta^\star)$ defined by $\varphi(s)(x)=C(x, s)$ is $H_{\mathfrak{a}\mathfrak{M}_{v^\star}}$. Thus we see as above that $H_{\mathfrak{a}}/H_{\mathfrak{a}\mathfrak{M}_{v^\star}}$ is an *abelian group of type* $(p, p, \cdots, p)$.

(i) Since $G$ is a *finite* group, the mappings $\mathfrak{a} \to G_{\mathfrak{a}}$, $\mathfrak{a} \to H_{\mathfrak{a}}$ take only a *finite* number of values. Let, for example, $G'$ be one of the values taken by $G_{\mathfrak{a}}$. If $\Phi$ denotes any set of ideals in $R_{v^\star}$ and we set

$$\mathfrak{b} = \bigcap_{\mathfrak{a} \in \Phi} \mathfrak{a}$$

we immediately verify that

$$G_{\mathfrak{b}} = \bigcap_{\mathfrak{a} \in \Phi} G_{\mathfrak{a}}.$$

Taking for $\Phi$ the set of all ideals $\mathfrak{a}$ for which $G_{\mathfrak{a}}=G'$, we deduce that this set has a *smallest element* $\mathfrak{a}(G')$. We obtain in this way a finite decreasing sequence

$$\mathfrak{a}_1 > \mathfrak{a}_2 > \cdots > \mathfrak{a}_q > (0)$$

such that the $G_{\mathfrak{a}_i}$ form a decreasing sequence of distinct subgroups of $G$. It follows from the construction that

$$\begin{aligned} G_{\mathfrak{a}} &= G_{\mathfrak{a}_1} \quad \text{for} \quad \mathfrak{a} \supset \mathfrak{a}_1 \\ G_{\mathfrak{a}} &= G_{\mathfrak{a}_2} \quad \text{for} \quad \mathfrak{a}_1 > \mathfrak{a} \supset \mathfrak{a}_2 \\ &\cdots\cdots \\ G_{\mathfrak{a}} &= G_{\mathfrak{a}_q} \quad \text{for} \quad \mathfrak{a}_{q-1} > \mathfrak{a} \supset \mathfrak{a}_q \\ G_{\mathfrak{a}} &= (1) \quad \text{for} \quad \mathfrak{a}_q > \mathfrak{a}. \end{aligned}$$

The ideals $\mathfrak{a}_1, \cdots, \mathfrak{a}_q$ are called the *ramification ideals* of $v^\star$ (and generalize the ramification numbers defined in Chapter V, § 10). An analogous sequence $\mathfrak{b}_1 > \mathfrak{b}_2 > \cdots > \mathfrak{b}_r > (0)$, with analogous properties, is defined by using the mapping $\mathfrak{a} \to H_{\mathfrak{a}}$ instead of $\mathfrak{a} \to G_{\mathfrak{a}}$.

**§ 13. Classical ideal theory and valuations.** Let $R$ be a UFD, and $K$ its quotient field. With every irreducible element $z$ in $R$, there is associated the $z$-*adic* valuation of $K$(§ 9, Example 1, p. 38). We have noticed already (§ 9, Example 2, p. 38) that the ring $R$ and the family $(F)$ of all $z$-adic valuations of $K$ enjoy the following properties:

($E_1$) *Every valuation $v$ in $(F)$ has rank* 1 *and is discrete.*
($E_2$) *The ring $R$ is the intersection of the valuation rings $R_v$ $(v \in (F))$.*
($E_3$) *For every $x \neq 0$ in $R$, we have $v(x) = 0$ for all $v$ in $(F)$ except a finite number of them* (we shall say "for almost all $v$ in $(F)$").
($E_4$) *For every $v$ in $(F)$, the valuation ring $R_v$ is equal to the quotient ring $R_{\mathfrak{p}(v)}$, where $\mathfrak{p}(v)$ is the center of $v$ on $R$.*

When we have a domain $R$ and a family $(F)$ of valuations of its quotient field $K$ which satisfy ($E_1$), ($E_2$), ($E_3$), ($E_4$), we say that $R$ is a *Krull domain* (or a *finite discrete principal order*), and that the family $(F)$ is a family of *essential valuations* of $R$. Property ($E_2$) shows that a Krull domain $R$ is *integrally closed.* The fact that every element of $K$ is a quotient of two elements of $R$ shows that condition ($E_3$) is equivalent with the seemingly stronger condition:

($E'_3$) *For every $x \neq 0$ in $K$, we have $v(x) = 0$ for almost all $v$ in $(F)$.*

Further examples of Krull domains may be given:

(a) *Dedekind domains.* A family of essential valuations in these domains is given by the set of all $\mathfrak{p}$-adic valuations (§ 9, Example 3, p. 38). A more general example is the following:

(b) *Integrally closed noetherian domains.* If $R$ is an integrally closed noetherian domain, then a family $(F)$ of essential valuations of $R$ is given by the $\mathfrak{p}$-adic valuations, where $\mathfrak{p}$ is any minimal prime ideal in $R$ (Theorem 16, Corollary 3, § 10).

REMARK. A Krull domain need not be noetherian; for example, polynomial rings in an infinite number of indeterminates, over a field, are non-noetherian UFD's.

The family $(F)$ of essential valuations of a Krull domain $R$ is uniquely determined by $R$. More precisely:

THEOREM 26. *Let $R$ be a Krull domain, and $(F)$ a family of essential valuations of $R$. Then the valuation rings $R_v$ $(v \in (F))$ are identical with the quotients rings $R_{\mathfrak{p}}$, where $\mathfrak{p}$ runs over the family of all minimal prime ideals in $R$.*

PROOF. Let $v \in (F)$, and let $\mathfrak{p}(v)$ denote its center on $R$. Since the quotient ring $R_{\mathfrak{p}(v)}$ is the valuation ring $(E_4)$ of a discrete, rank 1 valuation $(E_1)$, $\mathfrak{p}(v)R_{\mathfrak{p}(v)}$ is its unique proper prime ideal. Thus, taking into account the relations between prime ideals in $R$ and in $R_{\mathfrak{p}(v)}$ (Vol. I, Ch. IV, § 11, Theorem 19), $\mathfrak{p}(v)$ is a minimal prime ideal in $R$.

Conversely we have to show that every minimal prime ideal $\mathfrak{p}$ in $R$ is the center of some valuation $v$ in $(F)$. More generally we shall prove that every proper prime ideal $\mathfrak{p}$ in $R$ contains the center $\mathfrak{p}(v)$ of some valuation $v$ in $(F)$. Suppose this is not so. Take an element $x \neq 0$ in $\mathfrak{p}$. Since $\mathfrak{p} \neq R$, $x$ is not a unit in $R$. Hence $v\left(\frac{1}{x}\right) < 0$ for at least one valuation $v$ in $(F)(E_2)$. Denote by $v_1, \cdots, v_n$ the valuations $v$ in $(F)$ such that $v(x) > 0$ $(E_3)$. As was just pointed out, we must have $n \geqq 1$. Since no center $\mathfrak{p}(v_i)$ is contained in $\mathfrak{p}$, there exists an element $y_i \in \mathfrak{p}(v_i)$ such that $y_i \notin \mathfrak{p}$. Since the valuations $v_i$ have rank 1 and since $v_i(y_i) > 0$, there exists an integer $s(i)$ such that $v_i(y_i^{s(i)}) \geq v_i(x)$. Denoting by $y$ the product $\prod_i y_i^{s(i)}$, we have $v_i(y) \geq v_i(x)$ for all $i$, whence $v(y) \geq v(x)$ for all $v$ in $(F)$ since $v(x) = 0$ for every $v$ in $(F)$ distinct from $v_1, \cdots, v_n$. In other words, we have $v(y/x) \geq 0$ for all $v$ in $(F)$, whence $y/x \in R$ by $(E_2)$. But, since $\mathfrak{p}$ is a prime ideal, and since $y_i \notin \mathfrak{p}$, we have $y \notin \mathfrak{p}$, in contradiction with the fact that $y \in Rx \subset \mathfrak{p}$. Our theorem is thereby proved.

We now characterize UFD's and Dedekind domains among Krull domains. (From now on, all valuations have the additive group of integers as value group.)

THEOREM 27. *Let $R$ be a Krull domain, $(F)$ its family of essential valuations. In order for $R$ to be a* UFD, *it is necessary and sufficient that, for every $v$ in $(F)$, there exists an element $a_v$ in $R$ such that $v(a_v) = 1$ and $w(a_v) = 0$ for every $w \neq v$ in $(F)$.*

PROOF. For the necessity we observe that if $v$ is the $a$-adic valuation of a UFD $R$ ($a$ being an irreducible element in $R$), we have $v(a)=1$, and $w(a)=0$ for every other $b$-adic valuation $w$ of $R$ such that $w\neq v$. Conversely, suppose the existence of the elements $a_v$ in $R$. These elements are irreducible, since, from $a_v=xy$ with $x$ and $y$ in $R$, we deduce $v(x)+v(y)=1$ and $w(x)+w(y)=0$ for every $w\neq v$ in $(F)$, whence $w(x)=w(y)=0$ and either $v(x)=0$ and $v(y)=1$ or $v(x)=1$ and $v(y)=0$; therefore either $x$ or $y$ is a unit in $R$ since it has values 0 for all valuations in $(F)$ (use $(E_2)$). Secondly, for every element $x$ in $R$ we can write $x=u\cdot\prod_v a_v^{v(x)}$; from this we deduce that $v(u)=0$ for all $v$ in $(F)$, i.e., that $u$ is a unit in $R$ (since $u$ and $1/u$ belong to $R$ by $(E_2)$). Lastly such a representation $x=u\cdot\prod_v a_v^{n(v)}$ ($u$: unit in $R$; the $n(v)$ almost all zero) is necessarily unique, since $v(x)=v(u)+n(v)v(a_v)+\sum_{w\neq v} n(w)v(a_w)$ and since therefore $v(x)$ is equal to $n(v)$ by the hypothesis made on the elements $a_v$. These facts show that $R$ is a UFD.

THEOREM 28. *Let $R$ be a Krull domain, $(F)$ its family of essential valuations. In order for $R$ to be a Dedekind domain it is necessary and sufficient that the following equivalent conditions hold:*

(a) *Every proper prime ideal in $R$ is maximal.*
(b) *Every proper prime ideal in $R$ is minimal.*
(c) *Every non-trivial valuation of the quotient field of $R$ which is finite on $R$ is essential.*

PROOF. The equivalence of (a) and (b) is trivial. If (b) holds, then any non-trivial valuation $v$ of the quotient field $K$ of $R$ which is finite on $R$ has a minimal prime ideal $\mathfrak{p}$ as center, and its valuation ring contains the quotient ring $R_\mathfrak{p}$. As $R_\mathfrak{p}$ is the valuation ring of a rank 1 valuation (Theorem 26), it is a maximal proper subring of $K$ (§ 3, p. 10), thus proving that $R_\mathfrak{p}$ is the valuation ring of $v$, and that (c) holds. Conversely, if (c) holds, every proper prime ideal in $R$ is minimal by Theorem 26, since it is the center on $R$ of some non-trivial valuation (§ 4, Theorem 5).

We have already seen that condition (a) is necessary (Vol. I, Ch. V, § 6, Theorem 10). For proving the sufficiency of the equivalent conditions (a), (b), (c) we are going to prove first that *every proper prime* (therefore maximal) *ideal $\mathfrak{p}$ in $R$ is invertible.* We take an element $x\neq 0$ in $\mathfrak{p}$. For any prime ideal $\mathfrak{a}$ in $R$, we denote by $v_\mathfrak{a}$ the (essential) valuation having $\mathfrak{a}$ as center. Then $x^{-1}\prod_\mathfrak{a} \mathfrak{a}^{v_\mathfrak{a}(x)}$ (this product makes sense, by condition $(E_3)$) is a fractionary ideal $\mathfrak{b}$ such that $\min_{y\in\mathfrak{b}} v_\mathfrak{a}(y)=0$ for all $\mathfrak{a}$. Therefore

$\mathfrak{b}$ is an integral ideal, necessarily equal to $R$, for $\mathfrak{b}$ is not contained in any maximal ideal $\mathfrak{a}$. Consequently we have $Rx = \prod \mathfrak{a}^{v_{\mathfrak{a}}(x)}$, so each $\mathfrak{a}$ is invertible provided $v_{\mathfrak{a}}(x) > 0$ (Vol. I, Ch. V, § 6, lemma 4). In particular $\mathfrak{p}$ is invertible.

We now prove that every integral ideal $\mathfrak{a}$ in $R$ is invertible, and this will show that $R$ is a Dedekind domain by Theorem 12 of Vol. I, Ch. V, § 6. In fact, let us denote by $v_{\mathfrak{p}}(\mathfrak{a})$ the smallest value taken by $v_{\mathfrak{p}}$ on $\mathfrak{a}$, and consider the ideal $\mathfrak{a}' = \prod_{\mathfrak{p}} \mathfrak{p}^{v_{\mathfrak{p}}(\mathfrak{a})}$. It is clear that we have $\mathfrak{a} \subset \mathfrak{a}'$. Since $\mathfrak{a}'$ is invertible (as a product of invertible ideals), we can consider the ideal $\mathfrak{b} = \mathfrak{a}\mathfrak{a}'^{-1}$; this is an integral ideal since $\mathfrak{a} \subset \mathfrak{a}'$, and we have $\mathfrak{a} = \mathfrak{a}'\mathfrak{b}$. Since we have $v_{\mathfrak{p}}(\mathfrak{b}) = 0$ for every $\mathfrak{p}$, $\mathfrak{b}$ is necessarily equal to $R$, as it is not contained in any maximal ideal $\mathfrak{a}$. Therefore $\mathfrak{a} = \mathfrak{a}'$, and $\mathfrak{a}$ is invertible. Q.E.D.

We now study the behavior of normal domains under two simple types of extensions.

Given a field $K$ and a valuation $v$ of $K$, we consider the polynomial ring $K[X]$ in one indeterminate over K. If $P(X) = a_0 + a_1X + \cdots + a_nX^n$, $a_i \in K$, we set $v'(P(X)) = \min_{0 \leq i \leq n}(v(a_i))$. It is clear that we have $v'(P(X)+Q(X)) \geq \min\{v'(P(X)), v'(Q(X))\}$, and $v'(P(X)\cdot Q(X)) \geq v'(P(X)) + v'(Q(X))$. To prove the equality $v'(P(X)\cdot Q(X)) = v'(P(X)) + v'(Q(X))$, we consider, in $P(X) = a_0 + a_1X + \cdots + a_nX^n$ and in $Q(X) = b_0 + b_1X + \cdots + b_qX^q$, the smallest indices $i$, $j$ for which $v(a_i)$ and $v(b_j)$ reach their minima. Then the coefficient of $X^{i+j}$ in $P(X)Q(X)$ is the sum of $a_ib_j$ and of terms whose order for $v$ is strictly greater than $v(a_i) + v(b_j)$; the order of that coefficient is thus $v(a_i) + v(b_j) = v'(R) + v'(Q)$, showing that $v'(PQ) \leq v'(P) + v'(Q)$. It follows from Theorem 14 (§ 9) that $v'$ has a unique extension to a valuation of the rational function field $K(X)$. We shall also denote by $v'$ this valuation of $K(X)$, and we shall call it the *canonical extension* of $v$ to $K(X)$. We notice that $v$ and $v'$ have the same value group, hence also the same rank.

THEOREM 29. *Let $R$ be an integrally closed domain and $K$ its quotient field. Let $(F)$ be a family of valuations of $K$, the valuation rings of which have $R$ as intersection. Denote by $(F')$ the family of the canonical extensions $v'$ of elements $v \in (F)$ to the rational function field $K(X)$. Denote by $(G)$ the family of all $a(X)$-adic valuations of $K(X)$ ($a(X)$: irreducible polynomial in $K[X]$). Then*

(a) *The polynomial ring $R[X]$ is the intersection of all valuation rings $R_v$ where $v \in (F') \cup (G)$, and is therefore integrally closed.*

(b) *If $R$ is a Krull domain, and if $(F)$ is its family of essential valuations,* *then* $R[X]$ *is a Krull domain, and* $(F') \cup (G)$ *is its family of essential valuations.*

(c) *If* $R$ *is a* UFD, *then* $R[X]$ *is also a* UFD.

PROOF. (a) The intersection $\bigcap_{w \in (G)} R_w$ is the polynomial ring $K[X]$, by definition of the $a(X)$-adic valuations. Now, if a polynomial $P(X) = a_0 + a_1X + \cdots + a_nX^n$ ($a_i \in K$) satisfies the inequality $v'(P) \geq 0$ for every $v'$ in $(F')$, then we have $\min(v(a_i)) \geq 0$ for all $v$ in $(F)$, i.e., $v(a_i) \geq 0$ for every $v$ and every $i$, and this is equivalent to saying that $a_i \in R$ for every $i$. This proves (a).

(b) Suppose that $(F)$ is the family of essential valuations of the Krull domain $R$. We have to show that the set $(F') \cup (G)$ satisfies conditions $(E_1)$, $(E_2)$, $(E_3)$, $(E_4)$ with respect to the ring $R[X]$. Condition $(E_1)$ is trivial. Condition $(E_2)$ has been proved in (a). As for $(E_3)$, given a polynomial $P(X) = a_0 + a_1X + \cdots + a_nX^n$ there is only a finite number of $a(X)$-adic valuations $w$ in $(G)$ for which $w(P) > 0$, since $P$ has only a finite number of irreducible factors (in $K[X]$); on the other hand, if $a_i$ is a non-zero coefficient of $P(X)$, the valuations $v'$ in $(F')$ for which $v'(P) > 0$ are among those for which $v(a_i) > 0$, by definition of $v'$, and these latter valuations are finite in number according to $(E_3)$ as applied to $R$. It remains to show that $(E_4)$ holds.

Consider, first, an $a(X)$-adic valuation $w \in (G)$. Its center $\mathfrak{p}(w)$ in $R[X]$ is the set of all polynomials in $R[X]$ which are multiples of $a(X)$ (in $K[X]$). Since this prime ideal $\mathfrak{p}(w)$ does not contain any constant polynomial $\neq 0$, the quotient ring $(R[X])_{\mathfrak{p}(w)}$ contains $K[X]$. By the transitivity of quotient ring formations (Vol. I, Ch. IV, § 11, p. 231), this quotient ring is equal to $(K[X])_{\mathfrak{p}}$, where $\mathfrak{p}$ is the (prime) ideal generated by $\mathfrak{p}(w)$ in $K[X]$. But, since this ideal is the ideal generated by $a(X)$, the quotient ring we are dealing with is equal to $(K[X])_{(a(X))}$, and this latter ring is the valuation ring of $w$, by the structure of the $a(X)$-adic valuation.

Consider now a valuation $v'$ in $(F')$, extending canonically the valuation $v$ ($\in (F)$) of $K$. Its center $\mathfrak{p}(v')$ on $R[X]$ is the set of all polynomials $a_0 + a_1X + \cdots + a_nX^n$ for which $v(a_i) > 0$ for every $i$. Since the valuation ring $R_v$ of $v$ is a quotient ring of $R$, the quotient ring $(R[X])_{\mathfrak{p}(v')}$ contains $R_v$ and therefore contains also $R_v[X]$. If we denote by $a$ an element of $R_v$ such that $v(a) = 1$, and if we write every element of $K(X)$ under the form $a^qP(X)/Q(X)$ where $P$ and $Q$ are polynomials over $R_v$ such that $v'(P) = v'(Q) = 0$, the elements of the valuation ring of $v'$ are those for which $q \geq 0$. In other words, this valuation ring is $(R_v[X])_{\mathfrak{p}}$, where $\mathfrak{p}$ is the prime ideal in $R_v[X]$ generated by $a$. Now, this prime ideal $\mathfrak{p}$ is obviously the extension to $R_v[X]$ of the center $\mathfrak{p}(v')$ of $v'$ in $R[X]$. Thus, the valuation ring we are investigating, is, by the transitivity of quotient ring formations (Vol. I, Ch. IV, § 11, p. 231) equal to the quotient ring $(R[X])_{\mathfrak{p}(v')}$. The proof of (b) is now complete.

(c) We use the characterization of UFD's by Theorem 27. For $v'$ in $(F')$, we take an element $a_v$ in $R$ such that $v(a_v)=1$ and $u(a_v)=0$ for every $u \neq v$ in $(F)$. If we consider $a_v$ as a constant polynomial in $R[X]$, we have $v'(a_v)=1$, $u'(a_v)=0$ for every $u' \neq v'$ in $(F')$, and $w(a_v)=0$ for every $w$ in $(G)$, since $a_v$ is a constant polynomial. For the $a(X)$-adic valuation $w$ in $(G)$, we take for $a_w$ a constant multiple of $a(X)$, all the coefficients of which are in $R$ and are relatively prime; we then have $w(a_w)=1$, $u(a_w)=0$ for every $u \neq w$ in $(G)$, and $v'(a_w)=0$ for every $v'$ in $(F')$ since the coefficients of $a_w$ are relatively prime and cannot have strictly positive orders for $v$. Thus also (c) is proved.

REMARK. Observe that (c) has already been proved (Vol. I, Ch. I, § 17, Theorem 10) by elementary methods.

THEOREM 30. *Let $R$ be an integrally closed domain, $K$ its quotient field and $(F)$ a family of valuations of $K$, the valuation rings of which have $R$ as intersection. Let $K'$ be a finite algebraic extension of $K$, $R'$ the integral closure of $R$ in $K'$, and $(F')$ the family of all extensions to $K'$ of all valuations belonging to $(F)$. Then:*

(a) *$R'$ is the intersection of the valuation rings of the valuations belonging to $(F')$.*
(b) *If $R$ is a Krull domain, and if $(F)$ is its family of essential valuations, then $R'$ is a Krull domain and $(F')$ is its family of essential valuations.*
(c) *If $R$ is a Dedekind domain, so is $R'$.*

PROOF. (a) It is clear that $R'$ is contained in the intersection I of the valuation rings of the valuations belonging to $(F')$. Conversely consider an element $x$ of $K'$ such that $v'(x) \geq 0$ for all $v'$ in $(F')$. Let $K''$ denote the smallest normal extension of $K$ containing $K'$, and let $(F'')$ be the family of all extensions to $K''$ of valuations belonging to $(F)$. We obviously have $v''(x) \geq 0$ for all $v''$ in $(F'')$. Since $(F'')$ contains, together with $v''$, all the conjugates of $v''$ over $K$, we have $v''(x_i) \geq 0$ for every $v''$ in $(F'')$ and for every conjugate $x_i$ of $x$ over $K$. Now the coefficients $a_j$ of the minimal polynomial of $x$ over $K$ are sums of products of conjugates of $x$. Thus the valuation axioms show that we have $v''(a_j) \geq 0$ for all $v''$ in $(F'')$, i.e., $v(a_j) \geq 0$ for all $v$ in $(F)$. This means that the coefficients $a_j$ belong to $R$. Therefore the minimal polynomial of $x$ over $K$ yields an equation of integral dependence of $x$ over $R$, and assertion (a) is proved.

(b) If $v \in (F)$ is a discrete, rank 1 valuation, any extension $v'$ of $v$ to $K'$ is also discrete, of rank 1 (§ 11, Lemma 2, and Corollary); thus $(F')$ satisfies condition $(E_1)$. That $(F')$ verifies $(E_2)$ follows from assertion (a). Concerning $(E_3)$, consider an element $x \neq 0$ in $R'$ and an equation of integral dependence $x^n + a_{n-1}x^{n-1} + \cdots + a_0 = 0$ of $x$ over $R$. We may suppose $a_0 \neq 0$; otherwise we would divide by $x$. If we have $v'(x) > 0$ for $v'$ in $(F')$, we must have $v'(a_0) > 0$. But the valuations $v'$ in $(F')$ for which $v'(a_0) > 0$ are the extensions of the valuations $v$ in $(F)$ for which $v(a_0) > 0$ $(a_0 \in R)$. Since the latter are finite in number, by $(E_3)$ as applied to $(F)$, and since a valuation $v$ of $K$ has only a finite number of extensions to $K'$ (§ 7, Corollary 4 to Theorem 12), the number of valuations $v'$ in $(F')$ for which $v'(a_0) > 0$, is finite, whence also the number of valuations $v'$ in $(F')$ for which $v'(x) > 0$ is finite. Thus $(F')$ satisfies $(E_3)$.

We now check $(E_4)$. Let $v' \in (F')$ be an extension of $v \in (F)$, and denote by $\mathfrak{p}(v')$ and $\mathfrak{p}(v)$ the corresponding centers in $R'$ and $R$ respectively. The valuation ring $R_v$ of $v$ is the quotient ring $R_{\mathfrak{p}(v)} = R_M$, where $M$ denotes the complement of $\mathfrak{p}(v)$ in $R$. The integral closure $(R_v)'$ of $R_v = R_M$ in $K'$ is the quotient ring $R'_M$ (Vol. I, Ch. V, § 3, Example 2, p. 261). Since $\mathfrak{p}(v') \cap R = \mathfrak{p}(v)$, this integral closure is a subring of $R'_{\mathfrak{p}(v')}$. Now, the valuation ring of $v'$ is the quotient ring of $(R_v)' = R'_M$ with respect to the maximal ideal $\mathfrak{m}'$ which is the center of $v'$ in $(R_v)$ (§ 7, Theorem 12). By the transitivity of quotient ring formations (Vol. I, Ch. IV, § 10, p. 226), this valuation ring is therefore equal to $R'_{\mathfrak{p}(v')}$, and this completes the proof of (b).

(c) We use the characterization of Dedekind domains given in Theorem 28. If $R'$ contains two proper prime ideals $\mathfrak{p}'$, $\mathfrak{q}'$ such that $\mathfrak{p}' < \mathfrak{q}'$, then $\mathfrak{p}' \cap R$ and $\mathfrak{q}' \cap R$ are proper prime ideals in $R$ such that $\mathfrak{p}' \cap R < \mathfrak{q}' \cap R$ (Vol. I, Ch. V, § 2, Complement 1 to Theorem 2, p. 259). This contradicts the fact that $R$ is a Dedekind domain.

REMARK. Another proof of (c) has been given in a previous chapter (Vol. I, Ch. V, § 8, Theorem 19).

§ **14. Prime divisors in fields of algebraic functions.** We recall (Vol. I, Ch. II, § 13) that a field $K$, containing a ground field $k$, is said to be a field of algebraic functions over $k$, or, briefly, a function field over $k$, if it is finitely generated over $k$. In this section we shall study prime divisors of a function field $K/k$, i.e., the places or the valuations of $K/k$, which have dimension $r-1$ over $k$, where $r$ is the transcendence degree of $K/k$. For our immediate purpose it will be more convenient to treat prime divisors as valuations rather than as places.

We have already proven the existence of prime divisors; their existence is a special case of a more general theorem proven in §6 (Theorem 11 and its Corollary). Of considerable importance is the following theorem:

THEOREM 31. *Any prime divisor $v$ of a function field $K/k$ is a discrete valuation of rank* 1, *and the residue field $D_v$ of $v$ is itself a function field (of transcendence degree $r-1$ over $k$). Furthermore, the valuation ring $K_v$ of $v$ is the quotient ring of a finite integral domain $R$ (having $K$ as quotient field) with respect to a minimal prime ideal of $R$.*

PROOF. It is obvious that $v$ must have rank 1 since $v$ has maximum dimension $r-1$ and cannot therefore be composite with any other valuation of higher dimension (see § 3, Definition 1, Corollary 1, p. 10).

We fix $r-1$ elements $x_1, x_2, \cdots, x_{r-1}$ in $K$ whose $v$-residues in $D_v$ are algebraically independent over $k$. Then it is clear that these elements $x_i$ are also algebraically independent over $k$ (§ 6, Lemma 2; see also proof of Corollary 1 of that lemma). We extend $\{x_1, x_2, \cdots, x_{r-1}\}$ to a transcendence basis $\{x_1, \cdots, x_r\}$ of $K/k$ and we denote by $v'$ the restriction of $v$ to the field $k(x)$ $(=k(x_1, x_2, \cdots, x_r))$. Since $K$ is an algebraic extension of $k(x)$, it follows that $v$ and $v'$ have the same dimension (§ 6, Lemma 2, Corollary 1). Hence $v'$ is a prime divisor of $k(x)/k$. We first show that our theorem is true for $v'$ and for the purely transcendental extension field $k(x)$ $(=k(x_1, x_2, \cdots, x_r))$ of $k$. For this purpose we first observe that it is permissible to assume that $v'(x_r) \geqq 0$, since we can replace $x_r$ by $1/x_r$. Under this assumption, $v'$ is non-negative on the polynomial ring $R' = k[x_1, x_2, \cdots, x_r]$. If $\mathfrak{p}'$ is the center of $v'$ in $R'$, then the integral domain $k[x]/\mathfrak{p}'$ has transcendence degree $r-1$ over $k$ (since the $v$-residues of $x_1, x_2, \cdots, x_{r-1}$ are algebraically independent over $k$). If $\mathfrak{p}$ is a prime ideal in $R'$ such that $\mathfrak{p}' > \mathfrak{p}$ then, by Theorem 29 of Vol. I, Ch. II, § 12, we have tr.d. $R'/\mathfrak{p}' <$ tr.d. $R'/\mathfrak{p}$, i.e., $r-1 <$ tr.d. $R'/\mathfrak{p} \leqq r$, where all the transcendence degrees are relative to $k$. Hence tr.d. $R'/\mathfrak{p} = r =$ tr.d. $R'$, whence—again by the just cited theorem, $\mathfrak{p} = (0)$. Hence $\mathfrak{p}'$ is a minimal prime ideal in $R'$. Since $R'$ is noetherian and integrally closed, it follows that $R'_{\mathfrak{p}'}$ is a discrete valuation ring of rank 1 (§ 10, Theorem 16, Corollary 2). Since $R'_{\mathfrak{p}'}$ is contained in the valuation ring of $v'$ and since $R'_{\mathfrak{p}'}$ is a maximal subring of $k(x)$, it follows that $R'_{\mathfrak{p}'}$ is the valuation ring of $v'$. Thus $v'$ is discrete of rank 1, its residue field is the quotient field of the finite integral domain $k[x_1, x_2, \cdots, x_r]/\mathfrak{p}'$, and its valuation ring is the quotient ring of the polynomial ring $k[x_1, x_2, \cdots, x_r]$ with respect to the minimal prime ideal $\mathfrak{p}'$; so the theorem holds for $v'$. (Observe that $\mathfrak{p}'$ is a principal ideal $(f)$ in the UFD $k[x_1, x_2, \cdots, x_r]$ and that therefore $v'$ is merely the $f$-adic valuation of $k[x]$.)

The theorem can now easily be proved for $v$ and $K$ as follows: (1) since $K$ is a finite algebraic extension of $k(x)$ and $v$ is an extension of $v'$, also $v$ must be discrete (§ 11, Lemma 2, Corollary) and of rank 1 (§ 11, Lemma 2). (2) The residue field of $v$ is a finite algebraic extension of the residue field of $v'$ (§ 6, Lemma 2, Corollary 2) and is therefore also a finitely generated extension of $k$. (3) If $R$ denotes the integral closure of $k[x_1, x_2, \cdots, x_r]$ in $K$, then clearly $v$ is non-negative on $R$, the center $\mathfrak{p}$ of $v$ in $R$ is a prime ideal of dimension $r-1$ and is therefore a mimimal prime ideal in $R$; thus, since $R$ is a finite integral domain, hence noetherian, it follows, again by Theorem 16, Corollary 2 (§ 10) that $K_v = R_{\mathfrak{p}}$. This completes the proof.

We note the following consequence of our theorem:

COROLLARY. *If a valuation $v$ of a field $K/k$ of algebraic functions of $r$ independent variables has dimension $s$ and rank $r-s$, then $v$ is discrete, and its residue field $D_v$ is a field of algebraic functions of $s$ independent variables. In particular, every valuation of $K/k$ of maximum rank $r$ is discrete.*

For, let $v = v' \circ \bar{v}$, where $v'$ has rank $r-s-1$ and $\bar{v}$ is a rank 1 valuation of the residue field $D_{v'}$ of $v'$. The dimension of $v'$ is $\leqq r - \text{rank } v'$, i.e., $\dim v' \leqq s+1$, and since $\bar{v}$ is non-trivial it follows that $\dim v' = s+1$, while $\dim \bar{v} = s$. Using induction from $s+1$ to $s$, we may assume that $v'$ is discrete and that $D_{v'}$ is a field of algebraic functions of $s+1$ independent variables. Then $\bar{v}$ is a prime divisor of $D_{v'}/k$, hence also $\bar{v}$ and $v$ are discrete. If $v$ has rank $r$, then its dimension cannot exceed zero, and so $v$ must be discrete.

The converse of the last part of the theorem is also true, but before stating and proving it we must first prove a lemma which will be used several times in this section and which will form the cornerstone of the dimension theory developed in the next chapter (VII, § 7).

Let $R = k[x_1, x_2, \cdots, x_n]$ be a finite integral domain, of transcendence degree $r$, and let $\mathfrak{p}$ be a prime ideal in $R$, different from $R$. Then the canonical homomorphism $R \to R/\mathfrak{p}$ is an isomorphism on $k$, and we may therefore regard $k$ as a subfield of $R/\mathfrak{p}$. We define the *dimension* of the prime ideal $\mathfrak{p}$, in symbols: $\dim \mathfrak{p}$, as being the transcendence degree of $R/\mathfrak{p}$ over $k$.

By definition, we have always $\dim \mathfrak{p} \geqq 0$ if $\mathfrak{p} \neq R$. It is sometimes convenient to attach the dimension $-1$ to the unit ideal $R$. It is clear that *a prime ideal of dimension* 0 *is maximal.* The converse will be proved in the next chapter (VII, § 3, Lemma, p. 165).

If $\mathfrak{p}$ and $\mathfrak{p}'$ are two prime ideals in $R$, both different from $R$, and if $\mathfrak{p} < \mathfrak{p}'$, then the canonical homomorphism of $R/\mathfrak{p}$ onto $R/\mathfrak{p}'$ is proper and therefore the transcendence degree of $R/\mathfrak{p}$ is *greater* than the transcendence degree of $R/\mathfrak{p}'$ (Vol. I, Ch. II, § 12, Theorems 28 and 29). We have therefore proved that

$$(1) \qquad \text{“}\mathfrak{p} < \mathfrak{p}'\text{”} \Rightarrow \text{“}\dim \mathfrak{p} > \dim \mathfrak{p}'.\text{”}$$

In particular, since the prime ideal (0) has dimension $r$, it follows that *every proper prime ideal has dimension less than r and that every prime ideal of dimension $r-1$ is minimal.* The lemma which we wish to prove and which is fundamental in the dimension theory of finite integral domains is the converse of the second part of the last assertion:

LEMMA. *If $\mathfrak{p}$ is a minimal prime ideal in a finite integral domain $R=k[x_1, x_2, \cdots, x_n]$, of transcendence degree r, then $\mathfrak{p}$ has dimension $r-1$.*

PROOF. Assume first that $x_1, x_2, \cdots, x_n$ are algebraically independent over $k$, whence $r=n$ and $R$ is a polynomial ring in $n$ variables. Since $R$ is a unique factorization domain, $\mathfrak{p}$ is a principal ideal, say $\mathfrak{p}=Rf$, where $f$ is an irreducible element of $R$ (Vol. I, Ch. IV, § 14, statement following immediately the definition of minimal prime ideals, p. 238). The polynomial $f=f(x_1, x_2, \cdots, x_n)$ must have positive degree since $\mathfrak{p}\neq(1)$. Hence at least one of the elements $x_i$ actually occurs in the formal polynomial expression of $f$. Let, say, $x_n$ occur in $f$. Then $\mathfrak{p}$ contains no polynomial which is independent of $x_n$, since $\mathfrak{p}=Rf$. It follows that the $\mathfrak{p}$-residues of $x_1, x_2, \cdots, x_{n-1}$ are algebraically independent over $k$. This shows that $\dim \mathfrak{p}\geqq n-1$, whence $\dim \mathfrak{p}=n-1$ since $\mathfrak{p}\neq(0)$.

If $r<n$, we consider first the case in which the ground field $k$ is infinite. We use then the normalization theorem (Vol. I, Ch. V, § 4, Theorem 8) and we thus choose $r$ elements $z_1, z_2, \cdots, z_r$ in $R$ such that $R$ is integrally dependent on $R'=k[z_1, z_2, \cdots, z_r]$. We set $\mathfrak{p}'=\mathfrak{p}\cap R'$. Then $R'$ is a polynomial ring in $r$ variables. Since $R'$ is integrally closed and $\mathfrak{p}$ is minimal in $R$, $\mathfrak{p}'$ is necessarily minimal in $R'$ (Vol. I, Ch. V, § 3, Theorem 6) and hence, by the above proof, we have $\dim \mathfrak{p}'=r-1$. Consequently, by Vol. I, Ch. V, § 2, Lemma 1, $\dim \mathfrak{p}=r-1$.

If $k$ is a finite field we consider an algebraic closure $K$ of the field $k(x_1, x_2, \cdots, x_n)$ and we set $\bar{R}=\bar{k}[x_1, x_2, \cdots, x_n]$ where $\bar{k}$ is the algebraic closure of $k$ in $K$. Since $\bar{R}$ is integrally dependent over $R=k[x_1, x_2, \cdots, x_n]$, there exists at least one prime ideal in $\bar{R}$ which lies over $\mathfrak{p}$ (Vol. I, Ch. V, § 2, Theorem 3). Let $\bar{\mathfrak{p}}$ be such a prime ideal. Then also $\bar{\mathfrak{p}}$ is minimal in $\bar{R}$ (Vol. I, Ch. V, § 2, Complement 1 to Theorem 3, p. 259). Now, it is clear that the transcendence degree of $\bar{R}$ over $\bar{k}$ is the same as the transcendence degree of $R$ over $k$ (using a transcendence basis $\{z_1, z_2, \cdots, z_r\}$ of $R/k$, $z_i\in R$, and the transitivity of algebraic dependence, we see at once that $z_1, z_2, \cdots, z_r$ are algebraically independent over $\bar{k}$ and form a transcendence basis of $\bar{R}/\bar{k}$). Since $\bar{k}$ is an infinite field, we have, by the preceding case, that dim $\bar{\mathfrak{p}} = r-1$. Consequently, again by Lemma 1 of Vol. I, Ch. V, § 2, dim $\mathfrak{p} = r-1$, and the proof of the lemma is complete.

COROLLARY. *If $R$ is a finite integral domain (over a ground field $k$) and if a prime ideal $\mathfrak{p}$ in $R$ is such that the quotient ring $R_\mathfrak{p}$ is a valuation ring, then the associated valuation $v$ of the quotient field $K$ of $R$ is a prime divisor of $K/k$ and $\mathfrak{p}$ is a minimal prime ideal in $R$.*

For, since $R_\mathfrak{p}$ is noetherian, the valuation $v$ is discrete, of rank 1 (§ 10, Theorem 16) and $R_\mathfrak{p}$ is a maximal subring of $K$; therefore $\mathfrak{p}R_\mathfrak{p}$ is (not only a maximal but also) a minimal prime ideal of $R_\mathfrak{p}$, showing that $\mathfrak{p}$ is a minimal prime ideal in $R$. By the preceding lemma, we have therefore dim $\mathfrak{p} = r-1$, if $r$ is the transcendence degree of $R/k$, and hence $v$ is a prime divisor of $K/k$.

Let $V$ be an affine variety in an affine $n$-space, such that $V$ is defined and is irreducible over $k$ and $K$ is $k$-isomorphic with the function field $k(V)$ of $V/k$. We shall identify $K$ with $k(V)$. If $\mathscr{P}$ is a prime divisor† of $K/k$ which is finite on the coördinate ring $k[V]$ of $V$, then $\mathscr{P}$ has a center on $V$, and this center is a subvariety $W$ of $V$, defined and irreducible over $k$. The dimension of $W$ is at most equal to $r-1$.

THEOREM 32. *If $W$ is an $(r-1)$-dimensional irreducible subvariety of $V/k$, then the set of prime divisors of $K/k$ $(=k(V)/k)$ which have center $W$ on $V$ is finite and non-empty. If $\mathscr{P}$ is any prime divisor of $K/k$ having center $W$, then the residue field of $\mathscr{P}$ is a finite algebraic extension of the function field $k(W)$ of $W/k$.*

PROOF. There exist prime divisors of center $W$, since there exist non-trivial valuations of $K/k$ having center $W$ and since any such valuation must have dimension $r-1$ and must therefore be a prime divisor. We shall now show that there is only a finite number of prime divisors with center $W$.

Let $K = k(x_1, x_2, \cdots, x_n)$, where the $x_i$ are the non-homogeneous coördinates of the general point of $V/k$. Let $\mathfrak{q}$ be the prime ideal of $W$ in $k[x]$. Since dim $W = r-1$, we may assume that the $\mathfrak{q}$-residues of $x_1, x_2, \cdots, x_{r-1}$ are algebraically independent over $k$. Then $x_1, x_2, \cdots, x_{r-1}$ are also algebraically independent over $k$, and we may furthermore assume that $x_1, x_2, \cdots, x_r$ are algebraically independent over $k$. From our assumptions it follows that in the *polynomial ring* $k[x_1, x_2, \cdots, x_r]$ the prime ideal $\mathfrak{q}_0 = \mathfrak{q} \cap k[x_1, x_2, \cdots, x_r]$ is $(r-1)$-

† Without fear of confusion we are using here the same symbol $\mathscr{P}$ for prime divisors as was used for places in the beginning of the chapter.

dimensional, hence minimal. Let $\mathscr{P}_0$ denote the $\mathfrak{q}_0$-adic valuation of the field $k(x_1, x_2, \cdots, x_r)$; then $\mathscr{P}_0$ is the only valuation of $k(x_1, x_2, \cdots, x_r)/k$ which has center $\mathfrak{q}_0$ in $k[x_1, x_2, \cdots, x_r]$. Any prime divisor of $K/k$ which has center $W$ on $V$ has center $\mathfrak{q}$ in $k[x_1, x_2, \cdots, x_n]$, hence has center $\mathfrak{q}_0$ in $k[x_1, x_2, \cdots, x_r]$; in other words: any prime divisor $\mathscr{P}$ of $K/k$ with center $W$ on $V$ must be an extension of $\mathscr{P}_0$. Since $K$ is a finite algebraic extension of $k(x_1, x_2, \cdots, x_r)$, $\mathscr{P}_0$ has only a finite number of extensions to $K$, and this proves the finiteness of the set of prime divisors of $K/k$ having center $W$. If $\mathscr{P}$ is any prime divisor of that set, then the ring $k[x]/\mathfrak{q}$ can be canonically identified with a subring of the residue field $\Delta$ of $\mathscr{P}$. Hence, the quotient field of that ring, i.e., the field $k(W)$, is a subfield of $\Delta$. Since $\Delta/k$ is a function field, of transcendence degree $r-1$, and since also $k(W)/k$ has transcendence degree $r-1$, the theorem is proved.

There is an important case in which there is only one prime divisor of $K/k$ whose center is the given irreducible $(r-1)$-dimensional subvariety $W$ of $V/k$. If $W$ is an irreducible subvariety of $V/k$ and $\mathfrak{p}$ is the prime ideal of $W$ in the coördinate ring $R=k[x_1, x_2, \cdots, x_n]$ of $V/k$, then we mean by the *local ring of W (on V)* the quotient ring $R_\mathfrak{p}$. We denote this ring by $\mathfrak{o}(W; V)$. We say that $V/k$ is *normal* at $W$ if the local ring $\mathfrak{o}(W; V)$ is integrally closed (in this definition $W$ may be an irreducible subvariety of any dimension $\leqq r-1$). If $Q$ is any point of $V$ and $W$ is the irreducible subvariety of $V$ which has $Q$ as general point, we say that $V$ is *normal* at $Q$ if it is normal at $W$. That means then that the local ring $\mathfrak{o}(Q; V)$ is integrally closed. If $\mathfrak{f}$ denotes the conductor of the coördinate ring $R=k[x]$ in the integral closure of $R$ (Vol. I, Ch. V, § 5) and if $F$ is the (proper) subvariety of $V$ which is defined by the ideal $\mathfrak{f}$, then the irreducible subvarieties $W$ of $V/k$ such that $V/k$ is not normal at $W$ are precisely the subvarieties of $F$ (Vol. I, Ch. V, § 5, Corollary of Lemma). In particular, since $\dim F \leqq r-1$, *there is at most a finite number of irreducible $(r-1)$-dimensional subvarieties $W$ of $V/k$ such that $V/k$ is not normal at $W$.*

THEOREM 33. *If $W$ is an irreducible $(r-1)$-dimensional subvariety of $V/k$ such that $V/k$ is normal at $W$, then there is only one prime divisor of $K/k$ which has center $W$ on $V$. The valuation ring of that prime divisor coincides with the local ring $\mathfrak{o}(W; V)$, and its residue field coincides with the function field $k(W)$ of $W/k$.*

The proof is immediate: the ring $\mathfrak{o}(W; V)$ is an integrally closed, local domain which has only one proper prime ideal (since $W$ has dimension $r-1$, whence $\mathfrak{o}(W; V)=R_\mathfrak{p}$, where $\mathfrak{p}$ is a minimal prime ideal in $R$), and thus the theorem is a direct consequence of Theorem 16, Corollary 1 (§ 10).

Note that the first part of Theorem 33 is a special case of Theorem 16, Corollary 2 (§ 10), concerning minimal prime ideals in noetherian domains.

A variety $V/k$ is said to be *normal*, or *locally normal*, if it is normal at each of its points. It is clear that if the coördinate ring $k[x]$ of $V$ is integrally closed, then $V$ is normal. We shall prove now the converse:

THEOREM 34. *If an affine variety $V$ is normal then the coördinate ring $R$ of $V$ is integrally closed.*

This theorem is included in the following, more general and stronger result:

THEOREM 34'. *If $R$ is an integral domain and $M$ denotes the set of maximal prime ideals of $R$ then*

$$R = \bigcap_{\mathfrak{m} \in M} R_{\mathfrak{m}}.$$

For, the assumption that $V$ is normal signifies that $R_{\mathfrak{p}}$ is integrally closed for *any* prime ideal $\mathfrak{p}$ in the coördinate ring $R$ of $V$, and hence Theorem 34 is indeed a consequence of Theorem 34'. To prove Theorem 34' we first prove a lemma:

LEMMA. *Let $R$ be an integral domain, $\mathfrak{A}$ an ideal in $R$ and $x$ an element of $R$. If for every maximal ideal $\mathfrak{m}$ in $R$ it is true that $x$ belongs to the extended ideal $R_{\mathfrak{m}}\mathfrak{A}$, then $x \in \mathfrak{A}$.*

PROOF. Let $\mathfrak{m}$ be any maximal ideal in $R$. The assumption $x \in R_{\mathfrak{m}}\mathfrak{A}$ signifies that there exists an element $z_{\mathfrak{m}}$ (depending on $\mathfrak{m}$), *not in* $\mathfrak{m}$, such that $xz_{\mathfrak{m}} \in \mathfrak{A}$. In other words: $\mathfrak{A}:Rx \not\subset \mathfrak{m}$. The assumption that $x \in R_{\mathfrak{m}}\mathfrak{A}$ for all maximal ideals $\mathfrak{m}$ signifies therefore that the ideal $\mathfrak{A}:Rx$ is contained in no maximal ideal of $R$. Hence $\mathfrak{A}: Rx=(1)$, whence $x \in \mathfrak{A}$, as asserted.

REMARK. The lemma remains valid if $R$ is any ring with identity (and not an integral domain), provided the condition $x \in R_{\mathfrak{m}}\mathfrak{A}$, all $\mathfrak{m}$, is replaced by the condition $\varphi_{\mathfrak{m}}(x) \in R_{\mathfrak{m}} \cdot \varphi_{\mathfrak{m}}(\mathfrak{A})$, where $\varphi_{\mathfrak{m}}$ is the canonical homomorphism of $R$ into $R_{\mathfrak{m}}$ (see Vol. I, Ch. IV, § 9). The proof is similar to the one given above, and may be left to the reader.

Using the above lemma we can easily prove Theorem 34', as follows. We have only to prove the inclusion $\bigcap_{\mathfrak{m} \in M} R_{\mathfrak{m}} \subset R$, for the opposite inclusion is obvious. Let $z \in \bigcap_{\mathfrak{m} \in M} R_{\mathfrak{m}}$ and write $z$ in the form $z=x/y$, with $x, y \in R$. We have the assumption: $x \in R_{\mathfrak{m}} \cdot y$, for all $\mathfrak{m}$ in $M$. Hence, by the lemma (as applied to the ideal $\mathfrak{A}=Ry$) we conclude that $x \in Ry$, whence $z=x/y \in R$. Q.E.D.

A prime divisor $\mathscr{P}$ of $K/k$ which is finite† on the coördinate ring $k[V]$ of $V/k$ is said to be of the *first kind* or of the *second kind with respect to* $V$ according as the dimension of the center of $\mathscr{P}$ on $V$ is equal to $r-1$ or is less than $r-1$. This distinction between prime divisors of the first and of the second kind is classical. If $r>1$, then the prime divisors of $K/k$ which are of the first kind with respect to $V$ fall very short of exhausting the totality of prime divisors of $K/k$ which are finite on $k[V]$. We have in fact the following theorem:

THEOREM 35. *If $W$ is any proper subvariety of $V$, defined and irreducible over $k$, then there exist prime divisors of $K/k$ having center $W$ on $V$. If we denote by $M_W$ the set of all these prime divisors then*

$$(1) \qquad \bigcap_{\mathscr{P} \in M_W} K_{\mathscr{P}} = \text{integral closure of } \mathfrak{o}(W; V).$$

PROOF. If dim $W=r-1$, then everything has already been proved: $M_W$ is non-empty, by Theorem 32, and (1) follows from Theorem 8, § 5, since every valuation of $K/k$ with center $W$ is necessarily a prime divisor. If dim $W<r-1$, all the elements of $M_W$ are prime divisors of the second kind with respect to $V$, and our theorem asserts not only that $M_W$ is non-empty but also that the set $M_W$ is sufficiently ample as to insure that the intersection of the valuation rings $K_{\mathscr{P}}$, $\mathscr{P} \in M_W$, is the same as the intersection of *all* the valuation rings of valuations $v$ having center $W$ (this latter intersection being equal to the integral closure of the local ring $\mathfrak{o}(W, V)$, by Theorem 8, § 5).

Let $\mathfrak{p}$ be the prime ideal of $W$ in the ring $R=k[V]$, and let $\{w_1, w_2, \cdots, w_h\}$ be a basis of $\mathfrak{p}$. We consider the following $h$ rings $R'_i$:

$$R'_i = R\left[\frac{w_1}{w_i}, \frac{w_2}{w_i}, \cdots, \frac{w_h}{w_i}\right], \quad i = 1, 2, \cdots, h.$$

We note that $R'_i\mathfrak{p} = R'_i \cdot (w_1, w_2, \cdots, w_h) = R'_i \cdot w_i$.

We assert that for at least one value of $i$, $1 \leqq i \leqq h$, it is true that $R'_i w_i \cap R = \mathfrak{p}$. To see this we fix a valuation $v$ of $k(V)/k$ which has center $\mathfrak{p}$ in $R$, and we fix an index $i$ such that $v(w_i) = \min \{v(w_1), v(w_2), \cdots, v(w_h)\}$. Then the valuation ring $R_v$ contains $R'_i$. Let $\mathfrak{p}_1$ be the center of $v$ in $R'_i$. We have $R'_i w_i \subset \mathfrak{p}_1$ since $v(w_i) > 0$, and clearly $\mathfrak{p}_1 \cap R = \mathfrak{p}$. Since $R'_i w_i \cap R \supset \mathfrak{p}$, it follows that $R'_i w_i \cap R = \mathfrak{p}$, as asserted.

† Strictly speaking we should say "non-negative", since in our terminology a prime divisor is a valuation. However, in the present geometric context the term "finite" is more suggestive, since if the affine variety $V$ is thought of as part of a projective variety $V'$ then to say that $\mathscr{P}$ is non-negative on $k[V]$ is the same as saying that the center of $\mathscr{P}$ is not a subvariety at infinity (of $V'$) (see end of this section).

We give another (indirect) proof of the above assertion, which does not make use of the existence theorem for valuations. Assume that our assertion is false and that consequently there exists for each $i=1, 2, \cdots, h$ an element $\xi_i$ in $R$ such that $\xi_i \notin \mathfrak{p}$ and $\xi_i \in R'_i w_i$. Let $\xi = \xi_1 \xi_2 \cdots \xi_h$. Then also $\xi \notin \mathfrak{p}$ and $\xi \in R'_i w_i$, $i=1, 2, \cdots, h$. We can therefore write $\xi$ in the form $\xi = w_i \varphi_i(w_1, w_2, \cdots, w_h)/w_i^{\nu_i}$, where $\varphi_i$ is a form in $w_1, w_2, \cdots, w_h$, of degree $\nu_i$, with coefficients in $R$. Letting $\nu = \max \{\nu_i\}$, we have

$$\xi w_i^{\nu-1} = \psi_i(w_1, w_2, \cdots, w_h), \quad i = 1, 2, \cdots, h,$$

where $\psi_i$ is a form of degree $\nu$, with coefficients in $R$. It follows that the product of $\xi$ with every monomial $w_1^{\alpha_1} w_2^{\alpha_2} \cdots w_h^{\alpha_h}$ of degree $\alpha_1 + \alpha_2 + \cdots + \alpha_h = (\nu - 2)h + 1 = N$ is equal to a form of degree $N+1$ in $w_1, w_2, \cdots, w_h$, with coefficients in $R$. This implies that $\xi \mathfrak{p}^N \subset \mathfrak{p}^{N+1}$. Since $\xi \notin \mathfrak{p}$, this relation implies the relation $\mathfrak{m}^N = \mathfrak{m}^{N+1}$, where $\mathfrak{m}$ is the maximal ideal in the quotient ring $R_\mathfrak{p}$, in contradiction with Vol. I, Ch. IV, § 7, Theorem 12, Corollary 1, since $R$ is a noetherian integral domain and since $\mathfrak{p}$ is a proper prime ideal in $R$.

For simplicity of notations, assume that we have $R'_1 w_1 \cap R = \mathfrak{p}$. This relation implies at any rate that $w_1$ is a non-unit in $R'_1$ and that at least one isolated prime ideal $\mathfrak{p}'_1$ of $R'_1 w_1$ must contract to $\mathfrak{p}$ in $R$. By the principal ideal theorem (Vol. I, Ch. IV, § 14, Theorem 29), $\mathfrak{p}'_1$ is a minimal prime ideal in $R'_1$, and since $R'_1$ is a finite integral domain it follows that $\mathfrak{p}'_1$ has dimension $r-1$ (see Lemma). Consider now any valuation $v$ of $k(V)/k$ which is finite on $R'_1$ and has center $\mathfrak{p}'_1$. Then $v$ is necessarily a prime divisor since $\dim \mathfrak{p}'_1 = r-1$. *A fortiori*, $v$ is also finite on $R$. Its center in $R$ is clearly the prime ideal $\mathfrak{p}'_1 \cap R$, i.e., $\mathfrak{p}$. Thus $v$ is a prime divisor of $k(V)/k$ which has center $W$ on $V$, and this proves the first part of our theorem.

[The device used in the preceding proof, namely the transition from the ring $R$ to any of the rings $R'_i$, is frequently used in algebraic geometry; that device, interpreted geometrically, consists in applying to the variety $V$ a special birational transformation: a monoidal transformation of center $W$ (see Oscar Zariski, "Foundations of a general theory of birational correspondences," *Transactions of the American Mathematical Society*, vol. 53, p. 532).]

We now proceed to the proof of the second part of the theorem. Let $z$ be any element of $k(V)$ which is not contained in the integral closure of the quotient ring $R_\mathfrak{p}(=\mathfrak{o}(W; V))$. We set $y = 1/z$, $R' = R[y]$. Since $z$ does not belong to the integral closure of $R_\mathfrak{p}$, there exists a valuation $v$ of $k(V)/k$ which has center $\mathfrak{p}$ in $R$ and such that $v(z) < 0$ (§ 5, Theorem

8). Then $v(y) > 0$, $v$ is finite on $R'$, and if $\mathfrak{p}'$ denotes the center of $v$ in $R'$ then $y \in \mathfrak{p}'$ and $\mathfrak{p}' \cap R = \mathfrak{p}$. By the first part of the theorem, as applied to $R'$ and $\mathfrak{p}'$ instead of to $R$ and $\mathfrak{p}$, there exists a prime divisor $v^\star$ of $k(V)/k$ which is finite on $R'$ and has center $\mathfrak{p}'$. Then $v^\star$ is also finite on $R$, has center $\mathfrak{p}$ in $R$ (since $\mathfrak{p}' \cap R = \mathfrak{p}$) and furthermore $v^\star(z) < 0$ since $y \in \mathfrak{p}'$ and thus $v^\star(y) > 0$. Thus, we have found a prime divisor of center $W$ such that the valuation ring of that prime divisor does not contain $z$. This establishes (1) and completes the proof of the theorem.

We now go back to the prime divisors of $K/k$ which are of the first kind with respect to $V$. We denote by $S$ the set of these prime divisors. Let $\bar{R}$ be the integral closure of the coördinate ring $R = k[V]$ of $V/k$. Every prime divisor $v$ in $S$ is also finite on $\bar{R}$, the center of $v$ in $\bar{R}$ is a minimal prime $\bar{\mathfrak{p}}$ in $\bar{R}$, and the quotient ring $\bar{R}_{\bar{\mathfrak{p}}}$ is the valuation ring of $v$. Conversely, if $\bar{\mathfrak{p}}$ is any minimal prime ideal in $\bar{R}$, then $\bar{R}_{\bar{\mathfrak{p}}}$ is a discrete valuation ring of rank 1 (Theorem 16, Corollary 2, § 10) since $\bar{R}$ is noetherian, and if $v_{\bar{\mathfrak{p}}}$ is the associated valuation, then the center $\bar{\mathfrak{p}} \cap R$ of $v_{\bar{\mathfrak{p}}}$ in $R$ is a minimal prime ideal; in other words, the center of $v_{\bar{\mathfrak{p}}}$ on $V$ is of dimension $r-1$, and $v_{\bar{\mathfrak{p}}}$ is a prime divisor of the first kind with respect to $V$. Thus the set $S$ is given by the set of all $v_{\bar{\mathfrak{p}}}$ where $\bar{\mathfrak{p}}$ ranges over the set of all minimal prime ideals of $\bar{R}$. From Theorem 16, Corollary 3 (§ 10) we can now derive a number of consequences. In the first place, we have

$$\bigcap_{v \in S} K_v = \bar{R}. \tag{2}$$

If $w$ is any element of the function field $K$ of $V/k$, $w \neq 0$, then, for any $v$ in $S$, $v(w)$ is an integer, and there is only a finite number of prime divisors $v$ in $S$ such that $v(w) \neq 0$. We refer to $v(w)$ as the *order of $w$ at the prime divisor $v$*, and we say that $v$ is a prime *null divisor* or a prime *polar divisor* of $w$ according as $v(w) > 0$ or $v(w) < 0$. Any function $w$ in $K$, $w \neq 0$, has at most a finite number of prime null divisors and polar divisors in the set $S$, and the functions $w$ having no polar prime divisors of the first kind with respect to $V$ are those and only those functions which belong to the integral closure of the coördinate ring $R$ of $V/k$.

The situation is particularly simple if $V/k$ is a normal variety. In this case, every element $v$ of $S$ can be denoted without ambiguity by the symbol $v_W$, when $W$ is the center of $v$ on $V$, since $W$, which is of dimension $r-1$, uniquely determines the prime divisor $v_W$. We then introduce the free group $G$ generated by the irreducible $(r-1)$-dimensional subvarieties of $V/k$ and we call the elements of this group, *divisors*. A divisor $\Gamma$ on $V$ is therefore a formal finite sum $\Gamma = \Sigma m_i W_i$, where the $W_i$

are irreducible $(r-1)$-dimensional subvarieties of $V/k$ and the $m_i$ are integers. We write $\Gamma \succ 0$ if all the $m_i$ are non-negative, and we say then that $\Gamma$ is a non-negative divisor. We write $\Gamma > 0$ ($\Gamma$—a *positive* divisor) if $\Gamma \succ 0$ and $\Gamma \neq 0$. If now $w$ is any function in $K$, $w \neq 0$, then we can associate with $w$ a well-defined divisor on $V$, namely the divisor

$$(3) \qquad (w) = \Sigma v_W(w) \cdot W,$$

where the sum is extended to all the irreducible $(r-1)$-dimensional subvarieties of $V/k$ (the above sum is, of course, finite since the number of $W$'s for which $v_W(w) \neq 0$ is finite). The divisor $(w)$ defined in (3) is called *the divisor of the function* $w$. Then $(w) \succ 0$ if and only if $w \in \bar{R}$ and $(w) = 0$ if and only if $w$ is a unit in $\bar{R}$.

The above definitions refer to the *affine* variety $V$. That a function $w$ may have no polar prime divisors on $V$ without being a "constant" (i.e., without belonging to the ground field $k$ or to the algebraic closure of $k$ in $K$) is due precisely to the fact that our definitions refer to an affine variety $V/k$. In this frame of reference one loses track of the prime divisors "at infinity." The "correct" definitions are obtained if one deals with projective varieties. We shall do that in the next chapter (VII, § 4). However, even without introducing explicitly projective spaces and varieties in the projective space, we can arrive already here at the desired "correct" definition of the divisor of a function in the following fashion:

If $n$ is the dimension of the affine ambient space of our variety $V$, let $x_1, x_2, \cdots, x_n$ be the coördinates of the general point of $V/k$. We set

$$x_1^i = \frac{x_1}{x_i}, \quad x_2^i = \frac{x_2}{x_i}, \cdots, x_{i-1}^i = \frac{x_{i-1}}{x_i},$$

$$x_i^i = \frac{1}{x_i}, \quad x_{i+1}^i = \frac{x_{i+1}}{x_i}, \cdots, x_n^i = \frac{x_n}{x_i}$$

and we denote by $V_i$ the affine variety whose general point is $(x_1^i, x_2^i, \cdots, x_n^i)$. We set

$$R_0 = R = k[x_1, x_2, \cdots, x_n], \quad R_i = k[x_1^i, x_2^i, \cdots, x_n^i] = k[V_i].$$

The $n+1$ rings $R_i$ have $K$ as common quotient field (whence the $n+1$ varieties $V_i$ are birationally equivalent). We denote by $S_i$ the set of prime divisors of $K/k$ which are of the first kind with respect to $V_i$ (we set $V_0 = V$) and by $S^\star$ the union of the $n+1$ sets $S_i$. We note the following: *the only prime divisors $v$ in the set $S_i$, $i \neq 0$, which do not belong to $S_0$ are those at which $x_i^i$ has positive order* (or equivalently: $v(x_i) < 0$).

In fact, if $v(x_i) < 0$, then $v$ is not finite on $R_0$ and therefore $v \notin S_0$. On the other hand, if $v \in S_i$ and $v(x_i) \geqq 0$, then also $v(x_j) \geqq 0, j=1, 2, \cdots, n$, for $x_j = x_i \cdot x_j^i$ if $j \neq i$; whence $v$ is finite on $R_0$. Furthermore, we must now have $v(x_i)=0$ (since $v(x_i^i)$ is non-negative), and hence the $v$-residue of $x_i$ is different from zero. The relations $x_j = x_i \cdot x_j^i$ $(j \neq i)$ show therefore that the field generated over $k$ by $v$-residues of the $x_j^i (j=1, 2, \cdots, n)$ coincides with the field generated over $k$ by the residues of $x_1, x_2, \cdots, x_n$. This shows that the center of $v$ in $R$ also has dimension $r-1$, whence $v \in S_0$. We have therefore shown that there is only a finite number of prime divisors in $S^\star$ which do not belong to $S_0$. These are the prime divisors "at infinity" with respect to $V$.

We now can proceed as we did in the case of an affine variety, except that the set $S^\star$ now replaces the set $S_0(=S)$. *If now a function $w$ in $K$, $w \neq 0$, has no polar divisors, i.e., if we have $v(w) \geqq 0$ for all $v$ in $S^\star$, then $w$ must be a constant, i.e., $w$ is algebraic over $k$.* For, $w$ must then belong to the integral closure of each of the $n+1$ rings $R_i$. On the other hand, given *any* valuation $v$ of $K/k$, the valuation ring $K_v$ must contain at least one of the $n+1$ rings $R_i$: namely, if all $v(x_i)$ are $\geqq 0$ then $R_v \supset R_0$; otherwise if, say, $v(x_i) = \min \{v(x_1), \cdots, v(x_n)\}$ then $K_v \supset R_i$. It follows that $w$ belongs to all the valuation rings $K_v$ such that $K_v \supset k$, and hence $w$ must belong to the integral closure of $k$ in $K$, as asserted.

It would now be easy to develop the concept of a divisor and of the divisor of a function, with reference to the set of $n+1$ affine varieties $V_i$, especially if each $V_i$ is a normal variety. However, we shall postpone this to the next chapter (see VII, § $4^{\text{bis}}$).

**§ 15. Examples of valuations.** All the examples of valuations encountered in the preceding sections were discrete, of rank 1 (e.g., $\mathfrak{p}$-adic valuations of Dedekind domains, prime divisors of function fields, etc.). We shall give in this section a number of examples of valuations of various types, in particular examples of non-discrete valuations of rank 1. The algebraic function fields of transcendence degree $r > 1$, over a given ground field $k$, represent the best source of such illustrative material, and we shall in fact work exclusively with function fields in this section. As a matter of fact, we shall deal largely with pure transcendental extensions of a ground field $k$, for we know that if we extend a valuation $v$ of a field $K$ to a valuation of a finite algebraic extension of $K$, then the structure of the value group of $v$ (rank, rational rank, etc.) remains unaltered.

EXAMPLE 1. *Valuations of maximum rational rank.* Let $K = k(x_1, x_2, \cdots, x_r)$, where $x_1, x_2, \cdots, x_r$ are algebraically independent over $k$, and let $\alpha_1, \alpha_2, \cdots, \alpha_r$ be arbitrary, rationally independent real numbers. If $t$ is a parameter and we carry out the formal substitution $x_i \rightarrow t^{\alpha_i}$, then every monomial in $x_1, x_2, \ldots, x_r$ yields a power of $t$, and distinct monomials yield distinct powers of $t$ (since the $\alpha_i$ are rationally independent). If $f(x_1, x_2, \cdots, x_r)$ is any polynomial in $k[x_1, x_2, \cdots, x_r]$, then $f(t^{\alpha_1}, t^{\alpha_2}, \cdots, t^{\alpha_r})$ is a sum of powers of $t$, say, $ct^\beta +$ terms of degree $> \beta$, where $c \neq 0$, $\beta = n_1\alpha_1 + n_2\alpha_2 + \cdots + n_r\alpha_r$, and the $n_i$ are non-negative integers. If we set $v(f) = \beta$, then $v$ is a mapping of $k[x_1, x_2, \cdots, x_r]$ (the zero excluded) onto a group $\Gamma$ of real numbers, where $\Gamma = J\alpha_1 + J\alpha_2 + \cdots + J\alpha_r$ ($J$ = the additive group of integers). Note that *$\Gamma$ is the direct sum of the $r$ free cyclic groups $J\alpha_i$.* We have $v(fg) = v(f) + v(g)$, $v(f+g) \geqq \min \{v(f), v(g)\}$, and hence $v$ can be extended to a valuation $v$ of the field $K$ (§ 9, Theorem 14). The above group $\Gamma$ is the value group of $v$, and thus $v$ is non-discrete, of rank 1 and rational rank $r$. It is immediately seen that the residue field of $v$ is the ground field $k$, whence $v$ is zero-dimensional. If the $\alpha_i$ are all positive, then $v$ is non-negative on the polynomial ring $k[x_1, x_2, \cdots, x_r]$ and its center is the origin $x_1 = x_2 = \cdots = x_r = 0$ in the affine $r$-space.

We know that the rational rank of a rank 1 valuation of a field $K/k$, of transcendence degree $r$, is at most equal to $r$. In the above example this maximum $r$ of the rational rank is realized, and the value group turns out to be a direct sum of $r$ free cyclic groups. This is not accidental, for we have quite generally the following:

THEOREM 36. *If a valuation $v$ of a field $K/k$ of algebraic functions of $r$ independent variables has rational rank $r$ then the value group $\Gamma$ of $v$ is the direct sum of $r$ cyclic groups:*

$$\Gamma = J\tau_1 + J\tau_2 + \cdots + J\tau_r,$$

*where $J$ denotes the additive group of integers and $\tau_1, \tau_2, \cdots, \tau_r$ are rationally independent elements of $\Gamma$.*

PROOF. We fix in $\Gamma$ a set $\{\alpha_1, \alpha_2, \cdots, \alpha_r\}$ of rationally independent elements and then we fix in $K$ a set of elements $x_1, x_2, \cdots, x_r$ such that $v(x_i) = \alpha_i$. As in the preceding example one shows that the value group $\Gamma'$ of the restriction of $v$ to the field $k(x_1, x_2, \cdots, x_r)$ is then the group $\Gamma' = J\alpha_1 + J\alpha_2 + \cdots + J\alpha_r$, a direct sum of $r$ cyclic groups. If $n$ denotes the relative degree $[K: k(x_1, x_2, \cdots, x_r)]$ then we know that $\Gamma' \subset \Gamma \subset \frac{1}{n!}\Gamma'$ (§ 11, proof of Lemma 1). Now, the group $\frac{1}{n!}\Gamma'$ is a direct sum of $r$ cyclic groups and admits the basis elements $\alpha_1/n!$,

$\alpha_2/n!, \cdots, \alpha_r/n!$. Since $\Gamma$ is a subgroup, of finite index, of $\frac{1}{n!}\Gamma'$, also $\Gamma$ must possess a basis of $r$ elements $\tau_1, \tau_2, \cdots, \tau_r$, as asserted (Vol. I, Ch. IV, § 15, Lemma 2).

EXAMPLE 2. *Generalized power series expansions.* Consider formal power series $z(t) = a_0t^{y_0} + a_1t^{y_1} + \cdots + a_nt^{y_n} + \cdots$, where the coefficients $a_n$ are in $k$ and the exponents $y_n$ are real numbers such that $y_0 < y_1 < \cdots$, and $\lim y_n = \infty$. These power series with the usual formal rules for addition and multiplication form a field $k\{t\}$. This field admits a natural valuation $V$, of rank 1, defined by setting $V(z(t)) = y_0$, if $a_0 \neq 0$. Any isomorphism of $k(x_1, x_2, \cdots, x_r)$ into $k\{t\}$ will therefore yield a rank 1 valuation of $k(x_1, x_2, \cdots, x_r)$. Any such isomorphism is obtained by choosing for each variable $x_i$ a power series $z_i(t)$ in $k\{t\}$ *such that the r power series* $z_1(t), z_2(t), \cdots, z_r(t)$ *are algebraically independent.* The valuations thus obtained are all zero-dimensional and have $k$ as residue field. In particular, if the $z_1(t), z_2(t), \cdots, z_r(t)$ are power series with *integral* exponents, so that the "one-dimensional arc" $x_i = z_i(t)$ $(i = 1, 2, \cdots, r)$ is analytic and *does not lie on any proper algebraic subvariety of the affine r-space*, then we get a discrete zero-dimensional valuation of $k(x_1, x_2, \cdots, x_r)$, of rank 1. The condition that the "arc" $x_i = z_i(t)$ does not lie on any proper algebraic subvariety of the affine $r$-space is equivalent to our condition that the $r$ power series $z_i(t)$ be algebraically independent (over $k$). If this condition is not satisfied, then the $r$ power series $z_i(t)$ can be used to define valuations of rank $> 1$, as follows:

The polynomials $f(x_1, x_2, \cdots, x_r)$ in $k[x_1, x_2, \cdots, x_r]$ which give rise to true algebraic relations $f(z_1(t), z_2(t), \cdots, z_r(t)) = 0$ between the given power series $z_i(t)$ form a prime ideal $\mathfrak{p}$ in $k[x_1, x_2, \cdots, x_r]$. Let $v'$ be *any* valuation of $k(x_1, x_2, \cdots, x_r)$ which is non-negative on the polynomial ring $k[x_1, x_2, \cdots, x_r]$ and which has center $\mathfrak{p}$ in that ring. If $\bar{x}_i$ denotes the $\mathfrak{p}$-residue of $x_i$ then it is clear that the mapping $\bar{x}_i \to z_i(t)$, $i = 1, 2, \cdots, r$, defines a $k$-isomorphism of $k(\bar{x}_1 \bar{x}_2, \cdots, \bar{x}_r)$ into $k\{t\}$ and therefore also defines a rank 1 valuation $\bar{v}$ of the field $k(\bar{x}_1, \bar{x}_2, \cdots, \bar{x}_r)$. This latter field is a subfield of the residue field $\Delta_{v'}$ of $v'$, and the valuation $\bar{v}$ can be extended to a valuation of $\Delta_{v'}$ which has the same value group as $\bar{v}$. Denoting this extended valuation by the same letter $\bar{v}$, we have now a composite valuation $v = v' \circ \bar{v}$ of $k(x_1, x_2, \cdots, x_r)$, whose rank is one greater than the rank of $v'$. Note that this valuation is, in general, not uniquely determined by the "arc" $x_i = z_i(t)$; it depends on the choice of $v'$. The only case in which $v'$, and hence also $v$, is uniquely determined is the case in which the prime ideal $\mathfrak{p}$ is minimal, in which case $v'$ is necessarily the prime divisor of center $\mathfrak{p}$.

EXAMPLE 3. *Real valuations with preassigned value group.* Let

$$z_i(t) = a_{0i}t^{y_{0i}} + a_{1i}t^{y_{1i}} + \cdots + a_{ni}t^{y_{ni}} + \cdots,$$

where we assume that the power series $z_i(t)$ are algebraically independent. MacLane and Schilling have proved (see their joint paper "Zero-dimensional branches of rank 1 on algebraic varieties," Annals of Mathematics, v. 40 (1939), pp. 507–520) that if all the $a_{ni}$ are $\neq 0$, if $k$ is of characteristic zero and if the exponents $y_{ni}$ are rational linear combinations of $s+1$ given real numbers $1, \tau_1, \tau_2, \cdots, \tau_s$, then the field $k(t, t^{\tau_1}, t^{\tau_2}, \cdots, t^{\tau_s}, z_1(t), z_2(t), \cdots, z_r(t))$ has in the natural valuation $V$ a value group generated by $1, \tau_1, \tau_2, \cdots, \tau_r$ and all the exponents $y_{ni}$ of the given $r$ series $z_i(t)$. From this result one can easily obtain *the existence of a rank 1 valuation of $k(x_1, x_2, \cdots, x_r)$ with any preassigned value group $\Gamma$ of rational rank $s+1$ less than $r$.* For, let $1, \tau_1, \tau_2, \cdots, \tau_s$ be $s+1$ rationally independent elements of $\Gamma$ (we may assume, as we did, that one of these real numbers is 1). Since every element of $\Gamma$ is rationally dependent on $1, \tau_1, \tau_2, \cdots, \tau_s$, $\Gamma$ is a denumerable set. We can therefore find $r-s-1$ power series $z_i(t)$ in $k\{t\}$ such that the exponents of these power series generate the group $\Gamma$, and it is also possible to arrange the choice of these series in such a fashion that the $r$ series $t, t^{\tau_1}, \cdots, t^{\tau_s}, z_1(t), z_2(t), \cdots, z_{r-s-1}(t)$ be algebraically independent over $k$. By means of these $r$ series, and in view of the theorem of MacLane-Schilling cited above, we get a rank 1 valuation of $k(x_1, x_2, \cdots, x_r)$ with the preassigned value group $\Gamma$.

In particular, it follows that *if $r \geqq 2$ then any additive subgroup of the field of rational numbers is the value group of a suitable valuation of the field $k(x_1, x_2, \cdots, x_r)$ of rational functions of $r$ independent variables.* We shall illustrate this result by an example using a procedure which does not make use of the generalized formal power series. For simplicity, we shall restrict ourselves to the case $r=2$ and to the field $k(x_1, x_2)$. Let $\{m_1, m_2, \cdots\}$ be an arbitrary infinite sequence of positive integers such that $m_1m_2\cdots m_i \to +\infty$, and let $\{c_1, c_2, \cdots\}$ be a sequence of elements of $k$, where each $c_i$ is $\neq 0$. We define an infinite sequence of elements $u_i$ in $k(x_1, x_2)$, by induction, as follows: $u_1 = x_1$, $u_2 = x_2$, $u_{i+2} = (u_i - c_iu_{i+1}^{m_i})/u_{i+1}^{m_i}$, $i = 1, 2, \cdots$. We denote by $R$ the ring $k[u_1, u_2, \cdots, u_i, \cdots]$ and by $\mathfrak{q}$ the ideal generated in $R$ by the infinitely many elements $u_i$. Since every element of $R$ is congruent mod $\mathfrak{q}$ to an element of $k$, $\mathfrak{q}$ is either the unit ideal or is a maximal ideal in $R$. We prove that $\mathfrak{q} \neq R$. Assuming the contrary, there will exist an integer $h$ such that the ideal $\mathfrak{q}_h$ generated by $u_1, u_2, \cdots, u_h$ in $R_h = k[u_1, u_2, \cdots, u_h]$ is the unit ideal. Now, we have $u_i \in k[u_{i+1}, u_{i+2}]$, for all $i$, and furthermore $u_i$ belongs to the ideal generated by $u_{i+1}, u_{i+2}$ in $k[u_{i+1}, u_{i+2}]$. It follows that $R_h = k[u_{h-1}, u_h]$ and that $\mathfrak{q}_h = R_h(u_{h-1}, u_h)$. Since $u_{h-1}, u_h$ are algebraically independent over $k$, the relation $1 \in \mathfrak{q}_h$ is impossible.

Since $\mathfrak{q}$ is a proper (maximal) ideal, there exists a valuation $v$ of $k(x_1, x_2)$ such that $v$ is non-negative on $R$ and has center $\mathfrak{q}$ in $R$. Let $v$ be such a valuation. Since $v(u_{i+2}) > 0$, it follows that $v(u_i - c_i u_{i+1}^{mi}) > v(u_{i+1}^{mi})$, whence $v(u_i) = m_i v(u_{i+1})$. In particular,

$$(1) \qquad v(u_1) = m_1 m_2 \cdots m_s v(u_{s+1}).$$

Since $m_1 m_2 \cdots m_s \to \infty$, it follows that $v$ is non-discrete, therefore of rank 1, and necessarily of rational rank 1, for (1) shows that $v$ cannot be isomorphic with a direct product of two free cyclic groups. If we normalize the value group $\Gamma$ of $v$ by setting $v(u_1) = 1$, then (1) shows that $\Gamma$ contains all the rational numbers having denominator $m_1 m_2 \cdots m_s$, $s = 1, 2, \cdots$. *We shall now show that $\Gamma$ is actually the set of all rational numbers of the form* $\dfrac{n}{m_1 m_2 \cdots m_s}$, $s = 1, 2, \cdots$, *and that*

$$(2) \qquad K_v = \bigcup_{h=1}^{\infty} (R_h)_{\mathfrak{q}_h}.$$

To prove (2) we shall use the corollary of Theorem 10, § 5 (p. 21). We have $R_h \subset R_{h+1}$ and $\mathfrak{q}_{h+1} \cap R_h = \mathfrak{q}_h$; this last relation follows from the relations

$$(2') \qquad \mathfrak{q}_{h+1} = R_{h+1} \cdot (u_h, u_{h+1}), \quad u_{h-1} = u_h^{m_{h-1}}(u_{h+1} + c_{h-1}) \in \mathfrak{q}_{h+1}$$

and from the fact that $\mathfrak{q}_h$ is a maximal ideal in $R_h$. Hence, by the cited corollary of Theorem 10, (2) will be proved if we show that there exists no valuation of $K$ which, for every $h$, has center $\mathfrak{q}_h$ and is of the second kind with respect to $R_h$. Assume the contrary, and let $v'$ be such a valuation. Then $v'$ must have dimension 1 (since $K/k$ has transcendence degree 2), i.e., $v'$ must be a prime divisor, and the value group of $v'$ is therefore the additive group of integers. We must have $v'(u_h) > 0$, for all $h$, since $\mathfrak{q}_h$ is the center of $v'$ in $R_h$. On the other hand, we have also by (2'), $v'(u_{h-1}) = m_{h-1} v'(u_h)$, and in particular, $v'(u_1) = m_1 m_2 \cdots m_h v'(u_{h+1})$, for all $h$. This is in contradiction with the fact that all the numbers $v'(u_h)$ are positive integers, whereas $m_1 m_2 \cdots m_h \to +\infty$.

By (1), we have $v(u_{h+1}) = \dfrac{1}{m_1 m_2 \cdots m_h}$. Therefore, to prove our assertion concerning the value group $\Gamma$ we have only to show the following: if $f(u_1, u_2)$ is any polynomial in $k[u_1, u_2]$, then for $h$ sufficiently large we will have $f(u_1, u_2) = u_h^n f_h(u_h, u_{h+1})$, $f_h(0, 0) \neq 0$. To show this, we fix a positive integer $m$, sufficiently large, so as to satisfy the inequality $v(u_1^m) \geqq v(f(u_1, u_2))$, and we set $\xi = u_1^m / f(u_1, u_2)$. Then $\xi \in K_v$ and hence, by (2), $\xi \in (R_h)_{\mathfrak{q}_h}$, for large $h$, i.e., we have

$$(3) \qquad u_1^m / f(u_1, u_2) = A(u_h, u_{h+1}) / B(u_h, u_{h+1}), \quad B(0, 0) \neq 0.$$

Now, $u_1$, as a polynomial in $u_h$, $u_{h+1}$, has the form $u_1 = u_h^\rho \varphi(u_h, u_{h+1})$, where $\varphi(0, 0) \neq 0$. It follows then from (3) that

$$\begin{aligned} f(u_1, u_2) A(u_h, u_{h+1}) &= u_h^{\rho m} B(u_h, u_{h+1}) [\varphi(u_h, u_{h+1})]^m \\ &= u_h^{\rho m} C(u_h, u_{h+1}), \quad C(0, 0) \neq 0, \end{aligned}$$

and therefore, if $f(u_1, u_2)$ is expressed as a polynomial in $u_h$, $u_{h+1}$, its only irreducible factor which vanishes at $u_h = u_{h+1} = 0$ (if $f(u_1, u_2)$ has such a factor) must be $u_h$. In other words, $f$ must be of the form $u_h^n f_h(u_h, u_{h+1})$, $f_h(0, 0) \neq 0$, as asserted.

We thus see that we can take as $\Gamma$ any subgroup of the additive group of rational numbers. In particular, if $m_h = h$, then $\Gamma$ is the set of all rational numbers.

EXAMPLE 4. *Valuations of infinite relative degree.* If the algebraic closure $\bar{k}$ of the ground field $k$ has infinite relative degree over $k$, it is possible to construct zero-dimensional valuations of $k(x_1, x_2, \cdots, x_r)$, $r > 1$, having as residue field an infinite algebraic extension of $k$. We shall show this in the case $r = 2$. We assume for simplicity that the maximal separable extension of $k$ in $\bar{k}$ has already infinite relative degree over $k$. We fix in $\bar{k}$ an infinite sequence of elements $a_1, a_2, \cdots, a_n, \cdots$ which are separable over $k$ and such that the field $k(a_1, a_2, \cdots, a_n, \cdots)$ has infinite relative degree over $k$, and we consider in the $(x_1, x_2)$-plane the branch $x_2 = a_1 x_1 + a_2 x_1^2 + \cdots + a_n x_1^n + \cdots$. This branch determines a discrete zero-dimensional valuation $v$ of $k(x_1, x_2)$ which has center at the origin $(0, 0)$ and has rank 2 or 1 according as the branch is or is not algebraic (see second part of Example 2; we shall see in a moment that the above branch is in fact necessarily non-algebraic). *It will be sufficient to show that the residue field of $v$ coincides with the field $k(a_1, a_2, \cdots, a_n, \cdots)$.*

It is clear that the residue field of $v$ is contained in $k(a_1, a_2, \cdots, a_n, \cdots)$. It is also clear that $a_1$ belongs to the residue field of $v$, since $a_1$ is the $v$-residue of $x_2/x_1$. We assume that it has already been proved that $a_1, a_2, \cdots, a_{n-1}$ belong to the residue field of $v$. We set $w = a_1 x_1 + a_2 x_1^2 + \cdots + a_{n-1} x_1^{n-1}$ and we denote by $w_1 (= w), w_2, \cdots, w_g$ the conjugates of $w$ over $k(x_1)$: $w_i = a_1^{(i)}x_1 + a_2^{(i)}x_1^2 + \cdots + a_{n-1}^{(i)}x_1^{n-1}$. Let $f(t) = \prod_{i=1}^{g} (t - w_i)$, $\xi = f(x_2) = F(x_1, x_2) \in k[x_1, x_2]$. Then $F(x_1, z(x_1)) = a_n x_1^n \cdot (1 + \text{terms of degree} > 1) \cdot \prod_{i=2}^{g} \{(a_1 - a_1^{(i)})x_1 + \cdots + (a_{n-1} - a_{n-1}^{(i)})x_1^{n-1} + a_n x_1^n + \cdots\}$, which shows that the leading term of the power series $F(x_1, z(x_1))$ is of the form $a_n b x_1^h$, $h \geqq n$, where $b$ is an element of the field $k(a_1, a_2, \cdots, a_{n-1})$. Since $a_n b$ is the residue of $\xi/x_1^h$, it follows that $a_n$ belongs to the residue field of $v$.

[It now follows *a posteriori* that our branch is non-algebraic, since the residue field of a zero-dimensional, rank 2, valuation of $k(x_1, x_2)$ is always finitely generated over $k$, by Theorem 31, Corollary, § 14.]

In the following example, $k$ may be algebraically closed, and we are dealing with a function field $k(x_1, x_2, x_3)$ of three independent variables. In this case we can construct a 1-dimensional valuation whose residue field is not a finitely generated extension of $k$ (contrary to what happens in the case of prime divisors; see § 14, Theorem 31). We simply set, for instance: $x_3 = x_2 + \sqrt{x_1}x_2^2 + \cdots + \sqrt[n]{x_1}x_2^n + \cdots = z(x_2)$, i.e., we use the substitution $x_3 \to z(x_2)$ and we treat $k(x_1)$ as ground field. Then we get a discrete, rank 1 valuation of $k(x_1, x_2, x_3)$, whose residue field is $k(\sqrt{x_1}, \sqrt[3]{x_1}, \sqrt[4]{x_1}, \cdots)$.

EXAMPLE 5. *Prime divisors of the 2nd kind.* Consider the polynomial ring $k[x_1, x_2, \cdots, x_r]$ in $r$ independent variables, and for any polynomial $f$ in $k[x_1, x_2, \cdots, x_r]$ set $v(f) = m$ if $f$ has terms of degree $m$ but no terms of degree less than $m$. It is immediately seen that $v(fg) = v(f) + v(g)$ and that $v(f+g) \geqq \min\{v(f), v(g)\}$. Hence if we extend $v$ to the field $k(x_1, x_2, \cdots, x_r)$ by setting $v(f/g) = v(f) - v(g)$, we obtain a valuation $v$, discrete, of rank 1, which is non-negative on $k[x_1, x_2, \cdots, x_r]$ and whose center in this polynomial ring is the prime ideal $(x_1, x_2, \cdots, x_r)$. In other words, we are dealing with a valuation whose center, in this affine $r$-space, is the origin. On the other hand, *it is easily seen that $v$ is a prime divisor.* For, any non-zero polynomial $\xi$ in the ratios $x_2/x_1$, $x_3/x_1, \cdots, x_r/x_1$, with coefficients in $k$, is of the form $\dfrac{f(x_1, x_2, \cdots, x_r)}{x_1^m}$, where $f$ is a *form* of degree $m$. Hence $v(\xi) = 0$, since $v(f) = m$ and $v(x_1) = 1$, i.e., we have shown that the $v$-residues of the $r-1$ elements $x_2/x_1, \cdots, x_r/x_1$ are algebraically independent over $k$. Note that $v$ is also non-negative on the ring $k[x'_1, x'_2, \cdots, x'_r]$, where $x'_1 = x_1$, $x'_i = x_i/x_1$, $i = 2, 3, \cdots r$, and that the center of $v$ in that ring is the principal ideal $(x'_1)$. The valuation $v$ thus defined is, in some sense, the simplest prime divisor of $k(x_1, x_2, \cdots, x_r)$ whose center is the maximal ideal $\mathfrak{m}=(x_1, x_2, \cdots, x_r)$ and is sometimes referred to as the $\mathfrak{m}$-adic prime divisor. Our construction of $\mathfrak{m}$-adic prime divisors of the 2nd kind is merely a special case of a more general procedure which was used in § 14 in the construction of prime divisors of the second kind, having a preassigned center.

**§ 16. An existence theorem for composite centered valuations.** In the preceding section we have dealt exclusively with valuations of rank 1. By repeated applications of the procedures outlined in the case of rank 1 valuations, one obtains corresponding examples of valuations of higher rank. The arbitrary elements which one may wish to be able to preassign are the following: (1) the value groups; (2) the dimensions of the successive valuations with which the given valuation is to be composite; (3) the centers of these valuations. We shall devote this section to an existence theorem, for function fields, which bears on items (2) and (3) and which is a refinement of the theorem of existence of places with preassigned center (Theorem 5, § 4). Let $V/k$ be an irreducible variety, of dimension $r$, let $K=k(V)$ be the function field of $V/k$ and let $\mathcal{P}$ be a non-trivial place of $K/k$, of rank $m$, which has a center on $V$ (i.e., $\mathcal{P}$ is a place which is finite on the coördinate ring $k[V]$ of $V/k$). We have then a specialization chain for $\mathcal{P}$:

$$\mathcal{P}_{m-1} \to \mathcal{P}_{m-2} \to \cdots \to \mathcal{P}_1 \to \mathcal{P}, \tag{1}$$

where $\mathcal{P}_{m-j}$ is a place of $K/k$, of rank $j$. Necessarily each $\mathcal{P}_j$ has a center on $V$. Let $Q$ be the center of $\mathcal{P}$, $Q_j$ the center of $\mathcal{P}_j$ on $V, j=1, 2, \cdots, m-1$. Then also the points $Q_j$ form a specialization chain over $k$:

$$Q_{m-1} \xrightarrow{k} Q_{m-2} \xrightarrow{k} \cdots \xrightarrow{k} Q_1 \xrightarrow{k} Q. \tag{2}$$

If $s=\dim \mathcal{P}/k$, $s_j=\dim \mathcal{P}_j/k$, then

$$r-1 \geqq s_{m-1} > s_{m-2} > \cdots > s_1 > s \geqq 0 \tag{3}$$

and

$$s_j \geqq \dim Q_j/k. \tag{4}$$

The existence theorem which we wish to prove in this section is the following:

THEOREM 37. *Let $m$ be an integer such that $1 \leqq m \leqq r$ and let $s, s_1, \cdots, s_{m-1}$ be $m$ integers satisfying the inequalities* (3). *Let furthermore $Q, Q_1, \cdots, Q_{m-1}$ be $m$ points on $V$ such that* (2) *and* (4) *hold. Then there* *exists a specialization chain* (1) *of m places* $\mathscr{P}, \mathscr{P}_1, \cdots, \mathscr{P}_{m-1}$ *of* $K/k$ *such that the rank and the dimension of* $\mathscr{P}_j/k$ *are respectively* $m-j$ *and* $s_j$, *and that the center of* $\mathscr{P}_j$ *on* $V$ *is the point* $Q_j(\mathscr{P}_0=\mathscr{P}, s_0=s, Q_0=Q)$.

PROOF. We first consider the case $m=1$. Let $h=\dim Q/k$, whence $r-1 \geqq s \geqq h$, and let $\sigma=s-h$. We shall first achieve a reduction to the case $h=0$, as follows:

Let $x_1, x_2, \cdots, x_n$ be the non-homogeneous coördinates of the general point of $V/k$ and let $z_1, z_2, \cdots, z_n$ be the coördinates of $Q$. We may assume that $z_1, z_2, \cdots, z_h$ are algebraically independent over $k$, so that $z_{h+1}, z_{h+2}, \cdots, z_n$ are algebraically dependent over $k(z_1, z_2, \cdots, z_h)$. Then also $x_1, x_2, \cdots, x_h$ are algebraically independent over $k$, since the point $Q=(z_1, z_2, \cdots, z_n)$ is a specialization of the point $(x_1, x_2, \cdots, x_n)$ over $k$. It is clear that in the proof of our theorem it is permissible to replace $Q$ by any $k$-isomorphic point. Since the $k$-isomorphism of $k(z_1, z_2, \cdots, z_h)$ onto $k(x_1, x_2, \cdots, x_h)$, defined by $z_i \rightarrow x_i$, $i=1, 2, \cdots, h$, can be extended to an isomorphism of $k(z_1, z_2, \cdots, z_n)$ into the universal domain, we may assume that $x_i=z_i$, $i=1, 2, \cdots, h$. If we now extend our ground field $k$ to the field $k'=k(x_1, x_2, \cdots, x_h)$, our problem is to find a place $\mathscr{P}$ of $k'(x)$ *over* $k'$, of rank 1 and dimension $\sigma$, such that $x_i\mathscr{P}=z_i$, $i=h+1, h+2, \cdots, n$. This is the reduction to the case $h=0$, since the $z_i$ are algebraic over $k'$.

The case $m=1$ can now be divided into two sub-cases according as $\sigma=0$ or $\sigma>0$, i.e., according as $s=h$ or $s>h$. We consider first the case $\sigma=0$. In this case we may assume that we have originally $s=h=0$. We can carry out a second reduction to the case in which the ground field $k$ is algebraically closed. This reduction is straightforward, for if $\bar{k}$ is the algebraic closure of $k$ in the universal domain, then it is sufficient to construct a place $\bar{\mathscr{P}}$ of $\bar{k}(x_1, x_2, \cdots, x_n)/\bar{k}$, of rank 1 and dimension zero, such that $x_i\bar{\mathscr{P}}=z_i$ and to take for $\mathscr{P}$ the restriction of $\bar{\mathscr{P}}$ to $k(x_1, x_2, \cdots, x_n)$. We may therefore assume that $k$ is algebraically closed. In that case, upon replacing each $x_i$ by $x_i-z_i$ $(z_i \in k)$, we may also assume that $Q$ is the origin and that consequently the ideal $\tilde{\mathfrak{q}}$ in $k[x_1, x_2, \cdots, x_n]$ which is generated by $x_1, x_2, \cdots, x_n$ is not the unit ideal. By the normalization theorem (Vol. I, Ch. V, § 4, Theorem 8), we may also assume that $x_1, x_2, \cdots, x_r$ are algebraically independent over $k$ and that the ring $k[x_1, x_2, \cdots, x_n]$ is integrally dependent on $k[x_1, x_2, \cdots, x_r]$. Now, in § 15, Example 2, we have given general procedures for constructing places $\mathscr{P}_0$ of $k(x_1, x_2, \cdots, x_r)$, of rank 1 and dimension zero, which are finite on $k[x_1, x_2, \cdots, x_r]$ and have in that ring center $\mathfrak{q}$, where $\mathfrak{q}$ is the ideal generated by $x_1, x_2, \cdots, x_r$. Now, the ideal $\tilde{\mathfrak{q}}$ generated by $x_1, x_2, \cdots, x_n$ in the ring $k[x_1, x_2, \cdots, x_n]$, lies over $\mathfrak{q}$. Hence by Theorem 13, § 7, any place $\mathscr{P}_0$ such as above, has at least one extension $\mathscr{P}$ to $k(x_1, x_2, \cdots, x_n)$ whose center in $k[x_1, x_2, \cdots, x_n]$ is the prime ideal $\tilde{\mathfrak{q}}$. Since $\mathscr{P}$ and $\mathscr{P}_0$ have the same dimension and the same rank, our theorem is proved in the special case under consideration (case $m=1$, $s=h$).

Let now $m=1$ and $s>h$. By the first reduction achieved above we may assume that $h=0$, whence $s>0$. Let $\mathfrak{q}$ be the prime ideal of $Q$ in the ring $R=k[x_1, x_2, \cdots, x_n]=k[V]$. Since $Q$ is an algebraic point over $k$, $\mathfrak{q}$ is a maximal ideal in $R$. We pass to one of the rings $R'_i$ introduced in the course of the proof of Theorem 35 (§ 14, p. 95) (the ideal $\mathfrak{q}$ now plays the role of the prime ideal which in that proof was denoted by $\mathfrak{p}$). Using the same notations, we may assume that $R'_1 w_1 \cap R=\mathfrak{q}$. Let $\mathfrak{q}'_1$ be an isolated prime ideal of $R'_1 w_1$ such that $\mathfrak{q}'_1 \cap R=\mathfrak{q}$. Since $s \leqq r-1$, the ring $R'_1$ contains prime ideals of dimension $s$ which contain $\mathfrak{q}'_1$. We fix such a prime ideal $\mathfrak{q}'$ in $R'_1$. By the preceding part of the proof, there exists a place $\mathscr{P}$ of $k(V)$ of rank 1 and dimension $s$, such that $\mathscr{P}$ is finite on $R'_1$ and has center $\mathfrak{q}'$. Since $\mathfrak{q}$ is maximal in $R$, it follows from $\mathfrak{q}'_1 \cap R=\mathfrak{q}$ and $\mathfrak{q}'_1 \subset \mathfrak{q}'$ that $\mathfrak{q}' \cap R=\mathfrak{q}$, and hence $\mathfrak{q}$ is the center of $\mathscr{P}$ in $R$. This completes the proof in the case $m=1$.

For $m>1$, we shall use induction with respect to $m$. We therefore assume that there exists a specialization chain

$$\mathscr{P}_{m-1} \to \mathscr{P}_{m-2} \to \cdots \to \mathscr{P}_1$$

of $m-1$ places of $K/k$ such that $\mathscr{P}_j$ is of dimension $s_j$, of rank $m-j$, and has center $Q_j$ on $V (j=1, 2, \cdots, m-1)$. Let $\Sigma_1$ be the residue field of $\mathscr{P}_1$ and let $K_1=k(Q_1)$. We set

$$d = \max \{\dim Q_1/k+s-s_1, \dim Q/k\}. \tag{5}$$

Then $d$ is a non-negative integer, and we have

$$d \leqq \dim Q_1/k \tag{6}$$

since $s<s_1$ and $\dim Q/k \leqq \dim Q_1/k$, and

$$d \leqq s \tag{7}$$

since $\dim Q_1/k \leqq s_1$ and $\dim Q/k \leqq s$.

Now let $V_1/k$ be the irreducible variety having $Q_1$ as general point. Since $Q$ is a specialization of $Q_1$ over $k$, $Q$ is a point of $V_1$. From (5) and (6) it follows that $\dim Q/k \leqq d \leqq \dim Q_1/k$. If $d<\dim Q_1/k$, then, by the case $m=1$ of our theorem, there exists a place $\mathscr{P}'$ of $k(Q_1)/k$, of rank 1 and dimension $d$, such that the center of $\mathscr{P}'$ on $V_1$ is the point $Q$. If $d=\dim Q_1/k$, then it follows from (5) that necessarily $d=\dim Q/k$, for $\dim Q_1/k + s - s_1 < \dim Q_1/k$. Hence in this case, $Q$ and $Q_1$ have the same dimension over $k$ and are therefore $k$-isomorphic points. We then take for $\mathscr{P}'$ the $k$-isomorphism of the field $k(Q_1)$ which takes the point $Q_1$ into the point $Q$ ($\mathscr{P}'$—a trivial place of $k(Q_1)/k$, with center $Q$). In this case, $\mathscr{P}'/k$ still has dimension $d$, but the rank is zero.

By (5) and (7), we have

(8) $$0 \leqq s - d \leqq s_1 - \dim Q_1/k.$$

Since $s_1 - \dim Q_1/k$ is precisely the transcendence degree of $\Sigma_1/k(Q_1)$, it follows by Theorem 11, § 6 that there exists an extension $\bar{\mathscr{P}}$ of $\mathscr{P}'$ in $\Sigma_1$ which has relative dimension $s-d$. Note that in the case $d = \dim Q_1/k$, (in which case $\mathscr{P}'$ is a trivial place of $k(Q_1)/k$), we have $s-d<s_1-\dim Q_1/k$, and hence $\bar{\mathscr{P}}$ is not a trivial place of $\Sigma_1$. We set $\mathscr{P}=\mathscr{P}_1\bar{\mathscr{P}}$. Then $\mathscr{P}$ is a place of $K/k$, composite with $\mathscr{P}_1$, and it is clear that $Q$ is the center of $\mathscr{P}$ on $V$. We have $\dim \mathscr{P}/k = s$, since the residue field of $\bar{\mathscr{P}}$ has transcendence degree $s-d$ over the residue field of $\mathscr{P}'$, while the residue field of $\mathscr{P}'$ has transcendence degree $d$ over $k$. Now, if the extension $\bar{\mathscr{P}}$ of the place $\mathscr{P}'$ has exactly rank 1, then the rank of $\mathscr{P}$ is one greater than the rank of $\mathscr{P}_1$, i.e., the rank of $\mathscr{P}$ is $m$, and everything is proved. The rank of $\bar{\mathscr{P}}$ is certainly equal to 1 in the following case: $s_1 = \dim Q_1/k$. For, in that case we have $\dim Q/k \leqq s < s_1 = \dim Q_1/k$, whence $\mathscr{P}'$ is definitely a non-trivial place and hence has rank 1; and on the other hand, $\Sigma_1$ is now an algebraic extension of $k(Q_1)$, and therefore rank $\bar{\mathscr{P}}$ = rank $\mathscr{P}'$. The proof of the theorem is now therefore complete in the case $s_1 = \dim Q_1/k$. It follows that in order to complete the proof it will be sufficient to show the following: there exists a subring $R'$ of $k(x_1, x_2, \cdots, x_n)$ containing the ring $R = k[x_1, x_2, \cdots, x_n]$ and having the following properties: (1) $\mathscr{P}_1$ is finite on $R'$, and the center of $\mathscr{P}_1$ in $R'$ is a prime ideal $\mathfrak{q}'_1$ *which has dimension* $s_1$ (in other words: $\mathscr{P}_1$ is of the first kind with respect to $R'$); (2) $R'$ contains a prime ideal $\mathfrak{q}'$, of dimension $\leqq s$, such that $\mathfrak{q}' \cap R = \mathfrak{q}$ = prime ideal of $Q$ in $R$. For, if such a ring $R'$ exists, then by the preceding proof there will exist a place $\mathscr{P}$ of $K/k$, composite with $\mathscr{P}_1$ and having rank $m$, such that $\mathscr{P}$ has center $\mathfrak{q}'$ in $R'$ and has dimension $s$ over $k$. Then the center of $\mathscr{P}$ in $R$ will be necessarily $\mathfrak{q}$.

To show the existence of a ring $R'$ with the above properties, we fix a place $\tilde{\mathscr{P}} = \mathscr{P}_1\bar{\tilde{\mathscr{P}}}$ of $K/k$ which is composite with $\mathscr{P}_1$, has dimension $s$ over $k$, and has center $\mathfrak{q}$ in $R$ (the existence of such a place has just been shown above, independently of the condition $s_1 = \dim Q_1/k$). If $h = s_1 - \dim Q_1/k$, we fix $h$ elements $\bar{w}_1, \bar{w}_2, \cdots, \bar{w}_h$ in the residue field $\Sigma_1$ of $\mathscr{P}_1$ which are algebraically independent over $k(Q_1)$. We can also assume that these elements $\bar{w}_i$ belong to the valuation ring $\Sigma_{1\bar{\mathscr{P}}}$ of $\bar{\mathscr{P}}$. We then fix elements $w_1, w_2, \cdots, w_h$ in $k(x_1, x_2, \cdots, x_n)$ such that $\bar{w}_i = w_i\mathscr{P}_1$, and we set $R' = R[w_1, w_2, \cdots, w_h]$. It is immediately seen that this ring $R'$ satisfies our requirement (as prime ideal $\mathfrak{q}'$ we take the center of $\mathscr{P}$ in $R'$).

**§ 17. The abstract Riemann surface of a field.** Let $K$ be a field and $k$ a *subring* of $K$ (not necessarily a subfield). We denote by $S$ the set of all *non-trivial* valuations $v$ of $K$ which are non-negative on $k$, i.e., such that the valuation ring $R_v$ contains $k$. There is only one case in which $S$ is empty; it is the case in which $k$ is a field *and* $K$ is an algebraic extension of $k$ (Theorem 4, Corollary 1 and Theorem 5, Corollary 1, § 4). We shall exclude this case.

EXAMPLES: (1) *$k$ is a field.* In this case $S$ is the set of all non-trivial valuations of $K$ which are trivial on $k$. This is the case which occurs most frequently in algebraic geometry.

(2) *$k$ is a Dedekind domain.* In this case $S$ consists of valuations of two types: (a) valuations of $K$ which are trivial on the quotient field of $k$ and (b) valuations of $K$ which are extensions of the (discrete, rank 1) $\mathfrak{p}$-adic valuations of the quotient field of $k$, where $\mathfrak{p}$ is any proper prime ideal of $k$. The valuations of type (a) are missing if and only if $K$ is an algebraic extension of the quotient field of $k$; when they are present they have a residue field of the same characteristic as that of $K$. The characteristic of the residue field of a valuation of type (b) *may* be different from the characteristic of $K$: this case of *unequal characteristics* arises if and only if $K$ is of characteristic zero while the intersection of the prime ideal $\mathfrak{p}$ with the ring of (natural) integers is a prime ideal $(p)$ different from zero.

We shall now introduce a topology in the set $S$.

If $\mathfrak{o}$ is a subring of $K$, containing $k$, we denote by $E(\mathfrak{o})$ the set of all $v$ in $S$ such that $v$ is non-negative on $\mathfrak{o}$. We now let $\mathfrak{o}$ range over the family of all subrings of $K$ which contain $k$ *and are finitely generated over* $k$, and we take the family $E$ of corresponding sets $E(\mathfrak{o})$ as a basis of the open sets in $S$. We note that $E([\mathfrak{o}, \mathfrak{o}']) = E(\mathfrak{o}) \cap E(\mathfrak{o}')$, where $[\mathfrak{o}, \mathfrak{o}']$ denotes the ring generated by two given subrings $\mathfrak{o}, \mathfrak{o}'$ of $K$, and that $E(k) = S$. Therefore any finite intersection of basic open sets is itself a basic open set, and hence our choice of the basis $E$ defines indeed a topology in $S$. Note also that $\mathfrak{o} \subset \mathfrak{o}'$ implies $E(\mathfrak{o}) \supset E(\mathfrak{o}')$.

The topological space $S$ is called the *Riemann surface* of the field $K$ *relative to* $k$, or the Riemann surface of $K/k$.

We note that if $k'$ is the integral closure of $k$ in $K$ then the Riemann surface of $K/k'$ coincides with the Riemann surface $S$ of $K/k$ both as set *and* as topological space. The proof is straightforward.

We begin a study of the separation properties of $S$.

THEOREM 38. *The closure of an element $v$ of $S$* (i.e., *the closure of the set $\{v\}$ consisting of the single element $v$*) *is the set of all valuations $v' \in S$ which are composite with $v$.*

PROOF. Suppose that $v'$ is composite with $v$, so that we have for the corresponding valuation rings the inclusion $R_{v'} \subset R_v$. If $E(\mathfrak{o})$ is any basic open set such that $v$ belongs to the basic closed set $S-E(\mathfrak{o})$, then $R_v \not\supset \mathfrak{o}$, whence *a fortiori* $R_{v'} \not\supset \mathfrak{o}$, and thus $v' \in S-E(\mathfrak{o})$. Thus every basic closed set which contains $v$ necessarily contains $v'$, showing that $v'$ belongs to the closure of the set $\{v\}$. On the other hand, assume that $v'$ is not composite with $v$. We can then find an element $x$ of $K$ such that $v'(x)$ is non-negative while $v(x) < 0$. Then if we set $\mathfrak{o} = k[x]$ we will have $v \in S-E(\mathfrak{o})$, $v' \notin S-E(\mathfrak{o})$, and consequently $v'$ is not in the closure of the set $\{v\}$. This completes the proof.

We recall from topology that a topological space is said to be a $T_1$-space if every point of the space is a closed set. The following theorem will show that the Riemann surfaces which are $T_1$-spaces are, from an algebraic point of view, of a very special type.

THEOREM 39. *Let $k$ be an integrally closed subring of a field $K$. The Riemann surface $S$ of $K/k$ is a $T_1$-space if and only if one of the following two conditions is satisfied:*

(1) *$k$ is a field and $K/k$ has transcendence degree* 1; *or*

(2) *$k$ is a proper ring, $K$ is an algebraic extension of the quotient field of $k$, and for every proper prime ideal $\mathfrak{p}$ of $k$ it is true that the quotient ring $k_{\mathfrak{p}}$ is the valuation ring of a valuation of rank* 1.

PROOF. If condition (1) is satisfied then any valuation $v \in S$ has rank 1 (Corollary 1 of Definition 1, § 3). Hence, in this case $S$ is a $T_1$-space, by the preceding theorem.

Assume that condition (2) is satisfied, and let $v$ be any element of $S$. Since $v$ is non-trivial on $K$ and since $K$ is an algebraic extension of the quotient field of $k$, the center $\mathfrak{p}$ of $v$ in $k$ is not the zero ideal, hence $\mathfrak{p}$ is a proper prime ideal. If $v'$ is the restriction of $v$ to the quotient field of $k$ then $R_{v'} \supset k_{\mathfrak{p}}$, hence $R_{v'} = k_{\mathfrak{p}}$ since $k_{\mathfrak{p}}$ is a maximal subring of the quotient field of $k$ (p. 10). Thus $v'$, and hence also $v$, is of rank 1, whence again $S$ is a $T_1$-space.

Assume now that $S$ is a $T_1$-space. By the preceding theorem, every element $v$ of $S$ must be a valuation of rank 1. If $k$ is a field then the transcendence degree of $K/k$ cannot be greater than 1, for in the contrary case we can construct a valuation $v_0$ of $K/k$ whose residue field has positive transcendence degree over $k$, and compounding $v_0$ with a non-trivial valuation of the residue field of $v_0$ we would find a valuation of $K/k$ which has rank greater than 1. Suppose now that $k$ is a proper ring. Let $\mathfrak{p}$ be any proper prime ideal of $k$. If the quotient ring $k_{\mathfrak{p}}$ is not a valuation ring then there exists a valuation $v'$ of the quotient field of $k$ which has center $\mathfrak{p}$ in $k$ and which is of the second kind with respect to $k$ (Theorem 10, § 5). The residue field $\Delta$ of $v'$ is then of positive transcendence degree over the quotient field $k^\star$ of $k/\mathfrak{p}$. Compounding $v'$ with a non-trivial valuation of $\Delta/k^\star$ and extending the resulting composite valuation to a valuation of $K$ we find a valuation $v$ in $S$ which has rank $>1$, in contradiction with the preceding theorem. Hence $k_{\mathfrak{p}}$ is a valuation ring, and the corresponding valuation of the quotient field of $k$ must be of rank 1. Finally, $K$ must be an algebraic extension of the quotient field of $k$, for in the contrary case $S$ would contain valuations of rank $>1$, extensions of non-trivial valuations of the quotient field of $k$. This completes the proof.

Even in the special case in which $S$ is a $T_1$-space *it need not be a Hausdorff space.* Without attempting to give a complete classification of Hausdorff Riemann surfaces we shall make here only the following three observations:

(A) *In the case* (1) *of Theorem* 39 *the Riemann surface $S$ is never a Hausdorff space.* For, let $\mathfrak{o}=k[x_1, x_2, \cdots, x_n]$ and $\mathfrak{o}'=k[x'_1, x'_2, \cdots, x'_m]$ be two finitely generated subrings of $K$ and let $\mathfrak{o}^\star=k[x, x']=[\mathfrak{o}, \mathfrak{o}']$, whence $E(\mathfrak{o}^\star)$ is the intersection of $E(\mathfrak{o})$ and $E(\mathfrak{o}')$. If $\mathfrak{o}^\star$ is a proper ring then $E(\mathfrak{o}^\star)$ is non-empty. Assume that $\mathfrak{o}^\star$ is a field. From a result closely related to the Hilbert Nullstellensatz and proved in the next chapter it will follow that the generators $x_i$, $x'_j$ of $\mathfrak{o}^\star$ over $k$ are then necessarily algebraic over $k$ (see VII, § 3, Lemma, p. 165). Hence $K$ has positive transcendence degree over $\mathfrak{o}^\star$, and again $E(\mathfrak{o}^\star)$ is non-empty. We have thus shown that the intersection of any two non-empty basic open sets in $S$ is never empty. Hence $S$ is not a Hausdorff space.

Taking into account Theorem 39, *it follows that if $k$ is a field then $S$ is never a Hausdorff space.*

(B) Consider now the case (2) of Theorem 39. We may assume that *$k$ is integrally closed in $K$* (by a remark made earlier in this section). Then $K$ is the quotient field of $k$. If $S$ is a Hausdorff space then there must at least exist a pair of non-empty open sets in $S$, whence also a pair of non-empty basic open sets, having an empty intersection. In view of the relation $E(\mathfrak{o}) \cap E(\mathfrak{o}')=E([\mathfrak{o}, \mathfrak{o}'])$, it follows that *a necessary condition that $S$ be a Hausdorff space is that the field $K$ be a finitely generated ring extension of $k$.* It is obvious that in that case we have

$K=k[1/x]$, where $x$ is a suitable element of $k$, characterized by the property that it belongs to all the prime ideals $\mathfrak{p}$ of $k$, different from the zero ideal. However, the above condition may not be sufficient.

(C) *If $k$ is a proper ring of the type described in case* (2) *of Theorem* 39 *and if $K$ is any (finite or infinite) algebraic extension of the quotient field $L$ of $k$, then a sufficient condition for the Riemann surface of $K/k$ to be a Hausdorff space is that $k$ have only a finite number of prime ideals.* The statement is obvious if $K$ is a finite extension of $L$, for in that case the $T_1$-space $S$ has only a finite number of elements. In the infinite case, given two distinct elements $v'_1$ and $v'_2$ of $S$, there exists a field $F$ between $L$ and $K$, finite over $L$, such that the restrictions $v_1$ and $v_2$ of $v'_1$ and $v'_2$ to $F$ are distinct elements of the Riemann surface $S^\star$ of $F/k$. By the finite case, the elements $v_1$ and $v_2$ can be separated in $S^\star$ by two disjoint basic open sets. Taking the inverse images of these two open sets, under the restriction map $v' \to v =$ restriction of $v'$ in $F(v' \in S, v \in S^\star)$, we find in $S$ two basic open sets which are disjoint and separate $v'_1$ and $v'_2$.

Our next object is to prove the following theorem:

THEOREM 40. *The Riemann surface $S$ of $K/k$ is quasi-compact* (i.e., every open covering of $S$ contains a finite subcovering).

PROOF. Any valuation $v$ of $K$ is completely determined if one knows, for any element $x$ in $K$, whether $v(x)$ is positive, zero or negative. In other words, the elements $v$ in $S$ can be identified with certain mappings of $K$ into the set $Z$ consisting of the elements $-$, $0$, $+$. Using the customary notation $Z^K$ for the set of all mappings of a set $K$ into a set $Z$, we can therefore regard $S$ as a subset of $Z^K$. We now define a topology in $Z$ by taking as open subsets of $Z$ the empty set, the entire set $Z$ and the subset $\{0, +\}$, and introduce the corresponding usual topology in the product space $Z^K$. From the definition of the product topology it follows that in the induced topology on $S$ the basic open subsets are sets $E$ defined as follows: if $\{x_1, x_2, \cdots, x_n\}$ is any finite set of elements of $K$ then the set of all $v$ *in* $S$ such that $v(x_i) \in \{0, +\}$ is a set $E$. This agrees with our preceding definition of the topology of the Riemann surface $S$, and thus the latter is indeed a *subspace* of $Z^K$. To complete the proof we shall make *temporarily* two modifications in our definition of the space $S$:

(1) *We shall include in $S$ also the trivial valuation of $K$.* If we denote by $S^\star$ this enlarged set and define the topology of $S^\star$ in the same way as the topology of $S$ was defined, i.e., by means of subrings of $K$ which are finitely generated over $k$, we see at once that *every* basic open set in $S^\star$ contains the trivial valuation. Since $S$ is a subspace of $S^\star$, it

follows at once that $S$ is quasi compact if and only if $S^\star$ is quasi compact. We shall therefore prove the quasi compactness of $S^\star$. *In the rest of the proof we shall drop the asterisk*, so that temporarily (*until the end of the proof of the theorem*) it should be understood that $S$ contains the trivial valuation of $K$.

(2) *We shall also introduce in $Z$ a stronger topology* which will be Hausdorff, and we shall show that in the corresponding stronger topology of $Z^K$ the subset $S$ becomes a closed set. It will then follow, by Tychonoff's theorem, that in the induced stronger topology $S$ is compact (i.e., quasi compact and Hausdorff), whence *a fortiori* the Riemann surface $S$ is quasi compact (in its original weaker topology).

The stronger topology which we introduce in $Z$ shall be the discrete topology (every subset of $Z$ is open). For any $f$ in $Z^K$ the relation "$f \in S$" holds if and only if the following conditions are satisfied.

(a) The set of all $x$ in $K$ such that $f(x) \in \{0, +\}$ is closed under addition and multiplication.

(b) The above set contains $k$.

(c) If $f(x) \notin \{0, +\}$ (whence $x \neq 0$, by (b)) then $f(1/x) \in \{+\}$.

These conditions can be re-formulated as follows:

(a′) For any elements $x$, $y$ in $K$ we have either $f(x) = -$ or $f(y) = -$ or *both* $f(x+y)$ and $f(xy)$ are in $\{0, +\}$.

(b′) If $x \in k$ then $f(x) \in \{0, +\}$.

(c′) For any $x$ in $K$ either $f(x) \in \{0, +\}$ or $x \neq 0$ and $f(1/x) = +$.

For any $x$ in $K$ denote by $pr_x$ the mapping $f \to f(x)$ of $Z^K$ into $Z$. This is a continuous mapping. For any $x$ and $y$ in $K$ denote by $F_{x,y}$ the intersection of the following two subsets of $Z^K$.

$$pr_x^{-1}\{-\} \cup pr_y^{-1}\{-\} \cup pr_{x+y}^{-1}\{0, +\},$$
$$pr_x^{-1}\{-\} \cup pr_y^{-1}\{-\} \cup pr_{xy}^{-1}\{0, +\}.$$

The six sets which occur in the definition of $F_{x,y}$ are closed sets (since we have assigned to $Z$ the discrete topology). Hence $F_{x,y}$ is a closed set. Condition (a′) can now be written as follows:

(a″) $f$ belongs to the intersection of the sets $F_{x,y}$ ($x$ and $y$ arbitrary elements of $K$).

Similarly, conditions (b′) and (c′) can be written as follows:

(b″) $f$ belongs to the intersection of the sets $pr_x^{-1}\{0, +\}$, $x \in k$.

(c″) $f$ belongs to the intersection of the sets

$$pr_x^{-1}\{0, +\} \cup pr_{1/x}^{-1}\{+\}, \quad 0 \neq x \in K.$$

Thus $S$ is an intersection of closed sets and is therefore a closed set. This completes the proof of the theorem.

We shall now undertake a study of the Riemann surface $S$ from a different point of view. The objective of this study will be to show that $S$ can be regarded as the projective limit of an inverse system of certain topological spaces associated with finite subsets of $K$. The manner in which these spaces will be defined will be quite similar to that in which projective varieties are defined in algebraic geometry.

We mean by a *quasi-local ring* a commutative ring (noetherian or non-noetherian) with identity, in which the non-units form an ideal. Thus, every valuation ring is a quasi-local ring, and a quasi-local ring is a local ring (Vol. I, p. 228) if and only if it is noetherian. We consider the set $L$ of *all quasi-local rings* (noetherian or non-noetherian) between $k$ and $K$. For $P$ in $L$, we denote by $\mathfrak{m}(P)$ the (unique) maximal ideal of $P$. For $P, P' \in L$, we say that $P$ *dominates* $P'$ if $P' \subset P$ and $\mathfrak{m}(P') = P' \cap \mathfrak{m}(P)$. A subset $M$ of $L$ is said to be *irredundant* (resp., *complete*) if, for any valuation $v$ of $K/k$ (trivial or non-trivial), the valuation ring $R_v$ dominates at most one (resp., at least one) element of $M$. We say that a subset $M'$ of $L$ *dominates* a subset $M$ of $L$ and we write $M \leqq M'$ if every element of $M'$ dominates at least one element of $M$. This relation $M \leqq M'$ is obviously *transitive*. If we, furthermore, suppose that $M$ is irredundant, then, by the extension theorem (Theorem 5, § 4), the element $P$ of $M$ which is dominated by a given element $P'$ of $M'$ is unique; thus the transformation $P' \rightarrow P$ is a *mapping*, called the *domination mapping* and denoted by $d_{M',M}$. In the set of *irredundant* subsets of $L$, the relation $M \leqq M'$ defines a *partial ordering*; furthermore if $M, M', M''$ are irredundant subsets of $L$ such that $M \leqq M' \leqq M''$, then $d_{M'',M} = d_{M'',M'} d_{M',M}$. Notice, finally, that, if $M'$ dominates $M$ and if $M'$ is complete, then $M' d_{M',M}$ is complete.

We introduce in $L$ the following topology, which generalizes the topology we have defined on the Riemann-surface $S$ of $K/k$. If $\mathfrak{o}$ is any ring between $k$ and $K$, we denote by $L(\mathfrak{o})$ the subset of $L$ composed of all quasi-local rings $P$ containing $\mathfrak{o}$. We let $\mathfrak{o}$ range over the family of all subrings of $K$ which are *finitely generated over* $k$, and we take the family of corresponding sets $L(\mathfrak{o})$ as a basis for open sets in $L$. Since any finite intersection of sets $L(\mathfrak{o})$ ($\mathfrak{o}$ finitely generated) is a set of the same type, these sets constitute indeed a basis for open sets for a topology on $L$. When, in the sequel, a subset $M$ of $L$ is considered as a topological space, it is tacitly understood that its topology is induced by the topology of $L$.

The Riemann surface $S$ may be identified with a subset of $L$, and the topology on $S$ defined at the beginning of this section is obviously induced by the topology of $L$. Theorem 38 generalizes in the following way: *the closure of an element $P$ of $L$ is the set of all quasi-local rings $P'$ between $k$ and $P$*; the proof is similar to that of theorem 38.

For any ring $\mathfrak{o}$ between $k$ and $K$, we denote by $P(\mathfrak{o})$ the set of all prime ideals of $\mathfrak{o}$ which are $\neq 0$, and we assign to $P(\mathfrak{o})$ the following topology: a closed set is the set of all ideals $\mathfrak{p} \in P(\mathfrak{o})$ which contain a given ideal $\mathfrak{a}$; it is indeed clear that any intersection and any finite union of sets of this type is a set of the same type. We denote by $V(\mathfrak{o})$ the subset of $L$ composed of all quotient rings $\mathfrak{o}_{\mathfrak{p}}(\mathfrak{p} \in P(\mathfrak{o}))$.

LEMMA 1. *The mapping $f$ of $L(\mathfrak{o})$ into $P(\mathfrak{o})$ defined by $f(P) = \mathfrak{m}(P) \cap \mathfrak{o}$ is continuous. The restriction of $f$ to $V(\mathfrak{o})$ is a topological homeomorphism of $V(\mathfrak{o})$ onto $P(\mathfrak{o})$.*

PROOF. Any closed set in $P(\mathfrak{o})$ is an intersection of closed sets $F_x(x \in \mathfrak{o}, x \neq 0)$ of the following type: $F_x$ is the set of all prime ideals containing $x$. In order to prove that $f$ is continuous, it is sufficient to prove that $f^{-1}(F_x)$ is closed in $L(\mathfrak{o})$, i.e., that $f^{-1}(P(\mathfrak{o}) - F_x)$ is open. Now, for $P \in L(\mathfrak{o})$, the relations "$P \in f^{-1}(P(\mathfrak{o}) - F_x)$", "$x \notin \mathfrak{m}(P)$" and "$1/x \in P$" are equivalent, since $x \in \mathfrak{o} \subset P$; we thus have $f^{-1}(P(\mathfrak{o}) - F_x) = L(\mathfrak{o}) \cap L(k[1/x])$, which proves that the set is open.

Similarly, any basic open set in $V(\mathfrak{o})$ is a finite intersection of sets $U_x$ of the following type: $x$ is an element $\neq 0$ of the quotient field of $\mathfrak{o}$, and $U_x$ is the set of all $P \in V(\mathfrak{o})$ containing $x$. Since $f$ is a (1, 1) continuous mapping of $V(\mathfrak{o})$ onto $P(\mathfrak{o})$, to prove that $f$ is a homeomorphism it is therefore sufficient to prove that $f(V(\mathfrak{o}) - U_x)$ is closed. Now this follows from the fact that the relations "$\mathfrak{p} \in f(V(\mathfrak{o}) - U_x)$", "$x \notin \mathfrak{o}_{\mathfrak{p}}$" and "$\mathfrak{p}$ contains the ideal $\mathfrak{a}_x$ of all elements $d \in \mathfrak{o}$ such that $dx \in \mathfrak{o}$" are equivalent. Q.E.D.

For any ring $\mathfrak{o}$ between $k$ and $K$, the subset $V(\mathfrak{o})$ of $L$ is obviously irredundant. When $\mathfrak{o}$ is *finitely generated* over $k$, we say that $V(\mathfrak{o})$ is an *affine model* over $k$; the ring $\mathfrak{o}$, which is uniquely determined by $V(\mathfrak{o})$ since it is the intersection of all $P \in V(\mathfrak{o})$, is called the *defining ring* of the affine model $V(\mathfrak{o})$. A *model* $M$ over $k$ shall be by definition, any *irredundant* subset of $L$ which is a finite union $M = \bigcup_{i=1}^{n} V(\mathfrak{o}_i)$ of affine models over $k$.†

LEMMA 2. *For any model $M = \bigcup_{i=1}^{n} V(\mathfrak{o}_i)$ we have $M \cap L(\mathfrak{o}_i) = V(\mathfrak{o}_i)$, whence $V(\mathfrak{o}_i)$ is open in $M$. For a subset $H$ of $M$ to be open* (resp. *closed*) *in $M$, it is necessary and sufficient that $H \cap V(\mathfrak{o}_i)$ be open* (resp. *closed*) *in $V(\mathfrak{o}_i)$ for every $i$.*

† It may be easily proved that all the rings $\mathfrak{o}_i$ have then the same quotient field.

PROOF. The inclusion $V(\mathfrak{o}_i) \subset M \cap L(\mathfrak{o}_i)$ is obvious. Conversely, if $P \in M \cap L(\mathfrak{o}_i)$, $P$ contains $\mathfrak{o}_i$ and hence dominates the element $P' = (\mathfrak{o}_i)_{\mathfrak{p}'}$ of $V(\mathfrak{o}_i)$, where $\mathfrak{p}' = \mathfrak{m}(P) \cap \mathfrak{o}_i$. Since $M$ is irredundant, this implies $P = P'$, and proves the first assertion. The second assertion is now pure topology. The necessity of the condition is obvious. In the proof of the sufficiency it is enough to consider the case of open sets (replace $H$ by $M - H$). In this case, since $V(\mathfrak{o}_i)$ is open in $M$ and since $H \cap V(\mathfrak{o}_i)$ is open in $V(\mathfrak{o}_i)$, $H \cap V(\mathfrak{o}_i)$ is open in $M$, whence also $H$ is open in $M$, for $H$ is the union of the sets $H \cap V(\mathfrak{o}_i)$. Q.E.D.

LEMMA 3. *Let $M$ be a model and $M'$ a subset of $L$ which dominates $M$. Then the domination mapping $f = d_{M',M}$ is continuous.*

PROOF. Let $M = \bigcup_{i=1}^{n} V(\mathfrak{o}_i)$, where the $V(\mathfrak{o}_i)$'s are affine models, and let $U$ be an open set in $M$. We show that $f^{-1}(U)$ is open. Since $U$ is the union of the open sets $U \cap V(\mathfrak{o}_i)$ (Lemma 2), we may assume that $U$ is contained in some $V(\mathfrak{o}_i)$, say $V(\mathfrak{o}_1)$. Now, by Lemma 1, the mapping $g$ of $L(\mathfrak{o}_1)$ onto $V(\mathfrak{o}_1)$ defined by $g(P) = \mathfrak{o}_{1(\mathfrak{m}(P) \cap \mathfrak{o}_1)}$ is continuous. Since we obviously have $f^{-1}(U) = g^{-1}(U) \cap M'$, and since $L(\mathfrak{o}_1)$ is open in $L$, $f^{-1}(U)$ is open in $M'$. Q.E.D.

LEMMA 4. *Let $M$ be a complete model and let $f = d_{S,M}$ be the domination mapping of the Riemann surface $S$ into $M$. Then $f$ is continuous and closed.*

PROOF. The fact that $f$ is continuous is a particular case of Lemma 3. We thus have to prove that, for any closed set $F$ of $S$, $f(F)$ is closed in $M$. For any finite subset $I = \{x_1, \cdots, x_n\}$ of $K$, we denote by $F(I)$ the set of all valuations $v$ in $S$ such that $R_v$ does not contain $k[I]$; the sets $F(I)$ are the basic closed sets of $S$, whence $F$ is an intersection of such sets, say $F = \bigcap_{a \in A} F(I_a)$.

We first prove that, for any *finite intersection* $F'$ of basic closed sets of $S$, $f(F')$ is closed in $M$. We write $F' = \bigcap_{j=1}^{q} F(I_j)$, where $I_j = \{x_{j,1}, \cdots, x_{j,n(j)}\}$. Setting $F(x_{j,k}) = F(\{x_{j,k}\})$, we have $F(I_j) = F(x_{j,1}) \cup \cdots \cup F(x_{j,n(j)})$, whence

$$F' = \bigcap_{j=1}^{q} (F(x_{j,1}) \cup F(x_{j,2}) \cup \cdots \cup F(x_{j,n(j)})).$$

Using the distributivity of union with respect to intersection, we see that $F'$ is the union of the closed sets $G_s = F(x_{1,s(1)}) \cap F(x_{2,s(2)}) \cap \cdots \cap F(x_{q,s(q)})$, where $s$ ranges over the set $R$ of all integral valued mappings of $\{1, 2, \cdots, q\}$ such that $1 \leq s(j) \leq n(j)$ for $j = 1, \cdots, q$. Since $f(F') =$

$f(\bigcup_{s \in R} G_s) = \bigcup_{s \in R} f(G_s)$, and since $R$ is a finite set, it is sufficient to prove that each $f(G_s)$ is closed in $M$. To simplify notations we prove that, if $G = F(x_1) \cap F(x_2) \cap \cdots \cap F(x_q)(x_j \in K,\ x_j \neq 0)$, then $f(G)$ is closed. Notice that $G$ is the set of all valuations $v$ such that $v(x_j) < 0$ for every $j$, i.e., such that the valuation ideal $\mathfrak{M}_v$ contains all the elements $y_j = 1/x_j$.

For proving that $f(G)$ is closed in $M$, we use Lemma 2 and write $M = \bigcup_i V(\mathfrak{o}_i)$ where the $V(\mathfrak{o}_i)$ are affine models; it is sufficient to prove that $f(G) \cap V(\mathfrak{o}_i)$ is closed in $V(\mathfrak{o}_i)$ for any $i$. Let $\mathfrak{o}$ be any one of the rings $\mathfrak{o}_i$. We consider the ideal $\mathfrak{a} = \mathfrak{o} \cap \left(\sum_{j=1}^{q} y_j \cdot \mathfrak{o}[y_1, \cdots, y_q]\right)$ of $\mathfrak{o}$. If $P \in f(G) \cap V(\mathfrak{o})$, the prime ideal $\mathfrak{p} = \mathfrak{o} \cap \mathfrak{m}(P)$ is the center in $\mathfrak{o}$ of a valuation $v$ $(\in G)$ such that $\mathfrak{M}_v$ contains $y_1, \cdots, y_q$; then $\mathfrak{M}_v$ contains the ideal $\mathfrak{a}$, whence $\mathfrak{p}$ contains $\mathfrak{a}$. Conversely, if $\mathfrak{p}$ is a prime ideal in $\mathfrak{o}$ which contains $\mathfrak{a}$, it is easily seen that the ideal $\mathfrak{b}'$ of $\mathfrak{o}' = \mathfrak{o}[y_1, \cdots, y_q]$ generated by $\mathfrak{p}, y_1, \cdots, y_q$ contracts to $\mathfrak{p}$ in $\mathfrak{o}$. Thus the ideal $\mathfrak{b}' \cdot \mathfrak{o}'_{(\mathfrak{o}-\mathfrak{p})}$ is not the unit ideal of the quotient ring $\mathfrak{o}'_{(\mathfrak{o}-\mathfrak{p})}$ and is therefore contained in some maximal ideal $\mathfrak{M}'$ of $\mathfrak{o}'_{(\mathfrak{o}-\mathfrak{p})}$ (Vol. I, Ch. III, p. 151, Note I). By the extension theorem, $\mathfrak{M}'$ is the center *in* $\mathfrak{o}'_{(\mathfrak{o}-\mathfrak{p})}$ of some valuation $v$. The valuation ideal $\mathfrak{M}_v$ contains $y_1, \cdots, y_q$, whence $v \in G$; on the other hand $\mathfrak{p}$ is the center of $v$ in $\mathfrak{o}$. Therefore the quasi local ring $\mathfrak{o}_{\mathfrak{p}}$ belongs to $f(G) \cap V(\mathfrak{o})$. By Lemma 1, this proves that $f(G) \cap V(\mathfrak{o})$ is closed in $V(\mathfrak{o})$, as asserted.

To complete the proof, we have to pass to the case of an *infinite* intersection $F$ of basic closed sets, say $F = \bigcap_{a \in A} F(I_a)$ (where each $I_a$ is a finite subset of $K$). For every *finite subset* $B$ of the indexing set $A$, we denote by $F'_B$ the intersection of the sets $F(I_b)$, where $b$ ranges over $B$. We have $F = \bigcap_B F'_B$. The first part of the proof shows that $f(F'_B)$ is closed for every finite subset $B$ of $A$. It is therefore sufficient to prove that $f(F) = \bigcap_B f(F'_B)$. It is clear that the left-hand side of this relation is contained in the right-hand side. Conversely, let $P$ be an element of $M$ which belongs to $f(F'_B)$ for every $B$; this means that the subset $f^{-1}(P) \cap F'_B$ of $S$ is non-empty for every $B$. Since any finite intersection of sets $F'_B$ is itself a set of the same type, it follows that the family of sets $f^{-1}(P) \cap F'_B$ has the *finite intersection property*. Were the point $P$ of $M$ a closed set (equivalently: were $P$ a ring of quotients relative to a *maximal* ideal of one of the rings $\mathfrak{o}_i$ by which $M$ is defined) then all the sets of the above family would be closed in $S$ (since $f$ is continuous), and from the quasi-compactness of $S$ it would then follow that the sets of the collection have a non-empty intersection, i.e., $f^{-1}(P) \cap F$ is non-empty. Thus $P$ would belong to $f(F)$, and the proof would be complete. Bearing in mind this observation, we shall use the following device:

Let us denote by $k^\star$ the quasi-local ring $P$ and let $S^\star$ be the Riemann surface of $K/k^\star$. Then $S^\star$ is a subset of $S$ (since $k^\star \supset k$). If $\mathfrak{o} = k[z]$ is any finitely generated subring of $K$ and if we set $\mathfrak{o}^\star = k^\star[z]$, then $E^\star(\mathfrak{o}^\star) = E(\mathfrak{o}) \cap S^\star$, where $E^\star(\mathfrak{o}^\star)$ denotes the basic open set on $S^\star$ which is defined by $\mathfrak{o}^\star$. It follows that the topology of $S^\star$ is at least as strong as the topology induced on $S^\star$ by that of $S$.

We now set $\mathfrak{o}_i^\star = k^\star[\mathfrak{o}_i]$, $M^\star = \bigcup_i V^\star(\mathfrak{o}_i^\star)$, where the symbol $V^\star$ has the same meaning relative to the ring $k^\star$ as $V$ had relative to $K$. It is clear that $M^\star \subset M$, and since $S^\star \subset S$ and $M$ is irredundant, also $M^\star$ is irredundant. Since each $\mathfrak{o}_i^\star$ is finitely generated over $k^\star$, each $V^\star(\mathfrak{o}_i^\star)$ is an affine model over $k^\star$. *Therefore $M^\star$ is a model over $k^\star$.* If $\mathfrak{o}$ is one of the rings $\mathfrak{o}_i$ such that $P \supset \mathfrak{o}$ then $\mathfrak{o}^\star = P = k^\star$, the ideal $\mathfrak{m}(P)$ is a maximal ideal of $\mathfrak{o}^\star$ and therefore the point $P$ is a *closed* subset of $M^\star$. Now, it is obvious that if $f^\star$ is the domination mapping of $S^\star$ onto $M^\star$, then $f^{\star-1}(P) = f^{-1}(P)$. It follows that *$f^{-1}(P)$ is a closed subset of $S^\star$* and consequently also the sets $f^{-1}(P) \cap F_a^\star$ are closed subset of $S^\star$. Since $f^{-1}(P) \cap F_a^\star = f^{-1}(P) \cap F_a$, the sets $f^{-1}(P) \cap F_B'^\star$ coincide with the sets $f^{-1}(B) \cap F'_B$, and since the collection of the former has the finite intersection property, it follows, by the quasi-compactness of $M^\star$, that $f^{-1}(P) \cap F$ is non-empty. This completes the proof.

LEMMA 5. *If $M$ and $M'$ are two complete models such that $M'$ dominates $M$, then the domination mapping $d_{M',M}$ is both continuous and closed.*

PROOF. In fact, the continuity of $d_{M',M}$ follows from Lemma 3. On the other hand, if $F'$ is a closed subset of $M'$, we have $d_{M',M}(F') = d_{S,M}(d_{S,M'}^{-1}(F'))$, whence $d_{M',M}(F')$ is closed since $d_{S,M'}$ is continuous (Lemma 3) and since $d_{S,M}$ is closed (Lemma 4).

Among the complete models of $K$, we are going to single out a particularly interesting class of models, the *projective models*. Given a non-empty finite set $\{x_0, x_1, \cdots, x_n\}$ composed of *non-zero* elements of $K$, we set $\mathfrak{o}_i = k[x_0/x_i, x_1/x_i, \cdots, x_n/x_i]$ $(i = 0, 1, \cdots, n)$ and $M = \bigcup_{i=0}^{n} V(\mathfrak{o}_i)$. We prove that $M$ is a complete model.

(a) $M$ is *irredundant*. If fact, if $P$ and $P'$ are two elements of $M$ which are dominated by the same valuation ring $R_v$, $P$ and $P'$ cannot belong to the same affine model $V(\mathfrak{o}_i)$; so we have, for example, $P \in V(\mathfrak{o}_0)$

and $P' \in V(\mathfrak{o}_1)$. We set $\mathfrak{o}=k[\mathfrak{o}_0, \mathfrak{o}_1]$. The local rings $P$ and $P'$ are dominated by the quotient ring $\mathfrak{o}_\mathfrak{p}$ of $\mathfrak{o}$, where $\mathfrak{p}=\mathfrak{o} \cap \mathfrak{M}_v$. Since $\mathfrak{o}$ contains $x_1/x_0$ and $x_0/x_1$, these elements are units in $\mathfrak{o}$, hence also in $\mathfrak{o}_\mathfrak{p}$. Since $P$ contains $x_1/x_0$ and is dominated by $\mathfrak{o}_\mathfrak{p}$, it follows that $x_1/x_0$ is a unit in $P$; therefore, since $x_j/x_0 \in P$, we have $x_j/x_1=(x_j/x_0)/(x_1/x_0) \in P$ for every $j$, whence $P$ contains $\mathfrak{o}_1$ and consequently $\mathfrak{o}$. From the inclusions $\mathfrak{o} \subset P \subset \mathfrak{o}_\mathfrak{p}$ and from the fact that $\mathfrak{o}_\mathfrak{p}$ dominates $P$ we conclude that $\mathfrak{m}(P) \cap \mathfrak{o}=\mathfrak{p}$, whence the elements of $\mathfrak{o}-\mathfrak{p}$ are units in $P$. Therefore $P$ contains $\mathfrak{o}_\mathfrak{p}$, whence $P=\mathfrak{o}_\mathfrak{p}$. In a similar way, we see that $P'=\mathfrak{o}_\mathfrak{p}$. Consequently $P=P'$ and $M$ is irredundant.

(b) $M$ is *complete*. In fact, given any valuation $v$ of $K/k$, we choose an index $j$ for which $v(x_j)$ takes its least value. We then have $v(x_i/x_j) \geq 0$ for every $i$, whence $\mathfrak{o}_j \subset R_v$. Therefore the element $P=(\mathfrak{o}_j)_{(\mathfrak{o}_j \cap \mathfrak{M}_v)}$ of $M$ is dominated by $R_v$, and $M$ is complete.

From (a) and (b) it follows that $M$ is a *complete model*; we say that $M$ is the *projective model* over $k$ determined by $\{x_0, \cdots, x_n\}$.

We denote by $C$ (resp. $C'$) the set of all complete (resp. projective) models over $k$; it is clear that $C'$ is a subset of $C$. Both are ordered sets for the order relation $M \leq M'$.

LEMMA 6. *Let* $M=\bigcup_i V(\mathfrak{o}_i)$ *and* $M'=\bigcup_j V(\mathfrak{o}'_j)$ *be two models over* $k$. *We set* $\mathfrak{o}_{ij}=k[\mathfrak{o}_i, \mathfrak{o}'_j]$. *Then* $M''=\bigcup_{i,j} V(\mathfrak{o}_{ij})$ *is a model which dominates* $M$ *and* $M'$ *and is such that every subset* $N$ *of* $L$ *which dominates both* $M$ *and* $M'$ *dominates* $M''$. *If* $M$ *and* $M'$ *are affine* (resp. *complete, projective*), *so is* $M''$.

PROOF. We first show that $M''$ dominates both $M$ and $M'$. Given $P'' \in M''$, $P''$ belongs to some $V(\mathfrak{o}_{ij})$ whence contains some $\mathfrak{o}_i$; then $P''$ dominates the element $(\mathfrak{o}_i)_{(\mathfrak{m}(P'') \cap \mathfrak{o}_i)}$ of $M$; similarly for $M'$.

Now let $N$ be a subset of $L$ which dominates both $M$ and $M'$. Given $Q$ in $N$, $Q$ dominates some $P \in M$ and some $P' \in M'$; let $i$ and $j$ be indices such that $P \in V(\mathfrak{o}_i)$ and $P' \in V(\mathfrak{o}'_j)$. Then $Q$ contains both $\mathfrak{o}_i$ and $\mathfrak{o}'_j$, whence also $\mathfrak{o}_{ij}$. Consequently $Q$ dominates the element $(\mathfrak{o}_{ij})_{(\mathfrak{m}(Q) \cap \mathfrak{o}_{ij})}$ of $M''$.

In order to show that $M''$ is a model we have to show that it is irredundant. Let $P_1''$ and $P_2''$ be two elements of $M''$ which are dominated by the same valuation ring $R_v$ and let, for $s=1, 2$, $P''_s$ dominate $P_s \in M$ and $P'_s \in M'$. Since $P_1$ and $P_2$ are dominated by $R_v$, and since $M$ is a model, we have $P_1=P_2$; similarly $P'_1=P_2'$. If $i$ and $j$ are indices such that $P''_1$ belongs to $V(\mathfrak{o}_{ij})$, then we have seen that $P_1$ is a quotient ring of $\mathfrak{o}_i$ and $P'_1$ a quotient ring of $\mathfrak{o}'_j$. From the inclusion $\mathfrak{o}_{ij} \subset k[P_1, P'_1] \subset$

$P''_1$, and the fact that $P''_1$ is a quotient ring of $\mathfrak{o}_{ij}$, we deduce that $P''_1$ is a quotient ring of $k[P_1, P'_1]$, necessarily with respect to a prime ideal $\mathfrak{q}_1$; we obviously have $\mathfrak{q}_1 = k[P_1, P'_1] \cap \mathfrak{m}(P''_1)$, whence $\mathfrak{q}_1 = k[P_1, P'_1] \cap \mathfrak{M}_v$. Similarly $P''_2$ is also a quotient ring $k[P_1, P'_1]_{\mathfrak{q}_2}$, and we also have $\mathfrak{q}_2 = k[P_1, P'_1] \cap \mathfrak{M}_v$. Consequently $\mathfrak{q}_1 = \mathfrak{q}_2$, whence $P''_1 = P''_2$. This proves that $M''$ is irredundant.

We have thus proved that $M''$ is the least upper bound of $M$ and $M'$ in the ordered set of all models. This proves the *uniqueness* of $M''$; in particular $M''$ is *independent* of the representations of $M$ and $M'$ as finite unions of affine models.

Now, if $M$ and $M'$ are affine models, say $M = V(\mathfrak{o})$ and $M' = V(\mathfrak{o}')$, we have $M'' = V(k[\mathfrak{o}, \mathfrak{o}'])$, whence $M''$ is an affine model.

Let us now suppose that $M$ and $M'$ are projective models, respectively determined by $\{x_0, \cdots, x_n\}$ and $\{x'_0, \cdots, x'_q\}$. Setting $\mathfrak{o}_i = k[x_0/x_i, \cdots, x_n/x_i]$ and $\mathfrak{o}'_j = k[x'_0/x'_j, \cdots, x'_q/x'_j]$, the ring $\mathfrak{o}_{ij} = k[\mathfrak{o}_i, \mathfrak{o}'_j]$ is obviously equal to $k[x_0x'_0/x_ix'_j, \cdots, x_sx'_t/x_ix'_j, \cdots, x_nx'_q/x_ix'_j]$. Therefore $M''$ is the projective model determined by the set consisting of the $(n+1)(q+1)$ elements $x_sx'_t$.

Suppose finally that $M$ and $M'$ are complete. This means that the Riemann surface $S$ dominates both $M$ and $M'$. From what has been seen above, it follows that $S$ dominates $M''$, whence that $M''$ is complete. Q.E.D.

The model $M''$ defined in Lemma 6 is called the *join* of $M$ and $M'$ and is denoted by $J(M, M')$. The join of a finite number of models is defined inductively and enjoys the same properties as the join of two models. It is immediate that if $M'$ dominates $M$ then $J(M, M') = M'$. In particular, $J(M, M) = M$.

LEMMA 7 ("Chow's lemma"). *For any complete model $M$ there exists a projective model $M'$ which dominates $M$.*

PROOF. In fact, let us write $M = \bigcup_{i=1}^{q} V(\mathfrak{o}_i)$, where $\mathfrak{o}_i = k[x_{i,1}, \cdots, x_{i,n(i)}]$. We may assume that the elements $x_{i,j}$ are $\neq 0$. Let $M_i$ be the projective model determined by $\{1, x_{i,1}, \cdots, x_{i,n(i)}\}$. Then $V(\mathfrak{o}_i)$ is a subset of $M_i$. We take for $M'$ the join of all the projective models $M_i$ (whence $M'$ is a projective model, by Lemma 6). If $P'$ is any element of $M'$, then by Lemma 6, $P'$ dominates an element $P_i$ of $M_i$ for every $i$. Now let $R_v$ be a valuation ring which dominates $P'$. Since $M$ is complete, $R_v$ dominates some element $P$ of $M$; let $i$ be an index such that $P \in V(\mathfrak{o}_i)$. Since $P$ and $P_i$ are two elements of a model $M_i$ which are dominated by the same valuation ring $R_v$, they are equal. Therefore $P'$ dominates the element $P$ of $M$. Q.E.D.

It may be shown by examples that there exist complete models which are not projective (see M. Nagata, "Existence theorems for non-projective complete algebraic varieties," *Illinois J. of Mathematics*, Dec. 1958).

Lemma 6 shows that the ordered sets $C$ and $C'$ of all complete models and of all projective models respectively, are *directed sets*. Lemma 7 shows that $C'$ is a *cofinal* subset of $C$.

In view of these properties, the partially ordered set $C$ and the continuous mappings $d_{M',M}$ $(M, M' \in C, M \leqq M')$ give rise to an *inverse system* of topological spaces. The limit space of this inverse system, or the *projective limit* of the spaces $M \in C$ with respect to the mappings $d_{M',M}$ is then defined as the set $S(C)$ of all those points $P^0 = \{P_M; P_M \in M\}$ of the product $\prod_{M \in C} M$ which satisfy the relations $P_M = d_{M',M}(P_{M'})$, whenever $M \leq M'$; the topology in $S(C)$ is defined as the one induced in $S(C)$ by the usual product topology in the product space. We shall denote by $f_M$ the projection $P \to P_M$ of $S(C)$ into $M$. By definition of $S(C)$ we have $f_M = f_{M'} d_{M',M}$ whenever $M \leq M'$.

We define in an entirely similar way the projective limit $S(C')$ of the projective models $M \in C'$, and denote by $f'_M$ the natural mapping of $S(C')$ into $M$. Since $C'$ is a cofinal subset of $C$, the elementary theory of projective limits shows the existence of a natural homeomorphism of $S(C)$ onto $S(C')$. But we shall not need this elementary fact, as we are going to prove that both $S(C)$ and $S(C')$ are naturally homeomorphic to the Riemann surface $S$ of $K$.

In fact, given any element $v$ of $S$, the system of quasi-local rings $\{d_{S,M}(R_v)\}$ $(M \in C)$ is a point of $S(C)$ since we have $d_{S,M} = d_{S,M'} d_{M',M}$ whenever $M \leq M'$. We have thus a mapping $g$ of $S$ into $S(C)$, defined by $g(v) = \{d_{S,M}(R_v)\}$. Similarly, we obtain a mapping $g'$ of $S$ into $S(C')$.

THEOREM 41. *The mappings $g$ and $g'$ are topological homeomorphisms of $S$ onto $S(C)$ and $S(C')$ respectively.*

PROOF. We give the proof for $S(C')$, the proof for $S(C)$ being entirely analogous. Let $P^0 = \{P_M\}(M \in C')$ be a point of $S(C')$. Using the fact that $C'$ is a directed set we find that the union of the quasi local rings $P_M$ is a ring $\mathfrak{o}$, and that the union $\mathfrak{m}$ of their maximal ideals $\mathfrak{m}(P_M)$ is the ideal of non-units of $\mathfrak{o}$. Hence, there exists a valuation $v$ of $K$ such that $R_v$ dominates $\mathfrak{o}$. Therefore $R_v$ dominates each $P_M$; in other words, we have $g'(v) = P^0$. This shows that $g'$ maps $S$ *onto* $S(C')$.

Let $v$ and $v'$ be two distinct elements of $S$. We have either $R_v \not\subset R_{v'}$ or $R_{v'} \not\subset R_v$; thus there exists an element $x$ of $K$ which is contained in one and only one of the rings $R_v$ and $R_{v'}$. Then it is immediately seen that $v$ and $v'$ dominate distinct elements of the projective model $M$

determined by $\{1, x\}$. Consequently, $g'(v) \neq g'(v')$. Hence $g'$ is *one-to-one.*

Since all the mappings $d_{S,M}$ are continuous (Lemma 4), their "product mapping" $g': v \to \{d_{S,M}(R_v)\}$ is a continuous mapping of $S$ into $\prod_{M \in C'} M$, whence also a continuous mapping of $S$ onto the subspace $S(C')$.

It remains to be proved that $g'$ is *closed.* Let $F$ be a closed subset of $S$. We obviously have $g'(F) = S(C') \cap \left(\prod_{M \in C'} d_{S,M}(F)\right)$. By Lemma 4 each set $d_{S,M}(F)$ is closed, whence also the product of these sets is closed. Therefore $g'(F)$ is a closed subset of $S(C')$. Q.E.D.

NOTE: For further details concerning Riemann surfaces, and for applications of the compactness theorem 40 in Algebraic geometry (specifically, in the problem of local uniformization), see O. Zariski, "The compactness of the Riemann manifold of an abstract field of algebraic functions" (*Bull. Amer. Math. Soc.*, 1944) and "Local uniformization on algebraic varieties" (*Annals of Mathematics*, 1940).

**§ 18. Derived normal models.** Let $V/k$ be an affine variety (defined over a ground field $k$) in the affine $n$-space $A_n^K$ ($K$—a universal domain; see § 5$^{\text{bis}}$). Let $\mathfrak{o} = k[x_1, x_2, \cdots, x_n]$ be the coördinate ring of $V/k$; here $(x_1, x_2, \cdots, x_n)$ is a general point of $V/k$ and the $x_i$ may be assumed to belong to $K$ (since $K$ is a universal domain). Using the notations of § 5$^{\text{bis}}$ and of the preceding section, we have a natural mapping of $V$ onto the affine model $V(\mathfrak{o})$: to each point $Q$ of $V$ we let correspond its local ring $\mathfrak{o}(Q; V)$ on $V/k$. Two points of $V$ are then mapped into one and the same element of $V(\mathfrak{o})$ if and only if they are $k$-isomorphic points (§ 5bis). Thus, the affine model $V(\mathfrak{o})$ is obtained from the affine variety $V/k$ by identification of $k$-isomorphic points.

At the end of § 14 we have introduced implicitly (and we shall do that in more detail in VII, §§ 4 and 4$^{\text{bis}}$) the notion of a projective variety $V^*/k$, in the projective $n$-space $P_n^K$ over $K$, as the union of $n+1$ affine varieties $V_i$ $(i = 0, 1, \cdots, n)$ immersed in $P_n^K$. We start, namely, from a set of $n$ quantities $x_1, x_2, \cdots, x_n$ in $K$ and we define $V_i$ as the set of all points $(z_0, z_1, \cdots, z_{i-1}, 1, z_{i+1}, \cdots, z_n)$ in $P_n^K$ (the coördinates being homogeneous) such that the $n$-tuple $(z_0, z_1, \cdots, z_{i-1}, z_{i+1}, \cdots, z_n)$ is a specialization, over $k$, of the $n$-tuple

$$\left(\frac{x_0}{x_i}, \frac{x_1}{x_i}, \cdots, \frac{x_{i-1}}{x_i}, \frac{x_{i+1}}{x_i}, \cdots, \frac{x_n}{x_i}\right),$$

where $x_0 = 1$ (note for $i = 0$ this means the $n$-tuple $(x_1, x_2, \cdots, x_n)$). Thus $V_i$ lies in the affine space $P_n^K - H_i$, where $H_i$ is the hyperplane

$Y_i=0$, and if we take as non-homogeneous coördinates in that affine space the quotients $Y_0/Y_i, Y_1/Y_i, \cdots, Y_{i-1}/Y_i, Y_{i+1}/Y_i, \cdots, Y_n/Y_i$, then $(x_0/x_i, x_1/x_i, \cdots, x_{i-1}/x_i, x_{i+1}/x_i, \cdots, x_n/x_i)$ is a general point of $V_i/k$. It is then easily seen that there is a natural mapping of $V^\star$ onto the projective model $M$ determined by the set $\{x_0, x_1, \cdots, x_n\}$ and that, again, two points of $V^\star$ are mapped into one and the same point of $M$ if and only if they are $k$-isomorphic.

By analogy with our definition of normal varieties, given in § 14, we can define normality for the general models, over $k$, introduced in the preceding section ($k$ is now a ring, not necessarily a field). A model $M$ is normal if each element of $M$ is an *integrally closed* quasi-local domain. It is immediately seen that Theorem 34 of § 14 continues to be valid for these, more general models; we have, namely, that an affine model $V(\mathfrak{o})$ is normal if and only if $\mathfrak{o}$ is an integrally closed ring.

The concept of a derived normal model is of importance in algebraic geometry. We shall introduce this concept here with reference to the more general type of models considered in the preceding section. We shall find it convenient to denote the "ground ring" not by $k$ but by some other letter, and denote by $k$ the field of quotients of the ground ring. This will facilitate references to some theorems proved in volume I. We shall therefore denote the ground ring by $R$. Following Nagata ("A general theory of Algebraic Geometry over Dedekind domains," I, *American Journal of Mathematics*, vol. 58 (1956), p. 79 and p. 86), we will impose on $R$ the following conditions: (1) *$R$ is noetherian;* (2) *if $F$ is any finite algebraic extension of the quotient field of $R$ then the integral closure of $R$ in $F$ is a finite $R$-module.* We shall refer to an integral domain $R$ satisfying these two conditions as a *restricted domain.*

We note first of all that the "normalization lemma" proved in Volume I (Ch. V, § 4, Theorem 8) continues to be valid if the infinite field $k$ of that lemma is replaced by an infinite ground ring $R$, and the proof remains substantially the same. For the convenience of the reader we shall now restate the "normalization lemma" in the more general form in which it is now needed.

*Let $A=R[x_1, x_2, \cdots, x_n]$ be an integral domain, finitely generated over an infinite domain $R$, and let $d$ be the transcendence degree of the field of quotients of $A$ over the field of quotients $k$ of $R$. There exist $d$ linear combinations $y_1, y_2, \cdots, y_d$ of the $x_i$ with coefficients in $R$, such that $A$ is integral over $R[y_1, y_2, \cdots, y_d]$. If the field $k(x_1, x_2, \cdots, x_n)$ is separably generated over $k$, the $y_j$ may be chosen in such a way that $k(x_1, x_2, \cdots, x_n)$ is a separable extension of $k(y_1, y_2, \cdots, y_d)$.*

[Only the following modifications must be made in the proof of the normalization theorem as given in volume I: (a) It is permissible to assume that the polynomial $P(U, X_1, X_2, \cdots, X_n)$ has coefficients in $R$. (b) The elements $a_i$ $(i=1, 2, \cdots, n)$ must now be suitably chosen in $R$; this is possible, by Theorem 14 of Vol. I, Ch. I, § 18, since $R$ has infinitely many elements.]

With the aid of this generalized normalization theorem we can now also extend Theorem 9 of Vol. I, Ch. V, § 4 in the following form:

*Let $R$ be a restricted domain, $A=R[x_1, x_2, \cdots, x_n]$ an integral domain which is finitely generated over $R$, and let $F$ be a finite algebraic extension of the quotient field $k(x_1, x_2, \cdots, x_n)$ of $A$, where $k$ is the quotient field of $R$. Then the integral closure $A'$ of $A$ in $F$ is a finite $A$-module (and is therefore finitely generated over $R$).*

Again, the proof is substantially the same as that of the cited Theorem 9 of Vol. I, Ch. V, § 4. We shall give here only those extra steps or modifications in the proof that are needed for the complete proof of the above generalized statement.

(a) In the reduction to the case in which $F$ is the quotient field of $A$ we must take a basis $\{y_1, y_2, \cdots, y_q\}$ of $F$ over $k(x_1, x_2, \cdots, x_n)$ composed of elements which are integral over $A$ (and not merely over $k[x_1, x_2, \cdots, x_n]$). It is obvious that such a basis can be obtained by first finding a basis consisting of elements which are integral over $k[x_1, x_2, \cdots, x_n]$ and by multiplying each element of that basis by a suitable element of $R$.

(b) Assuming that we have already $F=$quotient field of $A$, we may furthermore replace $R$ by the integral closure $\bar{R}$ of $R$ and $A$ by $\bar{R}[x_1, x_2, \cdots, x_n]$. For, the algebraic closure, in $F$, of the quotient field $k$ of $R$, is a finite algebraic extension of $k$, and therefore $\bar{R}$ is a finite $R$-module ($R$ being a restricted domain). It is clear that $\bar{R}$ is also a restricted domain, and since the integral closure of $A$ in $F$ is the same as the integral closure of $\bar{R}[x_1, x_2, \cdots, x_n]$ in $F$, it is sufficient to prove that the integral closure in question is a finite module over $\bar{R}[x_1, x_2, \cdots x_n]$. We may therefore assume that $R$ is an integrally closed domain.

(c) In the next part of the proof the additional hypothesis is made to the effect that $R$ is an infinite domain and that $F\ (=k(x_1, x_2, \cdots, x_n))$ is separably generated over $k$ ($=$quotient field of $R$). Using the generalized normalization theorem, stated above, we find elements $z_1, z_2, \cdots z_d$ in $A$ such that $A$ is integral over the ring $B=R[z_1, z_2, \cdots, z_d]$ and such that $\{z_1, z_2, \cdots, z_d\}$ is a separating transcendence basis of $F/k(z_1, z_2, \cdots, z_d)$.

Then Corollary 1 of Theorem 7 (Vol. I, Ch. V, § 4) is applicable provided it is proved that $B$ is an *integrally closed* domain. We observe that $z_1, z_2, \cdots z_d$ *are algebraically independent over* $R$ *and that* $R$ *is integrally closed.* To prove that *this implies that* $R[z_1, z_2, \cdots, z_d]$ *is also integrally closed* it is sufficient to consider the case $d=1$. Let then $B=R[z]$, where $z$ is a transcendental over $R$, and let $\xi$ be an element of the integral closure of $R[z]$ (in the quotient field of $R[z]$). Then necessarily $\xi \in k[z]$. Let then $\xi=f(z)=a_0z^q+a_1z^{q-1}+\cdots+a_q$, where the $a_i$ are in $k$. The ring $B[\xi]$ is a finite $B$-module. Since $B[\xi]\subset k[z]$, the finiteness of the $B$-module $B[\xi]$ implies the existence of an element $d$ of $R$, $d\neq 0$, such that $dB[\xi]\subset B$.

In particular, $d\xi^i \in B$ for $i=1, 2, \cdots$. Since $z$ is transcendental over $R$ it follows from this that $da_0{}^i \in R$, for $i=1, 2, \cdots$. This implies that $a_0$ is integral over $R$, since $R$ is noetherian. Therefore $a_0 \in R$, $a_1z^{q-1}+\cdots+a_q$ is integral over $B$, and in a similar fashion it follows that $a_1, a_2, \cdots, a_q \in R$, which proves our assertion.†

Having settled these algebraic preliminaries, we now consider an affine model $V(\mathfrak{o})$, where $\mathfrak{o}$ is a ring between the ground ring $R$ and $K$, finitely generated over $R$. Let $F$ be a subfield of $K$ which is a finite algebraic extension of the quotient field of $\mathfrak{o}$, and let $\bar{\mathfrak{o}}$ be the integral closure of $\mathfrak{o}$ in $F$. Since we have just proved that $\bar{\mathfrak{o}}$ is a finite $\mathfrak{o}$-module (and hence is finitely generated over $R$), $\bar{\mathfrak{o}}$ is the defining ring of an affine model $V(\bar{\mathfrak{o}})$. This affine model is, of course, normal and is called *the derived normal model of* $V(\mathfrak{o})$ in $F$.

Let now $M=\bigcup_{i=1}^{n} V(\mathfrak{o}_i)$ be an arbitrary model over $R$. It has been pointed out in § 17 that the rings $\mathfrak{o}_i$ have necessarily the same quotient field. This field will be denoted by $R(M)$. Let $F$ be a subfield of $K$ which is a finite algebraic extension of $R(M)$, and let $\bar{\mathfrak{o}}_i$ be the integral closure of $\mathfrak{o}_i$ in $F$. We consider the finite union $M'=\bigcup_{i=1}^{n} V(\bar{\mathfrak{o}}_i)$ of affine models $V(\bar{\mathfrak{o}}_i)$. It is clear that $M'$ dominates $M$, for if $P'$ is any element of $M'$ and if, say, $P'=\bar{\mathfrak{o}}_{i,\bar{\mathfrak{p}}}$, where $\bar{\mathfrak{p}}$ is a prime ideal of $\bar{\mathfrak{o}}_i$, then $P'$ dominates the element $\mathfrak{o}_{i,\mathfrak{p}}$ of $M$, where $\mathfrak{p}=\bar{\mathfrak{p}}\cap\mathfrak{o}_i$. We now show that $M'$ is an *irredundant set, and is therefore a model over* $R$. Let $v$ be any valuation of $K/R$ such that the valuation ring $R_v$ dominates some element $P'$ of $M'$. Then $R_v$ dominates one and only one element $P$ of $M$ (since $M'\geq M$ and since $M$ is an irredundant subset of $L$). Let,

† We note that the assertion that $R[z]$ is integrally closed has already been proved earlier (§ 13, Theorem 29, part (a)) by valuation-theoretic methods, *without* the assumption that $R$ is noetherian.

say, $P' \in V(\bar{\mathfrak{o}}_i)$. Then $P' = \bar{\mathfrak{o}}_{i,\bar{\mathfrak{p}}}$ where $\bar{\mathfrak{p}}$ is a prime ideal in $\bar{\mathfrak{o}}_i$, and $P = \mathfrak{o}_{i,\mathfrak{p}}$ where $\mathfrak{p} = \bar{\mathfrak{p}} \cap \mathfrak{o}_i$. It is clear that $P'$ contains as subring the integral closure $\bar{P}$ of $P$ in $F$. Let $\mathfrak{P} = \mathfrak{m}(P') \cap \bar{P}$. The prime ideal $\mathfrak{P}$ in $\bar{P}$ is the center of $v$ in $\bar{P}$ and is thus uniquely determined by $v$. It is a *maximal* ideal in $\bar{P}$ since $\mathfrak{P} \cap P = \mathfrak{m}(P)$. We have $\mathfrak{m}(P') \cap \bar{P} = \mathfrak{P}$, whence $P'$ dominates the local ring $\bar{P}_{\mathfrak{P}}$. On the other hand, we have that $\bar{\mathfrak{o}}_i$ is a subring of $\bar{P}$ and that $\mathfrak{P} \cap \bar{\mathfrak{o}}_i = \bar{\mathfrak{p}}$ (since $\mathfrak{P}$ and $\bar{\mathfrak{p}}$ are the centers of $v$ in $\bar{P}$ and $\bar{\mathfrak{o}}_i$ respectively). Therefore $\bar{P}_{\mathfrak{P}}$ dominates $P'$. It follows that $P' = \bar{P}_{\mathfrak{P}}$, showing that $P'$ is uniquely determined and that $M'$ is therefore an irredundant subset of $L$.

The given model $M$ may possibly admit more than one representation as a finite union of affine models. However, the model $M'$ which we have just constructed, starting from a given representation of $M = \bigcup_{i=1}^{n} V(\mathfrak{o}_i)$, *depends only on $M$ and the field $F$*. For, the above proof of the irredundant character of $M'$ shows clearly that $M'$ is the set of all local rings $\bar{P}_{\mathfrak{P}}$, where $\bar{P}$ ranges over the set of integral closures, in $F$, of the elements $P$ of $M$, and where, for a given $\bar{P}$, $\mathfrak{P}$ ranges over the set of all maximal ideals of $\bar{P}$.

The model $M'$, constructed above, is called the *derived normal model* of $M$, in $F$, and will be denoted by $N(M, F)$. We repeat that $F$ must be assumed to be a finite algebraic extension of $R(M)$.

If $M$ and $M'$ are models over $R$ and $M'$ dominates $M$, we say that *$M'$ is complete over $M$* if every valuation ring $R_v$ ($v$—a valuation of $K/R$) which dominates an element of $M$ dominates also an element of $M'$. It is clear that *$N(M, F)$ is complete over $M$*. For, let $v$ be any valuation of $K/R$ such that $R_v$ dominates an element $P$ of $M$. Then $v$ has a center $\mathfrak{P}$ in the integral closure $\bar{P}$ of $P$ in $F$ (where $\mathfrak{P}$ is necessarily a maximal ideal in $\bar{P}$, since $\mathfrak{P} \cap P = \mathfrak{m}(P)$), and thus $R_v$ dominates the element $\bar{P}_{\mathfrak{P}}$ of $N(M, F)$.

In particular, it follows that if *$M$ is a complete model then also $N(M, F)$ is a complete model.*

THEOREM 41. *Let $M$ and $M'$ be two models over $R$ such that $M'$ is normal and such that the field $R(M')$ is a finite algebraic extension $F$ of the field $R(M)$. Then $M'$ is the derived normal model $N(M, F)$ of $M$ in $F$ if and only if the following condition is satisfied: if a normal model $M''$ dominates $M$ and is such that $R(M'') \supset F$, then $M''$ also dominates $M'$.*

PROOF. Let $M' = N(M, F)$, let $P''$ be any element of $M''$ and let $P$ be the element of $M$ which is dominated by $P''$. Since $R(M'') \supset F$ and $P''$ is integrally closed in its quotient field $R(M'')$, $P''$ contains the integral closure $\bar{P}$ of $P$ in $F$. We have $\mathfrak{m}(P'') \cap \bar{P} \cap P = \mathfrak{m}(P'') \cap P = \mathfrak{m}(P)$, showing that $\mathfrak{m}(P'') \cap \bar{P}$ is a maximal ideal $\mathfrak{P}$ of $\bar{P}$. Hence $P''$ dominates the element $\bar{P}_{\mathfrak{P}}$ of $M'$, showing that $M''$ dominates $M'$.

Conversely, assume that $M'$ satisfies the stated condition and denote by $M^\star$ the derived normal model $N(M, F)$. By our assumption, as applied to $M''=M^\star$, we have that $M^\star$ dominates $M'$. On the other hand, since $M'$ dominates $M$ and $R(M')=F$, it follows, from what we have just proved, that $M'$ dominates $M^\star$. Using the fact that both $M'$ and $M^\star$ are irredundant subsets of $L$ we conclude that $M'=M^\star$.

THEOREM 42. *If $M$ is a projective model, also $N(M, F)$ is a projective model.*

PROOF. Let $M$ be a projective model, over $R$, determined by $\{x_0, x_1, \cdots, x_n\}$, so that $M=\bigcup_{i=0}^{n} V(\mathfrak{o}_i)$, where $\mathfrak{o}_i=R[x_0/x_i, x_1/x_i, \cdots, x_n/x_i]$. Let $\bar{\mathfrak{o}}_i$ be the integral closure of $\mathfrak{o}_i$ in $F$. Then $N(M, F)=\bigcup_{i=0}^{n} V(\bar{\mathfrak{o}}_i)$. Let $\{w_{i1}, w_{i2}, \cdots,\}$ be a finite module basis of $\bar{\mathfrak{o}}_i$ over $\mathfrak{o}_i$. If $i$ and $j$ are any two indices in the set $(0, 1, \cdots, n)$ and if $w_i$ is any element of $\bar{\mathfrak{o}}_i$, then upon writing the relation of integral dependence of $w_i$ over $\mathfrak{o}_i$ we see at once that for all sufficiently high integers $q$ the elements $w_i x_i^q/x_j^q$ belong to $\bar{\mathfrak{o}}_j$. We can therefore choose a large integer $q$ such that $w_{i\nu_i} x_i^q/x_j^q \in \bar{\mathfrak{o}}_j$, for $i=0, 1, \cdots, n$ and for all $w_{i\nu_i}$ in the set $\{w_{i1}, w_{i2}, \cdots\}$. We denote by $z_0, z_1, \cdots, z_m$ the various monomials $x_0^{\alpha_0}x_1^{\alpha_1}\cdots x_n^{\alpha_n}$ of degree $q$, where we assume that $z_i=x_i^q$, $i=0, 1, \cdots, n$. We denote by $z_{m+1}, z_{m+2}, \cdots, z_N$ the various products $w_{i\nu_i}x_i^q$ $(i=0, 1, \cdots, n;\ \nu_i=1, 2, \cdots)$ and we consider the projective model $M'$ determined by the set $\{z_0, z_1, \cdots, z_N\}$. Let $\mathfrak{o}'_i=R\left[\frac{z_0}{z_i}, \frac{z_1}{z_i}, \cdots, \frac{z_N}{z_i}\right]$, $i=0, 1, \cdots, n$. We have $z_s/z_i \in \mathfrak{o}_i$ for $s=0, 1, 2, \cdots, m$ (since $z_i=x_i^q$ for $i=0, 1, \cdots, n$, and $z_s$ is a monomial in $x_0, x_1, \cdots, x_n$, of degree $q$, for $s=0, 1, 2, \cdots, m$). We also have $z_s/z_i \in \bar{\mathfrak{o}}_i$ for $s>m$, since $z_s/z_i$ is an element of the form $w_{j\nu_j}x_j^q/x_i^q$ for some $j=0, 1, \cdots, n$. Furthermore, the set of elements $z_s/z_i$, $s>m$, includes the basis $w_{i1}, w_{i2}, \cdots,$ of $\bar{\mathfrak{o}}_i$ over $\mathfrak{o}_i$. Hence $\mathfrak{o}'_i=\bar{\mathfrak{o}}_i$. Thus $M'\supset V(\mathfrak{o}'_i)=V(\bar{\mathfrak{o}}_i)$, $i=0, 1, \cdots, n$, and consequently $M'\supset N(M, F)$. Since $M'$ is irredundant and $N(M, F)$ is complete, it follows that $M'=N(M, F)$. This completes the proof.

Another proof of Theorem 42 will be given at the end of VII, § 4bis.

# VII. POLYNOMIAL AND POWER SERIES RINGS

Among commutative rings, the polynomial rings in a finite number of indeterminates enjoy important special properties and are frequently used in applications. As they are also of paramount importance in Algebraic Geometry, polynomial rings have been intensively studied. On the other hand, rings of formal power series have been extensively used in "algebroid geometry" and have many properties which are parallel to those of polynomial rings. In the first section of this chapter we shall define formal power series rings and we shall show that the main properties of polynomial rings which have been derived in previous chapters (see, in particular, Vol. I, Ch. I, §§ 16–18) hold also for formal power series rings. In the later sections of this chapter we shall give deeper properties of polynomial rings and, whenever possible, the parallel properties of power series rings.

§ **1. Formal power series.** Let $A$ be a (commutative) ring with element 1 and let $R=A[X_1, X_2, \cdots, X_n]$ be the polynomial ring in $n$ indeterminates over A. By a *formal power series in n indeterminates over A* we mean an infinite sequence $f=(f_0, f_1, \cdots, f_q, \cdots)$ of *homogeneous* polynomials $f_q$ in $R$, each polynomial $f_q$ being either 0 or of degree $q$. We define addition and multiplication of two power series $f=(f_0, f_1, \cdots, f_q, \cdots)$ and $g=(g_0, g_1, \cdots, g_q, \cdots)$ as follows:

(1) $$f+g = (f_0+g_0, f_1+g_1, \cdots, f_q+g_q, \cdots),$$

(2) $$fg = (h_0, h_1, \cdots, h_q, \cdots), \text{ where } h_q = \sum_{i+j=q} f_i g_j.$$

It is easily seen that with these definitions of addition and multiplication the set $S$ of all formal power series in $n$ indeterminates over $A$ becomes a *commutative ring*. This ring $S$, called the *ring of formal power series in n indeterminates over A*, shall be denoted by $A[[X_1, X_2, \cdots, X_n]]$. The *zero* of $S$ is the sequence $(0, 0, \cdots)$, and $(1, 0, 0, \cdots,)$ is the multiplicative identity of $S$.

Polynomials in $X_1, X_2, \cdots, X_n$, with coefficients in $A$, can be identified with formal power series, as follows: if $f \in A[X_1, X_2, \cdots, X_n]$ and $f = f_0 + f_1 + \cdots + f_m$, where each $f_i$ is a form which is either zero or of degree $i$, then we identify $f$ with the power series $(f_0, f_1, \cdots, f_m, 0, 0, \cdots)$. By this identification the polynomial ring $R = A[X_1, X_2, \cdots, X_n]$ becomes a subring of the power series ring $S = A[[X_1, X_2, \cdots, X_n]]$.

REMARK. If the ring $A$ is the field of real or complex numbers, then the power series $f$ which are *convergent* in a suitable neighborhood of the origin $X_1 = X_2 = \cdots = X_n = 0$ become an object of study. It can be shown that the convergent power series form a subring $S'$ of $S$ (this subring obviously contains all the polynomials). Most of the results proved in this section (in particular, the Weierstrass preparation theorem and its consequences) hold also for $S'$.

Let $f = (f_0, f_1, \cdots, f_q, \cdots)$ be a *non-zero* power series. The smallest index $q$ for which $f_q$ is different from zero will be called the *order* of $f$ and will be denoted by $\mathbf{o}(f)$. If $i = \mathbf{o}(f)$, then the form $f_i$ is called the *initial form* of $f$. We agree to attach the order $+\infty$ to the element 0 of $S$.

THEOREM 1. *If $f$ and $g$ are power series in $A[[X_1, X_2, \cdots, X_n]]$, then*

$$\mathbf{o}(f+g) \geqq \min\{\mathbf{o}(f), \mathbf{o}(g)\}, \tag{3}$$

$$\mathbf{o}(fg) \geqq \mathbf{o}(f) + \mathbf{o}(g). \tag{4}$$

*Furthermore, if $A$ is an integral domain then also $S$ is an integral domain and we have*

$$\mathbf{o}(fg) = \mathbf{o}(f) + \mathbf{o}(g). \tag{4'}$$

PROOF. The proofs of (3) and (4) are straightforward and are similar to the proofs given for polynomial rings in Vol. I, Ch. I (see, for instance, I, § 18, proof of Theorem 11; the only difference in the proof is that now we have to use the initial forms rather than the homogeneous components of highest degree). As to (4′), we observe that if $f \neq 0$ and $g \neq 0$ then the product $f_i g_j$ of the initial forms of $f$ and $g$ is different from zero (since the polynomial ring $A[X_1, X_2, \cdots, X_n]$ is an integral domain if $A$ is an integral domain) and is the initial form of $fg$.

The power series of *positive* order form an ideal in $S$. This ideal is generated by $X_1, X_2, \cdots, X_n$ and shall be denoted by $\mathfrak{X}$. For any integer $q \geqq 1$, the ideal $\mathfrak{X}^q$ consists of those power series which have order $\geqq q$. It follows that $\bigcap_{q=1}^{\infty} \mathfrak{X}^q = (0)$.

THEOREM 2. *If $f=(f_0, f_1, \cdots, f_q, \cdots)$ is a power series, then $f$ is a unit in $S$ if and only if the element $f_0$ of $A$ is a unit in $A$.*

PROOF. If $fg=1$, with $g=(g_0, g_1, \cdots, g_q, \cdots)$, then $f_0g_0=1$, and hence $f_0$ is a unit in $A$. Conversely, if $f_0$ is a unit in $A$, then we can find successively forms $g_0, g_1, \cdots, g_q, \cdots$, where $g_q$ is either zero or a form of degree $q$, such that $g_0f_0=1, g_1f_0+g_0f_1=0, \cdots, g_qf_0+g_{q-1}f_1+\cdots+g_0f_q=0, \cdots$. In fact, we have $g_0=f_0^{-1}$. Assuming that $g_0, g_1, \cdots, g_{q-1}$ have already been determined and that each $g_i$ is either zero or a form of degree $i$ $(0 \leqq i \leqq q-1)$, we set $g_q=-f_0^{-1}(g_{q-1}f_1+\cdots+g_0f_q)$, and it is clear that $g_q$ is then either zero or a form of degree $q$. If we now set $g=(g_0, g_1, \cdots, g_q, \cdots)$ then we find, by (2), that $fg=1$. This completes the proof.

COROLLARY 1. *If $k$ is a field, then the units of the power series ring $k[[X_1, X_2, \cdots, X_n]]$ are the power series of order 0. The ring $k[[X_1, X_2, \cdots, X_n]]$ is a local ring, and the ideal $\mathfrak{X}$ generated by $X_1, X_2, \cdots, X_n$ is its maximal ideal.*

Everything follows directly from Theorem 2 except the assertion (implicit in the statement that $k[[X_1, X_2, \cdots, X_n]]$ is a local ring) that $k[[X_1, X_2, \cdots, X_n]]$ is noetherian. This will be proved later on in this section (see Theorem 4).

COROLLARY 2. *If $k$ is a field and $S=k[[X]]$ is the power series ring in one indeterminate, then $\mathfrak{X}$ is the principal ideal $SX$, and every ideal in $S$ is a power of $\mathfrak{X}$. In other words, $S$ is a discrete valuation ring, of rank 1, and its non-trivial ideals are the ideals $SX^q$.*

Everything follows directly from Theorem 2 and from properties of $p$-adic valuations in unique factorization domains ($p$—an irreducible element; see VI, § 9, Examples of valuations, 2), by observing that if $f$ is a non-zero element of $k[[X]]$, of order $q$, then $f=X^qg$, where $g$ is a unit.

The valuation of which $k[[X]]$ is the valuation ring is the one in which the value of any non-zero element $f$ of $k[[X]]$ is the order $\mathbf{o}(f)$ of $f$. Now, Theorem 1 shows that, more generally, *if $A$ is an integral domain* and $S=A[[X_1, X_2, \cdots, X_n]]$ is the power series ring in any number of indeterminates over $A$, then the mapping $f \rightarrow \mathbf{o}(f)$ can be extended uniquely to a valuation of the quotient field of $S$ (in general, however, $S$ will *not* be the valuation ring of that valuation). If we denote by $\mathbf{o}$ that valuation, then it is clear that the center of $\mathbf{o}$ in $S$ (see Ch. VI, § 5) is the maximal ideal $\mathfrak{X}$ of $S$. We shall refer to this valuation $\mathbf{o}$ as the *$\mathfrak{X}$-adic valuation* of $S$ (or of the quotient field of $S$). It is clear that the valuation $\mathbf{o}$ is trivial on $A$, and hence we may assume that the residue field of $\mathbf{o}$ contains the quotient field of $A$.

THEOREM 3. *The quotients* $X_i/X_n$, $i=1, 2, \cdots, n-1$, *belong to the valuation ring of the* $\mathfrak{X}$*-adic valuation* $\mathbf{o}$. *If* $t_i$ *denotes the residue of* $X_i/X_n$ *in the valuation* $\mathbf{o}$, *then* $t_1, t_2, \cdots, t_{n-1}$ *are algebraically independent over* $A$, *and the residue field of* $\mathbf{o}$ *is* $k(t_1, t_2, \cdots, t_{n-1})$, *where* $k$ *is the quotient field of* $A$ ($A$, *an integral domain*).

PROOF. Since $\mathbf{o}(X_i)=1$, $i=1, 2, \cdots, n$, $\mathbf{o}(X_i/X_n)=0$, and the first assertion is proved. Let now $F(X_1, X_2, \cdots, X_{n-1})$ be any non-zero polynomial in $n-1$ indeterminates, with coefficients in $A$, and let $m$ be the degree of $F$. We set $g=g_m(X_1, X_2, \cdots, X_n)=X_n^m F(X_1/X_n, X_2/X_n, \cdots, X_{n-1}/X_n)$. Then $g$ is a *form* of degree $m$ in $X_1, X_2, \cdots, X_n$, with coefficients in $A$. We have $\mathbf{o}(g)=m=\mathbf{o}(X_n^m)$, hence the $\mathbf{o}$-residue of the quotient $g/X_n^m$ is different from zero. Since $g/X_n^m = F(X_1/X_n, X_2/X_n, \cdots, X_{n-1}/X_n)$ and since $\mathbf{o}$ is trivial on $A$, it follows that $F(t_1, t_2, \cdots, t_{n-1})\neq 0$, showing that $t_1, t_2, \cdots, t_{n-1}$ are algebraically independent over $A$.

The field $k(t_1, t_2, \cdots, t_{n-1})$ is contained in the residue field of $\mathbf{o}$, and it remains to show that these two fields coincide. Let $\xi$ be any element of the residue field of $\mathbf{o}$, $\xi\neq 0$, and let $f$ and $g$ be elements of $A[[X_1, X_2, \cdots, X_n]]$ such that $\xi$ is the $\mathbf{o}$-residue of $f/g$. Since $\xi\neq 0$, we must have $\mathbf{o}(f)=\mathbf{o}(g)$. Let $\mathbf{o}(f)=q$. Then both $f/X_n^q$ and $g/X_n^q$ have non-zero $\mathbf{o}$-residues, and the quotient of these two residues is $\xi$. It is therefore sufficient to show that the residues of $f/X_n^q$ and $g/X_n^q$ both belong to $k(t_1, t_2, \cdots, t_{n-1})$. Consider, for instance, $f/X_n^q$. Let $f_q$ be the initial form of $f$. Then $\mathbf{o}(f-f_q) > q$, whence the $\mathbf{o}$-residue of $f/X_n^q$ coincides with the $\mathbf{o}$-residue of $f_q/X_n^q$. Since $f_q(X_1, X_2, \cdots, X_n)/X_n^q=f_q(X_1/X_n, X_2/X_n, \cdots, X_{n-1}/X_n, 1)$, the $\mathbf{o}$-residue of $f_q/X_n^q$ is $f_q(t_1, t_2, \cdots, t_{n-1}, 1)$ and belongs therefore to $A(t_1, t_2, \cdots, t_{n-1})$. This completes the proof.

We note that the restriction of $\mathbf{o}$ to the polynomial ring $R=A[X_1, X_2, \cdots, X_n]$ is a prime divisor of the field $k(X_1, X_2, \cdots, X_n)$, with the same residue field as $\mathbf{o}$, and that if $n>1$ then this prime divisor is of the second kind with respect to the ring $R$, its center in $R$ being the point $X_1=X_2=\cdots=X_n=0$ (see Ch. VI, § 14).

We now go back to the general case, in which $A$ is an arbitrary ring. If we take the set of ideals $\mathfrak{X}^q$, $q=0, 1, 2, \cdots$, as a fundamental system of neighborhoods of the element 0 of $S$, then, by Theorem 1, $S$ becomes a *topological ring* (S. L. Pontrjagin, *Topological groups*, p. 172). [Elements "near" a given element $f_0$ of $S$ are those elements $f$ for which $f-f_0$ has high order. Since we have $\mathbf{o}((f+g)-(f_0+g_0))\geqq \min\{\mathbf{o}(f-f_0), \mathbf{o}(g-g_0)\}$, $\mathbf{o}(fg-f_0g_0)=\mathbf{o}(f(g-g_0)+g_0(f-f_0))\geqq \min\{\mathbf{o}(f)+\mathbf{o}(g-g_0), \mathbf{o}(g_0)+\mathbf{o}(f-f_0)\}$, and both $\mathbf{o}(f)$ and $\mathbf{o}(g_0)$ are non-negative integers, it follows that $f+g$ and $fg$ are near $f_0+g_0$ and $f_0g_0$ respectively provided $f$ and $g$ are sufficiently near $f_0$ and $g_0$; in other words, the ring operations in $S$ are indeed continuous.] Note that in view of the relation $\bigcap_{q=1}^{\infty} \mathfrak{X}^q=(0)$, $S$ is a Hausdorff space. As a matter of fact, the topology of $S$ can be induced by a suitable metric in $S$; namely, fix a real number $r>1$ and define the distance $d(f, g)$ between any two elements $f, g$ of $S$ by the formula $d(f, g)=r^{-q}$, where $q=\mathbf{o}(f-g)$.

*The space S is complete*, i.e., every Cauchy sequence $\{f^i\}$ of elements $f^i$ of $S$ converges in $S$. For let $f^i=(f_0{}^i, f_1{}^i, \cdots, f_q{}^i, \cdots)$. Since we are dealing with a Cauchy sequence, we must have $f_q{}^i=f_q{}^j$ for all $i, j \geqq n(q)$, where $n(q)$ is an integer depending on $q$. We set $f_q=f_q{}^i$ for $i=n(q)$ and $f=(f_0, f_1, \cdots, f_q, \cdots)$. Then $\mathbf{o}(f-f^i)>q$ if $i \geqq \max \{n(0), n(1), \cdots, n(q)\}$, showing that the sequence $\{f^i\}$ converges to $f$.

It follows in the usual way that if $\{f^i\}$ and $\{g^i\}$ are two Cauchy sequences, then

(5) $$\operatorname{Lim}(f^i+g^i) = \operatorname{Lim} f^i+\operatorname{Lim} g^i,$$

(5′) $$\operatorname{Lim} f^ig^i = \operatorname{Lim} f^i \cdot \operatorname{Lim} g^i.$$

Let now $\{h^i\}$ be an infinite sequence of power series satisfying the sole condition that $\mathbf{o}(h^i)$ *tends to* $\infty$ *with* $i$; in other words, $\{h^i\}$ is a Cauchy sequence whose limit is the element 0 of $S$. Then the partial sums $f^i=h^0+h^1+\cdots+h^i$ clearly form a Cauchy sequence. We express this by saying that the *infinite series* $h^0+h^1+\cdots+h^i+\cdots$ *is convergent* and we define the *infinite sum* $\sum_i h^i$ to be the limit $f$ of the sequence $\{f^i\}$:

$$\sum_{i=0}^{\infty} h^i = \lim_{i\to+\infty} (h^0+h^1+\cdots+h^i), \quad \text{if} \quad \mathbf{o}(h^i) \to +\infty.$$

It follows easily from the definition of $\sum_i h^i$ that this infinite sum is independent of the order in which the elements of the sequence $\{h^i\}$ are written. We have the usual rules of addition and multiplication of infinite series:

(6) $$\sum_i g^i+\sum_i h^i = \sum_i (g^i+h^i),$$

(6′) $$\sum_i g^i \cdot \sum_i h^i = \sum_i (g^0h^i+g^1h^{i-1}+\cdots+g^ih^0).$$

Relation (6) follows directly from (5). As to (6′), the left-hand side is, by (5′), the limit of the Cauchy sequence $\{\varphi^q\}$, where $\varphi^q=\sum_{i=0}^{q} g^i \cdot \sum_{i=0}^{q} h^i$

$= \sum_{i,j=0}^{q} g^i h^j$, while the right-hand side is the limit of the sequence $\{\psi^q\}$, where $\psi^q = \sum_{i+j \leqq q} g^i h^j$. Hence $\varphi^q - \psi^q$ is a sum of terms $g^i h^j$ in which at least one of the integers $i, j$ is $\geqq [q/2]$. Since $\mathbf{o}(g^i)$ and $\mathbf{o}(h^i)$ tend to $+\infty$ with $i$, it follows that the two sequences $\{\varphi^q\}$ and $\{\psi^q\}$ have the same limit, and this proves (6′).

We note that (6′) implies the distributive law

$$(6'') \qquad h \sum_i g^i = \sum_i hg^i.$$

We also note that if we have $h^i = 0$ for all sufficiently large values of $i$, say for $i > m$, so that the sequence $\{h^i\}$ is essentially a finite sequence, then the infinite sum $\sum_i h^i$ coincides with the sum of the elements $h^0, h^1, \cdots, h^m$ in the ring $S$.

We note that the inequality (3) generalizes to infinite sums, i.e., we have for any convergent series $\sum_i h^i$:

$$(7) \qquad \mathbf{o}\left(\sum_i h^i\right) \geqq \min_i\{\mathbf{o}(h^i)\}.$$

The notion of infinite sums allows us to write every power series $f = (f_0, f_1, f_2, \cdots, f_q, \cdots)$, where $f_q$ is a form of degree $q$ (or is zero), as an infinite sum; namely, we have

$$(8) \qquad f = \sum_{i=0}^{\infty} f_i, \quad \text{or} \quad f = f_0 + f_1 + \cdots + f_q + \cdots.$$

In this form, $f$ appears as an actual power series in $X_1, X_2, \cdots, X_n$. The partial sums $f^i$ are now polynomials $f_0 + f_1 + \cdots + f_i$. Each monomial which occurs in any of the forms $f_q$ will be called a *term* of the power series $f$.

In (8), every element $f$ of $S$ is represented as a limit of polynomials. Hence *$S$ is the closure of the polynomial ring $R = A[X_1, X_2, \cdots, X_n]$*, or—equivalently—*$R$ is everywhere dense in $S$*. The following characterization of subrings of $S$ which are everywhere dense in $S$ will be used in the sequel:

LEMMA 1. *A subring $L$ of $S$ is everywhere dense in $S$ if and only if $L$ has the following property: if $f_q$ is any form in $X_1, X_2, \cdots, X_n$, with coefficients in $A$, then $L$ contains at least one element whose initial form is $f_q$.*

PROOF. Assume that $L$ is everywhere dense in $S$ and let $f_q$ be a form, of degree $q$. If $n$ is an integer $> q$, $L$ must contain an element $f$ such that $\mathbf{o}(f - f_q) \geqq n$ (since $f_q$ must be the limit of a sequence of elements of $L$). Since $n>q$, the inequality $\mathbf{o}(f-f_q)\geqq n$ implies that $f_q$ is the initial form of $f$. Note that in this part of the proof we have not used the assumption that $L$ is a subring of $S$.

Conversely, assume that $L$ has the property stated in the lemma. Let $f$ be any element of $S$. We shall construct an infinite sequence $\{f^i\}$, $f^i\in L$, such that $\mathbf{o}(f-f^i)\geqq i$, whence $f=\operatorname{Lim} f^i$. For $i=0$ we simply set $f^0=0$. Let us assume that we have already defined the $n$ elements $f^0, f^1, \cdots, f^{n-1}$ in $L$ and that we have then $\mathbf{o}(f-f^i)\geqq i$ for $i=0, 1, \cdots, n-1$. If $\mathbf{o}(f-f^{n-1})\geqq n$ we set $f^n=f^{n-1}$. If $\mathbf{o}(f-f^{n-1})=n-1$, let $g_{n-1}$ be the initial form of $f-f^{n-1}$ and let $h^{n-1}$ be some element of $L$ whose initial form is $g_{n-1}$. If we set $f^n=f^{n-1}+h^{n-1}$, then $f^n\in L$, since $L$ is a subring of $S$, and we have $\mathbf{o}(f-f^n)=\mathbf{o}(f-f^{n-1}-h^{n-1})\geqq n$, since both $f-f^{n-1}$ and $h^{n-1}$ are of order $n-1$ and have the same initial form $g_{n-1}$. This completes the proof of the lemma.

We have seen in Vol. I, Ch. I that in any polynomial in $A[X_1, X_2, \ldots, X_n]$ one can substitute for the indeterminates elements of any overring of $A$ (see Vol. I, Ch. I, § 16, end of section). This operation of substitution cannot be performed for power series without further ado since infinite sums of power series have a meaning only if their partial sums form a Cauchy sequence (hence converge, in the formal sense explained above). Consider the power series ring $A[[Y_1, Y_2, \cdots, Y_m]]$ in $m$ indeterminates and $m$ power series $f^1(X_1, X_2, \cdots, X_n)$, $f^2(X_1, X_2, \cdots, X_n), \cdots, f^m(X_1, X_2, \cdots, X_n)$ in $n$ indeterminates, over $A$. *We assume that each of the $m$ power series $f^i$ is of order* $\geqq 1$. Under this assumption we proceed to define $g(f^1, f^2, \cdots, f^m)$, $g(Y_1, Y_2, \cdots, Y_m)$ being any power series in $A[[Y_1, Y_2, \cdots, Y_m]]$. Let $g=g_0+g_1+\cdots+g_q+\cdots$, $g_q$ being either zero or a form of degree $q$ in $Y_1, Y_2, \cdots, Y_m$, with coefficients in $A$. Then $g_q(f^1, f^2, \cdots, f^m)$ is defined as an element $\bar{g}^q$ of $A[[X_1, X_2, \cdots, X_n]]$. Furthermore, by Theorem 1, $\bar{g}^q$ is a power series of order $\geqq q$, since $g_q$ is a form of degree $q$ and since $\mathbf{o}(f^i)\geqq 1$, $1\leqq i\leqq m$. Hence the series $\sum_q \bar{g}^q$ is defined as an element of $A[[X_1, X_2, \cdots, X_n]]$. This power series $\sum_q \bar{g}^q$ in $A[[X_1, X_2, \cdots, X_n]]$ we call the *result of substitution of $f^1, f^2, \cdots f^m$ into $g(Y_1, Y_2, \cdots, Y_m)$, or the transform of $g(Y_1, Y_2, \cdots, Y_m)$ by the substitution $Y_i\to f^i$.* In symbols:

$$g(f^1, f^2, \cdots, f^m) = \sum_{q=0}^{\infty} g_q(f^1, f^2, \cdots, f^m) = \sum_{q=0}^{\infty} \bar{g}^q. \tag{9}$$

For fixed $f^1, f^2, \cdots, f^m$, (9) defines a mapping

$$g\to g(f^1, f^2, \cdots, f^m), \quad g\in A[[Y_1, Y_2, \cdots, Y_m]], \tag{10}$$

of $A[[Y_1, Y_2, \cdots, Y_m]]$ into $A[[X_1, X_2, \cdots, X_n]]$. We shall refer to (10) as the *substitution mapping* (relative to the substitution $Y_i \to f^i$). It follows easily from the rules (6) and (6′) of addition and multiplication of infinite sums, that *the substitution mapping* (10) *is a homomorphism.* Furthermore, the mapping (10) is *continuous* (with respect to the topology introduced earlier in power series rings). To see this it is sufficient to show that if $\mathfrak{Y}$ denotes the ideal generated in $A[[Y_1, Y_2, \cdots, Y_m]]$ by $Y_1, Y_2, \cdots, Y_m$, then the transform of $\mathfrak{Y}^i$ by (10) is contained in $\mathfrak{X}^{\rho(i)}$, where $\rho(i)$ tends to $\infty$ with $i$. This, however, is obvious, since from the definition of the substitution mapping it follows that if $g \in \mathfrak{Y}^i$ then $g(f^1, f^2, \cdots, f^m)$ belongs to $\mathfrak{X}^i$.

The image of the ring $A[[Y_1, Y_2, \cdots, Y_m]]$ under the substitution mapping (10) is a subring of $A[[X_1, X_2, \cdots, X_n]]$. We shall denote this subring by $A[[f^1, f^2, \cdots, f^m]]$.

It is not difficult to see that *any continuous homomorphism $\tau$ of $A[[Y_1, Y_2, \cdots, Y_m]]$ into $A[[X_1, X_2, \cdots, X_n]]$ is a substitution mapping.* For let $\tau(Y_i) = f^i$. The continuity of $\tau$ requires that high powers of $f^i$ belong to high powers of the ideal $\mathfrak{X}$. Hence $f^i \in \mathfrak{X}$, $i = 1, 2, \cdots, m$. Now, let $g = g_0 + g_1 + \cdots + g_q + \cdots$ be any power series in $Y_1, Y_2, \cdots, Y_m$. Since $\tau$ is a homomorphism we have $\tau(g_q) = g_q(f^1, f^2, \cdots, f^m)$ and $\tau\left(\sum_{q=0}^{i} g_q\right) = \sum_{q=0}^{i} g_q(f^1, f^2, \cdots, f^m)$. Since $g = \lim_{i \to \infty} \left(\sum_{q=0}^{i} g_q\right)$ and since $\tau$ is continuous, we must have

$$\tau(g) = \lim_{i \to \infty} \tau\left(\sum_{q=0}^{i} g_q\right) = \lim_{i \to \infty} \sum_{q=0}^{i} g_q(f^1, f^2, \cdots, f^m),$$

i.e., $\tau(g) = g(f^1, f^2, \cdots, f^m)$, in view of (9). This shows that $\tau$ is the substitution mapping relative to the substitution $Y_i \to f^i$.

In the special case $m = n$, the two rings $A[[Y_1, Y_2, \cdots, Y_m]]$ and $A[[X_1, X_2, \cdots, X_n]]$ coincide and we have $Y_i = X_i$. In this case, our substitution mapping defines a continuous homomorphism of the power series ring $A[[X_1, X_2, \cdots, X_n]]$ into itself. We now describe a case in which this homomorphism is an automorphism.

LEMMA 2. *Let $f^1, f^2, \cdots, f^n$ be $n$ power series in $A[[X_1, X_2, \cdots, X_n]]$ such that the initial form of $f^i$ is $X_i$ $(1 \leq i \leq n)$. Then the substitution mapping $\varphi$: $g(X_1, X_2, \cdots, X_n) \to g(f^1, f^2, \cdots, f^n)$ is an automorphism of the power series ring $A[[X_1, X_2, \cdots, X_n]]$.*

PROOF. We first show that the kernel of $\varphi$ is zero. Let $g$ be a non-zero power series in $A[[X_1, X_2, \cdots, X_n]]$ and let $g_s$ be its initial form. From (9) we find at once that $g(f^1, f^2, \cdots, f^n) - g_s \in \mathfrak{X}^{s+1}$.

Hence $g(f^1, f^2, \cdots, f^n) \neq 0$, and thus $g$ is not in the kernel of $\varphi$. Observe that we have shown here the following: *$g$ and $g(f^1, f^2, \cdots, f^n)$ have the same initial form.*

We next show that $\varphi$ maps $A[[X_1, X_2, \cdots, X_n]]$ *onto* itself, i.e., that $A[[f^1, f^2, \cdots, f^n]] = A[[X_1, X_2, \cdots, X_n]]$. If $g_s(X_1, X_2, \cdots, X_n)$ is any form, with coefficients in $A$, then we have just seen that $g_s(X_1, X_2, \cdots, X_n)$ is the initial form of the element $g_s(f^1, f^2, \cdots, f^n)$ of the ring $A[[f^1, f^2, \cdots, f^n]]$. It follows therefore from Lemma 1 that the ring $A[[f^1, f^2, \cdots, f^n]]$ is everywhere dense in $A[[X_1, X_2, \cdots, X_n]]$, and in order to prove the lemma we have only to show that $A[[f^1, f^2, \cdots, f^n]]$ is a closed subset of $A[[X_1, X_2, \cdots, X_n]]$. Assume then that we have an element $h$, such that $h = \lim_{i\to\infty} g^i(f^1, f^2, \cdots, f^n)$, where $g^i(X_1, X_2, \cdots, X_n)$ is in $A[[X_1, X_2, \cdots, X_n]]$. The order of $g^i(f^1, f^2, \cdots, f^n) - g^j(f^1, f^2, \cdots, f^n)$ is the same as the order of $g^i(X_1, X_2, \cdots, X_n) - g^j(X_1, X_2, \cdots, X_n)$. Hence $\{g^i(X_1, X_2, \cdots, X_n)\}$ must be a Cauchy sequence as well as $\{g^i(f^1, f^2, \cdots, f^n)\}$. Let $g = \lim g^i(X_1, X_2, \cdots, X_n)$. Since $\varphi$ is continuous, it follows that $h = \varphi(g) = g(f^1, f^2, \cdots, f^n)$, whence $h \in A[[f^1, f^2, \cdots, f^n]]$. Q.E.D.

COROLLARY 1. *Let $f^1, f^2, \cdots, f^m$ be $m$ power series in $A[[X_1, X_2, \cdots, X_n]]$, $m \leqq n$, such that the initial form of $f^i$ is $X_i$. Then the substitution $Y_i \to f^i$ defines an isomorphism $\varphi$: $g \to g(f^1, f^2, \cdots, f^m)$ of $A[[Y_1, Y_2, \cdots, Y_m]]$ into $A[[X_1, X_2, \cdots, X_n]]$.*

For the first part of the proof of Lemma 2 is independent of the assumption $m = n$.

COROLLARY 2. *Let $A$ be an integral domain and let $f^1, f^2, \cdots, f^m$ be $m$ power series in $A[[X_1, X_2, \cdots, X_n]]$, $m \leqq n$, such that the initial forms of the $f^i$ are linearly independent linear forms $f_1^1, f_1^2, \cdots f_1^m$. Then the substitution mapping $\varphi: g(Y_1, Y_2, \cdots, Y_m) \to g(f^1, f^2, \cdots, f^m)$ is an isomorphism of $A[[Y_1, Y_2, \cdots, Y_m]]$ into $A[[X_1, X_2, \cdots, X_n]]$. If, furthermore, $m = n$, $Y_i = X_i$, $i = 1, 2, \cdots, n$, and the determinant of the coefficients of the linear forms $f_1^1, f_1^2, \cdots, f_1^n$ is a unit in $A$ (in particular, if $A$ is a field and the above determinant is $\neq 0$), then $\varphi$ is an automorphism of $A[[X_1, X_2, \cdots, X_n]]$.*

If $g_s(Y_1, Y_2, \cdots, Y_m)$ is the initial form of a non-zero element $g(Y_1, Y_2, \cdots, Y_m)$ of $A[[Y_1, Y_2, \cdots, Y_m]]$, then we find, as in the case of the lemma, that $s$ is also the order of $\varphi(g)$, since $\varphi(g) - g_s(f_1^1, f_1^2, \cdots, f_1^m) \in \mathfrak{X}^{s+1}$ and since in the *integral domain* $A$ the linear independence of the linear forms $f_1^1, f_1^2, \cdots, f_1^m$ and the non-vanishing of the form $g_s$ imply that $g_s(f_1^1, f_1^2, \cdots, f_1^m)$ is different from zero and has an initial form of degree $s$ (to see this it is sufficient to pass to the quotient field of $A$).

If $m=n$ and if the determinant of the coefficients of the linear forms $f_1^1, f_1^2, \cdots, f_1^n$ is a unit in $A$, then, for each integer $q$, the *linear substitution* $X_i \to f_1^i$ maps *onto itself* the set of forms of degree $q$ in $X_1, X_2, \cdots, X_n$, with coefficients in $A$. It follows that also in the present case the ring $A[[f^1, f^2, \cdots, f^n]]$ has the property of containing power series with arbitrarily preassigned initial forms, with coefficients in $A$, and the rest of the proof of the lemma is now applicable without any change.

THEOREM 4. *If $A$ is a noetherian ring, then the power series ring $A[[X]]$ is also noetherian.*

PROOF. We give here a proof parallel to the second proof of Hilbert's basis theorem, cf. Vol. I, Ch. IV, § 1, i.e., a proof using the finite basis condition. Let $\mathfrak{A}$ be an ideal in $A[[X]]$. For any integer $i \geqq 0$ denote by $L_i(\mathfrak{A})$ the set of elements of $A$ consisting of 0 and of the coefficients of $X^i$ in all elements of $\mathfrak{A}$ which are of order $i$. Then $L_i(\mathfrak{A})$ is an ideal in $A$, and the ideals $L_i(\mathfrak{A})$ constitute an ascending sequence. Their union $L(\mathfrak{A})$ is the ideal in $A$ consisting of 0 and of the coefficients of the initial terms† of all non-zero elements of $\mathfrak{A}$. Since $A$ is noetherian, $L(\mathfrak{A})$ has a finite basis $\{a_1, \cdots, a_q\}$. We fix in $\mathfrak{A}$ a power series $F_i(X)$ whose initial term has $a_i$ as coefficient. Denote by $d$ the greatest integer among the orders of the series $F_i(X)$.

Now, for every $j<d$, let $\{b_{j1}, \cdots, b_{jn(j)}\}$ be a finite basis of the ideal $L_j(\mathfrak{A})$, and let $G_{jk_j}(X)$ be a power series in $\mathfrak{A}$ whose initial term is $b_{jk_j}X^j (1 \leq k_j \leq n(j))$. We shall prove that the ideal $\mathfrak{A}$ is generated by the series $F_i(X)$, $G_{jk_j}(X)$ $(1 \leq i \leq q;\ 0 \leq j < d;\ 1 \leq k_j \leq n(j))$. We prove this in two steps:

(a) Let $\mathfrak{A}'$ be the ideal $(G_{jk_j}(X))$ generated by the elements $G_{jk_j}(X)$. We have $\mathfrak{A}' \subset \mathfrak{A}$. Every element $P(X)$ of $\mathfrak{A}$ which has the order $j<d$ is congruent mod $\mathfrak{A}'$ to an element of $\mathfrak{A}$ which has order $\geq j+1$. In fact, the coefficient $c$ of the initial term $cX^j$ of $P(X)$ may be written in the form $c = \sum_{k_j=1}^{n(j)} c_{k_j} b_{jk_j}$ $(c_{k_j} \in A)$. Thus $P(X) - \sum_{k_j=1}^{n(j)} c_{k_j} G_{jk_j}(X)$ is of order $\geq j+1$. It follows by successive applications of this result that every element of order $j<d$ of $\mathfrak{A}$ is congruent mod $\mathfrak{A}'$ to an element of $\mathfrak{A}$ of order $\geq d$. It remains to prove that any element of order $\geq d$ of

† Since we are dealing now with power series in one variable, an initial form, of degree $i$, consists of just one term $cX^i$, $c \in A$.

$\mathfrak{A}$ is in the ideal $(F_i(X), G_{jk_j}(X))$. We will even prove that such an element is in the ideal $(F_1(X), \cdots, F_q(X))$.

(b) Let $P(X)$ be an element of $\mathfrak{A}$ of order $s \geq d$, and let $cX^s$ be its initial term. We may write $c = \sum_{i=1}^{q} c_i a_i$ $(c_i \in A)$. Thus $P(X) - \sum_{i=1}^{q} c_i X^{s-\mathbf{o}(F_i)} F_i(X)$ is an element of order $\geq s+1$ of $\mathfrak{A}$. By successive applications of this result we get $q$ sequences $\{c_i^n\}$ $(i=1, 2, \cdots, q;$ $n=s, s+1, \cdots;$ $c_i^s = c_i)$ of elements of $A$ such that, for every $n$, the power series

$$P(X) - \sum_{i=1}^{q} \left( \sum_{j=s}^{j=n} c_i^j X^{j-\mathbf{o}(F_i)} \right) F_i(X)$$

is of order $> n$. As the exponents $j - \mathbf{o}(F_i)$ tend to infinity with $j$, each of the infinite sums $\sum_{j=s}^{\infty} c_i^j X^{j-\mathbf{o}(F_i)}$ converges and represents an element $s_i(X)$ of $A[[X]]$. Since the order of the power series $P(X) - \sum_{i=1}^{q} s_i(X) F_i(X)$ is greater than $n$ for every $n$, this power series is 0, and we have $P(X) = \sum_{i=1}^{q} s_i(X) F_i(X)$. Q.E.D.

COROLLARY. *The power series ring $A[[X_1, \cdots X_n]]$ in $n$ indeterminates over a noetherian ring $A$ (in particular, over a field, or over the ring of integers) is noetherian.*

This follows from Theorem 4 by induction on $n$, since $A[[X_1, \cdots, X_n]]$ is isomorphic to $A[[X_1, \cdots, X_{n-1}]][[X_n]]$.

REMARK. A simple direct proof of the fact that $A[[X_1, \cdots, X_n]]$ is noetherian may be given if one uses the fact that the polynomial ring $A[X_1, \cdots, X_n]$ is noetherian. But, since this proof applies as well to a more general situation, we postpone it until the chapter on Local Algebra (see VIII, § 3, Example 1, p. 260). On the other hand we shall give later on in this section a proof that $k[[X_1, \cdots, X_n]]$ is noetherian ($k$, a field) using the Weierstrass' preparation theorem.

THEOREM 5. (Weierstrass preparation theorem) *Let $k$ be a field and let $F(X_1, \cdots, X_n)$ be a non-invertible power series (i.e., a non-unit in $k[[X_1, X_2, \cdots, X_n]]$) with coefficients in $k$. Suppose that $F(X_1, \cdots, X_n)$ contains terms of the form $aX_n^h$ with non-zero coefficient $a$, and denote by $s$ $(\geq 1)$ the smallest of all the exponents $h$ having this property. Then for every power series $G(X_1, \cdots, X_n)$ there exists a power series $U(X_1, \cdots,$*

*$X_n$) and $s$ power series $R_i(X_1, \cdots, X_{n-1})$ in $X_1, \cdots, X_{n-1}$ $(0 \leq i \leq s-1)$ such that*

$$(11) \quad G(X_1, \cdots, X_n) = U(X_1, \cdots, X_n)F(X_1, \cdots, X_n) + \sum_{i=0}^{s-1} R_i(X_1, \cdots, X_{n-1})X_n^i.$$

*The power series $U$ and $R_i$ are uniquely determined by $G$ and $F$.*

PROOF. For every power series $P(X_1, \cdots, X_n)$ denote by $r(P)$ the sum of all terms in $P$ which do not have $X_n^s$ as a factor, and by $h(P)$ the factor of $X_n^s$ in $P-r(P)$. In other words we have

$$(12) \qquad P = r(P)+X_n^s h(P),$$

where $r(P), h(P) \in k[[X_1, X_2, \cdots, X_n]]$ and where, furthermore, $r(P)$ is a polynomial in $X_n$, of degree $\leqq s-1$, with coefficients in $k[[X_1, X_2, \cdots, X_{n-1}]]$. Note that if the power series ring $k[[X_1, X_2, \cdots, X_n]]$ is thought of as a vector space over the field $k$, then both operations $r$ and $h$ are *linear transformations* in that vector space. By the definition of the integer $s$, $h(F)$ is a unit in $k[[X_1, X_2, \cdots, X_n]]$ (see Theorem 2), and $r(F)$, regarded as a polynomial in $X_n$, has all its coefficients in the maximal ideal of the ring $k[[X_1, X_2, \cdots, X_{n-1}]]$. We shall denote this maximal ideal by $\mathfrak{m}$.

The problem of finding power series $U$ and $R_0, R_1, \cdots, R_{s-1}$ such that (11) holds is equivalent to the problem of finding a power series $U$ such that the following relation holds:

$$(11a) \qquad h(G) = h(UF).$$

For if (11) holds, then $h(G-UF)=0$, whence (11a) holds by linearity of $h$. Conversely, assume that $U$ is a power series satisfying (11a). Then $h(G-UF)=0$, whence $G-UF=r(G-UF)$ (by (12)), i.e., $G-UF$ is a polynomial in $X_n$, of degree $\leqq s-1$, with coefficients in $k[[X_1, X_2, \cdots, X_{n-1}]]$, and so (11) holds.

We have $UF= Ur(F)+X_n^s Uh(F)$, and hence (11a) can be re-written as follows:

$$(11b) \qquad h(G) = h(Ur(F))+Uh(F),$$

and our problem is equivalent to finding a power series $U$ satisfying (11b). Since $h(F)$ is a unit in $k[[X_1, X_2, \cdots, X_n]]$ we shall try to construct the power series

$$(13) \qquad V = Uh(F).$$

We set

(14) $$M = -r(F)[h(F)]^{-1}.$$

Then, by (13), $Ur(F) = -MV$, and (11b) is equivalent to

(11c) $$h(G) = -h(MV) + V.$$

For every power series $P$, denote by $m(P)$ the power series $h(MP)$. Notice that $m$ is again a *linear operation* on power series. Furthermore, if $P$, considered as a power series in $X_n$ over $k[[X_1, \cdots, X_{n-1}]]$, has all its coefficients in some power $\mathfrak{m}^j$ of the maximal ideal $\mathfrak{m}$, then *$m(P)$ has all its coefficients in* $\mathfrak{m}^{j+1}$. For convenience we set $H = h(G)$. With these notations condition (11c) may be written as follows:

(11d) $$V = H + m(V).$$

Since $m$ is linear, condition (11d) implies that $V = H + m(H + m(V)) = H + m(H) + m^2(V)$, and, by successive applications:

(11e) $$V = H + m(H) + m^2(H) + \cdots + m^q(H) + m^{q+1}(V),$$
for any integer $q \geqq 0$.

The property of the operation $m$ which we have just pointed out above shows that $m^j(H)$ is at least of order $j$, and $m^{q+1}(V)$ is at least of order $q+1$. Thus the infinite sum $H + m(H) + m^2(H) + \cdots + m^q(H) + \cdots$ converges, and, *if* a power series $V$ satisfying (11d) exists, it must therefore be the series

(15) $$V = H + m(H) + m^2(H) + \cdots + m^q(H) + \cdots,$$

and this proves the *uniqueness* of $V$, whence of $U$ and of the $R_i$.

We now prove that the series $V$ given by (15) satisfies condition (11d). Let us write $V = H + m(H) + \cdots + m^q(H) + W_q$. The coefficients of $W_q$ ($W_q$ being considered as a power series in $X_n$) are all in $\mathfrak{m}^{q+1}$. Then, since $m$ is linear,

$$V - H - m(V) = H + \cdots + m^q(H) + W_q - H$$
$$- m(H) - \cdots - m^{q+1}(H) - m(W_q) = W_q - m^{q+1}(H) - m(W_q).$$

Thus all coefficients of $V - H - m(V)$ are in $\mathfrak{m}^{q+1}$. As this is true for every $q$, we have $V - H - m(V) = 0$, and condition (11d) holds. This proves the *existence* of $V$, whence also of $U$ and of the $R_i$.

REMARK. In the next chapter we shall give a somewhat shorter proof of the Weierstrass preparation theorem, based upon the properties of complete local rings. An advantage of the proof given here is that the questions of existence and unicity are treated simultaneously. A more substantial advantage is that the method of majorants is easily applicable to the resolving formula (11e), with the result that if $F$ and $G$ are convergent power series over the field of real or complex numbers, then the series $V$, $U$ and the $R_i$ are also convergent. To show this we open now a brief digression on *the preparation theorem for convergent power series.*

In the case of convergent power series over the field $k$ of real or complex numbers, the proof of the Weierstrass preparation theorem runs as follows. We recall† that a power series

$$F(X_1, \cdots, X_n) = \sum_q a_{q_1, \cdots, q_n} X_1^{q_n} \cdots X_n^{q_n}$$

is said to be *convergent* if there exists a neighborhood $N$ of the origin in $k^n$ such that the series $\sum_q a_{q_1 \cdots q_n} z_1^{q_1} \cdots z_n^{q_n}$ is absolutely convergent for every $(z_1, \cdots, z_n) \in N$. Then there exist positive real numbers $\mu$ and $\rho$ such that $|a_{q_1 \cdots q_n}| \leq \mu\rho^{-(q_1 + \cdots + q_n)}$. Conversely, the existence of two such real numbers implies that $\sum_q a_{q_1 \cdots q_n} z_1^{q_1} \cdots z_n^{q_1}$ converges in the neighborhood $N$ of 0 defined by $|z_i| < \rho$ $(i = 1, \cdots, n)$. It is easily seen that the convergent power series in $k[[X_1, \cdots, X_n]]$ form a *subring* of $k[[X_1, \cdots, X_n]]$, and that a convergent power series with a constant term $\neq 0$ admits as inverse a convergent power series. A series $\sum_q a_{q_1 \cdots q_n} X_1^{q_1} \cdots X_n^{q_n}$ with real positive coefficients is said to be a *majorant* of $\sum_q b_{q_1 \cdots q_n} X_1^{q_1} \cdots X_n^{q_n}$ if $|b_{q_1 \cdots q_n}| \leq a_{q_1 \cdots q_n}$ for all $q_1, \cdots, q_n$. It is clear that, in order to prove the convergence of a power series $F$, it is sufficient to prove that a majorant of $F$ converges. The inequality $|a_{q_1 \cdots q_n}| \leq \mu\rho^{-(q_1 + \cdots + q_n)}$ means that $\mu \Big/ \left(1 - \frac{X_1}{\rho}\right) \cdots \left(1 - \frac{X_n}{\rho}\right)$ is a majorant of $\sum a_{q_1 \cdots q_n} X_1^{q_1} \cdots X_n^{q_n}$.

In order to extend the Weierstrass preparation theorem to convergent power series, it is sufficient to prove that, if the series $M$ and $H$ are convergent, then

$$V = H + m(H) + \cdots + m^q(H) + \cdots,$$

is convergent (same notations as in the proof of Theorem 5). We notice that the coefficients of $V$ are polynomials with positive integral coefficients in the coefficients of $M$ and $H$. Thus, if we replace $M$ and

† See Bochner-Martin, "Several complex variables," Princeton (1948), Chap. II.

$H$ by majorants $M'$ and $H'$, and assuming that $M'$ is of positive order, then the power series

$$V' = H' + m'(H') + \cdots + m'^q(H') + \cdots$$

(where the operation $m'$ is defined by $m'(P) = h(M'P)$) is a majorant of $V$. We may take

$$H'_1 = \frac{\mu}{\left(1-\frac{X_1}{\rho}\right)\cdots\left(1-\frac{X_n}{\rho}\right)},$$

$$M' = \frac{\mu(X_1 + \cdots + X_{n-1})}{\left(1-\frac{X_1}{\rho}\right)\cdots\left(1-\frac{X_n}{\rho}\right)}.$$

(For the second one we write $M = N_1X_1 + \cdots + N_{n-1}X_{n-1}$ and we major separately each one of the series $N_i$.) Instead of $H'_1$, we take as majorant of $H$ the series

$$H' = \frac{\mu}{\left(1-\frac{X_1}{\rho}\right)\cdots\left(1-\frac{X_{n-1}}{\rho}\right)}\,\varphi\left(\frac{X_n}{\rho}\right),$$

where $\varphi(X)$ is a series in one variable, majoring $\frac{1}{1-X}$ and enjoying properties which we are going to describe.

We notice that the operation $m'$ is not only additive, but *linear over* $k[[X_1, \cdots, X_{n-1}]]$. We thus have

$$m'(H') = \frac{\mu^2(X_1 + \cdots + X_{n-1})}{\left(1-\frac{X_1}{\rho}\right)^2\cdots\left(1-\frac{X_{n-1}}{\rho}\right)^2}\,h\left(\varphi\left(\frac{X_n}{\rho}\right)\Big/\left(1-\frac{X_n}{\rho}\right)\right).$$

We set $X = \frac{X_n}{\rho}$. The series $V' = H' + m'(H') + \cdots + m'^q(H') + \cdots$ will be very easy to compute if $h\left(\frac{\varphi(X)}{1-X}\right)$ is a scalar multiple of the series $\varphi(X)$. By definition of the operation $h$, this is true if there exist a polynomial $P_{s-1}(X)$ of degree $\leq s-1$ and a real number $\lambda$ such that

$$\frac{\varphi(X)}{1-X} = P_{s-1}(X) + \lambda X^s\varphi(X).$$

Thus $\varphi(X)$ must be a rational function:

$$\varphi(X) = \frac{(1-X)P_{s-1}(X)}{1-\lambda X^s + \lambda X^{s+1}}.$$

We take $\lambda = 2^{s+1}$, and notice that the denominator $1-2^{s+1}X^s+2^{s+1}X^{s+1}$ factors into $(1-2X)(1+2X+2^2X^2+\cdots+2^{s-1}X^{s-1}-2^sX^s)$. The second factor takes the value 1 for $X=0$ and $-1$ for $X=1$. Therefore it admits a positive root $1/\alpha$ ($\alpha>1$). Thus the denominator $1-2^{s+1}X^s+2^{s+1}X^{s+1}$ may be written in the form $(1-2X)(1-\alpha X)$ $\bar{P}_{s-1}(X)$, where $\bar{P}_{s-1}(X)$ is a polynomial of degree $s-1$. We choose $P_{s-1}(X)$ to be just this polynomial $\bar{P}_{s-1}(X)$. We then have

$$\varphi(X) = \frac{1-X}{1-2X}\cdot\frac{1}{1-\alpha X},$$

and thus for this choice of $\varphi(X)$ we will have $h(\varphi(X)/(1-X)) = 2^{s+1}\varphi(X)$. As it is a rational function, this power series $\varphi(X)$ is convergent. Since

$$\frac{1-X}{1-2X} = \frac{1}{2}+\frac{1}{2}\frac{1}{1-2X}$$

the power series expansion of $\varphi(X)$ is

$$\frac{1}{2}(2+2X+4X^2+\cdots+2^nX^n+\cdots)(1+\alpha X+\cdots+\alpha^nX^n+\cdots).$$

Except for the constant term (which is equal to 1), the coefficient of $X^n$ is $\alpha^n+\alpha^{n-1}+2\alpha^{n-2}+\cdots+2^{n-1}$; since it is obviously $>1$, $\varphi(X)$ is a majorant of $1/(1-X) = 1+X+\cdots+X^n+\cdots$.

This being so, if we set $A = \mu\Big/\left(1-\frac{X_1}{\rho}\right)\cdots\left(1-\frac{X_{n-1}}{\rho}\right)$ and $B = \mu(X_1+\cdots+X_{n-1})\Big/\left(1-\frac{X_1}{\rho}\right)\cdots\left(1-\frac{X_{n-1}}{\rho}\right)$ and if we notice that, for every power series $\psi(X)$ (where $X = X_n/\rho$), we have

$$m'(\psi(X)) = h\left(B\frac{\psi(X)}{1-X}\right) = Bh\left(\frac{\psi(X)}{1-X}\right),$$

we get $m'(H') = m'(A\varphi(X)) = ABh(\varphi(X)/(1-X)) = 2^{s+1}AB\varphi(X)$. Hence, by repeated applications,

$$m'^2(H') = m'(2^{s+1}AB\varphi(X)) = 2^{2(s+1)}AB^2\varphi(X)$$

and $m'^q(H') = 2^{(s+1)q}AB^q\varphi(X)$ for every $q$. Then the computation of the infinite sum $V' = H'+m'(H')+\cdots+m'^q(H')+\cdots$, reduces to the computation of the sum of a geometric series:

$$V' = A\varphi(X)+A2^{s+1}B\varphi(X)+A2^{2(s+1)}B^2\varphi(X)+\cdots.$$

Hence

$$V' = A\varphi(X)\frac{1}{1-2^{s+1}B}$$

$$= \frac{\mu\varphi(X_n/\rho)}{\left(1-\frac{X_1}{\rho}\right)\cdots\left(1-\frac{X_{n-1}}{\rho}\right)\left(1-\frac{2^{s+1}\mu(X_1+\cdots+X_{n-1})}{(1-X_1/\rho)\cdots(1-X_{n-1}/\rho)}\right)}$$

$$= \frac{\mu\cdot\left(1-\frac{X_n}{\rho}\right)}{\left(1-2\frac{X_n}{\rho}\right)\left(1-\alpha\frac{X_n}{\rho}\right)\left[\left(1-\frac{X_1}{\rho}\right)\cdots\left(1-\frac{X_{n-1}}{\rho}\right)-2^{s+1}\mu(X_1+\cdots+X_{n-1})\right]}.$$

Since $V'$ is a rational function, this is a convergent power series. This proves the preparation theorem in the case of convergent power series.

A power series $F(X_1, X_2, \cdots, X_n)$ which contains a term $cX_n^s$ which is a power of $X_n$, with non-zero coefficient $c$, is said to be *regular* in $X_n$. To say that $F(X_1, X_2, \cdots, X_n)$ is regular in $X_n$ is equivalent to saying that $F(0, 0, \cdots, 0, X_n)$ is different from zero.

COROLLARY 1. *Let $F(X_1, X_2, \cdots, X_n)$ be a power series in $S=k[[X_1, X_2, \cdots, X_n]]$ which is regular in $X_n$ ($k$, a field) and let the order $s$ of the power series $F(0, 0, \cdots, 0, X_n)$ be $\geqq 1$ (in other words, it is assumed that $F$ is not a unit).*† *Then there exist power series $E(X_1, X_2, \cdots, X_n)$, $R_i(X_1, X_2, \cdots, X_{n-1})$ $(i=0, 1, \cdots, s-1)$ such that*

$$(16)\quad \begin{aligned} &F(X_1, X_2, \cdots, X_n) \\ &= E(X_1, X_2, \cdots, X_n)[X_n^s + R_{s-1}(X_1, X_2, \cdots, X_{n-1})X_n^{s-1} + \cdots \\ &\qquad + R_0(X_1, X_2, \cdots, X_{n-1})]. \end{aligned}$$

*The power series $E$, $R_i$ are uniquely determined by $F$; $E$ is a unit, and none of the $R_i$ is a unit.*

For if we apply Theorem 5 to the power series $G=-X_n^s$ we find

$$\begin{aligned} X_n^s + R_{s-1}(X_1, X_2, \cdots, X_{n-1})X_n^{s-1} + \cdots + R_0(X_1, X_2, \cdots, X_{n-1}) \\ = -U(X_1, X_2, \cdots, X_n)F(X_1, X_2, \cdots, X_n). \end{aligned}$$

Setting $X_1=X_2=\cdots=X_{n-1}=0$ in this identity we obtain on the right-hand side a power series in $X_n$ which has order $\geqq s$. Hence $R_i(0, 0, \cdots, 0)=0$, $0\leqq i\leqq s-1$, and no $R_i(X_1, X_2, \cdots, X_{n-1})$ is a unit. It follows at the same time that $U(0, 0, \cdots, 0, X_n)$ must be of order

† The corollary holds trivially also if $F$ is a unit; in (16) we have then $E=F$, while the expression in the square brackets is the element 1.

zero, whence $U(X_1, X_2, \cdots, X_n)$ is a unit. If we now set $E=U^{-1}$, we have (16). The unicity of $E$ and of the $R_i$ also follows from Theorem 5 in the special case $G=-X_n^s$.

The polynomial (in $X_n$)

$$(17)\quad F^{\star} = X_n^s + R_{s-1}(X_1, X_2, \cdots, X_{n-1})X_n^{s-1} + \cdots + R_0(X_1, X_2, \cdots, X_{n-1})$$

in (16) is called the *distinguished pseudo-polynomial* associated with $F$; it is defined only if $F$ is regular in $X_n$, and its degree $s$ (in $X_n$) is equal to the order of the power series $F(0, 0, \cdots, 0, X_n)$. The relation (16) shows that $F$ and $F^{\star}$ are associates in $S$.†

Note that $F^{\star}$ has the following two properties: (a) it is a monic polynomial in $X_n$; (b) its coefficients, other than the leading coefficient, are power series in $X_1, X_2, \cdots, X_{n-1}$ which belong to the maximal ideal of $k[[X_1, X_2, \cdots, X_{n-1}]]$. Before deriving other consequences of Weierstrass' preparation theorem, we point out the following consequence of (a) and (b): *if $S^{\star}$ denotes the ring*

$$k[[X_1, X_2, \cdots, X_{n-1}]][X_n],$$

*then*

$$(18)\qquad SF^{\star} \cap S^{\star} = S^{\star}F^{\star}.$$

We have to show the following: if $H^{\star}=hF^{\star}$, with $H^{\star} \in S^{\star}$ and $h \in S$, then $h \in S^{\star}$. Let $h=\sum_{q=0}^{\infty} h_q(X_1, X_2, \cdots, X_{n-1})X_n^q$ and let $s+m$ be the degree of $H^{\star}$ in $X_n$. Expressing the fact that $hF^{\star}$, regarded as a power series in $X_n$, is actually a polynomial of degree $s+m$, we find

$$(19)\qquad h_q + h_{q+1}R_{s-1} + \cdots + h_{q+s}R_0 = 0, \quad q > m.$$

Since the $R_i$ all belong to $\mathfrak{m}$, it follows from (19) that $h_q \in \mathfrak{m}$ if $q>m$. But then again (19) shows that $h_q \in \mathfrak{m}^2$ if $q>m$. By repeated application of this argument we find that $h_q \in \bigcap_{i=1}^{\infty} \mathfrak{m}^i$, whence $h_q=0$ for all $q>m$. Thus $h$ is a polynomial in $X_n$ (of degree $m$), showing that $h \in S^{\star}$.

Since $F$ and $F^{\star}$ are associates in $S$ we have $SF=SF^{\star}$. Then (18) shows that the residue class ring $S/SF$ contains $S^{\star}/S^{\star}F^{\star}$ as a subring.

COROLLARY 2. *The rings $S/SF$ and $S^{\star}/S^{\star}F^{\star}$ coincide.*

For if $G$ is any element of $S$ then Theorem 5 shows that $G$ is congruent *mod* $F$ to an element of $S^{\star}$.

† Note that the distinguished pseudo-polynomial of a unit $F$ is the element 1.

The following lemma shows that every non-zero power series in $k[[X_1, X_2, \cdots, X_n]]$ may be construed to be regular in $X_n$. More precisely, we have

LEMMA 3. *If* $F(X_1, X_2, \cdots, X_n)$ *is a non-zero power series in* $k[[X_1, X_2, \cdots, X_n]]$ *(k, a field), then there exists an automorphism* $\varphi$ *of* $k[[X_1, X_2, \cdots, X_n]]$ *such that* $\varphi(F)$ *is regular in* $X_n$.

PROOF. We assume first that $k$ is an infinite field. Let $f_q$ be the initial form of $F$. Since $k$ is infinite we can find elements $a_1, a_2, \cdots, a_{n-1}$ in $k$ such that $f_q(a_1, a_2, \cdots, a_{n-1}, 1) \neq 0$. Then we may use the linear substitution $X_i \to X_i + a_i X_n$ $(i = 1, 2, \cdots, n-1)$, $X_n \to X_n$ (compare with the normalization lemma of Vol. I, Ch. V, § 4, Theorem 8). By Lemma 2, Corollary 2, the corresponding substitution mapping $\varphi$ is an automorphism. Furthermore, the initial form of $\varphi(F)$ contains the term $f_q(a_1, a_2, \cdots, a_{n-1}, 1)X_n^q$. Hence $\varphi(F)$ is regular in $X_n$.

We now give a proof which is also valid for finite fields and which will show the existence of exponents $u_j$ $(j=1, \cdots, n-1)$ such that the automorphism $\varphi$ defined by $\varphi(X_n) = X_n$, $\varphi(X_j) = X_j + X_n^{u_j}$ has the required property, i.e., is such that $F(X_n^{u_1}, \cdots, X_n^{u_{n-1}}, X_n) \neq 0$. We order lexicographically the monomials which appear in $F$ with non-zero coefficients. Let $X_1^{a_1} \cdots X_n^{a_n}$ be the smallest one. Then, if $X_1^{b_1} \cdots X_n^{b_n}$ is another monomial which actually appears in $F$, we have, either $b_1 > a_1$, or $b_1 = a_1$ and $b_2 > a_2, \cdots$, or $b_1 = a_1, \cdots, b_{n-1} = a_{n-1}$ and $b_n > a_n$. The corresponding monomials in $F(X_n^{u_1}, \cdots, X_n^{u_{n-1}}, X_n)$ have $u_1a_1 + u_2a_2 + \cdots + u_{n-1}a_{n-1} + a_n$ and $u_1b_1 + u_2b_2 + \cdots + u_{n-1}b_{n-1} + b_n$ as exponents. If we take $u_{n-1} > a_n$, $u_{n-2} > u_{n-1}a_{n-1} + a_n, \cdots, u_1 > u_2a_2 + \cdots + u_{n-1}a_{n-1} + a_n$, then we get $u_1b_1 + \cdots + u_{n-1}b_{n-1} + b_n > u_1a_1 + \cdots + u_{n-1}a_{n-1} + a_n$: in fact, if the index $i$ is defined by the condition $a_1 = b_1, \cdots, a_{i-1} = b_{i-1}$, $a_i < b_i$, then the difference $u_1b_1 + \cdots + b_n - (u_1a_1 + \cdots + a_n)$ of the two above exponents is $u_i(b_i - a_i) + u_{i+1}(b_{i+1} - a_{i+1}) + \cdots + b_n - a_n$. The first term is $\geq u_i$, whereas the remainder is $\geq -(u_{i+1}a_{i+1} + \cdots + a_n)$, and thus the difference of the two exponents is $> 0$ since $u_i > u_{i+1}a_{i+1} + \cdots + a_n$. In other words, in $F(X_n^{u_1}, \cdots, X_n^{u_{n-1}}, X_n)$ the monomial with exponent $u_1a_1 + \cdots + u_n$ cannot be cancelled by any other, and hence $F(X_n^{u_1}, \cdots, X_n^{u_{n-1}}, X_n) \neq 0$.

COROLLARY. *Given any finite set of non-zero power series* $F_1, F_2, \cdots, F_h$ *in* $k[[X_1, X_2, \cdots, X_n]]$, *there exists an automorphism* $\varphi$ *of* $k[[X_1, X_2, \cdots, X_n]]$ *such that each of the h power series* $\varphi(F_i)$ *is regular in* $X_n$.

It is sufficient to apply the lemma to the product $F_1F_2 \cdots F_h$.

We give now a second proof of the fact that $k[[X_1, \cdots, X_n]]$ is noetherian. This proof can be applied *verbatim* to rings of convergent power series.

THEOREM 4′. *If $k$ is a field, the formal power series ring $k[[X_1, \cdots X_n]]$ is noetherian.*

We prove by induction on $n$ that every ideal $\mathfrak{A}$ in $k[[X_1, \cdots, X_n]]$ has a finite basis (the cases $n=0$ and $n=1$ being trivial). We may suppose that $\mathfrak{A} \neq (0)$. By replacing, if necessary, $\mathfrak{A}$ by an automorphic image $\varphi(\mathfrak{A})$, we may suppose that $\mathfrak{A}$ contains a power series $F$ which is regular in $X_n$ (Lemma 3). For every $G$ in $\mathfrak{A}$, we may write then $G = UF + \sum_{i=0}^{s-1} R_i X_n^{\,i}$ (Theorem 5). In other words, if we denote by $S'$ the power series ring $k[[X_1, \cdots, X_{n-1}]]$, we have $\mathfrak{A} = (F) + \mathfrak{A} \cap (S' + S'X_n + \cdots + S'X_n^{\,s-1})$. As $S'$ is a noetherian ring, by hypothesis, $\mathfrak{A} \cap (S' + S'X_n + \cdots + S'X_n^{\,s-1})$ is a finitely generated $S'$-module, since it is a submodule of the finitely generated $S'$-module $S' + S'X_n + \cdots + S'X_n^{\,s-1}$. A finite system of generators of $\mathfrak{A} \cap (S' + \cdots + S'X_n^{\,s-1})$ will thus constitute, together with $F$, a finite basis of $\mathfrak{A}$. Q.E.D.

We end this section with another application of the Weierstrass preparation theorem. The proof we will give can be applied almost *verbatim* to rings of convergent power series.

THEOREM 6. *If $k$ is a field, the formal power series ring $k[[X_1. \cdots, X_n]]$ is a unique factorization domain.*

PROOF. We proceed by induction on $n$, the cases $n=0$ and $n=1$ being trivial. Since $k[[X_1, \cdots, X_n]]$ is noetherian, we have to prove that, if $F$ is an irreducible power series, then the principal ideal $(F)$ is prime; in other words, we have to prove that, if $GH \in (F)$, then either $G$ or $H$ is a multiple of $F$. Let us write $GH = DF$. By replacing, if necessary, the series $F$, $G$, $H$, $D$ by automorphic images $\sigma(F)$, $\sigma(G)$, $\sigma(H)$, $\sigma(D)$, we may suppose that $F$, $G$, $H$, $D$ are regular in $X_n$ (corollary to Lemma 3). We denote by $F'$, $G'$, $H'$, $D'$ the distinguished pseudo-polynomials associated with $F$, $G$, $H$, $D$ (Corollary 1 to Theorem 5). Since the power series $G'H'$ differs from $GH$ by a unit only, and since it is a distinguished pseudo-polynomial of the right degree in $X_n$, it is the distinguished pseudo-polynomial associated with $GH$. Similarly $D'F'$ is the distinguished pseudo-polynomial associated with $DF$. As $DF = GH$, we have $D'F' = G'H'$, since the distinguished pseudo-polynomial associated with a given power series is unique.

Now, $F'$ is an irreducible element of $k[[X_1, X_2, \cdots, X_{n-1}]][X_n]$. In fact, assume that $g(X_1, X_2, \cdots, X_{n-1}; X_n)$ is a factor of $F'$ in $k[[X_1, X_2, \cdots, X_{n-1}]][X_n]$, not a unit in this latter ring. The leading coefficient of $g$ is a unit in $k[[X_1, X_2, \cdots, X_{n-1}]]$ since the leading coefficient of $F'$ is 1 (both $g$ and $F'$ being regarded as polynomials in $X_n$). Therefore $g$ must be of positive degree in $X_n$ (for $g$ is not a unit in $k[[X_1, X_2, \cdots, X_{n-1}]][X_n]$) and also $g(0, 0, \cdots, 0; X_n)$ must be of positive degree in $X_n$. Consequently $g(0, 0, \cdots, 0; X_n)$ is of the form $cX_n^h$, $h \geqq 1$, $c \in k$, $c \neq 0$, since $F'(0, 0, \cdots, 0; X_n)$ is also of this form. This shows that $g(0, 0, \cdots, 0; 0)=0$, i.e., that $g(X_1, X_2, \cdots, X_{n-1}, X_n)$ is a non-unit in $k[[X_1, X_2, \cdots, X_{n-1}, X_n]]$. Since $F$ is an irreducible element of $k[[X_1, X_2, \cdots, X_{n-1}, X_n]]$, we have proved that $F'$ is also an irreducible element of $k[[X_1, X_2, \cdots, X_{n-1}]][X_n]$. By the induction hypothesis, $k[[X_1, X_2, \cdots, X_{n-1}]]$ is a UFD, whence also $k[[X_1, X_2, \cdots, X_{n-1}]][X_n]$ is also a UFD. (Vol. I, Ch. I, § 18, Theorem 13.) Thus, from $D'F'=G'H'$ we deduce that either $G'$ or $H'$ is a multiple of $F'$ in $k[[X_1, X_2, \cdots, X_{n-1}]][X_n]$. Hence, *a fortiori*, either $G'$ or $H'$ is a multiple of $F'$ in $k[[X_1, X_2, \cdots, X_{n-1}, X_n]]$. Since $F'$, $G'$ and $H'$ differ from $F$, $G$ and $H$ only by unit factors in $k[[X_1, X_2, \cdots, X_{n-1}, X_n]]$, we conclude that either $G$ or $H$ is a multiple of $F$. This completes the proof.

COROLLARY. *If $F(X_1, X_2, \cdots, X_n)$ is a power series which is regular in $X_n$ and is an irreducible element of $k[[X_1, X_2, \cdots, X_n]]$, then the quotient field of the residue class ring $S/SF$ is a simple algebraic extension of the quotient field of $k[[X_1, X_2, \cdots, X_{n-1}]]$.*

This follows immediately from Corollary 2 of Theorem 5.

## § 2. Graded rings and homogeneous ideals.

Let $A$ be a ring and let $R=A[X_1, X_2, \cdots, X_n]$ be the polynomial ring over $A$, in $n$ indeterminates. Every element $F$ in $R$ can be written in the form of a finite sum $F=F_0+F_1+\cdots+F_j+\cdots$, where $F_j$ is either zero or a form of degree $j$. The form $F_j$ is called the *homogeneous component of degree $j$ of $F$*. The product of two *homogeneous* polynomials $f$ and $g$ is again homogeneous, and if $fg \neq 0$ then $\partial(fg)=\partial(f)+\partial(g)$ ($\partial=$degree). The homogeneous polynomials of a given degree $q$ form, together with zero, an additive group and a finite $A$-module $R_q$. We have

$$(1) \qquad R_q R_{q'} \subset R_{q+q'},$$

and $R$ is an infinite (weak) direct sum (see Vol. I, Ch. III, § $12^{\text{bis}}$) of the subgroups $R_q$:

$$(2) \qquad R = \sum_{q=-\infty}^{+\infty} R_q, \quad \text{the sum being direct,}$$

where, *in the present case of polynomial rings*, we have $R_q=(0)$ if $q<0$.

An ideal $\mathfrak{A}$ in $R$ is said to be *homogeneous* if the relation $F \in \mathfrak{A}$ implies that all homogeneous components of $F$ are in $\mathfrak{A}$.

In this section we shall derive a number of properties of homogeneous ideals. However, we shall not restrict ourselves to polynomial rings. We shall study homogeneous ideals in rings which are more general than polynomial rings, namely in *graded rings.*

DEFINITION. *A ring R is called a graded ring if it is a (weak) direct sum (in the sense of Vol. I, Ch. III, § 12*[bis]*) of additive subgroups* $R_q$ *of R satisfying relation* (1); *here q ranges over the set J of integers. An element of R is said to be homogeneous if it belongs to an* $R_q$, *and is said to be homogeneous of degree q if it belongs to* $R_q$ *and is different from zero.*

In a graded ring $R$ we have therefore the direct decomposition (2); it signifies that every non-zero element $F$ of $R$ can be written, in a unique way, as a finite sum of non-zero homogeneous elements of distinct degrees. These elements will be called the *homogeneous components* of $F$, and the homogeneous component of $F$ of least degree will be called the *initial component* of $F$.

If $S$ is a subring of $R$ we say that $S$ is *graded subring* of $R$ if $S$ is the (direct) sum of its subgroups $S_q = S \cap R_q$, i.e., if we have $S = \sum S_q$. It is clear that the sum is then necessarily direct and that $S$ is a graded ring.

We define homogeneous ideals in a graded ring in the same way as we have defined it above for polynomial rings. This definition can also be expressed by saying that an ideal $\mathfrak{A}$ in a graded ring $R$ is homogeneous if $\mathfrak{A}$ is also a graded subring of $R$.

Let $R$ and $R'$ be two graded rings: $R = \sum_q R_q$, $R' = \sum R'_q$. A homomorphism $\varphi$ of $R$ into $R'$ is said to be *homogeneous* of degree $s$ if $\varphi(R_q) \subset R'_{q+s}$ for all $q$.

LEMMA 1. (a) *If* $\varphi$ *is a homogeneous homomorphism of a graded ring R into a graded ring R', then the kernel* $\mathfrak{A}$ *of* $\varphi$ *is a homogeneous ideal in R, and the image of* $\varphi$ *is a graded subring of R'.* (b) *If* $\mathfrak{A}$ *is a homogeneous ideal in a graded ring R and* $\varphi$ *is the canonical homomorphism of R onto the ring* $R/\mathfrak{A}$, *then* $R/\mathfrak{A}$ *is a graded ring with respect to the decomposition* $R/\mathfrak{A} = \sum_q \varphi(R_q)$, *and the canonical homomorphism of R onto* $R/\mathfrak{A}$ *maps in* (1, 1) *fashion the set of homogeneous ideals of R containing* $\mathfrak{A}$ *onto the set of all homogeneous ideals of* $R/\mathfrak{A}$.

PROOF. Assume that $\varphi$ is a homogeneous homomorphism of $R$ into $R'$, of degree $s$. Let $F = \sum F_q (F_q \in R_q)$ be an element of the kernel $\mathfrak{A}$ of $\varphi$. We have $\sum \varphi(F_q) = 0$, with $\varphi(F_q) \in R'_{q+s}$, and therefore necessarily $\varphi(F_q) = 0$ for all $q$. This shows that all the homogeneous components of $F$ belong to $\mathfrak{A}$, whence $\mathfrak{A}$ is homogeneous.

Since $R=\sum R_q$ we find that $\varphi(R)=\sum \varphi(R_q)$, and since $\varphi(R_q)$ obviously coincides with $\varphi(R)\cap R'_{q+s}$, it follows at once that the image $S'=\varphi(R)$ is a graded subring of $R'$.

Now, let $\mathfrak{A}$ be a homogeneous ideal in a graded ring $R$ and let $\varphi$ be the canonical homomorphism of $R$ onto $R/\mathfrak{A}$. We set $S=R/\mathfrak{A}$, $S_q=\varphi(R_q)$. From $R=\sum R_q$ follows $S=\sum S_q$, and from $R_qR_{q'}\subset R_{q+q'}$ we deduce that $S_qS_{q'}\subset S_{q+q'}$. It remains to prove that the sum $\sum S_q$ is direct, or—equivalently—that if a finite sum $\bar{F}=\bar{F}_h+\bar{F}_{h+1}+\cdots$, with $\bar{F}_q\in S_q$, is zero, then each term $\bar{F}_q$ is zero. But this follows directly from our assumption that the ideal $\mathfrak{A}$ is homogeneous. The last statement of the lemma is obvious.

Of particular importance in this chapter will be those graded rings which contain a ring $A$ and are homomorphic images of polynomial rings $A[X_1, X_2, \cdots, X_n]$, with a *homogeneous* ideal in $A[X_1, X_2, \cdots, X_n]$ as kernel. We call such rings *finite homogeneous rings, over A.* More precisely: a ring $\bar{R}$, containing a ring $A$ and finitely generated over $A$, is *homogeneous* if there exists a homomorphism $\varphi$ of a polynomial ring $R=A[X_1, X_2, \cdots, X_n]$ onto $\bar{R}$ such that $\varphi$ is the identity on $A$ and such that the kernel of $\varphi$ is a homogeneous ideal in $R$. If we set $x_i=\varphi(X_i)$, then $\bar{R}=A[x_1, x_2, \cdots, x_n]$, and the homogeneity of the ring $\bar{R}$ signifies that every algebraic relation $F(x_1, x_2, \cdots, x_n)=0$ between the generators $x_i$, with coefficients in $A$, is a consequence of homogeneous relations. By the preceding lemma, a homogeneous ring $\bar{R}=A[x_1, x_2, \cdots, x_n]$ is a graded ring, the subgroup $\bar{R}_q$ of homogeneous elements of degree $q$ being the set of elements of the form $f(x_1, x_2, \cdots, x_n)$, where $f$ is a form of degree $q$, with coefficients in $A$. Note that a homogeneous ring $\bar{R}$ admits a set of generators $x_i$ which are *homogeneous and of the same degree.* It is not difficult to give examples of finitely generated graded rings (over a given ring $A$) which are not homogeneous. For instance, it can be shown (see end of this section) that the integral closure of a finite homogeneous integral domain, over a field $k$, is a finitely generated *graded* ring; however, this ring is not necessarily a homogeneous ring.

THEOREM 7. *In order that an ideal $\mathfrak{A}$ in a graded ring be homogeneous it is necessary and sufficient that $\mathfrak{A}$ possess a basis (finite or infinite) consisting of homogeneous elements.*

PROOF. Suppose that $\mathfrak{A}$ is homogeneous. If $\{F^{(\alpha)}\}$ is any basis of $\mathfrak{A}$, then all the homogeneous components $F_q^{(\alpha)}$ of all the $F^{(\alpha)}$ also belong to $\mathfrak{A}$ and obviously form a basis of $\mathfrak{A}$. Suppose, conversely, that an ideal $\mathfrak{A}$ possesses a basis $\{G^{(\lambda)}\}$ consisting of homogeneous elements. Let $F$ be any element of $\mathfrak{A}$ and let $\{F_q\}$ be the set of homogeneous components of $F$. We have then $F=\sum_\lambda P^{(\lambda)}G^{(\lambda)}$, $P^{(\lambda)} \in R$. If $P^{(\lambda)}=\sum P_q^{(\lambda)}$ is the decomposition of $P^{(\lambda)}$ into its homogeneous components, then $F=\sum_{\lambda, q} P_q^{(\lambda)}G^{(\lambda)}$, and in this sum the partial sum $\sum_{q+d(\lambda)=m} P_q^{(\lambda)}G^{(\lambda)}$, where $d(\lambda)$ denotes the degree of $G^{(\lambda)}$, is the homogeneous component $F_m$ of $F$, of degree $m$. Hence $F_m \in \mathfrak{A}$, and $\mathfrak{A}$ is homogeneous.

The class of homogeneous ideals in a graded ring $R$ is closed under the standard ideal-theoretic operations. More precisely:

THEOREM 8. *Let $\mathfrak{A}$ and $\mathfrak{B}$ be ideals in a graded ring.* (a) *If $\mathfrak{A}$ and $\mathfrak{B}$ are homogeneous, then $\mathfrak{A}+\mathfrak{B}$, $\mathfrak{A}\mathfrak{B}$, $\mathfrak{A}\cap\mathfrak{B}$ and $\mathfrak{A}:\mathfrak{B}$ are homogeneous.* (b) *If $\mathfrak{A}$ is homogeneous, then its radical $\sqrt{\mathfrak{A}}$ is homogeneous.*

PROOF. The assertions relative to $\mathfrak{A}+\mathfrak{B}$ and $\mathfrak{A}\mathfrak{B}$ are trivial, by Theorem 7. The assertion relative to $\mathfrak{A}\cap\mathfrak{B}$ results trivially from the definition. For $\mathfrak{A}:\mathfrak{B}$, take a basis $\{B^{(\lambda)}\}$ of $\mathfrak{B}$ consisting of homogeneous elements. If $F \in \mathfrak{A}:\mathfrak{B}$ and if $F=\sum_j F_j$ is the decomposition of $F$ into its homogeneous components, then we have $FB^{(\lambda)}=\sum_j F_jB^{(\lambda)} \in \mathfrak{A}$ for every $\lambda$. Since, for fixed $\lambda$, the products $F_jB^{(\lambda)}$ are homogeneous elements of different degrees, and since $\mathfrak{A}$ is homogeneous, we deduce that $F_jB^{(\lambda)} \in \mathfrak{A}$, for every $j$ and every $\lambda$. Therefore $F_j \in \mathfrak{A}:\mathfrak{B}$ for every $j$ (since $\{B^{(\lambda)}\}$ is a basis of $\mathfrak{B}$), and $\mathfrak{A}:\mathfrak{B}$ is homogeneous.

We now consider the radical $\sqrt{\mathfrak{A}}$ of a homogeneous ideal $\mathfrak{A}$. Let $F$ be an element of $\sqrt{\mathfrak{A}}$ and let $F=F_s+F_{s+1}+\cdots$ be the decomposition of $F$ into its homogeneous components, where $F_s$, then, is the initial component of $F$. We have $F^\rho=F_s^\rho+$ terms of degree $>s\rho$, and $F^\rho \in \mathfrak{A}$ for a suitable integer $\rho$. Since $\mathfrak{A}$ is homogeneous, it follows that $F_s^\rho \in \mathfrak{A}$, $F_s \in \sqrt{\mathfrak{A}}$. But then $F-F_s \in \sqrt{\mathfrak{A}}$ and therefore, by the same argument, also the initial component of $F-F_s$ belongs to $\sqrt{\mathfrak{A}}$. In this fashion we find that all the homogeneous components of $F$ belong to $\sqrt{\mathfrak{A}}$. Q.E.D.

COROLLARY. *If a primary ideal $\mathfrak{q}$ in a graded ring $R$ is homogeneous then its associated prime ideal is also homogeneous.*

Concerning prime homogeneous ideals the following useful remark can be made: in order to prove that a given *homogeneous* ideal $\mathfrak{p}$ is prime it is *sufficient* to verify that the property "$f \notin \mathfrak{p}$, $g \notin \mathfrak{p} \Rightarrow fg \notin \mathfrak{p}$" holds for homogeneous elements $f$ and $g$. In fact, assume that this property holds for homogeneous elements $f$ and $g$ and let $F$ and $G$ be two arbitrary elements of $R$ such that $F \notin \mathfrak{p}$, $G \notin \mathfrak{p}$. Let $F=F_r+F_{r+1}+\cdots$, $G=G_s+G_{s+1}+\cdots$ be the decompositions of $F$ and $G$ into homogeneous components. Let $F_{r+\rho}$ and $G_{s+\sigma}$ be the first homogeneous component of $F$ and $G$ respectively which does not belong to $\mathfrak{p}(\rho \geqq 0, \sigma \geqq 0)$. Then $F_{r+\rho}G_{s+\sigma} \notin \mathfrak{p}$, and therefore

$$[F-(F_r+F_{r+1}+ \cdots +F_{r+\rho-1})][G-(G_s+G_{s+1}+ \cdots +G_{s+\sigma-1})] \notin \mathfrak{p}$$

(since $\mathfrak{p}$ is homogeneous). Since $F_r+F_{r+1}+ \cdots +F_{r+\rho-1}$ and $G_s+G_{s+1}+ \cdots +G_{s+\sigma-1}$ belong to $\mathfrak{p}$, it follows that $FG \notin \mathfrak{p}$.

The above remark can be generalized to primary ideals:

LEMMA 2. *If a homogeneous ideal $\mathfrak{q}$ in a graded ring $R$ has the property that whenever a product $fg$ of two homogeneous elements belongs to $\mathfrak{q}$ and one factor, say $f$, does not belong to $\mathfrak{q}$, some power of the second factor $g$ belongs to $\mathfrak{q}$, then $\mathfrak{q}$ is a primary ideal.*

PROOF. The proof will be similar to the one given above for prime ideals, and we shall use the same notations. Assume that $F \notin \mathfrak{q}$ and that $FG \in \mathfrak{q}$. We have to show that $G \in \sqrt{\mathfrak{q}}$. In the proof we may assume that $F_r \notin \mathfrak{q}$, for we may replace $F$ by $F_{r+\rho}+F_{r+\rho+1}+ \cdots$ without affecting the conditions $F \notin \mathfrak{q}$ and $FG \in \mathfrak{q}$. The product $F_rG_s$ is either zero or is the initial component of $FG$, and hence $F_rG_s \in \mathfrak{q}$ since $\mathfrak{q}$ is homogeneous. Since $F_r \notin \mathfrak{q}$ it follows that $G_s \in \sqrt{\mathfrak{q}}$. Assume that it has already been proved that $G_s, G_{s+1}, \cdots, G_{s+m}$ belong to $\sqrt{\mathfrak{q}}$ and let $\mu$ be an integer such that $(G_s+G_{s+1}+ \cdots +G_{s+m})^\mu \in \mathfrak{q}$. Then $F(G-G_s- \cdots -G_{s+m})^\mu \in \mathfrak{q}$, and therefore, using again the fact that $F_r \notin \mathfrak{q}$, we find that $G_{s+m+1}^\mu \in \sqrt{\mathfrak{q}}$. Hence $G_{s+m+1} \in \sqrt{\mathfrak{q}}$. Q.E.D.

We shall use Lemma 2 and the next lemma for the study of primary decompositions of homogeneous ideals.

LEMMA 3. *Let $\mathfrak{A}$ be an ideal in a graded ring $R$ and let $\mathfrak{A}^\star$ denote the ideal generated by the homogeneous elements belonging to $\mathfrak{A}$. Then if $\mathfrak{A}$ is prime or primary, also $\mathfrak{A}^\star$ is prime or primary.*

PROOF. Let $F$ and $G$ be homogeneous elements such that $F \notin \mathfrak{A}^\star$ and $FG \in \mathfrak{A}^\star$. Then $F \notin \mathfrak{A}$. If $\mathfrak{A}$ is prime then $G \in \mathfrak{A}$; if $\mathfrak{A}$ is primary then $G^\rho \in \mathfrak{A}$, for some $\rho$. Since $G$ is homogeneous, it follows, by the definition of $\mathfrak{A}^\star$, that $G$ (or $G^\rho$) belongs to $\mathfrak{A}^\star$. Hence, by Lemma 2, the proof is complete.

We note that $\mathfrak{A}^\star$ is the greatest homogeneous ideal contained in $\mathfrak{A}$.

THEOREM 9. *Let $\mathfrak{A}$ be a homogeneous ideal in a graded ring $R$. If $\mathfrak{A}$ admits a primary representation $\mathfrak{A} = \cap\, \mathfrak{q}_j$, then it also admits a primary representation $\mathfrak{A} = \cap\, \mathfrak{q}^\star_j$ in which the $\mathfrak{q}^\star_j$ are primary homogeneous ideals.*

PROOF. We take for $\mathfrak{q}^\star_j$ the greatest homogeneous ideal contained in $\mathfrak{q}_j$. By Lemma 3, each $\mathfrak{q}^\star_j$ is a primary ideal, and we have $\cap\, \mathfrak{q}^\star_j \subset \mathfrak{A}$. On the other hand, since $\mathfrak{A}$ is homogeneous and $\mathfrak{A} \subset \mathfrak{q}_j$, it follows that $\mathfrak{A} \subset \mathfrak{q}^\star_j$, whence $\mathfrak{A} \subset \cap\, \mathfrak{q}^\star_j$. Thus, $\mathfrak{A} = \cap\, \mathfrak{q}^\star_j$, and the theorem is proved.

COROLLARY. *Let $\mathfrak{A}$ be a homogeneous ideal in a graded ring $R$ and assume that $\mathfrak{A}$ admits a primary representation. Then the isolated components of $\mathfrak{A}$ are homogeneous, and so are the associated prime ideals of $\mathfrak{A}$.*

This follows from Theorem 9 and from the uniqueness of the isolated primary components and of all the associated prime ideals of $\mathfrak{A}$.

Some of the direct components $R_q$ of a graded ring $R$ may be zero. An important case is the one in which $R_q=0$ for all negative integers $q$; that is so, for instance, if $R$ is a polynomial ring $A[X_1, X_2, \cdots, X_n]$ over a ring $A$. If $R_q=0$ for all negative $q$ then the ideal generated by the homogeneous elements of positive degree is given by $\sum_{q>0} R_q$ and is not the unit ideal unless $R_0=0$. This ideal shall be denoted by $\mathfrak{X}$. It is clear that if $R_0$ has no proper zero divisor, then $\mathfrak{X}$ is a prime ideal. A homogeneous ideal $\mathfrak{A}$ in $R$ shall be called *irrelevant* if $\mathfrak{X}\subset\sqrt{\mathfrak{A}}$. The consideration of the ideal $\mathfrak{X}$ is particularly useful if $R_0$ is a field or if $R$ is a polynomial ring $A[X_1, X_2, \cdots, X_n]$ over a ring $A$. In the first case, $\mathfrak{X}$ is a maximal ideal in $R$, it contains every proper homogeneous ideal, and every irrelevant ideal is either the unit ideal or is a primary ideal with $\mathfrak{X}$ as associated prime ideal (Vol. I, Ch. III, § 9, Theorem 13, Corollary 2). In the second case, $\mathfrak{X}$ is generated by $X_1, X_2, \cdots, X_n$.

The next two lemmas refer to finitely generated graded rings, i.e., to graded rings of the form $R=A[x_1, x_2, \cdots, x_n]$, where $A$ is a *noetherian* ring, $R_0=A$ and each $x_i$ is homogeneous of positive degree. These lemmas are useful in some applications. If $\mathfrak{B}$ is a homogeneous ideal in $R$ and $B=\cap\, \mathfrak{q}_i$ is a primary irredundant representation of $\mathfrak{B}$, the $\mathfrak{q}_i$ being homogeneous ideals, we denote by $\mathfrak{B}^\star$ the intersection of those primary components $\mathfrak{q}_j$ of $\mathfrak{B}$ which are non-irrelevant. Clearly $\mathfrak{B}^\star$ is uniquely determined by $\mathfrak{B}$, for the prime ideals $\sqrt{\mathfrak{q}_j}$ form an isolated system of prime ideals of $\mathfrak{B}$ (see Vol. I, Ch. IV, § 5, p. 212). For *any* ideal $\mathfrak{A}$ in $R$ we denote by $\mathfrak{A}_q$ the set $\mathfrak{A}\cap R_q$.

LEMMA 4. *If $\mathfrak{B}$ is a homogeneous ideal, then there exists an integer $s_0$ such that $\mathfrak{B}_s=\mathfrak{B}_s^\star$ for $s\geq s_0$ (in other words, $\mathfrak{B}$ and $\mathfrak{B}^\star$ coincide in the homogeneous elements of sufficiently high degree). Furthermore, $\mathfrak{B}^\star$ is the largest homogeneous ideal enjoying this property; in other words, if a homogeneous ideal $\mathfrak{B}'$ is such that there exists an integer $m$ such that $\mathfrak{B}'_s=\mathfrak{B}_s$ for $s\geqq m$, then $\mathfrak{B}'\subset\mathfrak{B}^\star$ and $\mathfrak{B}'^\star=\mathfrak{B}^\star$.*

PROOF. Let $\mathfrak{B}=\bigcap_{i=1}^{k}\mathfrak{q}_i$, where $\mathfrak{q}_i$ is non-irrelevant for $i=1, \cdots, h$, and is irrelevant for $i=h+1, \cdots, k$. We have $\mathfrak{B}_s=\bigcap_{i=1}^{k}\mathfrak{q}_{i,s}$. For

$i=h+1, \cdots, k$, $\mathfrak{q}_i$ contains a power of $\mathfrak{X}$, whence, for large $s$, $\mathfrak{q}_{i,s}$ is the entire group $R_s$.† Thus, for $s$ large, we have $\mathfrak{B}_s = \bigcap_{i=1}^{h} \mathfrak{q}_{i,s} = \mathfrak{B}_s^{\star}$, and this proves our first assertion. Suppose now that $\mathfrak{B}'$ is as indicated above. For $1 \leqq i \leqq h$, $\mathfrak{q}_i$ is non-irrelevant, whence its radical $\mathfrak{p}_i$ does not contain the ideal $\mathfrak{X}$. Therefore for any given $i$, $1 \leqq i \leqq h$, there exists an index $j$ depending on $i$ such that $x_j \notin \mathfrak{p}_i$. From this it follows that if $F \in \mathfrak{B}'$, then $F \in \mathfrak{q}_i$, since $x_j^m F \in \mathfrak{B} \subset \mathfrak{q}_i$. In other words, we have $\mathfrak{B}' \subset \mathfrak{B}^{\star}$. Applying the same result to the ideal $\mathfrak{B}'^{\star}$ (which also coincides with $\mathfrak{B}$ in the homogeneous elements of large degrees), we get $\mathfrak{B}'^{\star} \subset \mathfrak{B}^{\star}$, and, by exchanging $\mathfrak{B}$ and $\mathfrak{B}'$ we have $\mathfrak{B}^{\star} \subset \mathfrak{B}'^{\star}$. Hence $\mathfrak{B}^{\star} = \mathfrak{B}'^{\star}$ and all our assertions are proved.

LEMMA 5. *The ideal $\mathfrak{B}^{\star}$ is equal to $\mathfrak{B} : \mathfrak{X}^s$ for $s$ large enough.*

PROOF. The ideals $(\mathfrak{B} : \mathfrak{X}^s)$ form an ascending sequence; since $R$ is noetherian, this sequence stops increasing for large $s$: $(\mathfrak{B} : \mathfrak{X}^s) = (\mathfrak{B} : \mathfrak{X}^{s+1}) = \cdots$. With the notations of Lemma 4, we have $\mathfrak{q}_i : \mathfrak{X}^s = R$ for $h+1 \leqq i \leqq k$ and $s$ large enough, since $\mathfrak{q}_i$ contains all high powers of $\mathfrak{X}$. For $1 \leqq i \leqq h$, there exists an index $j(i)$ such that $x_{j(i)} \notin \sqrt{\mathfrak{q}_i}$, whence a relation such as $F x_{j(i)}^s \in \mathfrak{q}_i$ implies $F \in \mathfrak{q}_i$; in other words, we have $\mathfrak{q}_i = \mathfrak{q}_i : \mathfrak{X}^s$ for every $s$ and every $i$ such that $1 \leq i \leq h$. From this it follows that, for $s$ large, we have

$$\mathfrak{B} : \mathfrak{X}^s = \bigcap_{i=1}^{k} (\mathfrak{q}_i : \mathfrak{X}^s) = \bigcap_{i=1}^{h} (\mathfrak{q}_i : \mathfrak{X}^s) = \bigcap_{i=1}^{h} \mathfrak{q}_i = \mathfrak{B}^{\star}.$$

Our next theorem refers to a finite *homogeneous* ring $A[x_1, x_2, \cdots, x_n]$, where $A$ is now not necessarily noetherian.

THEOREM 10. *Let $\mathfrak{A}$ be an ideal in a finite homogeneous ring $A[x_1, x_2, \cdots, x_n]$ (all $x_i$ being homogeneous of the same degree). If $\mathfrak{A}$ is homogeneous then for every element $F(x_1, x_2, \cdots, x_n)$ in $\mathfrak{A}$ and for every $t$ in $A$ we have $F(tx_1, tx_2, \cdots, tx_n) \in \mathfrak{A}$. The converse is true if $A$ is an infinite field.*

PROOF. Let $F(x_1, x_2, \cdots, x_n) = \sum_j F_j(x_1, x_2, \cdots, x_n)$ be the decomposition of an element $F$ of $\mathfrak{A}$ into homogeneous components ($F_j$ stands for a form of degree $j$, with coefficients in $A$). We have

$$F(tx_1, tx_2, \cdots, tx_n) = \sum_j t^j F_j(x_1, x_2, \cdots, x_n).$$

If $\mathfrak{A}$ is homogeneous then $F_j(x_1, x_2, \cdots, x_n) \in \mathfrak{A}$, whence $F(tx_1, tx_2, \cdots, tx_n) \in \mathfrak{A}$. To prove the partial converse, we have only to

† If $h$ is an integer such that each $x_i$ is homogeneous of degree $\leqq h$ and if $\mathfrak{X}^q \subset \mathfrak{q}_i$ then $\mathfrak{q}_{i,s} \supset R_s$ as soon as $s \geqq hq$.

show that the (finite-dimensional) vector space $V$ which is spanned over $A$ by the homogeneous components $F_j$ of $F$ is also spanned by the family $\mathscr{F}$ of elements $F(tx_1, tx_2, \cdots, tx_n)$, $t \in A$. (It is clear that $\mathscr{F} \subset V$.) For that it is sufficient to show that any linear function $f$ on $V$ (with values in $A$) which is zero on $\mathscr{F}$ is also zero on $V$. Let $f(F_j) = c_j$. We have then $f(F(tx_1, tx_2, \cdots, tx_n)) = \sum c_j t^j = 0$ for all $t$ in $A$. Since $A$ is an infinite field, the vanishing of the polynomial $\sum c_j X^j$ for all values of $X$ in $A$ implies that all the coefficients $c_j$ of that polynomial must be zero. Hence $f = 0$ on $V$, as asserted.

REMARK. If $R = A[X_1, X_2, \cdots, X_n]$ is a polynomial ring and if $\nu$ denotes the degree of $F$, then the polynomial $\sum c_j X^j$ is of degree $\nu$, and the conclusion that $F$ belongs to $\mathfrak{A}$ would still be true in the case of a finite field $A$, provided $A$ has at least $\nu+1$ distinct elements $t_1, t_2, \cdots, t_{\nu+1}$. Another proof can be obtained by using the Vandermonde determinant $|t_i^j|$. The following is an example in which the second part of Theorem 10 fails to hold for a finite field $A$. Assume that $A$ is a field with two elements (0, 1). In this case, if $F(X_1, X_2, \cdots, X_n)$ is any polynomial *whose constant term is zero* then $F(tX_1, tX_2, \cdots, tX_n)$, $t \in A$, is either $F(X_1, X_2, \cdots, X_n)$ or 0. Thus, *every* ideal $\mathfrak{A}$ in $A[X_1, X_2, \cdots, X_n]$ which is contained in the maximal ideal $(X_1, X_2, \cdots, X_n)$ satisfies the condition "$F \in \mathfrak{A} \Rightarrow F(tX_1, tX_2, \cdots, tX_n) \in \mathfrak{A}$."

We shall conclude this section with the proof of a result which concerns the integral closure of a graded domain and which, in the special case of homogeneous finite integral domains, is of basic importance in the theory of normal varieties in the projective space (see § $4^{\text{bis}}$).

Let $R = \sum R_q$ be a graded *domain* and let $K$ be the quotient field of $R$. It is easy to see that *the element* 1 *of* $R$ *is a homogeneous element of degree zero*. For if $1 = \omega_m + \omega_{m+1} + \cdots + \omega_n$ ($\omega_q \in R_q$, $n \geqq m$, $\omega_m \neq 0$, $\omega_n \neq 0$), then $1 = \omega_m^2 + 2\omega_m\omega_{m+1} + \cdots + \omega_n^2 = \omega_m + \cdots \omega_n$. Since $\omega_m^2 \neq 0$ and $\omega_n^2 \neq 0$ it follows from the equality $\omega_m^2 + \cdots + \omega_n^2 = \omega_m + \cdots + \omega_n$ that $\omega_m^2 = \omega_m$ and $\omega_n^2 = \omega_n$. Since $\omega_m$ and $\omega_n$ are homogeneous, this implies that $m = n = 0$.

The group $R_0$ is obviously a ring, and is not the nullring since $1 \in R_0$.

An element $x$ of the quotient field $K$ will be said to be *homogeneous* if it is a quotient of homogeneous elements of $R$. If $x$ is a homogeneous element, and if, say, $x = \xi_q/\eta_r$, with $\xi_q \varepsilon R_q$ and $\eta_r \in R_r$, then it is immediately seen that the integer $q-r$ depends only on $x$. We say that $x$ is homogeneous of *degree* $q-r$. It is clear that the product and quotient of homogeneous elements are homogeneous and that the degree of a product is the sum of the degrees of the factors. Furthermore, the homogeneous elements of $K$, *of a given degree*, form, together with 0, a group. In particular, it follows that the homogeneous elements of $K$ which are of degree zero form a field. We shall denote this field by $K_0$.

More generally, we shall denote by $K_q$ the set of elements of $K$ which are homogeneous of degree $q$. As was pointed out above, we have $K_qK_{q'} \subset K_{q+q'}$ and hence the sum $\sum_{q \in J} K_q$ is a subring of $K$. Furthermore, it is easily seen that *the sum* $\sum K_q$ *is direct*. In fact, if we have a relation of the form $\xi_m + \xi_{m+1} + \cdots + \xi_n = 0$ $(\xi_q \in K_q,\ n \geqq m)$, then we express the $\xi_q$ as quotients of homogeneous elements of $R$, with the same denominator $\omega' \epsilon R_s$; say, $\xi_q = \omega_{s+q}/\omega'$, where $\omega_i \in R_i$. Then the above relation yields the relation $\omega_{s+m} + \omega_{s+m+1} + \cdots + \omega_{s+n} = 0$, and hence the $\omega_i$ are all zero, whence also the $\xi_q$ are all zero. We have shown therefore that the ring $\sum K_q$ is again a graded ring.

It is clear that the integers $q$ such that $K_q \neq 0$ form a subgroup $J'$ of the additive group $J$ of integers. Hence $J' = Jm$, where $m$ is some positive integer (we exclude the trivial case $R = R_0$). We may therefore assume that $J' = J$, for in the contrary case we may simply redefine the degree of the homogeneous elements of $R$ by assigning to any non-zero element of $R_q$ $(q \equiv 0 (\mathrm{mod}\ m))$ the degree $q/m$. We may therefore assume that there exist elements in $K$ which are homogeneous of degree 1.

Let $y \neq 0$ be a homogeneous element of degree 1. If $\xi$ is an element of $K_q$ then $\xi/y^q \in K_0$, $\xi \in K_0[y]$ if $q \geqq 0$ and $\xi \in K_0[1/y]$ if $q < 0$. Hence $R \subset K_0[y, 1/y]$, and therefore $K = K_0(y)$.

Note the relations

(3) $$K_q = K_0 \cdot y^q, \quad \sum_{q \in J} K_q = K_0[y, 1/y].$$

*We assert that $y$ is a transcendental over $K_0$.* For, assume that we have an algebraic relation $a_0y^n + a_1y^{n-1} + \cdots + a_n = 0$, $a_i \in K_0$. Then $a_iy^{n-i} \in K_{n-i}$, and therefore $a_iy^{n-i} = 0$, since $\sum K_q$ is a graded ring. Since $y \neq 0$, it follows that the $a_i$ are all zero, showing that $y$ is a transcendental over $K_0$.

Let $\bar{R}$ be the integral closure of $R$ in $K$. The theorem which we wish to prove is the following:

THEOREM 11. *The ring $\bar{R}$ is a graded subring of the ring $\sum_q K_q$. More precisely: if we set $\bar{R}_q = \bar{R} \cap K_q$, then $\bar{R} = \sum \bar{R}_q$. In the special case in which $R_q = 0$ for all negative integers $q$ also $\bar{R}_q = 0$ for negative $q$.*

PROOF. It was pointed out above that $R \subset K_0[y, 1/y]$. Now $K_0[y]$ is a polynomial ring over a field $K_0$ and is therefore integrally closed in its quotient field $K_0(y)(=K)$. The ring $K_0[y, 1/y]$ is the quotient ring of $K_0[y]$ with respect to the multiplicative system formed by the non-negative powers of $y$; this ring $K_0[y, 1/y]$ is therefore also integrally closed in $K$. (Vol. I, Ch. V, § 3, Example 2, p. 261.) Consequently $\bar{R} \subset K_0[y, 1/y] = \sum K_q$[by (3)]. *Every element of $\bar{R}$ is therefore a sum of homogeneous elements.* In particular, if $R_q = 0$ for all negative $q$, then $R \subset K_0[y]$ and therefore also $\bar{R} \subset K_0[y]$; thus in this special case, *every element of $\bar{R}$ is a sum of homogeneous elements of non-negative degree.*

Let

$$(4) \qquad \xi = \xi_s + \xi_{s+1} + \cdots + \xi_t$$

($\xi_q \in K_q$, $t \geqq s$) be an element of $\bar{R}$. To complete the proof of the theorem we have only to show that each $\xi_q$ ($q = s, s+1, \cdots, t$) is itself an element of $\bar{R}$.

We shall first consider the case in which the ring $R$ is noetherian. Since $\bar{R} \subset \sum K_q$, every element of $\bar{R}$ can be written as a quotient of two elements of $R$ such that the denominator is a homogeneous element. Since $\xi$ is integral over $R$, the ring $R[\xi]$ is a finite $R$-module. We can therefore find a *homogeneous element* $d$ in $R$, $d \neq 0$, such that $d.R[\xi] \subset R$. We have therefore, for every integer $i \geqq 0$, that $d\xi^i \in R$. If $\xi_s$ denotes, as in (4), the initial component of $\xi$, then the initial component of the element $d\xi^i$ of $R$ is $d\xi_s{}^i$. Hence $d\xi_s{}^i \in R$ for every integer $i \geqq 0$. We have therefore shown that all the non-negative powers of $\xi_s$ belong to the finite $R$-module $R.(1/d)$. Since we have assumed that $R$ is noetherian, it follows that the ring $R[\xi_s]$ itself is a finite $R$-module. *Therefore also $\xi_s$ is integral over $R$.* Then also $\xi - \xi_s = \xi_{s+1} + \cdots + \xi_t \in \bar{R}$, and in this fashion we can prove step by step that all the $\xi_q$, $q = s, s+1, \cdots, t$, belong to $\bar{R}$.

In the non-noetherian case we can achieve a reduction to the noetherian case, as follows:

Let

$$(5) \qquad \xi^n + a_1\xi^{n-1} + \cdots + a_n = 0, \quad a_i \in R,$$

be a relation of integral dependence for $\xi$ over $R$, and let $d \neq 0$ be a homogeneous element of $R$ such that $\xi_q d \in R$ for $q = s, s+1, \cdots, t$. We consider the following homogeneous elements of $R$: the element $d$, the products $\xi_q d$ ($q = s, s+1, \cdots, t$) and the homogeneous components of the coefficients $a_1, a_2, \cdots, a_n$ of the above relation (5). We denote these homogeneous elements, in some order, by $x_1, x_2, \cdots, x_N$, and we denote by $A$ the smallest subring of $R$ containing the elements $x_i$. Then $A=J[x_1, x_2, \cdots, x_N]$ if $R$ is of characteristic zero ($J$ = ring of integers) and $A=\mathfrak{J}_p[x_1, x_2, \cdots, x_N]$ if $R$ is of characteristic $p \neq 0$ ($\mathfrak{J}_p$ = prime subfield of $R$). In either case $A$ is a noetherian integral domain. If we set $A_q = A \cap R_q$ then it is immediately seen that $A = \sum A_q$ and that consequently $A$ is a graded subring of $R$. In fact, if $\eta$ is any element of $A$, let $\eta_q$ be the homogeneous component of $\eta$, of a given degree $q$, and let $\eta = f(x_1, x_2, \cdots, x_N)$, where $f(X_1, X_2, \cdots, X_N)$ is a polynomial with coefficients which are integers or integers mod the characteristic $p$ of $R$. If $q_i$ denotes the degree of the homogeneous element $x_i$ of $R$ and $f_q(X_1, X_2, \cdots, X_N)$ denotes the sum of terms $cX_1^{i_1} X_2^{i_2} \cdots X_N^{i_N}$ in $f$ such that $i_1q_1 + i_2q_2 + \cdots + i_Nq_N = q$ ($c \in J$ or $c \in \mathfrak{J}_p$), then it is clear that $\eta_q = f_q(x_1, x_2, \cdots, x_N)$ and hence $\eta_q \in A$.

Since the element $d$ and the products $\xi_q d$, $q = s, s+1, \cdots, t$, are included in the set $\{x_1, x_2, \cdots, x_N\}$, it follows that $\xi$ belongs to the quotient field of $A$. On the other hand, since also the homogeneous components of all the coefficients $a_i$ in (5) are also included in the set $\{x_1, x_2, \cdots, x_N\}$, it follows that $\xi$ is integrally dependent over $A$. Hence by the noetherian case, the homogeneous components $\xi_q$ of $\xi$ are integral over $A$, hence *a fortiori* also over $R$. This completes the proof of the theorem.

Theorem 11 can be generalized as follows:

Let $K'_0$ be an algebraic extension field of $K_0$ and let $K' = K'_0(y)$. We set $K'_q = K'_0 \cdot y^q$ ($q$—an integer), so that $\sum_q K'_q$ is obviously a graded ring. Then we have the following

COROLLARY. *Theorem* 11 *remains true if in the statement of that theorem we replace the field* $K$ *by the field* $K'$, *the graded ring* $\sum_q K_q$ *by the graded ring* $\sum_q K'_q$ *and the ring* $\bar{R}$ *by the integral closure* $\bar{R}'$ *of* $R$ *in* $K'$ (*in particular, we must write* $\bar{R}' = \sum \bar{R}'_q$, *where* $\bar{R}'_q = \bar{R}' \cap K'_q$).

The proof is immediate. For, the ring $\sum \bar{R}'_q$ (weak direct sum of the $\bar{R}'_q$) is obviously a graded ring, having $K'$ as quotient field, and $\bar{R}'$ is also the integral closure of this graded ring, in $K'$. Since $\bar{R}'_q$, by its very definition, consists of all the homogeneous elements of $K'$, of degree $q$, which are integral over the graded ring $\sum \bar{R}'_q$, it follows from Theorem 11 that $\bar{R}' = \sum \bar{R}'_q$.

REMARK. It is easily seen that *if* $z \in \bar{R}'_q$ *then* $z$ *satisfies an equation of the form*

$$(6)\quad z^n + a_q z^{n-1} + a_{2q} z^{n-1} + \cdots + a_{nq} = 0, \quad a_{iq} \in R_{iq},$$

*and that conversely, if an element* $z$ of $K'$ *satisfies such an equation (with the* $a_{iq}$ *in* $R_{iq}$*) then* $z \in \bar{R}'_q$. For, assume that $z \in \bar{R}'_q$ and let

$$z^n + b_1 z^{n-1} + b_2 z^{n-2} + \cdots + b_n = 0, \quad b_i \in R,$$

be an equation of integral dependence for $z$ over $R$. Each of the $n+1$ terms on the left-hand side of this equation belongs to the graded ring $\bar{R}'$. Therefore, if we denote by $a_{iq}$ the homogeneous component of $a_i$, of degree $iq$, then we find (6). Conversely, assume (6). Dividing (6) by $y^{nq}$ and observing that $a_{iq}/y^{iq} \in K_0$, we find that $z/y^q$ is algebraic over $K_0$ and therefore must belong to $K'_0$ (since $K'_0$ is the algebraic closure of $K_0$ in $K'$). Hence the element $z$ is homogeneous of degree $q$, and since it is integral over $R$ (in view of (6)) it must belong to $\bar{R}'_q$.

§ **3. Algebraic varieties in the affine space.** Let $k$ be a field and let $K$ be an algebraically closed extension of $k$. The field $k$ will be referred to as the *ground field*, while $K$ will be called the *coördinate domain*. Given an ideal $\mathfrak{A}$ in the polynomial ring $R = k[X_1, X_2, \cdots, X_n]$, we recall (VI, § 5bis) that *the variety of* $\mathfrak{A}$ in the affine space $A_n^K$ is the set $V$ of all points $(x) = (x_1, x_2, \cdots, x_n)$ $(x_i \in K)$ such that $f(x) = 0$ for all $f$ in $\mathfrak{A}$. We shall denote this variety by $\mathscr{V}(\mathfrak{A})$. The fact that $\mathfrak{A}$ is an ideal in the polynomial ring *over* $k$ is expressed by saying that $V$, the variety of $\mathfrak{A}$, is *defined* over $k$. Any point $(x)$ of $V$ is said to be a *zero* of the ideal $\mathfrak{A}$. For every subset $E$ of $A_n^K$ we denote by $\mathscr{I}(E)$ the set of all polynomials in $k[X_1, X_2, \cdots, X_n]$ which vanish at every point $(x)$ of $E$. Clearly, $\mathscr{I}(E)$ is an ideal. We shall denote by **I** the set of all ideals of the form $\mathscr{I}(E)$, $E \subset A_n^K$.

The set of points in $A_n^K$ which satisfy a finite set of equations $f_1 = 0, f_2 = 0, \cdots, f_q = 0$, where $f \in k[X_1, X_2, \cdots, X_n]$, is a variety, namely it is the variety of the ideal generated by the polynomials $f_1, f_2, \cdots, f_q$. Conversely, every variety can thus be defined by a finite system of polynomial equations, with coefficients in $k$, for every polynomial ideal has a finite basis.

We note the following relations:

(1) $$\mathfrak{A} \subset \mathfrak{B} \Rightarrow \mathscr{V}(\mathfrak{A}) \supset \mathscr{V}(\mathfrak{B}).$$

(1′) $$E \subset F \Rightarrow \mathscr{I}(E) \supset \mathscr{I}(F).$$

(2) $$\mathscr{V}\left(\sum_i \mathfrak{A}_i\right) = \bigcap_i \mathscr{V}(\mathfrak{A}_i).$$

(2′) $$\mathscr{I}\left(\bigcup_i E_i\right) = \bigcap_i \mathscr{I}(E_i).$$

(3) $$\mathscr{V}(\mathfrak{A} \cap \mathfrak{B}) = \mathscr{V}(\mathfrak{A}\mathfrak{B}) = \mathscr{V}(\mathfrak{A}) \cup \mathscr{V}(\mathfrak{B}).$$

(4) $$\mathscr{V}(\mathscr{I}(E)) \supset E.$$

(4′) $$\mathscr{I}(\mathscr{V}(\mathfrak{A})) \supset \mathfrak{A}.$$

(5) $$\mathscr{V}(\mathscr{I}(E)) = E \Leftrightarrow E \text{ is a variety.}$$

(5′) $$\mathscr{I}(\mathscr{V}(\mathfrak{A})) = \mathfrak{A} \Leftrightarrow \mathfrak{A} \in \mathbf{I}.$$

(6) $$\mathfrak{A} \in \mathbf{I} \Rightarrow \sqrt{\mathfrak{A}} = \mathfrak{A}.$$

All these relations, except (2), (3), (5) and (5′) are self-evident. In (2) the sum $\sum \mathfrak{A}_i$ is not meant to be necessarily finite. The inclusion $\mathscr{V}(\sum \mathfrak{A}_i) \subset \bigcap_i \mathscr{V}(\mathfrak{A}_i)$ follows from (1). The opposite inclusion follows from the definition of the ideal-theoretic sum $\sum \mathfrak{A}_i$ according to which every polynomial in $\sum \mathfrak{A}_i$ is a *finite* sum $\sum f_j$, each $f_j$ belonging to at least one of the ideals $\mathfrak{A}_i$; any such polynomial vanishes therefore on $\bigcap_i \mathscr{V}(\mathfrak{A}_i)$.

The inclusions $\mathscr{V}(\mathfrak{A}\mathfrak{B}) \supset \mathscr{V}(\mathfrak{A} \cap \mathfrak{B}) \supset \mathscr{V}(\mathfrak{A}) \cup \mathscr{V}(\mathfrak{B})$ again follow from (1) since $\mathfrak{A}\mathfrak{B} \subset \mathfrak{A} \cap \mathfrak{B}$. On the other hand, if $(x) \notin \mathscr{V}(\mathfrak{A}) \cup \mathscr{V}(\mathfrak{B})$, then these exist polynomials $f$ and $g$ such that $f \in \mathfrak{A}$, $g \in \mathfrak{B}$, $f(x)g(x) \neq 0$. Since $fg \in \mathfrak{A}\mathfrak{B}$, it follows that $(x) \notin \mathscr{V}(\mathfrak{A}\mathfrak{B})$. This shows that $\mathscr{V}(\mathfrak{A}\mathfrak{B}) \subset \mathscr{V}(\mathfrak{A}) \cup \mathscr{V}(\mathfrak{B})$, and (3) is proved.

The implication "$\mathscr{V}(\mathscr{I}(E)) = E \Rightarrow E$ is a variety" is self-evident. On the other hand, if $E$ is a variety, then $E = \mathscr{V}(\mathfrak{A})$, for some ideal $\mathfrak{A}$. We have, then, by (4′), $\mathscr{I}(E) \supset \mathfrak{A}$, whence $\mathscr{V}(\mathscr{I}(E)) \subset E$, and (5) now follows from (4). The proof of relation (5′) is quite similar (and is, in fact, dual to the proof of (5)).

From (2) and (3) it follows that intersections (finite or infinite) and finite unions of varieties are again varieties. The empty set (= variety of the unit ideal) and the whole space $A_n^K$ (= variety of the zero ideal) are varieties. It follows that $A_n^K$ becomes a *topological space* if the closed sets in $A_n^K$ are defined to be the algebraic varieties immersed in $A_n^K$. We have an induced topology on each variety $V$ immersed in $A_n^K$. Since intersections of varieties are again varieties, the closed subsets of $V$ are the algebraic varieties contained in $V$, i.e., the subvarieties of $V$.

If $E$ is any subset of $A_n^K$ then the closure of $E$ is, of course, the least variety containing $E$. If $V$ is any variety containing $E$, then $\mathscr{I}(V) \subset \mathscr{I}(E)$ and $V = \mathscr{V}(\mathscr{I}(V)) \supset \mathscr{V}(\mathscr{I}(E))$. *Hence* $\mathscr{V}(\mathscr{I}(E))$ *is the closure of* $E$. In particular, the closure of a point $P$ is the set of all points which are specializations of $P$ over $k$ (VI, § $5^{\text{bis}}$).

From (5) it follows that if $V_1$ and $V_2$ are distinct varieties, then $\mathscr{I}(V_1) \neq \mathscr{I}(V_2)$. Hence a *strictly descending chain* $V_1 > V_2 > \cdots >$

$V_i > \cdots$ *of varieties* gives rise to a strictly ascending chain of polynomial ideals $\mathscr{I}(V_1) < \mathscr{I}(V_2) < \cdots < \mathscr{I}(V_i) < \cdots$ and *is therefore necessarily finite.* This very special property of varieties shows that *every variety, with the above topology, is a quasi-compact space.*

A variety $V$ (defined over $k$) is said to be *reducible* (*over* $k$) if it can be decomposed into a sum of two varieties $V_1$ and $V_2$ which are defined over $k$ and are *proper* subsets of $V$. If such a decomposition does not exist, then $V$ is said to be *irreducible* (over $k$).

THEOREM 12. *A variety $V$ is irreducible if and only if its ideal $\mathscr{I}(V)$ is prime.*

PROOF. Assume that $V$ is irreducible and let $f_1, f_2$ be two polynomials such that $f_i \notin \mathscr{I}(V)$, $i=1, 2$. Let $W_i$ be the set of points of $V$ at which $f_i$ vanishes ($i=1, 2$). Then $W_i$ is a variety, and it is a proper subvariety of $V$, since $f_i \notin \mathscr{I}(V)$. Since $V$ is irreducible, also $W_1 \cup W_2$ is a proper subset of $V$. Let $(x)$ be a point of $V$, not in $W_1 \cup W_2$. Then $f_1(x) \neq 0$ *and* $f_2(x) \neq 0$, whence $f_1 f_2 \notin \mathscr{I}(V)$. This shows that $\mathscr{I}(V)$ is a prime ideal.

Conversely, assume that $\mathscr{I}(V)$ is a prime ideal. Let $V = V_1 \cup V_2$, where $V_i$ is a variety (defined over $k$), $i=1, 2$, and assume that $V_2 \neq V$. We shall show that $V_1 = V$ (and that therefore $V$ is irreducible). By (2′) we have $\mathscr{I}(V) = \mathscr{I}(V_1) \cap \mathscr{I}(V_2) \supset \mathscr{I}(V_1) \cdot \mathscr{I}(V_2)$. Since $\mathscr{I}(V_2) > \mathscr{I}(V)$ and $\mathscr{I}(V)$ is prime, it follows at once that $\mathscr{I}(V) \supset \mathscr{I}(V_1)$, whence $V = V_1$. Q.E.D.

THEOREM 13. *Every variety $V$ can be represented as a finite sum of irreducible varieties $V_i$:*

$$V = \bigcup_{i=1}^{h} V_i, \tag{7}$$

*and the decomposition* (7) *is unique* (*to within order of the* $V_i$) *if it is irredundant, i.e., if no $V_i$ is superfluous in* (7).

PROOF. The existence of a decomposition (7) into irreducible varieties follows easily by an indirect argument. Suppose, namely, that there exists a variety $V$ for which the existence assertion of the theorem is false. Then $V$ must be reducible, so that we can write $V = W \cup W'$, with $W < V$ and $W' < V$. Then the existence assertion of the theorem must be false for at least one of the two varieties $W$ or $W'$. What we have shown is that if the theorem is false for a given variety $V$ then there exists a *proper* subvariety $V_1$ of $V$ for which the theorem is still false. This conclusion leads to the existence of an infinite strictly descending chain $V > V_1 > V_2 > \cdots$ of varieties, in contradiction with a preceding result.

Suppose now that (7) is an irredundant decomposition of $V$ into irreducible varieties and let

$$V = \bigcup_{j=1}^{g} V'_j \tag{7'}$$

be another irredundant decomposition of $V$ into irreducible varieties. For any $V_i$, $1 \leqq i \leqq h$, we have $V_i = V \cap V_i = \bigcup_{j=1}^{g} (V'_j \cap V_i)$. Since $V_i$ is irreducible, at least one of the $g$ varieties $V'_j \cap V_i$ must coincide with $V_i$, i.e., we must have $V_i \subset V'_j$ for some $j$, $1 \leqq j \leqq g$. By the same argument we find $V'_j \subset V_s$ for some $s$, $1 \leqq s \leqq h$. We have then $V_i \subset V'_j \subset V_s$, and therefore $V_i = V'_j = V_s$ (since the proper inclusion $V_i < V_s$ would imply that $V_s$ is superfluous in (7)). We have shown that each one of the $h$ varieties $V_i$ coincides with one of the $g$ varieties $V'_j$; and conversely. This establishes the unicity assertion of the theorem.

The irreducible varieties $V_1, V_2, \cdots, V_h$ are called the *irreducible components* of $V$.

REMARK. In order to verify that a decomposition (7) into irreducible varieties is irredundant it is sufficient to verify that $V_i \not\subset V_j$ if $i, j = 1, 2, \cdots, h$ and $i \neq j$. For assume that we have a decomposition (7) into irreducible varieties which is not irredundant, and let, say, $V_1$ be superfluous. Then $V_1 \subset \bigcup_{i=2}^{h} V_i$, $V_1 = \bigcup_{i=2}^{h} (V_i \cap V_1)$. Since $V_1$ is irreducible, this implies that $V_1 = V_i \cap V_1$ for some $i \neq 1$, i.e., that $V_1 \subset V_i$ for some $i \neq 1$.

The above reasoning is similar to that which one uses to show that if a finite set of prime ideals $\{\mathfrak{p}_1, \mathfrak{p}_2, \cdots, \mathfrak{p}_h\}$ is such that $\mathfrak{p}_i \not\supset \mathfrak{p}_j$ for $i, j = 1, 2, \cdots, h$ and $i \neq j$, then no $\mathfrak{p}_i$ is superfluous in the intersection $\mathfrak{p}_1 \cap \mathfrak{p}_2 \cap \cdots \cap \mathfrak{p}_h$. (See Vol. I, Ch. IV, § 4, property A at end of section.)

COROLLARY 1. *If an irreducible variety $V$ has more than one point it is not a Hausdorff space.* [Compare with Theorem 39 of VI, § 17 and the observations (A), (B) and (C) following that theorem.]

If $V$ is irreducible, the union of two proper closed subsets of $V$ is never the entire variety $V$, or—equivalently—the intersection of two non-empty open subsets of $V$ is never empty, and hence $V$ is not a Hausdorff space.

COROLLARY 2. *Every ideal $\mathfrak{A}$ in the set* **I** *admits an irredundant representation as intersection of prime ideals:*

$$\mathfrak{A} = \mathfrak{p}_1 \cap \mathfrak{p}_2 \cap \cdots \cap \mathfrak{p}_h, \tag{8}$$

*The irredundant decomposition* (8) *is unique, each one of the h prime ideals* $\mathfrak{p}_i$ *is itself in the set* **I**, *and the h varieties* $V_i = \mathscr{V}(\mathfrak{p}_i)$ *are the irreducible components of the variety* $\mathscr{V}(\mathfrak{A})$.

Let $V = \mathscr{V}(\mathfrak{A})$. Since $\mathfrak{A} \in \mathbf{I}$, we have $\mathfrak{A} = \mathscr{I}(V)$. Let $V_1, V_2, \cdots, V_h$ be irreducible component of $V$. By the property (2′) we have $\mathfrak{A} = \mathfrak{p}_1 \cap \mathfrak{p}_2 \cap \cdots \cap \mathfrak{p}_h$, where $\mathfrak{p}_i = \mathscr{I}(V_i) \in \mathbf{I}$ is a prime ideal, by Theorem 12. Since $V_i \not\subset V_j$ for $i \neq j$, we have $\mathfrak{p}_i \not\supset \mathfrak{p}_j$ for $i \neq j$, and this shows that the representation $\mathfrak{p}_1 \cap \mathfrak{p}_2 \cap \cdots \cap \mathfrak{p}_h$ is irredundant. The unicity of the irredundant representation (8) of $\mathfrak{A}$ as an intersection of prime ideals follows from the general theorems on primary decompositions of ideals in noetherian rings (and could also be proved directly and in a straightforward fashion by an argument similar to the one employed in the proof of the second part of Theorem 13). We observe that the existence and unicity of an irredundant representation of $\mathfrak{A}$ as an intersection of prime ideals is an immediate consequence of the general decomposition theorems for ideals in noetherian rings and of the fact that $\mathfrak{A} = \sqrt{\mathfrak{A}}$ (see (6)). What is new in the above corollary is the assertion that the prime ideals $\mathfrak{p}_i$ in the decomposition (8) themselves belong to the set **I**.

We shall now prove the following important theorem:

THEOREM 14 (THE HILBERT NULLSTELLENSATZ): *The ideal* $\mathscr{I}(\mathscr{V}(\mathfrak{A}))$ *of the variety of an ideal* $\mathfrak{A}$ *in* $k[X_1, X_2, \cdots, X_n]$ *is the radical of* $\mathfrak{A}$. *Or equivalently:* if $F, F_1, F_2, \cdots, F_q$ *are polynomials in* $k[X_1, X_2, \cdots, X_n]$ *and if* $F$ *vanishes at every common zero of* $F_1, F_2, \cdots, F_q$ *(in an algebraically closed extension* $K$ *of* $k$*), then there exists an exponent* $\rho$ *and polynomials* $A_1, A_2, \cdots, A_q$ *in* $k[X_1, X_2, \cdots, X_n]$ *such that*

$$F^\rho = A_1F_1 + A_2F_2 + \cdots + A_qF_q. \tag{9}$$

PROOF. We first show that the following statement is equivalent to the Hilbert Nullstellensatz:

$$\text{If } \mathscr{V}(\mathfrak{A}) \text{ is empty then } \quad \mathfrak{A} = (1). \tag{10}$$

It is obvious that (10) is a consequence of the Hilbert Nullstellensatz, since the ideal of the empty variety is the unit ideal, and the only ideal $\mathfrak{A}$ whose radical is the unit ideal is the unit ideal itself. On the other hand, assume the truth of (10) and let $F, F_1, F_2, \cdots, F_q$ be polynomials in $k[X_1, X_2, \cdots, X_n]$ satisfying the conditions stated in the theorem. We introduce an additional indeterminate $T$. The polynomials $F_1, F_2, \cdots, F_q, 1 - TF$ have no common zero in $K$. Therefore, by (10), the ideal generated by these polynomials in $k[X_1, X_2,$

$\cdots, X_n, T]$ must be the unit ideal, and there exist then polynomials $B_i(X, T)$, $B(X, T)$ in $k[X_1, X_2, \cdots, X_n, T]$ such that

$$1 = B_1(X, T)F_1(X) + \cdots + B_q(X, T)F_q(X) + B(X, T)(1 - TF(X)).$$

Substituting $1/F(X)$ for $T$ in this identity and clearing denominators, we obtain a relation of the form (9).

Thus, in order to prove the Hilbert Nullstellensatz we have only to show the following: *if* $\mathfrak{A}$ *is an ideal different from* (1) *then* $\mathfrak{A}$ *has at least one zero in* $K$. Since every ideal different from (1) is contained in some proper prime ideal, it is sufficient to deal with the case of a prime ideal $\mathfrak{A} = \mathfrak{p}$, different from (1).

The proof that a prime ideal $\mathfrak{p}$, different from (1), has always a zero in $K$, is immediate *if* $K$ *is a universal domain* (see VI, § $5^{\text{bis}}$, p. 22). For in that case, one can always construct a $k$-isomorphism of the residue class ring

$$k[x_1, x_2, \cdots, x_n] = k[X_1, X_2, \cdots, X_n]/\mathfrak{p} \quad (x_i = \mathfrak{p}\text{-residue of } X_i)$$

into $K$, and if $\varphi$ is such an isomorphism then the point $(\varphi(x_1), \varphi(x_2), \cdots, \varphi(x_n))$ is a zero of $\mathfrak{p}$ in $K$. Thus, our proof of the Hilbert Nullstellensatz is complete if $K$ is a universal domain. The Nullstellensatz for the case of a universal domain is often referred to as the *weak Nullstellensatz*.

To prove the Nullstellensatz in all generality, it is sufficient to prove it in the case in which $K = \bar{k}$ = algebraic closure of $k$, for every algebraically closed extension of $k$ contains an algebraic closure $\bar{k}$ of $k$ and since, furthermore, the existence of a zero of $\mathfrak{p}$ in $A_n^{\bar{k}}$ will imply the existence of a zero of $\mathfrak{p}$ in every algebraically closed extension of $k$. We have therefore to show that "*every prime ideal* $\mathfrak{p}$ *in* $k[X_1, X_2, \cdots, X_n]$, *different from* (1), *has an algebraic zero*," *i.e., a zero* $(\xi_1, \xi_2, \cdots, \xi_n)$ *such that* $\xi_i \in \bar{k}$.

We shall give two proofs of this assertion.

FIRST PROOF. Since every prime ideal, different from (1), is contained in a maximal prime ideal, we may assume that $\mathfrak{p}$ is a maximal ideal. In that case, the residue class ring $k[x_1, x_2, \cdots, x_n] = k[X_1, X_2, \cdots, X_n]/\mathfrak{p}$ ($x_i = \mathfrak{p}$-residue of $X_i$) is a field, and the Hilbert Nullstellensatz results then as a consequence of the following lemma:

LEMMA. *If a finite integral domain* $k[x_1, x_2, \cdots, x_n]$ *over a field* $k$ *is a field, then the* $x_i$ *are algebraic over* $k$.

PROOF OF THE LEMMA. The lemma is obvious if $n = 1$, for if $x$ is a transcendental over $k$ then the polynomial ring $k[x]$ is definitely not a field (the polynomials of positive degree are non-units). We shall use induction with respect to $n$. The ring $S=k[x_1, x_2, \cdots, x_n]$, assumed to be a field, contains the field $k(x_1)$, and we have $S=k(x_1)[x_2, x_3, \cdots, x_n]$. Hence, by our induction hypothesis, the elements $x_2, x_3, \cdots, x_n$ are algebraic over $k(x_1)$. It remains to show that $x_1$ is algebraic over $k$.

Since each $x_i$, $2 \leqq i \leqq n$, is algebraic over $k(x_1)$, there exists a polynomial $a(X)$, with coefficients in $k$, such that $a(x_1) \neq 0$ and such that the $n-1$ products $a(x_1)x_i$, $2 \leqq i \leqq n$, are integral over $k[x_1]$. It follows that for any element $\xi = f(x_1, x_2, \cdots, x_n)$ of $S$ there exists an exponent $\rho$ (depending on $\xi$) such that $[a(x_1)]^\rho f(x_1, x_2, \cdots, x_n)$ is integral over $k[x_1]$. This holds, in particular, for every element $\xi$ of $k(x_1)$, since $k(x_1) \subset S$. Now, if $x_1$ were a transcendental over $k$, then $k[x_1]$ would be integrally closed in $k(x_1)$ and we would have, therefore, the absurd result that every element $\xi$ of $k(x_1)$ can be written as a quotient $A(x_1)/[a(x_1)]^\rho$ of two polynomials in $x_1$, with denominator equal to a power of a *fixed* polynomial $a(x_1)$, independent of $\xi$.

SECOND PROOF. This proof will be based on properties of integral dependence. We first of all achieve a reduction to the case in which $k$ is an infinite field. For this purpose we consider an algebraic closure $K$ of the field $k(X_1, X_2, \cdots, X_n)$ and in this field we consider the polynomial ring $\bar{k}[X_1, X_2, \cdots, X_n]$, where $\bar{k}$ is the algebraic closure of $k$ in $K$. If $\mathfrak{p}^e$ denotes the extension of $\mathfrak{p}$ to the ring $\bar{k}[X_1, X_2, \cdots, X_n]$, we have to show the existence of an algebraic zero $(\alpha_1, \alpha_2, \cdots, \alpha_n)$ of $\mathfrak{p}^e$, and thus, if we fix any prime ideal $\bar{\mathfrak{p}}$ in $\bar{k}[X_1, X_2, \cdots, X_n]$ such that $\bar{\mathfrak{p}} \supset \mathfrak{p}^e$, it will be sufficient to show the existence of an algebraic zero of $\bar{\mathfrak{p}}$. Thus we may replace in the proof the field $k$ by the field $\bar{k}$, and since $\bar{k}$ is an infinite field, we have the desired reduction. Assuming, then, that $k$ is infinite, we apply the normalization theorem (Vol. I, Ch. V, § 4, Theorem 8) to the integral domain $S=R/\mathfrak{p}=k[x_1, x_2, \cdots, x_n]$, and we thus get a set of $d$ algebraically independent elements $z_1, z_2, \cdots, z_d$ of $S/k$ ($d$=transcendence degree of $S/k$) such that $S$ is integral over $k[z_1, z_2, \cdots, z_d]$. We consider a specialization of $k[z]$ to $\bar{k}$ by assigning to $z_1, z_2, \cdots, z_d$ arbitrary values $a_1, a_2, \cdots, a_d$ in $\bar{k}$. The polynomials $f(z_1, z_2, \cdots, z_d)$ such that $f(a_1, a_2, \cdots, a_d)=0$ form a prime ideal $\mathfrak{q}_0$, in $k[z_1, z_2, \cdots, z_d]$, necessarily maximal, since $k[a_1, a_2, \cdots, a_d]$ is a field. Since $S$ is integral over $k[z_1, z_2, \cdots, z_d]$, there exists in $S$ a prime ideal $\mathfrak{q}$ lying over $\mathfrak{q}_0$ (Vol. I, Ch. V, § 2, Theorem 3). The residue class ring $S/\mathfrak{q}$ is integral over $k[z_1, z_2, \cdots, z_d]/\mathfrak{q}_0$ (Vol. I, Ch. V, § 2, Lemma 1), and this implies that the $\mathfrak{q}$-residues $\xi_i$ of the $x_i$ are algebraic over $k$. We have thus found an algebraic zero $(\xi_1, \xi_2, \cdots, \xi_n)$ of $\mathfrak{p}$.

A slight modification of the above proof makes it possible to avoid the use of the normalization theorem. For that purpose we fix an *arbitrary* transcendence basis $\{z_1, z_2, \cdots, z_d\}$ of the field $k(x_1, x_2, \cdots, x_n)/k$ such that the $z$'s belong to $k[x_1, x_2, \cdots, x_n]$ (for instance, we could take for $\{z_1, z_2, \cdots, z_d\}$ a suitable subset of $\{x_1, x_2, \cdots, x_n\}$). Each $x_i$ satisfies an equation of algebraic dependence, of the form $g_m(z_1, z_2, \cdots, z_d)x_i^m + \cdots + g_0(z_1, z_2, \cdots, z_d) = 0$, where the $g_\nu$ are polynomials with coefficients in $k$ and where we may assume that the leading coefficient $g_m$ is independent of $i$. Fix in the algebraic closure $\bar{k}$ of $k$ a set of elements $\bar{z}_1, \bar{z}_2, \cdots, \bar{z}_d$ such that $g_m(\bar{z}_1, \bar{z}_2, \cdots, \bar{z}_d) \neq 0$ (this is possible since $k$ is an infinite field). Let $\mathfrak{q}_0$ be the kernel of the $k$-homomorphism $k[z_1, z_2, \cdots, z_d] \rightarrow k[\bar{z}_1, \bar{z}_2, \cdots, \bar{z}_d]$ determined by the conditions $z_i \rightarrow \bar{z}_i$. We denote by $\mathfrak{o}$ the quotient ring of $k[z_1, z_2, \cdots, z_d]$ with respect to $\mathfrak{q}_0$, by $\mathfrak{o}^\star$ the integral closure of $\mathfrak{o}$ in $k(x_1, x_2, \cdots, x_n)$, and we fix a prime ideal $\mathfrak{q}^\star$ in $\mathfrak{o}^\star$ which lies over $\mathfrak{q}$. Since $g_m(\bar{z}_1, \bar{z}_2, \cdots, \bar{z}_d) \neq 0$, we have $g_m(z_1, z_2, \cdots, z_d) \notin \mathfrak{q}_0$ and hence $g_m(z_1, z_2, \cdots, z_d)$ is a unit in $\mathfrak{o}$. Consequently, each $x_i$ belongs to $\mathfrak{o}^\star$; the $\mathfrak{q}^\star$ residue $\bar{x}_i$ of each $x_i$ is algebraically dependent on $k[\bar{z}_1, \bar{z}_2, \cdots, \bar{z}_d]$ and thus $\bar{x}_i$ is algebraic over $k$. Since the mapping $k[x_1, x_2, \cdots, x_n] \rightarrow k[\bar{x}_1, \bar{x}_2, \cdots, \bar{x}_n]$ determined by the condition $x_i \rightarrow \bar{x}_i$ is a homomorphism (with kernel $\mathfrak{q}^\star \cap k[x_1, x_2, \cdots, x_n]$), $(\bar{x}_1, \bar{x}_2, \cdots, \bar{x}_n)$ is an algebraic zero of the prime ideal $\mathfrak{p}$.

Various consequences can be drawn from the Hilbert Nullstellensatz.

COROLLARY 1. *If $\mathfrak{p}$ is any prime ideal in $k[X_1, X_2, \cdots, X_n]$, then $\mathfrak{p}$ is the ideal of its own variety $\mathscr{V}(\mathfrak{p})$, and hence $\mathscr{V}(\mathfrak{p})$ is irreducible and* $\mathfrak{p} \in \mathbf{I}$.

For, $\sqrt{\mathfrak{p}} = \mathfrak{p}$, whence $\mathfrak{p} = \mathscr{I}(\mathscr{V}(\mathfrak{p})) \in \mathbf{I}$. The irreducibility of $\mathscr{V}(\mathfrak{p})$ follows from Theorem 12.

We have therefore a (1, 1) correspondence between the prime ideals $\mathfrak{p}$ in the polynomial ring $k[X_1, X_2, \cdots, X_n]$ and the varieties in $A_n^K$ which are defined and irreducible over $k$. The correspondence is such that if $\mathfrak{p}$ and $V$ are corresponding elements then $\mathfrak{p} = \mathscr{I}(V)$ and $V = \mathscr{V}(\mathfrak{p})$.

COROLLARY 2. *Every ideal which coincides with its own radical is the ideal of a variety and therefore belongs to the set* **I**. *This set* **I** *coincides therefore with the set of ideals $\mathfrak{A}$ such that $\mathfrak{A} = \sqrt{\mathfrak{A}}$; or equivalently,* **I** *is the set of all polynomial ideals which are finite intersections of prime ideals.*

For if $\mathfrak{A} = \sqrt{\mathfrak{A}}$ then $\mathfrak{A} = \mathscr{I}(\mathscr{V}(\mathfrak{A}))$, by the Hilbert Nullstellensatz. The rest of the corollary follows from relation (6) and from Theorem 13, Corollary 2.

COROLLARY 3. *If $\mathfrak{A}$ is a polynomial ideal and $\mathfrak{p}_1, \mathfrak{p}_2, \cdots, \mathfrak{p}_h$ are the* *isolated prime ideals of* $\mathfrak{A}$, *then the varieties* $\mathscr{V}(\mathfrak{p}_1), \mathscr{V}(\mathfrak{p}_2), \cdots, \mathscr{V}(\mathfrak{p}_h)$ *are the irreducible components of* $\mathscr{V}(\mathfrak{A})/k$.

Since $\mathfrak{p}_1 \cap \mathfrak{p}_2 \cap \cdots \cap \mathfrak{p}_h$ is an irredundant representation of $\sqrt{\mathfrak{A}}$ as intersection of prime ideals, the corollary follows from the irreducibility of $\mathscr{V}(\mathfrak{p}_i)$ and from Corollary 2 (since $\sqrt{\mathfrak{A}} \in \mathbf{I}$, by that corollary).

COROLLARY 4. *Let* $V$ *be a variety in* $A_n{}^K$, *defined over* $k$. *If a polynomial* $f$ *in* $k[X_1, X_2, \cdots, X_n]$ *vanishes at all the algebraic points of* $V$, *then* $f$ *vanishes at every point of* $V$.

Let $V_0$ be the set of algebraic points of $V$ and let $\mathfrak{A}$ be an ideal in $k[X_1, X_2, \cdots, X_n]$ such that $V = \mathscr{V}(\mathfrak{A})$. Then $V_0$ is the variety of the ideal $\mathfrak{A}$ in the affine space $A_n{}^{\bar{k}}$ over $\bar{k}$. By the Hilbert Nullstellensatz, as applied to the case $K = \bar{k}$, the vanishing of $f$ at every point of $V_0$ implies that $f \in \sqrt{\mathfrak{A}}$. Hence $f \in \mathscr{I}(V)$.

The last corollary shows that a variety $V$ in $A_n{}^K$ which is defined over $k$ is uniquely determined by the set of its algebraic points. Or, in topological terms: the set of all algebraic points of a variety $V$ is *everywhere dense* in $V$.

§ 4. **Algebraic varieties in the projective space.** Let $k$ be a ground field and let $K$ be an algebraically closed extension of $k$ ($K$= coördinate domain). The points of the *n-dimensional projective space* $P_n{}^K$ over $K$ are represented by ordered $(n+1)$-tuples $(y_0, y_1, \cdots, y_n)$ of elements of $K$, the $(n+1)$-tuple $(0, 0, \cdots, 0)$ being excluded and two $(n+1)$-tuples $(y_0, y_1, \cdots, y_n)$, $(y'_0, y'_1, \cdots, y'_n)$ representing the same point $P$ if and only if they are *proportional* (i.e., if there exists an element $t \neq 0$ in $K$ such that $y'_i = ty_i$, $i = 0, 1, \cdots, n$). The $(n+1)$-tuple $(y_0, y_1, \cdots, y_n)$ is called a set of *homogeneous coördinates* of the corresponding point. We shall often denote this point by $(y)$. If $(y)$ is a point $P$ in $P_n{}^K$, the field generated over $k$ by all the ratios $y_i/y_j$ such that $y_j \neq 0$ is independent of the choice of the set of homogeneous coördinates of $P$. This field will be denoted by $k(P)$. By the *dimension*, dim $P/k$, of $P$ (over $k$) we mean the transcendence degree of $k(P)/k$.

A set $(y_0, y_1, \cdots, y_n)$ of homogeneous coördinates of a point $P$ is called a set of *strictly homogeneous* coördinates of $P$ if the following condition is satisfied: the ideal of all polynomials $F(Y_0, Y_1, \cdots, Y_n)$ (*homogeneous or non-homogeneous*) such that $F(y_0, y_1, \cdots, y_n) = 0$ is homogeneous; or equivalently: *the ring* $k[y_0, y_1, \cdots, y_n]$ *is homogeneous* (in the sense of § 2).

LEMMA. *Let* $(y_0, y_1, \cdots, y_n)$ *be a set of homogeneous coördinates of a point P and let s be an index,* $0 \leqq s \leqq n$, *such that* $y_s \neq 0$. *Then* $(y_0, y_1, \cdots, y_n)$ *is a set of strictly homogeneous coördinates of P if and only if* $y_s$ *is a transcendental over the field* $k(P)$.

PROOF. Assume that $(y_0, y_1, \cdots, y_n)$ is a set of strictly homogeneous coördinates of $P$ and let $F(Z)$ be a non-zero polynomial in one indeterminate $Z$, with coefficients in $k(P)$. Since every element of $k(P)$ is a quotient of two forms in $k[y_0, y_1, \cdots, y_n]$, of like degree, we have

$$F(Z) = (\sum_i f^{(i)}(y_0, y_1, \cdots, y_n)Z^i)/f^{(0)}(y_0, y_1, \cdots, y_n),$$

where $f^{(0)}$ and the $f^{(i)}$ are forms in $k[Y_0, Y_1, \cdots, Y_n]$, of like degree $h$. We have $f^{(0)}(y_0, y_1, \cdots, y_n) \neq 0$, and $f^{(i)}(y_0, y_1, \cdots, y_n) \neq 0$ for some $i \neq 0$. Let $G(Y_0, Y_1, \cdots, Y_n) = \sum_i f^{(i)}(Y_0, Y_1, \cdots, Y_n)Y_s^i$. If, say, $f^{(\nu)}(y_0, y_1, \cdots, y_n) \neq 0$ and if we set $G_{\nu+h} = f^{(\nu)}(Y_0, Y_1, \cdots, Y_n)Y_s^\nu$, then $G_{\nu+h}$ is the homogeneous component of $G$, of degree $\nu + h$, and we have $G_{\nu+h}(y_0, y_1, \cdots, y_n) \neq 0$ since $y_s \neq 0$. Since the $y$'s are strictly homogeneous coördinates of $P$, it follows that $G(y_0, y_1, \cdots, y_n) \neq 0$, i.e., $F(y_s) \neq 0$. This shows that $y_s$ is a transcendental over $k(P)$. The proof of the converse is also straightforward and may be left to the reader.

COROLLARY. *If K has infinite transcendence degree over k every point of* $P_n^K$ *has sets of strictly homogeneous coördinates.*

Let $F(Y_0, Y_1, \cdots, Y_n)$ be a homogeneous polynomial over $k$ and let $P$ be a point of $P_n^K$. If some set of homogeneous coördinates $(y_0, y_1, \cdots, y_n)$ of $P$ satisfies the relation $F(y_0, y_1, \cdots, y_n) = 0$ then every set of homogeneous coördinates $(y'_0, y'_1, \cdots, y'_n)$ of $P$ will satisfy the relation $F(y'_0, y'_1, \cdots, y'_n) = 0$. We then say that the point $P$ is a *zero* of the form $F$ and that $F$ *vanishes at* $P$. If $\mathfrak{A}$ is a homogeneous ideal in $k[Y_0, Y_1, \cdots, Y_n]$, any common zero of the forms belonging to $\mathfrak{A}$ is called *a zero of the ideal* $\mathfrak{A}$, and the set of zeros of $\mathfrak{A}$ is called *the variety of* $\mathfrak{A}$ and is denoted by $\mathscr{V}(\mathfrak{A})$. An *algebraic (projective) variety* in $P_n^K$, *defined over k*, is any subset of $P_n^K$ which is the variety of some homogeneous ideal in $k[Y_0, Y_1, \cdots, Y_n]$. Only varieties defined over the given ground field $k$ will be considered, and the specification "defined over $k$" will be omitted.

If $E$ is any subset of $P_n^K$ then the set of forms in $k[Y_0, Y_1, \cdots, Y_n]$ which vanish at every point of $E$ is obviously the set of forms belonging to a well defined homogeneous ideal, namely to the ideal generated by these forms. This homogeneous ideal is called the *ideal of the set E* and will be denoted by $\mathscr{J}(E)$. We shall denote by **I** the set of all homogeneous ideals in $k[Y_0, Y_1, \cdots, Y_n]$ of the form $\mathscr{J}(E)$, $E \subset P_n^K$.

Note that if $\mathfrak{A}$ is an irrelevant ideal (§ 2, p. 154) then $\mathscr{V}(\mathfrak{A})$ is empty, for if $\mathfrak{A}$ is irrelevant then $Y_i^{\rho} \in \mathfrak{A}$ for some integer $\rho \geqq 1$ and for all $i$, showing that no point $(y_0, y_1, \cdots, y_n)$ (not all $y_i$ being zero) can be a zero of $\mathfrak{A}$.

As in the case of the affine space $A_n^K$, we have a natural topology in the projective space $P_n^K$ in which the algebraic (projective) varieties are the closed sets. The closure of any subset $E$ of $P_n^K$, i.e., the least variety containing $E$, is given by $\mathscr{V}(\mathscr{I}(E))$. By a *specialization* of a point $P$, *over* $k$, we mean any point $Q$ which belongs to the closure of the point $P$; in symbols: $P \xrightarrow{k} Q$.

These notations and terms are identical with those used in the preceding section for affine varieties. The formulas (1)–(6) continue to hold for projective varieties and homogeneous ideals, and there is no change whatsoever in the proofs except that whenever we use polynomials $f$, $g$, etc., we must now assume that $f, g, \cdots$ are forms. It is only necessary to bear in mind the fact that the set of homogeneous ideals is closed under all the basic ideal-theoretic operations (see § 2, Theorem 8). The definition of irreducible varieties can be repeated *verbatim* for projective varieties, and then Theorems 12 and 13 continue to hold, the proofs remaining the same (we need only recall, from § 2, that for a homogeneous ideal $\mathfrak{p}$ to be prime it is sufficient that the condition "$fg \in \mathfrak{p} \Rightarrow f \in \mathfrak{p}$ or $g \in \mathfrak{p}$" be satisfied for forms $f$ and $g$). Corollary 2 of Theorem 13 continues to hold, with the additional property that the prime ideals $\mathfrak{p}_1, \mathfrak{p}_2, \cdots, \mathfrak{p}_h$ in (8) are homogeneous. While going through the reasoning which was employed in the proof of that corollary the reader should bear in mind the fact proved in § 2 (Theorem 9, Corollary) that all the prime ideals of a homogeneous polynomial ideal (over a *field* $k$ of coefficients) are homogeneous.

In VI, § 5bis, we have introduced the notion of a general point of an irreducible affine variety and also the coördinate ring of such a variety. We shall now extend these definitions to varieties in the *projective* space $P_n^K$.

Let $V$ be a non-empty irreducible variety in $P_n^K$ and let $\mathfrak{p}$ be the homogeneous prime ideal of $V$ in $k[Y_0, Y_1, \cdots, Y_n]$. The residue class ring $k[Y_0, Y_1, \cdots, Y_n]/\mathfrak{p} = k[y_0, y_1, \cdots, y_n]$, where $y_i = \mathfrak{p}$-residue of $Y_i$, is called *the homogeneous coördinate ring* of $V$. It is clear that this ring is a finite *homogeneous integral domain* (over $k$), in the sense of the definition given in § 2 (p. 151).

Since $V$ is non-empty, not all the indeterminates $Y_i$ can belong to $\mathfrak{p}$. Hence not all $y_i$ are zero. However, the $y_i$ are not in general elements of $K$, and we cannot therefore, in general, regard $(y_0, y_1, \cdots, y_n)$ as a point of $P_n^K$. We do call, however, the $(n+1)$-tuple $(y_0, y_1, \cdots, y_n)$ *the general point of* $V$. Since the kernel $\mathfrak{p}$ of the canonical homomorphism $k[Y_0, Y_1, \cdots, Y_n] \to k[y_0, y_1, \cdots, y_n]$ is homogeneous, it follows that $(y_0, y_1, \cdots, y_n)$ *is a set of strictly homogeneous coördinates of the general point of* $V$.

If $K$ is a universal domain, there exist $k$-isomorphisms of $k(y_0, y_1, \cdots, y_n)$ into $K$. If $\sigma$ is such an isomorphism then the point $(\sigma(y_0), \sigma(y_1), \cdots, \sigma(y_n))$ is a point of $V$ and is also called a *general point* of $V$; the point $(y_0, y_1, \cdots y_n)$ may be singled out by referring to it as *the canonical general point of* $V$. Note that the set $(\sigma(y_0), \sigma(y_1), \cdots, \sigma(y_n))$ is a set of strictly homogeneous coördinates.

The quotient field of the homogeneous coördinate ring $k[Y]/\mathfrak{p}$ is *not* what is called the function field of $V$. We notice that $k[Y]/\mathfrak{p}$ is a graded ring (see § 2, Lemma 1, p. 150), whence we can talk about homogeneous elements of this ring. Then the set of all quotients $a/b$, where $a$ and $b$ are homogeneous elements of like degree in $k[Y]/\mathfrak{p}$ $(b \neq 0)$, is obviously a subfield of the quotient field of $k[Y]/\mathfrak{p}$. This subfield we call the *function field* of $V$, and we denote it by $k(V)$. The field $k(V)$ is generated over $k$ by all the ratios $y_i/y_j$ whose denominator is $\neq 0$; if $s$ is an index such that $y_s \neq 0$, we also have $k(V) = k(y_0/y_s, \cdots, y_n/y_s)$. The transcendence degree of $k(V)$ over $k$ is called the *dimension of* $V$ and also the *projective dimension* of the homogeneous prime ideal $\mathfrak{p}$. It is an integer between 0 and $n$. Since $y_0, y_1, \cdots, y_n$ are strictly homogeneous coördinates, it follows from the above lemma that $y_s$ is a transcendental over $k(V)$. Hence the transcendence degree of $k(y_0, y_1, \cdots, y_n)/k$ (= dimension of the prime ideal $\mathfrak{p}$) is one greater than the dimension of $V$ (or also, one greater than the projective dimension of $\mathfrak{p}$).

According to our preceding definitions, $k(V)$ is identical with $k(P)$, where $P$ is the canonical general point of $V$, and $\dim V = \dim P/k$.

From the Hilbert Nullstellensatz we can easily derive a corresponding Nullstellensatz for homogeneous polynomial ideals and projective varieties. We can see already that some modification will be necessary, for we have already pointed out that the projective variety $V$ of an *irrelevant* ideal $\mathfrak{A}$ is always empty, while in the non-homogeneous case the Hilbert Nullstellensatz tells us that only the unit ideal has the property that its (affine) variety is empty. Thus we cannot expect to have a *verbatim* extension of the Nullstellensatz. However, it turns out that the irrelevant ideals are the only exceptional ones:

THEOREM 15 (*Projective Nullstellensatz*): *If* $\mathfrak{A}$ *is a non-irrelevant homogeneous ideal in* $k[Y_1, \cdots, Y_n]$, *then* $\mathscr{V}(\mathfrak{A})$ *is non-empty and the ideal of the variety* $\mathscr{V}(\mathfrak{A})$ *is the radical of* $\mathfrak{A}$.

PROOF. We set $V=\mathscr{V}(\mathfrak{A})$ and we consider the *affine* variety $C(V)$ *in* $A_{n+1}{}^K$ which is the variety of the ideal $\mathfrak{A}$: it is the set of all points $(x_0, \cdots, x_n)$ in $A_{n+1}{}^K$ such that $F(x)=0$ for every $F$ in $\mathfrak{A}$. Since $\mathfrak{A}$ is a homogeneous ideal, the relation $(x_0, \cdots, x_n) \in C(V)$ implies $(tx_0, \cdots, tx_n) \in C(V)$ for every $t$ in $K$. Thus, *if* $V$ *is non-empty*, $C(V)$ is a union of straight lines containing the origin $(0, \cdots, 0)$.

It is furthermore clear that a point $(x_0, x_1, \cdots, x_n)$ of $A_{n+1}{}^K$, different from the origin, belongs to $C(V)$ if and only if the point of $P_n{}^K$ whose homogeneous coördinates are $x_0, x_1, \cdots, x_n$ belongs to $V$. The variety $C(V)$ is called the *representative cone* of $V$. Since $\mathfrak{A}$ is non-irrelevant, $C(V)$ is neither empty nor is it reduced to the origin (by the affine Nullstellensatz). Hence $V$ is non-empty.

Since $V$ is non-empty, it is clear that the (homogeneous) ideal of $V$ is contained in the ideal of $C(V)$. Conversely, if a polynomial $F(Y_0, \cdots, Y_n)$ vanishes on $C(V)$, we have, for every point $(x_0, \cdots, x_n)$ of $V$ and for every $t$ in $K$, $F(tx_0, \cdots, tx_n)=0$. Writing $F=\sum_{j=0}^{q} F_j$, where $F_j$ is either zero or a form of degree $j$, we get $F_0+tF_1(x)+\cdots+t^qF_q(x)=0$ for every $t$, whence $F_j(x)=0$ for every $j$ since the algebraically closed field $K$ is infinite. Therefore the homogeneous ideal of $V$ is equal to the ideal of $C(V)$. Theorem 15 now follows immediately from the affine Nullstellensatz. Q.E.D.

The four corollaries of Theorem 14 hold for projective varieties and homogeneous ideals with the following modifications:

In Corollary 1 it must be assumed that $\mathfrak{p}$ is a prime homogeneous ideal, different from the irrelevant ideal $\mathfrak{Y}$ which is generated by $(Y_0, Y_1, \cdots, Y_n)$.

Corollary 2 should read as follows: "Every ideal which coincides with its own radical and is not an irrelevant ideal is the ideal of a variety and therefore belongs to the set **I**. The set **I** is therefore the set of all polynomial ideals which are finite intersections of prime ideals and are different from the irrelevant prime ideal $(Y_0, Y_1, \cdots, Y_n)$."

In Corollary 3 it must be assumed that $\mathfrak{A}$ is not an irrelevant ideal.

In Corollary 4, $V$ is a projective variety in $P_n{}^K$, $f$ is in $k[Y_0, Y_1, \cdots, Y_n]$ and is a form. By an *algebraic* point in $P_n{}^K$ we mean a point whose homogeneous coördinates are proportional to elements of $\bar{k}$.

We note that the *existence* of algebraic points on every non-empty variety can also be proved by means of the existence theorem for algebraic places (VI, § 4, Theorem 5′, Corollary 2), as follows:

Let $(y_0, y_1, \cdots, y_n)$ be the canonical general point of an irreducible non-empty variety $V$ in $P_n^K$, and let $\mathscr{P}$ be an *algebraic* place of $k(y_0, y_1, \cdots, y_n)/k$. If $v$ is the corresponding valuation, we may assume that $v(y_0) \leqq v(y_i)$, $1 \leqq i \leqq n$. Let $\alpha_i = \mathscr{P}(y_i/y_0)$, where $\alpha_i$ is then different from $\infty$ and is algebraic over $k$. The point $(1, \alpha_1, \cdots, \alpha_n)$ is immediately seen to belong to $V$, and thus $V$ has an algebraic point, as asserted.

§ 4bis. **Further properties of projective varieties.** We shall begin by generalizing to projective varieties the notion of the center of a place and the notion of a divisor which have been given for affine varieties in the preceding chapter (VI, § 5bis and § 14).

Let $Q$ be a point $(z_0, z_1, \cdots, z_n)$ of $V$. We consider quotients $f(y_0, y_1, \cdots, y_n)/g(y_0, y_1, \cdots, y_n)$ of elements of the homogeneous coördinate ring of $V$, such that $f$ and $g$ are homogeneous, of *like degree*, and such that $g(z_0, z_1, \cdots, z_n) \neq 0$. These quotients form a ring, contained in the field $k(V)$, called *the local ring of $V$ at the point $Q$*, or, briefly, *the local ring of $Q$ (on $V$)*.

Without loss of generality we may assume that $z_0 \neq 0$. Then also $y_0 \neq 0$ since $Q$ is a specialization, over $k$, of the general point $(y_0, y_1, \cdots, y_n)$ of $V/k$. Set $x_i = y_i/y_0$, $a_i = z_i/z_0$. It is clear that the point $(a_1, a_2, \cdots, a_n)$ of the affine $n$-space is a specialization of the point $(x_1, x_2, \cdots, x_n)$. Therefore, if we consider the ring $k[x_1, x_2, \cdots, x_n]$ then the point $(a_1, a_2, \cdots, a_n)$ corresponds to a prime ideal $\mathfrak{p}$ of this ring, and the local ring of $Q$ is immediately seen to be equal to the quotient ring of the ring $k[x_1, x_2, \cdots, x_n]$ with respect to this prime ideal. The points of the projective space $P_n^K$ which do not lie in the hyperplane $Y_0 = 0$ form an affine space $A_n^K$. Denote by $V_a$ the intersection $V \cap A_n^K$. We have just seen that each point of $V_a$ is a specialization of $(x_1, x_2, \cdots, x_n)$ over $k$; also the converse is true and its proof is immediate. *Hence $V_a$ is an irreducible affine variety, with $(x_1, x_2, \cdots, x_n)$ as general point.* This connection between projective and affine varieties will be investigated in more detail in Section 6. For the moment we only wish to call attention to the fact which was established above, namely that *if $Q$ is any point of $V_a$ then the local ring of the projective variety $V$ at $Q$ is the same as the local ring of the affine variety $V_a$ at $Q$.* It is also clear that the function fields $k(V)$ and $k(V_a)$ coincide, both being given by the field $k(x_1, x_2, \cdots, x_n)$. We shall use these facts and notations in the remainder of the section.

Let $\mathscr{P}$ be a place of the function field $k(V)$ of $V$ and let us assume that the residue field of $\mathscr{P}$ is contained in the coördinate domain $K$ (this is no essential restriction on $\mathscr{P}$ if $K$ is a universal domain; see VI, § 5bis). If $v$ is the valuation determined by $\mathscr{P}$, then $v(y_i/y_j)$ is meaningful for any $i, j=0, 1, \cdots, n$, provided $y_j \neq 0$, since $y_i/y_j \in k(V)$. It is clear that there exists an index $s$ such that $v(y_i/y_s) \geqq 0$ for $i=0, 1, \cdots, n$. For such an index $s$ let $\mathscr{P}(y_i/y_s)=b_i \in K$. The $b_i$ are not all zero (since $b_s=1$) and thus determine a point $Q=(b_0, b_1, \cdots, b_n)$ of $P_n^K$. If $t$ is another index such that $v(y_i/y_t) \geqq 0$ for $i=0, 1, \cdots, n$ and if we set $\mathscr{P}(y_i/y_t)=c_i$, then $b_t c_s=1$, whence $b_t \neq 0$, $c_s \neq 0$, and furthermore, $c_i=\mathscr{P}(y_i/y_s \cdot y_s/y_t)=b_i c_s$, $i=0, 1, \cdots, n$. This shows that the point $Q$ above depends only on the place $\mathscr{P}$ and not on the choice of the index $s$. *It is easily seen that $Q$ belongs to $V$.* For if $f(Y_0, Y_1, \cdots, Y_n)$ is any form in the homogeneous ideal of $V$, then we have $f(y_0/y_s, y_1/y_s, \cdots, y_n/y_s)=0$, and since $\mathscr{P}$ is a $k$-homomorphism it follows that $f(b_0, b_1, \cdots, b_n)=0$, showing that $Q$ is on $V$. This point $Q$ is called *the center* of the place $\mathscr{P}$ on the variety $V$. *The properties* (1)–(6) *of the center of a place on an affine variety, given in VI, § 5bis, continue to hold for projective varieties.* The proofs are straightforward and may be left to the reader (it is best to prove property 5 and to use this property in the proof of the remaining properties).

In a similar way (i.e., by reduction to affine varieties) we can define the center $W$, on $V$, of any valuation of $k(V)/k$: $W$ will be a certain irreducible subvariety of $V$ (see VI, § 9, p. 38).

We now consider prime divisors of the function field $k(V)$ of $V$. Since $k(V)$ is a field of algebraic functions, namely $k(V)=k(x_1, x_2, \cdots, x_n)$, where $x_i=y_i/y_0$ (assuming that $y_0 \neq 0$), the results of VI, § 14 are applicable. In particular, every prime divisor of $k(V)$ is a discrete valuation, of rank 1. Furthermore, every irreducible $(r-1)$-dimensional subvariety $W$ of $V/k$ is the center of at least one and of at most a finite number of prime divisors. To see this, we have only to fix a general point $Q$ of $W/k$ and—assuming that $Q$ belongs to the affine variety $V_a$—observe that the prime divisors of $k(V)$ having center $W$ on $V$ coincide with the prime divisors of $k(V_a)$ $(=k(V))$ which have center $W_a$ on $V_a$, and then apply Theorem 32 of VI, § 14.

We say that our variety $V$ is *normal* at $W$ if the local ring $\mathfrak{o}(W; V)$ (i.e., the local ring of $V$ at the general point $Q$ of $W/k$) is integrally closed. Clearly, $V$ is normal at $W$ if and only if $V_a$ is normal at $W_a$ [since $\mathfrak{o}(W; V)=\mathfrak{o}(W_a, V_a)$]. We say that $V/k$ is *normal*, or *locally normal*, if it is normal at each of its points. Theorem 33 of VI, § 14 continues to hold for normal varieties in the projective space: if $V/k$ is normal at $W$ and $\dim W=r-1$, then there is only one prime divisor of $k(V)$ having center $W$. We denote this divisor by $v_W$. In particular, if $V/k$ is a normal variety then every irreducible $(r-1)$-dimensional subvariety $W$ of $V/k$ is the center of a unique prime divisor of $k(V)$.

We now assume that $V$ is normal and we introduce the free *group of divisors* on $V$, i.e., the group generated by the irreducible $(r-1)$-dimensional subvarieties $W$ of $V$. Using the notations of VI, § 14, p. 98, we can now define the divisor $(w)$ of any function $w \neq 0$ in $k(V)$:

$$(1) \qquad (w) = \sum v_W(w) \cdot W,$$

where the sum is extended to all the irreducible $(r-1)$-dimensional subvarieties of $V/k$. That the sum (1) is finite can be seen as follows:

In the first place, there is only a finite number of irreducible $(r-1)$-dimensional subvarieties $W$ of $V$ such that (a) $V_a$ contains the general point of $W_a/k$ and (b) $v_W(w) \neq 0$; this assertion concerns only the affine variety $V_a$ and has been proved in VI, § 14 (p. 97).

In the second place, since the intersection of $V$ with the hyperplane $Y_0=0$ is at most $(r-1)$-dimensional, there is only a finite number of $(r-1)$-dimensional irreducible subvarieties $W$ of $V$ which do not satisfy condition (a) above.

As has been proved in VI, § 14, p. 99, *if $w$ is not a constant, i.e., if $w$ is not algebraic over $k$*, then there exists at least one polar prime divisor of $w$, i.e., for at least one $W$ in (1) we must have $v_W(w) < 0$. Upon replacing $w$ by $1/w$ we see, under this same assumption, that we must also have $v_W(w) > 0$ for at least one $W$.

We now prove the following analogue of Theorem 34 of VI, § 14, for normal varieties in the projective space:

THEOREM 16. *Let $V/k$ be an irreducible variety in the projective space $P_n^K$ and let $R = k[y_0, y_1, \cdots, y_n]$ be the homogeneous coördinate ring of $V/k$. A necessary and sufficient condition that $V/k$ be normal is that the conductor of $R$ in the integral closure $\bar{R}$ of $R$ be an irrelevant ideal.*

PROOF. Assume that the conductor $\mathfrak{C}$ of $R$ in $\bar{R}$ is irrelevant and let $Q = (z_0, z_1, \cdots, z_n)$ be any point of $V$. We show that $V$ is normal at $Q$. Without loss of generality we may assume that $z_0 \neq 0$. We set $x_i = y_i/y_0$, $a_i = z_i/z_0$ and we call $V_a$ the affine variety consisting of those points of $V$ which do not lie in the hyperplane $Y_0=0$. The point $Q_a \equiv (a_1, a_2, \cdots, a_n)$ lies on $V_a$, and to say that $V$ is normal at $Q$ is the same as saying that $V_a$ is normal at $Q_a$. Now, the ring $k[x_1, x_2, \cdots, x_n]$ is the non-homogeneous coördinate ring of $V_a/k$. We shall show that *this ring is integrally closed*, whence it will follow that $V_a$ is a normal variety. Let $\xi$ be an element of the integral closure of $k[x_1, x_2, \cdots, x_n]$. Upon the substitution $x_i \to y_i/y_0$, and clearing denominators, an equation of integral dependence for $\xi$ over $k[x_1, x_2, \cdots, x_n]$ takes the form

(2) $$y_0^h\xi^\nu + f^{(1)}(y_0, y_1, \cdots, y_n)\xi^{\nu-1} + \cdots + f^{(\nu)}(y_0, y_1, \cdots, y_n) = 0,$$

where each $f^{(i)}$ is a form of degree $h$, with coefficients in $k$. Relation (2) implies that $y_0^h\xi \in \bar{R}$. Since the conductor $\mathfrak{C}$ of $R$ in $\bar{R}$ is irrelevant, it contains a power of each $y_i$. In particular, let, say, $y_0^N \in \mathfrak{C}$. Then $y_0^{N+h}\xi \in R$, and since $\xi$ is homogeneous of degree zero, we have $y_0^{N+h}\xi = g(y_0, y_1, \cdots, y_n)$, where $g$ is a form of degree $N+h$, with coefficients in $k$. Hence $\xi = g(1, x_1, x_2, \cdots, x_n) \in k[x_1, x_2, \cdots, x_n]$.

Conversely, assume that $V$ is normal. We have to show that there exists an integer $N$ such that $y_i^N\bar{R} \subset R$ for $i = 0, 1, \cdots, n$. Since $\bar{R}$ is a finite $R$-module (Vol. I, Ch. V, § 4, Theorem 9) and since each element of $\bar{R}$ is a sum of homogeneous elements of $\bar{R}$ (§ 2, Theorem 11), it is sufficient to show that for any homogeneous element $\omega$ of $\bar{R}$ there exists an integer $N$ (depending on $\omega$) such that $y_i^N\omega \in R$, for $i = 0, 1, \cdots, n$. Let us show, for instance, that $y_0^N\omega \in R$ for some $N$. Let $\nu$ be the degree of $\omega (\nu \geqq 0)$ and let $\omega$ satisfy an equation of integral dependence over $R$, of degree $g$ in $\omega$:

(3) $$\omega^g + A_1\omega^{g-1} + \cdots + A_g = 0, \quad A_i \in R.$$

Each coefficient $A_i$ is a sum of homogeneous elements of $R$, and thus the left-hand side of (3) is a sum of homogeneous elements of $\bar{R}$. Since $\bar{R}$ is a graded ring, the sum of terms having the same degree must vanish. In particular, the sum of terms of degree $\nu g$ must be zero. Hence we may assume that $A_i$ is a *form* in $y_0, y_1, \cdots, y_n$, of degree $\nu i$. But then (3) shows that $\omega/y_0^\nu$ is integral over $k[x_1, x_2, \cdots, x_n]$. Since $V$ is normal, also $V_a$ is normal, and *hence the ring* $k[x_1, x_2, \cdots, x_n]$ *is integrally closed* (VI, § 14, Theorem 34). Hence $\omega/y_0^\nu \in k[x_1, x_2, \cdots, x_n]$, i.e., $\omega/y_0^\nu = f(y_0, y_1, \cdots, y_n)/y_0^s$, where $f$ is a form of degree $s$. Hence $y_0^{s-\nu}\omega \in R$, as asserted. This completes the proof of the theorem.

A variety $V/k$ in $P_n^K$ is said to be *arithmetically normal* if its homogeneous coördinate ring $R = k[y_0, y_1, \cdots, y_n]$ is integrally closed. It follows from the preceding theorem that *an arithmetically normal variety is also normal.* The converse is not always true, as can be shown by examples.

For an arbitrary projective variety $V/k$, we consider a finite algebraic extension $F$ of the field $k(V)$ and we denote by $\bar{R}$ the integral closure, in $F(y_0)$, of the homogeneous coördinate ring $R$ of $V/k$. Since $y_0$ is a transcendental over $k(V)$ (we assume that $y_0 \neq 0$) it follows from

Corollary to Theorem 11 (§ 2) that $\bar{R}$ is a graded ring. Let $\bar{R}_q$ (respectively, $R_q$) be the set of homogeneous elements of $\bar{R}$ (respectively of $R$), of degree $q$. Since, for each $q \geqq 1$, $R_q$ is a finite dimensional vector space over $k$ and since $\bar{R}$ is a finite $R$-module (admitting an $R$-basis consisting of homogeneous elements), it follows at once that $\bar{R}_q$ is also a finite-dimensional vector space over $k$. Let $\{u_0, u_1, \cdots, u_m\}$ be a $k$-basis of $\bar{R}_q$ and let $\bar{V}_q$ be the projective variety whose general point is $(u_0, u_1, \cdots, u_m)$. A change of $k$-basis of $\bar{R}_q$ leaves $\bar{V}_q$ unchanged, up to projective equivalence. Thus $\bar{V}_q$ is uniquely determined for each integer $q \geqq 0$. We shall prove the following:

*If $q$ is sufficiently large then $\bar{V}_q$ is the derived normal model of $V/k$ in $F$,† and, moreover, $\bar{V}_q$ is an arithmetically normal variety provided $k$ is maximally algebraic in $F$.*

[The proof given below applies without modification to models over "restricted" domain (VI, § 18) and yields another proof of Theorem 42 of VI, § 18.]

Let $\mathfrak{o}_i = k[y_0/y_i, y_1/y_i, \cdots, y_n/y_i]$ and let $V_i$ be the affine model $V(\mathfrak{o}_i)$, so that $V$ is the union of $V_0, V_1, \cdots, V_n$. Let $V'_i$ be the derived normal model of $V_i$ in $F$, i.e., let $V'_i = V(\mathfrak{o}'_i)$, where $\mathfrak{o}'_i$ is the integral closure of $\mathfrak{o}_i$ in $F$. To prove that $\bar{V}_q$ is the derived normal model $N(V, F)$ of $V$ in $F$ it will be sufficient to show that $V'_i$ is a subset of $\bar{V}_q$ for $i=0, 1, \cdots, n$ (for, $N(V, F)$ is the union of the affine models $V'_0, V'_1, \cdots, V'_n$ and is a complete model, while $\bar{V}_q$, being a model, is an irredundant set; see VI, § 17). Let us show, for instance, that $V'_0 \subset \bar{V}_q$ if $q$ is sufficiently large.

Without loss of generality we may assume that the $k$-basis $\{u_0, u_1, \cdots, u_m\}$ of $\bar{R}_q$ includes the element $y_0^q$. Let, say, $u_0 = y_0^q$. Let $\bar{\mathfrak{o}}_0 = k[u_1/u_0, u_2/u_0, \cdots, u_m/u_0]$. Then the affine variety $V(\bar{\mathfrak{o}}_0)$ is a subset of $\bar{V}_q$. *We shall show that if $q$ is sufficiently large then $\mathfrak{o}'_0 = \bar{\mathfrak{o}}_0$.* This will establish the inclusion $V'_0 \subset \bar{V}_q$, for $q$ large.

Let $\xi$ be any element of $\bar{\mathfrak{o}}_0$. Then $\xi = \omega/u_0^h$, where $\omega$ is a form, of degree $h$, in $u_0, u_1, \cdots, u_m$, with coefficients in $k$, whence $\omega \in \bar{R}_{hq}$. The element $\omega$ satisfies a "homogeneous" relation of integral dependence over $R$, i.e., we have $\omega^g + a_1\omega^{g-1} + \cdots + a_g = 0$, where $a_i \in R_{ihq}$. Upon dividing this equation by $u_0^{hg}$ and observing that $a_i/u_0^{hi} = a_i/y_0^{hiq} \in \mathfrak{o}_0$, we see that $\xi \in \mathfrak{o}'_0$. We have therefore shown that $\bar{\mathfrak{o}}_0 \subset \mathfrak{o}'_0$ (for any $q$). To prove the opposite inclusion $\mathfrak{o}'_0 \subset \bar{\mathfrak{o}}_0$ (for large $q$) we first observe that the monomials $y_0^{q-1}y_i$ ($i=0, 1, \cdots, n$) belong to $R_q$, hence also to $\bar{R}_q$, and thus are linear combinations of $u_0, u_1, \cdots, u_m$,

† Here $V$ and $\bar{V}_q$ are regarded as models, i.e., as collections of local rings; see the opening paragraphs of VI, § 18.

with coefficients in $k$. Therefore $y_i/y_0 = y_i y_0^{q-1}/u_0 \in \bar{\mathfrak{o}}_0$. Thus $\mathfrak{o}_0 \subset \bar{\mathfrak{o}}_0$. On the other hand, if $\eta$ is any element of $\mathfrak{o}'_0$ then, upon writing an equation of integral dependence of $\eta$ on $\mathfrak{o}_0$, we see at once that for large $q$ the product $\eta y_0^q$ is integral over $R$ and therefore belongs to $\bar{R}_q$. Since $\mathfrak{o}'_0$ is a finite $\mathfrak{o}_0$-module, it follows therefore that if $\{\eta_1, \eta_2, \cdots, \eta_q\}$ is an $\mathfrak{o}_0$-basis of $\mathfrak{o}'_0$ and $q$ is sufficiently large, then all the products $\eta_\alpha y_0^q$ are linear combinations of $u_0, u_1, \cdots, u_m$, with coefficients in $k$, and therefore the $\eta_\alpha$ belong to $\bar{\mathfrak{o}}_0$. Since also $\mathfrak{o}_0$ is contained in $\bar{\mathfrak{o}}_0$, the inclusion $\mathfrak{o}'_0 \subset \bar{\mathfrak{o}}_0$ is proved, for all large $q$.

It now remains to prove that $\bar{V}_q$ is arithmetically normal, i.e., that the homogeneous coördinate ring $I = k[u_0, u_1, \cdots, u_m]$ of $\bar{V}_q/k$ is integrally closed (for large $q$). Let $I'$ be the integral closure of $I$ in its quotient field. Then $I'$ is a graded ring: $I' = I'_0 + I'_1 + \cdots + I'_h + \cdots$ (the degree $h$ of a homogeneous element of $I'$ being defined by stipulating that $u_0, u_1, \cdots, u_m$ are homogeneous elements of degree 1). Since $I \subset \bar{R}$, we have $I' \subset \bar{R}$ and hence $I'_h \subset \bar{R}_{hq}$. *We assert that* $I'_h = \bar{R}_{hq}$. To show this we first observe that $R$ is integral over $I$, since $y_i^q \in I$. Hence the elements of $\bar{R}_{hq}$, being integral over $R$, are also integral over $I$. Therefore, in order to show that $\bar{R}_{hq} \subset I'_h$ we have only to show that $\bar{R}_{hq}$ is contained in the quotient field of $I$. This, however, is obvious, since $\bar{R}_{hq} \subset F \cdot y_0^q \subset F(u_0)$ (assuming—as we may—that $u_0 = y_0^q$) and since $F(u_0)$ is precisely the quotient field of $I$, for large $q$ (we have just proved that if $q$ is large then $\bar{V}_q$ is a derived normal model of $V/k$ in $F$, whence—at any rate—$k(\bar{V}_q) = F$). We thus have shown that

$$(4) \qquad I' = \bar{R}_0 + \bar{R}_q + \bar{R}_{2q} + \cdots, \quad q\text{-large, say } q \geqq \sigma.$$

Since $\bar{R}$ is a finite $R$-module, we can write $\bar{R} = Rz_1 + Rz_2 + \cdots + Rz_t$, and we may assume that the $z_i$ are homogeneous elements of $\bar{R}$. Let $s_i$ be the degree of $z_i$ and let $\rho = \max(s_1, s_2, \cdots, s_t, \sigma)$, where $\sigma$ is defined in (4). We shall now show that if $q \geqq \rho$ then $\bar{V}_q$ is arithmetically normal.

If $q \geqq \max(s_1, s_2, \cdots, s_t)$ then we have clearly

$$(5) \qquad \bar{R}_q = R_{q-s_1}z_1 + R_{q-s_2}z_2 + \cdots + R_{q-s_t}z_t.$$

Let $j$ be any non-negative integer. Then:

$$\bar{R}_{q+j} = R_{q+j-s_1}z_1 + R_{q+j-s_2}z_2 + \cdots + R_{q+j-s_t}z_t = R_j\bar{R}_q.$$

Therefore, *a fortiori*, we have $\bar{R}_{q+j} = \bar{R}_j\bar{R}_q$. It follows that $\bar{R}_{hq} = (\bar{R}_q)^h$. If, now, also (4) holds, i.e., if we have $q \geqq \rho$, then we find $I' = \bar{R}_0 + \bar{R}_q + (\bar{R}_q)^2 + \cdots + (\bar{R}_q)^h + \cdots$. Recalling that $\bar{R}_q = ku_0 +$

$ku_1 + \cdots + ku_m$, we conclude that $I' \subset \bar{R}_0 + I$. Now, we have $R_0 = k$ and $\bar{R}_0 = k$ since we have assumed that $k$ is maximally algebraic in $F$ (and therefore also in $F(y_0)$). Therefore $I' \subset I$, i.e., $I' = I$, showing that $\bar{V}_q$ is arithmetically normal.

**§ 5. Relations between non-homogeneous and homogeneous ideals.** We consider the polynomial rings $R = k[X_1, \ldots, X_n]$ and ${}^hR = k[Y_0, Y_1, \cdots, Y_n]$ in $n$ and $n+1$ indeterminates, respectively, over the same field $k$. Our aim is to establish a natural correspondence between arbitrary ideals in $R$ and homogeneous ideals in ${}^hR$. Given any polynomial $F(X_1, \cdots, X_n)$ in $R$, different from zero, we first define its *homogenized polynomial* ${}^hF$ in ${}^hR$ as follows:

$$(1) \qquad {}^hF(Y_0, \cdots, Y_n) = Y_0^{\partial(F)} F(Y_1/Y_0, \cdots, Y_n/Y_0),$$

where $\partial(F)$ denotes, as usual, the (total) degree of $F$; the fact that ${}^hF$ is actually a polynomial, and not merely a rational function with denominator a power of $Y_0$, is clear. The homogenized polynomial ${}^hF$ is a form having the *same degree* as $F$. We leave to the reader the verification of the following formulas:

$$(2) \qquad {}^h(FG) = {}^hF \cdot {}^hG,$$

$$(3) \qquad Y_0^{\partial(F)+\partial(G)} \cdot {}^h(F+G) = Y_0^{\partial(F+G)} [Y_0^{\partial(G)} \cdot {}^hF + Y_0^{\partial(F)} \cdot {}^hG].$$

Note that (3) reduces to ${}^h(F+G) = {}^hF + {}^hG$ if $F$, $G$ and $F+G$ have the same degree and $F+G \neq 0$. Note also that ${}^hF$ is never a multiple of $Y_0$.

Conversely, with every polynomial $\varphi(Y_0, \cdots, Y_n)$ in ${}^hR$, we associate the polynomial ${}^a\varphi$ in $R$ defined as follows:

$$(1') \qquad {}^a\varphi(X_1, \cdots, X_n) = \varphi(1, X_1, \cdots, X_n).$$

Then it is clear that we have

$$(2') \qquad {}^a(\varphi\psi) = {}^a\varphi \cdot {}^a\psi$$

$$(3') \qquad {}^a(\varphi + \psi) = {}^a\varphi + {}^a\psi.$$

Actually we shall apply the operation ${}^a$ only to forms $\varphi$, so that from now on $\varphi$ will always denote a form (unless the opposite is stated explicitly). It is clear that if $Y_0^m$ is the highest power of $Y_0$ which divides $\varphi$, then the degree of ${}^a\varphi$ is equal to $\partial(\varphi) - m$.

We now study the *relations* between "$h$" and "$a$". It follows immediately from the definitions that we have

$$(4) \qquad {}^a({}^hF) = F.$$

On the other hand, we have $({}^{ha}\varphi)(Y_0, \cdots, Y_n) = Y_0^{\partial({}^a\varphi)} \cdot {}^a\varphi(Y_1/Y_0, \cdots, Y_n/Y_0) = Y_0^{\partial({}^a\varphi)} \cdot \varphi(1, Y_1/Y_0, \cdots, Y_n/Y_0)$. Hence

$$(5) \qquad {}^h({}^a\varphi) = Y_0^{\partial({}^a\varphi) - \partial(\varphi)} \cdot \varphi,$$

or, by the preceding observation,

$$(5') \qquad {}^h({}^a\varphi) = Y_0^{-m}\varphi,$$

where $Y_0^m$ is the highest power of $Y_0$ which divides $\varphi$. Thus ${}^h({}^a\varphi)$ is, in general, a divisor of $\varphi$. The inequality $\partial({}^a\varphi) < \partial(\varphi)$ can hold only if $\varphi$ is a *multiple of* $Y_0$, and ${}^h({}^a\varphi)$ is then the *form obtained from* $\varphi$ *by deleting the factor* $Y_0^m$ *contained in* $\varphi$. It follows that the homogeneous polynomials of the form ${}^hF$ in ${}^hR$ ($F \in R$) are exactly those polynomials which do not contain $Y_0$ as a factor.

We now extend the operations "$h$" and "$a$" to *ideals*. We shall denote ideals in $R$ by small German letters and ideals in ${}^hR$ by capital German letters. Given an ideal $\mathfrak{a}$ in $R$, the set of all forms ${}^hF$, $F \in \mathfrak{a}$, is *not* the set of all forms belonging to some homogeneous ideal, for this set does not contain any form which is divisible by $Y_0$. However, if we consider the set $S$ of all forms $Y_0^m \cdot {}^hF$ ($m \geqq 0$, $F \in \mathfrak{a}$), then it is easily seen that $S$ is the set of forms of a homogeneous ideal. To show this we have only to show that the difference of two forms in $S$, of like degree, is still in $S$, and that the product of any form in $S$ by an arbitrary form in ${}^hR$ also belongs to $S$. For, if this is shown, then it will follow that $S$ is the set of all forms which belong to the ideal generated by the elements of $S$. Now, all that will follow directly from the following characterization of $S$: *a form* $\varphi$ *belongs to* $S$ *if and only if* ${}^a\varphi \in \mathfrak{a}$. The proof is immediate and is as follows:

If ${}^a\varphi = F \in \mathfrak{a}$, then ${}^h({}^a\varphi) = {}^hF$, and thus, by (5′), $\varphi = Y_0^m \cdot {}^hF \in S$. Conversely, if $\varphi = Y_0^m \cdot {}^hF$, with $F$ in $\mathfrak{a}$, then ${}^a\varphi = {}^a({}^hF) = F \in \mathfrak{a}$, by (4).

We denote by ${}^h\mathfrak{a}$ the *homogeneous ideal* in ${}^hR$ which is generated by the forms belonging to $S$. Thus, *a form* $\varphi$ *belongs to* ${}^h\mathfrak{a}$ *if and only if* $\varphi$ *is of the type* $Y_0^m({}^hF)$, $m \geqq 0$, $F \in \mathfrak{a}$, *or, equivalently, if and only if* ${}^a\varphi \in \mathfrak{a}$.

THEOREM 17. *The operation* $\mathfrak{a} \to {}^h\mathfrak{a}$ *maps distinct ideals in* $R$ *into distinct ideals in* ${}^hR$; *it preserves inclusion and the usual ideal-theoretic operations, i.e., it has the following properties:*

[1] $\mathfrak{a} \supset \mathfrak{b} \Rightarrow {}^h\mathfrak{a} \supset {}^h\mathfrak{b}$.
[2] ${}^h(\mathfrak{a} + \mathfrak{b}) = {}^h\mathfrak{a} + {}^h\mathfrak{b}$.
[3] ${}^h(\mathfrak{a}\mathfrak{b}) = {}^h\mathfrak{a} \cdot {}^h\mathfrak{b}$.
[4] ${}^h(\mathfrak{a} \cap \mathfrak{b}) = {}^h\mathfrak{a} \cap {}^h\mathfrak{b}$.
[5] ${}^h(\mathfrak{a} : \mathfrak{b}) = {}^h\mathfrak{a} : {}^h\mathfrak{b}$.
[6] ${}^h(\sqrt{\mathfrak{a}}) = \sqrt{{}^h\mathfrak{a}}$.

*Furthermore:*

[7] *If* $\mathfrak{p}$ *is a prime ideal in* $R$, *then* ${}^h\mathfrak{p}$ *is also a prime ideal.*

[8] *If* $\mathfrak{p}$ *is prime and* $\mathfrak{q}$ *is an ideal primary for* $\mathfrak{p}$ *in* $R$, *then* ${}^h\mathfrak{q}$ *is primary for* ${}^h\mathfrak{p}$.

[9] *If* $\mathfrak{a}=\bigcap_i \mathfrak{q}_i$ *is an irredundant primary representation of an ideal* $\mathfrak{a}$ *in* $R$, *then* ${}^h\mathfrak{a}=\bigcap_i {}^h\mathfrak{q}_i$ *is an irredundant primary representation of* ${}^h\mathfrak{a}$.

PROOF: If $\mathfrak{a}$ is an ideal in $R$ then $\mathfrak{a}$ coincides with the set of all polynomials ${}^a\varphi$, $\varphi \in {}^h\mathfrak{a}$. This shows that if $\mathfrak{a}$ and $\mathfrak{b}$ are distinct ideals in $R$ then ${}^h\mathfrak{a} \neq {}^h\mathfrak{b}$.

[1] is obvious, and [2] follows from the fact that, for any ideal $\mathfrak{a}$, ${}^h\mathfrak{a}$ is generated by the forms ${}^hF$ where $F$ ranges over $\mathfrak{a}$. Similarly, [3] follows directly from (2) and from the definition of products of ideals. A form $\varphi$ belongs to ${}^h(\mathfrak{a} \cap \mathfrak{b})$ if and only if ${}^a\varphi \in \mathfrak{a} \cap \mathfrak{b}$, i.e., if and only if $\varphi$ belongs to ${}^h\mathfrak{a}$ *and* to ${}^h\mathfrak{b}$, and this proves [4]. The inclusion ${}^h(\mathfrak{a}:\mathfrak{b}) \subset {}^h\mathfrak{a}:{}^h\mathfrak{b}$ follows directly from [3] and [1] and from the definition of quotients of ideals. Conversely, let $\varphi \in {}^h\mathfrak{a}:{}^h\mathfrak{b}$ and let $F$ be any polynomial in $\mathfrak{b}$. Since ${}^hF \in {}^h\mathfrak{b}$ we have $\varphi \cdot {}^hF \in {}^h\mathfrak{a}$, whence ${}^a(\varphi \cdot {}^hF) \in \mathfrak{a}$. By (2′) and (4) we have ${}^a(\varphi \cdot {}^hF) = {}^a\varphi \cdot F$, and so the product ${}^a\varphi \cdot F$ belongs to $\mathfrak{a}$, for every $F$ in $\mathfrak{b}$. This implies that ${}^a\varphi \in \mathfrak{a}:\mathfrak{b}$, $\varphi \in {}^h(\mathfrak{a}:\mathfrak{b})$, showing ${}^h\mathfrak{a}:{}^h\mathfrak{b} \subset {}^h(\mathfrak{a}:\mathfrak{b})$.

Relation [6] follows from the following equivalences: $\varphi \in {}^h(\sqrt{\mathfrak{a}}) \Leftrightarrow {}^a\varphi \in \sqrt{\mathfrak{a}} \Leftrightarrow ({}^a\varphi)^m \in \mathfrak{a}$ for some integer $m \geqq 1 \Leftrightarrow {}^a(\varphi^m) \in \mathfrak{a} \Leftrightarrow \varphi^m \in {}^h\mathfrak{a} \Leftrightarrow \varphi \in \sqrt{{}^h\mathfrak{a}}$.

Let $\mathfrak{q}$ be a primary ideal in $R$ and let $\varphi$ and $\psi$ be two *forms* in ${}^hR$ such that $\varphi\psi \in {}^h\mathfrak{q}$, $\psi \notin {}^h\mathfrak{q}$. Then ${}^a\varphi \cdot {}^a\psi = {}^a(\varphi\psi) \in \mathfrak{q}$ and ${}^a\psi \notin \mathfrak{q}$. Consequently, $({}^a\varphi)^m \in \mathfrak{q}$ for some $m \geqq 1$, showing that $\varphi^m \in {}^h\mathfrak{q}$. It follows now from Lemma 2, § 2, that ${}^h\mathfrak{q}$ is primary. Similarly, it can be shown that if $\mathfrak{p}$ is prime then ${}^h\mathfrak{p}$ is prime, and this completes the proof of [7] and [8], in view of [6].

As to [9], the fact that $\bigcap_i {}^h\mathfrak{q}_i$ is a primary representation of ${}^h\mathfrak{a}$ follows from [4] and [8]. It remains to show that this representation is irredundant. If $j$ is any of the indices in the set $\{i\}$ then $\mathfrak{a} \neq \bigcap_{i \neq j} \mathfrak{q}_i$. Hence, by the first assertion of the theorem and by [4], we have ${}^h\mathfrak{a} \neq \bigcap_{i \neq j} {}^h\mathfrak{q}_i$. This completes the proof of the theorem.

Not every ideal in ${}^hR$ is of the form ${}^h\mathfrak{a}$, where $\mathfrak{a}$ is an ideal in $R$; in fact, no ideal of the form ${}^h\mathfrak{a}$, other than the unit ideal, can contain a power of $Y_0$. The question arises, therefore, of characterizing the class of ideals ${}^h\mathfrak{a}$, $\mathfrak{a} \subset R$. Before studying this question it will be convenient to extend the operation ${}^a$ to *ideals* in ${}^hR$.

Given a *homogeneous* ideal $\mathfrak{A}$ in ${}^hR$ the set of all polynomials of the form ${}^a\varphi$, where $\varphi$ ranges over the set of all *forms* in $\mathfrak{A}$, is easily seen to be an ideal by using the formulas (2′) and (3′) and by observing that: (a) if $\varphi$ and $\psi$ are forms in $\mathfrak{A}$ and $m = \partial(\psi) - \partial(\varphi) \geqq 0$, then the form $Y_0{}^m\varphi - \psi$ is in $\mathfrak{A}$ and we have ${}^a(Y_0{}^m\varphi - \psi) = {}^a\varphi - {}^a\psi$; (b) every polynomial in $R$ can be written in the form ${}^a\varphi$ with $\varphi$ a *form* in ${}^hR$ (see (4)). We denote this ideal by ${}^a\mathfrak{A}$.

We note the following properties of the composite operations ${}^{ah}$ and ${}^{ha}$:

(6) $$ {}^a({}^h\mathfrak{a}) = \mathfrak{a}, \quad \text{for any ideal } \mathfrak{a} \text{ in } R, $$

(7) $$ {}^h({}^a\mathfrak{A}) \supset \mathfrak{A}, \quad \text{for any homogeneous ideal } \mathfrak{A} \text{ in } {}^hR, $$

(7′) $$ Y_0{}^m({}^h({}^a\mathfrak{A})) \subset \mathfrak{A}, \quad \text{for some integer } m \geqq 1. $$

If $\mathfrak{a}$ is any ideal in $R$ then it follows from the definition of ${}^h\mathfrak{a}$ (and it has also been pointed out at the beginning of the proof of Theorem 17) that $\mathfrak{a}$ is the set of all polynomials ${}^a\varphi$, where $\varphi$ ranges over ${}^h\mathfrak{a}$. In other words, we have (6). Relation (7) is obvious, for if $\varphi \in \mathfrak{A}$ ($\varphi$, a form) then ${}^a\varphi \in {}^a\mathfrak{A}$ and ${}^h({}^a\varphi) \in {}^h({}^a\mathfrak{A})$, whence $\varphi \in {}^h({}^a\mathfrak{A})$, since $\varphi$ is a multiple of ${}^h({}^a\varphi)$, by (5′). On the other hand, if $\varphi$ is any form in ${}^h({}^a\mathfrak{A})$ then ${}^a\varphi = {}^a\psi$ for some form $\psi$ in $\mathfrak{A}$, and hence, by (5′), $\varphi$ and $\psi$ can differ only by a factor which is a power of $Y_0$. Thus, for every form $\varphi$ in ${}^h({}^a\mathfrak{A})$ there exists an integer $s = s(\varphi)$ such that $Y_0{}^s\varphi \in \mathfrak{A}$. Since ${}^h({}^a\mathfrak{A})$ has a finite basis, (7′) follows.

THEOREM 18. *The operation $\mathfrak{A} \to {}^a\mathfrak{A}$ maps the set of all homogeneous ideals in ${}^hR$ onto the set of all ideals in $R$; it preserves inclusion and the usual ideal-theoretic operations, i.e., it has the following properties ($\mathfrak{A}$ and $\mathfrak{B}$ are homogeneous ideals in ${}^hR$):*

{1} $\mathfrak{A} \supset \mathfrak{B} \Rightarrow {}^a\mathfrak{A} \supset {}^a\mathfrak{B}$.
{2} ${}^a(\mathfrak{A} + \mathfrak{B}) = {}^a\mathfrak{A} + {}^a\mathfrak{B}$.
{3} ${}^a(\mathfrak{A}\mathfrak{B}) = {}^a\mathfrak{A} \cdot {}^a\mathfrak{B}$.
{4} ${}^a(\mathfrak{A} \cap \mathfrak{B}) = {}^a\mathfrak{A} \cap {}^a\mathfrak{B}$.
{5} ${}^a(\mathfrak{A} : \mathfrak{B}) = {}^a\mathfrak{A} : {}^a\mathfrak{B}$.
{6} ${}^a(\sqrt{\mathfrak{A}}) = \sqrt{{}^a\mathfrak{A}}$.

Furthermore:

{7} *${}^a\mathfrak{A}$ is the unit ideal if and only if $\mathfrak{A}$ contains a power of $Y_0$.*
{8} *${}^a\mathfrak{A} = {}^a\mathfrak{B}$ if and only if $\mathfrak{A} : Y_0{}^s = \mathfrak{B} : Y_0{}^s$ for some integer $s$ (and hence also for all $s$ sufficiently large).*

{9} *If* $\mathfrak{P}$ *is a homogeneous prime ideal which is different from* (0) *and does not contain* $Y_0$ *then* ${}^a\mathfrak{P}$ *is a proper prime ideal.*

{10} *If* $\mathfrak{Q}$ *is a homogeneous primary ideal which is different from* (0) *and does not contain any power of* $Y_0$, *then* ${}^a\mathfrak{Q}$ *is a proper primary ideal, and if* $\mathfrak{P}$ *is the prime ideal of* $\mathfrak{Q}$ *then* ${}^a\mathfrak{P}$ *is the prime ideal of* ${}^a\mathfrak{Q}$.

{11} *If* $\mathfrak{A}=\bigcap_i \mathfrak{Q}_i$ *is an irredundant primary representation of* $\mathfrak{A}$, *all the* $\mathfrak{Q}_i$ *being homogeneous* (see § 2, Theorem 9), *then* ${}^a\mathfrak{A}=\bigcap_j {}^a\mathfrak{Q}_j$ *where the* $\mathfrak{Q}_j$ *are those primary components* $\mathfrak{Q}_i$ *of* $\mathfrak{A}$ *which do not contain any power of* $Y_0$, *and the representation* ${}^a\mathfrak{A}=\bigcap_j {}^a\mathfrak{Q}_j$ *is primary and irredundant.*

PROOF. We have $\mathfrak{a}={}^a({}^h\mathfrak{a})$, by (6), and this shows that the range of the operation $\mathfrak{A}\rightarrow {}^a\mathfrak{A}$ is the set of *all* ideals in $R$.

The relations {1}, {2} and {3} are obvious. The inclusion ${}^a(\mathfrak{A}\cap\mathfrak{B})\subset {}^a\mathfrak{A}\cap {}^a\mathfrak{B}$ follows from {1}. Conversely, let $F$ be any polynomial in ${}^a\mathfrak{A}\cap {}^a\mathfrak{B}$. Then $F={}^a\varphi={}^a\psi$, where $\varphi$ is a form in $\mathfrak{A}$ and $\psi$ is a form in $\mathfrak{B}$. It follows from (5′) that $\varphi=Y_0{}^m({}^hF)$ and $\psi=Y_0{}^{m'}({}^hF)$, and therefore, if say $m'\geqq m$, then $\psi=Y_0{}^{m'-m}\varphi$. Consequently $\psi\in\mathfrak{A}\cap\mathfrak{B}$, and $F={}^a\psi\in{}^a(\mathfrak{A}\cap\mathfrak{B})$, showing the opposite inclusion ${}^a\mathfrak{A}\cap{}^a\mathfrak{B}\subset{}^a(\mathfrak{A}\cap\mathfrak{B})$ and proving {4}.

The inclusion ${}^a(\mathfrak{A}:\mathfrak{B})\subset{}^a\mathfrak{A}:{}^a\mathfrak{B}$ follows from $\mathfrak{B}\cdot(\mathfrak{A}:\mathfrak{B})\subset\mathfrak{A}$, {3} and {1}. On the other hand, let $F$ be any polynomial in ${}^a\mathfrak{A}:{}^a\mathfrak{B}$ and let $m$ be an integer such that (7′) holds. Then we have:

$$(Y_0{}^m\cdot{}^hF)\mathfrak{B}\subset(Y_0{}^m\cdot{}^hF)({}^h({}^a\mathfrak{B}))=Y_0{}^m\cdot{}^h(F\cdot{}^a\mathfrak{B})\subset Y_0{}^m\cdot({}^h({}^a\mathfrak{A}))\subset\mathfrak{A},$$

and therefore $\varphi\in\mathfrak{A}:\mathfrak{B}$, where $\varphi=Y_0{}^m\cdot{}^hF$. Since ${}^a\varphi=F$, it follows that $F\in{}^a(\mathfrak{A}:\mathfrak{B})$, and this proves {5}.

The inclusion $(\sqrt{\mathfrak{A}})^\rho\subset\mathfrak{A}$, where $\rho$ is some integer $\geqq 1$, implies, by {1} and {3}, $[{}^a(\sqrt{\mathfrak{A}})]^\rho\subset{}^a\mathfrak{A}$, whence ${}^a(\sqrt{\mathfrak{A}})\subset\sqrt{{}^a\mathfrak{A}}$. On the other hand, if $F$ is any polynomial in $\sqrt{{}^a\mathfrak{A}}$ and $m$ is an integer satisfying (7′), then we have, for a suitable integer $\rho\geqq 1$: $(Y_0{}^m\cdot{}^hF)^\rho\in Y_0{}^m\,({}^h({}^a\mathfrak{A}))\subset\mathfrak{A}$, i.e., $\varphi\in\sqrt{\mathfrak{A}}$ where $\varphi=Y_0{}^m\cdot{}^hF$. Since ${}^a\varphi=F$, it follows therefore that $F\in{}^a(\sqrt{\mathfrak{A}})$. Hence $\sqrt{{}^a\mathfrak{A}}\subset{}^a(\sqrt{\mathfrak{A}})$, and this proves {6}.

If $Y_0{}^m\in\mathfrak{A}$, for some $m\geqq 1$, then ${}^a\mathfrak{A}=(1)$ since ${}^a(Y_0{}^m)=1$. Conversely, if ${}^a\mathfrak{A}=(1)$, then ${}^a\varphi=1$ for some form $\varphi$ in $\mathfrak{A}$, and by (5′), such a form is necessarily a power of $Y_0$. This proves {7}.

We have, for any integer $s\geqq 1$, $(\mathfrak{A}:Y_0{}^s)\cdot Y_0{}^s\subset\mathfrak{A}\subset(\mathfrak{A}:Y_0{}^s)$. Applying the operation ${}^a$ and using {1}, {3} and {7}, we find that ${}^a(\mathfrak{A}:Y_0{}^s)={}^a\mathfrak{A}$. Therefore, if $\mathfrak{A}:Y_0{}^s=\mathfrak{B}:Y_0{}^s$, for some $s$, then ${}^a\mathfrak{A}={}^a\mathfrak{B}$. Conversely, assume that ${}^a\mathfrak{A} = {}^a\mathfrak{B}$. Since $\mathfrak{A} : Y_0^i \subset \mathfrak{A} : Y_0^{i+1}$ and similarly for $\mathfrak{B}$, there exists an integer $s$ such that $\mathfrak{A} : Y_0^s = \mathfrak{A} : Y_0^{s+1} = \cdots$, $\mathfrak{B} : Y_0^s = \mathfrak{B} : Y_0^{s+1} = \cdots$. We have ${}^a(\mathfrak{A} : Y_0^s) = {}^a\mathfrak{A} = {}^a\mathfrak{B} = {}^a(\mathfrak{B} : Y_0^s)$, hence

$$(8) \qquad {}^{ha}(\mathfrak{A} : Y_0^s) = {}^{ha}(\mathfrak{B} : Y_0^s).$$

On the other hand, we have, by (7′): $Y_0^m({}^{ha}(\mathfrak{A} : Y_0^s)) \subset \mathfrak{A} : Y_0^s$, for some integer $m \geqq 1$, and therefore, by our choice of $s$, ${}^{ha}(\mathfrak{A} : Y_0^s) \subset \mathfrak{A} : Y_0^s$. Consequently, by (7), we have ${}^{ha}(\mathfrak{A} : Y_0^s) = \mathfrak{A} : Y_0^s$. Similarly we obtain ${}^{ha}(\mathfrak{B} : Y_0^s) = \mathfrak{B} : Y_0^s$. Therefore $\mathfrak{A} : Y_0^s = \mathfrak{B} : Y_0^s$, by (8), and this establishes {8}.

Let $\mathfrak{Q}$ be a homogeneous primary ideal in ${}^hR$, different from (0) and not containing any power of $Y_0$, and let $F$ and $G$ be two polynomials in $R$ such that $FG \in {}^a\mathfrak{Q}$, $F \notin {}^a\mathfrak{Q}$. From (7′), and from the fact that $FG \in {}^a\mathfrak{Q}$ it follows that $Y_0^m({}^hF)({}^hG) \in \mathfrak{Q}$, for some $m \geqq 1$. Since $\mathfrak{Q}$ is primary and $F \notin {}^a\mathfrak{Q}$, $Y_0^m \notin {}^a\mathfrak{Q}$, it follows that ${}^hG \in \sqrt{\mathfrak{Q}}$, whence $G = {}^a({}^hG) \in {}^a(\sqrt{\mathfrak{Q}}) = \sqrt{{}^a\mathfrak{Q}}$. This shows that ${}^a\mathfrak{Q}$ is primary, and thus the first part of {10} is established. In a similar fashion one proves {9}. The second part of {10} follows from {6}.

The first part of {11} follows from {4} and {7}. That all the $\mathfrak{Q}_j$ are primary follows from {10}. To prove the assertion of irredundancy, let $\nu$ be any one of the indices $j$ and let $\mathfrak{A}_\nu = \bigcap_{i \neq \nu} \mathfrak{Q}_i$. We have $\mathfrak{A}_\nu \not\subset \mathfrak{Q}_\nu$ since $\bigcap_i \mathfrak{Q}_i$ is an irredundant representation. *A fortiori*, $\mathfrak{A}_\nu : (Y_0^s) \not\subset \mathfrak{Q}_\nu$, for all $s \geqq 1$. On the other hand, since no power of $Y_0$ belongs to $\mathfrak{Q}_\nu$ we have $\mathfrak{Q}_\nu : (Y_0^s) = \mathfrak{Q}_\nu$ and hence $\mathfrak{A} : (Y_0^s) \subset \mathfrak{Q}_\nu : (Y_0^s) = \mathfrak{Q}_\nu$, for all $s$. Consequently, $\mathfrak{A}_\nu : (Y_0^s) \neq \mathfrak{A} : (Y_0^s)$, for all $s$. It follows then by {8} that ${}^a\mathfrak{A}_\nu \neq {}^a\mathfrak{A}$, i.e., $\bigcap_{j \neq \nu} {}^a\mathfrak{Q}_j \neq \bigcap_j {}^a\mathfrak{Q}_j$. This completes the proof of {11} and of Theorem 18.

COROLLARY. *If $\mathfrak{A}$ is any homogeneous ideal in ${}^hR$ then, with the same notations as in part {11} of Theorem 18 we have:*

$$(9) \qquad {}^h({}^a\mathfrak{A}) = \bigcap \mathfrak{Q}_j.$$

*In particular, we have $\mathfrak{A} = {}^h({}^a\mathfrak{A})$ if and only if no prime ideal of $\mathfrak{A}$ contains $Y_0$. The set of ideals of the form ${}^h\mathfrak{a}$, where $\mathfrak{a}$ is an ideal in $R$, coincides with the set of homogeneous ideals $\mathfrak{A}$ in ${}^hR$ no prime ideal of which contains $Y_0$.*

Relation (9) follows immediately from part {11} of Theorem 18, part [4] of Theorem 17, and from relations (7) and (7′). The last assertion of the corollary follows by observing that if $\mathfrak{A} = {}^h\mathfrak{a}$ then ${}^a\mathfrak{A} = {}^a({}^h\mathfrak{a}) = \mathfrak{a}$ and hence ${}^h({}^a\mathfrak{A}) = {}^h\mathfrak{a} = \mathfrak{A}$.

Note that the preceding corollary shows that for any homogeneous ideal $\mathfrak{A}$ the ideal $^h(^a\mathfrak{A})$ can be characterized as the greatest homogeneous ideal $\mathfrak{B}$ such that $^a\mathfrak{B} = {^a\mathfrak{A}}$.

REMARK. Another method for studying the operation $\mathfrak{A} \rightarrow {^a\mathfrak{A}}$ is to notice the existence of two different ways for passing from $^hR = k[Y_0, \cdots, Y_n]$ to $R = k[X_1, \cdots, X_n]$:

(1) The mapping $\varphi \rightarrow {^a\varphi}$, where now $\varphi$ is not necessarily a form, is, by formulas (2′) and (3′), a *homomorphism* of $^hR$ into $R$, and formula (4) shows that it maps $^hR$ *onto* $R$. Its kernel is obviously the ideal $\mathfrak{K} = (Y_0 - 1)$. If we identify $R$ with $^hR/\mathfrak{K}$, the ideal $^a\mathfrak{A}$ gets identified with $(\mathfrak{A} + \mathfrak{K})/\mathfrak{K}$. In other words: the passage from $^hR$ to $R$ may be regarded as a *residue class ring formation.* This proves, for example, assertions {2} and {3} in Theorem 18.

(2) Another way of looking at the mapping $\mathfrak{A} \rightarrow {^a\mathfrak{A}}$ is to imbed $R = k[X_1, \cdots, X_n]$ in $k(Y_0, \cdots, Y_n)$ by setting $X_1 = Y_1/Y_0, \cdots, X_n = Y_n/Y_0$. Then $R$ is contained in the *quotient ring* $S = k[Y_0, \cdots, Y_n]_M$, where $M$ is the multiplicative system formed by the powers of $Y_0$; and we have

$$(10) \qquad R = k[Y_1/Y_0, \cdots, Y_n/Y_0] = S \cap k(Y_1/Y_0, \cdots, Y_n/Y_0).$$

In fact, the inclusion $R \subset S \cap k(Y_1/Y_0, \cdots, Y_n/Y_0)$ is clear. Conversely, if a rational function $P(Y_0, \cdots, Y_n)/Y_0^q$ ($P$ = polynomial) belongs to $k(Y_1/Y_0, \cdots, Y_n/Y_0)$, it remains invariant if we multiply the variables $Y_0, \cdots, Y_n$ by one and the same quantity, whence $P$ is a homogeneous polynomial of degree $q$, and our rational function belongs to $k[Y_1/Y_0, \cdots, Y_n/Y_0]$.

By this identification the polynomial $(^a\varphi)(X_1, \cdots, X_n)$ corresponding to a *form* $\varphi$ of degree $q$ becomes $\varphi(Y_0, \cdots, Y_n)/Y_0^q = \varphi(1, Y_1/Y_0, \cdots, Y_n/Y_0)$. Thus if $\mathfrak{A}$ is a *homogeneous ideal,* $^a\mathfrak{A}$ becomes the ideal generated in $R$ by (and—in fact—consisting of) the elements $\varphi(Y_0, \cdots, Y_n)/Y_0^q$ where $\varphi$ is a form in $\mathfrak{A}$ and where $q$ is its degree. It is clear that this ideal is contained in $\mathfrak{A}k[Y_0, \cdots, Y_n]_M \cap k[Y_1/Y_0, \cdots, Y_n/Y_0]$. Conversely, if a polynomial $P(Y_1/Y_0, \cdots, Y_n/Y_0)$ belongs to the ideal $\mathfrak{A}k[Y_0, Y_1, \cdots, Y_n]_M$, it may be written in the form $A(Y_0, \cdots, Y_n)/Y_0^q$, where $A(Y) \in \mathfrak{A}$; as $P(Y_1/Y_0, \cdots, Y_n/Y_0) = \varphi(Y_0, \cdots, Y_n)/Y_0^r$ where $\varphi$ is a form of degree $r$, not a multiple of $Y_0$, this implies that $q \geqq r$ and that $A = Y_0^{q-r}\varphi$. Hence $A$ is a *form* of degree $q$, and $P(X_1, \cdots, X_n) = A(1, X_1, \cdots, X_n)$ is an element of $^a\mathfrak{A}$. Hence

$$(10') \qquad {^a\mathfrak{A}} = \mathfrak{A}k[Y_0, \cdots, Y_n]_M \cap k(Y_1/Y_0, \cdots, Y_n/Y_0).$$

We have already mentioned above that the representation of ${}^a\mathfrak{A}$ as $(\mathfrak{A}+\mathfrak{K})/\mathfrak{K}$ proves immediately assertions {2} and {3} in Theorem 18. On the other hand, formula (10′) proves immediately the assertions {4}, {6}, {9} and {10} in Theorem 18, provided one takes into account the behavior of intersections, radical, prime ideals and primary ideals under quotient ring formation and under contraction (Vol. I, Ch. IV, §§ 8 and 10).

We shall end this section with a discussion of the extension of the preceding results to arbitrary finite integral domains $S=k[x_1, x_2, \cdots, x_n]$. Guided by the imbedding (10) of the polynomial ring $R=k[X_1, X_2, \cdots, X_n]$ in the field $k(Y_0, Y_1, \cdots, Y_n)$, we proceed as follows:

We adjoin a transcendental $y_0$ to the quotient field of $S$, we set $y_i=y_0x_i$, $i=1, 2, \cdots, n$, and we denote by ${}^hS$ the ring $k[y_0, y_1, \cdots, y_n]$. It is immediately seen that ${}^hS$ is a *homogeneous* ring over $k$ (compare with the proof of the lemma in § 4). For every homogeneous element $\alpha=\varphi(y_0, y_1, \cdots, y_n)$ of degree $q$, where $\varphi$ is a form, we set ${}^a\alpha=\varphi(1, x_1, x_2, \cdots, x_n)=\alpha/y_0^q$. Since $q$ is determined by $\alpha$, ${}^a\alpha$ depends only on $\alpha$. If we attempt now to define ${}^ha$ for any element $a$ in $S$ by analogy with the definition given in the case of polynomial rings, we meet a difficulty arising from the fact that there are in general infinitely many polynomials $F(X_1, X_2, \cdots, X_n)$ with the property that $F(x_1, x_2, \cdots, x_n)=a$. Were we to agree to take for $F$ a polynomial of smallest possible degree, say $\nu$, and then define ${}^ha$ to be $y_0^\nu F(y_1/y_0, y_2/y_0, \cdots, y_n/y_0)$, we would find that the relation ${}^h(ab)={}^ha{}^hb$ is not necessarily satisfied. However, we do not need a definition of the operation ${}^h$ for *elements* of $S$; what we need is only to define that operation for *ideals* in $S$. The definition is the same as in the case of polynomial rings, namely: if $\mathfrak{a}$ is an ideal in $S$, ${}^h\mathfrak{a}$ is the ideal generated by the homogeneous elements $\varphi$ of ${}^hS$ such that ${}^a\varphi \in \mathfrak{a}$. On the other hand, if $\mathfrak{A}$ is any *homogeneous* ideal in ${}^hS$ we define ${}^a\mathfrak{A}$ as the ideal consisting of all elements of $S$ of the form ${}^a\varphi$, where $\varphi$ is any homogeneous element of $\mathfrak{A}$. With these definitions, *Theorems* 17 *and* 18 *remain valid if $R$, ${}^hR$ and $Y_0$ are replaced by $S$, ${}^hS$ and $y_0$ respectively.* Similarly, formulas (6), (7) and (7′) as well as the corollary to Theorem 18 remain valid. We shall briefly prove this assertion.

Let $\tau$ be the $k$-homomorphism of the polynomial ring $R=k[X_1, X_2, \cdots, X_n]$ onto $S=k[x_1, x_2, \cdots, x_n]$, such that $X_i\tau=x_i$, and let $\mathfrak{n}$ be the kernel of $\tau$. We can extend $\tau$ to a homomorphism (which we shall continue to denote by the same letter $\tau$) of $k[Y_0, Y_1, \cdots, Y_n]$ by setting $Y_0\tau=y_0$. Then $\tau$ induces a homomorphism of the *subring* ${}^hR$ of

$k[Y_0, X_1, \cdots, X_n]$ (note that $Y_i = Y_0 X_i$) onto ${}^hS$ such that $Y_i\tau = y_i$. The kernel of this homomorphism of ${}^hR$ onto ${}^hS$ is easily seen to be the ideal ${}^h\mathfrak{n}$. From now on we shall identify $S$ with $R/\mathfrak{n}$ and ${}^hS$ with ${}^hR/{}^h\mathfrak{n}$:

$$S = R/\mathfrak{n}, \quad {}^hS = {}^hR/{}^h\mathfrak{n}. \tag{11}$$

We note that, by (6), we can also write

$$ {}^hS = {}^hR/\mathfrak{N}, \quad S = {}^a({}^hR)/{}^a\mathfrak{N}, \tag{11'}$$

where $\mathfrak{N} = {}^h\mathfrak{n}$.

Now the canonical homomorphism of $R$ onto $R/\mathfrak{n}$ maps in (1, 1) fashion the set of all ideals of $R$ which contain the kernel $\mathfrak{n}$ onto the set of all ideals in $S$, and this mapping preserves inclusion and all the usual ideal-theoretic operations (see Vol. I, Ch. III, § 7, formulae (11)–(16)); this mapping also sends prime and primary ideals into prime and primary ideals respectively (see Vol. I, Ch. III, § 8, Theorem 11 and III, § 9, Theorem 14), and transforms irredundant primary representations into irredundant primary representations (Vol. I, Ch. IV, § 5, Remark at the end of the section). A similar statement holds for the canonical homomorphism of ${}^hR$ onto ${}^hS$ and for the induced mapping of the set of all homogeneous ideals of ${}^hR$ which contain the kernel $\mathfrak{N} = {}^h\mathfrak{n}$ onto the set of all homogeneous ideals of ${}^hS$. In view of these facts, it is seen at once that the validity of Theorems 17 and 18 for $R$ and ${}^hR$ implies their validity for $S$ and ${}^hS$.

§ 6. **Relations between affine and projective varieties.** With every point $P = (x_1, x_2, \cdots, x_n)$ of $A_n^K$ we associate the point $\varphi(P)$ of $P_n^K$ having $\{1, x_1, x_2, \cdots, x_n\}$ as a set of homogeneous coördinates. The mapping $P \to \varphi(P)$ of $A_n^K$ into $P_n^K$ is one to one, for if two points $\varphi(P) = (1, x_1, x_2, \cdots, x_n)$ and $\varphi(P') = (1, x'_1, x'_2, \cdots, x'_n)$ coincide, then we must have, for some $t$ in $K$, $1 = t \cdot 1$, $x'_i = tx_i$, showing that $t = 1$, $x'_i = x_i$, $P' = P$. This mapping is not onto, for no point of the form $(y_0, y_1, \cdots, y_n)$ with $y_0 = 0$ can be in $\varphi(A_n^K)$. However, every other point of $P_n^K$ is in $\varphi(A_n^K)$, for if $y_0 \neq 0$ and if we set $x_i = y_i/y_0$, then the point $(y_0, y_1, \cdots, y_n)$ in $P_n^K$ is the $\varphi$-image of the point $(x_1, x_2, \cdots, x_n)$ of $A_n^K$. Thus, *the mapping* $\varphi$ *identifies the affine space* $A_n^K$ *with the complement of the hyperplane* $Y_0 = 0$ *in the projective space* $P_n^K$. We think of having carried out this identification and we shall regard therefore the affine space $A_n^K$ as a subset of the projective space $P_n^K$. The hyperplane $Y_0 = 0$ is called the *hyperplane at infinity* (for the above identification), and the points or varieties which are contained in the hyperplane at infinity are said to be *at infinity*. The points not in the hyperplane $Y_0=0$ are said to be *at finite distance*.

In this section we shall denote algebraic varieties in the projective space by capital letters such as $V, W, \cdots$, while algebraic varieties in the affine space will be denoted by small letters such as $v, w, \cdots$. Similarly, capital German letters $\mathfrak{A}, \mathfrak{B}, \cdots$ will be used to denote homogeneous ideals in $k[Y_0, Y_1, \cdots, Y_n]$, while small German letters $\mathfrak{a}, \mathfrak{b}, \cdots$, will denote ideals (homogeneous or non-homogeneous) in $k[X_1, X_2, \cdots, X_n]$. If $V$ is a variety in $P_n{}^K$ we shall denote by ${}^aV$ *the intersection of* $V$ *with* $A_n{}^K$ and we shall call ${}^aV$ *the affine restriction of* $V$:

$$(1) \qquad {}^aV = V \cap A_n{}^K.$$

The fact that ${}^aV$ is also a variety (an affine variety) is included in the following relation: If $\mathfrak{A}$ is a homogeneous ideal in $k[Y_0, Y_1, \cdots, Y_n]$ then

$$(2) \qquad {}^a(\mathscr{V}(\mathfrak{A})) = \mathscr{V}({}^a\mathfrak{A}).$$

(It is understood that in (2) the operator $\mathscr{V}$ has two different meanings according as it is applied to a homogeneous or non-homogeneous ideal: $\mathscr{V}(\mathfrak{A})$ means the projective variety of the homogeneous ideal $\mathfrak{A}$, while $\mathscr{V}({}^a\mathfrak{A})$ stands for the affine variety of the ideal ${}^a\mathfrak{A}$.) The proof of (2) is straightforward: a point $P=(1, x_1, x_2, \cdots, x_n)$ of $P_n{}^K$ belongs to $\mathscr{V}(\mathfrak{A})$ if and only if $\varphi(1, x_1, x_2, \cdots, x_n)=0$ for all forms $\varphi(Y_0, Y_1, \cdots, Y_n)$ in the ideal $\mathfrak{A}$, and since ${}^a\mathfrak{A}$ consists of all the polynomials $\varphi(1, X_1, X_2, \cdots, X_n)$ such that $\varphi(Y_0, Y_1, \cdots, Y_n)$ is a form in $\mathfrak{A}$, we see that a point $P$ of $A_n{}^K$ belongs to $\mathscr{V}(\mathfrak{A})$ if and only if the $n$-tuple $(x_1, x_2, \cdots, x_n)$ of its non-homogeneous coördinates is a zero of the ideal ${}^a\mathfrak{A}$, and this proves (2) and shows that ${}^a(\mathscr{V}(\mathfrak{A}))$ is an affine variety.

If $v$ is any affine variety we denote by ${}^hv$ *the least algebraic* (projective) *variety containing* $v$, or equivalently, using the topology in $P_n{}^K$ introduced in § 4:

$$(3) \qquad {}^hv = \text{closure of } v \text{ in } P_n{}^K.$$

We call ${}^hv$ the *projective extension* of $v$.

THEOREM 19. *If* $v$ *is an affine variety in* $A_n{}^K$ *then*

$$(4) \qquad {}^{ah}v = v.$$

*The mapping* $v \rightarrow {}^hv$ *maps in* (1, 1) *fashion the set of all affine varieties in* $A_n{}^K$ *onto the set of all projective varieties having no irreducible components at infinity. If* $v$ *is irreducible, so is* ${}^hv$, *and if* $\bigcup_i v_i$ *is the irredun-*

*dant decomposition of an affine variety* $v$ *into irreducible components, then* $\bigcup_i {}^h v_i$ *is the irredundant decomposition of* ${}^h v$ *into irreducible components. If* $V$ *is an irreducible projective variety, not at infinity, then* ${}^a V$ *is irreducible and*

$$(5) \qquad {}^{ha}V = V.$$

PROOF. We first observe that if $\mathfrak{a}$ is any ideal in $k[X_1, X_2, \cdots, X_n]$ then

$$(6) \qquad {}^a(\mathscr{V}({}^h\mathfrak{a})) = \mathscr{V}(\mathfrak{a}).$$

This follows immediately by setting $\mathfrak{A} = {}^h\mathfrak{a}$ in (2) and by recalling that ${}^{ah}\mathfrak{a} = \mathfrak{a}$ (§ 5, (6)). Now, let $v$ be a given affine variety. Then $v = \mathscr{V}(\mathfrak{a})$ for some ideal $\mathfrak{a}$ in $k[X_1, X_2, \cdots, X_n]$. Formula (6) shows that there exists a projective variety $V$ such that ${}^aV = v$ (namely, the variety $\mathscr{V}({}^h\mathfrak{a})$). Since ${}^hv$ is the smallest projective variety containing $v$, it follows *a fortiori* that ${}^{ah}v = v$, which proves (4). Formula (4) also shows that if $v_1$ and $v_2$ are distinct affine varieties then ${}^hv_1 \neq {}^hv_2$, for ${}^a({}^hv_1) = v_1 \neq v_2 = {}^a({}^hv_2)$. Hence the mapping $v \rightarrow {}^hv$ is (1, 1).

Let $v$ be an irreducible non-empty variety and let ${}^hv = V_1 \cup V_2$ where $V_1$ and $V_2$ are projective varieties. We have, by (4): $v = {}^a(V_1 \cup V_2) = {}^aV_1 \cup {}^aV_2$. Since $v$ is irreducible, either ${}^aV_1$ or ${}^aV_2$ coincides with $v$. Let, say, ${}^aV_1 = v$. Then $V_1 \supset v$ and hence $V_1 \supset {}^hv$ (by definition of ${}^hv$), i.e., $V_1 = {}^hv$. This shows that ${}^hv$ is irreducible. Note that since $v$ is non-empty, ${}^hv$ is not at infinity.

Let $v$ be an arbitrary affine variety and let $v = \bigcup_i v_i$ be the irredundant decomposition of $v$ into irreducible varieties. We know that each variety ${}^hv_i$ is irreducible, and it is clear that ${}^hv = \bigcup_i {}^hv_i$ (the closure of a finite union of sets is the union of the closures). It remains to show that the representation $\bigcup_i {}^hv_i$ is irredundant. If it were not irredundant, say if ${}^hv_1$ were superfluous, then we would have ${}^hv_1 \subset {}^hv_i$ for some $i \neq 1$ (see § 3, Remark following the proof of Theorem 13) and hence, by (4), $v_1 \subset v_i$, which is impossible.

Let $V$ be an irreducible projective variety in $P_n{}^K$, not at infinity. By (4) we have ${}^{aha}V = {}^aV$, i.e., the two projective varieties ${}^{ha}V$ and $V$ differ only by points at infinity. If, then, we denote by $L_\infty$ the hyperplane at infinity, then $({}^{ha}V) \cup L_\infty \supset V$. Since $V$ is irreducible and $L_\infty \not\supset V$, it follows that ${}^{ha}V \supset V$. On the other hand, $V$ contains ${}^aV$ and therefore $V \supset {}^{ha}V$, which proves (5). The irreducibility of ${}^aV$ follows from (5), from the irreducibility of $V$ and the preceding part of the theorem.

We have just shown that every irreducible projective variety $V$, not at infinity, is the map of some affine variety under the operation ${}^h$, namely of ${}^aV$. This, and the other assertions of the theorem which have already been established, show that the mapping $v \to {}^hv$ maps the set of affine varieties onto the set of projective varieties having no irreducible components at infinity. This completes the proof of the theorem.

COROLLARY. *The mapping $V \to {}^aV$ maps the set of all projective varieties in $P_n^K$ onto the set of all affine varieties in $A_n^K$. If $V = \bigcup_i V_i$ is the irredundant decomposition of $V$ into irreducible components, then ${}^aV = \bigcup {}^aV_j$ where the $V_j$ are those irreducible components of $V$ which are not at infinity, and $\bigcup {}^aV_j$ is the irredundant decomposition of ${}^aV$ into irreducible components. If $V$ and $V'$ are two projective varieties then ${}^aV = {}^aV'$ if and only if $V$ and $V'$ differ at most by irreducible components which are at infinity.*

The first assertion of the corollary follows from (4). It is clear that ${}^aV = \bigcup_j {}^aV_j$ if $V = \bigcup V_i$ (the notations being as in the corollary), for ${}^aV_i$ is empty for any $V_i$ which is not a $V_j$. From Theorem 19 we know that each ${}^aV_j$ is an irreducible variety. If $j'$ and $j''$ are two distinct indices $j$ then neither of the two varieties $V_{j'}$ and $V_{j''}$ contains the other. Hence, by (5), neither of the two varieties ${}^aV_{j'}$ and ${}^aV_{j''}$ contains the other. This shows that the decomposition ${}^aV = \bigcup_j {}^aV_j$ is irredundant. The last part of the corollary now follows immediately.

In addition to formula (2) the following ideal-theoretic relations may be pointed out:

$$(7) \qquad {}^a(\mathscr{I}(V)) = \mathscr{I}({}^aV),$$

$$(8) \qquad {}^h(\mathscr{I}(v)) = \mathscr{I}({}^hv),$$

$$(9) \qquad \mathscr{V}({}^a\mathfrak{A}) = {}^a(\mathscr{V}(\mathfrak{A})),$$

$$(10) \qquad \mathscr{V}({}^h\mathfrak{a}) = {}^h(\mathscr{V}(\mathfrak{a})).$$

(For the sake of symmetry we have reproduced here in (9) the formula (2).)

In these relations, $V$ and $v$ denote arbitrary varieties, projective and affine respectively, $\mathfrak{A}$ is an arbitrary homogeneous ideal in $k[Y_0, Y_1, \cdots, Y_n]$ and $\mathfrak{a}$ is an arbitrary ideal in $k[X_1, X_2, \cdots, X_n]$. The symbols $\mathscr{I}(V)$ and $\mathscr{I}({}^hv)$ refer to the homogeneous ideals of $V$ and ${}^hv$ in $k[Y_0, Y_1, \cdots, Y_n]$ while $\mathscr{I}(v)$ and $\mathscr{I}({}^aV)$ denote the ideals of the affine varieties $v$ and ${}^aV$ in $k[X_1, X_2, \cdots, X_n]$.

To prove (7), we observe that if $F(X_1, X_2, \cdots, X_n)$ is any polynomial in $k[X_1, X_2, \cdots, X_n]$ and if $\varphi = {}^hF$ (see § 5), then $F \in \mathscr{I}({}^aV)$ if and only if the form $\varphi(Y_0, Y_1, \cdots, Y_n)$ vanishes at each point of ${}^aV$, or—equivalently, if and only if $\psi = Y_0\varphi \in \mathscr{I}(V)$. We thus see that $F \in \mathscr{I}({}^aV)$ if and only if there exists a form $\psi$ in $\mathscr{I}(V)$ such that ${}^a\psi = F$, i.e., if and only if $F \in {}^a(\mathscr{I}(V))$.

If a form $\varphi(Y_0, Y_1, \cdots, Y_n)$ is in $\mathscr{I}({}^hv)$ then $\varphi(1, X_1, X_2, \cdots, X_n) \in \mathscr{I}(v)$, whence $\varphi(Y_0, Y_1, \cdots, Y_n) \in {}^h(\mathscr{I}(v))$. On the other hand, if a form $\varphi(Y_0, Y_1, \cdots, Y_n)$ belongs to ${}^h(\mathscr{I}(v))$, then it is clear that this form vanishes at every point of $v$, i.e., at all points of ${}^hv$ which are at finite distance (since ${}^hv \cap A_n{}^K = v$). Since ${}^hv$ has no irreducible components at infinity, it follows at once that $\varphi$ vanishes on ${}^hv$, and this proves (8).

As to (10), let $\mathfrak{p}_1, \mathfrak{p}_2, \cdots, \mathfrak{p}_s$ be the isolated prime ideals of $\mathfrak{a}$. By Theorem 17 (§ 5), ${}^h\mathfrak{p}_1, {}^h\mathfrak{p}_2, \cdots, {}^h\mathfrak{p}_s$ are the isolated prime ideals of ${}^h\mathfrak{a}$. If we set $v_i = \mathscr{V}(\mathfrak{p}_i)$ and $V_i = \mathscr{V}({}^h\mathfrak{p}_i)$, then $v_1, v_2, \cdots, v_s$ are the irreducible components of $\mathscr{V}(\mathfrak{a})$, while $V_1, V_2, \cdots, V_s$ are the irreducible components of $\mathscr{V}({}^h\mathfrak{a})$. Since $\mathfrak{p}_i = {}^{ah}\mathfrak{p}_i$, it follows from (9) that $v_i = {}^aV_i$. Therefore, by (5), $V_i = {}^hv_i$ and consequently, by Theorem 19, the irreducible components of ${}^h(\mathscr{V}(\mathfrak{a}))$ are also $V_1, V_2, \cdots, V_s$. This establishes (10). Note that in the proof of (10) we used implicitly the Hilbert Nullstellensatz (or equivalent consequences).

We conclude this section by comparing corresponding irreducible varieties in $P_n{}^K$ and $A_n{}^K$. Let $W$ be an *irreducible* variety in $P_n{}^K$, not contained in the hyperplane at infinity, and let $\mathfrak{P}$ be its prime ideal in $k[Y_0, \cdots, Y_n]$. Then its affine restriction $w = {}^aW$ is an *irreducible* affine variety in $A_n{}^K$ (Theorem 19). We have, as was seen above:

$$w = {}^aW, \quad W = {}^hw,$$

and every irreducible affine variety may be written as ${}^aW$, with $W$ irreducible.

For studying the *prime ideal* $\mathfrak{p}$ of $w$, we use formula (10′) in § 5 which gives (after identifying $X_i$ with $Y_i/Y_0$):

$$(11) \qquad \mathfrak{p} = \mathfrak{P}\cdot k[Y_0, \cdots, Y_n]_M \cap k\left(\frac{Y_1}{Y_0}, \cdots, \frac{Y_n}{Y_0}\right),$$

where $M$ denotes the multiplicative system $\{1, Y_0, \cdots, Y_0{}^q, \cdots\}$. Let $y_j$ be the $\mathfrak{P}$-residue of $Y_j$. The ring $k[Y_0, \cdots, Y_n]_M/\mathfrak{P}\cdot k[Y_0, \cdots, Y_n]_M$ is, (Vol. I, Ch. IV, § 10, form. 1) (and since $\mathfrak{p} \cap M = \emptyset$), isomorphic to the quotient ring $k[y_0, \cdots, y_n]_{M'}$ where $M'$ is the multiplicative system $\{1, y_0, \cdots, y_0{}^q, \cdots,\}$ in the homogeneous coördinate ring $k[y_0, \cdots, y_n]$. By formula (11) the affine coördinate ring $k[x_1, \cdots,$

$x_n]$ of $w = {}^aW$ (which is $k[X_1, \cdots, X_n]/\mathfrak{p} = k[Y_1/Y_0, \cdots, Y_n/Y_0]/\mathfrak{p})$ is a subring of $k[y_0, \cdots, y_n]_{M'}$. More precisely, it is the intersection of this ring with the field $k(x_1, \cdots, x_n) = k(y_1/y_0, \cdots, y_n/y_0)$. We have therefore

$$(12) \qquad k[x_1, \cdots, x_n] = k[y_0, \cdots, y_n]_{M'} \cap k(y_1/y_0, \cdots, y_n/y_0).$$

It follows from this that the *function field of* $w = {}^aW$ *is equal to the function field of* $W$. Thus, in particular, an irreducible projective variety (not contained in $Y_0 = 0$) has the *same dimension* as its affine restriction; and an irreducible affine variety has the same dimension as its projective extension.

If the homogeneous coördinate ring of $W$ is *integrally closed*, then $k[y_0, \cdots, y_n]_{M'}$ is integrally closed (Vol. I, Ch. V, § 3, Example 2, p. 261). It follows then from (12) that the coördinate ring of ${}^aW$ is also integrally closed. In other words, if $W$ is arithmetically normal then ${}^aW$ is normal. This result is included in the results proved in § 4[bis].

**§ 7. Dimension theory in finite integral domains.** The basic theorems of dimension theory in finite integral domains are essentially included in, or are easy consequences of, two general theorems on noetherian rings: the lemma on minimal prime ideals proved in VI, § 14 (p. 91) and the "principal ideal theorem" proved in Vol. I, Ch. IV, § 14 (Theorem 29). To derive the main facts of dimension theory in finite integral domains from these two general theorems will be our first object in this section. The proofs in this theory are, as a rule, of inductive character, and the induction is carried out by passage to residue class rings modulo a prime ideal. It is therefore not feasible to deal with the dimension theory of polynomial rings separately, outside the general framework of dimension theory of arbitrary finite integral domains. It is for this reason that we do not confine ourselves in this section to polynomial rings.

Our second object in this section will be to derive the dimension theory *ab initio*, without presupposing the two general theorems cited above, but rather proving again these two theorems (in the special case of finite integral domains) by using special properties of finite integral domains. The special properties which play a particular role are those expressed by the "normalization theorem" (Theorem 25) and by the lemma preceding that theorem.

Let $k$ be an arbitrary ground field and let $R = k[x_1, x_2, \cdots, x_n]$ be a finite integral domain over $k$ (the $x_i$ are not necessarily algebraically independent over $k$). We denote by $r$ the transcendence degree of $R$

over $k$. We recall our definition of the dimension of a prime ideal $\mathfrak{p}$ in $R$ (VI, § 14, p. 90): the dimension of $\mathfrak{p}$ is the transcendence degree of $R/\mathfrak{p}$ over $k$ (it is tacitly assumed that $\mathfrak{p} \neq R$ and that consequently $R/\mathfrak{p}$ contains $k$; if $\mathfrak{p} = R$, we may define the dimension of $R/\mathfrak{p}$ as $-1$). The lemma proved in VI, § 14, p. 91, states that *if* $\mathfrak{p}$ *is a minimal prime ideal in* $R$ *then* $\dim \mathfrak{p} = r-1$. We recall also from Vol. I, Ch. IV, § 14, p. 240, that if $h = h(\mathfrak{p})$, resp. $d = d(\mathfrak{p})$, is the height, resp. the depth, of $\mathfrak{p}$, then there exists at least one strictly ascending chain

$$\mathfrak{p}_0 < \mathfrak{p}_1 < \cdots < \mathfrak{p}_{h-1} < \mathfrak{p}_h = \mathfrak{p}, \tag{1}$$

resp. at least one strictly descending chain

$$R > \mathfrak{p}_0 > \mathfrak{p}_1 > \cdots > \mathfrak{p}_{d-1} > \mathfrak{p}_d = \mathfrak{p} \tag{1'}$$

of prime ideals and there does not exist such a chain with more than $h+1$ (resp. $d+1$) prime ideals. We note that in the case of an integral domain $R$, the first term $\mathfrak{p}_0$ in the above *ascending* chain is necessarily the ideal (0).

In all that follows it is necessary to bear in mind that if $\mathfrak{p}$ and $\mathfrak{p}'$ are two prime ideals in $R$ then

$$\text{“}\mathfrak{p} < \mathfrak{p}'\text{” implies} \begin{cases} \dim \mathfrak{p} > \dim \mathfrak{p}', \\ h(\mathfrak{p}) < h(\mathfrak{p}'), \\ d(\mathfrak{p}) \gtrsim d(\mathfrak{p}'). \end{cases} \tag{2}$$

The first inequality follows from the fact that the canonical homomorphism of $R/\mathfrak{p}$ onto $R/\mathfrak{p}'$ is proper (cf. Vol. I, Ch. II, § 12, Theorems 28 and 29), the last two inequalities are self-evident. In particular, it follows that every *proper* prime ideal has dimension $< r$.

The main theorem of dimension theory is the following:

THEOREM 20. *If* $\mathfrak{p} \neq R$ *is a prime ideal of dimension* $s$, *then*

$$h(\mathfrak{p}) = r - s, \tag{3}$$

$$d(\mathfrak{p}) = s. \tag{3'}$$

PROOF. We shall prove (3) by induction from $s+1$ to $s$, since (3) is trivial if $s = r$ (in which case $\mathfrak{p} = (0)$). We assume $\mathfrak{p} \neq (0)$. From (1) it follows that $r > \dim \mathfrak{p}_1 > \cdots > \dim \mathfrak{p}_{h-1} > \dim \mathfrak{p} = s$, and hence $h(\mathfrak{p}) \leqq r-s$ (note that $\mathfrak{p}_0 = (0)$, whence $\dim \mathfrak{p}_0 = r$). At any rate, $h(\mathfrak{p})$ is finite, and we can therefore find a prime ideal $\mathfrak{p}'$ in $R$ such that $\mathfrak{p}' < \mathfrak{p}$ (since $\mathfrak{p} \neq (0)$) and such that there are no prime ideals between $\mathfrak{p}'$ and $\mathfrak{p}$. Then $\mathfrak{p}/\mathfrak{p}'$ is a minimal prime ideal in the finite integral domain $\bar{R} = R/\mathfrak{p}'$, and hence, by the cited lemma of VI, § 14, we have $1 + \dim \mathfrak{p}/\mathfrak{p}' = \dim \mathfrak{p}'$, i.e., $\dim \mathfrak{p}' = s+1$ (since $\dim \mathfrak{p}/\mathfrak{p}' = \dim \mathfrak{p}$). By our induction hypothesis we have $h(\mathfrak{p}')=r-s-1$, and hence $h(\mathfrak{p})\geqq r-s$. This establishes (3).

We shall prove (3′) by induction from $s-1$ to $s$ since (3′) is trivial if $s=0$ (in which case $\mathfrak{p}$ is necessarily a maximal ideal). From (1′) it follows that $0\leqq \dim \mathfrak{p}_0<\dim \mathfrak{p}_1<\cdots<\dim \mathfrak{p}_{d-1}<\dim \mathfrak{p}=s$, and hence $d(\mathfrak{p})\leqq s$. We now consider the finite integral domain $\bar{R}=R/\mathfrak{p}$, and we fix a minimal prime ideal $\bar{\mathfrak{p}}'$ in $\bar{R}$. Let $\mathfrak{p}'$ be the prime ideal in $R$ which contains $\mathfrak{p}$ and is such that $\mathfrak{p}'/\mathfrak{p}=\bar{\mathfrak{p}}'$. Since $\bar{R}$ has transcendence degree $s$, we have $\dim \mathfrak{p}'=\dim \bar{\mathfrak{p}}'=s-1$ (by the cited lemma of VI, § 14). By our induction hypothesis we have therefore $d(\mathfrak{p}')=s-1$, and consequently $d(\mathfrak{p})\geqq s$. This establishes (3′) and completes the proof of the theorem.

COROLLARY 1. *If $\mathfrak{p}$ and $\mathfrak{p}'$ are two prime ideals in $R$ such that $\mathfrak{p}<\mathfrak{p}'$ and if $s$ and $s'$ are the dimensions of $\mathfrak{p}$ and $\mathfrak{p}'$ respectively, then there exists at least one strictly ascending chain of $s-s'+1$ prime ideals connecting $\mathfrak{p}$ and $\mathfrak{p}'$:*

$$(4)\qquad \mathfrak{p}<\mathfrak{p}_1<\mathfrak{p}_2<\cdots<\mathfrak{p}_{s-s'-1}<\mathfrak{p}',$$

*and there does not exist any such chain with more than $s-s'+1$ ideals. Furthermore, any strictly ascending chain of $q+1$ prime ideals connecting $\mathfrak{p}$ and $\mathfrak{p}'$, $q<s-s'$, can be refined to a chain* (4) *of maximum length.*

The first assertion of the corollary follows from the fact that in the ring $R/\mathfrak{p}$, which has transcendence degree $s$ over $k$, the prime ideal $\mathfrak{p}'/\mathfrak{p}$, which has dimension $s'$, must have height $s-s'$. The second part of the corollary follows by applying the first part to each pair of consecutive members of the given chain of $q+1$ prime ideals.

COROLLARY 2. *If $\mathfrak{p}$ and $\mathfrak{p}'$ are prime ideals in $R$ such that $\mathfrak{p}<\mathfrak{p}'$ and such that no prime ideal can be inserted between $\mathfrak{p}$ and $\mathfrak{p}'$, then the dimensions of $\mathfrak{p}$ and $\mathfrak{p}'$ differ by unity, and so do their heights and depths.*

Obvious.

COROLLARY 3. *In a finite integral domain $R$ of transcendence degree $r>0$ there exist proper prime ideals of all dimensions* $0, 1, 2, \cdots, r-1$.

This follows from the theorem, in the special case $\mathfrak{p}=(0)$.

COROLLARY 4. *Let ${}^hR=k[y_0, y_1, \cdots, y_n]$ be a homogeneous finite integral domain and let $\mathfrak{P}$ and $\mathfrak{P}'$ be prime ideals in ${}^hR$, of dimension $s+1$ and $s'+1$ respectively, such that $\mathfrak{P}<\mathfrak{P}'$. If $\mathfrak{P}$ and $\mathfrak{P}'$ are homogeneous, then there exists at least one strictly ascending chain of $s-s'+1$ prime homogeneous ideals connecting $\mathfrak{P}$ and $\mathfrak{P}'$.*

Assuming that $y_0\neq 0$ we set $x_i=y_i/y_0$, $i=1, 2, \cdots, n$, and we consider the integral domain $R=k[x_1, x_2, \cdots, x_n]$. We apply the results proved in § 5 in regard to the relationship between homogeneous ideals in ${}^hR$ and arbitrary ideals in $R$. We set $\mathfrak{p} = {}^a\mathfrak{P}$ and $\mathfrak{p}' = {}^a\mathfrak{P}'$. Then $\mathfrak{p}$ and $\mathfrak{p}'$ are prime ideals of dimensions $s$ and $s'$ respectively, and we have $\mathfrak{p} < \mathfrak{p}'$. By Corollary 1 we have a chain (4) of $s - s' + 1$ prime ideals connecting $\mathfrak{p}$ and $\mathfrak{p}'$. Then the chain $\mathfrak{P} < {}^h\mathfrak{p}_1 < {}^h\mathfrak{p}_2 < \cdots < {}^h\mathfrak{p}_{s-s'-1} < \mathfrak{P}'$ satisfies the required conditions.

COROLLARY 5. *In a homogeneous finite integral domain ${}^hR$ of transcendence degree $r+1$ there exist proper homogeneous ideals of all dimensions* $0, 1, 2, \cdots, r$.

We note, in regard to Corollary 5, that the irrelevant prime ideal in ${}^hR$ is the only prime homogeneous ideal of dimension 0, since every homogeneous ideal (different from the unit ideal) is contained in the irrelevant prime ideal. We recall also from § 4 that if a homogeneous prime ideal $\mathfrak{p}$ in ${}^hR$ has dimension $s+1$ then its projective dimension is $s$, and $s$ is also the dimension of the variety $\mathscr{V}(\mathfrak{p})$ of $\mathfrak{p}$.

The preceding results have an immediate geometric interpretation in terms of algebraic varieties, in view of the (1, 1) correspondence that exists between the homogeneous prime ideals in the polynomial ring $k[Y_0, Y_1, \cdots, Y_n]$ and the irreducible algebraic varieties in the projective space $P_n{}^K$ (see § 4). Thus the lemma proved in VI, § 14, concerning minimal prime ideals, signifies, geometrically speaking, that *every maximal irreducible (proper) subvariety of an r-dimensional irreducible variety has dimension* $r-1$. This is true for both projective and affine varieties. Corollary 3 implies that every irreducible affine variety $V$, of dimension $r > 0$, carries points of all dimension $0, 1, \cdots, r-1$, and in particular $V$ carries algebraic points. This yields another proof of the Hilbert Nullstellensatz. As a matter of fact, Corollary 3 implies that if a finite integral domain $R$ is a field (hence has no proper prime ideals), then its transcendence degree is 0, and this is precisely the lemma which we have proved in § 3 and from which we were able to derive the Nullstellensatz in a straightforward manner.

Other important consequences of the lemma proved in VI, § 14 are obtained by making use of the "principal ideal theorem" proved in Vol. I, Ch. IV, § 14 (Theorem 29). We have namely

THEOREM 21. *If $R$ is a finite integral domain, of transcendence degree $r$, and $f \neq 0$ is a non-unit in $R$, then every isolated prime ideal of $Rf$ has dimension* $r-1$.

This is simply a re-statement of the "principal ideal theorem" in which use is made of the knowledge that every minimal prime ideal in $R$ has dimension $r-1$.

COROLLARY. If *${}^hR$ is a homogeneous finite integral domain, of transcendence degree $r+1$, and $f$ is a homogeneous element of ${}^hR$, different* *from zero and of positive degree, then every isolated prime ideal of* ${}^hR\cdot f$ *has projective dimension* $r-1$.

Obvious, since $f$ is a non-unit, ${}^hR\cdot f$ is a homogeneous ideal and every prime ideal of ${}^hR\cdot f$ is homogeneous (§ 2, Theorem 9).

In geometric terms, the above corollary may be stated as follows: *if $V$ is an irreducible $r$-dimensional algebraic variety in the projective space $P_n^K$ and $f(Y_0, Y_1, \cdots, Y_n)$ is a form in $k[Y_0, Y_1, \cdots, Y_n]$, of positive degree, such that the hypersurface $H: f(Y)=0$ does not contain $V$, then each irreducible component of the intersection $V \cap H$ has dimension $r-1$.* To see this we have only to take for ${}^hR$ the homogeneous coördinate ring $k[y_0, y_1, \cdots, y_n]$ of the variety $V$ (§ 4, p. 170), observe that our assumption concerning the form $f(Y_0, Y_1, \cdots, Y_n)$ signifies that the element $f(y_0, y_1, \cdots, y_n)$ of ${}^hR$ is different from zero and finally recall (§ 3, Theorem 14, Corollary 3) that the irreducible components of $H \cap V$ are the varieties of the isolated prime ideals of the principal ideal generated in ${}^hR$ by $f(y_0, y_1, \cdots, y_n)$.

In the same way as Theorem 21 represents a re-statement of the "principal ideal theorem" (Vol. I, Ch. IV, § 14, Theorem 29), the following generalization of Theorem 21 is a re-statement of Theorem 30 of IV, § 14:

THEOREM 22. *If $R$ is a finite integral domain, of transcendence degree $r$, and $\mathfrak{A}$ is a proper ideal in $R$ which admits a basis of $s$ elements, then every isolated prime ideal of $\mathfrak{A}$ has dimension $\geqq r-s$.*

COROLLARY. *If ${}^hR$ is a homogeneous finite integral domain, of transcendence degree $r+1$, and $\mathfrak{A}$ is a proper homogeneous ideal in $R$ which admits a basis of $s$ elements, then every isolated prime ideal of $\mathfrak{A}$ has projective dimension $\geqq r-s$.*

The maximum of the dimensions of the isolated prime ideals of an ideal $\mathfrak{A}$ in a finite integral domain $R$ is called the *dimension of* $\mathfrak{A}$. If $R$ is a homogeneous finite integral domain and $\mathfrak{A}$ is a homogeneous ideal in $R$, then one uses preferably the *projective* dimension of $\mathfrak{A}$, which is defined as the dimension of $\mathfrak{A}$ diminished by unity. Theorem 22 and its corollary assert that, under the conditions stated, the dimension, resp. the projective dimension of $\mathfrak{A}$, is not less than $r-s$.

An ideal $\mathfrak{A}$ in a finite integral domain $R$ is said to be *unmixed* (or *equidimensional*) if all its prime ideals have the same dimension. It is clear that an unmixed ideal has no imbedded prime ideals. A previous theorem on principal ideals in noetherian integrally closed domains (Vol. I, Ch. V, § 6, Theorem 14) permits us to strengthen Theorem 22 in the case in which $R$ is integrally closed (in particular, then, if $R$ is a polynomial ring):

THEOREM 23. *If $R$ is an integrally closed finite integral domain, of transcendence degree $r$, then every proper principal ideal in $R$ is unmixed, of dimension $r-1$. If, in addition, $R$ is also a unique factorization domain (in particular, if $R$ is a polynomial ring), then also the converse is true, i.e., every unmixed ideal of dimension $r-1$ is principal.*

PROOF. The first part of the theorem is a restatement of Theorem 14 in Vol. I, Ch. V, § 6. Conversely, if $\mathfrak{A}$ is an unmixed ideal, of dimension $r-1$, then all its prime ideals $\mathfrak{p}_i$ are minimal, and if $R$ is a UFD then each $\mathfrak{p}_i$ is a principal ideal $(G_i)$ and each primary ideal having $\mathfrak{p}_i$ as associated prime ideal is a power $(G_i^{\nu_i})$ of $\mathfrak{p}_i$. We have therefore, for suitable integers $\nu_i$, $\mathfrak{A}=\bigcap_i (G_i^{\nu_i})$. Since $R$ is a UFD and the $G_i$ are two by two relatively prime, it follows that $\mathfrak{A}$ coincides with the principal ideal generated by $\prod_i G_i^{\nu_i}$. Q.E.D.

For *polynomial rings* we have the following special result concerning zero-dimensional (whence maximal) prime ideals.

THEOREM 24. *Every zero-dimensional prime ideal $\mathfrak{p}$ in a polynomial ring $R=k[X_1, X_2, \cdots, X_n]$ in $n$ indeterminates ($k$, a field) has a basis of $n$ elements.*

(NOTE. By Theorem 22, $\mathfrak{p}$ can have no basis of less than $n$ elements.)

PROOF. We shall proceed by induction with respect to $n$, the case $n=0$ being trivial. Since $\mathfrak{p}$ is zero-dimensional, each $X_i$ is algebraic over $k$ modulo $\mathfrak{p}$, i.e., $\mathfrak{p} \cap k[X_i] \neq (0)$. Consider $\mathfrak{p}_1=\mathfrak{p} \cap k[X_1]$. The ideal $\mathfrak{p}_1$ is principal, say $(f_1(X_1))$, where $f_1(X_1)$ is necessarily an irreducible polynomial in $k[X_1]$. Let $\bar{R}=R/R\cdot\mathfrak{p}_1=k[\alpha_1, \bar{X}_2, \cdots, \bar{X}_n]$, where $\alpha_1$ is the $\mathfrak{p}_1$-residue of $X_1$, and $\bar{X}_2, \bar{X}_3, \cdots, \bar{X}_n$ are the $R\cdot\mathfrak{p}_1$-residues of $X_2, X_3, \cdots, X_n$ respectively. Since $\alpha_1$ is algebraic over $k$ (being a root of $f_1(X_1)$), we have $k[\alpha_1]=k(\alpha_1)$ is a field. It is clear that $\bar{X}_2, \bar{X}_3, \cdots, \bar{X}_n$ are algebraically independent over $k(\alpha_1)$, since $R\mathfrak{p}_1$ is the principal ideal generated by $f_1(X_1)$. *Hence $\bar{R}$ is a polynomial ring in $n-1$ variables $\bar{X}_2, \bar{X}_3, \cdots, \bar{X}_n$, over the field $k(\alpha_1)$.* By our induction hypothesis, the zero-dimensional prime ideal $\bar{\mathfrak{p}}=\mathfrak{p}/R\mathfrak{p}_1$ in $\bar{R}$ has a basis of $n-1$ elements, say $\{\bar{f}_2, \bar{f}_3, \cdots, \bar{f}_n\}$. If, then, $f_i$ is any element in $R$ whose $R\mathfrak{p}_1$ residue is $\bar{f}_i$ $(i=2, 3, \cdots, n)$ then $\{f_1, f_2, \cdots, f_n\}$ is a basis of $\mathfrak{p}$.

REMARK. It follows from the above proof that the prime ideal $\mathfrak{p}$ has a basis consisting of $n$ polynomials of the form

$$f_1(X_1),\ f_2(X_1, X_2), \cdots, f_n(X_1, X_2, \cdots, X_{n-1}, X_n).$$

If $\alpha_1, \alpha_2, \cdots, \alpha_n$ are the $\mathfrak{p}$-residues of $X_1, X_2, \cdots X_n$ respectively, then we can take for $f_i(X_1, X_2, \cdots, X_i)$ any polynomial in $k[X_1, X_2,$

$\cdots, X_i]$ such that $f_i(\alpha_1, \alpha_2, \cdots, \alpha_{i-1}, X_i)$ is the minimal polynomial of $\alpha_i$ over $k(\alpha_1, \alpha_2, \cdots, \alpha_{i-1})$ $(=k[\alpha_1, \alpha_2, \cdots, \alpha_{i-1}])$. The following additional conditions *determine* the polynomials $f_i$ uniquely and lead to a *canonical* basis of $\mathfrak{p}$ (relative to the ordering $X_1, X_2, \cdots, X_n$ of the variables):

(1) Each $f_i$ is of degree $\nu_i$ in $X_i$, where $\nu_i = [k(\alpha_1, \alpha_2, \cdots, \alpha_i):k(\alpha_1, \alpha_2, \cdots, \alpha_{i-1})]$, and is monic as a polynomial in $X_i$.

(2) Each $f_i$ is of degree $<\nu_j$ in $X_j$, for $j=1, 2, \cdots, i-1$.

If $k$ is algebraically closed, then the $\alpha_i$ are in $k$, and $\mathfrak{p}$ has the following basis:

$$X_1 - \alpha_1, X_2 - \alpha_2, \cdots, X_n - \alpha_n.$$

Before proceeding any further in dimension theory we shall show how all the preceding results can be obtained without recourse to general theorems on noetherian rings. It is clear that Theorem 21 is the key result, from which all the theorems proved in this section (and also the two earlier results, namely the lemma of VI, § 14 and Theorem 29 of Vol. I, Ch. IV, § 14, in the special case of finite integral domains) follow as immediate consequences. We shall therefore show how Theorem 21 can be proved directly by using special properties of finite integral domains. Actually, we shall find it essential, in this new treatment, to deal only with *homogeneous* finite integral domains. Therefore, what we shall prove directly is Theorem 21 *in the case of homogeneous domains R*, i.e., we shall prove the *corollary* of Theorem 21. In view of the relationship between ideals in $R$ and ${}^hR$, established in § 5, Theorem 21 in the general case is an immediate consequence of the "homogeneous" formulation given in the corollary of that theorem.

We first prove a general lemma on finitely generated homogeneous rings. Let $R=A[y_0, y_1, \cdots, y_n]$ be such a ring, where $A$ is an arbitrary commutative ring (with element 1). A set of *homogeneous* elements $z_0, z_1, \cdots, z_m$ of $R$, of positive degrees, is said to be a *homogeneous system of integrity*, if the ring $R$ is integral over the ring $S= A[z_0, z_1, \cdots, z_m]$.

LEMMA. *In order that a set $\{z_0, z_1, \cdots, z_m\}$ of homogeneous elements of R, of positive degrees, be a homogeneous system of integrity it is necessary and sufficient that the ideal $\mathfrak{Z}$ generated in R by $z_0, z_1, \cdots, z_m$ be irrelevant.*

PROOF. We first observe that it is sufficient to prove the lemma under the assumption that the $m+1$ elements $z_i$ are of the same degree. For if, say, $\nu_i$ is the degree of $z_i$ and if we set $u_i = z_i^{\nu/\nu_i}$, where $\nu = \nu_0\nu_1 \cdots \nu_m$, then the $m+1$ elements $u_0, u_1, \cdots, u_m$ of $R$ are homogeneous, of the same degree $\nu$. It is clear that the ideal generated by $z_0, z_1, \cdots, z_m$ in $R$ is irrelevant if and only if the ideal generated by $u_0, u_1, \cdots, u_m$ is irrelevant. On the other hand, since $u_i \in S$ and since the $z_i$ are integral over the ring $A[u_0, u_1, \cdots, u_m]$, it follows that $R$ is integral over $S$ if and only if $R$ is integral over $A[u_0, u_1, \cdots, u_n]$. We shall therefore assume that the $z_i$ have the same degree $\nu$.

Assume that $\{z_0, z_1, \cdots, z_m\}$ is a homogeneous system of integrity. Each of the elements $y_i$ is then integral over the ring $S = A[z_0, z_1, \cdots, z_m]$. Consider one of the $y_i$'s, say $y_0$. Let

$$(5) \qquad \psi(y_0; z_0, z_1, \cdots, z_m) = y_0^s + \varphi_{s-1}(z_0, z_1, \cdots, z_m)y_0^{s-1} + \cdots + \varphi_0(z_0, z_1, \cdots, z_m) = 0$$

be an equation of integral dependence for $y_0$ over $S$. Here each $\varphi_j$ is a polynomial with coefficients in $A$. We have $z_i = F_i(y_0, y_1, \cdots, y_n)$, where $F_i$ is a *form*, of degree $\nu$, with coefficients in $A$. We set $G(Y_0, Y_1, \cdots, Y_n) = \psi(Y_0, F_0(Y_0, Y_1, \cdots, Y_n), F_1(Y_0, Y_1, \cdots, Y_n), \cdots, F_m(Y_0, Y_1, \cdots, Y_n))$.

We have then $G(y_0, y_1, \cdots, y_n) = 0$. Since $R$ is a homogeneous ring it follows that we must have $G_\mu(y_0, y_1, \cdots, y_n) = 0$, for every homogeneous component $G_\mu$ of the polynomial $G(Y_0, Y_1, \cdots, Y_n)$. In particular, we have $G_s(y_0, y_1, \cdots, y_n) = 0$. If we denote by $\varphi_{j,\rho}(Z_0, Z_1, \cdots, Z_m)$ the homogeneous component of degree $\rho$ of $\varphi_j$ ($j = s-1, s-2, \cdots, 0$), then we find that $G_s(Y_0, Y_1, \cdots, Y_n) = Y_0^s + \sum_{\rho=1}^{s'} \varphi_{s-\rho\nu,\rho}(F_0(Y), F_1(Y), \cdots, F_m(Y))\, Y_0^{s-\rho\nu}$, where $s' = [s/\nu]$. Hence

$$(6) \qquad y_0^s + \varphi_{s-\nu,1}(z_0, z_1, \cdots, z_m)y_0^{s-\nu} + \varphi_{s-2\nu,2}(z_0, z_1, \cdots, z_m)y_0^{s-2\nu} + \cdots = 0.$$

This is still a relation of integral dependence for $y_0$ over $S$, but now the coefficients $\varphi_{s-\nu,1}, \varphi_{s-2\nu,2}, \cdots$ are *homogeneous* in the $z_i$'s, of *positive* degrees $1, 2, \cdots$, respectively. It follows from (6) that $y_0^s$ belongs to the ideal $\mathfrak{Z}$. Similarly, some power of each of the elements $y_1, y_2, \cdots, y_n$ belongs to $\mathfrak{Z}$. Hence $\mathfrak{Z}$ is an irrelevant ideal.

Conversely, let us assume that the ideal $\mathfrak{Z}$ is irrelevant. To prove that $R$ is integral over $S$ it will be sufficient to prove that $R$ is a finite $S$-module (see Vol. I, Ch. V, § 1, condition ($c''$); actually, these two properties of $R$ are equivalent, in view of Vol. I, Ch. V, § 1, Theorem 1). Let $t$ be an exponent such that $\mathfrak{Y}^t \subset \mathfrak{Z}$, where $\mathfrak{Y}$ is the prime irrelevant ideal generated by $y_0, y_1, \cdots, y_n$, and let $\{\omega_j\}$ be the (finite) set of monomials $y_0^{\alpha_0}y_1^{\alpha_1} \cdots y_n^{\alpha_n}$ of degree $\alpha_0 + \alpha_1 + \cdots + \alpha_n < t$. We shall show that *every monomial* $\xi$ *in the* $y$'*s can be written* in the form $\sum_j a_j\omega_j$, with $a_j \in S$, and thus establish that the monomials $\omega_j$ form a basis of $R$ over $S$.

We proceed by induction with respect to the degree $\partial(\xi)$ of the monomial $\xi$, for the assertion is trivial if $\partial(\xi)<t$. We therefore assume that $\partial(\xi)\geqq t$ and we write $\xi=MG$, where $M$ is a monomial of degree $t$ and $G$ is a monomial of degree $\partial(\xi)-t$. Since $M \in \mathfrak{Y}^t \subset \mathfrak{Z}$, we can write $M=\sum_i H_i z_i$, where we may assume (since $R$ is a homogeneous ring) that each $H_i$ is a homogeneous element of $R$, of degree $t-\partial(z_i)$. Substituting into the above expression of $\xi$ we find $\xi=\sum_i GH_i z_i$. The coefficient $GH_i$ of $z_i$ is homogeneous, of degree $\partial(\xi)-t+t-\partial(z_i)=\partial(\xi)-\partial(z_i)<\partial(\xi)$, and is therefore a linear combination of monomials $y_0^{\alpha_0}y_1^{\alpha_1}\cdots y_n^{\alpha_n}$, of degree $<\partial(\xi)$, with coefficients in $A$. By our induction hypothesis we have therefore $GH_i=\sum_j \sigma_{i,j}\omega_j$, with $\sigma_{i,j}\in S$. Hence $\xi=\sum_j \sigma_j\omega_j$, where $\sigma_j=\sum_i \sigma_{i,j}z_i \in S$. This completes the proof of the lemma.

The following application of the preceding lemma is an extension of the normalization theorem which was proved in Vol. I, Ch. V, § 4 (Theorem 8) only for infinite ground fields:

THEOREM 25. (*Normalization theorem.*) *If* $R=k[x_1, x_2, \cdots, x_n]$ *is a finite integral domain over a ground field* $k$ *and* $r$ *is the transcendence degree of* $R$ *over* $k$, *then there exist sets of* $r$ *elements* $z_1, z_2, \cdots, z_r$ *in* $R$ *such that* $R$ *is integral over the ring* $k[z_1, z_2, \cdots, z_r]$. *If* $R$ *is homogeneous, then the* $z_i$ *can be chosen so as to be homogeneous, and for one of the* $z_i$ *we can choose an arbitrary homogeneous element of positive degree.*

PROOF. We first consider the case in which $R$ is a homogeneous ring. By the preceding lemma we have only to show that there exists a set of homogeneous elements $z_1, z_2, \cdots, z_r$ in $R$, of positive degree, such that the ideal $\mathfrak{Z}$ generated by these elements is irrelevant. To say that a *proper homogeneous* ideal $\mathfrak{Z}$ is irrelevant is the same as saying that it is of dimension zero, for the irrelevant prime ideal $(x_1, x_2, \cdots, x_n)$ is the only zero-dimensional prime homogeneous ideal in $R$ and the prime ideals of a homogeneous ideal are all homogeneous (§ 2, Theorem 9). To prove that there exist homogeneous elements $z_1, z_2, \cdots, z_r$, of positive degree, having the property that the ideal $\mathfrak{Z}$ is irrelevant, it will be sufficient to show that we can choose the elements $z_1, z_2, \cdots z_r$ in such a way as to satisfy the following condition: *if* $\mathfrak{Z}_i$ *is the ideal generated by* $z_1, z_2, \cdots, z_i$, $i=1, 2, \cdots, r$, *and* $\mathfrak{Z}_0$ *denotes the zero ideal, then*

$$\dim \mathfrak{Z}_0 > \dim \mathfrak{Z}_1 > \cdots > \dim \mathfrak{Z}_r, \tag{7}$$

where $\mathfrak{Z}_r$ is, of course, the ideal $\mathfrak{Z}$. For if (7) holds then it follows that $\dim \mathfrak{Z}_i = r - i$, $i = 1, 2, \cdots, r$, since $\dim \mathfrak{Z}_0 = r$ and $\dim \mathfrak{Z}_r \geqq 0$ (the elements $z_i$ being all of positive degree, the ideal $\mathfrak{Z}$ is not the unit ideal).

We choose for $z_1$ an arbitrary homogeneous element of positive degree and different from zero. Then, of course, we will have $r > \dim \mathfrak{Z}_1$. Assume that we have already found elements $z_1, z_2, \cdots, z_i$ such that $r > \dim \mathfrak{Z}_1 > \dim \mathfrak{Z}_2 > \cdots > \dim \mathfrak{Z}_i$. Let $\mathfrak{p}_1, \mathfrak{p}_2, \cdots, \mathfrak{p}_h$ be the isolated prime ideals of $\mathfrak{Z}_i$. For any $j = 1, 2, \cdots, h$ we can find a homogeneous element $u_j$ which belongs to $\bigcap_{\nu \neq j} \mathfrak{p}_\nu$ but does not belong to $\mathfrak{p}_j$. Upon replacing each of the $h$ elements $u_1, u_2, \cdots, u_h$ by a suitable power of that element, we can arrange matters so that the elements $u_j$ are all of the same degree. Then the element $z_{i+1} = u_1 + u_2 + \cdots + u_h$ does not belong to any of the ideals $\mathfrak{p}_j$. Let $\rho$ be the dimension of the ideal $\mathfrak{Z}_{i+1}$ generated by $z_1, z_2, \cdots, z_{i+1}$, and let $\mathfrak{p}$ be a prime ideal of $\mathfrak{Z}_{i+1}$ which has dimension $\rho$. The ideal $\mathfrak{p}$ contains also $\mathfrak{Z}_i$ and thus contains at least one of the ideals $\mathfrak{p}_1, \mathfrak{p}_2, \cdots, \mathfrak{p}_h$. Let, say, $\mathfrak{p} \supset \mathfrak{p}_1$. Since $z_{i+1} \in \mathfrak{p}$ and $z_{i+1} \notin \mathfrak{p}_1$, we have $\mathfrak{p} > \mathfrak{p}_1$. Therefore $\rho = \dim \mathfrak{Z}_{i+1} = \dim \mathfrak{p} < \dim \mathfrak{p}_1 \leqq \dim \mathfrak{Z}_i$. This completes the proof of the theorem in the homogeneous case.

In the general case we adjoin a transcendental $y_0$ to the field $k(x_1, x_2, \cdots, x_n)$, we set $y_i = y_0 x_i$ and we consider the homogeneous finite integral domain ${}^hR = k[y_0, y_1, \cdots, y_n]$. This domain has transcendence degree $r+1$. By the preceding case, there exists a homogeneous system of integrity $\{u_0, u_1, \cdots, u_r\}$ in ${}^hR$ consisting of $r+1$ elements, and we can take as one of these elements an arbitrary non-zero homogeneous element of positive degree. We take for $u_0$ the element $y_0$. If $\nu_i$ is the degree of $u_i$ we replace $\{u_0, u_1, \cdots, u_r\}$ by the following homogeneous system of integrity $\{v_0, v_1, \cdots, v_r\}$: $v_0 = y_0^\nu$, $v_i = u_i^{\nu/\nu_i}$, where $\nu = \nu_1 \nu_2 \cdots \nu_r$. The elements $v_j$ of this new system of integrity are all of the same degree $\nu$. For each element $y_i$ $(i = 1, 2, \cdots, n)$ there exists a relation of integral dependence over $k[v_0, v_1, \cdots, v_r]$ which has the form:

$$(8) \quad y_i^s + \varphi_1(v_0, v_1, \cdots, v_r) y_i^{s-\nu} + \varphi_2(v_0, v_1, \cdots, v_r) y_i^{s-2\nu} + \cdots = 0,$$

where $\varphi_q$ is a form of degree $q$ (see proof of Lemma, equation (6)). If we set $z_j = v_j/v_0$, $j = 1, 2, \cdots, r$, and divide (8) by $y_0^s$, relation (8) yields the following relation:

$$x_i^s + \varphi_1(1, z_1, \cdots, z_r) x_i^{s-\nu} + \varphi_2(1, z_1, z_2, \cdots, z_r) x_i^{s-2\nu} + \cdots = 0,$$

and this is an equation of integral dependence for $x_i$ over $k[z_1, z_2, \cdots, z_r]$. This completes the proof.

We shall now proceed to our stated objective of giving a self-contained proof of Theorem 21 based on the above lemma and on the normalization theorem just proved. For reasons explained earlier we shall deal only with the homogeneous case of Theorem 21, i.e., with the corollary to Theorem 21. We shall denote by $R$ (instead of by ${}^hR$, as in the corollary) our homogeneous ring $k[y_0, y_1, \cdots, y_n]$, but we continue to denote by $r+1$ the transcendence degree of $R/k$. Let $f$ be a non-zero homogeneous element of $R$, of positive degree. We set $z_0=f$ and we choose $r$ other homogeneous elements $z_1, z_2, \cdots, z_r$ in $R$, of positive degree, such that the set $\{z_0, z_1, \cdots, z_r\}$ is a homogeneous system of integrity of $R$ (Theorem 25). Note that since $R$ has transcendence degree $r+1$ over $k$ and the $y_i$ are integral over $k[z_0, z_1, \cdots, z_r]$, the $r+1$ elements $z_i$ are algebraically independent over $k$.

Let $\mathfrak{p}$ be any isolated prime ideal of the principal ideal $Rf$ $(=Rz_0)$ and let $\bar{z}_i$ be the $\mathfrak{p}$-residue of $z_i$. *We shall prove that* $\bar{z}_1, \bar{z}_2, \cdots, \bar{z}_r$ *are algebraically independent over* $k$. This will establish the fact that $\mathfrak{p}$ has (affine) dimension $r$ (and projective dimension $r-1$) and will settle Theorem 21.

Let $h(Z_1, Z_2, \cdots, Z_r)$ be any non-zero polynomial in $r$ indeterminates, with coefficients in $k$. We have to prove that $h(\bar{z}_1, \bar{z}_2, \cdots, \bar{z}_r) \neq 0$, or—equivalently—that $h(z_1, z_2, \cdots, z_r) \notin \mathfrak{p}$. Since this has to be shown to be true for any isolated prime ideal $\mathfrak{p}$ of $Rz_0$, we see that what we are asserting is equivalent to the assertion that *the element* $h(z)=h(z_1, z_2, \cdots, z_r)$ *is prime to the radical* $\sqrt{Rz_0}$, i.e., that we have:

$$\sqrt{Rz_0}:R\cdot h(z) = \sqrt{Rz_0}. \tag{9}$$

Let $\eta$ be any element of $\sqrt{Rz_0}:R\cdot h(z)$. We have then, upon denoting by $\xi$ a suitable power $\eta^\rho$ of $\eta$:

$$[h(z)]^\rho \xi = uz_0, \quad (u \in R,\ \xi = \eta^\rho). \tag{10}$$

Let

$$u^s + a_1(z_0, z_1, \cdots, z_r)u^{s-1} + \cdots + a_s(z_0, z_1, \cdots, z_r) = 0 \tag{11}$$

be the equation of least degree which $u$ satisfies over the field $k(z_0, z_1, \cdots, z_r)$. Since $u$ is integral over $k[z_0, z_1, \cdots, z_r]$ and since $z_0, z_1, \cdots, z_r$ are algebraically independent over $k$ (whence $k[z_0, z_1, \cdots, z_r]$ is an integrally closed domain), it follows that (11) must be an equation of integral dependence for $u$ over $k[z_0, z_1, \cdots, z_r]$ (Vol. I, Ch. V, § 3, Theorem 4), i.e., the $a_i(z_0, z_1, \cdots, z_r)$ are polynomials, with coefficients in $k$. From (10) and (11) we deduce that the equation of least degree that $\xi$ satisfies over $k(z_0, z_1, \cdots, z_r)$ is the following:

$$(12) \quad \xi^s + \frac{a_1(z_0, z_1, \cdots, z_r)z_0}{[h(z)]^p} \cdot \xi^{s-1} + \cdots + \frac{a_s(z_0, z_1, \cdots, z_r)z_0^s}{[h(z)]^{ps}} = 0.$$

Since also $\xi$ is integrally dependent over $k[z_0, z_1, \cdots, z_r]$, it follows again by the cited Theorem 4 of Vol. I, Ch. V, § 3, that the quotients $a_j(z_0, z_1, \cdots, z_r)/[h(z)]^{pj}$ must be polynomials (in the algebraically independent elements $z_0, z_1, \cdots, z_r$). Then (12) shows at once that $\xi^s \in R \cdot z_0$, i.e., $\xi \in \sqrt{R \cdot z_0}$. Therefore also $\eta$ belongs to $\sqrt{R \cdot z_0}$ (since $\xi = \eta^p$). This proves (9) and completes the proof of Theorem 21.

**§ 8. Special dimension-theoretic properties of polynomial rings.** In this section we are going to prove two special results of dimension-theoretic nature which hold in polynomial rings and which do not extend to arbitrary finite integral domains.

THEOREM 26 (MACAULAY). *Let $\mathfrak{A}$ be an ideal in $R = k[X_1, \cdots, X_n]$, of dimension $n-h$. If $\mathfrak{A}$ is generated by $h$ elements $F_1, \cdots, F_h$, then $\mathfrak{A}$ is unmixed.*

PROOF. We proceed by induction on $h$, the case $h = 0$ being trivial (and the case $h = 1$ having already been treated in Theorem 23, § 7). We have to show that every associated prime ideal $\mathfrak{p}$ of $\mathfrak{A}$ is $(n-h)$-dimensional. Let $d$ be the dimension of $\mathfrak{p}$. Since $\mathfrak{A}$ is $(n-h)$-dimensional, we already know that $d \leqq n-h$. Since $k[X_1, \cdots, X_n]/\mathfrak{p}$ has transcendence degree $d$, $d$ of the variables $X_i$, say $X_1, \cdots, X_d$, are algebraically independent mod $\mathfrak{p}$. In other words we have $\mathfrak{p} \cap k[X_1, \cdots, X_d] = (0)$, whence the multiplicatively closed set $M$ of non-zero elements of $k[X_1, \cdots, X_d]$ has no element in common with $\mathfrak{p}$.

We consider the quotient ring $R_M$ and the extended ideals $\mathfrak{A}R_M$ and $\mathfrak{p}R_M$. It has been seen in Vol. I, Ch. IV, § 10 (Theorems 16 and 17) that $\mathfrak{p}R_M$ is an associated prime ideal of $\mathfrak{A}R_M$. Now $R_M$ is obviously the polynomial ring

$$k(X_1, \cdots, X_d)[X_{d+1}, \cdots, X_n].$$

Since $R/\mathfrak{p}$ and $R_M/\mathfrak{p}R_M$ have the same quotient field, $R_M/\mathfrak{p}R_M$ has transcendence degree $d$ over $k$, whence it has transcendence degree 0 over $k(X_1, \cdots, X_d)$. In other words, $\mathfrak{p}R_M$ is a maximal ideal. In order to show that $d = n-h$, i.e., that $h = n-d$, we argue by contradiction and suppose that $h < n-d$. Then we have, in the polynomial ring $R_M$ in $n-d$ variables, an ideal $\mathfrak{A}R_M$ generated by $h < n-d$ elements and admitting a zero-dimensional associated prime ideal. Furthermore, it is easily seen that the ideal $\mathfrak{A}R_M$ has dimension $n-d-h$. For, the prime ideals of $\mathfrak{A}R_M$ are the ideals of the form $\mathfrak{o}R_M$, where $\mathfrak{o}$ is any prime ideal of $\mathfrak{A}$ which does not intersect $M$. If $s$ is the dimension of $\mathfrak{o}$ then the above argument, given for the ideal $\mathfrak{p}$, shows that $\mathfrak{o}R_M$ has dimension $s-d \leqq n-h-d$, since $s \leqq n-h$. Hence dim $\mathfrak{A}R_M \leqq n-d-h$, and since $\mathfrak{A}R_M$ is generated by $h$ elements we must have dim $\mathfrak{A}R_M = n-d-h$, in view of Theorem 22 (§ 7).

We shall show, however, coming back to the notations of Theorem 26, that if $h<n$ and $\mathfrak{p}$ is any prime zero-dimensional ideal then $\mathfrak{p}$ is not an associated prime ideal of $\mathfrak{A}$. This result, when applied to the polynomial ring $R_M$, will contradict our assumption that $h<n-d$ and will complete the proof of the theorem.

We consider the ideal $\mathfrak{B}=(F_1, \cdots, F_{h-1})$. Since it is generated by $h-1$ elements, its dimension is at least $n-h+1$ (Theorem 22, § 7). If the dimension of $\mathfrak{B}$ were greater than $n-h+1$, $\mathfrak{B}$ would admit an isolated prime ideal $\mathfrak{v}$ of dimension $>n-h+1$. Then all the isolated prime ideals of $(\mathfrak{v}, F_h)$ would be of dimension $>n-h$ (Theorem 21, § 7, applied to $k[X_1, \cdots, X_n]/\mathfrak{v}$), and this would contradict the fact that they contain $\mathfrak{A}$. Thus the dimension of $\mathfrak{B}$ is $n-h+1$, and our induction hypothesis implies that all the associated prime ideals $\mathfrak{v}_1, \cdots, \mathfrak{v}_r$ of $\mathfrak{B}$ have dimension $n-h+1$. We denote by $\mathfrak{v}_{r+1}, \cdots, \mathfrak{v}_{r'}$ the associated prime ideals of $\mathfrak{A}$ which are of dimension $n-h$. Since $h<n$, none of the ideals $\mathfrak{v}_j$ $(1 \leqq j \leqq r')$ is maximal. We are going to construct an element $D$ of the maximal ideal $\mathfrak{p}$, of a particular type, which does not belong to any $\mathfrak{v}_j$. For this we need a lemma:

LEMMA 1. *Given a finite family* $\{\mathfrak{v}_1, \cdots, \mathfrak{v}_{r'}\}$ *of non-maximal prime ideals in* $k[X_1, \cdots, X_n]$, *there exists an index* $t$ *and a polynomial* $\varphi(X_1, \cdots, X_{t-1})$ *such that the* $\mathfrak{v}_j$*-residue of* $Y_t = X_t + \varphi(X_1, \cdots, X_{t-1})$ *is transcendental over* $k$ *for every* $j$.

PROOF OF LEMMA 1. We renumber the $\mathfrak{v}_j$ in such a way that:

(a) The $\mathfrak{v}_j$-residue of $X_1$ is transcendental over $k$ for $j=1, \cdots, r(1)$ and algebraic over $k$ for $j>r(1)$;

(b) The $\mathfrak{v}_j$-residue of $X_2$ is transcendental over $k$ for $j=r(1)+1, \cdots, r(2)$, and algebraic over $k$ for $j>r(2)$;†

and so on. Since, for every $j$, one at least of the elements $X_1, \cdots, X_n$ has a transcendental $\mathfrak{v}_j$-residue ($\mathfrak{v}_j$ having dimension $>0$), all the ideals $\mathfrak{v}_j$ are included in our renumbering; in other words: there exists an index $t \leqq n$ such that, for $j=r(t-1)+1, \cdots, r'$, the $\mathfrak{v}_j$-residue of $X_t$

† If the $\mathfrak{v}_j$-residue of $X_1$ is algebraic for all $j (1 \leqq j \leqq r')$ then, of course, the set of indices $1, 2, \cdots, r(1)$ is empty. A similar remark applies to the set of indices $r(1)+1, \cdots, r(2)$ introduced in (b).

is transcendental over $k$ and the $\mathfrak{v}_j$-residues of $X_1, \cdots, X_{t-1}$ are algebraic over $k$. Thus, for any polynomial $\varphi(X_1, \cdots, X_{t-1})$, and for $j=r(t-1)+1, \cdots, r'$, the $\mathfrak{v}_j$-residue of $X_t+\varphi(X_1, \cdots, X_{t-1})$ is transcendental over $k$. We take now $j$ so as to satisfy the inequality $r(t-2)+1 \leqq j \leqq r(t-1)$. Then, since $X_{t-1}$ is transcendental over $k$, mod $\mathfrak{v}_j$, there is at most *one* exponent $a$ such that $X_t+X_{t-1}{}^a$ is algebraic over $k$, mod $\mathfrak{v}_j$ (otherwise some differences $X_{t-1}{}^a-X_{t-1}{}^b$ $(a \neq b)$ would be algebraic over $k$, mod $\mathfrak{v}_j$, whence also $X_{t-1}$ would be algebraic over $k$, mod $\mathfrak{v}_j$). We can thus find an exponent $a(t-1)$ such that $X_t+X_{t-1}{}^{a(t-1)}$ is transcendental over $k$, mod $\mathfrak{v}_j$ for $j=r(t-2)+1, \cdots, r(t-1)$, and also for $j=r(t-1)+1, \cdots, r'$ in view of what has been seen above about $X_t+\varphi(X_1, \cdots, X_{t-1})$.

Since $X_1, \cdots, X_{t-2}$ are algebraic over $k$, mod $\mathfrak{v}_j$ for $j>r(t-2)$, it follows that, for any polynomial $\psi(X_1, \cdots, X_{t-2})$ and for any $j>r(t-2)$, $X_t+X_{t-1}{}^{a(t-1)}+\psi(X_1, \cdots, X_{t-2})$ is transcendental over $k$, mod $\mathfrak{v}_j$. As above, for every $j$ such that $r(t-3)+1 \leqq j \leqq r(t-2)$, there exists at most *one* exponent $a$ such that $X_t+X_{t-1}{}^{a(t-1)}+X_{t-2}{}^a$ is algebraic over $k$ mod $\mathfrak{v}_j$. Thus we can find an exponent $a(t-2)$ such that $X_t+X_{t-1}{}^{a(t-1)}+X_{t-2}{}^{a(t-2)}$ is transcendental over $k$ mod $\mathfrak{v}_j$ for $r(t-3)+1 \leqq j \leqq r(t-2)$, and also for $j>r(t-2)$ from what has been seen above. Continuing in the same manner, we get a polynomial $Y_t=X_t+X_{t-1}{}^{a(t-1)}+ \cdots +X_1{}^{a(1)}$ whose $\mathfrak{v}_j$-residue is transcendental over $k$ for every $j$. This completes the proof of the lemma.

*We now return to the proof of Theorem* 26. From the structure of the polynomial $Y_t$ we immediately see that

$$k[X_1, \cdots, X_n] = k[X_1, \cdots, X_{t-1}, Y_t, X_{t+1}, \cdots, X_n].$$

Since the ideal $\mathfrak{p}$ is maximal, the $\mathfrak{p}$-residue $y_t$ of $Y_t$ is algebraic over $k$. Let $D=f(Y_t)$ be the minimal polynomial of $y_t$ over $k$. We have $D \in \mathfrak{p}$. Since the $\mathfrak{v}_j$-residue of $Y_t$ is transcendental over $k$ for every $j$, $D$ does *not* belong to any $\mathfrak{v}_j$. It is clear that the residue class ring $k[X_1, \cdots, X_n]/(D)$ is isomorphic with the polynomial ring $k(y_t)[X_1, \cdots, X_{t-1}, X_{t+1}, \cdots, X_n]$.

Let now $a$ be an element of $R=k[X_1, \cdots, X_n]$ such that $a\mathfrak{p} \subset \mathfrak{A} = (F_1, \cdots, F_h)$. We then have $aD \in \mathfrak{A} = (\mathfrak{B}, F_h)$ and there exists an element $b$ in $R$ such that $aD-bF_h \in \mathfrak{B}$. Thus $bF_h \in (\mathfrak{B}, D)$.

Now, $F_h$ *does not belong to any associated prime ideal* $\mathfrak{B}$ *of the ideal* $(\mathfrak{B}, D)$. In fact, since $(\mathfrak{B}, D)/(D)$ is generated by $h-1$ elements in $R/(D)$ (namely by the classes of $F_1, \cdots, F_{h-1}$, mod $D$), since its dimension is $n-h$ (for $D$ has been chosen outside of the prime ideals of $\mathfrak{B}$) and since $R/(D)$ is a polynomial ring in $n-1$ variables, the induction hypothesis shows that every associated prime ideal $\mathfrak{B}/(D)$ of $(\mathfrak{B}, D)/(D)$, whence also every associated prime ideal $\mathfrak{B}$ of $(\mathfrak{B}, D)$, is of dimension $n-h$. If such a prime ideal $\mathfrak{B}$ were to contain $F_h$, it would contain $\mathfrak{A}$, in contradiction with the fact that $D$ has been chosen outside of the $(n-h)$-dimensional prime ideals of $\mathfrak{A}$.

This being so, the relation $bF_h \in (\mathfrak{B}, D)$ implies $b \in (\mathfrak{B}, D)$ (Vol. I, Ch. IV, § 6, Theorem 11). Thus there exists an element $c$ in $R$ such that $b-cD \in \mathfrak{B}$. This relation, together with the relation $aD-bF_h \in \mathfrak{B}$, implies that $(a-cF_h)D \in \mathfrak{B}$. But, since $D$ has been chosen outside all the associated prime ideals of $\mathfrak{B}$, we deduce that $a-cF_h \in \mathfrak{B}$, and that consequently $a \in (\mathfrak{B}, F_h)=\mathfrak{A}$. This shows that $(\mathfrak{A}:\mathfrak{p})=\mathfrak{A}$, whence $\mathfrak{p}$ cannot be an associated prime ideal of $\mathfrak{A}$. This concludes the proof of Macaulay's Theorem.

Before proving an important result about the dimension of the sum of two ideals, we need a lemma about unmixed ideals:

LEMMA 2. *Let $\mathfrak{A}$ be an ideal in $R=k[X_1, \cdots, X_n]$, different from $R$, and let $\{z_1, \cdots, z_d\}$ be a finite set of algebraically independent elements of $R/\mathfrak{A}$ over $k$ such that $R/\mathfrak{A}$ is integral over $k[z]$. Then $\mathfrak{A}$ has dimension $d$, and a necessary and sufficient condition that $\mathfrak{A}$ be unmixed is that no element of $k[z]$, different from zero, be a zero divisor in $R/\mathfrak{A}$.*

PROOF. Let $\mathfrak{p}$ be any prime ideal of $R$ containing $\mathfrak{A}$, and let $\bar{\mathfrak{p}}=\mathfrak{p}/\mathfrak{A}$. Then $R/\mathfrak{p}$ is integral over $k[z]/(\bar{\mathfrak{p}} \cap k[z])$, whence the dimension of $\mathfrak{p}$ is $\leqq d$. On the other hand, there exists a prime ideal $\bar{\mathfrak{p}}$ of $R/\mathfrak{A}$ which contracts to $(0)$ in $k[z]$ (Vol. I, Ch. V, § 2, Theorem 3), and for such an ideal $\bar{\mathfrak{p}}$ the corresponding prime ideal $\mathfrak{p}$ of $R$ (i.e., the ideal $\mathfrak{p}$ such that $\mathfrak{p} \supset \mathfrak{A}$ and $\bar{\mathfrak{p}}=\mathfrak{p}/\mathfrak{A}$) has dimension $d$. This proves that $\mathfrak{A}$ has dimension $d$, and, moreover, that the associated prime ideals $\mathfrak{p}$ of $\mathfrak{A}$ which are of dimension $d$ are those for which $(\mathfrak{p}/\mathfrak{A}) \cap k[z]=(0)$.

Now, in $R/\mathfrak{A}$, the set of zero divisors is $\bigcup_i (\mathfrak{p}_i/\mathfrak{A})$, where the $\mathfrak{p}_i$ are the associated prime ideals of $\mathfrak{A}$ (Vol. I, Ch. IV, § 6, Theorem 11, Corollary 3). Thus the condition that no element of $k[z]$, different from zero, is a zero divisor in $R/\mathfrak{A}$, is equivalent to the condition that we have $(\mathfrak{p}_i/\mathfrak{A}) \cap k[z]=(0)$ for every $i$, i.e., that all the ideals $\mathfrak{p}_i$ be of dimension $d$. The proof of the lemma is now complete.

Let $\mathfrak{A}$ and $\mathfrak{B}$ be two ideals in an arbitrary finite integral domain $R$. If $\mathfrak{A}$ is a *prime* ideal of dimension $a$, and $\mathfrak{B}$ an ideal of dimension $b$ *generated by $n-b$ elements*, then the application of Theorem 22 (§ 7) to $R/\mathfrak{A}$ shows that if $\mathfrak{A}+\mathfrak{B} \neq (1)$, then all the isolated prime ideals of $\mathfrak{A}+\mathfrak{B}$ have dimension $\geqq a-(n-b)=a+b-n$. This result continues to be true if $\mathfrak{A}$ is not a prime ideal, provided we suppose that all the isolated prime ideals $\mathfrak{p}_i$ of $\mathfrak{A}$ have dimension $a$: in fact every isolated prime ideal $\mathfrak{q}$ of $\mathfrak{A}+\mathfrak{B}$ contains some $\mathfrak{p}_i$ and is therefore an isolated prime ideal of $\mathfrak{p}_i+\mathfrak{B}$. Therefore the dimension of every isolated prime ideal of $\mathfrak{A}+\mathfrak{B}$ is at least $a+b-n$. This fact will be useful in the proof of the next theorem where we show that in the special case of a polynomial ring this same result remains valid without the assumption that $\mathfrak{B}$ is generated by exactly $n-b$ elements:

THEOREM 27. *Let $\mathfrak{A}$ and $\mathfrak{B}$ be two prime ideals in $R=k[X_1, \cdots, X_n]$ of dimensions $a$ and $b$. If $\mathfrak{A}+\mathfrak{B}$ is not the unit ideal, then all the isolated prime ideals of $\mathfrak{A}+\mathfrak{B}$ have dimension $\geqq a+b-n$.*

PROOF. We introduce a second copy $k[X'_1, \cdots, X'_n]$ of $k[X_1, \cdots, X_n]$ and denote by $\mathfrak{B}'$ the ideal in $k[X'_1, \cdots, X'_n]$ corresponding to $\mathfrak{B}$. In the polynomial ring in $2n$ variables $k[X_1, \cdots, X_n, X'_1, \cdots, X'_n]$, we consider the ideal $\mathfrak{U}$ generated by $\mathfrak{A}$ and $\mathfrak{B}'$, and the ideal $\mathfrak{D}$ generated by $X'_1-X_1, \cdots, X'_n-X_n$. We first prove that there is a $1-1$ correspondence between the isolated prime ideals of $\mathfrak{A}+\mathfrak{B}$ (in $k[X]$) and the isolated prime ideals of $\mathfrak{U}+\mathfrak{D}$ (in $k[X, X']$), and that this correspondence preserves dimensions.

With every prime ideal $\mathfrak{p}$ in $k[X]$, we shall associate the ideal $\bar{\mathfrak{p}}=(\mathfrak{p}, X'_1-X_1, \cdots, X'_n-X_n)$ in $k[X, X']$. The $k$-homomorphism of $k[X, X']$ onto $k[X]$ defined by $\varphi(X_i)=\varphi(X'_i)=X_i$ obviously admits $\mathfrak{D}$ as kernel. Since $\varphi^{-1}(\mathfrak{p})=\bar{\mathfrak{p}}$, $\bar{\mathfrak{p}}$ is a prime ideal and has the same dimension as $\mathfrak{p}$. Furthermore, the inverse image $\varphi^{-1}(\mathfrak{A}+\mathfrak{B})$ contains $\mathfrak{A}$, $\mathfrak{B}'$ and $\mathfrak{D}$, whence it also contains $\mathfrak{U}+\mathfrak{D}$. Conversely, if $F(X, X') \in \varphi^{-1}(\mathfrak{A}+\mathfrak{B})$ we have $F(X, X) \in \mathfrak{A}+\mathfrak{B}$, and we may write $F(X, X)= A(X)+B(X)$, $(A(X) \in \mathfrak{A}, B(X) \in \mathfrak{B})$. Since $F(X, X')\equiv F(X, X)$ mod $\mathfrak{D}$, and since $B(X)\equiv B(X')$ mod $\mathfrak{D}$, we have $F(X, X')\equiv A(X)+B(X')$ mod $\mathfrak{D}$, i.e., $F(X, X') \in \mathfrak{U}+\mathfrak{D}$. Therefore $\mathfrak{U}+\mathfrak{D}$ is the inverse image of $\mathfrak{A}+\mathfrak{B}$ under $\varphi$, and this proves that *$\varphi^{-1}(\mathfrak{p})$ is an isolated prime ideal of $\mathfrak{U}+\mathfrak{D}$, if and only if $\mathfrak{p}$ is an isolated prime ideal of* $\mathfrak{A}+\mathfrak{B}$ (Vol. I, Ch. IV, Remark at the end of § 5, p. 213).

The consideration of $\varphi$ proves also that $\mathfrak{D}$ is a prime ideal of dimension $n$. Since it is generated by $n$ elements (e.g., by $X'_1-X_1, \cdots, X'_n-X_n$), the remark preceding Theorem 27 shows that Theorem 27 will be proved if we prove that *all the isolated prime ideals* of $\mathfrak{U}$ have dimension $a+b$: in fact, every isolated prime ideal $\bar{\mathfrak{p}}$ of $\mathfrak{U}+\mathfrak{D}$ will have then dimension $\geqq a+b+n-2n=a+b-n$, and hence also every isolated prime ideal of $\mathfrak{A}+\mathfrak{B}$ will have dimension $\geqq a+b-n$.

We shall even prove that $\mathfrak{U}$ is an unmixed ideal of dimension $a+b$, and for this purpose we shall use Lemma 2 and the results on tensor products established in Vol. I, Ch. III, § 14.

We set $k[x]=k[X]/\mathfrak{A}$, where $X$ stands for $\{X_1, X_2, \cdots, X_n\}$, $x$ stands for $\{x_1, x_2, \cdots, x_n\}$ and $x_i$ denotes the $\mathfrak{A}$-residue of $X_i$. Similarly, we denote by $k[x']$ the ring $k[X']/\mathfrak{B}'$. We know that $k[X, X']$ is a tensor product of $k[X]$ and $k[X']$, over $k$ (Vol. I, Ch. III, § 14, p. 184). It follows from Theorem 35 of Vol. I, Ch. III, § 14, that the rings $k[x] \otimes k[x']$ and $k[X, X']/\mathfrak{U}$ are $k$-isomorphic and that there exists a $k$-isomorphism $f$ of the first ring onto the second such that if $F(X)$ is any element of $k[X]$ then $f$ sends $F(x)$ into the $\mathfrak{U}$-residue of $F(X)$ and $F(x')$ into the $\mathfrak{U}$-residue of $F(X')$.

By the normalization theorem we can find $a$ algebraically independent elements $z_1, z_2, \cdots, z_a$ in $k[x]$ and $b$ algebraically independent elements $z'_1, z'_2, \cdots, z'_b$ in $k[x']$ such that $k[x]$ is integral over $k[z]$ and $k[x']$ is integral over $k[z']$ (§ 7, Theorem 25). Then, by the linear disjointness of $k[x]$ and $k[x']$ in $k[x] \otimes k[x']$ (after identification of the rings $k[x]$ and $k[x']$ with the corresponding subrings in the tensor product; see Vol. I, Ch. III, § 14, p. 183), the $a+b$ elements $z$, $z'$ are algebraically independent over $k$, and it is clear that the ring $k[x] \otimes k[x']$ is integrally dependent over $k[z, z']$. Since $\mathfrak{A}$ and $\mathfrak{B}$ are prime ideals, the rings $k[x]$ and $k[x']$ are integral domains. Hence by Theorem 36 of Vol. I, Ch. III, § 14 it follows that no element of $k[z, z']$ $(=k[z] \otimes k[z'])$, different from zero, is a zero divisor in $k[x] \otimes k[x']$. If we now carry over these conclusions to the ring $k[X, X']/\mathfrak{U}$, by means of the isomorphism $f$, and if we use Lemma 2, we find at once that the ideal $\mathfrak{U}$ is unmixed and has dimension $a+b$. Q.E.D.

Theorem 27 has the following geometric application. Let $V$ and $W$ be two irreducible varieties in affine space $A_n{}^K$. The dimensions of their prime ideals $\mathfrak{A}$, $\mathfrak{B}$ in $k[X_1, \cdots, X_n]$ are $\dim(V)$ and $\dim(W)$, respectively. We have seen that $V \cap W$ is the variety of the ideal $\mathfrak{A}+\mathfrak{B}$ (§ 3, formula (2)), and that the irreducible components of this variety are the varieties of the isolated prime ideals of $\mathfrak{A}+\mathfrak{B}$ (§ 3, Theorem 14, Corollary 3). Therefore, Theorem 27 may be translated into:

THEOREM 27′. *If $V$ and $W$ are two irreducible varieties in $A_n{}^K$ and if the intersection $V \cap W$ is non-empty, then every irreducible component of $V \cap W$ has a dimension* $\geqq \dim(V)+\dim(W)-n$.

REMARK. Theorem 27 does not extend to arbitrary finite integral domains. In other words, if $V$ and $W$ are two subvarieties of an ambient variety $Z$, then it is not necessarily true that, for every irreducible component $C$ of $V \cap W$, we have

$$\dim(C) \geqq \dim(V)+\dim(W)-\dim(Z). \tag{1}$$

For example, we take for $Z$ the cone in $A_4^K$, with equation $X_1X_2 - X_3X_4 = 0$. The planes $V(X_1 = X_3 = 0)$ and $W(X_2 = X_4 = 0)$ are subvarieties of $Z$, and their intersection $C$ is reduced to the origin. We have $\dim(Z) = 3$, $\dim(V) = \dim(W) = 2$, $\dim(C) = 0$, and the above inequality is not verified. However, it can be proved that this inequality holds for every irreducible component $C$ of $V \cap W$ which is *simple* on $Z$.

For non-empty varieties $V$, $W$ in *projective* space $P_n^K$, *the inequality* (1) *is still valid for every irreducible component $C$ of $V \cap W$.* In fact, the homogeneous prime ideals $\mathfrak{A}$, $\mathfrak{B}$ of $V$ and $W$ in $k[Y_0, Y_1, \cdots, Y_n]$ have projective dimensions $\dim(V)$ and $\dim(W)$, whence their ordinary dimensions are $\dim(V)+1$ and $\dim(W)+1$, respectively. *Furthermore,* $\mathfrak{A}+\mathfrak{B} \neq (1)$, *for both* $\mathfrak{A}$ *and* $\mathfrak{B}$ *are contained in the irrelevant ideal* $(Y_0, Y_1, \cdots, Y_n)$. Hence, if $C$ is an irreducible component of $V \cap W$, its prime ideal $\mathfrak{p}$, being an isolated prime ideal of $\mathfrak{A}+\mathfrak{B}$, has an ordinary dimension $\geqq (\dim(V)+1)+(\dim(W)+1)-(n+1) = \dim(V) + \dim(W) - n + 1$, and a projective dimension $\geqq \dim(V) + \dim(W) - n$. Note that in the projective case we have established the inequality (1) without *assuming* that $V \cap W$ is non-empty. Hence in the projective case we have the following result (which has no affine analogue):

*If* $\dim(V)+\dim(W) \geqq n$, *then* $V \cap W$ *is non-empty.*

Again we note that this last result is not generally true if the ambient variety of $V$ and $W$ is an arbitrary variety $Z$. In other words, if the dimensions $a$ and $b$ of two subvarieties $V$ and $W$ of an $n$-dimensional irreducible projective variety $Z$ satisfy the inequality $a+b \geqq n$, then it is not necessarily true that $V \cap W$ is non-empty (even if $Z$ is a variety free from singularities). The simplest example is the following: $Z$ is a ruled irreducible quadric surface ($n=2$) and $V$, $W$ are straight lines on $Z$ ($a=b=1$) which belong to the same ruling of $Z$ (and are therefore skew lines).

**§ 9. Normalization theorems.** In this section we intend to give a new version of the "normalization theorem" proved in § 7 (Theorem 25), together with a systematic treatment of the so-called "normalization methods." These methods partly reduce the study of arbitrary ideals in a polynomial ring (or in a power series ring) to the study of the ideals generated by a certain number of the variables. The treatment we give, as well as the treatment given in the second half of § 7, is independent of the dimension theory developed in the first part of § 7. Moreover in our present treatment we shall deal simultaneously with polynomial rings *and* power series rings. In the next section we shall apply the normalization methods to the study of the dimension theory of power series rings.

Let $R$ denote either the polynomial ring $A[X_1, \cdots, X_n]$ or the power series ring $A[[X_1, \cdots, X_n]]$ in $n$ variables over an arbitrary commutative ring $A$ (with unit element, as usual). We recall (§ 7, p. 198) that a system of $n$ *forms* $(F_1, \cdots, F_n)$, with strictly positive degrees, is said to be a *homogeneous system of integrity* in $R = A[X_1, X_2, \cdots, X_n]$ if the elements $X_1, \cdots, X_n$ are *integral* over the ring $S = A[F_1, \cdots, F_n]$.

In the case of a power series ring $R = A[[X_1, X_2, \cdots, X_n]]$ we modify this definition by requiring that $X_1, X_2, \cdots, X_n$ be integral over the ring $S = A[[F_1, F_2, \cdots, F_n]]$. By replacing the forms $F_i$ by suitable powers, we can always reduce the study of a homogeneous system of integrity to that of a homogeneous system of integrity consisting of forms of *like degrees*. The following lemma is useful.

LEMMA 1. *If $\{F_1, \cdots, F_n\}$ are n forms of like degree d which constitute a homogeneous system of integrity in R, then each indeterminate $X_i$ satisfies a relation of the form*

$$X_i^{sd} + \varphi_{i,s-1}(F_1, \cdots, F_n)X_i^{(s-1)d} + \cdots + \varphi_{i,0}(F_1, \cdots, F_n) = 0,$$

*where $\varphi_{i,j}$ is a form of degree $s-j$ and s is a suitable integer.*

PROOF. We consider a relation of integral dependence

$$X_i^t + \psi_{t-1}(F)X_i^{t-1} + \cdots + \psi_0(F) = 0 \tag{1}$$

satisfied by $X_i$ over $S$. We single out the terms on the left-hand side which are of degree $t$ in $X_1, X_2, \cdots, X_n$, and obtain

$$X_i^t + \psi^\star_{t-1}(F)X_i^{t-1} + \cdots + \psi_0^\star(F) = 0. \tag{2}$$

In this relation, $\psi_j^\star$ must be a form, and $\psi_j^\star(F)$ must be either zero or a form of degree $t-j$ in $X_1, \cdots, X_n$. Since $F_1, \cdots, F_n$ are forms of degree $d$, the degree $t-j$ of $\psi_j^\star(F)$ (considered as a form in $X_1, \cdots, X_n$) must be a multiple of $d$, say $t-j=dm$ (unless $\psi_j^\star(F)$ is zero). Hence relation (2) may be written as follows:

$$X_i^t + \varphi_1^\star(F)X_i^{t-d} + \cdots + \varphi_m^\star(F)X_i^{t-md} + \cdots = 0, \tag{3}$$

where $\varphi_j^\star = \psi^\star_{t-jd}$ is a form of degree $j$. Factoring out a suitable power of $X_i$, we see that we may assume that $t$ is a multiple $sd$ of $d$, and this proves Lemma 1.

Lemma 1 has the following easy *consequences*:

(1) The notion of homogeneous system of integrity is *the same for polynomials and power series*. More precisely, a system of $n$ forms

$\{F_1, \cdots, F_n\}$ is a homogeneous system of integrity in $A[[X_1, \cdots, X_n]]$ if and only if it is a homogeneous system of integrity in $A[X_1, \cdots, X_n]$.

(2) It follows from Lemma 1 that, *if* $\{F_1, \cdots F_n\}$ *is a homogeneous system of integrity, then the ideal* $(F_1, \cdots, F_n)$ *is irrelevant*: in fact, since the forms $\varphi_{i,j}$ have strictly positive degrees, the relation given in Lemma 1 implies that $X_i^{sd} \in (F_1, \cdots, F_n)$. This result, together with its converse, has already been proved in the lemma of § 7 (see p. 198) for the case of polynomial rings (and also, more generally, for the case of finitely generated homogeneous rings). The converse result holds also in the case of power series, i.e., we have that if the ideal $(F_1, \cdots, F_n)$ is irrelevant (the $F_i$ being forms), then every $X_i$ is integral over $A[[F_1, \cdots, F_n]]$. In fact, we can even prove the following result (which is *not* a trivial consequence of the integrity of $X_1, \cdots, X_n$ over $A[[F_1, \cdots, F_n]]$, in contrast with the case of polynomials): *If* $F_1, \cdots, F_n$ *are forms such that the ideal* $(F_1, \cdots, F_n)$ *is irrelevant, then* $R = A[[X_1, \cdots, X_n]]$ *is a finite module over* $S = A[[F_1, \cdots, F_n]]$. For the proof we first observe that the proof of the above cited lemma (§ 7, p. 198) provides us with a finite system $\{\omega_j\}$ of monomials in $X_1, \cdots, X_n$ such that every monomial $m_\alpha(X)$ can be written in the form $m_\alpha(X) = \sum_j a_{\alpha j}\omega_j$ with $a_{\alpha j}$ in $S$. Hence, in order to prove our assertion it suffices to show that the coefficients $a_{\alpha j}$ may be chosen in such a way that their orders (in $A[[X_1, \cdots, X_n]]$) tend to infinity with the order of $m_\alpha(X)$. We may assume (since we can replace the $F_i$'s by suitable powers) that the forms $F_i$ have the same degree $d$. Then, as in the proof of Lemma 1, we write $a_{\alpha j}$ as a power series $\psi_{\alpha j}(F)$, and we decompose $\psi_{\alpha j}$ into an infinite sum $\psi_{\alpha j} = \sum_q \varphi_{\alpha j}{}^{(q)}$ of forms ($\varphi_{\alpha j}{}^{(q)}$ having degree $q$). Singling out the terms of degree $\partial(m_\alpha(X))$ in the sum $\sum_j a_{\alpha j}\omega_j$, we get

$$m_\alpha(X) = \sum_{j,q} \varphi_{\alpha j}{}^{(q)}(F)\omega_j,$$

the summation being extended to all pairs $(j, q)$ of integers such that $\partial(m_\alpha(X)) = dq + \partial(\omega_j)$. This proves our assertion.

The last result, together with its analogue for polynomials, may be generalized in such a way as to make the notion of homogeneous system of integrity useful also for questions which arise in a non-homogeneous set-up.

THEOREM 28. *Let* $\{G_1, \cdots, G_n\}$ *be a system of* $n$ *elements in* $R$ *having a homogeneous system of integrity* $\{F_1, \cdots, F_n\}$ *as system of initial forms in the power series case (of highest degree forms in the* *polynomial case). Then $R$ is a finite module over $T=A[[G_1, \cdots, G_n]]$ (or over $T=A[G_1, \cdots, G_n]$).*

PROOF. Upon replacing $G_1, \cdots, G_n$ by suitable powers, we may assume that the forms $F_1, \cdots, F_n$ have like degree $d$. We have just proved the existence of a finite set of monomials $\omega_j(X)$ which is a basis of $R$ over $S=A[[F_1, \cdots, F_n]]$ (or $A[F_1, \cdots, F_n]$). Furthermore, we have seen that every form $u(X)$ in $R$ may be written as

$$u(X) = \sum_j \varphi_j(F)\omega_j(X) \tag{4}$$

where $\varphi_j$ is a form such that $d\cdot\partial(\varphi_j)+\partial(\omega_j)=\partial(u)$. Now, if $v(X)$ is any element of $R$, we apply (4) to each monomial $u(X)$ which occurs in $v(X)$, and by addition of terms we find a relation of the form

$$v(X) = \sum_j \psi_j(F)\omega_j(X), \tag{4'}$$

where the $\psi_j(F)$ are power series in $F_1, F_2, \cdots, F_n$ (or polynomials in $F_1, F_2, \cdots, F_n$), and where

$$d\cdot\mathbf{o}(\psi_j) \geqq \mathbf{o}(v)-\mathbf{o}(\omega_j) \text{ in the power series case,} \tag{5}$$

with equality for at least one value of $j$;

$$d\cdot\partial(\psi_j) \leqq \partial(v)-\partial(\omega_j) \text{ in the polynomial case,} \tag{5'}$$

again with equality for at least one value of $j$.

We prove that $\{\omega_j(X)\}$ is also a basis of $R$ *over* $T$, and thus will prove Theorem 28. For every $v(X)$ in $R$, we consider the difference $v(X)-\sum_j \psi_j(G)\omega_j(X)$, where $v(X)=\sum_j \psi_j(F)\omega_j(X)$ (formula (4')). In the power series case, replacing the $F_i$'s by the $G_i$'s leaves unchanged the initial form of the element $\sum_j \psi_j(F)\omega_j(X)$ of $R$, for the equality sign holds in (5) for at least one value of $j$. We therefore have

$$\mathbf{o}(v(X)-\sum_j \psi_j(G)\omega_j(X)) > \mathbf{o}(v(X)). \tag{6}$$

Suppose, by induction on $s$, that we have found power series $\psi_{js}(G)$ such that $\mathbf{o}(v(X)-\sum_j \psi_{js}(G)\omega_j(X)) \geqq s$. We can then write, by (4')

$$v(X)-\sum_j \psi_{js}(G)\omega_j(X) = \sum_j \chi_{js}(F)\omega_j(X),$$

where, by (5), the order $\mathbf{o}(\chi_{js})$ satisfies the inequality $d\cdot\mathbf{o}(\chi_{js}) \geqq s-\mathbf{o}(\omega_j)$. If we set $\psi_{j,s+1}=\psi_{j,s}+\chi_{js}$, and replace the $F_i$'s by the $G_i$'s, we get

$$\mathbf{o}(v(X)-\sum \psi_{j,s+1}(G)\omega_j(X)) \geqq s+1.$$

Furthermore the inequalities $d \cdot \mathbf{o}(\chi_{js}) \geqq s - \mathbf{o}(\omega_j)$ show that the sequences $\{\psi_{j,s}\}$ are *Cauchy sequences* in the power series ring in $n$ variables. Let $\mu_j$ be the limit of $\{\psi_{j,s}\}$. By passage to the limit we obviously have $v(X) = \sum_j \mu_j(G)\omega_j(X)$, and this proves our assertion in the power series case.

In the polynomial case we proceed by induction on the degree $s$ of $v(X)$. The case $s=0$ is straightforward, if care has been taken to include the monomial 1 among the $\omega_j(X)$'s. As above, we write $v(X) = \sum_j \psi_j(F)\omega_j(X)$, and consider the differences $v'(X) = v(X) - \sum_j \psi_j(G)\omega_j(X)$. Since the replacement of the $F_i$'s by the $G_i$'s leaves the highest degree form of $\sum_j \psi_j(F)\omega_j(X)$ unchanged, we have $\partial(v') < \partial(v)$, and our induction hypothesis shows that $v'(X) = \sum_j \chi_j(G)\omega_j(X)$ with suitable polynomials $\chi_j$. Therefore $v(X)$ is also a linear combination of the monomials $\omega_j(X)$, with coefficients in $A[G_1, \cdots, G_n]$. Q.E.D.

A system of $n$ elements $G_1, \cdots, G_n$ of $R$ satisfying the conditions of Theorem 28 is called a *normal system of integrity of R*.

THEOREM 29. *Let A be an integral domain and let $G_1, \cdots, G_n$ be a normal system of integrity in $R = A[[X_1, \cdots, X_n]]$ (or $R = A[X_1, \cdots, X_n]$). Then the elements $G_1, \cdots, G_n$ are analytically* (*or algebraically*) *independent over A.*

PROOF. We first treat the polynomial case, which is quite simple. Let $K$ be the quotient field of $A$. Since every $X_i$ is integral over $A[G_1, \cdots, G_n]$, the field $K(X_1, \cdots, X_n)$ is algebraic over $K(G_1, \cdots, G_n)$. As the former has transcendence degree $n$ over $K$, it follows that $\{G_1, \cdots, G_n\}$ is a transcendence basis of this field (Vol. I, Ch. II, § 12, Theorem 25). Therefore these elements are algebraically independent over $K$, and, *a fortiori*, over $A$.

In the power series case, suppose that we have a non-zero power series $\varphi(Y_1, \cdots, Y_n)$ in $n$ variables over $A$ such that $\varphi(G_1, \cdots, G_n) = 0$. We denote by $F_i$ the initial form of $G_i$, and by $d_i$ its degree. With every monomial $\mu = Y_1^{s_1} \cdots Y_1^{s_n}$ appearing with a non-zero coefficient in $\varphi$, we associate the integer $w(\mu) = s_1 d_1 + \cdots + s_n d_n$. The monomials $\mu$ for which $w(\mu)$ takes its smallest value $q$ are finite in number. Thus the sum of the corresponding terms of $\varphi$ is a polynomial $\psi \neq 0$, and the difference $\varphi - \psi$ contains only monomials $\mu$ for which $w(\mu) > q$. From this it follows that $\varphi(G_1, \cdots, G_n) - \psi(F_1, \cdots, F_n)$, considered as a power series in $X_1, \cdots, X_n$, contains only terms of degree $> q$.

Now, since $\{F_1, \cdots, F_n\}$ is a homogeneous system of integrity, the first part of the proof shows that the element $\psi(F_1, \cdots, F_n)$ of $A[X_1, \cdots, X_n]$ is different from 0. Since it is a form of degree $q$, it follows that $\varphi(G_1, \cdots, G_n)$ is a power series of order $q$ in $X_1, X_2, \cdots, X_n$, in contradiction with the hypothesis that $\varphi(G_1, \cdots, G_n)=0$. Therefore the relation $\varphi(G_1, \cdots, G_n)=0$ implies $\varphi=0$, and this proves our assertion.

REMARK. The conclusion in Theorem 29 remains valid if we suppose only that $A$ is a ring without nilpotent elements.

The key to the so-called normalization methods is the following theorem, which is contained in the "normalization theorem" proved in § 7 (Theorem 25). We, however, give here a proof which is independent of dimension theory.

THEOREM 30. *Let $k$ be a field, and let $F$ be a non-constant form in $R=k[X_1, \cdots, X_n]$. Then there exists a homogeneous system of integrity $\{F_1, \cdots, F_n\}$ in $R$ such that $F_1=F$.*

PROOF. We first study the case in which $k$ is an *infinite* field, in which case the proof is a mere repetition of the proof of the normalization lemma given in Vol. I, Ch. V, § 4 (Theorem 8). Namely, we choose elements $a_2, \cdots, a_n$ of $k$ such that $F(1, a_2, \cdots, a_n)\neq 0$. Then, if we set $G(X_1, \cdots, X_n)=F(X_1, X_2+a_2X_1, \cdots, X_n+a_nX_1)$ and $d=\partial F$, the coefficient of $X_1{}^d$ in $G$ is $F(1, a_2, \cdots, a_n)\neq 0$. The relation $F(X_1, \cdots, X_n)=G(X_1, X_2-a_2X_1, \cdots, X_n-a_nX_1)$ shows that the ideal $\mathfrak{A}$ generated by $F, X_2-a_2X_1, \cdots, X_n-a_nX_1$ contains $X_1{}^dF(1, a_2, \cdots, a_n)$ $(=G(X_1, 0, \cdots, 0))$, and hence it contains also $X_1{}^d$, since $F(1, a_2, \cdots, a_n)\neq 0$. Since $\mathfrak{A}$ contains $X_1{}^d$ and $X_j-a_jX_1$ for any $j\geqq 2$, it contains $X_j{}^d$. Hence $\mathfrak{A}$ is irrelevant, and $\{F, X_2-a_2X_1, \cdots, X_n-a_nX_1\}$ is a homogeneous system of integrity.

If $k$ is a *finite field*, we proceed by induction on the number $n$ of variables. For $n=1$ our assertion is trivial. The case $n=2$ requires a special proof (which does not make use of the finiteness of $k$). We set $X_2/X_1=T$ and write $F(X_1, X_2)=X_1{}^dF(1, T)$, where $d=\partial F$. The polynomial $F(1, T)$ is $\neq 0$. We choose a polynomial $G(T)$ in $k[T]$ which is relatively prime to $F(1, T)$, and denote by $r$ its degree. It is well known that every polynomial $H(T)$ of degree $\leqq d+r-1$ may be written in the form

$$H(T) = A(T)F(1, T)+B(T)G(T),$$

where $\partial A<\partial G=r$ and $\partial B<\partial(F(1, T))\leqq d$. In particular, for every $q$ such that $0\leqq q\leqq d+r-1$, we have

$$T^q = A_q(T)F(1, T)+B_q(T)G(T),$$

where $\partial A_q < r$ and $\partial B_q < d$. Upon multiplication by $X_1^{d+r-1}$, we get

$$X_1^{d+r-1-q}X_2^q = (X_1^{r-1}A_q(T))F(X_1, X_2) + (X_1^{d-1}B_q(T))\cdot(X_1^rG(T)).$$

As $X_1^{r-1}A_q(T)$, $X_1^{d-1}B_q(T)$ and $X_1^rG(T)$ are forms in $X_1$ and $X_2$, this shows that the ideal generated by $F(X_1, X_2)$ and $X_1^rG(T)$ contains all the monomials of degree $d+r-1$. It is therefore irrelevant, and our assertion is proved for $n=2$.

We now study the passage from $n-1$ to $n$, under the assumptions that $n$ is $\geqq 3$ and that $k$ is a *field of characteristic* $p \neq 0$ (this is implied by the hypothesis that $k$ is finite). For every power $q$ of $p$, the $q$-th power $F(X_1, \cdots, X_n)^q$ is a polynomial in $X_1^q, \cdots, X_n^q$, which we shall denote by $F_q(X_1^q, \cdots, X_n^q)$. By renumbering the variables, we may assume that $X_n$ actually occurs in $F$. We assert that there exists a power $q$ of $p$ and a form $G(X_1, \cdots, X_{n-1})$ of degree $q$ such that $F_q(X_1^q, \cdots, X_{n-1}^q, G(X_1, \cdots, X_{n-1})) \neq 0$. In fact, $F(X_1, \cdots, X_n)$, considered as a polynomial in $X_n$ over $k(X_1, \cdots, X_{n-1})$, has only a finite number of roots in the algebraic closure $K$ of $k(X_1, \cdots, X_{n-1})$. On the other hand, since $n-1 \geqq 2$, there exists, for any power $q$ of $p$, a form $G^{(q)}(X_1, \cdots, X_{n-1})$ of degree $q$ which is not the $p$-th power of any element of $k(X_1, \cdots, X_{n-1})$. Thus the elements $\sqrt[q]{G^{(q)}(X_1, \cdots, X_{n-1})}$ of $K$ are all distinct, since their minimal equations $T^q - G^{(q)} = 0$ have distinct degrees. Therefore one of them must be distinct from the roots of the polynomial $F(X_1, \cdots, X_n)$ (regarded as a polynomial in $X_n$). If we take for $G$ this polynomial $G^{(q)}$, we have $F_q(X_1^q, \cdots, X_{n-1}^q, G(X_1, \cdots, X_{n-1})) \neq 0$, as asserted.

This being so we denote by $H$ the form $F_q(X_1^q, \cdots, X_{n-1}^q, G(X_1, \cdots, X_{n-1}))$. By our induction hypothesis there exist $n-2$ forms $H_2, \cdots, H_{n-1}$ in $k[X_1, \cdots, X_{n-1}]$ such that the ideal generated by $H, H_2, \cdots, H_{n-1}$ is irrelevant in $k[X_1, \cdots, X_{n-1}]$. On the other hand $H$ is congruent to $F^q$ modulo $X_n^q - G(X_1, \cdots, X_{n-1})$, whence $H$ belongs to the ideal generated by $F$ and $X_n^q - G$. Therefore the ideal generated by $F, X_n^q - G, H_2, \cdots, H_{n-1}$ is irrelevant in $k[X_1, \cdots, X_n]$, and this proves our assertion.

REMARK. In characteristic 0, we can still find an exponent $q$ and a form $G(X_1, \cdots, X_{n-1})$ of degree $q$ such that the $q$-th roots of $G$ are distinct from the roots of $F$ (considered as a polynomial in $X_n$). Then the last part of the proof may be extended to the case of characteristic 0 if we take for $H(X_1, \cdots, X_{n-1})$ the so-called "resultant" of the elimination of $X_n$ between $F(X_1, \cdots, X_n)$ and $X_n^q - G(X_1, \cdots, X_{n-1})$.

COROLLARY. *Let $k$ be a field. Any non-constant (resp. non-invertible)* *element $F$ of $R=k[X_1, \cdots, X_n]$ (resp. $k[[X_1, \cdots, X_n]]$) may be included is some normal system of integrity of $R$.*

We apply Theorem 30 to the highest (resp. lowest) degree form of $F$, and then use Theorem 28.

As our last topic in this section, we now define the notion of *system of integrity*. A system of $n$ elements $F_1, \cdots, F_n$ in $R=A[X_1, \cdots, X_n]$ (resp. $A[[X_1, \cdots, X_n]]$) is called a system of integrity of $R$ if $R$ is a *finite module over* $A[F_1, \cdots, F_n]$ (resp. $A[[F_1, \cdots, F_n]]$). It follows from Theorem 28 that a normal system of integrity (in particular, a homogeneous system of integrity) is actually a system of integrity. The two theorems which are given below give the existence of systems of integrity which are "adapted" to the study of a given chain of ideals. As these theorems will mostly be used for studying the dimension theory of power series rings, and in order to avoid tedious repetitions, these theorems will only be stated and proved in the power series case. Statements and proofs in the polynomial case are entirely analogous.

THEOREM 31. *Let $k$ be a field and $\mathfrak{A}$ a proper ideal in $R=k[[X_1, \cdots, X_n]]$. There exists a system of integrity $\{F_1, \cdots, F_n\}$ of $R$, such that, if we set $S=k[[F_1, \cdots, F_n]]$, then the ideal $\mathfrak{A} \cap S$ is generated by $F_1, \cdots, F_d$, where $d$ is an integer, $1 \leqq d \leqq n$. The classes $f_{d+1}, \cdots, f_n$ of $F_{d+1}, \cdots, F_n$* mod $\mathfrak{A}$ *are analytically independent over $k$, and $R/\mathfrak{A}$ is a finite module over $k[[f_{d+1}, \cdots, f_n]]$.*

PROOF. Among the finite subsets of $\mathfrak{A}$ which are contained in systems of integrity of $R$, we choose one with the greatest possible number $d$ of elements. Let $\{F_1, \cdots, F_d\}$ be such a subset and let $\{F_1, \cdots, F_n\}$ be a corresponding system of integrity. We assert that

$$\mathfrak{A} \cap k[[F_{d+1}, \cdots, F_n]] = (0). \tag{7}$$

In fact, assuming the contrary we could introduce an element $P_{d+1} \neq 0$ of this intersection in some normal system of integrity $\{P_{d+1}, \cdots, P_n\}$ of $k[[F_{d+1}, \cdots, F_n]]$ (Corollary to Theorem 30), since, by Theorem 29, this ring is a power series ring. Then $k[[F_{d+1}, \cdots, F_n]]$ would be a finite module over $k[[P_{d+1}, \cdots, P_n]]$, whence also $k[[F_1, \cdots, F_n]]$ would be a finite module over $k[[F_1, \cdots, F_d, P_{d+1}, \cdots, P_n]]$, and therefore $k[[X_1, \cdots, X_n]]$ would be a finite module over $k[[F_1, \cdots, F_d, P_{d+1}, \cdots, P_n]]$. Hence $\{F_1, \cdots, F_d, P_{d+1}, \cdots, P_n\}$ would be a system of integrity of $R$ such that $\{F_1, \cdots, F_d, P_{d+1}\} \subset \mathfrak{A}$, in contradiction with the maximality of $d$.

Since the ideal $\mathfrak{A} \cap S$ contains $F_1, \cdots, F_d$ and since relation (7) holds, it follows that $\mathfrak{A} \cap S$ is generated by $F_1, \cdots, F_d$. Thus $R/\mathfrak{A}$ contains $S/(F_1, \cdots, F_d)=k[[f_{d+1}, \cdots, f_n]]$ as a subring, and relation

(7) proves that $f_{d+1}, \cdots, f_n$ are analytically independent over $k$. Finally, since $R$ is a finite module over $S$, $R/\mathfrak{A}$ is obviously a finite module over $k[[f_{d+1}, \cdots, f_n]]$. Q.E.D.

The analogue of Theorem 31 in the polynomial case contains the normalization theorem of § 7 (Theorem 25).

THEOREM 32. *Let* $\mathfrak{A}_1 \subset \mathfrak{A}_2 \subset \cdots \subset \mathfrak{A}_q$ *be a sequence of proper ideals in* $R = k[[X_1, \cdots, X_n]]$. *Then there exists a system of integrity* $\{F_1, \cdots, F_n\}$ *of* $R$ *and* $q$ *integers* $1 \leqq d_1 \leqq d_2 \leqq \cdots \leqq d_q \leqq n$ *such that, if we set* $S = k[[F_1, \cdots, F_n]]$, *then* $\mathfrak{A}_j \cap S$ *is generated by* $F_1, \cdots, F_{d_j}$. *The classes* $f_{d_j+1}, \cdots, f_n$ *of* $F_{d_j+1}, \cdots, F_n$ *module* $\mathfrak{A}_j$ *are analytically independent over* $k$, *and* $R/\mathfrak{A}_j$ *is a finite module over* $k[[f_{d_j+1}, \cdots, f_n]]$.

PROOF. In case $q = 1$, Theorem 32 reduces to Theorem 31. We proceed by induction on $q$, and suppose that we have a system of integrity $\{G_1, \cdots, G_n\}$ of $R$ and $q-1$ integers $d_1 \leqq d_2 \leqq \cdots \leqq d_{q-1}$ such that, in the ring $T = k[[G_1, \cdots, G_n]]$, the ideal $T \cap \mathfrak{A}_j$ is generated by $G_1, \cdots, G_{d_j}$ for $j = 1, \cdots, q-1$.

The ideal $\mathfrak{A}_q \cap T$ contains $G_1, \cdots, G_{d_{q-1}}$ since $\mathfrak{A}_q \supset \mathfrak{A}_{q-1}$. We consider the subring $T' = k[[G_{d_{q-1}+1}, \cdots, G_n]]$ of $T$, which is a power series ring, by Theorem 29. If $\mathfrak{A}_q \cap T' = (0)$, we take $d_q = d_{q-1}$, $F_i = G_i$, $S = T$, and then $S \cap \mathfrak{A}_q$ is generated by $F_1, \cdots, F_{d_q}$ as asserted. Otherwise we apply Theorem 31 to the ring $T'$ and to the ideal $\mathfrak{A}_q \cap T'$: there exists an integer $d_q$ such that $1 \leqq d_q \leqq n - d_{q-1}$ and a system of integrity $\{H_{d_{q-1}+1}, \cdots, H_{d_q}, \cdots, H_n\}$ of $T'$ such that $\mathfrak{A}_q \cap T' \cap k[[H_{d_{q-1}+1}, \cdots, H_n]]$ is generated by $H_{d_{q-1}+1}, \cdots, H_{d_q}$. Then it is easily seen that $\{G_1, \cdots, G_{d_{q-1}}, H_{d_{q-1}+1}, \cdots, H_n\}$ is a system of integrity of $R$, and that for this system of integrity all the assertions about the ideals $S \cap \mathfrak{A}_j$ $(j = 1, 2, \cdots, q)$ are satisfied. The other assertions are easily verified, as in Theorem 31. Q.E.D.

§ 10. **Dimension theory in power series rings.** As was shown in § 7, the normalization methods provide a smooth treatment of dimension theory of finite integral domains. In this section we shall give a similar treatment of dimension theory in power series rings, over a field. However, since, in this case, the elementary methods of § 7, based upon the notion of transcendence degree, are not available, it will be necessary to use some deeper results of the general theory of prime ideals in noetherian domains.

We first consider the situation described in Theorem 32 (§ 9), in the case in which $\mathfrak{A}_1, \cdots, \mathfrak{A}_q$ are *distinct* (*proper*) *prime ideals*. Since $R = k[[X_1, \cdots, X_n]]$ is integral over $S = k[[F_1, \cdots, F_n]]$, the ideals

$\mathfrak{A}_j \cap S$ must be distinct (Vol. I, Ch. V, § 2, Theorem 3, Complement 1). Therefore we must have $d_1 < d_2 < \cdots < d_q$. Since the integers $d_i$ all lie between 0 and $n$, it follows that $q \leqq n$. As also (0) is a prime ideal in $R$, we have therefore:

THEOREM 33. *In $R = k[[X_1, \cdots, X_n]]$, any chain of prime ideals distinct from $R$ has at most $n+1$ terms.*

The existence of maximal chains of prime ideals of $R$, i.e., of chains of prime ideals ($\neq R$) with $n+1$ terms, is immediately proved by the example of $(0) < (X_1) < (X_1, X_2) < \cdots < (X_1, \cdots, X_n)$. Now we have a more precise result:

THEOREM 34. *Any chain $\mathfrak{p}_1 < \mathfrak{p}_2 < \cdots < \mathfrak{p}_q$ of prime ideals (distinct from $R$) in $R = k[[X_1, \cdots, X_n]]$ can be refined into a chain of $n+1$ prime ideals (distinct from $R$).*

PROOF. We again use Theorem 32 (§ 9): there exists a system of integrity $\{F_1, \cdots, F_n\}$ of $R$ and a sequence of integers $d_1 \leqq d_2 \leqq \cdots \leqq d_q$ such that, in $S = k[[F_1, \cdots, F_n]]$, the ideal $\mathfrak{p}_j \cap S$ is generated by $F_1, \cdots, F_{d_j}$. As was pointed out above, the integers $d_j$ are distinct. Let $\mathfrak{F}_i (0 \leqq i \leqq n)$ be the prime ideal in $S$ generated by $F_1, \cdots, F_i$ (we set $\mathfrak{F}_0 = (0)$). To prove the theorem it will be sufficient to show that, given any index $i$ $(0 \leqq i \leqq n)$ distinct from $d_1, \cdots, d_q$, there exists at least one prime ideal $\mathfrak{p}$ in $R$ such that $\mathfrak{p} \cap S = \mathfrak{F}_i$ and such that the family of ideals $\{\mathfrak{p}, \mathfrak{p}_1, \cdots, \mathfrak{p}_q\}$ is still totally ordered by inclusion.

We assume that there exists an index $r$ such that $d_r < i < d_{r+1}$ (the cases $i < d_1$ and $d_q < i$ are treated in a similar, and even simpler, manner). We consider the factor ring $R' = R/\mathfrak{p}_r$, its subring $S' = S/\mathfrak{p}_r \cap S = S/\mathfrak{F}_{d_r}$, the prime ideals $\mathfrak{F}' = \mathfrak{F}_{d_{r+1}}/\mathfrak{F}_{d_r}$ and $\mathfrak{F}'' = \mathfrak{F}_i/\mathfrak{F}_{d_r}$ in $S'$ and the prime ideal $\mathfrak{p}' = \mathfrak{p}_{r+1}/\mathfrak{p}_r$ in $R'$. The ring $R'$ is an integral domain, integral over $S'$, and we have $\mathfrak{F}'' < \mathfrak{F}'$ and $\mathfrak{p}' \cap S' = \mathfrak{F}'$. Since $S'$ is integrally closed (as it is a power series ring over a field; see Theorem 6, § 1), we may apply the "going down Theorem" (Vol. I, Ch. V, § 3, Theorem 6): there exists a prime ideal $\mathfrak{p}''$ in $R'$ such that $\mathfrak{p}'' < \mathfrak{p}'$ and $\mathfrak{p}'' \cap S' = \mathfrak{F}''$. If we set $\mathfrak{p}'' = \mathfrak{p}/\mathfrak{p}_r$ we deduce from this that $\mathfrak{p} \cap S = F_i$ and that $\mathfrak{p}_r < \mathfrak{p} < \mathfrak{p}_{r+1}$. Q.E.D.

COROLLARY 1. *Given any prime ideal $\mathfrak{p}$ in $R = k[[X_1, \cdots, X_n]]$, its height and its depth satisfy the relation $h(\mathfrak{p}) + d(\mathfrak{p}) = n$.*

Consider, in fact, two chains of prime ideals $\mathfrak{p}_0 = (0) < \mathfrak{p}_1 < \cdots < \mathfrak{p}_{h(\mathfrak{p})}$ where $\mathfrak{p}_{h(\mathfrak{p})} = \mathfrak{p}$ and $\mathfrak{q}_0 = \mathfrak{p} < \mathfrak{q}_1 < \cdots < \mathfrak{q}_{d(\mathfrak{p})}$ where $\mathfrak{q}_{d(\mathfrak{p})}$ is a maximal ideal (actually the unique maximal ideal of $R$). Their reunion is a chain with $h(\mathfrak{p}) + d(\mathfrak{p}) + 1$ terms, which cannot be refined any more. Thus $h(\mathfrak{p}) + d(\mathfrak{p}) + 1 = n + 1$.

COROLLARY 2. *If $\mathfrak{p}$ and $\mathfrak{p}'$ are two prime ideals in $R = k[[X_1, \cdots, X_n]]$*

*such that* $\mathfrak{p} < \mathfrak{p}'$ *and such that no prime ideal can be inserted between* $\mathfrak{p}$ *and* $\mathfrak{p}'$, *then their heights differ by unity, and so do their depths.*

In fact there exists a chain $(c)$ of $n+1$ prime ideals in $R$ admitting $\mathfrak{p}$ and $\mathfrak{p}'$ as terms. Since $h(\mathfrak{p})+d(\mathfrak{p})=n$ and $h(\mathfrak{p}')+d(\mathfrak{p}')=n$, the ideals $\mathfrak{p}$ and $\mathfrak{p}'$ must necessarily be the $(h(\mathfrak{p})+1)$-st and the $(h(\mathfrak{p}')+1)$-st terms of this chain. Since their indices in $(c)$ differ by unity, our assertions follow.

REMARK (1). The depth of a prime ideal $\mathfrak{p}$ of $R=k[[X_1, \cdots, X_n]]$ is sometimes called its *dimension.* Thus the unique prime ideal of dimension 0 of $R$ is its maximal ideal. On the other hand, Theorem 34 shows that the $(n-1)$-dimensional prime ideals of $R$ are its minimal prime ideals; they are principal since $R$ is a unique factorization domain (§ 1, Theorem 6).

REMARK (2). It follows from the proof of Theorem 34 that, if $\{F_1, \cdots, F_n\}$ is a system of integrity such that $\mathfrak{p} \cap k[[F_1, \cdots, F_n]]$ is generated by $F_1, \cdots, F_q$ (cf. Theorem 31, § 9), then the *dimension of* $\mathfrak{p}$ *is* $n-q$. This shows that, in the case of a prime ideal, the integer $q$ is independent of the chosen system of integrity $\{F_1, \cdots, F_n\}$. We also see immediately that the factor ring $k[[X_1, \cdots, X_n]]/\mathfrak{p}$ is integral over a power series ring in $d(\mathfrak{p})$ variables; the integer $d(\mathfrak{p})$ is also called the *dimension of the ring* $k[[X_1, \cdots, X_n]]/\mathfrak{p}$; this notion shall be generalized in VIII, § 9 (in the framework of the dimension theory of local rings).

REMARK (3). Conversely, if $\mathfrak{p}$ is a prime ideal in $R=k[[X_1, \cdots, X_n]]$ such that $R/\mathfrak{p}$ is *integral* over a power series ring $S'$ in $d$ variables, then $d$ is the dimension $d(\mathfrak{p})$ of $\mathfrak{p}$. In fact, a chain of $d(\mathfrak{p})+1$ distinct prime ideals in $R/\mathfrak{p}$ gives, by contraction, a chain of $d(\mathfrak{p})+1$ distinct prime ideals in $S'$, whence $d(\mathfrak{p}) \leqq d$ by Theorem 33. On the other hand, a chain of $d+1$ prime ideals in $S'$ gives, by application of the "going up Theorem" (Vol. I, Ch. V, § 2, Theorem 3, Corollary) a chain of $d+1$ prime ideals in $R/\mathfrak{p}$, whence $d \leqq d(\mathfrak{p})$.

REMARK (4). Remarks (2) and (3) give a characterization of the dimension $d(\mathfrak{p})$ of a prime ideal $\mathfrak{p}$ in $R=k[[X_1, \cdots, X_n]]$: it is the number of variables of any power series ring over which $R/\mathfrak{p}$ is *integral.* Here stops the analogy with the polynomial case. In fact, the following sentence "the maximum number of elements of $R/\mathfrak{p}$ which are analytically independent over $k$" *cannot* be taken as a convenient definition of the dimension of $\mathfrak{p}$, since this number is always *infinite* as soon as the depth of $\mathfrak{p}$ is larger than 1. For proving this it suffices to show the existence of infinitely many analytically independent power series in the power series ring in two variables $k[[x, y]]$. It is even sufficient to prove the existence of *three* analytically independent power series

$a, b, c$ in $k[[x, y]]$, since, as $k[[a, b]]$ contains three analytically independent power series $u, v, w$, the power series $u, v, w, c$ are also analytically independent; by repeated applications we then get infinitely many analytically independent power series in $k[[x, y]]$.

For constructing three analytically independent power series in $k[[x, y]]$, we first notice that, *if the power series $s_1(x), \cdots, s_n(x)$ are algebraically independent over $k$, then the power series $ys_1(x), \cdots, ys_n(x)$ are analytically independent over $k$.* For, if $\varphi$ is a power series such that $\varphi(ys_1(x), \cdots, ys_n(x)) = 0$, and if we write $\varphi = \sum_{j=0}^{\infty} \varphi_j$, $\varphi_j$ denoting a form of degree $j$, we get $\sum_{j=0}^{\infty} y^j \varphi_j(s_1(x), \cdots, s_n(x)) = 0$, whence $\varphi_j(s_1(x), \cdots, s_n(x)) = 0$ and $\varphi_j = 0$ since the series $s_1(x), \cdots, s_n(x)$ are algebraically independent over $k$.

It is therefore sufficient to prove the existence of three algebraically independent power series in $k[[x]]$, for example 1, $x$ and $s(x)$, where $s(x)$ is transcendental over $k(x)$. The existence of such a transcendental power series may be proved by various methods, some of these using cardinality arguments, others (valid only in characteristic 0) using the existence of transcendental analytic functions like $e^x$ or $\sin x$. We give here a third method, inspired by Liouville's construction of transcendental numbers, and prove that the series

$$s(x) = 1 + x + x^{2!} + x^{3!} + \cdots + x^{n!} + \cdots,$$

is transcendental (over $k(x)$). Suppose that $s(x)$ is a root of a polynomial $F(T)$ of degree $q$: $F(T) = a_0(x) + a_1(x)T + \cdots + a_q(x)T^q$, with $a_i(x) \in k[x]$, and let $d$ be the maximum of the degrees of the polynomials $a_i(x)$. We may assume that $F(T)$ is irreducible over $k(x)$. For any polynomial $p(x)$, the series $s(x) - p(x)$ is a root of the polynomial $G(T) = F(T + p(x))$. We set

$$G(T) = b_0(x) + b_1(x)T + \cdots + b_q(x)T^q,$$

where $b_0(x) = a_0(x) + a_1(x)p(x) + \cdots + a_q(x)p(x)^q$. We have $b_0(x) \neq 0$ since $G(T)$ is irreducible over $k(x)$. On the other hand, we have $\partial b_0 \leqq d + q \cdot \partial p$, where $\partial$ denotes, as usual, the degree of a polynomial. We take for $p(x)$ the polynomial $1 + x + x^{2!} + x^{3!} + \cdots + x^{(n-1)!}$, where $n$ is an integer such that $n > d+1$ and $n > q+1$. We then have $\partial p = (n-1)!$ and $\partial b_0 \leqq d + q \cdot \partial p < (n-1) + (n-1)(n-1)! \leqq (n-1)! + (n-1)(n-1)! = n!$, whence $\partial b_0 < n!$. On the other hand, in the relation

$$G(s(x) - p(x)) = b_0(x) + b_1(x)(x^{n!} + x^{(n+1)!} + \cdots) + \cdots + b_q(x)(x^{n!} + \cdots)^q = 0,$$

all the terms, except those of $b_0(x)$, have $x^{n!}$ as a factor. This contradicts the facts that $b_0(x) \neq 0$ and that $\partial b_0 < n!$.

**§ 11. Extension of the ground field.** Let $k$ be a field and let $K$ be an extension field of $k$. The polynomial ring $R = k[X] = k[X_1, X_2, \cdots, X_n]$ may be considered as a subring of the polynomial ring $S = K[X] = K[X_1, X_2, \cdots, X_n]$. We shall study in this section the *extension* to $S$ of ideals in $R$. Most of the results of this section can also be derived from properties of tensor products and free joins (Vol. I, Ch. III, §§ 14, 15). However, *on the whole* we shall deal with our present topic *ab initio*, for the following reasons: (1) in view of the special importance of polynomial ideals and their extensions it seems desirable to have a self-contained treatment which can be given at an early stage, without having to develop first the machinery of tensor products; (2) most of the results concerning the behavior of polynomial ideals under ground field extensions admit direct and simple proofs. However, we shall constantly emphasize the connection between the results of this section and those of Sections 14 and 15 of Chapter III. This connection is based on the following two facts: (1) the polynomial ring $S$ is a tensor product of $K$ and the polynomial ring $R$, over $k$; (2) if $\mathfrak{a}$ is an ideal in $R$, then the extension $\mathfrak{a}^e$ of $\mathfrak{a}$ to $S$ may be viewed as the ideal generated in the tensor product $R \otimes K$ by the ideal $\mathfrak{a}$ of $R$ and the ideal (0) of $K$, and hence Theorem 35 of Vol. I, Ch. III, § 14 is applicable. In other words, we have that the residue class ring $S/\mathfrak{a}^e$ is $k$-isomorphic with the tensor product $R/\mathfrak{a} \otimes K$.

The notational conventions will be the same as in Vol. I, Ch. IV, § 8. Ideals in $R$ and in $S$ will be denoted respectively by small and capital German letters. All the formulas (1)–(8) concerning extensions and contraction of ideals, given in Vol. I, Ch. IV, § 8, naturally continue to hold in the present case. However, some of the *inclusions* given there can now be improved to *equalities*. Namely, we now have the following equalities:

(1) $$\mathfrak{a}^{ec} = \mathfrak{a};$$

(2) $$(\mathfrak{a} \cap \mathfrak{b})^e = \mathfrak{a}^e \cap \mathfrak{b}^e$$

(3) $$(\mathfrak{a} : \mathfrak{b})^e = \mathfrak{a}^e : \mathfrak{b}^e,$$

whereas in the general case treated in Vol. I, Ch. IV, § 8, we could only assert that $\mathfrak{a}^{ec} \supset \mathfrak{a}$, $(\mathfrak{a} \cap \mathfrak{b})^e \subset \mathfrak{a}^e \cap \mathfrak{b}^e$ and $(\mathfrak{a} : \mathfrak{b})^e \subset \mathfrak{a}^e : \mathfrak{b}^e$. We shall now prove relations (1)–(3).

We fix once and for always a basis $\{u_i\}$ of $K$ over $k$. This basis may of course have infinitely many elements, and we agree to include the element 1 of $k$ in the basis; let, say, $u_1 = 1$. It is clear that the $u_i$ are also linearly independent over the polynomial ring $R = k[X_1, X_2, \cdots, X_n]$, and every element of $S$ has a unique expression of the form $\sum u_i f_i(X)$, where the $f_i(X)$ belong to $R$ and all but a finite number of the $f_i(X)$ are zero.

Since $S = \sum Ru_i$, it follows that $\mathfrak{a}^e = \sum \mathfrak{a}u_i$. Hence if $z$ is any element of $\mathfrak{a}^e$ then $z = \sum u_i z_i$, $z_i \in \mathfrak{a}$. Now if $z \in R$ then the relation $(z_1 - z) + \sum_{i \neq 1} u_i z_i = 0$ and the linear independence of the $u_i$ over $R$ implies $z = z_1 \in \mathfrak{a}$. We have thus proved that $\mathfrak{a}^{ec} \subset \mathfrak{a}$, and this establishes (1).

To prove (2), let $z \in \mathfrak{a}^e \cap \mathfrak{b}^e = \sum \mathfrak{a}u_i \cap \sum \mathfrak{b}u_i$. Then in the unique expression $z = \sum u_i z_i$ of $z$ as a linear combination of the $u_i$ with coefficients in $R$, the $z_i$ belong both to $\mathfrak{a}$ and to $\mathfrak{b}$. This shows that $z \in (\mathfrak{a} \cap \mathfrak{b})^e$ and establishes (2).

Finally, let $z \in \mathfrak{a}^e : \mathfrak{b}^e$, $z = \sum u_i z_i$, $z_i \in R$. If $b$ is any element of $\mathfrak{b}$ then we must have $zb = \sum u_i z_i b \in \mathfrak{a}^e$, whence $z_i b \in \mathfrak{a}$, $z_i \in \mathfrak{a} : \mathfrak{b}$, showing that $z \in (\mathfrak{a} : \mathfrak{b})^e$. This proves (3).

We observe that relation (1) is also a consequence of the above cited Theorem 35 of Vol. I, Ch. III, § 14. In fact, according to that theorem we have $(\mathfrak{a}, \mathfrak{b}) \cap R = \mathfrak{a}$, where $\mathfrak{b}$ is now the ideal (0) in $K$ and $(\mathfrak{a}, \mathfrak{b})$ is therefore the ideal $\mathfrak{a}^e$.

In view of relation (1), the ring $R/\mathfrak{a}$ may be regarded as a subring of $S/\mathfrak{a}^e$. We shall assume from now on that $\mathfrak{a} \neq R$. In that case we may also regard the field $K$ as being contained in $S/\mathfrak{a}^e$. Furthermore, since $S$ is generated by $X_1, X_2, \cdots, X_n$ and the elements of $K$, and since the $\mathfrak{a}^e$-residues of the $X_i$ belong to $R/\mathfrak{a}$, it follows that $S/\mathfrak{a}^e$ is generated by its two subrings $K$ and $R/\mathfrak{a}$. By the cited Theorem 35 of Vol. I, Ch. III, § 14 the ring $S/\mathfrak{a}^e$ must be a tensor product of $R/\mathfrak{a}$ and $K$, over $k$; in other words, $K$ and $R/\mathfrak{a}$ are linearly disjoint over $k$. This can be verified directly as follows:

Let $v_1, v_2, \cdots, v_q$ be elements of $K$ which are linearly independent over $k$, and assume that we have a relation of the form $\sum v_j z_j \in \mathfrak{a}^e$, where the $z_j$ are in $R$. We have to prove that the $z_j$ belong to the ideal $\mathfrak{a}$. This time we fix a basis $\{u_i\}$ of $K/k$ which includes the elements $v_j$: say, $u_j = v_j$, for $j = 1, 2, \cdots, q$. Then we have $\sum v_j z_j = \sum u_i x_i$, $x_i \in \mathfrak{a}$, and from the linear independence of the $u_i$ over $R$ we deduce that $z_j = x_j \in \mathfrak{a}$, $j = 1, 2, \cdots, q$, as asserted.

Before we proceed with the general case of an arbitrary extension field $K$ of $k$ we need a result concerning the special case in which $K$ is a pure transcendental extension of $k$. For convenience, we adopt from now on the following notation: if $\mathfrak{p}$ is a prime ideal in ring $R$, different from $R$, then we denote by $F(\mathfrak{p})$ the quotient field of $R/\mathfrak{p}$.

THEOREM 35. *Let $\mathfrak{p}$ be a prime ideal in $R$ and $\mathfrak{q}$ a primary ideal in $R$ having $\mathfrak{p}$ as associated prime ideal. If $K$ is a pure transcendental extension of $k$, then $\mathfrak{p}^e$ is a prime ideal, $\mathfrak{q}^e$ is a primary ideal having $\mathfrak{p}^e$ as associated prime ideal, and $\mathfrak{p}^e$ has the same dimension as $\mathfrak{p}$. Furthermore, if $\{t_i\}$ is a set of algebraically independent generators of $K/k$ then $F(\mathfrak{p}^e)=F(\mathfrak{p})(\{t_i\})$ and the $t_i$ are also algebraically independent over $F(\mathfrak{p})$.*

PROOF. Assume that the theorem has already been proved in the case in which $K$ has finite transcendence degree over $k$. It is then easy to see that the theorem holds in the general case. Namely, to prove that $\mathfrak{p}^e$ is prime, assume that we have $F(X)G(X) \in \mathfrak{p}^e$, where $F$ and $G$ belong to $S$. We write $F(X)G(X)=\sum A_i(X)\varphi_i(X)$, where $A_i(X) \in S$ and $\varphi_i(X) \in \mathfrak{p}$. The coefficients of the polynomials $F$, $G$ and $A_i$ belong already to some intermediate field $K'$ between $k$ and $K$ which has finite transcendence degree over $k$ and is itself a pure transcendental extension of $k$. If we use the superscript $e'$ to denote extension of ideals to $R'=K'[X_1, X_2, \cdots, X_n]$, we have then that $F(X)G(X) \in \mathfrak{p}^{e'}$. Hence, by the finite case, either $F$ or $G$ belong to $\mathfrak{p}^{e'}$ and therefore also to $\mathfrak{p}^e$. This shows that $\mathfrak{p}^e$ is prime.

Similarly, to show that $\mathfrak{q}^e$ is primary and has $\mathfrak{p}^e$ as associated prime ideal, we have only to show that if we have a relation of the form $F(X)G(X) \in \mathfrak{q}^e$, where $F$ and $G$ belong to $S$, and if $F(X) \notin \mathfrak{p}^e$ then $G(X) \in \mathfrak{q}^e$ (since the relations $\mathfrak{q}^e \subset \mathfrak{p}^e$ and $\mathfrak{p}^e \subset \sqrt{\mathfrak{q}^e}$ are obvious). Now, this assertion follows again easily by considering a suitable intermediate field $K'$ between $k$ and $K$, having finite transcendence degree over $k$. In a similar way one deals with the other parts of the theorem.

We may therefore assume that $K$ has finite transcendence degree over $k$. This allows us to use induction with respect to the transcendence degree of $K/k$ and reduce the proof of the theorem to the case in which $K$ *is a simple transcendental extension* of $k$. Let then $K=k(u)$, $u$ being a transcendental over $k$.

Let $x_1, x_2, \cdots, x_n$ denote the $\mathfrak{p}$-residues of $X_1, X_2, \cdots, X_n$. We have $S/\mathfrak{p}^e=k(u)[x_1, x_2, \cdots, x_n]$, and $k[x_1, x_2, \cdots, x_n]$ is an integral domain. Hence in order to prove that $\mathfrak{p}^e$ is prime, i.e., that $S/\mathfrak{p}^e$ is an integral domain, we have only to show that $u$ is a transcendental over $k(x_1, x_2, \cdots, x_n)$. In other words, we have to show that if we have a relation of the form $\sum z_iu^i \in \mathfrak{p}^e$, where $z_i \in R$, then $z_i \in \mathfrak{p}$. But this is obvious, since the powers of $u$ are linearly independent over $R$ and since $\mathfrak{p}^e=\sum \mathfrak{p}v_j$, where $\{v_j\}$ is any basis of $K/k$ (choose a basis $\{v_j\}$ which includes the powers of $u$).

We have therefore that $u$ is a transcendental over $F(\mathfrak{p})=k(x_1, x_2, \cdots, x_n)$ and that $F(\mathfrak{p}^e)=F(\mathfrak{p})(u)$. It follows that tr.d. $k(x_1, x_2, \cdots, x_n, u)/k(u)=$ tr.d. $k(x_1, x_2, \cdots, x_n)/k$. This proves that $\dim \mathfrak{p}^e=\dim \mathfrak{p}$ and that $\mathfrak{p}^e$ is a prime ideal.

Let $\mathfrak{P}$ be a prime ideal of $\mathfrak{q}^e$. To complete the proof of the theorem we have only to show that $\mathfrak{P}=\mathfrak{p}^e$. It is clear that $\mathfrak{P}\supset\mathfrak{p}^e$, since $\mathfrak{P}\supset\mathfrak{q}$ and therefore $\mathfrak{P}\supset\mathfrak{p}$. We shall now show that $\mathfrak{P}\subset\mathfrak{p}^e$.

Let $F(X)$ be any element of $\mathfrak{P}$:

$$F(X) = (y_m u^m + y_{m-1}u^{m-1} + \cdots + y_0)/f(u),$$

where $y_i \in R$ and $f(u) \in k[u]$. Since $\mathfrak{P}$ is a prime ideal of $\mathfrak{q}^e$, $\mathfrak{q}^e$ is a proper subset of the ideal $\mathfrak{q}^e : SF$. Let $G(X)$ be an element of this ideal, not contained in $\mathfrak{q}^e$:

$$G(X) = (z_\mu u^\mu + z_{\mu-1}u^{\mu-1} + \cdots + z_0)/g(u),$$

where $z_j \in R$ and $g(u) \in k[u]$. At least one of the polynomials $z_j$ does not belong to $\mathfrak{q}$. If the leading polynomial $z_\mu$ belongs to $\mathfrak{q}$, then we replace $G(X)$ by $G_1(X)=G(X)-z_\mu u^\mu/g(u)$ and observe that also $G_1(X)$ belongs to $\mathfrak{q}^e:SF$ and does not belong to $\mathfrak{q}^e$. We therefore may assume that $z_\mu \notin \mathfrak{q}$. From $F(X)G(X) \in \mathfrak{q}^e$ follows that $y_m z_\mu \in \mathfrak{q}$ and hence $y_m \in \mathfrak{p}$ since $z_\mu \notin \mathfrak{q}$. Since $\mathfrak{p}\subset\mathfrak{P}$, it follows that also the polynomial

$$F_1(X) = (y_{m-1}u^{m-1} + \cdots + y_0)/f(u)$$

belongs to $\mathfrak{P}$ and hence we conclude, as before, that $y_{m-1} \in \mathfrak{p}$. Continuing in this fashion we conclude that all $y_i$ are in $\mathfrak{p}$, whence $F(X) \in \mathfrak{p}^e$. This concludes the proof of the theorem.

Those assertions of the theorem which concern the prime ideal $\mathfrak{p}$ and its extension $\mathfrak{p}^e$ are easy consequences of Vol. I, Ch. III, § 14, Theorem 36 and Corollary. In fact, if we denote by $K'$ the quotient field of $F(\mathfrak{p})$ then, by the cited theorem, the ring $S/\mathfrak{p}^e$ is a subring of the tensor product $K \otimes K'$, and by the corollary to that theorem (Vol. I, Ch. III, § 14, p. 186) the generators $t_i$ are also algebraically independent over $K'$. It follows that $K \otimes K'$ is an integral domain, that the quotient field of $K \otimes K'$ is the purely transcendental extension $K'(\{t_i\})$ of $K'$ and that the transcendence degree of $K'/k$ is the same as the transcendence degree of $K'(\{t_i\})/k(\{t_i\})$. Since it is obvious that the ring $K \otimes K'$ and its subring $S/\mathfrak{p}^e$ have the same quotient field, everything is proved.

We now go back to arbitrary extensions $K$ of $k$ and we prove

THEOREM 36. *If $\mathfrak{q}$ is a primary ideal in $R$ and $\mathfrak{p}=\sqrt{\mathfrak{q}}$ then the prime* *ideals of* $\mathfrak{q}^e$ *are those and only those prime ideals* $\mathfrak{P}$ *in* $S$ *which satisfy the conditions* $\mathfrak{P}^c = \mathfrak{p}$ *and* $\dim \mathfrak{P} = \dim \mathfrak{p}$.

PROOF. Let $\mathfrak{P}$ be a prime ideal of $\mathfrak{q}^e$. We have $\mathfrak{P}^c \supset \mathfrak{q}$, whence $\mathfrak{P}^c \supset \mathfrak{p}$ since $\mathfrak{P}^c$ is prime. We also have $\mathfrak{q}^e : \mathfrak{P} > \mathfrak{q}^e$, and hence *a fortiori* $\mathfrak{q}^e : \mathfrak{P}^{ce} > \mathfrak{q}^e$, since $\mathfrak{P}^{ce} \subset \mathfrak{P}$. By (3) we can write $\mathfrak{q}^e : \mathfrak{P}^{ce} = (\mathfrak{q} : \mathfrak{P}^c)^e$. Hence we have $(\mathfrak{q} : \mathfrak{P}^c)^e > \mathfrak{q}^e$, and therefore taking contractions in $R$ and using (1) we find $\mathfrak{q} : \mathfrak{P}^c > \mathfrak{q}$. Therefore $\mathfrak{P}^c \subset \mathfrak{p}$, showing that $\mathfrak{P}^c = \mathfrak{p}$.

Let $K'$ be an intermediate field between $k$ and $K$ such that $K'$ is a pure transcendental extension of $k$ and $K$ is an algebraic extension of $K'$. We denote by $R'$ the polynomial ring $K'[X_1, X_2, \cdots, X_n]$ and by $\mathfrak{p}'$, $\mathfrak{q}'$ the extended ideals $R'\mathfrak{p}$ and $R'\mathfrak{q}$. The ideal $\mathfrak{q}^e$ is also the extension of $\mathfrak{q}'$ to $S$. Since, by the preceding theorem, $\mathfrak{q}'$ is primary and $\mathfrak{p}' = \sqrt{\mathfrak{q}'}$, it follows, by the preceding part of proof, that $\mathfrak{P} \cap R' = \mathfrak{p}'$. Since $K$ is algebraic over $K'$, $S$ is integral over $R'$. Hence $\dim \mathfrak{P} = \dim \mathfrak{p}'$ (Vol. I, Ch. V, § 2, Lemma 1). Since, by the preceding theorem, we have $\dim \mathfrak{p}' = \dim \mathfrak{p}$, we conclude that $\dim \mathfrak{P} = \dim \mathfrak{p}$.

Conversely, assume that $\mathfrak{P}$ is a prime ideal in $S$ such that $\mathfrak{P}^c = \mathfrak{p}$ and $\dim \mathfrak{P} = \dim \mathfrak{p}$. Since $\mathfrak{P} \supset \mathfrak{p} \supset \mathfrak{q}$, we have $\mathfrak{P} \supset \mathfrak{q}^e$ and therefore $\mathfrak{P}$ must contain at least one prime ideal of $\mathfrak{q}^e$. However, if $\mathfrak{P}_1$ is a prime ideal of $\mathfrak{q}^e$ contained in $\mathfrak{P}$, then $\mathfrak{P}_1$ must coincide with $\mathfrak{P}$, since $\dim \mathfrak{P}_1 = \dim \mathfrak{P}$ ($\dim \mathfrak{P}_1 = \dim \mathfrak{p}$, by preceding part of the proof, and $\dim \mathfrak{P} = \dim \mathfrak{p}$, by hypothesis). This completes the proof of the theorem.

COROLLARY 1. *If* $\mathfrak{a}$ *is an unmixed ideal in* $R$, *then also* $\mathfrak{a}^e$ *is an unmixed ideal, of the same dimension as* $\mathfrak{a}$.

If $\mathfrak{a} = \mathfrak{q}$ is a primary ideal, then all the prime ideals of $\mathfrak{q}^e$ have the same dimension, equal to the dimension of $\mathfrak{p} = \sqrt{\mathfrak{q}}$. Thus $\mathfrak{q}^e$ is unmixed, of the same dimension as $\mathfrak{q}$. If now $\mathfrak{a}$ is an arbitrary ideal, the corollary follows from relation (2).

COROLLARY 2. *If* $K$ *is a purely inseparable extension of* $k$ *and* $\mathfrak{q}$ *is a primary ideal in* $R$, *then also* $\mathfrak{q}^e$ *is primary.*

For let $\mathfrak{P}$ be a prime ideal of $\mathfrak{q}^e$. If $F \in \mathfrak{P}$, then for some integer $f \geqq 0$ the polynomial $F^{p^f}$ is contained in $R$ and therefore belongs to $\mathfrak{P}^c$, i.e., to $\mathfrak{p}$, where $\mathfrak{p} = \sqrt{\mathfrak{q}}$. Conversely, if $F$ is a polynomial in $S = K[X_1, X_2, \cdots, X_n]$ such that $F^{p^f}$ belongs to $\mathfrak{p}$ for some integer $f \geqq 0$, then $F^{p^f} \in \mathfrak{P}$ and hence $F \in \mathfrak{P}$, since $\mathfrak{P}$ is prime. Hence $\mathfrak{P}$ is uniquely characterized as the set of all polynomials $F$ in $S$ such that $F^{p^f} \in \mathfrak{p}$ for some integer $f \geqq 0$. Thus $\mathfrak{q}^e$ has only one prime ideal and is therefore primary.

COROLLARY 3. *If* $\mathfrak{q}$ *is a primary ideal in* $R$ *and* $\mathfrak{p} = \sqrt{\mathfrak{q}}$ *then the prime ideals of* $\mathfrak{q}^e$ *coincide with the prime ideals of* $\mathfrak{p}^e$.

Obvious.

We shall now study in more detail the behavior of a prime ideal $\mathfrak{p}$ in $R$ under extension to $S$. We give the following definitions:

(1) $\mathfrak{p}$ *splits* in $S$ if $\mathfrak{p}^e$ is not a primary ideal.

(2) $\mathfrak{p}$ is *unramified* in $S$ if $\mathfrak{p}^e$ is an intersection of prime ideals (or, equivalently, if $\mathfrak{p}^e = \sqrt{\mathfrak{p}^e}$). In the contrary case $\mathfrak{p}$ is said to be *ramified* in $S$.

(3) $\mathfrak{p}$ is *absolutely prime* if for *every* extension $K$ of $k$ the ideal $\mathfrak{p}^e$ is prime. In other words: $\mathfrak{p}$ is absolutely prime if it is unramified and does not split, for any extension $K$ of $k$.

(4) $\mathfrak{p}$ is *quasi-absolutely prime* if $\mathfrak{p}^e$ is a primary ideal for *any* extension $K$ of $k$.

(5) $\mathfrak{p}$ is *absolutely unramified* if $\mathfrak{p}$ is unramified for any extension $K$ of $k$.

Since the ring $S/\mathfrak{p}^e$ is the tensor product $K \otimes R/\mathfrak{p}$ over $k$, we can state the following lemma:

LEMMA. *If $\mathfrak{p}$ is a prime ideal in $R$ then*

(1) *$\mathfrak{p}$ does not split in $S$ if and only if every zero divisor in $K \otimes R/\mathfrak{p}$ is nilpotent (or—equivalently—if and only if the zero ideal in the tensor product $K \otimes R/\mathfrak{p}$ is primary);*

(2) *$\mathfrak{p}$ is unramified in $S$ if and only if zero is the only nilpotent element in $K \otimes R/\mathfrak{p}$;*

(3) *$\mathfrak{p}^e$ is a prime ideal if and only if $K \otimes R/\mathfrak{p}$ is an integral domain.*

In Vol. I, Ch. III, § 15 (Theorem 39) we have proved that if $K$ and $K'$ are two integral domains containing a field $k$ and if the quotient field of one of these domains is separable over $k$, then $K \otimes_k K'$ has no proper nilpotent elements. This yields at once the following consequence of the above lemma:

COROLLARY. *If either $K$ or the quotient field $F(\mathfrak{p})$ of $R/\mathfrak{p}$ is separable over $k$, then $\mathfrak{p}$ is unramified in $S$. In particular, if $F(\mathfrak{p})$ is separable over $k$, then $\mathfrak{p}$ is an absolutely unramified prime ideal. If $k$ is a perfect field (in particular, if $k$ is a field of characteristic zero) then every prime ideal in the polynomial ring $R = k[X_1, X_2, \cdots, X_n]$ is absolutely unramified.*

The sufficient condition for absolutely unramified prime ideals, given in the above corollary, is actually also a necessary condition. We have therefore the following

THEOREM 37. *A necessary and sufficient condition for a prime ideal $\mathfrak{p}$ in the polynomial ring $R = k[X_1, X_2, \cdots, X_n]$ to be absolutely unramified is that the quotient field of $R/\mathfrak{p}$ be separable over $k$.*

PROOF. If $\mathfrak{p}$ is absolutely unramified we take for $K$ the field $k^{p^{-1}}$ (we may assume that $p \neq 0$). Let $x_1, x_2, \cdots, x_n$ be the $\mathfrak{p}$-residues of

$X_1 X_2, \cdots, X_n$ respectively. We have $R/\mathfrak{p}=k[x_1, x_2, \cdots, x_n]$ and $S/\mathfrak{p}^e=K[x_1, x_2, \cdots, x_n]=K \otimes k[x_1, x_2, \cdots, x_n]$.

By assumption, $\mathfrak{p}^e$ is an intersection of prime ideals. By Corollary 2 to Theorem 36, $\mathfrak{p}^e$ is a primary ideal. Hence $\mathfrak{p}^e$ is a prime ideal, and $K[x_1, x_2, \cdots, x_n]$ is an integral domain. By the definition of tensor products, $R/\mathfrak{p}$ and $K$ are linearly disjoint in $K \otimes R/\mathfrak{p}$. We have therefore that the quotient field of $R/\mathfrak{p}$ and the field $k^{p^{-1}}$ are linearly disjoint, over $k$, in their common overfield $k^{p^{-1}}(x_1, x_2, \cdots, x_n)$, and the theorem now follows from the definition of separability (Vol. I, Ch. II, § 15, p. 113). Q.E.D.

We now characterize the prime ideals which are quasi-absolutely prime. If $K$ is a subfield of a field $\Omega$, we say, as in Vol. I, Ch. III, § 15 (p. 196), that $K$ is quasi-maximally algebraic in $\Omega$ if every element of $\Omega$ which is separable algebraic over $K$ belongs to $K$ (or equivalently: if every element of $\Omega$ which is algebraic over $K$ is purely inseparable over $K$). We say that $K$ is *maximally algebraic* in $\Omega$ if $K$ is algebraically closed *in* $\Omega$, i.e., if every element of $\Omega$ which is algebraic over $K$ belongs to $K$.

THEOREM 38. *If $\mathfrak{p}$ is a prime ideal in a polynomial ring $R=k[X_1, X_2, \cdots, X_n]$, then $\mathfrak{p}$ is quasi-absolutely prime if and only if $k$ is q.m.a. in the field $F(\mathfrak{p})$ (=quotient field of $R/\mathfrak{p}$).*

PROOF. Assume that $\mathfrak{p}$ is quasi-absolutely prime and let $\alpha$ be an element of $F(\mathfrak{p})$ which is separable algebraic over $k$. We shall show that $\alpha \in k$.

Let $G(T)=T^q+a_{q-1}T^{q-1}+\cdots+a_0$, $a_i \in k$, be the minimal polynomial of $\alpha$ over $k$. We take for $K$ a normal extension of $k$ such that $G(T)$ factors completely in linear factors over $K$:

$$G(T) = (T-c'_1)(T-c'_2)\cdots(T-c'_q), \quad c'_i \in K. \tag{4}$$

Since $\alpha \in F(\mathfrak{p})$, there exist polynomials $A(X)$, $B(X)$ in $R$ such that $\alpha=A(x)/B(x)$, where $x_1, x_2, \cdots, x_n$ are the $\mathfrak{p}$-residues of $X_1, X_2, \cdots, X_n$ and $B(x)\neq 0$. Upon substitution in (4) and after clearing denominators, the equation $G(\alpha)=0$ yields the relation

$$\prod_{i=1}^{q}(A(x)-c'_iB(x)) = 0.$$

This is to be viewed as a relation in the ring $K[X]/\mathfrak{p}^e$ and is therefore equivalent to

$$\prod_{i=1}^{q}(A(X)-c'_iB(X)) \in \mathfrak{p}^e. \tag{5}$$

By assumption, $\mathfrak{p}^e$ is a primary ideal, and its radical $\mathfrak{P}$ is therefore a prime ideal. Hence, at least one of the $q$ factors on the left-hand side of (5) must belong to $\mathfrak{P}$. Now, $\mathfrak{p}^e$, and therefore also $\mathfrak{P}$, is invariant under all the $k$-automorphisms of $K$ (more precisely: under all the $k[X]$ – automorphism of $K[X]$ which are extensions of $k$-automorphisms of $K$), and the $k$-automorphisms of $K$ act transitively on the $q$ roots $c'_i$ of $G(T)$. *Hence all the $q$ factors $A(X)-c'_iB(X)$ belong to $\mathfrak{P}$.* Now, the $q$ roots $c'_i$ are distinct, since $G(T)$ is a separable polynomial. If $q$ were greater than 1, it would then follow that $B(X)$ belongs to $\mathfrak{P}$. Then some power of $B(X)$ would belong to $\mathfrak{p}^e$ and hence also to $\mathfrak{p}$, since $\mathfrak{p}^{ec}=\mathfrak{p}$. Hence $B(X)$ itself would belong to $\mathfrak{p}$, in contradiction with the fact that $B(x)\neq 0$. Hence $q$ must be equal to 1, and this proves that $\alpha \in k$.

We now assume that $k$ is q.m.a. in $F(\mathfrak{p})$. We consider an arbitrary extension $K$ of $k$ and we must prove that the extended ideal $\mathfrak{p}^e$ of $\mathfrak{p}$ in $K[X_1, X_2, \cdots, X_n]$ is primary. Let $K'$ be an intermediate field between $k$ and $K$ such that $K'$ is a pure transcendental extension of $k$ and $K$ is an algebraic extension of $K'$. By Theorem 35, the ideal $\mathfrak{p}'=\mathfrak{p}\cdot K'[X]$ is prime. Furthermore, the field $F(\mathfrak{p}')$ (=quotient field of $K'[X]/\mathfrak{p}'$) is a pure transcendental extension of the field $F(\mathfrak{p})$, and if say $\{t_i\}$ is a set of generators of $K'$ over $k$ consisting of algebraically independent elements over $k$, then $F(\mathfrak{p}')=F(\mathfrak{p})(\{t_i\})$, and the $t_i$ are also algebraically independent over $F(\mathfrak{p})$ (Theorem 35). Hence, by the lemma proved in Vol. I, Ch. III, § 15 (p. 196), the field $K'$ is q.m.a. in $F(\mathfrak{p}')$. Since $\mathfrak{p}^e$ is also the extension of $\mathfrak{p}'$ to $K[X]$, we see that we have now achieved a reduction to the case of ground fields $K'$ and $K$ in which the bigger ground field $K$ is an algebraic extension of the smaller one, $K'$. We may therefore assume that *$K$ is an algebraic extension of $k$.*

We fix a prime ideal $\mathfrak{P}$ of $\mathfrak{p}^e$. To show that $\mathfrak{p}^e$ is primary we have only to show that $\mathfrak{P}=\sqrt{\mathfrak{p}^e}$. It will be sufficient to show that $\mathfrak{P}\subset\sqrt{\mathfrak{p}^e}$, since the opposite inclusion is obvious. Let $F(X)$ be any element of $\mathfrak{P}$ and let $\xi_i$ denote the $\mathfrak{P}$-residue of $X_i$. Then $\xi_i$ is also the $\mathfrak{p}$-residue of $X_i$, since $\mathfrak{P}^c=\mathfrak{p}$ (see Theorem 36), and we have $F(\xi)=0$. We fix a suitable power $p^s$ of the characteristic $p$ such that the coefficients of the polynomial $G(X)=[F(X)]^{p^s}$ are separable algebraic over $k$. Let $\alpha$ be a primitive element of the field generated by the coefficients of $G(X)$ over $k$ and let $f(T)$ be the minimal polynomial of $\alpha$ over $k$. If $q$ is the degree of $f(T)$, then we can write $G(X)$ in the form

$$(6) \qquad G(X) = \sum_{j=0}^{q-1} G_j(X)\alpha^j, \quad G_j(X) \in k[X].$$

We have $G(\xi)=0$, i.e.,

$$(7) \qquad \sum_{j=0}^{q-1} G_j(\xi)\alpha^j = 0.$$

If the coefficients $G_j(\xi)$ in (7) are not all zero, then (7) is a relation of algebraic dependence for $\alpha$ over $F(\mathfrak{p})$ $(=k(\xi))$ and it is of degree $<q$. However, *the assumption that $k$ is q.m.a. in $F(\mathfrak{p})$ implies that the polynomial $f(T)$ remains irreducible in $F(\mathfrak{p})[T]$.* This follows from the fact that $f(T)$ is a separable polynomial. In fact, if $f_1(T)$ is a factor of $f(T)$ in $F(\mathfrak{p})[T]$ and if we assume that the leading coefficient of $f_1(T)$ is 1, then the coefficients of $f_1(T)$ are elements of $F(\mathfrak{p})$ which are separable algebraic over $k$ (since these coefficients belong to a decomposition field of the separable polynomial $f(T)$) and therefore must belong to $k$. Thus $f_1(T)$ must divide $f(T)$ already in $k[T]$ and therefore must coincide with $f(T)$. From the irreducibility of $f(T)$ over $F(\mathfrak{p})$ follows that all the coefficients $G_j(\xi)$ in (7) must be zero. That signifies that the polynomials $G_j(X)$ belong to $\mathfrak{p}$. Hence, by (6), $G(X) \in \mathfrak{p}^e$, and consequently $F(X) \in \sqrt{\mathfrak{p}^e}$, since $G(X)$ is the $p^s$-th power of $F(X)$. This completes the proof of Theorem 38.

The preceding theorem can also be derived from two basic theorems on free joins of integral domains, namely Theorems 38 and 40 of Vol. I, Ch. III, § 15. We first observe that by the above lemma and by Theorem 38 of Vol. I, Ch. III, § 15 and its corollary 2 (Vol. I, p. 195) it follows that $\mathfrak{p}$ does not split in $S$ if and only if $R/\mathfrak{p}$ and $K$ are quasi-linearly disjoint over $k$. Hence by Theorem 40 of Vol. I, Ch. III, § 15, it follows at once that if $k$ is q.m.a. in $F(\mathfrak{p})$ then $\mathfrak{p}$ is quasi-absolutely prime. Conversely, if $\mathfrak{p}$ is quasi-absolutely prime and if $\alpha$ is an element of $F(\mathfrak{p})$ which is separable algebraic over $k$, then we take for $K$ the field $k(\alpha)$ and we then conclude, by Theorem 40 of Vol. I, Ch. III, § 15, that $F(\mathfrak{p})$ and $k(\alpha)$ must be quasi-linearly disjoint over $k$. Now suppose that $\alpha$ does not belong to $k$. Then 1 and $\alpha$ are linearly independent over $k$. Since $\alpha$ is separable algebraic over $k$, it follows that for any integer $s$ the elements 1 and $\alpha^{p^s}$ are also linearly independent over $k$ (see Vol. I, Ch. II, § 23, Theorem 8). By the quasi-linear disjointness of $k(\alpha)$ and $F(\mathfrak{p})$ over $k$ it would then follow that 1 and $\alpha$ are also linearly independent over $F(\mathfrak{p})$, and this is in contradiction with the fact that $\alpha$ belongs to $F(\mathfrak{p})$.

From the preceding results we obtain at once a characterization of absolutely prime ideals. Let us say that a field $F$ is a *regular extension* of a subfield $k$ of $F$ if the following two conditions are satisfied: (1) $F$ is a separable extension of $k$ and (2) $k$ is maximally algebraic in $F$. We observe that in the presence of condition (1), condition (2) can be replaced by the weaker condition (2′) that $k$ be q.m.a. in $F$. For, if (1) and (2′) hold and $\alpha$ is any element of $F$ which is algebraic over $k$, then also it follows from the very definition of separability in terms of linear disjointness (Vol. I, Ch. II, § 15, p. 113) that $k(\alpha)$, as a subfield of $F$, is also separable over $k$. Hence $\alpha$ is separable algebraic over $k$ and thus belongs to $k$. This shows that (2) holds.

THEOREM 39. *Let $\mathfrak{p}$ be a prime ideal in a polynomial ring $R=k[X_1, X_2, \cdots, X_n]$. A necessary and sufficient condition that $\mathfrak{p}$ be absolutely prime is that the field $F(\mathfrak{p})$ be a regular extension of $k$.*

PROOF. It is clear that $\mathfrak{p}$ is absolutely prime if and only if $\mathfrak{p}$ is both quasi-absolutely prime and absolutely unramified. Our theorem is therefore a direct consequence of Theorems 37 and 38, in view of the remark just made above in regard to the equivalence of the conditions (2) and (2′) (in presence of condition (1)).

REMARK. The results derived in this section give us information not only about the behavior of a given prime ideal $\mathfrak{p}$ under various extensions $K$ of the ground field but also about the behavior of the various prime ideals $\mathfrak{p}$ in $R$ under a *fixed* extension $K$ of $k$. Thus we have shown that (1) if $K$ is a pure transcendental extension of $k$ then $\mathfrak{p}^e$ is prime for every $\mathfrak{p}$ (Theorem 35); (2) if $K$ is a separable extension of $k$ then $\mathfrak{p}^e$ is an intersection of prime ideals for every $\mathfrak{p}$ (Corollary of Lemma); (3) and finally, if $K$ is a pure inseparable extension of $k$ then $\mathfrak{p}^e$ is primary for every $\mathfrak{p}$. To these results we can now add the following: (4) *If $k$ is q.m.a. in $K$ then $\mathfrak{p}^e$ is primary for every $\mathfrak{p}$;* (5) If $K$ is a regular extension of $k$ then $\mathfrak{p}^e$ is prime for every $\mathfrak{p}$. (4) follows directly from Theorem 40 of Vol. I, Ch. III, § 15 (but could also be derived from the results established in this section).

## § 12. Characteristic functions of graded modules and homogeneous ideals.

Let $R$ be a graded ring (§ 2). We recall that if $R_q$ denotes the set of all homogeneous elements of $R$, of degree $q$, we have $R=\sum_{-\infty}^{\infty} R_q$, where the sum is direct. In this section we restrict ourselves to graded rings for which we have $R_q=(0)$ for $q<0$. We also recall that $R_qR_r\subset R_{q+r}$.

A *graded module* $M$ over $R$ is a module $M$, together with a *direct* sum decomposition

$$M = \sum_{-\infty}^{+\infty} M_q$$

of the additive group of $M$, such that, for every pair of integers $(q, r)$, we have

$$R_r M_q \subset M_{q+r}.$$

The elements of $M_q$ are said to be *homogeneous of degree q*. Given any element $x \in M$, we can write, in a unique way,

$$x = \sum_{-\infty}^{+\infty} x_q,$$

where $x_q \in M_q$ and where all terms, except a finite number, are zero. The element $x_q$ is called the *homogeneous component of degree q* of $x$. The notation $M_q$ for the additive group of homogeneous elements of degree $q$ of $M$ will be used without further warning.

A submodule $N$ of $M$ is said to be *homogeneous* if the relation $x \in N$ implies that all the homogeneous components of $x$ belong to $N$.

The homogeneous submodules of $R$, where $R$ is considered as a module over itself, are obviously its homogeneous *ideals* (§ 2). As in the case of ideals (Theorem 7, § 2) one proves that, in order for a submodule $N$ of a graded module $M$ to be homogeneous, it is necessary and sufficient that $N$ be generated by homogeneous elements of $M$. It is a straightforward matter to verify that a homogeneous submodule $N$ of a graded module $M$ is itself a *graded module*, and that the difference module $M-N$ is also a graded module. The proof is the same as that of Lemma 1, part (b), § 2.

Given two graded $R$-modules $M$ and $M'$ and an integer $d$, a homomorphism $\theta$ of $M$ into $M'$ is said to be *homogeneous of degree d* if $\theta(M_q) \subset M'_{d+q}$ for every $q$ (i.e., if the image of any homogeneous element of degree $q$ of $M$ is a homogeneous element of degree $d+q$ in $M'$). For example, if $a_d$ is a homogeneous element of degree $d$ in $R$, the mapping $x \rightarrow a_d x$ of $M$ into itself is a homogeneous homomorphism of degree $d$.

If $\theta$ is a *homogeneous homomorphism* of degree $d$ of the graded module $M$ into the graded module $M'$, then the *kernel* $\theta^{-1}(0)$ of $\theta$ is a homogeneous submodule of $M$, and the *image* $\theta(M)$ is a homogeneous submodule of $M'$. In fact, if $\theta(\sum x_q)=0$ $(x_q \in M_q)$, we have $\sum \theta(x_q)=0$. As the $\theta(x_q)$ are homogeneous elements of distinct degrees, this relation implies $\theta(x_q)=0$ for every $q$, whence $x_q \in \theta^{-1}(0)$, and the kernel $\theta^{-1}(0)$ is homogeneous. Similarly, if $y \in \theta(M)$, then $y=\theta(\sum x_q)$ $(x_q \in M_q)$, $y=\sum \theta(x_q)$ is the decomposition of $y$ in homogeneous components, and these components belong to $\theta(M)$. Thus the image $\theta(M)$ is homogeneous. The difference module $M'-\theta(M)$ is called the *cokernel* of $\theta$; it is also a graded module.

THEOREM 40. *Let $A$ be a ring and $M$ a graded module over the polynomial ring $R=A[X_1, \cdots, X_n]$. Let $M=\sum M_q$ be the direct sum decomposition of $M$. Then each $M_q$ is an $A$-module. If, furthermore, $M$ is a finitely generated $R$-module, then each $M_q$ is a finitely generated $A$-module.*

PROOF. Since $A$ coincides with the set $R_0$ of elements of degree 0 of $R$, the first assertion is clear. If, now, $M$ is finitely generated, $M$ admits a finite set $\{y_1, \cdots, y_s\}$ of *homogeneous* generators, since the homogeneous components of the elements of any basis of $M$ themselves generate $M$. This being so, each element $y$ of $M_q$ may be written in the form $y=\sum P_j(X)\cdot y_j$, where $P_j(X)$ is a form of degree $q-d^0(y_j)$ where $d^0$ denotes the degree of a homogeneous element. Then the elements $m_\alpha(X)y_j$ ($m_\alpha(X)$: monomials of degree $q-d^0(y_j)$) constitute a finite set of generators of the $A$-module $M_q$.

THEOREM 41 (Hilbert-Serre). *Let $A$ be a ring satisfying the descending chain condition (d.c.c.), $M$ a finitely generated graded module over $R=A[X_1, \cdots, X_n]$ and $M=\sum M_q$ the direct sum decomposition of $M$. Then $M_q$, considered as an $A$-module, has a finite length $\varphi_M(q)$. For sufficiently large $q$, the function $\varphi_M(q)$ is a polynomial in $q$ whose degree is at most $n-1$.*†

PROOF. The fact that the length $\varphi_M(q)$ of $M_q$ is finite follows immediately from Theorem 40 and from the fact that $A$ is a ring with d.c.c. In order to prove that $\varphi_M(q)$ is a polynomial in $q$ for $q$ large enough, we proceed by induction on the number $n$ of variables.

For $n=0$, $R$ is reduced to $A=R_0$. Since $M$ admits a finite system of homogeneous generators $\{y_1, \cdots, y_s\}$, the non-zero homogeneous elements of $M$ can only be of degree $d^0(y_i)$ for some $i$. Thus, for $q>\max(d^0(y_i))$, we have $M_q=(0)$, whence $\varphi_M(q)=0$. This proves our assertion for $n=0$.

In the general case, consider the homomorphism $\theta: y \to X_n \cdot y$ of $M$ into $M$. It is a homogeneous homomorphism, of degree 1. Let $N=\theta^{-1}(0)$ be its kernel, and let $P=M-\theta(M)$ (the difference module) be its co-kernel. Both $N$ and $P$ have $X_n$ in their *orders* (see Vol. I, Ch. III, § 6). They can therefore be construed as graded modules *over* $K[X_1, \cdots X_{n-1}]$ (see Vol. I, Ch. III, § 6) and the induction hypothesis may be applied to them. We write $N=\sum N_q$, $P=\sum P_q$.

Consider the following sequence of modules and homomorphisms:

$$0 \to N \xrightarrow{i} M \xrightarrow{\theta} M \xrightarrow{j} P \to 0, \tag{1}$$

† For the purposes of this theorem we attach to the zero polynomial the degree $-1$. The proof of the Hilbert theorem given below is essentially due to Serre, at least in its cohomological formulation.

where all homomorphisms, except the one in the middle, are natural homomorphisms. ($i$ is the inclusion isomorphism into $M$, $j$ the canonical homomorphism of $M$ onto the difference module $P$.) In this sequence, the image of each homomorphism is equal to the kernel of the following one. In the terminology of cohomological algebra this is expressed by saying that the sequence (1) is *exact*. If we start from the homogeneous elements of degree $q$ of $N$ or $M$, we get the exact sequence

$$(2) \qquad 0 \rightarrow N_q \xrightarrow{i} M_q \xrightarrow{\theta} M_{q+1} \xrightarrow{j} P_{q+1} \rightarrow 0.$$

We now use the following lemma (a proof of this lemma will be given immediately after the proof of the theorem):

LEMMA. *Let $0 \rightarrow E_1 \rightarrow E_2 \rightarrow \cdots \rightarrow E_n \rightarrow 0$ be an exact sequence of $A$-modules, having finite lengths $\ell(E_i)$. Then the alternating sum $\ell(E_1) - \ell(E_2) + \ell(E_3) - \cdots + (-1)^{n-1}\ell(E_n)$ of the lengths of these modules is equal to* 0.

In our particular case, and with the notation which has been introduced before, the lemma gives the relation

$$(3) \qquad \varphi_M(q+1) - \varphi_M(q) = \varphi_P(q+1) - \varphi_N(q).$$

By the induction hypothesis, for $q$ large enough $\varphi_P(q+1)$ and $\varphi_N(q)$ are polynomials in $q$, of degree at most $n-2$. Hence the first difference $\varphi_M(q+1) - \varphi_M(q)$ is, for $q$ large, a polynomial of degree at most $n-2$ in $q$, and this polynomial takes integral values for all large values of $q$.

We now observe that since $q^s = s!\binom{q}{s} +$ a polynomial in $q$, of degree $s-1$, it follows that every polynomial $f(q)$ in $q$, of degree less than or equal than a given integer $d$, can be written in the form

$$f(q) = c_0\binom{q}{d} + c_1\binom{q}{d-1} + \cdots + c_{d-1}\binom{q}{1} + c_d,$$

with suitable coefficients $c_i$. Now, we assert that if $f(q)$ takes integral values for all large values of $q$ then the coefficients $c_i$ are integers. We prove this by induction with respect to $d$ since the case $d = 0$ is trivial. If we make use of the identity

$$(4) \qquad \binom{h+1}{s} - \binom{h}{s} = \binom{h}{s-1}$$

we find the following expression for the first difference $f(q+1) - f(q)$:

$$f(q+1) - f(q) = c_0\binom{q}{d-1} + c_1\binom{q}{d-2} + \cdots + c_{d-1}.$$

Since the first difference also takes integral values for large $q$ and since it is a polynomial of degree at most $d-1$, it follows from our induction hypothesis that $c_0, c_1, \cdots, c_{d-1}$ are integers. Since the binomial coefficients $\binom{q}{s}$ are integers for all $q$ and $s$ it follows from the above expression of $f(q)$ that also $c_d$ is an integer, and this proves our assertion.

Applying this result to the first difference $\varphi_M(m)-\varphi_M(m-1)$, where $m$ is a sufficiently high integer, say $m \geqq N$, we can write

$$(5) \quad \varphi_M(m)-\varphi_M(m-1) = a_0\binom{m-1}{n-2}+a_1\binom{m-1}{n-3}+\cdots+a_{n-2}, \quad (m \geqq N)$$

where the $a_i$ are integers. Let us also write

$$(6) \quad \varphi_M(m)-\varphi_M(m-1) = a_0\binom{m-1}{n-2}+a_1\binom{m-1}{n-3}+\cdots+a_{n-2}+c_m, \qquad m = 2, 3, \cdots, N-1,$$

$$\varphi_M(1) = c_1,$$

where we set $\binom{t}{s}=0$ if $t<s$ and where $c_1, c_2, \cdots, c_{N-1}$ are integers. If we add relations (4) for $h=q-1, q-2, \cdots, s-1$ we find the identity

$$\binom{q}{s} = \binom{q-1}{s-1}+\binom{q-2}{s-1}+\cdots+\binom{s}{s-1}+1,$$

and using this identity we find, by adding the relations (5) for $m=q, q-1, \cdots, N$ and the $N-1$ relations (6):

$$\varphi_M(q) = a_0\binom{q}{n-1}+a_1\binom{q}{n-2}+\cdots+a_{n-2}\binom{q}{1}+a_{n-1}, \quad (q \geqq N)$$

where $a_{n-1}$ is a suitable constant, necessarily an integer, since $a_0, a_1, \cdots, a_{n-2}$ are integers and since $\varphi_M(q)$ takes integral values for all large $q$. This completes the proof of the theorem.

We now give a proof of the lemma. We consider the homomorphism $f_i\colon E_i \to E_{i+1}$ $(i=1, 2, \cdots, n)$ where $E_{n+1}$ is the module (0) and $f_n$ is the zero homomorphism. Since $f_i(E_i)$ is isomorphic with $E_i/f_i^{-1}(0)$, we have the relation

$$\ell(E_i) = \ell(f_i(E_i))+\ell(f_i^{-1}(0)), \quad i = 1, 2, \cdots, n.$$

Since the sequence is exact, this relation may also be written as follows:

$$\ell(E_i) = \ell(f_{i+1}^{-1}(0))+\ell(f_i^{-1}(0)), \quad i = 1, 2, \cdots, n-1.$$
$$\ell(E_n) = \ell(f_n^{-1}(0)).$$

Thus the alternating sum $\ell(E_1)-\ell(E_2)+\cdots+(-1)^{n-1}\ell(E_n)$ is equal to $\ell(f_1^{-1}(0))$. Since $f_1$ is an isomorphism, we have $f_1^{-1}(0)=0$, and this completes the proof of the lemma.

REMARK (1). The most important case in which Theorem 41 may be applied is the one in which $A$ is a *field* $k$ and $M$ is a residue class ring $k[X_1, \cdots, X_n]/\mathfrak{A}$ of $k[X_1, \cdots, X_n]$ modulo a *homogeneous ideal* $\mathfrak{A}$. Then the function $\varphi_M(q)$ is denoted by $\chi(\mathfrak{A}; q)$ and is called the *characteristic function* of the ideal $\mathfrak{A}$. The integer $\chi(\mathfrak{A}; q)$ is the greatest number of forms of degree $q$ in $k[X_1, X_2, \cdots X_n]$ which are linearly independent modulo $\mathfrak{A}$ over $k$.

REMARK (2). Let $E$ and $F$ be two homogeneous submodules of the graded module $M$. Since $(E+F)_q = E_q + F_q$ and $(E \cap F)_q = E_q \cap F_q$, the $\mathfrak{A}$-modules $(E+F)_q/E_q$ and $F_q/(E \cap F)_q$ are isomorphic. Therefore we have for every $q$, the relation:

$$\varphi_E(q) + \varphi_F(q) = \varphi_{E+F}(q) + \varphi_{E \cap F}(q). \tag{7}$$

In the case of two homogeneous ideals $\mathfrak{A}$, $\mathfrak{B}$ in the polynomial ring $R$, (7) gives the relation

$$\chi(\mathfrak{A}; q) + \chi(\mathfrak{B}; q) = \chi(\mathfrak{A}+\mathfrak{B}; q) + \chi(\mathfrak{A} \cap \mathfrak{B}; q). \tag{8}$$

REMARK (3). Let $E$ be a homogeneous submodule of a graded module $M$. Since $E_q \subset M_q$, we have the relation

$$\varphi_E(q) \leqq \varphi_M(q). \tag{9}$$

In the case of two homogeneous ideals $\mathfrak{A}$, $\mathfrak{B}$ such that $\mathfrak{A} \subset \mathfrak{B}$, relation (9) gives:

$$\chi(\mathfrak{A}; q) \geqq \chi(\mathfrak{B}; q). \tag{10}$$

REMARK (4). It is often necessary to distinguish the characteristic function $\varphi_E(q)$ (or $\chi(\mathfrak{A}; q)$) from the polynomial which is equal to this function for $q$ large enough. In such a case we denote this polynomial by $\bar{\varphi}_E$ (or $\bar{\chi}(\mathfrak{A}; q)$). We call this polynomial the *characteristic polynomial* of $E$ (or $\mathfrak{A}$).

The degree of the characteristic polynomial of a homogeneous ideal $\mathfrak{A}$ is closely related to the dimension of $\mathfrak{A}$. More precisely, we have the following theorem:

THEOREM 42. *Let $\mathfrak{A}$ be a homogeneous ideal in $k[X_1 X_2, \cdots, X_n]$. Then the degree of the characteristic polynomial $\bar{\chi}(\mathfrak{A}; q)$ of $\mathfrak{A}$ is equal to the projective dimension of $\mathfrak{A}$* (see § 4).

PROOF. Theorem 42 is a particular case of:

THEOREM 42′. *Let $E$ be a finitely generated graded module over $R = k[X_1, \cdots, X_n]$ and $F$ a homogeneous submodule of $E$. Then the degree of $\bar{\varphi}_{E/F}(q)$ is equal to the greatest projective dimension of the associated prime ideals $\mathfrak{p}_i$ of the submodule $F$* (see Vol. I, Ch. IV, Appendix).

PROOF. We recall that the radical $\mathfrak{w}$ of the submodule $F$ is the set of all elements $a \in R$ for which there exists an exponent $s$ such that $a^s E \subset F$ (Vol. I, Ch. IV, Appendix). As in Theorem 8, § 2, it is easily seen that $\mathfrak{w}$ is a *homogeneous* ideal in $R$. It follows from Vol. I, Ch. IV, Appendix, that the isolated prime ideals $\mathfrak{p}_i$ of $F$ are the (necessarily isolated) prime ideals of $\mathfrak{w}$. These ideals are therefore homogeneous (Theorem 9, Corollary, § 2). Let $d-1$ be the greatest integer among their projective dimensions. By normalization (§ 9, Theorem 31) there exists a homogeneous system of integrity $\{G_1, \cdots, G_n)$ composed of forms of like degree $h$ in $k[X_1, \cdots, X_n]$ such that $\mathfrak{w} \cap k[G_1, \cdots, G_n] = (G_{d+1}, \cdots, G_n)$. Since $\mathfrak{w}$ is the radical of $F$, there exists an exponent $h'$ such that $G_j^{h'}E \subset F$ for $j = d+1, \cdots, r$. Thus, if we set $F_i = G_i^{h'}$, $k[F_1, \cdots, F_n]$ is a homogeneous subring of $k[X_1, \cdots, X_n]$, and, since $k[X_1, \cdots, X_n]$ is a finite module over $k[F_1, \cdots, F_n]$, $E$ is also a finite graded module over $k[F_1, \cdots, F_n]$. Since $E/F$ is annihilated by the ideal $(F_{d+1}, \cdots, F_n)$, $E/F$ is actually a graded module over $k[F_1, \cdots, F_d]$. Then Theorem 41 shows that the degree of $\bar{\varphi}_{E/F}(q)$ is *at most* $d-1$.

On the other hand, no non-zero element of $k[F_1, \cdots, F_d]$ is in the radical of the submodule (0) of $E/F$. Let $(\alpha_1, \cdots, \alpha_s)$ be a finite basis of $M = E/F$ over $S = k[F_1, \cdots, F_d]$, composed of homogeneous elements. The radical of (0) in $S\alpha_i$, i.e., the set $\mathfrak{A}_i$ of elements $x \in S$ such that $x^e\alpha_i = 0$ for some $e$, is an ideal in $S$ (and even a homogeneous one). Since $\bigcap \mathfrak{A}_i$ is obviously contained in the radical of (0) in $M$, we have $\bigcap \mathfrak{A}_i = (0)$, and this implies that some $\mathfrak{A}_i$, say $\mathfrak{A}_1$, is the ideal (0) (as $S$ is an integral domain). In particular, we have $x\alpha_1 \neq 0$ for every $x \neq 0$ in $S$. Thus $S\alpha_1$ is a *free* submodule of $M = E/F$. If we denote by $t$ the degree of $\alpha_1$, the vector space $(E/F)_{t+hh'q}$ contains, as subspace, the set of all elements $f(F_1, \cdots, F_d)\alpha_1$, where $f$ is a form of degree $q$ (remember that $F_j$ is a form of degree $hh'$). Since the space of forms of degree $q$ in $d$ variables has dimension $\psi(q) = \binom{d+q-1}{d-1}$, which is a polynomial *of degree* $d-1$ in $q$, and since we have the inequality

$$\bar{\varphi}_{E/F}(t+hh'q) \geq \psi(q), \text{ for large } q,$$

it follows that the degree of $\bar{\varphi}_{E/F}$ is *at least* $d-1$, and this proves Theorem 42′.

If $\mathfrak{A}$ is a homogeneous ideal in $k[X_1, X_2, \cdots, X_n]$, of projective dimension $r$, we have

$$\bar{\chi}(\mathfrak{A}; q) = a_0\binom{q}{r} + a_1\binom{q}{r-1} + \cdots + a_{r-1}\binom{q}{1} + a_r,$$

where the coefficients $a_0, a_1, \cdots, a_r$ are integers (see Theorem 41). Here $a_0$ is necessarily a positive integer, since $\bar{\chi}(\mathfrak{A}; q)$ is positive for large positive integers $q$. This coefficient $a_0$ is called *the degree of the ideal* $\mathfrak{A}$. The integer $p_a(\mathfrak{A}) = (-1)^r(a_r - 1)$ is called the *arithmetic genus* of $\mathfrak{A}$. If $\mathfrak{A}$ is a prime ideal, the degree and the arithmetic genus of $\mathfrak{A}$ correspond to well-defined geometric characters of the irreducible $r$-dimensional variety $V = \mathscr{V}(\mathfrak{A})$. Thus, if $k$ is algebraically closed, then $a_0$ is *the order of the variety* $V$, i.e., the number of intersections of $V$ with a general $(n-r)$-dimensional subspace of $S_n$. If $V$ is a curve

$(r=1)$ without singular points, then $p_a(\mathfrak{A})$ is the ordinary genus of the curve.

**§ 13. Chains of syzygies.** In this section $A$ denotes a noetherian ring, and all the $A$-modules are tacitly assumed to be unitary and finitely generated.

Given any finite $A$-module $M$, and any finite basis $\{x_1, \cdots, x_q\}$ of $M$, we may consider $M$ as a *difference module* $F(M)-S(M)$ of the *free module* $F(M)=A^q$ generated over $A$ by $q$ basic elements. (We may take for $A^q$ the $q$-fold direct sum $A\oplus A\oplus \cdots \oplus A$.) The submodule $S(M)$ of $A^q$ is called the *module of the relations* satisfied by the elements $x_1, \cdots, x_q$: in fact its elements are the "vectors"† $\{a_1, \cdots, a_q\} \in A^q$ which satisfy $a_1x_1+ \cdots +a_qx_q=0$. One says also that $S(M)$ is *the first module of syzygies of* $M$ (with respect to the basis $\{x_1, \cdots, x_q\}$).

Since $S(M)$ is a submodule of the finite $A$-module $A^q=F(M)$, it is a finite $A$-module. Let $\{y_1, \cdots, y_h\}$ be any finite basis of $S(M)$. We can then consider $S(M)$ as a difference module $F(S(M))-S(S(M))$ of a free module by the first module of syzygies of $S(M)$. We set $F(S(M))=F_2(M)$, $S(S(M))=S_2(M)$. This procedure can be continued, and we set, inductively, $F_{n+1}(M)=F(S_n(M))$ and $S_{n+1}(M)=S(S_n(M))$. Thus: $S_n(M)=F_{n+1}(M)-S_{n+1}(M)$. The module $S_n(M)$ is called the *n*th-*module of syzygies of* $M$. Notice that this module *depends on the choice* of the bases in $M, S_1(M), \cdots, S_{n-1}(M)$. Here $F_1(M)$ and $S_1(M)$ stand respectively for $F(M)$ and $S(M)$.

The situation we have just described may be conveniently described in terms of an *exact sequence*

$$(1)\qquad F_{n+1}(M) \xrightarrow{\varphi_n} F_n(M) \xrightarrow{\varphi_{n-1}} F_{n-1}(M) \rightarrow \cdots \rightarrow F_1(M) \xrightarrow{\varphi_0} M \rightarrow 0.$$

Here $\varphi_n$ is the natural homomorphism of $F_{n+1}(M)$ onto $S_n(M)$ and may be considered as a homomorphism of $F_{n+1}(M)$ into $F_n(M)$. Its image is $S_n(M)$, and, by definition, its kernel is $S_{n+1}(M)$. Thus the image of $\varphi_n$ is *equal* to the kernel of $\varphi_{n-1}$. This proves the exactness of our sequence, since the homomorphism $\varphi_0$ is onto.

The exact sequence (1) is called a *chain of syzygies* of the module $M$. We say that a chain of syzygies

$$\cdots \rightarrow F_n \xrightarrow{\varphi_{n-1}} F_{n-1} \rightarrow \cdots \rightarrow F_1 \rightarrow M \rightarrow 0$$

of an $A$-module $M$ *terminates at the* n-*th term* if the module of syzygies

† For convenience, we shall use in this section the term "vector" in a wider sense than in Vol. I, Ch. I, § 21, i.e., also if $A$ is not a field.

$S_{n-1}$ is a free module. We can then complete the exact sequence by setting $F_n = S_{n-1}$, $F_{n+1} = 0$:

$$0 \to F_n \to F_{n-1} \to \cdots \to F_1 \to M \to 0.$$

We now study the influence of the *choice of bases* upon the structure of the modules of syzygies. Let $\{x_1, \cdots, x_n\}$, $\{y_1, \cdots, y_q\}$ be two bases of the $A$-module $M$, and let $M = F - S$ and $M = F' - S'$ be the representations of $M$ as difference module of free modules deduced from these bases. We may take $\{x_1, \cdots, x_n, y_1, \cdots, y_q\}$ as basis of $M$ and write $M = A^{n+q} - T$. The syzygy module $T$ is the set of vectors $\{a_i, b_j\} \in A^{n+q}$ such that $\sum a_i x_i + \sum b_j y_j = 0$. It contains a submodule which can be identified with $S$, namely the set of vectors $\{a_i, 0\}$ such that $\sum a_i x_i = 0$. Now, since $\{x_i\}$ is a basis of $M$, we can write $y_j = \sum c_{ji} x_i$, and the vector $t_j = \{c_{j1}, \cdots, c_{jn}, 0, \cdots, 0, -1, 0, \cdots, 0\}$ belongs to $T$; here the number of zeros which precede $-1$ is equal to $j-1$. These $q$ vectors $t_j$ are obviously linearly independent, over $A$, mod $A^n$, whence *a fortiori* mod $S$. Furthermore, $T$ is generated by $S$ and by the vectors $t_j$: if $\{a_i, b_j\} \in T$, we have $0 = \sum a_i x_i + \sum b_j y_j = \sum a_i x_i + \sum b_j c_{ji} x_i = \sum_i (a_i + \sum_j b_j c_{ji}) x_i = 0$; thus the vector $\{a_i, b_j\} + \sum_j b_j t_j = \{a_i + \sum_j b_j c_{ji}, 0\}$ belongs to $S$. This proves that $T$ is the *direct sum of $S$ and a free module.* Similarly $T$ is also the direct sum of $S'$ and a free module. If we call *equivalent* two $A$-modules $S$, $S'$ for which there exist *free* $A$-modules $L$, $L'$ such that the direct sums $S \oplus L$, $S' \oplus L'$ are *isomorphic*, then we have proved:

LEMMA 1. *Two first modules of syzygies $S$, $S'$ of an $A$-module $M$ with respect to two bases of $M$ are equivalent.*

In order to prove that *all* the modules of syzygies of $M$ are *uniquely determined up to equivalence* we need only to observe that the notion of equivalent modules is actually an equivalence relation and to prove the following:

LEMMA 2. *If $M$ and $M'$ are equivalent modules, and if $S$ and $S'$ are two first modules of syzygies of $M$* and *$M'$, then $S$ and $S'$ are equivalent.*

PROOF. We have, by assumption, $M \oplus L \simeq M' \oplus L'$, where $L$ and $L'$ are free modules. If $\{x_1, x_2, \cdots, x_m\}$ is a basis of $M$ with respect to which $S$ is derived and if $\{z_1, z_2, \cdots, z_h\}$ is a free basis of $L$, then $\{x_1, x_2, \cdots, x_m, z_1, z_2, \cdots, z_h\}$ is a basis of $M \oplus L$. Since any relation $\sum a_i x_i + \sum b_j z_j =$ implies $\sum a_i x_i = 0$ and $b_j = 0$, it follows that the first module of syzygies of $M \oplus L$, relative to the basis $\{x_1, x_2, \cdots, x_m, z_1, z_2, \cdots, z_h\}$, is isomorphic with $S$. Similarly, $S'$ is isomorphic with the first module of syzygies of $M' \oplus L'$, relative to a suitable basis.

Since two first modules of syzygies of *isomorphic* modules $M \oplus L$ and $M' \oplus L'$ are equivalent by Lemma 1, it follows that also $S$ and $S'$ are equivalent. Q.E.D.

In the case of a *graded* module $M$ over a *graded* ring $A$ (see § 12) we shall restrict ourselves to *graded modules of syzygies*. They are constructed in the following way: we take a finite basis $\{x_1, \cdots, x_q\}$ of $M$ composed of *homogeneous* elements and denote by $d_i$ the degree of $x_i$. Let $F$ be the free $A$-module generated by $q$ elements $X_1, \cdots, X_q$, $X_i$ being considered as having degree $d_i$ (whence the additive group of homogeneous elements of degree $n$ of $F$ is $\sum A_{n-d_i} X_i$). The homomorphism $\varphi$ of $F$ onto $M$ defined by $\varphi(X_i) = x_i$ is homogeneous of degree 0. Its kernel $S$, which is the first module of syzygies of $M$ with respect to $\{x_1, \cdots, x_q\}$, is therefore a graded module. We apply the same procedure to $S$, etc. Thus, in the exact sequence

$$F_{n+1} \xrightarrow{\varphi_n} F_n \xrightarrow{\varphi_{n-1}} F_{n-1} \to \cdots \to F_1 \xrightarrow{\varphi_0} M \to 0$$

all the homomorphisms are now homogeneous, of degree 0.

From now on we make one of the following assumptions

(a) *Either $A$ is a graded ring $\sum_{i=0}^{\infty} A_i$, with $A_0$ a field, and $M$ is a graded $A$-module*, in which case we tacitly limit ourselves to *graded* modules of syzygies;

(b) *or $A$ is a local ring.*

We denote by $\mathfrak{m}$ the ideal $\sum_{i=1}^{\infty} A_i$ in case (a), the maximal ideal of $A$ in case (b). In both cases, $A/\mathfrak{m}$ is a *field*. We have stipulated earlier in this section that, given a chain of syzygies

$$F_n \xrightarrow{\varphi_{n-1}} F_{n-1} \to \cdots \to F_1 \xrightarrow{\varphi_0} M \to 0$$

for the module $M$, the assertion that it stops at the $n$-th step means that the module $S_{n-1} = \varphi_{n-1}(F_n) = \varphi_{n-2}{}^{-1}(0)$ is *free*. We prove that, in either case (a) or (b), this property is *independent of the choice of the chain of $M$*. In fact, in another chain of syzygies, the $(n-1)$-th module $S'_{n-1}$ of syzygies is equivalent to $S_{n-1}$, by Lemma 2. We thus have to show that *a module which is equivalent to a free module is itself a free module.* In view of the definition of equivalence, this assertion will follow from the following lemma:

LEMMA 3. *Under hypotheses (a) or (b), if a module $E$ and a free module $F$ are such that the direct sum $E \oplus F$ is a free module $G$, then $E$ is free.*

PROOF. It is sufficient to prove the lemma in the case in which $F$ is generated by a single element $x$. Let $\{g_1, \cdots, g_q\}$ be a linearly independent basis of $G$, the $g_i$ and $x$ being homogeneous in case (a). We set $g_i = e_i + a_i x$ ($e_i \in E$, $a_i \in A$, both $e_i$ and $a_i$ being uniquely determined). For any element $u$ of $E$ we can write $u = \sum c_i g_i$, whence $u = \sum c_i e_i + (\sum c_i a_i)x$. Since $E \cap Ax = (0)$, it follows that $\sum c_i a_i = 0$. Therefore the module $E$ is *generated* by the elements $e_1, \cdots, e_q$.

On the other hand, we may write $x = \sum b_i g_i$ ($b_i \in A$), whence $x = \sum b_i e_i + (\sum b_i a_i)x$. This implies $\sum b_i e_i = 0$ and $\sum b_i a_i = 1$. If $A$ is a graded ring, the elements $a_i$, $b_i$ of $A$ are homogeneous, and the relation $\sum b_i a_i = 1$ implies that at least one of the $b_i$, say $b_1$, is different from 0 and is of degree 0. If $A$ is a local ring, $\sum b_i a_i = 1$ implies that at least one of the $b_i$ say $b_1$, is outside the maximal ideal $\mathfrak{m}$. In both cases $b_1$ is a unit, and the relation $\sum b_i e_i = 0$ shows that $E$ is generated by $e_2, \cdots, e_q$.

We now show that these $q-1$ elements $e_2, e_3, \cdots, e_q$ are linearly independent. Given any relation $\sum_{i=1}^{q} c_i e_i = 0$ ($c_i \in A$), we have

$$0 = \sum_i c_i(g_i - a_i x) = \sum_i c_i\Big(g_i - a_i \sum_j b_j g_j\Big) = \sum_i \Big(c_i - b_i \sum_j a_j c_j\Big) g_i,$$

whence $c_i = \Big(\sum_j a_j c_j\Big) b_i$ for every $i$. Thus every relation satisfied by $e_1, \cdots, e_q$ is proportional to $\sum b_i e_i = 0$. Since $b_1$ is invertible, only the trivial relation does not contain $e_1$. Q.E.D.

We now strengthen our assumptions. Namely, we shall assume that

(a)′ *either the ring $A$ is a polynomial ring in $n$ variables over a field $k$, or*

(b)′ *the ring $A$ is a regular local ring of dimension $n$* (see VIII, § 11).

In both cases, there exist $n$ elements $\xi_1, \cdots, \xi_n$ of $A$ such that

[1] *The ideal $\mathfrak{m}$ is generated by $\xi_1, \cdots, \xi_n$;*

[2] *If $\mathfrak{m}_j$ denotes the ideal $(\xi_1, \cdots, \xi_j)$, then $(\mathfrak{m}_{j-1} : \mathfrak{m}_j) = \mathfrak{m}_{j-1}$.* In fact, in case (a)′ we take for $\xi_1, \cdots, \xi_n$ the variables in $A$ (and these elements are homogeneous). In case (b)′ we take for $\{\xi_1, \cdots, \xi_n\}$ a regular system of parameters of $A$ (VIII, § 11). Then

THEOREM 43 (Hilbert). *Under hypothesis (a)′ or (b)′, any chain of syzygies of any $A$-module $M$ terminates at the $(n+1)$-st step. If $M$ is a submodule of a free module, any chain of syzygies of $M$ terminates at the $n$-th step.*

PROOF. The first assertion follows from the second, since the first module of syzygies of any module is a submodule of a free module. We thus suppose that $M$ is a submodule of a free module $F_0$, we set $M=S_0$, and we consider a chain of syzygies

$$\cdots \to F_k \xrightarrow{\varphi_{k-1}} F_{k-1} \to \cdots \to F_1 \xrightarrow{\varphi_0} M \to 0$$

of $M$; as usual we set $S_k=\varphi_k\ (F_{k+1})=\varphi_{k-1}^{-1}(0)$. We now prove a lemma:

LEMMA 4.

*For* $0 \leqq j \leqq n$ and $k>j$ *we have*

$$S_k \cap \mathfrak{m}_j F_k = \mathfrak{m}_j S_k. \tag{2}$$

*If $M$ is a submodule of a free module then the equality* (2) *holds also for* $j=k$.

PROOF. The assertion is trivial for $j=0$, $\mathfrak{m}_0$ denoting the zero ideal. We proceed by induction with respect to $j$. We thus assume that (2) is true for given $j \geqq 0$ for every $k>j$, and for every $k \geqq j$ if $M$ is a submodule of a free module, and we proceed to prove that $S_r \cap \mathfrak{m}_{j+1}F_r = \mathfrak{m}_{j+1}S_r$ if $r>j+1$, and also if $r=j+1$ provided $M$ is a submodule of a free module. We have only to prove that any element $d$ of $S_r \cap \mathfrak{m}_{j+1}F_r$ belongs to $\mathfrak{m}_{j+1}S_r$. Let $d=\xi_1 a_1+\xi_2 a_2+\cdots+\xi_{j+1}a_{j+1} (a_i \in F_r)$. Since $d \in S_r$ we have

$$0=\varphi_{r-1}(d)=\xi_1\varphi_{r-1}(a_1)+\xi_2\varphi_{r-1}(a_2)+\cdots+\xi_{j+1}\varphi_{r-1}(a_{j+1}).$$

If $r-1>0$, $F_{r-1}$ is a free module. Therefore $\mathfrak{m}_jF_{r-1}:(\xi_{j+1})=\mathfrak{m}_jF_{r-1}$, in view of property [2] of the ideals $\mathfrak{m}_j$, and consequently

$$\varphi_{r-1}(a_{j+1}) \in \mathfrak{m}_jF_{r-1}. \tag{3}$$

Now, if $r>j+1$, then $r-1>0$. Thus (3) holds unconditionally if $r>j+1$. Assume, however, that $r=j+1$ and $r-1=0$, whence $j=0$, $r=1$. In that case, $d=\xi_1 a_1$, $\xi_1\varphi_0(a_1)=0$. If $M$ is a submodule of a free module, the relation $\xi_1\varphi_0(a_1)=0$ implies $\varphi_0(a_1)=0$, and hence (3) still holds in this case.

Using (3) we now set $\varphi_{r-1}(a_{j+1})=\xi_1 v_1+\cdots+\xi_j v_j$, with $v_i \in F_{r-1}$. As $\varphi_{r-1}(a_{j+1}) \in S_{r-1}$, our induction hypothesis shows that we may assume that $v_i \in S_{r-1}$. We set $v_i=\varphi_{r-1}(b_i)$, $b_i \in F_r$.

Consider the elements $a'_i=a_i+\xi_{j+1}b_i$ $(i=1,\cdots,j)$, and $a'_{j+1}=a_{j+1}-\xi_1 b_1-\cdots-\xi_j b_j$ of $F_r$. It is clear that $d=\xi_1 a'_1+\cdots+\xi_j a'_j+\xi_{j+1}a'_{j+1}$. On the other hand, we have $\varphi_{r-1}(a'_{j+1})=\varphi_{r-1}(a_{j+1})-\sum_{i=1}^{j}\xi_i\varphi_{r-1}(b_i)=\varphi_{r-1}(a_{j+1})-\sum_{i=1}^{j}\xi_i v_i=0$, whence $a'_{j+1} \in S_r$. If we apply the induction hypothesis to the element $d-\xi_{j+1}a'_{j+1}=\xi_1 a'_1+\cdots+$

$\xi_j a'_j$ of $S_r \cap \mathfrak{m}_j F_r$, we see that this element belongs to $\mathfrak{m}_j F_r$. Therefore $d$ belongs to $\mathfrak{m}_{j+1} F_r$.

This completes the proof of Lemma 4, and we now continue with the proof of the theorem. Let us suppose that we have chosen, for constructing $S_{n+1}$, a basis $(u_1, \cdots, u_q)$ of $S_n$ in which no element is a linear combination of the others (the $u_i$ being, of course, homogeneous in case (a)'). Then, in any relation $\sum \alpha_i u_i = 0$, all the elements $\alpha_i$ belong to $\mathfrak{m}$, otherwise one of them would be invertible (in the graded case we must decompose first $\sum \alpha_i u_i = 0$ into homogeneous components). In other words, we have $S_{n+1} \subset \mathfrak{m} F_{n+1} = \mathfrak{m}_n F_{n+1}$. By Lemma 4 we know that $S_{n+1} \cap \mathfrak{m}_n F_{n+1} = \mathfrak{m}_n S_{n+1}$. Hence $S_{n+1} = \mathfrak{m}_n S_{n+1} = \mathfrak{m} S_{n+1}$. This same reasoning and Lemma 4 show that if $M$ is a submodule of a free module, then $S_n = \mathfrak{m} S_n$.

Now, the relation $S_i = \mathfrak{m} S_i$ (where $i$ is either $n$ or $n+1$) implies $S_i = (0)$. In the graded case, to see this we need only to consider a homogeneous element $\alpha \neq 0$ of $S_i$, of smallest degree. In the local case, we take a finite basis $\{z_1, \cdots, z_r\}$ of $S_i$, write $z_j = \sum_\nu \mu_{j\nu} z_\nu$ with $\mu_{j\nu} \in \mathfrak{m}$, i.e., $\sum (\delta_{j\nu} - \mu_{j\nu}) z_\nu = 0$; this implies $d z_\nu = 0$, where $d = \det(\delta_{j\nu} - \mu_{j\nu})$; and since $d \equiv 1 \pmod{\mathfrak{m}}$, $d$ is invertible, whence $z_\nu = 0$ for every $\nu$, and $S_i = (0)$.

The fact that $S_i = (0)$ signifies that $S_{i-1}$ is *free*, and this proves Theorem 43.

From now on we suppose that *hypothesis* (a') *or* (b') *holds.*

The smallest integer $d$ such that any chain of syzygies of the $A$ module $M$ terminates at the $(d+1)$-th step is called the *cohomological dimension of* $M$, and is denoted by $\delta(M)$. We set $\delta(0) = -1$, by convention. For $M \neq 0$ to be free, it is necessary and sufficient that $\delta(M) = 0$. *If $M$ is a factor module $F/S$ of a free module $F$ and is not itself free, then*

$$\delta(M) = 1 + \delta(S). \tag{4}$$

For comparing cohomological dimensions of modules, submodules and factor modules the following lemma is useful.

LEMMA 5. *Let $M$ be an $A$-module, $M'$ a submodule of $M$, $M''$ the factor module $M/M'$, $S'$ a first module of syzygies of $M'$ and $S''$ a first module of syzygies of $M''$. Then $M$ has a first module of syzygies $S$ admitting $S'$ as submodule and $S''$ as corresponding factor module.*

PROOF. Let $\{x_i\}$, $\{\bar{y}_j\}$ be systems of generators of $M'$ and $M''$ giving rise to $S'$ and $S''$: $S'$ is the kernel of the homomorphism $\varphi'$ of the free module $F' = \sum AX_i$ onto $M'$ defined by $\varphi'(X_i) = x_i$, and $S''$ is the kernel of the homomorphism $\varphi''$ of the free module $F'' = \sum AY_j$ onto $M''$ defined by $\varphi''(Y_j)=\bar{y}_j$. Choose any element $y_j$ of $M$ in the residue class $\bar{y}_j$, and define a homomorphism $\varphi$ of the direct sum $F=F'\oplus F''$ into $M$ by setting $\varphi(X_i)=x_i$, $\varphi(Y_j)=y_j$. It is easily verified that $\varphi$ maps $F$ *onto* $M$. The kernel $S$ of $\varphi$ contains all the pairs $(s', 0) \in F$ such that $s' \in S'$, and $S'$ may therefore be identified with a submodule of $S$. On the other hand the canonical homomorphism $\pi$ of $F$ onto $F''$ maps $S$ onto $S''$, and the kernel $\pi^{-1}(0) \cap S$ is exactly $S'$ (the proofs are straightforward, and we leave them to the reader). Q.E.D.

If a module $T$ contains a submodule $T'$ such that $T'$ and the corresponding factor module $T/T'$ are both free, then $T$ is also free. It follows therefore from Lemma 5 that, *if $M'$ is a submodule of $M$, then*

$$\delta(M) \leqq \max(\delta(M'), \delta(M/M')). \tag{5}$$

Similarly, if $\delta(M)$ and $\delta(M/M')$ are $\leqq q$, the $q$-th module of syzygies $S_q$ of $M$ is free and admits a submodule $S'_q$ (i.e., the $q$-th module of syzygies of $M'$) such that $S_q/S'_q$ is free. From the fact that $S_q/S'_q$ is free follows that $S'_q$ is a direct summand of $S_q$, and, by Lemma 3, $S'_q$ is free. Therefore:

$$\delta(M') \leqq \max(\delta(M), \delta(M/M')). \tag{6}$$

Finally, if $\delta(M)$ and $\delta(M')$ are $\leqq q$, then we may assume that the $(q+1)$-th module of syzygies $S'_{q+1}$ of $M'$ is reduced to 0. Then, since $S_{q+1}$ is free, a $(q+1)$-th module of syzygies $S_{q+1}/S'_{q+1}$ of $M/M'$ is free. Therefore

$$\delta(M/M') \leqq 1+\max(\delta(M), \delta(M')). \tag{7}$$

LEMMA 6. *Let $L$ be a free module $\neq 0$, $M$ a submodule of $L$ such that $M \subset \mathfrak{m}L$, and let $a$ be a non-invertible element $\neq 0$ of the ring $A$ such that $M:Aa=M$. Then $\delta(M+aL)=1+\delta(M)$.*

PROOF. The hypothesis $M=M:Aa$ is equivalent with the relation $M \cap aL=aM$. Therefore the module $(M+aL)/aL$ is isomorphic to $M/(M \cap aL)=M/aM$. Since $A$ is an integral domain, $aM$ is isomorphic to $M$, whence $\delta(M/aM) \leq 1+\delta(M)$ by (7). Since $aL$ is free and $\neq 0$, we have $\delta(aL)=0$, whence, by (5), $\delta(M+aL) \leq \max(0, 1+\delta(M)) \leq 1+\delta(M)$. We now prove, by induction on $\delta(M)$, that we have the *equality*

$$\delta(M+aL) = 1+\delta(M). \tag{a}$$

This is true for $\delta(M)=-1$, since, then, $M=0$ and $\delta(M+aL)=0$. We first show that, if $M\neq 0$, then *$M+aL$ is not free*. If $M+aL$ is free, it admits a linearly independent basis $(y_i)_{1\leq i\leq n}$, where $y_i=m_i+ax_i$ with

$m_i \in M$ and $x_i \in L$. For any $m \in M$, we can write $m = \sum_{i=1}^{n} b_i y_i (b_i \in A)$, i.e., $m = \sum_{i=1}^{n} b_i m_i + a\left(\sum_{i=1}^{n} b_i x_i\right)$, whence $\sum_{i=1}^{n} b_i x_i \in M$ since $M:Aa = M$; setting $M' = \sum_{i=1}^{n} Am_i$, we thus see that $M \subset M' + aM$, whence $M/M' \subset \mathfrak{m}(M/M')$, and $M/M' = 0$ as at the end of proof of Theorem 43; in other words, the $m_i$ generate the module $M$. Similarly, for any $x$ in $L$, we can write $ax = \sum_{i=1}^{n} b_i y_i = \sum_{i=1}^{n} b_i m_i + a\left(\sum_{i=1}^{n} b_i x_i\right)$ $(b_i \in A)$, whence $x - \sum_{i=1}^{n} b_i x_i \in M \subset \mathfrak{m}L$; as above, we deduce that $L$ is generated by the elements $x_i$. Since their number $n$ is equal to the maximum number of linearly independent elements in $M + aL$, they are linearly independent (remember that $L$ may be imbedded in a vector space over the quotient field of $A$). Now, for every $i$, we can write $ax_i = \sum_{j=1}^{n} b_{ij}(m_j + ax_j)$ $(b_{ij} \in A)$, whence $x_i - \sum_{j=1}^{n} b_{ij} x_j \in M \subset \mathfrak{m}L$; since $(x_i)$ is a linearly independent basis of $L$, this implies $b_{ij} \equiv \delta_{ij} \pmod{\mathfrak{m}}$, whence the matrix $(b_{ij})$ is invertible. Hence, from $ax_i = \sum_{j=1}^{n} b_{ij}(m_j + ax_j)$, we deduce that $m_j + ax_j \in aL$, whence $m_j \in aL$, and therefore $M \subset aL$ since $M$ is generated by the elements $m_j$. Taking into account the relation $M \cap aL = aM$, it follows that $M = aM \subset \mathfrak{m}M$, and, as above, that $M = 0$.

This being so, relation (a) is true for $\delta(M) = 0$, since we know that $\delta(M + aL) \leq 1$ and that $M + aL$ is not free. We thus assume that $\delta(M) \geq 1$. We represent $M$ as a factor module $F/S$ of a free module $F$; as at the end of the proof of Theorem 43, we may assume that $S \subset \mathfrak{m}F$; since $M$ is not free, we have $\delta(S) = \delta(M) - 1$ (by (4)). Any relation of the form $ax \in S$ $(x \in F)$ implies that $a\bar{x} = 0$ ($\bar{x}$ = image of $x$ in $M$), whence $\bar{x} = 0$ and $x \in S$, since $M$ is a submodule of a free module and since $a \neq 0$; in other words, we have $S:Aa = S$. Our induction hypothesis shows that $\delta(S + aF) = 1 + \delta(S) = \delta(M)$. Since $M/aM$ is isomorphic to $F/(S + aF)$, it admits $S + aF$ as first module of syzygies. Now, we have seen that $(M + aL)/aL$ and $M/aM$ are isomorphic; since $aL$ is free and thus admits $0$ as first module of syzygies, Lemma 5 shows that $M + aL$ admits a first module of syzygies isomorphic to $S + aF$. As $M + aL$ is not free, (4) shows that we have $\delta(M + aL) = 1 + \delta(S + aF)$ $= 1 + \delta(M)$. Q.E.D.

In particular, if we take $k$ non-invertible elements $y_1, \cdots, y_k$ of $A$ such that, for every $i$, $y_i$ is outside of all the associated prime ideals of $(y_1, \cdots, y_{i-1}) = \mathfrak{a}_{i-1}$ (where we set $\mathfrak{a}_{-0} = (0)$), then $\mathfrak{a}_{i-1} : Ay_i = \mathfrak{a}_{i-1}$, and successive applications of Lemma 6 show that

$$\delta(\mathfrak{a}_k) = k-1. \tag{8}$$

Our hypothesis is satisfied, for instance, by the ideals $\mathfrak{m}_j$ introduced earlier in this section. In the polynomial case (case (a)′) the theorem of Macaulay shows that every ideal $\mathfrak{a}$ which is of dimension $n-k$ and is generated by $k$ elements, satisfies our hypothesis. In general, any ideal $(y_1, \cdots, y_k)$ satisfying the conditions $(y_1, \cdots, y_{i-1}) : Ay_i = (y_1, \cdots, y_{i-1})$ $(i = 1, 2, \cdots, k)$ is said to belong to *the principal class.*

LEMMA 7. *Let $E$ be an $A$-module $\neq (0)$, and $a_1, \cdots, a_q$ non-invertible elements of $A$ such that, for every $i \leqq q$, we have $(a_1E + \cdots + a_{i-1}E) : Aa_i = a_1E + \cdots + a_{i-1}E$. Then we have $\delta(E/(a_1E + \cdots + a_qE)) = q + \delta(E)$.*

PROOF. If we set $E_i = E/(a_1E + \cdots + a_{i-1}E)$ it suffices to prove that $\delta(E_{i+1}) = 1 + \delta(E_i)$. We have $E_{i+1} = E_i/a_iE_i$, and, by hypothesis, the submodule $(0)$ of $E_i$ satisfies the condition $(0) : Aa_i = (0)$ (in other words: "$a_i$ is not a zero divisor in the module $E_i$"). We represent $E_i$ as a factor module $L/S$ of a free module $L$. We may assume that $S \subset \mathfrak{m}L$. Then $E_i/a_iE_i$ is isomorphic with $L/(S + a_iL)$, and the relation $(0) : Aa_i = (0)$ (in $E_i$) implies $S : Aa_i = S$ (in $L$). Thus Lemma 6 shows that $\delta(S + a_iL) = 1 + \delta(S)$, whence $\delta(E_{i+1}) = 1 + \delta(E_i)$, since $\delta(E_i) = 1 + \delta(S)$ and $\delta(E_{i+1}) = 1 + \delta(S + a_iL)$.

COROLLARY. *With the hypotheses and notations of Theorem* 43, *we have* $\delta(A/\mathfrak{m}) = n$.

THEOREM 44. *Under hypothesis* (a)′ *or* (b)′, *let $M$ be an $A$-module, $(0) = \bigcap_i N_i$ a reduced primary representation of the submodule* (0) *in $M$, and $\mathfrak{p}_i$ the associated prime ideal of the primary submodule $N_i$* (Vol. I, Ch. IV, Appendix). *Then the cohomological dimension $\delta(M)$ of $M$ is greater than or equal to* max $(h(\mathfrak{p}_i))$, *where $h(\mathfrak{p}_i)$ denotes the height of the prime ideal $\mathfrak{p}_i$* (Vol. I, Ch. IV, § 14, p. 240).

PROOF. Let us denote by $h(M)$ the integer max $(h(\mathfrak{p}_i))$; we have to prove the inequality $h(M) \leqq \delta(M)$ for every $A$-module $M$. We first prove it in the case $h(M) = n$. In that case one of the ideals $\mathfrak{p}_i$, say $\mathfrak{p}_1$, is the ideal $\mathfrak{m}$. We take an element $y \neq 0$ in the intersection $\bigcap_{i \geqslant 2} N_i$. There exists then an exponent $r$ such that $\mathfrak{m}^r y = (0)$. Taking for $r$ the smallest exponent such that $\mathfrak{m}^r y = (0)$, and denoting by $x$ any non-zero element of $\mathfrak{m}^{r-1}y$, we have $\mathfrak{m}x = (0)$. As the submodule $M' = Ax$ is annihilated by $\mathfrak{m}$, it is, in a natural way, a vector space over $A/\mathfrak{m}$. Since $M'$ is the direct sum of a certain number of copies of $A/\mathfrak{m}$, we have $\delta(M')=n$ (corollary to Lemma 7). We have to show that $\delta(M)=n$. Suppose this is not the case, i.e., that $\delta(M)<n$ (Theorem 43). We set $M''=M/M'$, and we consider some $(n-1)$-st modules of syzygies $S, S', S''$ of $M, M', M''$; by repeated applications of Lemma 5, we may assume that $S'$ is a submodule of $S$, and that $S''$ is $S/S'$. The assumption that $\delta(M)<n$ means that $S$ is free. Hence $S'$ is a first module of syzygies of $S''$, whence an $n$-th module of syzygies of $M''$, and is therefore free (Theorem 43). This implies $\delta(M')\leqq n-1$, in contradiction with $\delta(M')=n$.

We now prove the inequality $h(M)\leqq\delta(M)$ by induction on $n-h(M)$. If $h(M)<n$, none of the ideals $\mathfrak{p}_i$ is equal to $\mathfrak{m}$, whence, since these ideals are prime, there exists an element $a$ of $\mathfrak{m}$ such that $a\notin\mathfrak{p}_i$ for every $i$; this element may be assumed to be homogeneous in case (a)'. Then the submodule (0) of $M$ satisfies the relation $(0):Aa=(0)$ (Vol. I, Ch. IV, Appendix), and we therefore have $\delta(M/aM)=1+\delta(M)$ (Lemma 7). If we show that $h(M/aM)\geqq h(M)+1$, our proof will be complete, since, by the induction hypothesis, we have the inequality $h(M/aM)\leqq \delta(M/aM)$.

We thus show that $h(M/aM)\geqq h(M)+1$. Let $\mathfrak{p}$ be an associated prime ideal of (0) in $M$ such that $h(\mathfrak{p})=h(M)$, i.e., let $\mathfrak{p}$ be one of the prime ideals $\mathfrak{p}_i$ having the greatest possible height. Since $a\notin\mathfrak{p}$, it is sufficient to show the existence of an associated prime ideal $\mathfrak{p}'$ of (0) in $M/aM$ such that $\mathfrak{p}'\supset\mathfrak{p}+Aa$. Suppose this is not so. Then the union $\bigcup\mathfrak{p}'_j$ of the associated prime ideals of (0) in $M/aM$ does not contain $\mathfrak{p}+Aa$, and there exists an element $b$ in $\mathfrak{p}+Aa$ such that $b\notin\mathfrak{p}'_j$ for all $j$ (Vol. I, Ch. IV, § 6, *Remark*, p. 215). Then the submodule (0) of $M/aM$ satisfies the relation $(0):Ab=(0)$, whence we have $aM:Ab=aM$. If we write $b=c+da$ ($c\in\mathfrak{p}$, $d\in A$), the relation $cx\in aM$ (where $x\in M$) implies $(c+da)x\in aM$, whence we have $x\in aM$; in other words we have $aM:Ac=aM$. We shall show that we have $(0):Ac=(0)$ (in $M$), and this will contradict the fact that $c\in\mathfrak{p}$ and terminate the proof. In fact, the relation $cx=0$ with $x\in M$ implies $cx\in aM$, whence $x\in aM$ and $x = ax_1$ with $x_1\in M$; then $cx=0$ gives $acx_1=0$, whence $cx_1=0$; by repeated applications we get $x_1=ax_2$ with $x_2\in M$ and $cx_2=0$, and so on, whence $x=a^nx_n$ with $x_n\in M$ for every $n$. This is impossible unless $x=0$ in the polynomial case (a)', since $a$ is then a homogeneous element of positive degree. In the local ring case (b)', this also implies $x=0$: we have $x\in\bigcap_{n=1}^{\infty}a^nM$ and $\bigcap_{n=1}^{\infty}a^nM=(0)$, by the generalization of Krull's theorem (Vol. I, Ch. IV, Appendix), since the element $1+a$ is invertible in $A$. Q.E.D.

We terminate this section by showing how *Hilbert's theorem on characteristic polynomials* may be deduced from *Hilbert's theorem on syzygies*. We restrict ourselves to the case of a graded module $M$ over a polynomial ring $A=K[X_1, \cdots, X_n]$ over a *field* $K$. We consider a chain of syzygies of $M$:

$$\text{(S)} \qquad 0 \to F_j \to F_{j-1} \to \cdots \to F_1 \to M \to 0,$$

where the sequence is exact, and where $j \leqq n+1$. Denote by $d_{i1}, \cdots, d_{is(i)}$ the degrees of the generators of the free $A$-module $F_i$. For $q \geqq \max_{1 \leqq j \leqq s(i)}(d_{ij})$ the vector space $F_i^{(q)}$ of elements of degree $q$ in $F_i$ has dimension $\varphi_i(q) = \sum_{j=1}^{s(i)} \binom{n+q-d_{ij}-1}{n-1}$, and this is a polynomial of degree $n-1$ in $q$. Since the exact sequence $(S)$ induces an exact sequence

$$0 \to F_j^{(q)} \to F_{j-1}^{(q)} \to \cdots \to F_1^{(q)} \to M^{(q)} \to 0$$

in the homogeneous components of degree $q$, then, for $q \geqq \max_{i,j} (d_{ij})$, we have:

$$\dim_K (M^{(q)}) = -\varphi_1(q) + \varphi_2(q) + \cdots + (-1)^j \varphi_j(q)$$

by the result about alternating sums of dimensions in an exact sequence (§ 12, Lemma 1). Thus, for $q \geqq \max_{i,j} (d_{ij})$, $\dim_K (M^{(q)})$ is a polynomial of degree at most $n-1$ in $q$.

Notice that we have only used the fact that a chain of syzygies of $M$ stops somewhere, and not the more precise inequality $j \leqq n+1$.

# VIII. LOCAL ALGEBRA

§ 1. **The method of associated graded rings.** Let $A$ be a ring with element 1, $\mathfrak{m}$ an ideal in $A$ ($\mathfrak{m} \neq A$) and $E$ an $A$-module. The ideals $\mathfrak{m}^n$ (where we set $\mathfrak{m}^0 = A$) form a descending sequence of ideals in $A$, and the modules $\mathfrak{m}^n E$ form a descending sequence of submodules of $E$. We consider the direct sums

$$G(A) = \sum_{n=0}^{\infty} \mathfrak{m}^n/\mathfrak{m}^{n+1}, \quad G(E) = \sum_{n=0}^{\infty} \mathfrak{m}^n E/\mathfrak{m}^{n+1}E.$$

These are *graded* abelian groups, the elements of $\mathfrak{m}^n/\mathfrak{m}^{n+1}$ or $\mathfrak{m}^n E/\mathfrak{m}^{n+1}E$ being considered as homogeneous elements of degree $n$.

We are going to define a *multiplication* between elements of $G(A)$ and $G(E)$. It is sufficient to define the product $\bar{a}\bar{x}$ of *homogeneous* elements, where, say, $\bar{a}$ belongs to $\mathfrak{m}^n/\mathfrak{m}^{n+1}$ and $\bar{x}$ belongs to $\mathfrak{m}^q E/\mathfrak{m}^{q+1}E$. We fix representatives $a$ and $x$ of $\bar{a}$ and $\bar{x}$ respectively, where $a \in \mathfrak{m}^n$ and $x \in \mathfrak{m}^q E$. We have $ax \in \mathfrak{m}^{n+q}E$, and the class of $ax$ mod $\mathfrak{m}^{n+q+1}E$ is easily seen to depend only on $\bar{a}$ and $\bar{x}$. We denote by $\bar{a}\bar{x}$ this element of $\mathfrak{m}^{n+q}E/\mathfrak{m}^{n+q+1}E$. We have $\partial(\bar{a}\bar{x}) = \partial(\bar{a}) + \partial(\bar{x})$, where $\partial$ denotes the degree of a homogeneous element.

Taking $E = A$ we get, in particular, a multiplication in $G(A)$. One verifies, in a straightforward manner, that this multiplication is associative, commutative, and distributive with respect to the addition. Thus $G(A)$ is a *graded ring*, called the *associated graded ring* of $A$ with respect to the ideal $\mathfrak{m}$, and sometimes denoted by $G_{\mathfrak{m}}(A)$.

On the other hand, a straightforward verification shows that with respect to the multiplication $\bar{a}\bar{x}$ ($\bar{a} \in G(A)$, $\bar{x} \in G(E)$) defined above, the group $G(E)$ is a graded $G(A)$-module. This module is called the *associated graded module* of $E$, with respect to the ideal $\mathfrak{m}$; it is sometimes denoted by $G_{\mathfrak{m}}(E)$.

Suppose that the ideal $\mathfrak{m}$ admits a *finite basis* $\{m_1, \cdots, m_q\}$. As the monomials of degree $n$ in the $m_j$'s constitute a basis of $\mathfrak{m}^n$, the ring $G(A)$ is generated, over the ring $A/\mathfrak{m}$, by the classes $\bar{m}_j$ of the $m_j$'s mod. $\mathfrak{m}^2$ (this follows from the above definition of multiplication in $G(A)$, as applied to the elements $\bar{m}_j$). We can therefore write $G(A) = (A/\mathfrak{m})[\bar{m}_1, \cdots, \bar{m}_q]$. If we introduce $q$ indeterminates $x_1, \cdots, x_q$, we see that *the graded ring $G(A)$ is isomorphic to a residue class ring $(A/\mathfrak{m})[x_1, \cdots, x_q]/\mathfrak{I}$ of the polynomial ring $(A/\mathfrak{m})[x_1, \cdots, x_q]$ modulo a homogeneous ideal $\mathfrak{I}$.* In particular, $G(A)$ is a *noetherian* graded ring, if $A/\mathfrak{m}$ is noetherian and if $\mathfrak{m}$ is finitely generated. Similarly, if $E$ admits a finite basis $\{e_1, \cdots, e_r\}$, $\mathfrak{m}^n E$ is generated by the products $b_\lambda e_j$ where the $b_\lambda$'s are the monomials of degree $n$ in $m_1, \cdots, m_q$. Therefore $G(E)$, considered as a $G(A)$-module, *is generated by the residue classes $\bar{e}_1, \cdots, \bar{e}$ of $e_1, \cdots, e_r$* mod $\mathfrak{m}E$, and is therefore a *finite* $G(A)$-module.

Given any element $x$ of $E$, we denote by $v(x)$ the largest integer $n$ such that $x \in \mathfrak{m}^n E$. For $x \in \bigcap_{n=0}^{\infty} \mathfrak{m}^n E$, we set $v(x) = +\infty$. Then we have, if $v(x)$ is finite:

$$x \in \mathfrak{m}^{v(x)} E, \quad x \notin \mathfrak{m}^{v(x)+1} E. \tag{1}$$

The function $v$ is called the *order function* on the module $E$. This definition applies also to the particular case $E = A$. For $x$, $y$ in $E$ and $a$, $b$ in $A$ we obviously have:

$$v(x+y) \geqq \min(v(x), v(y)), \quad v(a+b) \geqq \min(v(a), v(b)); \tag{2}$$

$$v(ax) \geqq v(a) + v(x), \quad v(ab) \geqq v(a) + v(b). \tag{3}$$

Note that $v$ is not, in general, a valuation of $A$.

Given an element $x$ of $E$ which does not belong to $\bigcap_{n=0}^{\infty} \mathfrak{m}^n E$, we call the *initial form* of $x$ and denote by $G(x)$ the residue class of $x$ in $\mathfrak{m}^{v(x)} E/\mathfrak{m}^{v(x)+1} E$. For $x$ in $\bigcap_{n=0}^{\infty} \mathfrak{m}^n E$ we set $G(x) = 0$. This definition applies also to the particular case $E = A$.

The definition of the multiplication in $G(A)$ shows that the relations $ab \in \mathfrak{m}^{v(a)+v(b)+1}$, $v(ab) > v(a) + v(b)$ and $G(a)G(b) = 0$ are equivalent. Therefore we can state:

THEOREM 1. *Let $A$ be a ring and $\mathfrak{m}$ an ideal in $A$. If the associated graded ring $G(A)$ is a domain, then $A' = A/\bigcap_{n=0}^{\infty} \mathfrak{m}^n$ is also a domain, and the order function in $A'$ is a valuation of $A'$.*

Let $F$ be a submodule of $E$. We have $\mathfrak{m}^n(E/F) = (\mathfrak{m}^n E + F)/F$. Therefore $\mathfrak{m}^n(E/F)/\mathfrak{m}^{n+1}(E/F)$ is canonically isomorphic to $(\mathfrak{m}^n E + F)/(\mathfrak{m}^{n+1} E + F)$, hence also to $\mathfrak{m}^n E/\{(\mathfrak{m}^n E) \cap (\mathfrak{m}^{n+1} E + F)\}$. Since $(\mathfrak{m}^n E) \cap (\mathfrak{m}^{n+1} E + F)$ contains $\mathfrak{m}^{n+1} E$, the factor module $\mathfrak{m}^n(E/F)/\mathfrak{m}^{n+1}(E/F)$ may be considered as a factor module of $\mathfrak{m}^n E/\mathfrak{m}^{n+1} E$ (the corresponding submodule being $\{(\mathfrak{m}^n E) \cap (\mathfrak{m}^{n+1}E+F)\}/\mathfrak{m}^{n+1}E)$. It follows that $G(E/F)$ is canonically isomorphic to a *factor module* of $G(E)$, the corresponding submodule of $G(E)$ being $\sum_{n=0}^{\infty} \{(\mathfrak{m}^n E) \cap (\mathfrak{m}^{n+1}E+F)\}/\mathfrak{m}^{n+1}E$, i.e., the homogeneous submodule of $G(E)$ which is generated by the *initial forms of the elements of* $F$. This submodule is called the *leading submodule* of $F$. In the particular case where $E=A$ and where $F$ is an ideal $\mathfrak{a}$ in $A$, the leading submodule of $\mathfrak{a}$ is a homogeneous ideal in $G(A)$, and is called the *leading ideal* of $\mathfrak{a}$. As was pointed out above, the group $\mathfrak{m}^n(A/\mathfrak{a})/\mathfrak{m}^{n+1}(A/\mathfrak{a})$ is canonically isomorphic to $(\mathfrak{m}^n+\mathfrak{a})/(\mathfrak{m}^{n+1}+\mathfrak{a})$; here $A/\mathfrak{a}$ is viewed as an $A$-module. Now, if we call $\bar{\mathfrak{m}}$ the ideal $(\mathfrak{m}+\mathfrak{a})/\mathfrak{a}$ which corresponds to $\mathfrak{m}$ in the residue class ring $A/\mathfrak{a}$, then $(\mathfrak{m}^n+\mathfrak{a})/(\mathfrak{m}^{n+1}+\mathfrak{a})$ is canonically isomorphic to $\bar{\mathfrak{m}}^n/\bar{\mathfrak{m}}^{n+1}$. If we now apply Theorem 1 to the ring $A/\mathfrak{a}$ and to the ideal $\bar{\mathfrak{m}}$, we find the following result:

THEOREM 2. *Let $A$ be a ring and let $\mathfrak{m}$ and $\mathfrak{a}$ be two ideals in $A$. If $\bar{\mathfrak{m}}$ denotes the ideal $(\mathfrak{m}+\mathfrak{a})/\mathfrak{a}$ in the ring $A/\mathfrak{a}$, then the associated graded module of the $A$-module $A/\mathfrak{a}$, with respect to $\mathfrak{m}$, is canonically isomorphic to the associated graded ring of $A/\mathfrak{a}$ with respect to $\bar{\mathfrak{m}}$. Furthermore, if the leading ideal of $\mathfrak{a}$, in the associated graded ring of $A$ with respect to $\mathfrak{m}$, is prime, then the ideal $\bigcap_{n=0}^{\infty} (\mathfrak{a}+\mathfrak{m}^n)$ is also prime.*

We now give a sufficient condition for a ring $A$ to be an *integrally closed* domain. A domain $R$ is said to be *completely integrally closed* if it satisfies the following condition:

(c) *Every element $x$ of the quotient field $K$ of $R$, for which there exists an element $d \neq 0$ in $R$ such that $dx^n \in R$ for every $n \geqq 0$, is an element of $R$.*

Since every element $x$ of $K$ which is integral over $R$ satisfies the hypothesis of condition (c), a completely integrally closed domain is integrally closed. The converse is true if $R$ is noetherian since, then, every element $x$ of $K$ which satisfies the hypothesis in (c) is integral over $R$ (as $R[x]$ is then contained in the finite $R$-module $d^{-1}R$).

THEOREM 3. *Let $A$ be a ring, and $\mathfrak{m}$ an ideal in $A$ such that $\bigcap_{n=0}^{\infty} (Ac+\mathfrak{m}^n)=Ac$ for every $c$ in $A$. If the associated graded ring $G_{\mathfrak{m}}(A)$ is a completely integrally closed domain, then $A$ itself is a completely integrally closed domain.*

PROOF. Our hypothesis implies, in particular, that $\bigcap_{n=0}^{\infty} \mathfrak{m}^n=(0)$. Thus, by Theorem 1, $A$ is a domain, and the order function $v$ is a valuation of $A$. Let $x$ be an element of the quotient field $K$ of $A$ for which there exists an element $d \neq 0$ in $A$ such that $dx^n \in A$ for every $n \geqq 0$. Let us write $x = a/b$ $(a, b \in A)$. We have to prove that $a \in Ab$. Since $Ab = \bigcap_{n=0}^{\infty} (Ab + \mathfrak{m}^n)$ by hypothesis, we are reduced to proving that $a \in Ab + \mathfrak{m}^n$ for every $n \geqq 0$. This we prove by induction on $n$, the case $n = 0$ being trivial ($\mathfrak{m}^0$ being the unit ideal).

Suppose that we have $a \in Ab + \mathfrak{m}^n$. We have to prove that $a \in Ab + \mathfrak{m}^{n+1}$. We write $a = ub + w$ $(u \in A, w \in \mathfrak{m}^n)$. Since $dx^q \in A$ for every $q$, we have $d(x-u)^q \in A$ for every $q$, or—since $x = a/b = u + w/b$—$dw^q \in Ab^q$. We can thus write $dw^q = w_q b^q$ with $w_q \in A$, for every $q$. Since the order function in $A$ is a valuation, the passage to initial forms preserves products, whence $G(d)G(w)^q = G(w_q)G(b)^q$ for every $q$. Since $G_{\mathfrak{m}}(A)$ is completely integrally closed, this implies that $G(w)/G(b) \in G(A)$. Setting $G(w) = G(b)G(u')$ with $u'$ in $A$, the definition of the multiplication in $G(A)$ shows that $w \equiv bu'$ $(\mathfrak{m}^{n+1})$ (since $w \in \mathfrak{m}^n$). Thus $a$ is congruent to $b(u+u')$ mod $\mathfrak{m}^{n+1}$, and therefore $a$ belongs to $Ab + \mathfrak{m}^{n+1}$. Q.E.D.

§ **2. Some topological notions. Completions.** We assume that the reader is familiar with the elementary notions concerning topological spaces, metric spaces and completion of metric spaces.

A ring $A$ in which a topology is given, is said to be a *topological ring* (with respect to the given topology) if the ring operations in $A$ are continuous, i.e., if the mappings $(a, b) \to a - b$ and $(a, b) \to ab$ of the topological space $A \times A$ into the topological space $A$ are continuous.

Let $A$ be a topological ring. An $A$-module $E$, in which a topology is given, is said to be a *topological $A$-module*, if the mapping $(x, y) \to x - y$ of $E \times E$ into $E$ and the mapping $(a, x) \to ax$ $(a \in A, x \in E)$ of $A \times E$ into $E$ are both continuous. Thus, a topological $A$-module is first of all a topological (additive) group, and, furthermore, the multiplication of elements of $A$ by elements of $E$ is continuous. In particular, a topological ring $A$ is also a topological $A$-module.

Let $E$ be a topological $A$-module and $\Sigma(E)$ a system of open sets in $E$ which contain the zero 0 of $E$ and satisfy the following condition: (1) Any open set in $E$ containing 0 contains a set of the system $\Sigma(E)$ (in other words: $\Sigma(E)$ is a local open basis at 0). Then we have: (2) The system of sets of the form $x + U$, where $x \in E$ and $U \in \Sigma(E)$, is an open basis of $E$. Such a set $\Sigma(E)$ is called a *basis of neighborhoods of* 0 for the *topological module E*.

Let $A$ be a topological ring and let $\Sigma(A)$ be a basis of neighborhoods of the zero of $A$, in the sense of the above definition. It is easily verified that the system $\Sigma(A)$ enjoys the following properties:

(a) The intersection of any two sets of the system $\Sigma(A)$ contains a third set of that system.

(b) If $U$ is any set in the system $\Sigma(A)$ then there exists a set $W$ in $\Sigma(A)$ such that $W-W\subset U$ and $W^2\subset U$ (here $W-W$ and $W^2$ denote respectively the sets of all elements $a-b$ and $ab$, where $a$ and $b$ are in $W$).

(c) If $U$ is any set in the system $\Sigma(A)$, $a$ any element of $U$ and $b$ any element of $A$, then there exists a set $W$ in $\Sigma(A)$ such that $W+a\subset U$ and $Wb\subset U$.

It can be shown that if $A$ is a ring and $\Sigma(A)$ is a system of subsets of $A$ satisfying conditions (a), (b) and (c), then there exists one and only one topology in $A$ such that $A$ is a topological ring with respect to that topology and $\Sigma(A)$ is a basis of neighborhoods of 0 of the topological ring $A$.

Let $A$ be a topological ring and $E$ a topological $A$-module. Let $\Sigma(A)$ be a basis of neighborhoods of the zero of $A$ and let $\Sigma(E)$ be a basis of neighborhoods of the zero of $E$. It is easily verified that the system $\Sigma(E)$ enjoys the following properties (similar to the above properties (a), (b) and (c)):

(a′) The intersection of any two sets in the system $\Sigma(E)$ contains a third set of that system.

(b′) If $U'$ is any set in $\Sigma(E)$ then there exists a set $W'$ in $\Sigma(E)$ and a set $W$ in $\Sigma(A)$ such that $W'-W'\subset U'$ and $WW'\subset U'$.

(c′) If $U'$ is any set in $\Sigma(E)$, $x$ any element of $U'$, $y$ any element of $E$, and $b$ any element of $A$, then there exists a set $W'$ in $\Sigma(E)$ and a set $W$ in $\Sigma(A)$ such that $W'+x\subset U'$, $bW'\subset U'$ and $Wy\subset U'$.

It can be shown that if $\Sigma(A)$ is a basis of neighborhoods of the zero of a topological ring $A$ and if $\Sigma(E)$ is a system of subsets of an $A$ module $E$ such that conditions (a′), (b′) and (c′) are satisfied, then there exists one and only one topology in $E$ such that with respect to that topology $E$ is a topological $A$-module and $\Sigma(E)$ is a basis of neighborhoods of the zero of the topological module $E$.

The proofs of the preceding assertions are similar to the proofs of the similar assertions concerning topological groups, and for these proofs the reader is referred to Pontrjagin's "Topological Groups."

According to the above definitions, a topological ring or a topological module need not be a Hausdorff space. It is well known that *if the zero of a topological module $E$ is a closed set then $E$ is a Hausdorff space.* (*Proof:* If $x$, $y$ are *distinct* elements of $E$, let $V$ be a neighborhood of $y-x$ which does not contain the zero of $E$, and let $U=x-y+V$. Then $U$ is a neighborhood of zero such that $x-y\notin U$. Let $W$ be another neighborhood of zero such that $W - W \subset U$; then $x + W$ and $y + W$ are disjoint neighborhoods of $x$ and $y$.) The above proof gives also the following result: *if $\Sigma(E)$ is a basis of neighborhoods of the zero of a topological A-module, then E is a Hausdorff space if and only if the intersection of the sets of the system $\Sigma(E)$ consists only of the zero of E.*

We shall be concerned primarily with topologies in $A$ which can be defined by using *powers of ideals in A*, in the following fashion:

If $\mathfrak{m}$ is an ideal in $A$, the powers $\mathfrak{m}^n$ $(r = 0, 1, 2, \cdots)$ form a system $\Sigma(A)$ satisfying the conditions (a), (b) and (c). We have in fact: (a) $\mathfrak{m}^n \cap \mathfrak{m}^{n'} = \mathfrak{m}^n$ if $n \geqq n'$; (b) $\mathfrak{m}^n - \mathfrak{m}^n = \mathfrak{m}^n$ and $(\mathfrak{m}^n)^2 \subset \mathfrak{m}^n$ since the powers $\mathfrak{m}^n$ are ideals; (c) $\mathfrak{m}^n + a \subset \mathfrak{m}^n$ and $\mathfrak{m}^n b \subset \mathfrak{m}^n$, if $a \in \mathfrak{m}^n$ and $b \in A$. We define the $\mathfrak{m}$-*topology* of $A$ as being the one in which the ideals $\mathfrak{m}^n$ constitute a basis of neighborhoods of the zero of $A$. In a similar fashion, if $E$ is an $A$-module we define the $\mathfrak{m}$-*topology* of $E$ as being the one in which the submodules $\mathfrak{m}^n E$ constitute a basis of neighborhoods of the zero of $E$ (these submodules are easily seen to satisfy the conditions (a′), (b′) and (c′), the system $\Sigma(A)$ being the system of ideals $\mathfrak{m}^n$). With respect to this $\mathfrak{m}$-topology, the module $E$ is a Hausdorff space if and only if $\bigcap_{n=1}^{\infty} \mathfrak{m}^n E = (0)$.

LEMMA 1. *The closure $\bar{S}$ of a subset $S$ of $E$ is equal to $\bigcap_{n=0}^{\infty} (S + \mathfrak{m}^n E)$.*

PROOF. If $x \in \bar{S}$, there exists, for every $n$, a point $s_n$ of $S$ such that $s_n \in x + \mathfrak{m}^n E$. Hence $x \in s_n + \mathfrak{m}^n E \subset S + \mathfrak{m}^n E$ for every $n$. Conversely, if $x \in \bigcap_{n=0}^{\infty} (S + \mathfrak{m}^n E)$, there exists, for every $n$, a point $s_n$ of $S$ such that $x \in s_n + \mathfrak{m}^n E$, whence $s_n \in x + \mathfrak{m}^n E$ and $x \in \bar{S}$.

In particular, the closure of a submodule $F$ is the *submodule* $\bigcap_{n=0}^{\infty} (F + \mathfrak{m}^n E)$. A *closed* submodule $F$ is a submodule such that $F = \bigcap_{n=0}^{\infty} (F + \mathfrak{m}^n E)$.

If a submodule $F$ of $E$ is *open*, it contains some basic neighborhood $\mathfrak{m}^s E$. Conversely, if a submodule $F$ contains some $\mathfrak{m}^s E$, we have $x + \mathfrak{m}^s E \subset F$ for every $x$ in $F$, whence $F$ contains a neighborhood of each of its points, and is therefore open. Since the relation $\mathfrak{m}^s E \subset F$ implies $\mathfrak{m}^n E + F = F$ for every $n \geqq s$, it follows from Lemma 1 that *every open submodule of E is closed.*

Denoting by $v$ the order function in $E$ (see § 1), the $\mathfrak{m}$-topology of $E$ can be defined by the *distance*

$$d(x, y) = e^{-v(x-y)}, \quad e\text{—real, } e > 1. \tag{1}$$

By formula (2) (§ 1) this distance satisfies the "strong triangle inequality":

(2) $$d(x, z) \leqq \max\{d(x, y), d(y, z)\}.$$

Naturally, this distance function does not define a metric in $E$, in the usual sense, unless $E$ is a Hausdorff space; we have namely $d(x, y)=0$ if and only if $x-y \in \bigcap_{n=0}^{\infty} \mathfrak{m}^n E$. Nevertheless we can speak of Cauchy sequences $\{x_n\}$ in $E$: they are the sequences such that $x_n - x_{n+i} \in \mathfrak{m}^{N(n)}E$ for all $i \geqq 0$, where $N(n) \to +\infty$ as $n \to +\infty$. In view of the strong triangle inequality (2) it is seen at once that $\{x_n\}$ is a Cauchy sequence if and only if $d(x_n, x_{n+1}) \to 0$. A *null sequence* $\{x_n\}$ is one for which $d(x_n, 0) \to 0$. A *limit* of a sequence $\{x_n\}$ is any element $y$ of $E$ such that $\{x_n - y\}$ is a null sequence. If $\{x_n\}$ has a limit $y$, then $y'$ is also a limit of $\{x_n\}$ if and only if $y' - y \in \bigcap_{n=0}^{\infty} \mathfrak{m}^n E$. The module $E$ is *complete* if every Cauchy sequence in $E$ converges in $E$ (i.e., has a limit in $E$). In view of the strong triangle inequality, if $E$ is complete then the convergent series $\sum_{n=0}^{\infty} z_n$ are those whose general term $z_n$ tends to zero.

Let $F$ now be a submodule of $E$. The factor $A$-module $E/F$ admits a unique topology such that the canonical mapping $f: E \to E/F$ is both open and continuous: it is the topology defined by taking as basis of neighborhoods of the zero of $E/F$ the $f$-images of the basic neighborhoods $\mathfrak{m}^n E$ of the zero of $E$. The basic neighborhoods of the zero in $E/F$ are therefore the submodules $\mathfrak{m}^n(E/F) = (\mathfrak{m}^n E + F)/F$; in other words: *the natural topology of the factor module* $E/F$ (regarded as an $A$-module) *is again the* $\mathfrak{m}$-*topology of* $E/F$. We note that since both $f$ and $f^{-1}$ are open, it follows that the topological space $E/F$ is obtained from $E$ by *topological identification.*

A *submodule* $F$ of $E$ admits two topologies: the *induced* topology defined by the neighborhoods $\mathfrak{m}^n E \cap F$, and its own $\mathfrak{m}$-*topology* defined by the neighborhoods $\mathfrak{m}^n F$. As $\mathfrak{m}^n E \cap F \supset \mathfrak{m}^n F$, the latter is *stronger* than the former (i.e., it has more open sets; or, equivalently, the natural mapping of $F$ in $E$ is continuous for the $\mathfrak{m}$-topologies). These two topologies coincide in one important case:

THEOREM 4. *If $A$ is a noetherian ring and $E$ a finite $A$-module, then, for every submodule $F$ of $E$, the $\mathfrak{m}$-topology of $F$ is induced by the $\mathfrak{m}$-topology of $E$.*

PROOF. In the appendix to Chapter IV (Vol. I) we have proved that, given any ideal $\mathfrak{b}$ in $A$, there exists an integer $s$ and a submodule $F'$

of $E$ containing $\mathfrak{b}^sE$ such that $\mathfrak{b}F=F\cap F'$; thus $\mathfrak{b}F\supset F\cap\mathfrak{b}^sE$. In particular, any basic neighborhood $\mathfrak{m}^nF$ for the $\mathfrak{m}$-topology contains some basic neighborhood $F\cap\mathfrak{m}^{ns}E$ of the induced topology. Q.E.D.

Another proof of Theorem 4 may be deduced from the following result, due to E. Artin and D. Rees:

THEOREM 4′. *Let $A$ be a noetherian ring, $E$ a finite $A$-module, $F$ a submodule of $A$, and $\mathfrak{m}$ an ideal in $A$. There exists an integer $k$, depending only on $A$, $E$, $F$ and $\mathfrak{m}$, such that $\mathfrak{m}^nE\cap F=\mathfrak{m}^{n-k}(\mathfrak{m}^kE\cap F)$ for every $n\geqq k$.*

PROOF. The fact that Theorem 4′ implies Theorem 4 is clear. For proving Theorem 4′ we introduce an indeterminate $X$, and consider the set $A'$ of polynomials $\sum_i m_iX^i$ with $m_i\in\mathfrak{m}^i$; this set is clearly a subring of $A[X]$, and even a *noetherian* ring, for, if $\{a_1,\cdots,a_q\}$ is a finite basis of the ideal $\mathfrak{m}$, we have $A'=A[a_1X,\cdots,a_qX]$. We consider also the set $E'$ of formal sums $z_0+z_1X+\cdots+z_jX^j$ where $z_i\in\mathfrak{m}^iE$; it is an additive group for coefficientwise addition, and even an $A'$-module if we set $(m_iX^i)(z_jX^j)=m_iz_jX^{i+j}$ and extend this multiplication by linearity (it may be observed that $E'$ is isomorphic with the tensor product $A'\otimes E$ (Vol. I, Ch. III, § 14), but we shall not use this). If we make the convention that an element $uX^j$ of $A'$, or $E'$, is homogeneous of degree $j$, then $E'$ becomes a graded module over the graded ring $A'$. Finally $E'$ is a *finite* $A'$-module, for, if $\{y_1,\cdots,y_r\}$ is a basis of the $A$-module $E$, then it is clearly also a basis of the $A'$-module $E'$.

This being so, we notice that the set $F'$ of formal sums $z_0+z_1X+\cdots+z_jX^j$ such that $z_i\in\mathfrak{m}^iE\cap F$ is a homogeneous submodule of the graded $A'$-module $E'$. Thus $F'$ is generated, as an $A'$-module, by a finite number of homogeneous elements, say $u_1X^{n(1)},\cdots,u_qX^{n(q)}$ $(u_i\in\mathfrak{m}^{n(i)}E\cap F)$. Let $k$ be the greatest of the integers $n(i)$. We consider an element $z$ of $\mathfrak{m}^nE\cap F$, where $n\geqq k$. The element $zX^n$ of $F'$ may thus be written in the form $zX^n=\sum(a_iX^{n-n(i)})(u_iX^{n(i)})$, where $a_i\in\mathfrak{m}^{n-n(i)}$. Since $n-n(i)\geqq n-k$, we have $a_iu_i\in\mathfrak{m}^{n-k}\mathfrak{m}^{k-n(i)}u_i$, whence $a_iu_i\in\mathfrak{m}^{n-k}(\mathfrak{m}^kE\cap F)$, for $\mathfrak{m}^{k-n(i)}u_i$ is contained in

$$\mathfrak{m}^{k-n(i)}(\mathfrak{m}^{n(i)}E\cap F)\subset\mathfrak{m}^kE\cap F.$$

Therefore we have $z\in\mathfrak{m}^{n-k}(\mathfrak{m}^kE\cap F)$ and we have proved the inclusion $\mathfrak{m}^nE\cap F\subset\mathfrak{m}^{n-k}(\mathfrak{m}^kE\cap F)$. Since the opposite inclusion $\mathfrak{m}^{n-k}(\mathfrak{m}^kE\cap F)\subset\mathfrak{m}^nE\cap F$ is obvious, Theorem 4′ is proved. Q.E.D.

An important case in which Theorem 4 may be applied is the one in which we are given a noetherian ring $A$, an ideal $\mathfrak{m}$ in $A$, and an overring

$B$ of $A$ which is a finite $A$-module. Then, since $\mathfrak{m}^n B$ is the ideal $(\mathfrak{m}B)^n$ of $B$, the $\mathfrak{m}$-topology of $B$ ($B$ being considered as an $A$-module) coincides with the $(\mathfrak{m}B)$-topology of *the ring* $B$. Thus $A$, with its $\mathfrak{m}$-topology, is a *topological subspace* of $B$, when $B$ is considered with its $(\mathfrak{m}B)$-topology.

It may be noticed that, if $\mathfrak{m}$ and $\mathfrak{m}'$ are two ideals of a ring $A$ for which there exist exponents $a$ and $b$ such that $\mathfrak{m} \supset \mathfrak{m}'^a$ and $\mathfrak{m}' \supset \mathfrak{m}^b$ (i.e., two ideals with the same radical, in the noetherian case), then the $\mathfrak{m}$-topology and the $\mathfrak{m}'$-topology coincide on every $A$-module.

Let $A$ be a ring, $\mathfrak{m}$ an ideal in $A$, and $E$ an $A$-module. We suppose that $A$ and $E$ are Hausdorff spaces for their $\mathfrak{m}$-topologies, i.e., that $\bigcap_{n=0}^{\infty} \mathfrak{m}^n = (0)$ and that $\bigcap_{n=0}^{\infty} \mathfrak{m}^n E = (0)$. As metric spaces, $A$ and $E$ may be completed; call $\hat{A}$ and $\hat{E}$ their completions. The uniformly continuous mappings $(a, b) \to a+b$, $(x, y) \to x+y$, $(a, b) \to ab$, $(a, x) \to ax$ from $A \times A$, $E \times E$, $A \times A$, $A \times E$ into $A$, $E$, $A$, $E$, respectively, may be extended in a unique way, by continuity, to uniformly continuous mappings from $\hat{A} \times \hat{A}$, $\hat{E} \times \hat{E}$, $\hat{A} \times \hat{A}$ and $\hat{A} \times \hat{E}$ into $\hat{A}$, $\hat{E}$, $\hat{A}$, $\hat{E}$. We write these extended mappings additively and multiplicatively, as the old ones. Since algebraic identities are preserved by passage to the limit, these mappings define in $\hat{A}$ and $\hat{E}$ the structure of a *topological ring* and a *topological $\hat{A}$-module*, respectively. We shall often say that $\hat{A}$ (or $\hat{E}$) is *the $\mathfrak{m}$-adic completion of* $A$ (or $E$).

We emphasize that we have defined the completions $\hat{A}$ (or $\hat{E}$) *only if $A$ and $E$ are Hausdorff spaces* (in their $\mathfrak{m}$-topologies).

THEOREM 5. *Let $A$ be a ring, $\mathfrak{m}$ an ideal in $A$ and $E$ a finite $A$-module. If $A$ and $E$ are Hausdorff spaces for their $\mathfrak{m}$-topologies, then the completion $\hat{E}$ of $E$ is, as an $\hat{A}$-module, generated by $E$,* i.e., *we have $\hat{E} = \hat{A}E$.*

PROOF. Let $\{x_1, \cdots, x_q\}$ be an $A$-basis of $E$. Any element $y$ of $\hat{E}$ is the limit of a Cauchy sequence $\{y_n\}$ of elements of $E$. We have that $y_{n+1} - y_n$ belongs to $\mathfrak{m}^{s(n)}E$, where $s(n) \to \infty$ as $n \to \infty$. We can therefore write: $y_{n+1} - y_n = \sum_{j=1}^{q} a_{nj} x_j$, with $a_{nj} \in \mathfrak{m}^{s(n)}$. We set $y_1 = \sum b_{1j} x_j$, with $b_{1j} \in A$, and define inductively $b_{n+1,j}$ as being $b_{nj} + a_{nj}$. Then we have, by induction, $y_n = \sum_{j=1}^{q} b_{nj} x_j$, and, furthermore, the $q$ sequences $\{b_{1j}, b_{2j}, \cdots\}$ are Cauchy sequences in $A$. Let $b_j$ denote the limit of the sequence $\{b_{nj}\}$ in $\hat{A}$. In the equality

$$y - \sum_{j=1}^{q} b_j x_j = y - y_n + \sum_{j=1}^{q} (b_{nj} - b_j) x_j,$$

the right-hand side tends to 0 as $n$ tends to infinity. Hence $y = \sum_j b_j x_j$, and our assertion is proved.

COROLLARY 1. *If, in addition to the assumptions made in Theorem 5, we also assume that the ideal* $\mathfrak{m}$ *admits a finite basis, then the closures of* $\mathfrak{m}^n E$ *in* $\hat{E}$ *and of* $\mathfrak{m}^n$ *in* $\hat{A}$ *are* $\hat{A}\mathfrak{m}^n E = (\hat{A}\mathfrak{m})^n E$ *and* $\hat{A}\mathfrak{m}^n = (\hat{A}\mathfrak{m})^n$ *respectively. We have* $\mathfrak{m}^n E = (\hat{A}\mathfrak{m})^n E \cap E$ *and* $\mathfrak{m}^n = (\hat{A}\mathfrak{m})^n \cap A$. *The topologies of* $\hat{E}$ *and* $\hat{A}$ *considered as completions of* $E$ *and* $A$ *are their* $(\hat{A}\mathfrak{m})$*-topologies.*

In fact, since $\mathfrak{m}^s \mathfrak{m}^n E = \mathfrak{m}^{n+s} E \cap \mathfrak{m}^n E$ for every $s$, the $\mathfrak{m}$-topology of $\mathfrak{m}^n E$ is induced by the $\mathfrak{m}$-topology of $E$. Thus the closure of $\mathfrak{m}^n E$ in $\hat{E}$ may be identified, as a topological $\hat{A}$-module, with the $\mathfrak{m}$-adic completion of $\mathfrak{m}^n E$. Since our hypotheses imply that $\mathfrak{m}^n E$ is a finite $A$-module, Theorem 5 shows that this completion is $\hat{A}\mathfrak{m}^n E = (\hat{A}\mathfrak{m})^n E$. In particular, the closure of $\mathfrak{m}^n$ in $\hat{A}$ is $(\hat{A}\mathfrak{m})^n$. Taking into account the fact that the module $\mathfrak{m}^n E$ is closed in $E$ (as it is an open submodule), the second part of the corollary follows from the well-known topological fact that, given a metric space $S$ and a subset $T$ of $S$, the intersection of $S$ and of the closure of $T$ in $\hat{S}$ is the closure of $T$ in $S$. The last part of the corollary follows from the well-known topological fact that, given a metric space $S$ and a point $x$ of $S$, a basis of neighborhoods of $x$ in $\hat{S}$ is formed by the closures in $\hat{S}$ of the neighborhoods of $x$ in $S$. Q.E.D.

COROLLARY 2. *Let* $F$ *be a submodule of* $E$. *If, in addition to the assumptions made in Theorem 5, we also assume that* $A$ *is noetherian, then the closure of* $F$ *in* $\hat{E}$ *is* $\hat{A}F$, *and the closure of* $F$ *in* $E$ *is* $\hat{A}F \cap E$. *If* $F$ *is closed in* $E$, *then* $F = \hat{A}F \cap E$.

For, $A$ being noetherian, $F$ is a finite $A$-module, and hence, by Theorem 4, the closure of $F$ in $\hat{E}$ coincides with the $\mathfrak{m}$-adic completion of $F$. The first assertion of the corollary follows then from Theorem 5. The remaining assertions are topologically trivial.

THEOREM 6. *Let* $A$ *be a noetherian ring,* $\mathfrak{m}$ *an ideal in* $A$, $E$ *a finite* $A$*-module, and* $F$ *a submodule of* $E$ *which is closed with respect to the* $\mathfrak{m}$*-topology of* $E$. *If* $A$ *and* $E$ *are Hausdorff spaces in their* $\mathfrak{m}$*-topologies, then* $\hat{E}/\hat{A}F$ *and the completion of* $E/F$ *are canonically isomorphic as topological* $\hat{A}$*-modules.*

PROOF. By Corollary 2 to Theorem 5 we have $F = \hat{A}F \cap E$, whence the group $E/F$ may be algebraically identified with a subgroup of $\hat{E}/\hat{A}F$. The identification topology of $\hat{E}/\hat{A}F$ admits the subgroups $(\hat{A}F + \hat{A}\mathfrak{m}^n E)/\hat{A}F$ as basic neighborhoods and hence induces on $E/F$ the $\mathfrak{m}$-topology, since $(\hat{A}F + \hat{A}\mathfrak{m}^n E) \cap E = \hat{A}(F + \mathfrak{m}^n E) \cap E = F + \mathfrak{m}^n E$, by Corollary 2 to Theorem 5 (this corollary is applicable since $F + \mathfrak{m}^n E$ is open and therefore closed). Hence the *topological space* $E/F$ is a subspace of $\hat{E}/\hat{A}F$. Since $\hat{E}/\hat{A}F$ is a factor group of a complete metric group, topology shows that it is complete. For completing the proof it remains to be observed that $E/F$ is dense in $\hat{E}/\hat{A}F$, and this is obvious since $E$ is dense in $\hat{E}$. Q.E.D.

COROLLARY 1. *Let $A$ be a noetherian ring, $\mathfrak{m}$ an ideal in $A$ and $E$ a finite $A$-module. If $A$ and $E$ are Hausdorff spaces with respect to their $\mathfrak{m}$-topologies, then the associated graded rings of $A$ with respect to $\mathfrak{m}$ and of $\hat{A}$ with respect to $\hat{A}\mathfrak{m}$ are canonically isomorphic. More generally, the associated graded modules of $E$ with respect to $\mathfrak{m}$ and of $\hat{E}$ with respect to $\hat{A}\mathfrak{m}$ are canonically isomorphic.*

In fact, $\mathfrak{m}^n E$ is closed, since it is open, and $\mathfrak{m}^{n+1}E$ is an open and closed submodule of $\mathfrak{m}^n E$. Thus $\mathfrak{m}^n E/\mathfrak{m}^{n+1}E$ is discrete for its $\mathfrak{m}$-topology. Therefore it is identical to its completion, which is isomorphic to $\hat{A}\mathfrak{m}^n E/\hat{A}\mathfrak{m}^{n+1}E$ by Theorem 6. This proves our assertion.

COROLLARY 2. *Let $A$ be a noetherian ring, $\mathfrak{m}$ and $\mathfrak{a}$ two ideals of $A$ such that $\mathfrak{a}$ is closed in the $\mathfrak{m}$-topology. Then the completion of $A/\mathfrak{a}$ (for its $(\mathfrak{m}+\mathfrak{a})/\mathfrak{a}$-topology) is canonically isomorphic to $\hat{A}/\hat{A}\mathfrak{a}$.*

In fact the $(\mathfrak{m}+\mathfrak{a})/\mathfrak{a}$-topology of $A/\mathfrak{a}$ coincides with the $\mathfrak{m}$-topology of $A/\mathfrak{a}$ considered as an $A$-module, and we thus have a special case of Theorem 6, with $E=A$ and $F=\mathfrak{a}$.

We terminate this section by introducing a useful notation. Let $A$ be a ring which is a complete Hausdorff space for its $\mathfrak{m}$-topology, let $\{x_1, \cdots, x_q\}$ be a finite system of elements *of* $\mathfrak{m}$, and let $F(X_1, \cdots, X_q)$ be a *formal power series* with coefficients in a subring $B$ of $A$. We write $F$ as an infinite sum of forms $F=\sum_{n=0}^{\infty} F_n$, $F_n$ being a form of degree $n$. Then, since $F_n(x_1, \cdots, x_q) \in \mathfrak{m}^n$, the series $\sum_{n=0}^{\infty} F_n(x_1, \cdots, x_q)$ converges in $A$ as $A$ is complete. The sum of that series, which is uniquely determined since $A$ is a Hausdorff space, is denoted by $F(x_1, \cdots, x_q)$. The mapping $F \to F(x_1, \cdots, x_q)$ is obviously a homomorphism $\varphi$ of $B[[X_1, \cdots, X_q]]$ into $A$ (cf. Chapter VII, § 1), and is continuous if one takes $B[[X_1, \cdots, X_q]]$ with its natural topology (i.e., with its $(X_1, \cdots, X_q)$-topology). The image of this homomorphism $\varphi$ is a subring of $A$, which is denoted by $B[[x_1, \cdots, x_q]]$. If $\varphi$ is one to one, we say that $x_1, \cdots, x_q$ are *analytically independent* over $B$, and in that case $B[[x_1, \cdots, x_q]]$ is isomorphic to the power series ring in $q$ variables over $B$.

**§ 3. Elementary properties of complete modules.** In this section we study some finiteness properties of complete rings and of modules over complete rings.

THEOREM 7. *Let $A$ be a ring, $\mathfrak{m}$ an ideal in $A$, $E$ an $A$-module and $F$ a submodule of $E$. Suppose that $A$ is a complete Hausdorff space for its $\mathfrak{m}$-topology, and that $E$ is a Hausdorff space for its $\mathfrak{m}$-topology. Let $\{x_1, \cdots, x_q\}$ be a finite system of elements of $F$ such that their initial forms $G(x_i)$ generate (over $G_{\mathfrak{m}}(A)$) the leading submodule of $F$ in $G_{\mathfrak{m}}(E)$. Then the elements $\{x_1, \cdots, x_q\}$ generate $F$.*

PROOF. Let $y$ be any element of $F$. We are going to show inductively the existence, for every $n \geqq 0$, of elements $a_{ni}$ of $A$ such that

$$(1) \qquad y \equiv \sum_{i=1}^{q} a_{ni}x_i \bmod \mathfrak{m}^n E.$$

This is obvious for $n=0$. Suppose (1) holds for a given integer $n$ and for suitable elements $a_{ni}$ in $A$. If the element $y-\sum_{i=1}^{q} a_{ni}x_i$ is in $\mathfrak{m}^{n+1}E$, we take $a_{n+1,i}=a_{ni}$. If not, then the initial form $G\left(y-\sum_{i=1}^{q} a_{ni}x_i\right)$ is an element of degree $n$ in the leading submodule $F'$ of F. As $F'$ is generated by the homogeneous elements $G(x_i)$ we can write $G\left(y-\sum_{i=1}^{q} a_{ni}x_i\right)=\sum_{i=1}^{q} G(c_{ni})G(x_i)$, where the $c_{ni}$ are elements of $A$ such that $\partial(G(c_{ni}))=n-\partial(G(x_i))$. By the definition of initial forms, we have $y-\sum_{i=1}^{q} a_{ni}x_i \equiv \sum_{i=1}^{q} c_{ni}x_i \pmod{\mathfrak{m}^{n+1}E}$. We take, in this case, $a_{n+1,i}=a_{ni}+c_{ni}$.

The choice of the elements $c_{ni}$ shows that $\{a_{ni}\}$ is a Cauchy sequence for every $i$. Since $A$ is complete, this sequence admits a limit $a_i \in A$. In the equality

$$y-\sum_{i=1}^{q} a_i x_i = y-\sum_{i=1}^{q} a_{ni}x_i + \sum_{i=1}^{q} (a_{ni}-a_i)x_i,$$

the right-hand side tends to 0 as $n \to \infty$. Since $E$ is a Hausdorff space, this implies $y=\sum_{i=1}^{q} a_i x_i$. Q.E.D.

COROLLARY 1. *$A$, $\mathfrak{m}$ and $E$ being as in Theorem 7, suppose that $G_{\mathfrak{m}}(E)$ is a finite $G_{\mathfrak{m}}(A)$-module. Then $E$ is a finite $A$-module.*

We apply Theorem 7 to the case $F=E$.

COROLLARY 2. *$A$, $\mathfrak{m}$ and $E$ being as in Theorem 7, suppose that $E/\mathfrak{m}E$ is a finite $(A/\mathfrak{m})$-module. Then $E$ is a finite $A$-module. If the classes of $x_1, \cdots, x_q$ mod $\mathfrak{m}E$ generate $E/\mathfrak{m}E$, the elements $x_i$ generate $E$.*

In fact, the $G(A)$-module $G(E)$ is generated by $E/\mathfrak{m}E$ since every element of $\mathfrak{m}^nE$ may be written as a sum of elements of the form $m_1 \cdots m_n x$ ($m_i \in \mathfrak{m}$, $x \in E$), and since, if such an element is not in

$\mathfrak{m}^{n+1}E$, its initial form is $G(m_1) \cdots G(m_n)G(x)$, with $G(m_i) \in \mathfrak{m}/\mathfrak{m}^2$ and $G(x) \in E/\mathfrak{m}E$. It follows that $G(E)$ is a finite $G(A)$-module. Thus Corollary 2 follows from Corollary 1 and Theorem 7.

COROLLARY 3. *$A$, $\mathfrak{m}$ and $E$ being as in Theorem 7, suppose that $G(E)$ is a noetherian $G(A)$-module. Then $E$ is a noetherian $A$-module.*

In fact, for every submodule $F$ of $E$, the leading submodule of $F$ is finitely generated. By Theorem 7, $F$ itself is then finitely generated.

COROLLARY 4. *Let $A$ be a ring, and $\mathfrak{m}$ an ideal in $A$ such that $A$ is a complete Hausdorff space for its $\mathfrak{m}$-topology. If $\mathfrak{m}$ is finitely generated and if $A/\mathfrak{m}$ is noetherian, then $A$ is noetherian.*

In fact, we have seen in § 1 that, under these conditions, $G(A)$ is a noetherian ring. Thus Corollary 4 follows from Corollary 3.

COROLLARY 5. *Let $A$ be a noetherian ring, and $\mathfrak{m}$ an ideal in $A$. If $A$ is a Hausdorff space in its $\mathfrak{m}$-topology, then $\hat{A}$ is a noetherian ring.*

In fact, we have seen in § 1 that $G(A)$ is a noetherian ring. Since $G(A)$ and $G(\hat{A})$ are isomorphic (§ 2, Corollary 1 to Theorem 6), $G(\hat{A})$ is noetherian. Thus Corollary 5 follows from Corollary 3.

EXAMPLES:

(1) We give a second proof of the fact that, *if $R$ is a noetherian ring, then the power series ring $A = R[[X_1, \cdots, X_n]]$ is noetherian.* If we denote by $\mathfrak{M}$ the ideal $(X_1, \cdots, X_n)$, it is easily seen (see Chapter VII, § 1) that $R[[X_1, \cdots, X_n]]$ is a complete Hausdorff space for its $\mathfrak{M}$-topology. Since $\mathfrak{M}$ is finitely generated, and since $A/\mathfrak{M} = R$ is noetherian, Corollary 4 shows that $A$ is noetherian.

It may be observed that $R[[X_1, \cdots, X_n]]$ is the *completion* of the polynomial ring $R[X_1, \cdots, X_n]$ for the $(X_1, \cdots, X_n)$-topology of this latter ring. Thus our assertion follows also from Corollary 5.

Notice also that the *associated graded ring* of $R[[X_1, \cdots, X_n]]$ is the polynomial ring $R[X_1, \cdots, X_n]$. Thus, by Theorem 3, § 1, if $R$ is a noetherian integrally closed ring, then $R[[X_1, \cdots, X_n]]$ is also integrally closed.†

† Here we use the fact that, if $R$ is integrally closed then so is $R[X_1, \cdots, X_n]$. This may be proved as follows. By induction on $n$, we are reduced to proving that $R[X]$ is integrally closed. Let $K$ be the quotient field of $R$. If $z \in K(X)$ is integrally dependent on $R[X]$ then $z \in K[X]$, as $K[X]$ is integrally closed. We write $z = \sum_{i=0}^{q} a_i X^i$ with $a_i \in K$. We consider an equation of integral dependence for $z$ over $R[X]$ and substitute for $X$, in that relation, $q+1$ distinct elements $u_j$ of an algebraic closure of the prime subfield of $K$ which are integral over $R$. This shows that the $a_i$ are integral over $R$, whence $z \in R[X]$. Another proof is implicitly contained in the proof of Theorem 11 of VII, § 2, where we replace $R$ by $R[X]$ (whence $K_0$ by $K$): it follows from that proof that each term $a_i X^i$ is integrally dependent on $R[X]$ and this easily leads to the desired conclusion. [See also VI, § 13, Theorem 29, for a proof using valuations.]

(2) We now give a second proof of the "existence" part in the *Weierstrass preparation theorem* (Chapter VII, § 1, Theorem 5). We are given a power series $F$ in $R=K[[X_1, \cdots, X_n]]$ which is regular in $X_n$; more precisely the coefficient $c_s$ of $X_n^s$ in $F$ is an invertible element of $R'=K[[X_1, \cdots, X_{n-1}]]$, and $c_j$ is not invertible for $j<s$. We have to prove that every element $G$ of $R$ may be written in the form $G=UF+\sum_{j=0}^{s-1} S_jX_n^j$, where $U\in R$ and $S_j\in R'$. The hypothesis about $F$ implies that the ring $R/(X_1, \cdots, X_{n-1}, F)=K[[X_n]]/(F(0, \cdots, 0, X_n))$ is isomorphic to $K[[X_n]]/(X_n^s)$, whence this ring admits $\{1, x_n, \cdots, x_n^{s-1}\}$ as a linear basis over $K$ ($x_n$ denoting the residue class of $X_n$). Therefore, by Theorem 7, Corollary 2 (applied with $A$, $\mathfrak{m}$ and $E$ being replaced by $K[[X_1, \cdots, X_{n-1}, F]]$, $(X_1, \cdots, X_{n-1}, F)$ and $K[[X_1, \cdots, X_n]]$) $\{1, X_n, \cdots, X_n^{s-1}\}$ is a basis of $R$, $R$ being considered as a module over $K[[X_1, \cdots, X_{n-1}, F]]$. In other words, we can write

$$G=\sum_{j=0}^{s-1} \varphi_j(X_1, \cdots, X_{n-1}, F)X_n^j.$$

By putting in evidence the term $S_j(X_1, \cdots, X_{n-1})$ of $\varphi_j$ which does not contain $F$, and by factoring out $F$ in the other terms, we see that we can write $G=UF+\sum_{j=0}^{s-1} S_j(X_1, \cdots, X_{n-1})X_n^j$, as asserted.

§ 4. **Zariski rings.**† We are going to study the pairs $(A, \mathfrak{m})$, formed by a noetherian ring $A$ and an ideal $\mathfrak{m}$ in $A$, such that every submodule $F$ of every finite $A$-module $E$ is closed for the $\mathfrak{m}$-topology of $E$.

THEOREM 8. *Let $A$ be a noetherian ring, $\mathfrak{m}$ an ideal in $A$, $E$ a finite $A$-module, and $F$ a submodule of $E$. For $F$ to be closed in the $\mathfrak{m}$-topology of $E$, it is necessary and sufficient that $\mathfrak{p}_i+\mathfrak{m}\neq A$ for every associated prime ideal $\mathfrak{p}_i$ of $F$.*

PROOF. The assertion that $F$ is closed is equivalent to the relation $\bigcap_{n=0}^{\infty} (F+\mathfrak{m}^nE)=F$ (Lemma 1, § 2). By Krull's theorem (Vol. I, Ch. IV, Appendix) applied to $E/F$ this relation is equivalent to the following property of $F$: for every $a\equiv 1 \pmod{\mathfrak{m}}$ and for every $x\in E$, $x\notin F$, we have $ax\notin F$. This means that every element $a\equiv 1 \pmod{\mathfrak{m}}$ is outside all the associated prime ideals $\mathfrak{p}_i$ of $F$ (Vol. I, Ch. IV, Appendix), i.e.,

† These rings, which have been first studied by the senior author in his paper "Generalized semi-local rings" (*Summa Brasiliensis Mathematicae*), have been so designated by the junior author in his monograph "Algèbre locale" (*Mémorial des Sciences Mathématiques*, fasc. CXXIII, 1953).

that for every $i$, no element of $\mathfrak{p}_i$ is congruent to 1 mod $\mathfrak{m}$. This is obviously equivalent to the necessary and sufficient condition given in the theorem.

COROLLARY. *Let $A$ be a noetherian ring, $\mathfrak{m}$ an ideal in $A$, $E$ a finite $A$-module and $F$ a submodule of $E$. Let $F=\bigcap_i F_i$ be a primary representation of $F$, and $\mathfrak{p}_i$ the radical of the primary module $F_i$. Then the closure of $F$ in $E$ for the $\mathfrak{m}$-topology is the intersection $\bigcap_j F_j$ of those primary components $F_j$ of $F$ for which $\mathfrak{p}_j+\mathfrak{m}\neq A$.*

In fact, each $F_j$ is closed by Theorem 8, and hence also $\bigcap_j F_j$ is closed. It remains to be proved that $F$ *is dense in* $\bigcap_j F_j$. Let $x$ be any element of $\bigcap_j F_j$. For every index $\nu$ such that $\mathfrak{p}_\nu+\mathfrak{m}=A$ we choose an exponent $s(\nu)$ such that $\mathfrak{p}_\nu^{s(\nu)}E\subset F_\nu$. Since $\mathfrak{p}_\nu$ and $\mathfrak{m}$ are comaximal, $\mathfrak{p}_\nu^{s(\nu)}$ and $\mathfrak{m}^n$ are comaximal for every $n$, and there exist elements $p_{\nu_n}$ of $\mathfrak{p}_\nu^{s(\nu)}$ and $m_{\nu_n}$ of $\mathfrak{m}^n$ such that $p_{\nu_n}+m_{\nu_n}=1$. The element $y=\left(\prod_\nu p_{\nu_n}\right)x$ is in every $F_\nu$ since $\mathfrak{p}_\nu^{s(\nu)}x\subset F_\nu$, whence $y$ is in $F$ since $x\in\bigcap_j F_j$. On the other hand, we have $y\equiv x(\text{mod } \mathfrak{m}^nE)$ since $\prod_\nu(1-m_{\nu_n})\equiv 1(\mathfrak{m}^n)$. Thus every neighborhood of $x$ has points in common with $F$, and this proves our assertion.

THEOREM 9. *Let $A$ be a noetherian ring, and $\mathfrak{m}$ an ideal in $A$. The following conditions are equivalent:*

(a) *For every finite $A$-module $E$ and every submodule $F$ of $E$, $F$ is closed for the $\mathfrak{m}$-topology of $E$* $\left(\text{i.e., } F = \bigcap_{n=0}^{\infty} (F+\mathfrak{m}^nE)\right)$.

(a′) *$A$ is a Hausdorff space in its $\mathfrak{m}$-topology, and for every finite $A$-module $E$ and every submodule $F$ of $E$ we have $F=\hat{A}F\cap E$.*

(b) *Every finite $A$-module $E$ (in particular, $A$ itself) is a Hausdorff space in its $\mathfrak{m}$-topology.*

(c) *Every ideal in $A$ is closed in the $\mathfrak{m}$-topology of $A$.*

(d) *The ideal $\mathfrak{m}$ is contained in the intersection of all the maximal ideals of $A$.*

(e) *Every element of* $1+\mathfrak{m}$ *is invertible in $A$.*

(f) *For every finite $A$-module $E$ the relation $E=\mathfrak{m}E$ implies $E=(0)$.*

PROOF. We shall give a cyclic proof (a) ⇒ (b) ⇒ (c) ⇒ (d) ⇒ (e) ⇒ (f) ⇒ (a), and in the course of the proof we shall also establish the equivalence of (a) and (a′). For $F=(0)$, (a) implies $\bigcap_{n=0}^{\infty}\mathfrak{m}^nE=(0)$, i.e., (a) implies (b). *Therefore,* (a) *also implies* (a′), for if both $A$ and $E$ are

Hausdorff spaces in their $\mathfrak{m}$-topologies, then, by Corollary 2 to Theorem 5 (§ 2), $\hat{A}F \cap E$ is the closure of $F$ in $E$.

Assume (b), and let $\mathfrak{a}$ be any ideal in $A$. The $A$-module $A/\mathfrak{a}$ is finite (it has the $\mathfrak{a}$-residue of 1 as a basis), and hence $\bigcap_{n=0}^{\infty} \mathfrak{m}^n(A/\mathfrak{a})=(0)$. This signifies that $\bigcap_{n=0}^{\infty} (\mathfrak{m}^n+\mathfrak{a})=\mathfrak{a}$ and thus (b) implies (c). If (c) holds and if $\mathfrak{p}$ is maximal ideal in $A$, we cannot have $\mathfrak{p}+\mathfrak{m}^n=A$ for every $n$ (otherwise $\mathfrak{p}$ would not be closed). Since $\mathfrak{p}+\mathfrak{m}=A$ implies $\mathfrak{p}+\mathfrak{m}^n=A$ for every $n$, we conclude that $\mathfrak{p}+\mathfrak{m}\neq A$. Hence $\mathfrak{p}\supset\mathfrak{m}$ since $\mathfrak{p}$ is maximal, and therefore (d) holds.

If (d) holds and if $1+m$ ($m\in\mathfrak{m}$) is an element of $1+\mathfrak{m}$, we have $1+m\notin\mathfrak{p}$ for every maximal ideal $\mathfrak{p}$ since $m\in\mathfrak{m}\subset\mathfrak{p}$. Thus the principal ideal $(1+m)$ must be the unit ideal, and $1+m$ is invertible in $A$.

If (e) holds and if $\{x_1, \cdots, x_q\}$ is a finite basis of a module $E$ such that $E=\mathfrak{m}E$, we have relations $x_i=\sum m_{ij}x_j$, with $m_{ij}\in\mathfrak{m}$. If we set $d=\det(\delta_{ij}-m_{ij})$ (where the $\delta_{ij}$ are the Kronecker symbols), this implies $dx_i=0$ for every $i$. Since $d$ belongs to $1+\mathfrak{m}$, it is invertible, whence $x_i=0$ and $E=(0)$.

Suppose that (f) holds. If $\mathfrak{a}$ is an ideal such that $\mathfrak{a}+\mathfrak{m}=A$, and if we set $E=A/\mathfrak{a}$, we have $\mathfrak{m}E=(\mathfrak{a}+\mathfrak{m}A)/\mathfrak{a}=(\mathfrak{a}+\mathfrak{m})/\mathfrak{a}=A/\mathfrak{a}=E$, whence $E=(0)$ and $\mathfrak{a}=A$. In particular, we have $\mathfrak{p}+\mathfrak{m}\neq A$ for every prime ideal $\mathfrak{p}$ of $A$ distinct from $A$. Thus Theorem 8 proves that (a) holds. Q.E.D.

Finally, if (a') holds, then, in the special case $E=A$, it follows from Corollary 2 to Theorem 5 (§ 2) that every ideal in $A$ is a closed set in the $\mathfrak{m}$-topology of $A$, and hence (a') implies (c).

COROLLARY. *Let $A$ be a noetherian ring and $\mathfrak{m}$ an ideal in $A$ such that every element of $1+\mathfrak{m}$ is invertible in $A$. Then, if $E$ is a finite $A$-module and $F$ a submodule of $E$ whose leading submodule* (§ 1, p. 250) *is equal to $G_{\mathfrak{m}}(E)$, then $E=F$.*

In fact the associated graded module of $E/F$ is $G(E)/G(F)=(0)$ (§ 1, p. 250). Therefore we have $\mathfrak{m}(E/F)=E/F$, whence $E/F=(0)$ since (e) implies (f).

DEFINITION. *A noetherian ring $A$ is said to be a Zariski ring with respect to an ideal $\mathfrak{m}$ in $A$ if $A$ and $\mathfrak{m}$ satisfy the equivalent conditions listed in Theorem* 9.

We shall often simply say "$A$ is a Zariski ring" when the nature of the ideal $\mathfrak{m}$ is clear from the context. Notice that $\mathfrak{m}$ may be replaced by any ideal having the same radical as $\mathfrak{m}$.

*Examples of Zariski rings:*

(1) A noetherian *local* ring, with respect to its maximal ideal (by (d)).

(2) A noetherian ring $A$ admitting only a finite number of maximal ideals $\mathfrak{m}_i$, with respect to their intersection $\mathfrak{m} = \bigcap_i \mathfrak{m}_i$ (by (d)). Such a ring is said to be *semi-local.*

(3) A noetherian ring $A$ which is a *complete* Hausdorff space in its $\mathfrak{m}$-topology. In fact, every element $1 - m$ ($m \in \mathfrak{m}$) of $1 + \mathfrak{m}$ is invertible, since it admits $1 + m + m^2 + \cdots + m^n + \cdots$ as an inverse. In particular, if $A$ is a noetherian ring and $\mathfrak{m}$ an ideal in $A$ such that $A$ is a Hausdorff space in its $\mathfrak{m}$-topology, then $\hat{A}$ is a Zariski ring, since $\hat{A}$ is noetherian (Corollary 5 to Theorem 7, § 3).

(4) A *factor ring* $A/\mathfrak{a}$ of a Zariski ring, with respect to the ideal $(\mathfrak{m} + \mathfrak{a})/\mathfrak{a}$.

THEOREM 10. *Let $A$ be a Zariski ring with respect to the ideal $\mathfrak{m}$. In order that $A$ be a semi-local (local) ring, it is necessary and sufficient that $A/\mathfrak{m}$ be a ring satisfying the descending chain condition (a ring satisfying the* d.c.c., *with only one prime ideal).*

PROOF. Suppose that $A$ is semi-local. Then the radical of $\mathfrak{m}$ is the intersection of the maximal ideals of $A$. Hence $A/\mathfrak{m}$ is a noetherian ring in which every prime ideal different from (1) is maximal, i.e., $A/\mathfrak{m}$ is a ring satisfying the d.c.c. (Vol. I, Ch. IV, § 2, Theorem 2). Similarly if $A$ is local. Conversely, if $A/\mathfrak{m}$ satisfies the d.c.c., there is only a finite number of prime ideals $\mathfrak{p}_i/\mathfrak{m}$ in $A/\mathfrak{m}$, and they are maximal (Vol. I, Ch. IV, § 2, Theorem 2). Since all the maximal ideals in $A$ contain $\mathfrak{m}$ (Theorem 9, (d)), $A$ has only a finite number of maximal ideals, whence $A$ is semi-local. Similarly, if $A/\mathfrak{m}$ has only one prime ideal different from (1), $A$ has only one maximal ideal, and is a local ring. Q.E.D.

COROLLARY. *The completion $\hat{A}$ of a semi-local (local) ring $A$ is a semi-local (local) ring.*

In fact, we have seen that $\hat{A}$ is a Zariski ring with respect to $\hat{A}\mathfrak{m}$. Since $\hat{A}/\hat{A}\mathfrak{m}$ is isomorphic to $A/\mathfrak{m}$ (§ 2, Theorem 6, Corollary 1), our assertion follows from Theorem 10.

Let $A$ be a Zariski ring with respect to the ideal $\mathfrak{m}$. If $f$ is a *linear mapping* of an $A$- module $E$ into an $A$-module $F$, $f$ is *uniformly continuous* for the $\mathfrak{m}$-topologies, since $f(\mathfrak{m}^n E) \subset \mathfrak{m}^n F$. Thus $f$ can be extended by continuity, and in a unique way, to a mapping $\hat{f}$ of $\hat{E}$ into $\hat{F}$. By passage to the limit it is easily seen that $\hat{f}$ is *$\hat{A}$-linear.*

THEOREM 11. *Let $A$ be a Zariski ring, and $E \xrightarrow{f} F \xrightarrow{g} G$ be an exact sequence of finite $A$-modules and of $A$-linear mappings. Then the sequence $\hat{E} \xrightarrow{\hat{f}} \hat{F} \xrightarrow{\hat{g}} \hat{G}$ is exact.*

PROOF. Our hypothesis signifies that $f(E)=g^{-1}(0)$ and implies that $g(f(x))=0$ for every $x$ in $E$. Hence, by continuity, we have $\bar{g}(\bar{f}(\xi))=0$ for every $\xi$ in $\hat{E}$. Thus the kernel $\bar{g}^{-1}(0)$ of $\bar{g}$ contains the image $\bar{f}(\hat{E})$ of $\bar{f}$. We have to prove that these two submodules of $\hat{F}$ are equal, i.e., that every element $\eta$ of $\hat{F}$ such that $\bar{g}(\eta)=0$ is in $\bar{f}(\hat{E})$.

The submodule $G'=g(F)$ of $G$ has the $\mathfrak{m}$-topology as induced topology (§ 2, Theorem 4). Thus its completion $\hat{G}'$ is identical with its closure in $\hat{G}$. By continuity $\bar{g}$ maps $\hat{F}$ into $\hat{G}'$, and, since $\bar{g}(\hat{F})$ is a closed submodule of $\hat{G}'$ (as $\hat{A}$ is a Zariski ring) which contains $g(F)=G'$, we have $\bar{g}(\hat{F})=\hat{G}'$.

Consider now an element $\eta$ of $\hat{F}$ such that $\bar{g}(\eta)=0$. We approximate $\eta$ by an element $y_n$ of $F$ such that $\eta-y_n \in \hat{A}\mathfrak{m}^n\hat{F}=\mathfrak{m}^n\hat{F}$. Then, since $\bar{g}(\eta)=0$, we have $g(y_n) \in \mathfrak{m}^n\bar{g}(\hat{F}) \cap g(F)=\mathfrak{m}^n\hat{G}' \cap G'=\mathfrak{m}^nG'$ (by Theorem 9, (a'))$=\mathfrak{m}^ng(F)=g(\mathfrak{m}^nF)$. In other words, there exists an element $y'_n$ of $\mathfrak{m}^nF$ such that $g(y_n)=g(y'_n)$. Since $g(y_n-y'_n)=0$, the fact that $f(E)=g^{-1}(0)$ implies that $y_n-y'_n \in f(E)$. Hence $y_n \in f(E)+\mathfrak{m}^nF$, and $\eta \in f(E)+\mathfrak{m}^n\hat{F}$. Since this holds for every $n$, it follows that $\eta$ is in the closure of $f(E)$ in $\hat{F}$. Since the submodule $\bar{f}(\hat{E})$ of $\hat{E}$ is closed and contains $f(E)$, we conclude that $\eta \in \bar{f}(\hat{E})$. Q.E.D.

REMARK. We have seen, in the course of the proof, that $\bar{g}(\hat{F})$ is the closure of $g(F)$. For the same reason $\bar{f}(\hat{E})$ is the closure of $f(E)$.

COROLLARY 1. *Let $A$ be a Zariski ring, $E$ a finite $A$-module, and $\{x_1, \cdots, x_q\}$ a finite family of elements of $E$. Then every linear relation $\sum_i \alpha_i x_i=0$, with coefficients $\alpha_i$ in $\hat{A}$, satisfied by the $x_i$ in $\hat{E}$, is a linear combination (with coefficients in $\hat{A}$) of relations $\sum_i a_{ji}x_i=0$ with coefficients in $A$.*

Consider the free module $F=\sum_{i=1}^{q} AX_i$ with $q$ generators over $A$, and the homomorphism $g$ of $F$ into $E$ defined by $g(X_i)=x_i$. Let $R$ be the kernel of $g$, i.e., let $R$ be the module of relations satisfied by the $x_i$ over $A$. The sequence

$$0 \to R \xrightarrow{i} F \xrightarrow{g} E$$

is exact ($i$ denoting the natural mapping of $R$ into $F$). By Theorem 11, we get an exact sequence

$$0 \to \hat{R} \xrightarrow{\bar{i}} \hat{F} \xrightarrow{\bar{g}} \hat{E},$$

which shows that $\hat{R}$ is isomorphic to the kernel of $\bar{g}$. Since $\hat{F}$ is obviously the free module $\sum_i \hat{A}X_i$, this means that $\hat{R}$ is isomorphic to the module of relations satisfied by the $x_i$ over $\hat{A}$. As $\hat{R}=\hat{A}R$ (§ 2, Theorem 5), our assertion is proved.

We point out that Corollary 1 together with Theorem 5, § 2, imply that the completion $\hat{E}$ is isomorphic to the *tensor product* $\hat{A}\otimes_A E$. The preservation of exactness proved in Theorem 11 is *not* a general property of tensor products; the fact that exactness is preserved in the present case means that the torsion functor $\mathrm{Tor}_1{}^A(\hat{A}, E)$ is 0 for every finite $A$-module $E$.

COROLLARY 2. *Let $A$ be a Zariski ring, and let $F$ and $G$ be two submodules of a finite $A$-module $E$. Then $\hat{A}(F\cap G)=\hat{A}F\cap\hat{A}G$.*

We consider the mapping $g$ of the direct sum $F\oplus G$ into $E$ defined by $g(x, y)=x-y$ $(x\in F, y\in G)$. The kernel $K$ of $g$ is the set of elements $(x, x)$ with $x\in F$ and $x\in G$, and is therefore isomorphic to $F\cap G$. From the exact sequence

$$0\to K\xrightarrow{i}F\oplus G\xrightarrow{g}E$$

we deduce, by Theorem 11, the exact sequence

$$0\to\hat{K}\xrightarrow{\bar{i}}\hat{F}\oplus\hat{G}\xrightarrow{\bar{g}}\hat{E},$$

where $\bar{g}$ is defined by $\bar{g}(\xi, \eta)=\xi-\eta(\xi\in\hat{F}, \eta\in\hat{G})$. Thus $\hat{K}$ may be identified with $\hat{F}\cap\hat{G}$, i.e., with $\hat{A}F\cap\hat{A}G$. Since $\hat{K}=\hat{A}K=\hat{A}(F\cap G)$, the corollary follows.

In particular, if $\mathfrak{a}$ and $\mathfrak{b}$ are *ideals* in $A$, we have $\hat{A}(\mathfrak{a}\cap\mathfrak{b})=\hat{A}\mathfrak{a}\cap\hat{A}\mathfrak{b}$.

COROLLARY 3. *Let $A$ be a Zariski ring, $E$ and $F$ two finite $A$-modules, $f$ a linear mapping of $E$ into $F$, and $F'$ a submodule of $F$. Then $\hat{A}f^{-1}(F')=\bar{f}^{-1}(\hat{A}F')$.*

We denote by $g$ the linear mapping of $E$ into $F/F'$ defined by $g(x)=$ residue class of $f(x)$ mod $F'$. We have the exact sequence

$$0\to f^{-1}(F')\xrightarrow{i}E\xrightarrow{g}F/F',$$

from which we deduce

$$0\to\hat{A}f^{-1}(F')\xrightarrow{\bar{i}}\hat{E}\xrightarrow{\bar{g}}\hat{F}/\hat{A}F'.$$

If $\xi$ is any element of $\hat{E}$, then $\bar{g}(\xi)$ is the residue class of $\bar{f}(\xi)$ mod $\hat{A}F'$ ($\bar{g}$ being the composition of $\bar{f}$ and the canonical mapping of $\hat{F}$ onto $\hat{F}/\hat{A}F'$, since $g$ is the composition of $f$ and the canonical mapping of $F$ onto $F/F'$). Thus the kernel $\hat{A}f^{-1}(F')$ of $\bar{g}$ is $\bar{f}^{-1}(\hat{A}F')$.

COROLLARY 4. *Let $A$ be a Zariski ring, $a$ an element of $A$, $E$ a finite $A$-module, and $G$ a submodule of $E$. Then $\hat{A}(G:Aa)=\hat{A}G:\hat{A}a$.*

In fact, $G:Aa$ is the submodule of all elements $x$ in $E$ such that $ax \in G$. To obtain the corollary, it suffices to apply Corollary 3 to the case $F=E$, $F'=G$ and to take for $f$ the mapping $x \to ax$ of $E$ into $E$.

COROLLARY 5. *Let $A$ be a Zariski ring, $E$ a finite $A$-module, $E'$ a submodule of $E$ and $z$ an element of $E$. Then $\hat{A}(E':Az)=\hat{A}E':Az$.*

We recall that $E':Az$ is the ideal of all elements $a$ in $A$ such that $az \in E'$. We apply Corollary 3 to the case $E=A$, $F=E$, $F'=E'$, and take for $f$ the mapping $a \to az$.

In particular, if $\mathfrak{b}$ is an ideal in $A$ and $a$ an element of $A$, we have $\hat{A}(\mathfrak{b}:Aa) = \hat{A}\mathfrak{b}:\hat{A}a$.

COROLLARY 6. *Let $A$ be a Zariski ring. If an element $c$ of $A$ is not a zero divisor in $A$, it is not a zero divisor in the completion $\hat{A}$.*

In fact $(0):\hat{A}c=\hat{A}((0):Ac)=(0)$ by Corollary 4 (or 5).

COROLLARY 7. *Let $A$ be a Zariski ring, $E$ a finite $A$-module, $F$ a submodule of $E$ and $\mathfrak{a}$ an ideal in $A$. Then $\hat{A}(F:\mathfrak{a})=\hat{A}F:\hat{A}\mathfrak{a}$.*

Let $\{a_1, \cdots, a_q\}$ be a finite basis of $\mathfrak{a}$. We have $F:\mathfrak{a}=\bigcap_{j=1}^{q} (F:Aa_j)$. Thus Corollary 7 follows from Corollaries 2 and 4.

COROLLARY 8. *Let $A$ be a Zariski ring, $E$ a finite $A$-module, $F$ and $G$ two submodules of $E$. Then $\hat{A}(F:G)=\hat{A}F:\hat{A}G$.*

We take a finite basis $\{z_1, \cdots, z_q\}$ of $G$, we observe that $F:G= \bigcap_{j=1}^{q} (F:Az_j)$, and apply Corollaries 2 and 5.

In particular, if $\mathfrak{a}$ and $\mathfrak{b}$ are two ideals in $A$, we have $\hat{A}(\mathfrak{a}:\mathfrak{b})=\hat{A}\mathfrak{a}:\hat{A}\mathfrak{b}$.

Let us now study more closely the relations between a noetherian ring $A$ (not necessarily a Zariski ring) and its completion $\hat{A}$ with respect to the $\mathfrak{m}$-topology. It will be convenient to include in this study (at least at the initial stage) also those rings $A$ which are not Hausdorff spaces. However, we have not yet defined the completion $\hat{A}$ of a ring $A$, with respect to its $\mathfrak{m}$-topology, if $A$ is not a Hausdorff space. We shall do so now. It is clear that if we set $A'=A/\bigcap_{n=1}^{\infty} \mathfrak{m}^n$ and $\mathfrak{m}'= \mathfrak{m}/\bigcap_{n=1}^{\infty} \mathfrak{m}^n$, then $A'$ is a Hausdorff space in its $\mathfrak{m}'$-topology. *We define $\hat{A}$ to be the completion $\hat{A}'$ of $A'$, with respect to the $\mathfrak{m}'$-topology of $A'$.* If $\bigcap_{n=1}^{\infty} \mathfrak{m}^n \neq (0)$, then $A$ is not any more a subring of $\hat{A}$, but we have the canonical homomorphism $A \to A' \to \hat{A}'=\hat{A}$ of $A$ into $\hat{A}$, and this homomorphism is a continuous mapping.

As $\hat{A}$ is a Zariski ring, every element of $1+\hat{A}\mathfrak{m}$, and in particular the image of every element of $1+\mathfrak{m}$, is invertible in $\hat{A}$. Since $S=1+\mathfrak{m}$ is a multiplicatively closed set in $A$, we are led to study the *quotient ring* $A_S$.

The kernel $\mathfrak{n}$ of the canonical homomorphism $\varphi: A \rightarrow A_S$ is (Vol. I, Ch. IV, § 9) the set of all elements $b$ in $A$ for which there exists an element $s=1-m$ in $S (m \in \mathfrak{m})$ such that $bs=0$. This last relation implies $b=bm=bm^2=\cdots=bm^n$, hence $\mathfrak{n}$ is contained in $\bigcap_{n=0}^{\infty} \mathfrak{m}^n$. Conversely, if $b$ is an element of $\bigcap_{n=1}^{\infty} \mathfrak{m}^n$, then there exists an integer $q$ such that the ideal $\mathfrak{m}b$ contains $\mathfrak{m}^q \cap Ab$ (Vol. I, Ch. IV, § 7, Lemma 1) and therefore is equal to $Ab$ (since $b \in \mathfrak{m}^q$, all $q$); we thus have $b \in \mathfrak{m}b$, i.e., $b=mb$ for some $m$ in $\mathfrak{m}$, whence $b(1-m)=0$ and $b$ belongs to $\mathfrak{n}$. Since the ideal $\bigcap_{n=1}^{\infty} \mathfrak{m}^n$, which is the closure of 0 in $A$, is also the kernel of the homomorphism of $A$ into its completion $\hat{A}$, it follows that *the quotient ring $A_S$ may be identified with a subring of the completion $\hat{A}$.*

From now on we simplify matters by replacing $A$ by $A'$, i.e., by assuming that $A$ is a Hausdorff space in its $\mathfrak{m}$-topology. In that case we have $\bigcap_{n=0}^{\infty} \mathfrak{m}^n=(0)$ and hence no element of $S$ is a zero-divisor in $A$. Therefore $A$ is a subring of $A_S$. We consider the $\mathfrak{m}$-topology on $A_S$ (considered as an $A$-module), i.e., the topology defined by the powers of the ideal $\mathfrak{m}A_S$. It is clear (since $\mathfrak{m}^n \subset \mathfrak{m}^n A_S \cap A$) that the $\mathfrak{m}$-topology of $A$ is stronger than the topology induced in $A$ by the $\mathfrak{m}$-topology of $A_S$. On the other hand, if an element $a$ of $A$ belongs to $\mathfrak{m}^n A_S$, we have $a(1+m) \in \mathfrak{m}^n$ for some $m$ in $\mathfrak{m}$, whence $a \in \mathfrak{m}^n + a\mathfrak{m}$; this implies $a \in \mathfrak{m}^n + (\mathfrak{m}^n + a\mathfrak{m})\mathfrak{m} = \mathfrak{m}^n + a\mathfrak{m}^2$, whence, by successive applications, $a \in \mathfrak{m}^n + a\mathfrak{m}^n = \mathfrak{m}^n$. We have therefore shown that $\mathfrak{m}^n = \mathfrak{m}^n A_S \cap A$, and hence *the $\mathfrak{m}$-topology of $A$ is induced by the $(\mathfrak{m}A_S)$-topology of $A_S$.*

It follows that $\hat{A}$ is also the *completion* of $A_S$. We now remark that $A_S$ is a *Zariski ring*, i.e., that every element $y=1+\frac{m}{1+m'}$ $(m, m' \in \mathfrak{m})$ of $1+\mathfrak{m}A_S$ is invertible. In fact, we have $y=\frac{1+m+m'}{1+m'}$, and, since $1+m+m'$ $(\in 1+\mathfrak{m})$ has an inverse $x$ in $A_S$, the element $x(1+m')$ is the inverse of $y$.

Since the passage from a Zariski ring to its completion has been extensively described by Theorem 11 and its corollaries, and since the passage from $A$ to $A_S$ has been described in detail in Vol. I, Ch. IV, § 11, we have now a certain amount of information about the passage from $A$ to $\hat{A}$. As an illustration we prove

THEOREM 12. *Let* $\mathfrak{m}$ *be an ideal in a noetherian ring* $A$, *such that* $A$ *is a Hausdorff space in its* $\mathfrak{m}$-*topology (i.e., such that* $\bigcap_{n=0}^{\infty} \mathfrak{m}^n = (0)$). *Let* $\mathfrak{a}$ *be a closed ideal in* $A$, *let* $\hat{A}\mathfrak{a} = \mathfrak{q}_1^\star \cap \mathfrak{q}_2^\star \cap \cdots \cap \mathfrak{q}_n^\star$ *be an irredundant primary representation of* $\hat{A}\mathfrak{a}$, *and let* $\mathfrak{p}_i^\star$ *be the prime ideal which is the radical of* $\mathfrak{q}_i^\star$. *Then* $\mathfrak{a} = \bigcap_i (\mathfrak{q}_i^\star \cap A)$ *is a primary representation of* $\mathfrak{a}$ *and* $\mathfrak{p}_i^\star \cap A$ *is the associated prime ideal of* $\mathfrak{q}_i^\star \cap A$ *and is contained in an associated prime ideal of* $\mathfrak{a}$.

PROOF. Consider the quotient ring $A_S$ where $S = 1 + \mathfrak{m}$. As $\mathfrak{a}$ is closed, we have $\hat{A}\mathfrak{a} \cap A = \mathfrak{a}$ (Corollary 2 to Theorem 5, § 2), whence $\mathfrak{a}A_S \cap A = \mathfrak{a}$. By using properties of quotient rings, we see that it would be sufficient to prove Theorem 12 for $\mathfrak{m}A_S$, $A_S$ and $\mathfrak{a}A_S$, instead of for $\mathfrak{m}$, $A$ and $\mathfrak{a}$ respectively. In other words, we may assume that $A$ is a Zariski ring.

Any element $x$ of $\mathfrak{p}_i^\star$ is a zero divisor mod $\hat{A}\mathfrak{a}$. Since any regular element in $A/\mathfrak{a}$ is also regular in $\hat{A}/\hat{A}\mathfrak{a}$ (Corollary 6 to Theorem 11), every element of $\mathfrak{p}_i^\star \cap A$ is a zero divisor mod $\mathfrak{a}$ and therefore belongs to some associated prime ideal of $\mathfrak{a}$. On the other hand, $\mathfrak{q}_i^\star \cap A$ is obviously a primary ideal admitting $\mathfrak{p}_i^\star \cap A$ as radical. Therefore, from $\hat{A}\mathfrak{a} = \bigcap_i \mathfrak{q}_i^\star$ and from $\hat{A}\mathfrak{a} \cap A = \mathfrak{a}$, we deduce that $\mathfrak{a} = \bigcap_i (\mathfrak{q}_i^\star \cap A)$. This is a (not necessarily irredundant) primary representation of $\mathfrak{a}$. Q.E.D.

COROLLARY 1. *If, furthermore,* $\mathfrak{a}$ *is a prime ideal, we have* $\mathfrak{a} = \mathfrak{q}_i^\star \cap A = \mathfrak{p}_i^\star \cap A$ *for every* $i$.

In fact $\mathfrak{p}_i^\star \cap A$ is contained in an associated prime ideal of $\mathfrak{a}$, i.e., $\mathfrak{p}_i^\star \cap A$ is contained in $\mathfrak{a}$.

COROLLARY 2. *With the same assumptions on* $A$ *and* $\mathfrak{m}$ *as in Theorem* 12, *assume furthermore that the closed ideal* $\mathfrak{a}$ *admits an irredundant primary representation* $\mathfrak{a} = \mathfrak{Q}_1 \cap \mathfrak{Q}_2 \cap \cdots \cap \mathfrak{Q}_h$ *such that none of the prime ideals* $\mathfrak{P}_j = \sqrt{\mathfrak{Q}_j}$ *is embedded. Then* $\hat{A}\mathfrak{a} = \hat{A}\mathfrak{Q}_1 \cap \hat{A}\mathfrak{Q}_2 \cap \cdots \cap \hat{A}\mathfrak{Q}_h$.

As in Theorem 12, let $\hat{A}\mathfrak{a} = \mathfrak{q}_1^\star \cap \mathfrak{q}_2^\star \cap \cdots \cap \mathfrak{q}_n^\star$ be an irredundant primary prepresentation of $\hat{A}\mathfrak{a}$ and let $\mathfrak{q}_i = \mathfrak{q}_i^\star \cap A$, $\mathfrak{p}_i = \mathfrak{p}_i^\star \cap A = \sqrt{\mathfrak{q}_i}$. It is clear that $\mathfrak{P}_1, \mathfrak{P}_2, \cdots, \mathfrak{P}_h$ are among the prime ideals $\mathfrak{p}_1, \mathfrak{p}_2, \cdots, \mathfrak{p}_n$ and that each $\mathfrak{p}_i$ contains one of the prime ideals $\mathfrak{P}_1, \mathfrak{P}_2, \cdots, \mathfrak{P}_h$. On the other hand, by Theorem 12, each $\mathfrak{p}_i$ is contained in one of the prime ideals $\mathfrak{P}_1, \mathfrak{P}_2, \cdots, \mathfrak{P}_h$. Since no $\mathfrak{P}_j$ is embedded, it follows that the set $\{\mathfrak{p}_1, \mathfrak{p}_2, \cdots, \mathfrak{p}_n\}$ coincides with the set $\{\mathfrak{P}_1, \mathfrak{P}_2, \cdots, \mathfrak{P}_h\}$ (the $n$ prime ideals $\mathfrak{p}_i$ are, however, not necessarily distinct). Hence each of the ideals $\mathfrak{Q}_j$ is the intersection of those $\mathfrak{q}_i$ for which $\mathfrak{p}_i=\mathfrak{P}_j$. If, say, $\mathfrak{Q}_1=\mathfrak{q}_1\cap\mathfrak{q}_2\cap\cdots\cap\mathfrak{q}_s$ then $\hat{A}\mathfrak{Q}_1\subset\hat{A}\mathfrak{q}_1\cap\hat{A}\mathfrak{q}_2\cap\cdots\hat{A}\mathfrak{q}_s$, and similarly for the ideals $\hat{A}\mathfrak{Q}_j$. Hence $\bigcap_{j=1}^{h}\hat{A}\mathfrak{Q}_j\subset\bigcap_{i=1}^{n}\hat{A}\mathfrak{q}_i\subset\bigcap_{i=1}^{n}\mathfrak{q}_i^{\star}=\hat{A}\mathfrak{a}$, and since the opposite inclusions $\hat{A}\mathfrak{a}\subset\bigcap_{j=1}^{h}\hat{A}\mathfrak{Q}_j$ is obvious, the corollary is proved.

§ 5. **Comparison of topologies in a noetherian ring.** Let $A$ be a noetherian ring. One is led to consider on $A$, not only the $\mathfrak{m}$-topologies (where $\mathfrak{m}$ is an ideal in $A$), but also topologies of a more general type. For example, if $\mathfrak{p}$ is a prime ideal in $A$, one may construct the local ring $A_{\mathfrak{p}}$, consider its natural topology (defined by the powers of the maximal ideal $\mathfrak{p}A_{\mathfrak{p}}$) and the induced topology on $A$. In this topology, the *symbolic* powers $\mathfrak{p}^{(n)}$ $(=(\mathfrak{p}^n)^{ec}$; Vol. I, Ch. IV, § 12) constitute a basis of neighborhoods of 0; notice that we have $\mathfrak{p}^{(n)}\cdot\mathfrak{p}^{(q)}\subset\mathfrak{p}^{(n+q)}$ (Vol. I, Ch. IV, § 12, Theorem 23).

More generally, given a noetherian ring $A$ and a *descending sequence* $(\mathfrak{a}_n)$ *of ideals of* $A$ such that

$$\mathfrak{a}_n\mathfrak{a}_q\subset\mathfrak{a}_{n+q},\tag{1}$$

we define the $(\mathfrak{a}_n)$-topology of $A$ as being the topology in which the ideals $\mathfrak{a}_n$ constitute a basis of neighborhoods of 0, the basic neighborhoods of any other element $a$ of $A$ being the cosets $a+\mathfrak{a}_n$. With respect to this topology, $A$ is a topological ring, and as in § 2, this topology is induced in $A$ by a metric, satisfying the strong triangular inequality. This space is Hausdorff if and only if $\bigcap_{n=0}^{\infty}\mathfrak{a}_n=(0)$.

In the case of a complete *semi-local ring* $A$, the next theorem gives an "extremal" property of the natural topology of $A$.

THEOREM 13 (CHEVALLEY). *Let $A$ be a complete semi-local ring*, $\mathfrak{m}$ *the intersection of its maximal ideals, and* $(\mathfrak{a}_n)$ *a descending sequence of ideals of $A$ such that* $\bigcap_{n=0}^{\infty}\mathfrak{a}_n=(0)$. *Then there exists an integral valued function $s(n)$ which tends to infinity with $n$, such that* $\mathfrak{a}_n\subset\mathfrak{m}^{s(n)}$.

PROOF. We shall use an indirect argument. Suppose that there exists an integer $s$ such that $\mathfrak{a}_n\not\subset\mathfrak{m}^s$ for every $n$. Since the ring $A/\mathfrak{m}^s$ satisfies the d.c.c. (§ 4, Theorem 10), and since in this ring the ideals $(\mathfrak{a}_n+\mathfrak{m}^s)/\mathfrak{m}^s$ form a descending sequence of ideals $\neq(0)$, their intersection is $\neq(0)$, and there exists an element $x_s\notin\mathfrak{m}^s$ such that $x_s\in\mathfrak{a}_n+\mathfrak{m}^s$ for every $n$.

We now define, by induction on $t \geq s$, a *Cauchy sequence* of elements $x_t$ of $A$ such that

(2) $$x_t \equiv x_s \pmod{\mathfrak{m}^s}$$

(3) $$x_t \in \mathfrak{a}_n + \mathfrak{m}^t \text{ for every } n.$$

We suppose that $x_t$ is already constructed, and proceed to construct $x_{t+1}$. The relation $x_t \in \mathfrak{a}_n + \mathfrak{m}^t$ implies that the ideal $\mathfrak{a}_n$ has a non-empty intersection with coset $x_t + \mathfrak{m}^t$. We pass to the ring $A/\mathfrak{m}^{t+1}$ and we denote by $\bar{x}_t$ the coset $x_t + \mathfrak{m}^{t+1}$. By (3), the set $\bar{x}_t + (\mathfrak{m}^t/\mathfrak{m}^{t+1})$ has a non-empty intersection with each one of the ideals $(\mathfrak{a}_n + \mathfrak{m}^{t+1})/\mathfrak{m}^{t+1}$. As $A/\mathfrak{m}^{t+1}$ satisfies the d.c.c., the intersection $\bigcap_{n=0}^{\infty} (\mathfrak{a}_n + \mathfrak{m}^{t+1})/\mathfrak{m}^{t+1}$ coincides with one of the ideals $(\mathfrak{a}_n + \mathfrak{m}^{t+1})/\mathfrak{m}^{t+1}$, and hence there exists an element $\bar{x}_{t+1}$ of the set $\bar{x}_t + (\mathfrak{m}^t/\mathfrak{m}^{t+1})$ which lies in all the ideals $(\mathfrak{a}_n + \mathfrak{m}^{t+1})/\mathfrak{m}^{t+1}$. We take for $x_{t+1}$ a representative of $\bar{x}_{t+1}$ in $A$. We have then $x_{t+1} \in \mathfrak{a}_n + \mathfrak{m}^{t+1}$ for every $n$, and $x_{t+1} \equiv x_t \pmod{\mathfrak{m}^t}$. The latter consequence, together with (2), implies that $x_{t+1} \equiv x_s \pmod{\mathfrak{m}^s}$. Thus $x_{t+1}$ satisfies conditions (2) and (3). On the other hand, the relation $x_{t+1} \equiv x_t \pmod{\mathfrak{m}^t}$ implies that $(x_t)$ is a Cauchy sequence.

Since $A$ is complete, the Cauchy sequence $(x_t)$ has a *limit* $x \in A$. From (2) we deduce, since $\mathfrak{m}^s$ is closed, that $x \equiv x_s \pmod{\mathfrak{m}^s}$, whence $x \notin \mathfrak{m}^s$ (since $x_s \notin \mathfrak{m}^s$). The relations $x_{t+1} \equiv x_t \pmod{\mathfrak{m}^t}$ imply that $x \equiv x_t \pmod{\mathfrak{m}^t}$, whence, by using (3), it follows that $x$ belongs to $\mathfrak{a}_n + \mathfrak{m}^t$ for every $n$ and every $t$. From $x \in \bigcap_{t=s}^{\infty} (\mathfrak{a}_n + \mathfrak{m}^t)$, and from the fact that ideals in $A$ are closed sets, we deduce that $x \in \mathfrak{a}_n$ for every $n$. Since $\bigcap_{n=0}^{\infty} \mathfrak{a}_n = (0)$ by hypothesis, we deduce that $x = 0$, in contradiction with $x \notin \mathfrak{m}^s$. Q.E.D.

In topological terms, Theorem 13 signifies that the natural topology of the complete semi-local ring $A$ is *weaker* than any other $(\mathfrak{a}_n)$-topology of $A$ for which $A$ is a Hausdorff space. This resembles a classical property of *compact* spaces whereby a compact space possesses no *Hausdorff* topologies which are *strictly weaker* than the given topology of the compact space. As a matter of fact, the complete semi-local ring $A$, without being in general compact in its $\mathfrak{m}$-topology (we have compactness if and only if $A/\mathfrak{m}$ is a ring with a finite number of elements), is however *linearly compact*† in the sense that, given a family $\mathfrak{a}_\alpha$

† See S. Lefshetz, "Algebraic Topology", p. 78 (*Amer. Math. Soc. Coll. Publ.*, vol. 27, 1942) for the theory of linearly compact vector spaces. The theory of linearly compact modules is analogous, without any significant changes.

of ideals in $A$ and a family of *cosets* $\mathfrak{c}_\alpha = x_\alpha + \mathfrak{a}_\alpha$ with the finite intersection property (i.e., such that $\bigcap_i \mathfrak{c}_{\alpha_i} \neq \emptyset$ for every finite family $\alpha_1, \cdots, \alpha_n$ of indices), then $\bigcap_\alpha \mathfrak{c}_\alpha$ is non-empty. For verifying this property one first proves, by using the d.c.c. in $A/\mathfrak{m}^n$, that $\bigcap_\alpha (\mathfrak{c}_\alpha + \mathfrak{m}^n)$ is $\neq \emptyset$; this being established, one constructs a Cauchy sequence $\{x_n\}$ such that $x_n \in \bigcap_\alpha (\mathfrak{c}_\alpha + \mathfrak{m}^n)$ for every $n$, and it is easily seen that $x = \lim_n x_n$ is an element of $\bigcap_\alpha \mathfrak{c}_\alpha$. In more sophisticated terms this amounts to proving that each $A/\mathfrak{m}^n$ is linearly compact, and that $A$ is the *inverse limit* of the factor rings $A/\mathfrak{m}^n$.

COROLLARY 1. *Let $A$ be a noetherian ring and $\mathfrak{m}$ an ideal in $A$ such that $A$ is a Hausdorff space in its $\mathfrak{m}$-topology. If $c \in A$ is not a zero divisor, then $c$ is not a zero divisor in the completion $\hat{A}$, and we have $\mathfrak{m}^n : Ac \subset \mathfrak{m}^{s(n)}$ where $s(n) \to \infty$ with $n$.*

PROOF. We first consider the case in which $A$ is a semi-local ring and $\mathfrak{m}$ the intersection of the maximal ideals of $A$. Since $c$ is not a zero divisor in $\hat{A}$ (§ 4, Corollary 6 to Theorem 11), we have $\bigcap_{n=0}^{\infty} (\hat{A}\mathfrak{m}^n : \hat{A}c) = \left(\bigcap_{n=0}^{\infty} \hat{A}\mathfrak{m}^n\right) : \hat{A}c = (0) : \hat{A}c = (0)$. Hence, by Theorem 13, we have $\hat{A}\mathfrak{m}^n : \hat{A}c \subset \hat{A}\mathfrak{m}^{s(n)}$, where $s(n) \to \infty$ with $n$, and from this we deduce that $\mathfrak{m}^n : Ac \subset \mathfrak{m}^{s(n)}$ (§ 4, Corollary 4 to Theorem 11).

Let now $\mathfrak{p}$ be a prime ideal in $A$. By applying what has just been proved to the local ring $A_\mathfrak{p}$, and denoting by $\bar{c}$ the image of $c$ in $A_\mathfrak{p}$, we see that $\mathfrak{p}^n A_\mathfrak{p} : A_\mathfrak{p}\bar{c} \subset \mathfrak{p}^{s(n)} A_\mathfrak{p}$, provided $\bar{c}$ is not a zero divisor in $A_\mathfrak{p}$. If we denote by $\mathfrak{n}$ the kernel of the homomorphism $A \to A_\mathfrak{p}$ (i.e., the set of all elements $x$ of $A$ for which there exists an element $s \notin \mathfrak{p}$ such that $sx = 0$), then $\bar{c}$ is not a zero divisor if and only if $cx \notin \mathfrak{n}$ for any $x \notin \mathfrak{n}$. Now, if $x \notin \mathfrak{n}$, we have $xs \neq 0$ for all $s \notin \mathfrak{p}$, and since $c$ is not a zero divisor, it follows that $cxs \neq 0$ for all $s \notin \mathfrak{p}$, whence $cx \notin \mathfrak{n}$. We have therefore shown that $\bar{c}$ is indeed a regular element. Coming back from $A_\mathfrak{p}$ to $A$, we deduce from $\mathfrak{p}^n A_\mathfrak{p} : A_\mathfrak{p}\bar{c} \subset \mathfrak{p}^{s(n)} A_\mathfrak{p}$ that $\mathfrak{p}^{(n)} : Ac \subset \mathfrak{p}^{(s(n))}$.

We consider now an arbitrary power $\mathfrak{m}^j$ of $\mathfrak{m}$ and a primary representation $\mathfrak{m}^j = \cap\, \mathfrak{q}_i$. If $\mathfrak{p}_i$ denotes the radical of $\mathfrak{q}_i$, there is an exponent $t(i)$ such that $\mathfrak{p}_i^{t(i)} \subset \mathfrak{q}_i$ and consequently also $\mathfrak{p}_i^{(t(i))} \subset \mathfrak{q}_i$ since $\mathfrak{q}_i$ is primary for $\mathfrak{p}_i$. By what has been proved above, there exists an exponent $r(i)$ such that $\mathfrak{p}_i^{(r(i))} : Ac \subset \mathfrak{p}_i^{(t(i))}$. We will have then $\mathfrak{p}_i^{(r(i))} : Ac \subset \mathfrak{q}_i$. Denoting by $r$ the greatest of the exponents $r(i)$, we deduce that $\mathfrak{p}_i^{(r)} : Ac \subset \mathfrak{m}^j$, and therefore that $\mathfrak{p}_i^r : Ac \subset \mathfrak{m}^j$. Now, since $\mathfrak{m}^j$ is contained in $\mathfrak{p}_i$, we have $\mathfrak{m}^{jr} \subset \mathfrak{p}_i^r$, whence $\mathfrak{m}^{jr} : Ac \subset \mathfrak{m}^j$. This proves that $\mathfrak{m}^n : Ac$ is contained in $\mathfrak{m}^j$ for $n$ large enough. In other words (since $j$ may be taken to be arbitrarily large), we have $\mathfrak{m}^n : Ac \subset \mathfrak{m}^{s(n)}$, where $s(n) \to \infty$ with $n$.

Finally, suppose that $\alpha$ is an element of $\hat{A}$ such that $c\alpha = 0$. We approximate $\alpha$ by $a_n$ in $A : \alpha - a_n \in \hat{A}\mathfrak{m}^n$. Then $ca_n$ belongs to $\hat{A}\mathfrak{m}^n \cap A = \mathfrak{m}^n$ (§ 2, Corollary 1 to Theorem 5). We thus have $a_n \in \mathfrak{m}^{s(n)}$ with $s(n) \to \infty$. This proves that the limit $\alpha$ of the sequence $\{a_n\}$ is necessarily 0. Q.E.D.

REMARKS:

(a) Notice that the hypothesis $\bigcap_{n=0}^{\infty} \mathfrak{m}^n = (0)$ has only been used in the last part of the proof. The relation $\mathfrak{m}^n : Ac \subset \mathfrak{m}^{s(n)}$ holds without this hypothesis.

(b) A part of Corollary 1 may be strengthened by using the Theorem of Artin and Rees (Theorem 4′, § 2). Let $A$ be a noetherian ring, $\mathfrak{m}$ any ideal in $A$, and $c$ an element of $A$. Then there exists an integer $k$ such that, for $n \geq k$, we have

$$\mathfrak{m}^n : Ac \subset \mathfrak{m}^{n-k} + ((0) : Ac). \tag{4}$$

In fact Theorem 4′ (§ 2) proves the existence of an integer $k$ such that $\mathfrak{m}^n \cap Ac = \mathfrak{m}^{n-k}(\mathfrak{m}^k \cap Ac)$ for every $n \geq k$. Thus, if $x \in \mathfrak{m}^n : Ac$, we have $xc \in \mathfrak{m}^n \cap Ac$, whence $xc \in \mathfrak{m}^{n-k}(\mathfrak{m}^k \cap Ac) \subset \mathfrak{m}^{n-k}c$. Hence we can write $xc = x'c$ with $x' \in \mathfrak{m}^{n-k}$. Therefore $x$ belongs to $\mathfrak{m}^{n-k} + ((0) : Ac)$ since $x = x' + (x - x')$. This proves formula (4). If $c$ is not a zero divisor in $A$, we have $(0) : Ac = (0)$, whence

$$\mathfrak{m}^n : Ac \subset \mathfrak{m}^{n-k} \text{ for every } n \geq k. \tag{5}$$

COROLLARY 2. *Let $A$ be a complete semi-local ring, $B$ an overring of $A$ and $\mathfrak{M}$ an ideal in $B$ such that $\bigcap_{n=0}^{\infty} \mathfrak{M}^n = (0)$. If the ideal $\mathfrak{M} \cap A$ admits the intersection $\mathfrak{m}$ of the maximal ideals of $A$ as radical, then the $\mathfrak{m}$-topology of $A$ is induced by the $\mathfrak{M}$-topology of $B$.*

In fact, the induced topology of $A$ is defined by the ideals $\mathfrak{a}_n = \mathfrak{M}^n \cap A$. Since $\bigcap_{n=0}^{\infty} \mathfrak{a}_n = (0)$, we have $\mathfrak{a}_n \subset \mathfrak{m}^{s(n)}$ (Theorem 13). On the other hand, since there exists an exponent $q$ such that $\mathfrak{m}^q \subset \mathfrak{M} \cap A$, we have $\mathfrak{m}^{qn} \subset \mathfrak{M}^n \cap A = \mathfrak{a}_n$. Thus the ideals $\mathfrak{a}_n$ and $\mathfrak{m}^n$ define the same topology on $A$.

REMARK. The conclusion of Corollary 2 does not necessarily hold if $A$ is a non-complete semi-local ring. However, in that case, it is still true that $\mathfrak{m}^{q^n} \subset \mathfrak{M}^n$, i.e., that the identity mapping $\varphi$ of $A$ into $B$ is uniformly continuous for the $\mathfrak{m}$-topology of $A$ and the $\mathfrak{M}$-topology of $B$. Thus $\varphi$ may be extended, by continuity, to a homomorphism $\bar{\varphi}$ of $\hat{A}$ into $\hat{B}$. If $\bar{\varphi}$ is *one to one* then $A$ is a *topological subspace* of $B$: in fact, Corollary 2 shows then that $\hat{A}$ is a topological subspace of $\hat{B}$. (The converse is also true: If $A$ is a topological subspace of $B$, then the identity mapping $\varphi$ of $A$ into $B$ admits as extension the identity mapping of $\hat{A}$ into $\hat{B}$.) In some important cases, the dimension theory of local rings permits to prove that $\bar{\varphi}$ is a one to one mapping.†

COROLLARY 3. *Let $A$ be a noetherian ring, $\mathfrak{m}$ a maximal ideal in $A$, and $\mathfrak{q}$ a prime ideal in $A$ which is contained in $\mathfrak{m}$. Then, if the $\mathfrak{m}$-adic completion of $A$ is an integral domain, the $(\mathfrak{q}^{(n)})$-topology of $A$ is stronger than its $\mathfrak{m}$-topology.*

It is clear that the $(\mathfrak{q}^n)$-topology of $A$ (with ordinary powers instead of symbolic ones) is stronger than its $\mathfrak{m}$-*topology*, since $\mathfrak{q}^n \subset \mathfrak{m}^n$. Let $\hat{A}$ be the $\mathfrak{m}$-adic completion of $A$. Since $A/\mathfrak{m}$ is a field, $\hat{A}$ is a local ring (Theorem 10, § 4). We first prove that $\bigcap_{n=0}^{\infty} \hat{A}\mathfrak{q}^{(n)} = (0)$. Let $\mathfrak{q}^\star$ be any isolated prime ideal of $\hat{A}\mathfrak{q}$. Since $\mathfrak{q}$ is closed (§ 4, Theorem 8), Corollary 1 to Theorem 12 (§ 4) may be applied, and we have $\mathfrak{q}^\star \cap A = \mathfrak{q}$. By definition of symbolic powers there exists, for every $n$, an element $c_n$ of $A$, $c_n \notin \mathfrak{q}$, such that $c_n\mathfrak{q}^{(n)} \subset \mathfrak{q}^n$. Therefore $c_n\hat{A}\mathfrak{q}^{(n)} \subset (\hat{A}\mathfrak{q})^n \subset \mathfrak{q}^{\star n}$, and, since $c_n \notin \mathfrak{q}$ and $\mathfrak{q}^\star \cap A = \mathfrak{q}$, it follows that $\hat{A}\mathfrak{q}^{(n)} \subset \mathfrak{q}^{\star(n)}$. Now, since $\hat{A}$ is a domain, the intersection of the symbolic powers of any prime ideal in $\hat{A}$ is (0) (Vol. I, Ch. IV, § 12). Thus $\bigcap_{n=0}^{\infty} \mathfrak{q}^{\star(n)} = (0)$, whence, *a fortiori*, $\bigcap_{n=0}^{\infty} \hat{A}\mathfrak{q}^{(n)} = (0)$.

This being so, Theorem 13 shows that the $(\hat{A}\mathfrak{q}^{(n)})$-topology of $\hat{A}$ is stronger than its natural local ring topology. Since $\mathfrak{q}^{(n)}$ is closed in $A$ (§ 4, Theorem 8), we have $\hat{A}\mathfrak{q}^{(n)} \cap A = \mathfrak{q}^{(n)}$ (§ 2, Corollary 2 to Theorem 5), whence the $(\mathfrak{q}^{(n)})$-topology of $A$ is induced by the $(\hat{A}\mathfrak{q}^{(n)})$-topology of $\hat{A}$. Thus the $(\mathfrak{q}^{(n)})$-topology of $A$ is stronger than its $\mathfrak{m}$-topology.

COROLLARY 4. *Let $R$ be a noetherian domain, $\mathfrak{p}$ and $\mathfrak{q}$ two prime ideals in $A$ such that $\mathfrak{p} \supset \mathfrak{q}$. If the $(\mathfrak{p}^{(n)})$-completion of $A$ has no zero divisors, then the $(\mathfrak{q}^{(n)})$-topology of $A$ is stronger than its $(\mathfrak{p}^{(n)})$-topology. Furthermore, if for each prime ideal $\mathfrak{m}$ containing $\mathfrak{p}$ it is true that the $(\mathfrak{m}^{(n)})$-completion of $A$ has no zero divisors, then the $(\mathfrak{q}^{(n)})$-topology of $A$ is stronger than its $\mathfrak{p}$-topology.*

† See O. Zariski, "A simple analytical proof of a fundamental property of birational transformations," Proc. Nat. Acad. Sci. USA, v. 35 (1949), pp. 62–66.

We set $A' = A_{\mathfrak{p}}$, $\mathfrak{p}' = \mathfrak{p}A_{\mathfrak{p}}$, $\mathfrak{q}' = \mathfrak{q}A_{\mathfrak{p}}$. Since the $\mathfrak{p}'$-topology of $A'$ induces the $(\mathfrak{p}^{(n)})$-topology of $A$, the hypotheses of Corollary 3 are satisfied by $A'$, $\mathfrak{p}'$, $\mathfrak{q}'$. Therefore the $(\mathfrak{q}'^{(n)})$-topology of $A'$ is stronger than its $\mathfrak{p}'$-topology. Since the former induces the $(\mathfrak{q}^{(n)})$-topology on $A$ (as $\mathfrak{q}'^{(n)} \cap A = \mathfrak{q}^{(n)}$ by Vol. I, Ch. IV, § 11, Theorem 19), and the latter induces the $(\mathfrak{p}^{(n)})$-topology, our first assertion is proved. As to the second assertion, we decompose $\mathfrak{p}^n$ into primary components: $\mathfrak{p}^n = \mathfrak{p}^{(n)} \cap \mathfrak{a}_1 \cap \cdots \cap \mathfrak{a}_s$, where $\mathfrak{a}_j$ is primary for a prime ideal $\mathfrak{m}_j > \mathfrak{p}$. Then there exists an exponent $u_j$ such that $\mathfrak{m}_j^{(u_j)} \subset \mathfrak{a}_j$, whence, by the first part of the corollary there exists an exponent $i(j, n)$ such that $\mathfrak{p}^{(i(j,n))} \subset \mathfrak{m}_j^{(u_j)} \subset \mathfrak{a}_j$. Setting $i(n) = \max(n, i(1, n), \cdots, i(s, n))$, we therefore have $\mathfrak{p}^{(i(n))} \subset \mathfrak{p}^n$. Hence, again by the first part of the corollary, there exists an exponent $t(n)$ such that $\mathfrak{q}^{(t(n))} \subset \mathfrak{p}^{(i(n))} \subset \mathfrak{p}^n$. This proves our second assertion.

In the course of the proof of Corollary 4, we have proved:

COROLLARY 5. *Let $A$ be a noetherian integral domain, $\mathfrak{p}$ a prime ideal in $A$ such that for every prime ideal $\mathfrak{m} > \mathfrak{p}$, the $(\mathfrak{m}^{(n)})$-completion of $A$ is a domain. Then the $(\mathfrak{p}^{(n)})$-topology of $A$ coincides with its $(\mathfrak{p}^n)$-topology.*

In fact, we have seen that under these assumptions high symbolic powers of $\mathfrak{p}$ are contained in high ordinary powers of $\mathfrak{p}$. The converse being obvious, our assertion is proved.

COROLLARY 6. *Let $A$ be a complete semi-local ring, $\mathfrak{m}$ the intersection of its maximal ideals, $B$ a commutative ring, $(\mathfrak{b}_n)$ a descending sequence of ideals in $B$ such that $\mathfrak{b}_p\mathfrak{b}_q \subset \mathfrak{b}_{p+q}$, $\bigcap_{q=0}^{\infty} \mathfrak{b}_q = (0)$, and $\varphi$ a continuous homomorphism of $A$ (considered with its $\mathfrak{m}$-topology) into $B$ (considered with its $(\mathfrak{b}_n)$-topology). Then $\varphi(A)$ is a closed subring of $B$.*

In fact, we have two topologies on $\varphi(A)$: the topology $T$ induced by that of $B$, and the topology $T'$ obtained by identifying $\varphi(A)$ to the factor ring $A/\varphi^{-1}(0)$ of $A$. The fact that $\varphi$ is continuous signifies that $T'$ is stronger than $T$. By Theorem 6, § 2, $\varphi(A)$ is complete for the topology $T'$, and is obviously a semi-local ring. Since $\varphi(A)$ is a Hausdorff space for $T$, it follows from Theorem 13 that $T' = T$. Therefore $\varphi(A)$, considered as a subspace of $B$, is complete, hence closed.

This, again, is a property which may be compared to a well-known property of compact spaces: a continuous image of a compact space $A$ in a Hausdorff space $B$ is a closed subset of $B$.

THEOREM 14. *Let $A$ be a noetherian ring, $\mathfrak{a}$ and $\mathfrak{b}$ two ideals in $A$ such that $\mathfrak{b} \subset \mathfrak{a}$ and such that $A$ is complete and Hausdorff in its $\mathfrak{a}$-topology. Then $A$ is complete in its $\mathfrak{b}$-topology.*

PROOF. Let $(b_n)$ be a Cauchy-sequence for the $\mathfrak{b}$-topology of $A$.

Since $\mathfrak{b} \subset \mathfrak{a}$, $(b_n)$ is also a Cauchy-sequence for the $\mathfrak{a}$-topology, whence it admits a limit $b \in A$. We have then $b_n \equiv b \pmod{\mathfrak{a}^{s(n)}}$ ($s(n) \to \infty$ with $n$). We now use more explicitly the fact that $(b_n)$ is a Cauchy-sequence in the $\mathfrak{b}$-topology. For every index $j \geq n$, we have $b_j - b_n \in \mathfrak{b}^{t(n)}$ where $t(n) \to \infty$ with $n$. From this and from $b_j - b \in \mathfrak{a}^{s(j)}$ we deduce that $b_n - b \in \mathfrak{b}^{t(n)} + \mathfrak{a}^{s(j)}$ for every $j \geq n$. Since $A$ is Hausdorff and is complete in its $\mathfrak{a}$-topology, it is a Zariski ring, whence $\mathfrak{b}^{t(n)}$ is closed in the $\mathfrak{a}$-topology. This means that $\mathfrak{b}^{t(n)} = \bigcap_{j=0}^{\infty} (\mathfrak{b}^{t(n)} + \mathfrak{a}^{s(j)})$, and hence $b_n - b \in \mathfrak{b}^{t(n)}$. This proves that $b$ is also the limit of the sequence $(b_n)$ for the $\mathfrak{b}$-topology. Q.E.D.

COROLLARY. *Let $A$ be a noetherian ring, $\mathfrak{a}$ and $\mathfrak{b}$ two ideals of $A$ such that $\mathfrak{b} \subset \mathfrak{a}$. Denote by $A'$ ($A''$) the ring $A$ considered with its $\mathfrak{b}$-topology (with its $\mathfrak{a}$-topology). Then the identity mapping $\varphi: A' \to A''$ is uniformly continuous, and if $A''$ is a Zariski ring then the extension $\bar{\varphi}: \hat{A}' \to \hat{A}''$ of $\varphi$ to $\hat{A}'$ is one to one.*

The fact that $\varphi$ is uniformly continuous follows immediately from the relation $\mathfrak{b} \subset \mathfrak{a}$. Theorem 14 shows that $\hat{A}''$ is complete for its $(\mathfrak{b}\hat{A}'')$-topology. If $A''$ is a Zariski ring, then we have $\mathfrak{b}^n \hat{A}'' \cap A = \mathfrak{b}^n$, whence the $(\mathfrak{b}\hat{A}'')$-topology of $\hat{A}''$ induces on $A$ its $\mathfrak{b}$-topology. Thus the completion $\hat{A}'$ of $A$ (for its $\mathfrak{b}$-topology) is canonically isomorphic to the closure of $A$ in $\hat{A}''$ considered with its $(\mathfrak{b}\hat{A}'')$-topology. In other words, $\hat{A}'$ is canonically isomorphic to a subring of $\hat{A}''$. This proves our assertion.

REMARK. It follows from the corollary that, if a Cauchy sequence of elements of $A'$ tends to zero in $A''$, then it also tends to zero in $A'$.

§ 6. **Finite extensions.** THEOREM 15. *Let $A$ be a noetherian ring, $\mathfrak{m}$ an ideal in $A$, and $B$ a ring containing $A$ which is a finite $A$-module. Then the $\mathfrak{m}$-topology of the $A$-module $B$ is identical with the $(\mathfrak{m}B)$-topology of the ring $B$ and induces on $A$ the $\mathfrak{m}$-topology. Furthermore*

(a) *For $B$ to be a Hausdorff space, it is necessary and sufficient that no element of $1 + \mathfrak{m}$ be a zero divisor in $B$.*

(b) *If $A$ is a Zariski ring, so is $B$.*

(c) *If $A$ is complete, so is $B$.*

(d) *If $A$ is semi-local, and if $\sqrt{\mathfrak{m}}$ is the intersection of the maximal ideals of $A$, then $B$ is semi-local and $\sqrt{\mathfrak{m}B}$ is the intersection of the maximal ideals of $B$.*

PROOF. The two parts of the first assertion follow respectively from the relation $(\mathfrak{m}B)^n = \mathfrak{m}^n B$ and from Theorem 4 (§ 2). Assertion (a) is a restatement of Krull's theorem for modules (Vol. I, Ch. IV, Appendix).

Concerning (b) we notice that every finite $B$-module $E$ is a finite $A$-module and hence is a Hausdorff space for its $\mathfrak{m}$-topology, since $A$ is a Zariski ring (see § 4, Theorem 9, property (b)). Since the $\mathfrak{m}$-topology of $E$ coincides with its $(\mathfrak{m}B)$-topology (in view of $\mathfrak{m}^n E = \mathfrak{m}^n BE$), it follows that every finite $B$-module $E$ is a Hausdorff space for its $(\mathfrak{m}B)$-topology. Hence $B$ is a Zariski ring, by Theorem 9, property (b).

If $A$ is complete, then Theorem 5, § 2 shows that $\hat{B} = \hat{A} \cdot B = B$. This proves (c). If $A$ is semi-local, then it is a Zariski ring, whence $B$ is also a Zariski ring, by (b). We have that $B/\mathfrak{m}B$ is a finite module over $A/(\mathfrak{m}B \cap A)$. On the other hand, since $\mathfrak{m}B \cap A \supset \mathfrak{m}$, the ring $A/(\mathfrak{m}B \cap A)$ is a homomorphic image of $A/\mathfrak{m}$ and therefore satisfies the d.c.c. (Theorem 10, § 4). Consequently $B/\mathfrak{m}B$ also satisfies the d.c.c., and Theorem 10, § 4 shows that $B$ is semi-local. Q.E.D.

REMARK. Assertion (b) proves that, if every element of $1+\mathfrak{m}$ is invertible in $A$, then every element of $1+\mathfrak{m}B$ is invertible in $B$.

THEOREM 16. *Let $A$ be a noetherian ring, $\mathfrak{m}$ an ideal in $A$, and $B$ a ring containing $A$. Suppose that $B$ is a finite $A$-module and that $A$ is a Zariski ring. Then:*

(a) *The closure of $A$ in $\hat{B}$ is the completion $\hat{A}$ of $A$, $\hat{B}$ is a finite $\hat{A}$-module, isomorphic to $\hat{A} \otimes_A B$ (here $\hat{B}$ is defined by considering the $\mathfrak{m}$-topology of $B$).*

(b) *If no element $\neq 0$ in $A$ is a zero divisor in $B$, then every element $\alpha$ of $\hat{A}$ which is a zero divisor in $\hat{B}$ is already a zero divisor in $\hat{A}$.*

PROOF. Assertion (a) has already been proved; the stronger statement about $\hat{A} \otimes_A B$ may be found in the remark following Corollary 1 to Theorem 11, § 4.

Assume now that no element of $A$, different from zero, is a zero-divisor in $B$. There exists an element $d \neq 0$ in $A$ and a finite family $\{b_j\}$ of elements of $B$ which are linearly independent over $A$ such that $dB \subset \sum_j Ab_j$. By completion we have $d\hat{B} \subset \sum \hat{A}b_j$, and the elements $b_j$ are still linearly independent over $\hat{A}$ (Corollary 1 to Theorem 11, § 4). If $\alpha$ is an element of $\hat{A}$ such that $\alpha\beta = 0$ for some $\beta \neq 0$ in $\hat{B}$, we write $d\beta = \sum_j \alpha_j b_j$ $(\alpha_j \in \hat{A})$. The relation $\alpha\beta = 0$ yields $d\alpha\beta = \sum_j \alpha\alpha_j b_j = 0$, whence $\alpha\alpha_j = 0$ for every $j$. It is impossible that all the $\alpha_j$'s be equal to 0, since this would imply $d\beta = 0$ in contradiction with the fact that the element $d$ of $A$ is not a zero divisor in $B$, whence also in $\hat{B}$ (Corollary 6 to Theorem 11, § 4). Therefore $\alpha$ is a zero divisor in $\hat{A}$.

It follows from the proof that the conclusion of (b) continues to hold if, instead of assuming that *all* the elements $\neq 0$ of $A$ are regular elements in $B$, we only assume that *there exists* an element $d \neq 0$ in $A$, which is not a zero divisor in $B$, and a finite family, $\{b_j\}$ of linearly independent elements of $B$ over $A$, such that $dB \subset \sum Ab_j$. In particular, the conclusion of (b) is true if $B$ is a *free* $A$-module, since we can take $d=1$ in this case.

**§ 7. Hensel's lemma and applications.** Let $A$ be a ring which is complete for its $\mathfrak{m}$-topology, where $\mathfrak{m}$ is an ideal in $A$. We intend to show how certain relations occurring in the ring $A/\mathfrak{m}$ (i.e., congruences mod $\mathfrak{m}$) may be "lifted" to analogous relations (not congruences) occurring in the ring $A$ itself. The completeness of $A$ is essential for this purpose. Historically, the completion of the ring $J$ of integers with respect to its $(Jp)$-topology ($p$, a prime number), which is called the *ring of p-adic integers*, was the first striking example of the theories developed in this section. The $p$-adic integers have been introduced by Hensel with the explicit purpose of deducing from congruences modulo $p$ actual equalities holding in some ring containing $J$.

For technical reasons it will be convenient to prove first a lemma which is in a sense a generalization to modules of the classical Hensel's lemma:

LEMMA ("BILINEAR LEMMA"). *Let $A$ be a ring, $\mathfrak{m}$ an ideal in $A$, $E$, $E'$, $F$ three finite $A$-modules. We suppose that $F$ is a Hausdorff space for its $\mathfrak{m}$-topology and that $A$ is complete. Let $f$ be a bilinear mapping of $E \times E'$ into $F$; denote by $\bar{f}$ the bilinear mapping of $(E/\mathfrak{m}E) \times (E'/\mathfrak{m}E')$ into $F/\mathfrak{m}F$ canonically deduced from $f$. Suppose we are given elements $y \in F$, $\alpha \in E/\mathfrak{m}E$, $\alpha' \in E'/\mathfrak{m}E'$ such that*

(1) *The class $\bar{y}$ of $y$* mod $\mathfrak{m}F$ *is equal to $\bar{f}(\alpha, \alpha')$.*
(2) *$F/mF = \bar{f}(\alpha, E'/\mathfrak{m}E') + \bar{f}(E/\mathfrak{m}E, \alpha')$.*

*Then there exist elements $a$ and $a'$ in $E$ and $E'$ respectively, such that $\alpha$ is the residue classes of $a$* mod $\mathfrak{m}E$, *$\alpha'$ is the residue class of $a'$* mod $\mathfrak{m}E'$, *and such that $y = f(a, a')$.*

PROOF. We prove, by induction on $n$, the existence of elements $a_n$ and $a'_n$ in $E$ and $E'$ respectively, having $\alpha$ and $\alpha'$ as residue classes, and such that $y \equiv f(a_n, a'_n) \pmod{\mathfrak{m}^n F}$. This is true in the case $n=1$, by assumption (1). We now go from $n$ to $n+1$. Since $y - f(a_n, a'_n) \in \mathfrak{m}^n F$, we may write $y - f(a_n, a'_n) = \sum_j m_j z_j$ with $m_j \in \mathfrak{m}^n$, $z_j \in F$. By assumption (2) there exists an element $w_j$ in $E$ and an element $w'_j$ in $E'$ such that $z_j \equiv f(a_n, w'_j) + f(w_j, a'_n) \pmod{\mathfrak{m}F}$. Thus the element

$$\begin{aligned} y - f(a_n + \sum_j m_j w_j, a'_n + \sum_j m_j w'_j) &= y - f(a_n, a'_n) - \sum_j m_j z_j \\ &+ \sum_j m_j \{z_j - f(a_n, w'_j) - f(w_j, a'_n)\} - \sum_{i,j} m_i m_j f(w_i, w'_j) \end{aligned}$$

belongs to $\mathfrak{m}^{n+1}F + \mathfrak{m}^{2n}F = \mathfrak{m}^{n+1}F$ (since $n \geq 1$). We can thus take $a_{n+1} = a_n + \sum_j m_j w_j$, $a'_{n+1} = a'_n + \sum_j m_j w'_j$.

This choice shows that $a_{n+1} \equiv a_n$ $(\mathfrak{m}^n E)$; thus $(a_n)$, and similarly $(a'_n)$, is a Cauchy sequence. Since $E$ and $E'$ are finite modules over a complete ring $A$, they are complete for their $\mathfrak{m}$-topologies (the proof of this assertion is similar to that of Theorem 5, § 2). Thus the sequences $(a_n)$, $(a'_n)$ admit limits $a \in E$, $a' \in E'$. Their residue classes mod $\mathfrak{m}E$ and $\mathfrak{m}E'$ are obviously $\alpha$ and $\alpha'$. Since $y - f(a_n, a'_n)$ tends to 0, we have $y = f(a, a')$ since $f$ is continuous and since $F$ is a Hausdorff space. Q.E.D.

THEOREM 17 (Hensel's lemma). *Let $A$ be a complete local ring, $\mathfrak{m}$ its maximal ideal, $f(X) \in A[X]$ a monic polynomial of degree $n$ over $A$. For every polynomial $h(X) \in A[X]$, we denote by $\bar{h}(X)$ the polynomial over $A/\mathfrak{m}$ obtained from $h(X)$ on replacing its coefficients by their $\mathfrak{m}$-residues. If $\alpha(X)$ and $\alpha'(X)$ are relatively prime monic polynomials over $A/\mathfrak{m}$ of degrees $r$ and $n-r$ such that $\bar{f}(X) = \alpha(X)\alpha'(X)$, then there exist two monic polynomials $g(X)$, $g'(X)$ over $A$, of degrees $r$ and $n-r$, such that $\bar{g}(X) = \alpha(X)$, $\bar{g}'(X) = \alpha'(X)$ and $f(X) = g(X)g'(X)$.*

PROOF. We apply the "bilinear lemma," taking for $E$, $E'$, $F$ the modules of polynomials over $A$, of degrees respectively $\leqq r$, $\leqq n-r$ and $\leqq n$, and for the bilinear mapping $f$ the multiplication of polynomials. We take $\alpha(X)$ for $\alpha$, $\alpha'(X)$ for $\alpha'$, and $f(X)$ for $y$. Assumption (1) in the "bilinear lemma" is verified. As to assumption (2), we note that, since $\alpha(X)$ and $\alpha'(X)$ are relatively prime, every polynomial $\beta(X)$ over $A/\mathfrak{m}$ may be written as a linear combination

$$\beta(X) = \lambda(X)\alpha(X) + \lambda'(X)\alpha'(X)(\lambda(X), \lambda'(X) \in (A/\mathfrak{m})[X]);$$

furthermore, if $\partial(\beta) \leq n$, we may choose $\lambda$ and $\lambda'$ in such a way that $\partial(\lambda) \leq n-r$, and $\partial(\lambda') \leq r$ (this follows easily from the euclidean algorithm in $(A/\mathfrak{m})[X]$). Thus the bilinear lemma proves that there exist polynomials $h(X)$, $h'(X)$ of degrees $r$ and $n-r$, such that $\bar{h}(X) = \alpha(X)$, $\bar{h}'(X) = \alpha'(X)$, $h(X)h'(X) = f(X)$.

For completing the proof, it suffices to show that $h(X)$ and $h'(X)$ may be replaced by monic polynomials. The highest degree terms of these polynomials are of the form $(1+m)X^r, (1+m)^{-1}X^{n-r}$ with $m \in \mathfrak{m}$ (since $\bar{h}(X)$ and $h(X)h'(X)$ are monic). It is thus sufficient to divide $h(X)$ by $1+m$, and to multiply $h(X)$ by $1+m$. Q.E.D.

COROLLARY 1. *Let $A$ be a complete local ring, $\mathfrak{m}$ its maximal ideal, and $f(X)$ a monic polynomial over $A$. Suppose that $\bar{f}(X)$ admits a simple root $\alpha \in A/\mathfrak{m}$. Then there exists an element $a$ of $A$, having $\alpha$ as $\mathfrak{m}$-residue, and such that $f(a) = 0$; furthermore the root $a$ of $f(X)$ is simple.*

In fact we can write $\bar{f}(X)=(X-\alpha)\varphi(X)$, where $\varphi(X)$ is prime to $X-\alpha$. Theorem 17 shows the existence of a monic polynomial $X-a$ which divides $f(X)$ and such that $\bar{a}=\alpha$. If $a$ were a multiple root of $f(X)$, $\alpha$ would be a multiple root of $\bar{f}(X)$, in contradiction with our assumption.

EXAMPLES:

(1) The polynomial $X^2+1$ has two simple roots in the prime field $GF(5)$, namely the classes of 2 and 3. Thus it admits two roots in the ring of 5-adic integers. Similarly for $X^2-2$ in the ring of 7-adic integers.

(2) Let $A$ be a complete local domain whose residue field $A/\mathfrak{m}$ is the finite field $GF(q)$. Since the equation $X^{q-1}=1$ admits $q-1$ simple roots in $GF(q)$, the ring $A$ contains all the $(q-1)$-th roots of unity. These roots form a multiplicative subgroup $V$ of $A$. If $A$ has characteristic $p\neq 0$ (where $q=p^b$), $V$ is even the multiplicative group of a subfield of $A$, since the $(p^b-1)$-th roots of unity, in a field of characteristic $p$, constitute the set of non-zero elements of a subfield. This subfield is canonically isomorphic to $A/\mathfrak{m}=GF(q)$.

(3) *Theorem of implicit functions.* Let $A$ be the power series ring $K[[x_1, \cdots, x_m]]$ in $m$ variables over a field $K$, and let

$$P(z) = z^n+a_{n-1}(x)z^{n-1}+ \cdots +a_1(x)z+a_0(x)$$

be a monic polynomial over $A$. Suppose that the polynomial

$$z^n+a_{n-1}(0)z^{n-1}+ \cdots +a_1(0)z+a_0(0)$$

admits a simple root $\alpha \in K$. Then there exists a power series $g(x)$ such that $g(0)=\alpha$ and such that $P(g(x))=0$. In particular, if $d$ is an integer which is prime to the characteristic of $K$, and if $f(x)$ is a power series whose constant term is $\neq 0$ and is a $d$-th power in $K$, $f(x)$ itself is a $d$-th power in $K[[x_1, \cdots, x_m]]$ (use the polynomial $P(z)=z^d-f(x)$).

COROLLARY 2. *Let $A$ be a complete local ring having the same characteristic $p$ as its residue field $A/\mathfrak{m}$. Then there exists a subfield $L$ of $A$ such that $A/\mathfrak{m}$ is purely inseparable over the image of $L$ in $A/\mathfrak{m}$.*

Let us denote by $\varphi$ the canonical mapping of $A$ into $A/\mathfrak{m}$. We first prove that $A$ contains at least one field. In the first place, the ring $A$ contains the "prime ring" $R_0$ formed by the integral multiples $n.1$ of $1$ $(n=0, \pm 1, \pm 2, \cdots)$. If $p\neq 0$, $R_0$ is a field. If $p=0$, $R_0$ is the ring of integers, and, since $A/\mathfrak{m}$ has characteristic 0, every integer $\neq 0$ is outside of $\mathfrak{m}$, and is therefore a unit in $A$, thus proving that $A$ contains the field of rational numbers. This being so, the family $\Phi$ of all subfields of $A$, ordered by inclusion, admits, by Zorn's lemma, a maximal element $L$. If $A/\mathfrak{m}$ were transcendental over $\varphi(L)$, we could find an element $x$ of $A$ such that $\varphi(x)$ is transcendental over $\varphi(L)$; then all the non-zero elements of the polynomial ring $L[x]$ would be outside of $\mathfrak{m}$, therefore units, and $A$ would then contain the quotient field of $L[x]$, in contradiction with the maximality of $L$. Thus $A/\mathfrak{m}$ is algebraic over $\varphi(L)$. Suppose that $A/\mathfrak{m}$ contains an element $\eta$ which is separable algebraic over $\varphi(L)$. Let $Y^n+\beta_{n-1}Y^{n-1}+\cdots+\beta_1 Y+\beta_0$ be the minimal polynomial of $\eta$ over $\varphi(L)$, and let $b_j$ be the representative of $\beta_j$ lying in $L$. The polynomial $f(Y)=Y^n+b_{n-1}Y^{n-1}+\cdots+b_1 Y+b_0$ over $A$ is such that $\eta$ is a simple root of $\bar{f}(Y)$; thus, by Corollary 1, $f(Y)$ admits a simple root $y \in A$ such that $\varphi(y)=\eta$. Since $\varphi$ induces an isomorphism of $L$ onto $\varphi(L)$ which carries $f(Y)$ to $\bar{f}(Y)$, the polynomial $f(Y)$ is irreducible over $L$, and the ring $L[y]$ is isomorphic to $\varphi(L)[\eta]$, which is a field. In view of the maximality of $L$, this implies that $y \in L$, whence $\eta \in \varphi(L)$. Therefore $A/\mathfrak{m}$ is purely inseparable over $\varphi(L)$.

REMARK. If $A/\mathfrak{m}$ is a field of characteristic 0, then $A$ itself has characteristic 0, and Corollary 2 shows the existence of a subfield $L$ of $A$ which is a *field of representatives* for the residue classes mod $\mathfrak{m}$. It can be proved that such a field of representatives exists whenever $A$ is complete and has the same characteristic as $A/\mathfrak{m}$ (see § 12, Theorem 27). However, already from the proof of Corollary 2 it follows that such a field of representatives exists under the additional assumption that $A/\mathfrak{m}$ admits a separating transcendence basis over its prime field (in particular if $A/\mathfrak{m}$ is a finitely generated extension of its prime field).

The bilinear lemma may also be used for showing that a complete semi-local ring is isomorphic to a direct product of complete *local* rings. We prove a slightly more general result.

THEOREM 18. (Decomposition theorem). *Let $A$ be a ring, and $\mathfrak{m}$ an ideal in $A$ such that $A$ is a complete Hausdorff space for its $\mathfrak{m}$-topology. If $A/\mathfrak{m}$ is the direct sum of two ideals $\mathfrak{v}/\mathfrak{m}$ and $\mathfrak{v}'/\mathfrak{m}$, then $A$ is the direct sum of the ideals $\mathfrak{n}=\bigcap_{n=0}^{\infty}\mathfrak{v}^n$ and $\mathfrak{n}'=\bigcap_{n=0}^{\infty}\mathfrak{v}'^n$. The $\mathfrak{m}$-topology of $\mathfrak{n}$ (considered as an $A$-module), its $(\mathfrak{m}\mathfrak{n})$-topology, and the topology induced on $\mathfrak{n}$ by the $\mathfrak{m}$-topology of $A$ are all identical, and $\mathfrak{n}$ is a complete Hausdorff space for this topology. Similarly for $\mathfrak{n}'$. The rings $\mathfrak{n}/\mathfrak{m}\mathfrak{n}$, $A/\mathfrak{v}'$ and $\mathfrak{v}/\mathfrak{m}$ (or $\mathfrak{n}'/\mathfrak{m}\mathfrak{n}'$, $A/\mathfrak{v}$ and $\mathfrak{v}'/\mathfrak{m}$) are isomorphic.*

PROOF. Let $\varepsilon$ and $\varepsilon'$ ($\varepsilon \in \mathfrak{v}/\mathfrak{m}$, $\varepsilon' \in \mathfrak{v}'/\mathfrak{m}$) be the orthogonal idempotents corresponding to the decomposition $A/\mathfrak{m}=(\mathfrak{v}/\mathfrak{m}) \oplus (\mathfrak{v}'/\mathfrak{m})$ (Vol. I, Ch. III, § 13): we have $\varepsilon+\varepsilon'=1$, $\varepsilon\varepsilon'=0$. We apply the bilinear lemma to the case in which $E=E'=F=A$, $f$ is the multiplication in $A$, and

$\alpha = \varepsilon$, $\alpha' = \varepsilon'$, $y = 0$. Conditions (1) and (2) are satisfied. There thus exist elements $a$, $a'$ of $A$, admitting $\varepsilon$ and $\varepsilon'$ respectively as $\mathfrak{m}$-residues, and such that $aa' = 0$. We have $a + a' \equiv 1 \pmod{\mathfrak{m}}$, whence $a + a'$ is an invertible element of $A$ since $A$ is complete (use formal expansion of $1/(1-T)$). Then the elements $e = a/(a+a')$, $e' = a'/(a+a')$ satisfy the relations $ee' = 0$, $e + e' = 1$, and are therefore two orthogonal idempotents. We therefore have a direct sum decomposition $A = Ae \oplus Ae'$ (Vol. I, Ch. III, § 13).

We now proceed to prove the assertions about the ideals $\mathfrak{v}$ and $\mathfrak{v}'$. Since $e$ admits $\varepsilon$ as $\mathfrak{m}$-residue, it belongs to $\mathfrak{v}$, whence to every power $\mathfrak{v}^n$, since $e$ is an idempotent. We therefore have $Ae \subset \mathfrak{n} = \bigcap_{n=0}^{\infty} \mathfrak{v}^n$, and similarly $Ae' \subset \mathfrak{n}' = \bigcap_{n=0}^{\infty} \mathfrak{v}'^n$. Since $\mathfrak{v}$ and $\mathfrak{v}'$ are comaximal, $\mathfrak{v}^n$ and $\mathfrak{v}'^n$ are comaximal, and we have $\mathfrak{v}^n \cap \mathfrak{v}'^n = \mathfrak{v}^n\mathfrak{v}'^n = (\mathfrak{v}\mathfrak{v}')^n = (\mathfrak{v} \cap \mathfrak{v}')^n = \mathfrak{m}^n$ (Vol. I, Ch. III, § 13, Theorem 31). Since $\bigcap_{n=0}^{\infty} \mathfrak{m}^n = (0)$, this implies that $\mathfrak{n} \cap \mathfrak{n}' = (0)$. From $Ae \subset \mathfrak{n}$, $Ae' \subset \mathfrak{n}'$ and $Ae + Ae' = A$, we deduce that $\mathfrak{n} + \mathfrak{n}' = A$, whence $A$ is the direct sum of the ideals $\mathfrak{n}$, $\mathfrak{n}'$. The relation $\mathfrak{n} = \mathfrak{n}(Ae + Ae') \subset Ae + (0)$ (since $\mathfrak{n}e' \subset \mathfrak{n}\mathfrak{n}' = (0)$) proves that $\mathfrak{n} = Ae$; similarly $\mathfrak{n}' = Ae'$. Both the $\mathfrak{m}$-topology and the $(\mathfrak{m}\mathfrak{n})$-topology of $\mathfrak{n}$ admit the ideals $\mathfrak{m}^n\mathfrak{n}$ as basic neighborhoods of 0, since, on the one hand, we have $\mathfrak{m}^n\mathfrak{n} = \mathfrak{m}^n\mathfrak{n}^n$ (as $\mathfrak{n}$ is an idempotent ideal), and on the other hand, we have $\mathfrak{m}^n = \mathfrak{m}^n e \oplus \mathfrak{m}^n e'$, and hence $\mathfrak{m}^n \cap \mathfrak{n} = \mathfrak{m}^n \cap Ae = Ae\mathfrak{m}^n = \mathfrak{m}^n\mathfrak{n}$. This shows that the $(\mathfrak{m}\mathfrak{n})$-topology of $\mathfrak{n}$ is induced by the $\mathfrak{m}$-topology of $A$. If $\{x_i\}$ is a Cauchy sequence of elements of $\mathfrak{n}$ and if $x$ is the limit of that sequence *in* $A$, then we observe that $xe$ is the limit of the sequence $\{x_i e\}$, i.e., of $\{x_i\}$. Since $A$ is a Hausdorff space it follows that $x = xe \in \mathfrak{n}$, showing that also $\mathfrak{n}$ is a complete (Hausdorff) space. Finally, since we have proved above that $\mathfrak{m}\mathfrak{n} = \mathfrak{m} \cap \mathfrak{n}$, it follows that the ring $\mathfrak{n}/\mathfrak{m}\mathfrak{n}$ is isomorphic to $\mathfrak{n}/(\mathfrak{m} \cap \mathfrak{n}) = (\mathfrak{m} + \mathfrak{n})/\mathfrak{m} = (\mathfrak{m} + Ae)/\mathfrak{m} = \varepsilon \cdot (A/\mathfrak{m}) = \mathfrak{v}/\mathfrak{m} \simeq A/\mathfrak{v}'$. This completes the proof of Theorem 18.

COROLLARY 1. *Let $A$ be a ring, $\mathfrak{m}$ an ideal in $A$ such that $A$ is a complete Hausdorff space for its $\mathfrak{m}$-topology. If $A/\mathfrak{m}$ is a direct sum of $q$ ideals $\mathfrak{v}_j/\mathfrak{m}$, then $A$ is the direct sum of the ideals $\mathfrak{n}_j = \bigcap_{n=0}^{\infty} \mathfrak{v}_j^n$. We have $\mathfrak{m}^n\mathfrak{n}_j = \mathfrak{m}^n \cap \mathfrak{n}_j$, and the $(\mathfrak{m}\mathfrak{n}_j)$-topology of the ring $\mathfrak{n}_j$ coincides with the topology induced on $\mathfrak{n}_j$ by the $\mathfrak{m}$-topology of $A$. The rings $\mathfrak{n}_j/\mathfrak{m}\mathfrak{n}_j$, $\mathfrak{v}_j/\mathfrak{m}$ and $A/\sum_{i \neq j} \mathfrak{n}_i$ are isomorphic.*

We proceed by induction on the number $q$ of the ideals $\mathfrak{n}_j$, the case

$q=2$ being covered by Theorem 18. We notice that $A/\mathfrak{m}$ is the direct sum of $\mathfrak{v}_1/\mathfrak{m}$ and of $\sum_{j\neq 1} \mathfrak{v}_j/\mathfrak{m}$, and we apply Theorem 18 and the induction hypothesis. Further details are left to the reader.

COROLLARY 2. *A complete semi-local ring A is a direct sum of complete local rings. In particular, if A is a domain, it is a local ring.*

Let $\mathfrak{m}$ be the intersection of the maximal ideals $\mathfrak{p}_i$ of $A$. Then $A/\mathfrak{m}$ is the direct sum of the ideals $\mathfrak{v}_j/\mathfrak{m}$, where $\mathfrak{v}_j=\prod_{i\neq j} \mathfrak{p}_i=\bigcap_{i\neq j} \mathfrak{p}_i$. Since $\mathfrak{p}_j$ is maximal, $\mathfrak{v}_j/\mathfrak{m}\cong A/\mathfrak{p}_j$ is a field. Thus the direct summand $\mathfrak{n}_j=\bigcap_{n=0}^{\infty} \mathfrak{v}_j^n$ of $A$ which is such that $\mathfrak{n}_j/\mathfrak{m}\mathfrak{n}_j=\mathfrak{v}_j/\mathfrak{m}$ is a field and which is a complete Zariski ring for its $(\mathfrak{m}\mathfrak{n}_j)$-topology (§ 4, Example 3), is a local ring (Theorem 10, § 4). The second assertion follows from the fact that a direct sum of $h$ ideals, $h>1$, is always a ring with zero-divisors.

REMARK. Let $A$ be a (not necessarily complete) semi-local ring, and $\mathfrak{m}$ the intersection of its maximal ideals $\mathfrak{p}_i$. We consider $\hat{A}$; the ideal $\hat{A}\mathfrak{p}_i$ is maximal, and $\hat{A}$ is the direct sum of the ideals $\mathfrak{n}_j=\bigcap_{n=0}^{\infty}\left(\bigcap_{i\neq j} \hat{A}\mathfrak{p}_i\right)^n$. Denote by $\varphi_j$ the "projection" of $\hat{A}$ onto $\mathfrak{n}_j$; there exists an idempotent $e$ such that $\varphi_j(x)=ex$ for any $x$ in $\hat{A}$ and such that $1-e\in\bigcap_{n=0}^{\infty} \hat{A}\mathfrak{p}_j^n$. Consider the restriction of $\varphi_j$ to $A$. For every element $y\in A$, $y\notin\mathfrak{p}_j$, $\varphi_j(y)$ is invertible, since it does not belong to the maximal ideal $\hat{A}\mathfrak{m}\cap\mathfrak{n}_j$ of $\mathfrak{n}_j$. On the other hand, the kernel of $\varphi_j$ is $\bigcap_{n=0}^{\infty} \hat{A}\mathfrak{p}_j^n$, whence the kernel of its restriction to $A$ is $A\cap\bigcap_{n=0}^{\infty} \hat{A}\mathfrak{p}_j^n=\bigcap_{n=0}^{\infty}(A\cap\hat{A}\mathfrak{p}_j^n)=\bigcap_{n=0}^{\infty}\mathfrak{p}_j^n$ (Theorem 9, § 4). We notice that $\bigcap_{n=0}^{\infty}\mathfrak{p}_j^n$ is also the kernel of the canonical homomorphism of $A$ into the quotient ring $A_{\mathfrak{p}_j}$ (Vol. I, Ch. IV, Theorems 19 and 20). Therefore the subring $\varphi_j(A)_{\varphi_j(\mathfrak{p}_j)}$ of $\mathfrak{n}_j$ is isomorphic to $A_{\mathfrak{p}_j}$. This subring is a local ring, its local ring topology is obviously induced by the topology of $\mathfrak{n}_j$, and the ring is dense in $\mathfrak{n}_j$ since $A$ is dense in $\hat{A}$, and since therefore $\varphi_j(A)$ is also dense in $\mathfrak{n}_j$. Therefore *the direct summand $\mathfrak{n}_j$ is isomorphic with the completion of the quotient ring $A_{\mathfrak{p}_j}$.*

§ 8. **Characteristic functions.** Let $A$ be a semi-local ring, $\mathfrak{m}$ the intersection of its maximal ideals. We recall (see § 2) that an ideal $\mathfrak{v}$ of $A$ is *open* if and only if it contains some power of $\mathfrak{m}$. For any integer $n\geqq 1$, the $\mathfrak{m}$-topology of $A$ is identical with the $\mathfrak{m}^n$-topology of $A$.

Since $A$ is a semi-local, it follows from Theorem 10, § 4 (where we replace $\mathfrak{m}$ by $\mathfrak{m}^n$) that the ring $A/\mathfrak{m}^n$ satisfies the d.c.c. If $\mathfrak{v}$ contains a power $\mathfrak{m}^n$ of $\mathfrak{m}$, then *a fortiori* also the ring $A/\mathfrak{v}$ satisfies the d.c.c., since this ring is a homomorphic image of $A/\mathfrak{m}^n$. Conversely, if $A/\mathfrak{v}$ satisfies the d.c.c., we must have $\mathfrak{m}^n+\mathfrak{v}=\mathfrak{m}^{n+1}+\mathfrak{v}$ for all large $n$, and therefore $\mathfrak{v}\supset\mathfrak{m}^n$ if $n$ is large, since $\mathfrak{v}$ is a closed set and since therefore $\mathfrak{v}$ is the intersection of the ideals $\mathfrak{m}^n+\mathfrak{v}$. Thus, the condition that $\mathfrak{v}$ is open is equivalent to the condition that the ring $A/\mathfrak{v}$ satisfies the d.c.c.

Let $\{\mathfrak{p}_j\}$ be the set of maximal ideals of $A$. Then the primary representation of $\mathfrak{v}$ is

(1) $$\mathfrak{v} = \bigcap_j \mathfrak{q}_j = \prod_j \mathfrak{q}_j,$$

where $\mathfrak{q}_j$ is either primary for $\mathfrak{p}_j$ or is equal to $A$. The ideal $\mathfrak{v}^n$ is equal to $\prod_j \mathfrak{q}_j^n = \bigcap_j \mathfrak{q}_j^n$ since the ideals $\mathfrak{q}_j^n$ are pairwise comaximal. The length of the ideal $\mathfrak{v}^n$ is therefore *finite* (Vol. I, Ch. IV, § 13, Theorem 24). We call *characteristic function of the ideal* $\mathfrak{v}$ and denote by $P_\mathfrak{v}(n)$ the length $\lambda(\mathfrak{v}^n)$. Since the ideals $\mathfrak{q}_j^n$ are pairwise comaximal, and since $\mathfrak{v}^n=\bigcap_j \mathfrak{q}_j^n$, the ring $A/\mathfrak{v}^n$ is isomorphic to the direct sum of the rings $A/\mathfrak{q}_j^n$. Hence

(2) $$P_\mathfrak{v}(n) = \sum_j \lambda(\mathfrak{q}_j^n).$$

Since the ideal $\mathfrak{q}_j^n$ is primary for $\mathfrak{p}_j$, the length $\lambda(\mathfrak{q}_j^n)$ is equal to the length of the ideal $\mathfrak{q}_j^n A_{\mathfrak{p}_j}$ in the local ring $A_{\mathfrak{p}_j}$; in other words, (2) may be written as

(2′) $$P_\mathfrak{v}(n) = \sum_j P_{\mathfrak{q}_j A_{\mathfrak{p}_j}}(n) = \sum_j P_{\mathfrak{v} A_{\mathfrak{p}_j}}(n).$$

We now prove that the characteristic function $P_\mathfrak{v}(n)$ is a *polynomial* for $n$ large enough. More generally:

THEOREM 19. *Let $A$ be a semi-local ring, $\mathfrak{v}$ an open ideal in $A$, and $E$ a finite $A$-module. Then the length of the $A$-module $E/\mathfrak{v}^nE$ is finite, and is a polynomial $n$ for $n$ large enough. In particular, the characteristic function $P_\mathfrak{v}(n)$ is a polynomial in $n$ for $n$ large enough.*

PROOF. We consider the associated graded module $G(E)=\sum_{n=0}^{\infty} \mathfrak{v}^nE/\mathfrak{v}^{n+1}E$. It is a finite module over the associated ring $G(A)=\sum_{n=0}^{\infty} \mathfrak{v}^n/\mathfrak{v}^{n+1}$ (§ 1), and $G(A)$ is a factor ring $(A/\mathfrak{v})[X_1, \cdots, X_q]/\mathfrak{S}$ of the polynomial ring $R=(A/\mathfrak{v})\,[X_1, \cdots, X_q]$ modulo a homogeneous ideal

$\mathfrak{S}$. Hence $G(E)$ is a finite graded module over $R$. Since $A/\mathfrak{v}$ is a ring which satisfies the d.c.c., we may apply Hilbert's theorem (VII, § 12, Theorem 41), which tells us that the length of $\mathfrak{v}^nE/\mathfrak{v}^{n+1}E$ is a polynomial in $n$, for $n$ large enough. From this it follows also that the length $\ell(E/\mathfrak{v}^nE) = \sum_{j=0}^{n-1} \ell(\mathfrak{v}^jE/\mathfrak{v}^{j+1}E)$ is a polynomial for $n$ large enough. Q.E.D.

The polynomial to which $P_{\mathfrak{v}}(n)$ is equal for large values of $n$ is called the *characteristic polynomial* of the ideal $\mathfrak{v}$. We shall denote it by $\bar{P}_{\mathfrak{v}}(n)$, and sometimes by $P_{\mathfrak{v}}(n)$ when we are dealing with large values of $n$. As in Chapter VII, § 12 (see p. 233), we find also here that, if $\bar{P}_{\mathfrak{v}}(n)$ is a polynomial of degree $d$, its coefficients are *integral multiples* of $1/d!$.

We first prove some *simple results* about characteristic functions.

LEMMA 1. *Let $\mathfrak{v}$ be an open ideal in a semi-local ring $A$. Then*

$$P_{\hat{A}\mathfrak{v}}(n) = P_{\mathfrak{v}}(n).$$

PROOF. In fact, $A/\mathfrak{v}^n$ and $\hat{A}/\hat{A}\mathfrak{v}^n$ are isomorphic (§ 2, Theorem 6).

LEMMA 2. *If $\mathfrak{v}$ and $\mathfrak{v}'$ are open ideals in a semi-local ring $A$ and if $\mathfrak{v} \subset \mathfrak{v}'$, then $P_{\mathfrak{v}}(n) \geq P_{\mathfrak{v}'}(n)$ for all $n$.*

PROOF. Obvious.

LEMMA 3. *Let $\mathfrak{v}$ be an open ideal in a semi-local ring $A$ and let $x$ be an element of $\mathfrak{v}$. Then*

$$P_{\mathfrak{v}/Ax}(n) = P_{\mathfrak{v}}(n) - \lambda(\mathfrak{v}^n : Ax).$$

PROOF. In fact, $(A/Ax)/(\mathfrak{v}/Ax)^n$ is isomorphic to $A/(\mathfrak{v}^n + Ax)$, whence

$$P_{\mathfrak{v}}(n) - P_{\mathfrak{v}/Ax}(n) = \ell(A/\mathfrak{v}^n) - \ell(A/(\mathfrak{v}^n + Ax)) = \ell((\mathfrak{v}^n + Ax)/\mathfrak{v}^n)$$
$$= \ell(Ax/(\mathfrak{v}^n \cap Ax)) = \ell(Ax/(Ax \cdot (\mathfrak{v}^n : Ax))) = \ell(A/(\mathfrak{v}^n : Ax)) = \lambda(\mathfrak{v}^n : Ax).$$

(Notice that the kernel of $y \to yx$ is contained in $\mathfrak{v}^n : Ax$.)

Lemma 3 is useful in the following way. Let $s$ be the greatest exponent such that $x \in \mathfrak{v}^s$ (whence $x \notin \mathfrak{v}^{s+1}$). It is clear that $\mathfrak{v}^n : Ax \supset \mathfrak{v}^{n-s}$. If we can prove that $\mathfrak{v}^n : Ax$ is "not too different" from $\mathfrak{v}^{n-s}$, we can deduce from Lemma 3 that $P_{\mathfrak{v}/Ax}(n)$ is not very different from $P_{\mathfrak{v}}(n) - P_{\mathfrak{v}}(n-s)$, a circumstance which is useful for devising proofs by induction (see below). We thus introduce the following notion: an element $x$ of $A$ is said to be *superficial of order* $s$ for $\mathfrak{v}$ if $x \in \mathfrak{v}^s$ and if there exists an integer $c$ such that

$$(3) \qquad (\mathfrak{v}^n : Ax) \cap \mathfrak{v}^c = \mathfrak{v}^{n-s}$$

for every large enough $n$. It follows from Lemma 3 that:

LEMMA 4. *Let $\mathfrak{v}$ be an open ideal in a semi-local ring $A$, and $x$ be an element of $A$ which is superficial of order $s$ for $\mathfrak{v}$. Then there exists an integer $c$ such that*

$$P_{\mathfrak{v}}(n)-P_{\mathfrak{v}}(n-s) \leq P_{\mathfrak{v}/Ax}(n) \leq P_{\mathfrak{v}}(n)-P_{\mathfrak{v}}(n-s)+P_{\mathfrak{v}}(c),$$

*for $n$ large enough.*

PROOF. In fact, (3) implies that we have

$$\begin{aligned} \ell((\mathfrak{v}^n:Ax)/\mathfrak{v}^{n-s}) &= \ell((\mathfrak{v}^n:Ax)/((\mathfrak{v}^n:Ax)\cap\mathfrak{v}^c)) \\ &= \ell((\mathfrak{v}^c+(\mathfrak{v}^n:Ax))/\mathfrak{v}^c) \leq \ell(A/\mathfrak{v}^c) = P_{\mathfrak{v}}(c), \end{aligned}$$

whence $0\leq P_{\mathfrak{v}}(n-s)-\lambda(\mathfrak{v}^n:Ax)\leq P_{\mathfrak{v}}(c)$. Using now Lemma 3, we easily get the double inequality in Lemma 4.

It follows from Lemma 4 that the *characteristic polynomial* $\bar{P}_{\mathfrak{v}/Ax}(n)$ differs from the polynomial $\bar{P}_{\mathfrak{v}}(n)-\bar{P}_{\mathfrak{v}}(n-s)$ *only by its constant term.*

We are now led to the question whether, given an open ideal $\mathfrak{v}$ in a semi-local ring $A$, there exist superficial elements for $\mathfrak{v}$. The following result gives a partial answer:

LEMMA 5. *Let $\mathfrak{v}$ be an open ideal in a semi-local ring $A$. There exist an integer $s$ and an element $x$ of $A$ such that $x$ is superficial of order $s$ for $\mathfrak{v}$.*

PROOF. Let $s$ be an integer and $x$ an element of $\mathfrak{v}^s$ such that $x\notin\mathfrak{v}^{s+1}$. The relation $\mathfrak{v}^n:Ax=\mathfrak{v}^{n-s}$ is valid if the initial form $\bar{x}$ of $x$ in the associated graded ring $G(A)=\sum_{n=0}^{\infty}\mathfrak{v}^n/\mathfrak{v}^{n+1}$ is not a zero divisor in $G(A)$ (§ 1, p. 249). The relation $(\mathfrak{v}^n:Ax)\cap\mathfrak{v}^c=\mathfrak{v}^{n-s}$ is true if every homogeneous element $\alpha\in G(A)$ such that $\alpha\bar{x}=0$ has a degree $<c$. This being so, we consider in $G(A)$ the associated prime ideals $\mathfrak{P}_j$ of (0). We assume that, for $1\leqq j\leqq h$, $\mathfrak{P}_j$ does not contain the ideal $\mathfrak{X}=\sum_{n=1}^{\infty}\mathfrak{v}^n/\mathfrak{v}^{n+1}$ of elements of positive degree in $G(A)$, and that $\mathfrak{P}_j\supset\mathfrak{X}$ for $h+1\leqq j\leqq k$. *It is easily seen that there exists a homogeneous element $\bar{x}$ of positive degree*, say $s$, *such that $\bar{x}\notin\mathfrak{P}_j$ for $1\leqq j\leqq h$.*† To prove this, we may replace the set $\{\mathfrak{P}_j\}$, $1\leqq j\leqq h$, by the set $\{\mathfrak{P}_\mu\}$ of maximal elements in that set. Since $\mathfrak{X}\not\subset\mathfrak{P}_\mu$, there exists a homogeneous element $\beta_\mu$ of $\mathfrak{X}$ such that $\beta_\mu\notin\mathfrak{P}_\mu$. On the other hand, for $\mu'\neq\mu$ there exists a homogeneous element $\chi_{\mu\mu'}$ in $\mathfrak{P}_{\mu'}$, which does not belong to $\mathfrak{P}_\mu$. We set $\chi_\mu=(\beta_\mu\prod_{\mu'\neq\mu}\chi_{\mu\mu'})^{n(\mu)}$, the exponents $n(\mu)$ being chosen in such a way that the elements $\chi_\mu$ have the same degree. We then have $\chi_\mu\notin\mathfrak{P}_\mu$ and $\chi_\mu\in\mathfrak{P}_{\mu'}$ for $\mu'\neq\mu$. Hence the element $\bar{x}=\sum_\mu\chi_\mu$ satisfies the conditions $\bar{x}\notin\mathfrak{P}_\mu$ for $1\leqq\mu\leqq h$.

† This result generalizes to homogeneous ideals and homogeneous elements, a result proved in Vol. I, Ch. IV, § 6 (*Remark*, p. 215).

Let $\mathfrak{Q}_j$ be a primary component of (0) corresponding to $\mathfrak{P}_j$. For $h+1 \leq j \leq k$, $\mathfrak{Q}_j$ contains some power of $\mathfrak{X}$. Let $c$ be an exponent such that $\mathfrak{X}^c \subset \bigcap_{j=h+1}^{k} \mathfrak{Q}_j$. Suppose now that $\alpha$ is a homogeneous element of $G(A)$ such that $\bar{x}\alpha = 0$. From $\bar{x} \notin \mathfrak{P}_j$ for $1 \leqq j \leqq h$, we deduce that $\alpha \in \mathfrak{Q}_j$ for $1 \leqq j \leqq h$. If, furthermore, the degree of $\alpha$ is $\geqq c$, we deduce from $\mathfrak{X}^c \subset \bigcap_{j=h+1}^{k} \mathfrak{Q}_j$, that $\alpha$ belongs to all the primary components of (0), whence that $\alpha = 0$. Therefore, if $\alpha$ is different from 0, its degree is $< c$, and $\bar{x}$ has the required property.

REMARKS ABOUT LEMMA 5.

(1) The proof of Lemma 5 shows that, given a finite family $(\mathfrak{J}_m)$ of homogeneous ideals of $G(A)$ such that no $\mathfrak{J}_m$ contains any power of $\mathfrak{X}$, the homogeneous element $\bar{x}$ may be chosen as to satisfy the condition $x \notin \mathfrak{J}_m$ for every $m$: in fact, for every $m$, we add to the family $\mathfrak{P}_j$ $(1 \leqq j \leqq h)$ of homogeneous prime ideals, an associated prime ideal $\mathfrak{P}'_m$ of $\mathfrak{J}_m$ which does not contain $\mathfrak{X}$. It follows that, *given a finite family* $\{\mathfrak{b}_m\}$ *of ideals of* $A$, *none of which is open, there exists an integer s and an element x which is superficial of order s for* $\mathfrak{v}$ *and which does not belong to any* $\mathfrak{b}_m$. In fact, we take for $\mathfrak{J}_m$ the leading ideal of $\mathfrak{b}_m$ (§ 1), and notice that $\mathfrak{J}_m$ does not contain any power of $\mathfrak{X}$, for if, say, $\mathfrak{J}_m \supset \mathfrak{X}^t$, then $\mathfrak{v}^t \subset \mathfrak{b}_m + \mathfrak{v}^{t+1}$ and therefore $\mathfrak{v}^t \subset \mathfrak{b}_m + \mathfrak{v}^{t+n}$ for every $n$; since $\mathfrak{b}_m$ is closed, i.e., since $\bigcap_{n=0}^{\infty} (\mathfrak{b}_m + \mathfrak{v}^n) = \mathfrak{b}_m$, this would imply $\mathfrak{v}^t \subset \mathfrak{b}_m$, in contradiction with the fact that $\mathfrak{b}_m$ is not open.

(2) Superficial elements of a *given* order do not necessarily exist (for example, the maximal ideal of $K[[X, Y]]/(XY(X+Y))$, where $K$ is a field with two elements, has no superficial elements of order 1). However, such a circumstance is due to the finiteness of the residue field. In fact, we now prove that, given *a local ring A whose residue field* $A/\mathfrak{m}$ *is infinite and given an open ideal* $\mathfrak{v}$ *of A* (*i.e., an ideal which is primary for the maximal ideal* $\mathfrak{m}$), *then for any finite family* $\mathfrak{b}_m$ *of non-open ideals of A and any integer* $s > 0$, *there exists an element x of A which is superficial of order s for* $\mathfrak{v}$ *and which does not belong to any* $\mathfrak{b}_m$. In order to prove this assertion we denote by $\mathfrak{J}_m$ the leading ideal of $\mathfrak{b}_m$, and by $\mathfrak{P}_j$ the prime ideals of (0) in $G(A)$ which do not contain $\mathfrak{X}$; then, as in Remark 1, none of the ideals $\mathfrak{J}_m$, $\mathfrak{P}_j$ contains any power of $\mathfrak{X}$. Since $\mathfrak{X}^s$ is generated by $\mathfrak{v}^s/\mathfrak{v}^{s+1}$, none of the ideals $\mathfrak{J}_m$, $\mathfrak{P}_j$ contains $\mathfrak{v}^s/\mathfrak{v}^{s+1}$. Thus, in the $(A/\mathfrak{v})$-module $E = \mathfrak{v}^s/\mathfrak{v}^{s+1}$ we have a finite number of submodules $F_i$ distinct from $E$ (namely, the intersections of $E$ with the ideals $\mathfrak{J}_m$ and $\mathfrak{P}_j$), and we have to find an element of $E$ which does not belong to any

$F_i$. We first notice that, if we denote by $\mathfrak{m}'$ the maximal ideal $\mathfrak{m}/\mathfrak{v}$ of $A/\mathfrak{v}$, we have $F_i + \mathfrak{m}'E \neq E$, for if $E = F_i + \mathfrak{m}'E$ then $E = F_i + \mathfrak{m}'(F_i + \mathfrak{m}'E)$ $= F_i + \mathfrak{m}'^2E = \cdots = F_i + \mathfrak{m}'^nE$ for every $n$, and hence (since $\mathfrak{m}'$ is a nilpotent ideal) $E = F_i$, a contradiction. Therefore, in $E/\mathfrak{m}'E$, which is a vector space over the infinite field $A/\mathfrak{m}$, the subspaces $(F_i + \mathfrak{m}'E)/\mathfrak{m}'E$ are distinct from the entire space, whence there exists an element $\bar{\xi}$ of $E/\mathfrak{m}'E$ such that $\bar{\xi} \notin (F_i + \mathfrak{m}'E)/\mathfrak{m}'E$, for every $i$. If we take for $\bar{x}$ a representative of $\bar{\xi}$ in $E$, we have $\bar{x} \notin F_i$, for every $i$, and we may take for $x$ any element of $\mathfrak{v}^s$ having $\bar{x}$ as initial form.

## § 9. Dimension theory. Systems of parameters.

Let $A$ be a semi-local ring, $\mathfrak{m}$ the intersection of its maximal ideals, $\mathfrak{v}$ and $\mathfrak{v}'$ two ideals in $A$ admitting $\mathfrak{m}$ as radical. Then $\mathfrak{v}$ and $\mathfrak{v}'$ are open sets, and the characteristic functions $P_{\mathfrak{v}}(n)$ and $P_{\mathfrak{v}'}(n)$ are defined (§ 8). Furthermore, there exist integers $a$ and $b$ such that

$$\mathfrak{v} \supset \mathfrak{v}'^a \quad \text{and} \quad \mathfrak{v}' \supset \mathfrak{v}^b.$$

Thus it follows from Lemma 2, § 8, that

$$P_{\mathfrak{v}}(n) \leq P_{\mathfrak{v}'}(an) \quad P_{\mathfrak{v}'}(n) \leq P_{\mathfrak{v}}(bn).$$

These inequalities imply that the polynomials $\tilde{P}_{\mathfrak{v}}(n)$ and $\tilde{P}_{\mathfrak{v}'}(n)$ have *the same degree d*. This degree is called the *dimension of the semi-local ring A*, and denoted by dim $(A)$. If we denote by $\mathfrak{p}_j$ the maximal ideals of $A$, formula (2') in § 8 (p. 284) shows that

$$(1) \qquad \qquad \dim(A) = \max_j (\dim(A_{\mathfrak{p}_j})).$$

Since $A_{\mathfrak{p}_j}$ is a local ring, expression (1) of dim $A$ allows, in many dimension-theoretic questions, a reduction to the case of local rings. It follows from Lemma 1, § 8, that the *completion $\hat{A}$ of $A$* has the same dimension as $A$. If $\mathfrak{a}$ is an ideal in a semi-local ring $A$, the dimension of the semi-local ring $A/\mathfrak{a}$ is called the *dimension of the ideal $\mathfrak{a}$*.

THEOREM 20. *Let $A$ be a local ring, $\mathfrak{m}$ its maximal ideal. The following integers are equal:*

(a) *The dimension $d$ of $A$.*

(b) *The height $h$ of the prime ideal $\mathfrak{m}$ (Vol. I, Ch. IV, § 14).*

(c) *The smallest integer $r$ for which there exist $r$ elements of $A$ which generate an ideal which is primary for $\mathfrak{m}$.*

PROOF. The equality of the integers defined by (b) and (c) has been proved in Vol. I, Ch. IV, § 14, Theorems 30 and 31. We prove that $d = r$. More generally we prove that, *given a semi-local ring $A$, its dimension d is equal to the smallest integer r for which there exist r elements of $A$*

*which generate an ideal having as radical the intersection* $\mathfrak{m}$ *of the maximal ideals of* $A$.

We first prove that, if an ideal $\mathfrak{v}$ has $\mathfrak{m}$ as radical then it cannot be generated by less than $d$ $(=\dim A)$ elements. In fact, suppose that $\mathfrak{v}$ may be generated by $s$ elements. Then the associated graded ring $\sum_{n=0}^{\infty} \mathfrak{v}^n/\mathfrak{v}^{n+1}$ is isomorphic to a factor ring of the polynomial ring $(A/\mathfrak{v})[X_1, \cdots, X_s]$ by a homgeneous ideal $\mathfrak{J}$. Since the module of polynomials of degree $\leq n$ in $(A/\mathfrak{v})[X_1, \cdots, X_s]$ has a length equal to $\ell(A/\mathfrak{v})\binom{n+s}{s}$, it follows that the length of $\sum_{i=0}^{n-1} \mathfrak{v}^i/\mathfrak{v}^{i+1}$, i.e., $\ell(A/\mathfrak{v}^n)$, is $\leq \ell(A/\mathfrak{v})\binom{n+s}{s}$. As $\binom{n+s}{s}$ is a polynomial of degree $s$ in $n$, it follows that the degree $d$ of $\bar{P}_{\mathfrak{v}}(n)$ is $\leq s$.

It remains to be proved that there exists an ideal $\mathfrak{v}$ generated by $d$ elements and admitting $\mathfrak{m}$ as radical. For the proof we proceed by induction on $d$. If $d=0$, $P_{\mathfrak{m}}(n)$ remains constant for $n \geq n_0$, whence $\mathfrak{m}^{n_0}=(0)$ since $\bigcap_{n=0}^{\infty} \mathfrak{m}^n=0$; we may then take $\mathfrak{v}=(0)$, since we agree that $(0)$ is generated by the empty set of elements of $A$. Suppose now that $A$ has dimension $d>0$, and that our assertion has been proved for every semi-local ring of dimension $d-1$. By Lemma 5, § 8, there exists an integer $q$ and an element $x_d$ of $\mathfrak{m}$ which is superficial of order $q$ for $\mathfrak{m}$. Then (Lemma 4, § 8) the characteristic polynomial $\bar{P}_{\mathfrak{m}/Ax_d}(n)$ differs from $\bar{P}_m(n)-\bar{P}_m(n-q)$ only by its constant term. It follows that $\bar{P}_{\mathfrak{m}/Ax_d}(n)$ is a polynomial of degree $d-1$ (note that we have assumed that $d>0$), i.e., that $A/Ax_d$ has dimension $d-1$. By our induction hypothesis there exist elements $\bar{x}_1, \cdots, \bar{x}_{d-1}$ of $\bar{A}=A/Ax_d$ such that the radical of $\sum_{i=1}^{d-1} \bar{A}\bar{x}_i$ is the intersection $\mathfrak{m}/Ax_d$ of the maximal ideals of $\bar{A}=A/Ax_d$. Taking representatives $x_1, \cdots, x_{d-1}$ of $\bar{x}_1, \cdots, \bar{x}_{d-1}$ in $A$, we see immediately that $\mathfrak{m}$ is the radical of $\sum_{i=1}^{d} Ax_i$. This completes the proof.

In the above proof of Theorem 20 we have used from Vol. I, Ch. IV, § 14, Theorem 31 which is quite elementary, but we have also used Theorem 30 which is rather difficult and uses the "Principal ideal theorem" (Theorem 29, Vol. I, Ch. IV, § 14). We shall give here a second proof of Theorem 20 which is independent of the cited Theorem 30 but uses the lemma of Artin-Rees (§ 2, Theorem 4'; more specifically formula (5) of § 5) and the properties of characteristic polynomials. In this second proof we shall establish the inequalities $r \leqq h$ and $h \leqq d$ (the inequality $d \leqq r$ has already been established in the first part of the preceding proof).

The inequality $r \leq h$ follows from Vol. I, Ch. IV, § 14, Theorem 31. In fact, if $h$ is the height of $\mathfrak{m}$, there exists an ideal $\mathfrak{q}$ generated by $h$ elements and admitting $\mathfrak{m}$ as an isolated prime ideal. As $\mathfrak{m}$ is the maximal ideal of a local ring $A$, this implies that $\mathfrak{q}$ is primary for $\mathfrak{m}$. We therefore have $r \leqq h$.

We finally prove the inequality $h \leq d$. We proceed by induction on $d$. The case $d = 0$ is easy, since $P_{\mathfrak{m}}(n)$ is then constant for $n \geq n_0$, whence $\mathfrak{m}^{n_0} = (0)$, and $A$ is a primary ring with $\mathfrak{m}$ as unique prime ideal. Now we suppose that $A$ has dimension $d$, and we consider a maximal chain $\mathfrak{p}_0 < \mathfrak{p}_1 < \cdots < \mathfrak{p}_h = \mathfrak{m}$ of prime ideals in $A$. Since the length of $A/(\mathfrak{p}_0 + \mathfrak{m}^n)$ is not greater than the length of $A/\mathfrak{m}^n$, we have $P_{\mathfrak{m}/\mathfrak{p}_0}(n) \leq P_{\mathfrak{m}}(n)$, whence the dimension $d'$ of $A' = A/\mathfrak{p}_0$ is $\leq d$. We choose an element $x' \neq 0$ in the ideal $\mathfrak{p}_1/\mathfrak{p}_0$ of $A'$, and denote by $\mathfrak{m}'$ the maximal ideal $\mathfrak{m}/\mathfrak{p}_0$ of the local ring $A'$. By formula (5) of § 5 there exists an integer $k$ such that $\mathfrak{m}'^n : A'x' \subset \mathfrak{m}'^{n-k}$ for every $n \geq k$. Since $\mathfrak{m}'^n : A'x'$ obviously contains $\mathfrak{m}'^{n-1}$, we have the double inequality $P_{\mathfrak{m}'}(n-k) \leq \lambda(\mathfrak{m}'^n : A'x') \leq P_{\mathfrak{m}'}(n-1)$. Thus, by Lemma 3 (§ 8), we have $P_{\mathfrak{m}'}(n) - P_{\mathfrak{m}'}(n-1) \leq P_{\mathfrak{m}'/A'x'}(n) \leq P_{\mathfrak{m}'}(n) - P_{\mathfrak{m}'}(n-k)$, from which it follows that the degree of $P_{\mathfrak{m}'/A'x'}(n)$, i.e., the dimension of $A'/A'x'$, is equal to $d'-1$. Since $d'-1 \leq d-1$, the induction hypothesis shows that the length of any chain of prime ideals in $A'/A'x'$ is $\leq d-1$. In particular the chain $\mathfrak{p}_1/(\mathfrak{p}_0 + Ax) < \cdots < \mathfrak{p}_h/(\mathfrak{p}_0 + Ax)$ ($x$: element of $\mathfrak{p}_1$ admitting $x'$ as $\mathfrak{p}_0$-residue) has at most $d$ terms. Therefore $h \leq d$, and our assertion is proved. Q.E.D.

REMARKS:

(1) We can even easily deduce Theorem 30 of Vol. I, Ch. IV, § 14 from the present Theorem 20. In fact, if $R$ is a noetherian ring, $\mathfrak{a}$ an ideal in $R$ admitting a basis of $r$ elements and $\mathfrak{p}$ an isolated prime ideal of $\mathfrak{a}$, then the dimension of the local ring $R_{\mathfrak{p}}$ is $\leq r$ by Theorem 20 (c), whence the height of the maximal ideal $\mathfrak{p}R_{\mathfrak{p}}$ (i.e., the height of $\mathfrak{p}$; see Vol. I, Ch. IV, § 11, Theorem 19) is at most $r$ and this proves the cited Theorem 30. Furthermore, the principal ideal theorem (Vol. I, Ch. IV, § 14, Theorem 29), which is a particular case of Theorem 30 (Vol. I, Ch. IV, § 14), is also an easy consequence of Theorem 20. We have therefore sketched an alternative treatment of the theory of prime ideals in noetherian rings, which is smoother than the one given in Vol. I, Ch. IV, § 14, but which is less elementary since it essentially uses the theory of characteristic polynomials.

(2) Concerning the existence of an ideal $\mathfrak{v}$ generated by $d$ elements and admitting $\mathfrak{m}$ as radical, we give now another proof which makes use of properties of polynomial ideals established in the preceding chapter.

Let $\sum_{n=0}^{\infty} \mathfrak{m}^n/\mathfrak{m}^{n+1} = k[X_1, X_2, \cdots, X_m]/\mathfrak{a}$, where $k = A/\mathfrak{m}$ and $\mathfrak{a}$ is a homogeneous ideal. The length of $\mathfrak{m}^n/\mathfrak{m}^{n+1}$ is given by $P_{\mathfrak{m}}(n+1) - P_{\mathfrak{m}}(n)$, for large $n$, and is therefore equal to a polynomial of degree $d-1$. On the other hand, that length is also equal to $\bar{\chi}(\mathfrak{a}; n)$, where $\bar{\chi}$ is the characteristic function of the ideal $\mathfrak{a}$. Hence the projective dimension of $\mathfrak{a}$ is equal to $d-1$ (VII, § 12, Theorem 42). Let $\{\varphi_i(X)\}$ be a set of $d$ forms in $k[X]$ such that the ideal generated by $\mathfrak{a}$ and the forms $\varphi_i(X)$ is irrelevant. For each $i$ fix an element $z_i$ in $A$ whose initial form is the $\mathfrak{a}$-residue of $\varphi_i$, and denote by $\mathfrak{z}$ the ideal generated in $A$ by the $d$ elements $z_i$. Then the leading ideal of $\mathfrak{z}$ contains a power of the leading ideal of $\mathfrak{m}$, say the $q$-th power. From this follows in the usual way that $\mathfrak{m}^q$ is contained in the intersection of the ideals $\mathfrak{z} + \mathfrak{m}^i$, i.e., $\mathfrak{m}^q$ is contained in $\mathfrak{z}$, and hence the $z_i$ generate an ideal having $\mathfrak{m}$ as radical.

COROLLARY 1. *Let $A$ be a semi-local ring and $x$ an element which is superficial for some ideal $\mathfrak{v}$ admitting $\mathfrak{m}$ as radical. Then*

$$\dim (A/Ax) = \dim (A) - 1.$$

This has already been established in the course of the proof of Theorem 20.

COROLLARY 2. *Let $A$ be a local ring and $x$ an element of $A$ which is not a zero divisor in $A$. Then*

$$\dim (A/Ax) = \dim (A) - 1.$$

In fact, it follows from relation (1) and from Theorem 20 that dim $(A)$ is the height $h(\mathfrak{m})$ of the maximal ideal $\mathfrak{m}$ of $A$ and that $\dim (A/Ax) = h(\mathfrak{m}/Ax)$. Since a maximal chain of prime ideals in $A/Ax$ corresponds to a chain $\mathfrak{p}_1 < \mathfrak{p}_2 < \cdots < \mathfrak{p}_{q+1} = \mathfrak{m}$ in $A$ in which $\mathfrak{p}_1$ is an isolated prime ideal of $Ax$ and therefore (Vol. I, Ch. IV, § 14, Corollary 2 to Theorem 29) a prime ideal of height 1, we have $q = \dim (A/Ax) \leqq \dim A - 1$. On the other hand, let $\{\bar{z}_1, \bar{z}_2, \cdots, \bar{z}_q\}$ be a system of parameters in $A/Ax$ and let $z_i$ be a representative of $\bar{z}_i$ in $A$. Then the elements $z_1, z_2, \cdots, z_q, x$ generate in $A$ an ideal which is primary for $\mathfrak{m}$. Hence $q + 1 \geqq \dim A$.

COROLLARY 3. *Let $A$ be a semi-local ring, and $B$ an overring of $A$ which is a finite $A$-module. Then* $\dim (A) = \dim (B)$.

If $\{\mathfrak{P}_j\}$ is a chain of distinct prime ideals of $B$, then the ideals $A \cap \mathfrak{P}_j$ are distinct, since $B$ is integral over $A$ (Vol. I, Ch. V, § 2, Complement 1 to Theorem 3), and therefore $\dim B \leq \dim A$. If $\mathfrak{p}_1 < \mathfrak{p}_2 < \cdots < \mathfrak{p}_h$ is a chain of prime ideals of $A$, then the corollary to Theorem 3 of § 2, Vol. I, Ch. V, shows successively the existence of prime ideals $\mathfrak{P}_1, \cdots, \mathfrak{P}_h$ in $B$ such that $\mathfrak{P}_1 \cap A = \mathfrak{p}_1$, $\mathfrak{P}_2 \cap A = \mathfrak{p}_2$ and $\mathfrak{P}_2 > \mathfrak{P}_1, \cdots, \mathfrak{P}_h \cap A = \mathfrak{p}_h$ and $\mathfrak{P}_h > \mathfrak{P}_{h-1}$. Since the ideals $\mathfrak{P}_j$ are obviously distinct, we have $\dim A \leq \dim B$. Q.E.D.

Given a *local ring* $A$ of dimension $d$, a system $\{x_1, \cdots, x_d\}$ of $d$ elements of $A$ which generates a primary ideal for the ideal $\mathfrak{m}$ of non-units of $A$ is called a *system of parameters* of $A$. Theorem 20 shows the existence of systems of parameters in any local ring. It is clear that $\{x_1, \cdots, x_d\}$ is also a system of parameters of the completion $\hat{A}$ of $A$. Notice that if $\{x_1, \cdots, x_d\}$ is a system of parameters of $A$, then the *dimension of* $A/(Ax_1 + \cdots + Ax_j)$ *is* $d-j$. In fact, more generally, given any $j$ elements $y_1, \cdots, y_j$ of a local ring $A$, we have

$$d' = \dim (A/(Ay_1 + \cdots + Ay_j)) \geq \dim (A) - j,$$

since, given a system of parameters $\{\bar{z}_1, \cdots, \bar{z}_{d'}\}$ in $A/(\sum Ay_i)$, the elements $\{y_1, \cdots, y_j, z_1, \cdots, z_{d'}\}$ ($z_i$: a representative of $\bar{z}_i$ in $A$) generate an open ideal in $A$, whence $d' + j \geq \dim (A)$. Furthermore, if $\{x_1, \cdots, x_j\}$ is a subset of a system of parameters $\{x_1, \cdots, x_d\}$, the ideal generated by the residue classes of $x_{j+1}, \cdots, x_d \bmod (Ax_1 + \cdots + Ax_j)$ is open, whence $\dim (A/(Ax_1 + \cdots + Ax_j)) \leq d - j$, and this proves our assertion.

We intend to study the "relations" between the elements of a system of parameters:

THEOREM 21. *Let $A$ be a local ring, $\{x_1, \cdots, x_d\}$ a system of parameters of $A$, $\mathfrak{m}$ the maximal ideal of $A$, and $F(X_1, \cdots, X_d)$ a homogeneous polynomial of degree $s$ over $A$. Let $\mathfrak{q}$ be the primary ideal $\sum_{i=1}^{d} Ax_i$. If $F(x_1, \cdots, x_d) \in \mathfrak{m}\mathfrak{q}^s$, then all the coefficients of $F$ are in $\mathfrak{m}$.*

PROOF. Consider the direct sum $\sum_{n=0}^{\infty} \mathfrak{q}^n/\mathfrak{m}\mathfrak{q}^n = k[X_1, \cdots, X_d]/\mathfrak{a}$, where $k = A/\mathfrak{m}$ and $\mathfrak{a}$ is a homogeneous ideal in $k[X]$. We have to show that $\mathfrak{a}$ is the zero ideal. Suppose the contrary is true. Then the dimension theory of polynomial rings tells us that we can choose $d-1$ forms $\varphi_i(X)$, of positive degrees, such that the ideal generated by $\mathfrak{a}$ and the forms $\varphi_i$ is irrelevant. The $\mathfrak{a}$-residues of the $\varphi_i$ will therefore be a homogeneous system of integrity in the ring $k[X]/\mathfrak{a}$ (see the Lemma in VII, § 7). We may assume that the $\varphi_i$ are of like degree $t > 0$. We choose elements $z_1, z_2, \cdots, z_{d-1}$ of $\mathfrak{q}^t$ whose $\mathfrak{m}\mathfrak{q}^t$-residues are precisely the $\mathfrak{a}$-residues of the forms $\varphi_i$, and we denote by $\mathfrak{b}$ the ideal generated by the $z_i$. Expressing the fact that the $\mathfrak{m}\mathfrak{q}$-residue of each $x_j$ satisfies an equation of integral dependence over the ring generated over $k$ by the $\mathfrak{a}$-residues of the $\varphi_i$, one finds at once that there exists an integer $h$ such that $x_j^h \in \mathfrak{b} + \mathfrak{m}\mathfrak{q}^h, j = 1, 2, \cdots, d$. From this it follows easily that $\mathfrak{q}^n \subset \mathfrak{b} + \mathfrak{m}\mathfrak{q}^n$ for all large $n$, whence $\mathfrak{q}^n \subset \mathfrak{b}$, for large $n$; a contradiction, since $\mathfrak{b}$ is generated by only $d-1$ elements.

COROLLARY 1. *Let $A$ be a local ring, $K$ a subfield of $A$, and $\{x_1, \cdots, x_d\}$ a system of parameters of $A$. Then the elements $x_1, \cdots, x_d$ are algebraically independent over $K$.*

Let $G(X_1, \cdots, X_d)$ be a non-zero polynomial over $K$ such that $G(x_1, \cdots, x_d) = 0$. Denote by $F(X_1, \cdots, X_d)$ the lowest degree form of $G$, and by $s$ the degree of $F$. From $G(x_1, \cdots, x_d) = 0$, we deduce that $F(x_1, \cdots, x_d) \in \mathfrak{q}^{s+1}$, where $\mathfrak{q} = \sum Ax_i$, whence $F(x_1, \cdots, x_d) \in \mathfrak{m}\mathfrak{q}^s$. Then Theorem 21 shows that all the coefficients of $F$ are in $\mathfrak{m}$. Since they are all in the subfield $K$ of $A$, this implies that they are all 0, in contradiction with the fact that $F$ is the lowest degree form of a non-zero polynomial.

COROLLARY 2. *Let $A$ be a complete local ring, $K$ a subfield of $A$, and $\{x_1, \cdots, x_d\}$ a system of parameters of $A$. Then the elements $x_1, \cdots, x_d$ are analytically independent over $K$* (cf. § 2, p. 258).

As in Corollary 1, we consider a non-zero power series $G(X_1, \cdots, X_d)$ over $K$ such that $G(x_1, \cdots, x_d) = 0$, and we write $G = \sum_{j=s}^{\infty} F_j$ where $F_j$ is a form of degree $j$ and where $F_s \neq 0$. The relation $F_s(x_1, \cdots, x_d) = -\sum_{j=s+1}^{\infty} F_j(x_1, \cdots, x_d) \in \mathfrak{q}^{s+1}$ (where $\mathfrak{q} = \sum_{i=1}^{d} Ax_i$) implies, as in Corollary 1, that all the coefficients of $F_s$ are equal to 0, in contradiction with $F_s \neq 0$.

REMARK. Let $A$ be a complete local ring containing a field $K$ such that $A/\mathfrak{m}$ ($\mathfrak{m}$: maximal ideal of $A$) is a finite algebraic extension of the canonical image of $K$ in $A/\mathfrak{m}$. Then, *if $\{x_1, \cdots, x_d\}$ is a system of parameters of $A$, $A$ is a finite module over $B = K[[x_1, \cdots, x_d]]$.* In fact, $A$ is a module over $B$, and, if we denote by $\mathfrak{X}$ the ideal $(x_1, \cdots, x_d)$ of $B$, the natural topology of $A$ is its $\mathfrak{X}$-topology (since $A\mathfrak{X}$ is primary for $\mathfrak{m}$). Furthermore $A/A\mathfrak{X}$ is a finite dimensional vector space over $K$, since $A\mathfrak{X}$ contains a power of $\mathfrak{m}$ and since $A/\mathfrak{m}$, $\mathfrak{m}/\mathfrak{m}^2$, $\mathfrak{m}^2/\mathfrak{m}^3$, etc., are finite-dimensional vector spaces over $K$. Since $K = B/\mathfrak{X}$, and since $B$ is complete, our assertion follows from Corollary 2 to Theorem 7 (§ 3).

§ 10. **Theory of multiplicities.** Let $A$ be a semi-local ring of dimension $d$, and $\mathfrak{q}$ an open ideal of $A$, admitting the intersection $\mathfrak{m}$ of the maximal ideals $\mathfrak{p}_j$ of $A$ as radical. Then the characteristic polynomial $\bar{P}_{\mathfrak{q}}(n)$ is of degree $d$, by the definition of the dimension of $A$ (§ 9). Its leading term has the form

$$e(\mathfrak{q})n^d/d!,$$

where $e(\mathfrak{q})$ is an integer (cf. VII, § 12). The integer $e(\mathfrak{q})$ is called the *multiplicity of the ideal* $\mathfrak{q}$. The integer $e(\mathfrak{m})$ is called the *multiplicity of the semi-local ring A*.

If all the quotient rings $A_{\mathfrak{p}_j}$ have dimension $d$, then it is clear that

$$(1) \qquad e(\mathfrak{q}) = \sum_j e(\mathfrak{q}A_{\mathfrak{p}_j}).$$

Denoting by $\mathfrak{q}_j$ the primary component of $\mathfrak{q}$ relative to $\mathfrak{p}_j$, we deduce from (1) that we have also

$$(2) \qquad e(\mathfrak{q}) = \sum_j e(\mathfrak{q}_j).$$

In an important case it is possible to reduce the study of multiplicities to the case of ideals generated by systems of parameters.

THEOREM 22. *Let A be a local ring, m its maximal ideal,* $\mathfrak{q}$ *an ideal which is primary for* $\mathfrak{m}$. *If* $A/\mathfrak{m}$ *is an infinite field, there exists an ideal* $\mathfrak{q}'\subset\mathfrak{q}$, *generated by a system of parameters and such that* $e(\mathfrak{q}')=e(\mathfrak{q})$.

PROOF. We proceed by induction on the dimension $d$ of $A$. For $d=0$, every proper ideal $\mathfrak{q}$ of $A$ is nilpotent, whence all the characteristic polynomials $\bar{P}_{\mathfrak{q}}(n)$ are equal to the constant $\ell(A)$ (=length of the $A$-module $A$); we may thus take $\mathfrak{q}'=(0)$, since we agree that (0) is generated by the empty set, which is thus the only system of parameters of $A$.

We now pass to the case $d=1$. Note that, if $d=1$, then $\mathfrak{m}$ is the only prime ideal in $A$ which is not an isolated prime ideal of (0). We take an element $x$ of $\mathfrak{q}$ which is superficial of order 1 for $\mathfrak{q}$ and which lies outside all isolated prime ideals of (0) (Remark 2 to Lemma 5, § 8). Everything is quite simple if $x$ is not a zero divisor in $A$ (or, equivalently, if $\mathfrak{m}$ is not a prime ideal, necessarily imbedded, of (0)). In fact the relations $(\mathfrak{q}^n:Ax)\cap\mathfrak{q}^c=\mathfrak{q}^{n-1}$ (§ 8, relation (3), p. 285) and $\mathfrak{q}^n:Ax\subset\mathfrak{q}^{s(n)}$ with $s(n)\to\infty$ (§ 5, Corollary 1 to Theorem 13; here is where we use the assumption that $x$ is not a zero divisor) show that $\mathfrak{q}^n:Ax=\mathfrak{q}^{n-1}$ for $n$ large enough, whence $\bar{P}_{\mathfrak{q}/Ax}(n)=\bar{P}_{\mathfrak{q}}(n)-\bar{P}_{\mathfrak{q}}(n-1)$ (Lemma 3, § 8). Since $\bar{P}_{\mathfrak{q}}(n)$ is of degree 1, the right-hand side is $e(\mathfrak{q})$. On the other hand, since $x$ is not contained in any of the isolated prime ideals of (0), $Ax$ is primary for $\mathfrak{m}$, $\mathfrak{m}/Ax$ is the only prime ideal of $A/Ax$, whence the local ring $A/Ax$ has dimension 0. Consequently, for $n$ large, $\bar{P}_{\mathfrak{q}/Ax}(n)$ is the length of $A/Ax$. This is also the length of $Ax^{n-1}/Ax^n$ since $A/Ax$ and $Ax^{n-1}/Ax^n$ are isomorphic under the mapping $z+Ax \rightarrow zx^{n-1}+Ax^n$ ($x$ not being a zero divisor). We therefore have $e(\mathfrak{q})=e(Ax)$ in this case.

Still in case $d=1$, we now assume that $\mathfrak{m}$ is an imbedded prime ideal of (0) (i.e., that all the elements of $\mathfrak{m}$ are zero-divisors in $A$). Then the annihilator $\mathfrak{a}$ of $x$ is an ideal which, if considered as an $A$-module, has a finite length $s$ (since it is annihilated by $Ax$ and since, by Corollary 1 of Theorem 20, § 9, $A/Ax$ has dimension zero). We consider the factor ring $A^\star=A/\mathfrak{a}$. For every open ideal $\mathfrak{v}$ of $A$, we have $\mathfrak{v}^n \cap \mathfrak{a}=(0)$ for $n$ large since $\bigcap_{n=0}^{\infty} (\mathfrak{v}^n \cap \mathfrak{a})=(0)$ and since $\mathfrak{a}$ is an $A$-module of finite length. Thus, if we set $\mathfrak{v}^\star=(\mathfrak{v}+\mathfrak{a})/\mathfrak{a}$, we have $P_{\mathfrak{v}^\star}(n)=\ell(A^\star/\mathfrak{v}^{\star n})=\ell(A/(\mathfrak{a}+\mathfrak{v}^n))=\ell(A/\mathfrak{v}^n)-\ell((\mathfrak{a}+\mathfrak{v}^n)/\mathfrak{v}^n)=\ell(A/\mathfrak{v}^n)-\ell(\mathfrak{a}/(\mathfrak{a}\cap\mathfrak{v}^n))$ $=\ell(A/\mathfrak{v}^n)-s$ for $n$ large enough, since $\mathfrak{v}^n \cap \mathfrak{a}=(0)$ for large $n$. In other words, we have $\bar{P}_{\mathfrak{v}^\star}(n)=\bar{P}_{\mathfrak{v}}(n)-s$, whence $e(\mathfrak{v})=e(\mathfrak{v}^\star)$, since $\bar{P}_{\mathfrak{v}}$ is a polynomial of degree 1. In particular, we have $e(\mathfrak{q})=e(\mathfrak{q}^\star)$ and $e(Ax)=e(A^\star x^\star)$ ($\mathfrak{q}^\star=(\mathfrak{q}+\mathfrak{a})/\mathfrak{a}$, $x^\star=\mathfrak{a}$-residue of $x$; note that the ideal $Ax$ is primary, with $\mathfrak{m}$ as associated prime ideal). Since $x^\star$ is superficial of order 1 for $\mathfrak{q}^\star$ and is not a zero-divisor in $A^\star$, the first part of the proof shows that $e(\mathfrak{q}^\star)=e(A^\star x^\star)$. Therefore $e(\mathfrak{q})=e(Ax)$.

Now, in the passage from $d-1$ to $d$ for $d\geq 2$ no complications will be caused any more by zero-divisors. Let $d$ be the dimension of $A$. We take again an element $x$ which is superficial of order 1 for $\mathfrak{q}$ (Remark 2 to Lemma 5, § 8), and set $A^\star=A/Ax$, $\mathfrak{q}^\star=\mathfrak{q}/Ax$. Then the polynomials $\bar{P}_{\mathfrak{q}^\star}(n)$ and $\bar{P}_{\mathfrak{q}}(n)-\bar{P}_{\mathfrak{q}}(n-1)$ differ only by their constant term (Lemma 4, § 8). Since they are of degree $d-1\geq 1$, they have the same leading term, whence $e(\mathfrak{q}^\star)n^{d-1}/(d-1)!=e(\mathfrak{q})n^{d-1}/(d-1)!$ (since $n^d-(n-1)^d$ has $dn^{d-1}$ as leading term) and therefore $e(\mathfrak{q}^\star)=e(\mathfrak{q})$. By the induction hypothesis there exists a system of parameters $\{x^\star_1, \cdots, x^\star_{d-1}\}$ ($x^\star_j \in \mathfrak{q}^\star$) of $A^\star$ such that $e(\mathfrak{q}^\star)=e\left(\sum_{j=1}^{d-1} A^\star x^\star_j\right)$. Then if $x_j$ denotes a representative of $x^\star_j$ in $\mathfrak{q}$, $\{x_1, \cdots, x_{d-1}, x\}$ is obviously a system of parameters in $A$; let $\mathfrak{q}'$ be the ideal generated by this system in $A$. By Lemma 3, § 8, we have $P_{\mathfrak{q}'/Ax}(n)=P_{\mathfrak{q}'}(n)-\lambda(\mathfrak{q}'^n:Ax)\geq P_{\mathfrak{q}'}(n)-P_{\mathfrak{q}'}(n-1)$ since $\mathfrak{q}'^n:Ax\supset\mathfrak{q}'^{n-1}$. Therefore $e(\mathfrak{q}'/Ax)=e\left(\sum_{j=1}^{d-1} A^\star x^\star_j\right)\geq e(\mathfrak{q}')$. From this inequality and from the relation $e(\mathfrak{q}^\star)=e\left(\sum_{j=1}^{d-1} A^\star x^\star_j\right)$ we deduce that $e(\mathfrak{q})=e(\mathfrak{q}^\star)\geq e(\mathfrak{q}')$. Since $\mathfrak{q}'\subset\mathfrak{q}$, we have also $e(\mathfrak{q})\leq e(\mathfrak{q}')$. Therefore $e(\mathfrak{q}')=e(\mathfrak{q})$, and our theorem is proved.

REMARK. If the ideal $\mathfrak{q}$ is generated by a system of parameters $\{y_1, \cdots, y_d\}$, it is also generated by a system of parameters $\{x_1, \cdots, x_d\}$ such that

(3) $$e(\mathfrak{q}) = e(\mathfrak{q}/Ax_1) = \cdots = e(\mathfrak{q}/(Ax_1 + \cdots + Ax_{d-1})).$$

In fact, there exists, by a reasoning similar to the one given in Remark 2 to Lemma 5 (§ 8), an element $x_1$ of $\mathfrak{q}$ which is superficial of order 1 for $\mathfrak{q}$ and which may be written in the form $x_1 = uy_1 + a_2y_2 + \cdots + a_dy_d$ with $u \notin \mathfrak{m}$. (Observe that $[\mathfrak{m}y_1 + (y_2, \cdots, y_d) + \mathfrak{q}^2]/\mathfrak{q}^2$ is a proper submodule of $\mathfrak{q}/\mathfrak{q}^2$.) Then, by the last part of the proof of Theorem 22, we have $e(\mathfrak{q}) = e(\mathfrak{q}/Ax_1)$, and $\mathfrak{q}$ is also generated by $\{x_1, y_2, \cdots, y_d\}$. We operate in the same way in $A/Ax_1$, $A/(Ax_1 + Ax_2)$, etc., up to $x_{d-1}$. We take $x_d$ to be $y_d$. Now we consider the one-dimensional local ring $B = A/(Ax_1 + \cdots + Ax_{d-1})$ and the residue class $x^\star$ of $x_d$ in $B$. We have $e(\mathfrak{q}) = e(Bx^\star)$ by relation (3), and $x^\star$ is a parameter in $B$. If $x^\star$ is not a *zero-divisor in B* (i.e., if the maximal ideal of $B$ is not entirely composed of zero divisors) then the modules $B/Bx^\star$ and $Bx^{\star n}/Bx^{\star n+1}$ are isomorphic (under the mapping $z + Bx^\star \to zx^{\star n} + Bx^{\star n+1}$). Then $e(Bx^\star)$ which is equal to the length of $Bx^{\star n}/Bx^{\star n+1}$ (since $B$ has dimension 1), is equal to $e(B/Bx^\star)$, i.e., to $\ell(A/\mathfrak{q})$. Therefore, if $x^\star$ is not a zero divisor in $B$, *the multiplicity of $\mathfrak{q}$ is equal to its length.* The condition that $x^\star$ is not a zero-divisor in $B$ is fulfilled if $\mathfrak{m}$ is not an imbedded prime ideal of $Ax_1 + \cdots + Ax_{d-1}$, and, in particular, if this ideal is *unmixed*; this is the case if $A$ is a *regular local ring* (see Cohen's extension of Macaulay's theorem in § 12, Theorem 29).

In general, we have the following relation between lengths and multiplicities:

THEOREM 23. *Let $A$ be a local ring, $\{x_1, \cdots, x_d\}$ a system of parameters of $A$, $\mathfrak{q}$ the ideal $\sum_{i=1}^{d} Ax_i$. Then $e(\mathfrak{q}) \leq \ell(A/\mathfrak{q})$. If $e(\mathfrak{q}) = \ell(A/\mathfrak{q})$, then the associated graded ring $G_\mathfrak{q}(A) = \sum_{n=0}^{\infty} \mathfrak{q}^n/\mathfrak{q}^{n+1}$ is isomorphic to the polynomial ring $B = (A/\mathfrak{q})\ [X_1, \cdots, X_d]$; and conversely.*

PROOF. In fact $G_\mathfrak{q}(A)$ is isomorphic to $B/\mathfrak{I}$, where $\mathfrak{I}$ is a homogeneous ideal (§ 1). Denoting by $\mathfrak{X}$ the ideal $\sum_{i=1}^{d} BX_i$, we have $\ell(A/\mathfrak{q}^n) = \ell(B/(\mathfrak{X}^n + \mathfrak{I})) \leq \ell(B/\mathfrak{X}^n) = \ell(A/\mathfrak{q})\binom{n+d-1}{d}$. Since $\ell(A/\mathfrak{q}^n)$ is a polynomial of degree $d$ in $n$ for $n$ large, this implies the inequality $e(\mathfrak{q}) \leq \ell(A/\mathfrak{q})$.

If the ideal $\mathfrak{I}$ is $\neq (0)$ then it contains a form $F(X) \neq 0$, say of degree $q$, whence also all the products of $F(X)$ by the monomials of degree

$<n-q$. These products generate an $(A/\mathfrak{q})$-module whose length is at least $\binom{n-q+d-1}{d}$. Then the formula $\ell(A/\mathfrak{q}^n)=\ell(B/(\mathfrak{X}^n+\mathfrak{J}))$ implies that $P_{\mathfrak{q}}(n)\leq \ell(A/\mathfrak{q})\binom{n+d-1}{d}-\binom{n-q+d-1}{d}$, whence $e(\mathfrak{q})<\ell(A/\mathfrak{q})$. This proves our second assertion. The converse is obvious.

Notice that, if $e(\mathfrak{q})=\ell(A/\mathfrak{q})$, the function $P_{\mathfrak{q}}(n)$ is a polynomial in $n$ for *all* values of $n$ (and not only for the large ones). Furthermore, since the initial form $X_i$ of $x_i$ is not a zero divisor in $G_{\mathfrak{q}}(A)=B$, we have $\mathfrak{q}^n:x_i=\mathfrak{q}^{n-1}$ for every $n$ and every $i$. As noticed in the remark following Theorem 22, this always happens if $A$ is a regular local ring.

We conclude this section with the proof of a theorem which not only can be used in certain cases for the computation of multiplicities, but also gives information on the behavior of multiplicities under finite integral extensions. This theorem is the algebraic counterpart of the projection formula for intersection cycles in Algebraic Geometry:

THEOREM 24. *Let $A$ be a local ring, $\mathfrak{m}$ its maximal ideal, $\mathfrak{q}$ an ideal in $A$ which is primary for $\mathfrak{m}$, and $B$ an overring of $A$ which is a finite $A$-module. Then $B$ is a semi-local ring, and $B\mathfrak{q}$ is an open ideal in $B$. Let $\{\mathfrak{p}_i\}$ be the set of maximal ideals of $B$ and let $\mathfrak{q}_i$ be the primary component of $B\mathfrak{q}$ relative to $\mathfrak{p}_i$. If no element $\neq 0$ in $A$ is a zero divisor in $B$, then the polynomials $[B:A]\bar{P}_{\mathfrak{q}}(n)$ and $\sum_i [B/\mathfrak{p}_i:A/\mathfrak{m}]\bar{P}_{\mathfrak{q}_i}(n)$ have the same degree and the same leading term.*†

PROOF. The assertions that $B\mathfrak{q}$ is an open ideal and that $B$ is semi-local follow from Theorem 15, part (d), § 6. For $n$ large enough the integer $\bar{P}_{\mathfrak{q}_i}(n)$ is the length of $B/\mathfrak{q}_i^n$ considered as a $B$-module. We first prove that $[B/\mathfrak{p}_i:A/\mathfrak{m}]\bar{P}_{\mathfrak{q}_i}(n)$ is the length of $B/\mathfrak{q}_i^n$ considered as *an $A$-module.* In fact, since $\mathfrak{q}_i$ is primary for $\mathfrak{p}_i$, there exists a chain of ideals

$$B > B\mathfrak{p}_i = \mathfrak{v}_1 > \mathfrak{v}_2 > \cdots > \mathfrak{v}_{d_i-1} > \mathfrak{v}_{d_i} = \mathfrak{q}_i^n,$$

such that $d_i=\bar{P}_{\mathfrak{q}_i}(n)$ and such that $\mathfrak{v}_j/\mathfrak{v}_{j+1}$ is a one-dimensional vector space over $B/\mathfrak{p}_i$ (Vol. I, Ch. IV, § 13, Theorem 28, Corollary 2). Therefore $\mathfrak{v}_j/\mathfrak{v}_{j+1}$ is, in a natural way, a $[B/\mathfrak{p}_i:A/\mathfrak{m}]$-dimensional vector space over $A/\mathfrak{m}$ and is therefore an $A$-module of length $[B/\mathfrak{p}_i:A/\mathfrak{m}]$. By addition we see that $[B/\mathfrak{p}_i:A/\mathfrak{m}]\bar{P}_{\mathfrak{q}_i}(n)$ is the length of $B/\mathfrak{q}_i^n$ considered as an $A$-module.

Furthermore, since $B\mathfrak{q}^n=\bigcap_i \mathfrak{q}_i^n$, and since the ideals $\mathfrak{q}_i^n$ are pairwise

† We denote by $[B:A]$ the maximum number of elements of $B$ which are linearly independent over $A$. It is equal to the dimension of the total quotient ring of $B$ considered as a vector space over the quotient field of $A$.

comaximal, $B/\mathfrak{q}^n$ is isomorphic to the direct sum of the rings $B/\mathfrak{q}_i^n$. Thus $\sum_i [B/\mathfrak{p}_i : A/\mathfrak{m}]\bar{P}_{\mathfrak{q}_i}(n)$ is, for $n$ large, the length of the *A-module* $B/B\mathfrak{q}^n$.

From now on we use the notation $\ell(E)$ for denoting the length of an *A-module E*. We have to compare $[B:A]\ell(A/\mathfrak{q}^n)$ and $\ell(B/B\mathfrak{q}^n)$. Since $B$ is a finite $A$-module, we can find in $B$ a maximal system $\{b_1, \cdots, b_r\}$ $(r=[B:A])$ of elements which are linearly independent over $A$. Then there exists an element $a \neq 0$ in $A$ such that

$$(4) \qquad aB \subset E = \sum_{i=1}^{r} Ab_i \subset B.$$

The kernel of the canonical mapping of $E$ onto $(E+\mathfrak{q}^nB)/\mathfrak{q}^nB$ obviously contains $\mathfrak{q}^nE$. We thus have a canonical mapping of $E/\mathfrak{q}^nE$ *onto* $(E+\mathfrak{q}^nB)/\mathfrak{q}^nB$, whence the length of the latter module is at most equal to the length of the former. Since $E$ is a free $A$-module with $r$ generators it follows that

$$(5) \quad r\ell(A/\mathfrak{q}^n) = \ell(E/\mathfrak{q}^nE) \geq \ell((E+\mathfrak{q}^nB)/\mathfrak{q}^nB) \geq \ell((aB+\mathfrak{q}^nB)/\mathfrak{q}^nB),$$

i.e.,

$$(5') \qquad r\ell(A/\mathfrak{q}^n) \geq \ell(B/\mathfrak{q}^nB) - \ell(B/(aB+\mathfrak{q}^nB)).$$

On the other hand, the kernel of the canonical mapping of $aB$ onto $(aB+\mathfrak{q}^nE)/\mathfrak{q}^nE$ obviously contains $\mathfrak{q}^naB$. We therefore have a canonical mapping of $aB/\mathfrak{q}^naB$ *onto* $(aB+\mathfrak{q}^nE)/\mathfrak{q}^nE$. Since, by assumption, $a$ is not a zero divisor in $B$, the modules $aB/\mathfrak{q}^naB$ and $B/\mathfrak{q}^nB$ are isomorphic. Thus we deduce, as above, the inequality

$$\ell(B/\mathfrak{q}^nB) \geq \ell((aB+\mathfrak{q}^nE)/\mathfrak{q}^nE) = \ell(E/\mathfrak{q}^nE) - \ell(E/(aB+\mathfrak{q}^nE)),$$

and since $E$ is a free $A$-module with $r$ generators, this yields the inequalities

$$(6) \quad r\ell(A/\mathfrak{q}^n) \leq \ell(B/\mathfrak{q}^nB) + \ell(E/(aB+\mathfrak{q}^nE)) \leq \ell(B/\mathfrak{q}^nB) + \ell(E/(aE+\mathfrak{q}^nE)).$$

Again, since $E$ is a free $A$-module with $r$ generators, it follows from (6) that we have

$$(6') \qquad r\ell(A/\mathfrak{q}^n) \leq \ell(B/\mathfrak{q}^nB) + r\ell(A/(aA+\mathfrak{q}^nA)).$$

We set $B' = B/Ba$. Then $B'$ is an $A$-module and we have $B'\mathfrak{q} = (B\mathfrak{q}+Ba)/Ba$. With this notation, inequalities (5′) and (6′) yield (for $n$ large)

$$(7) \qquad \ell(B/B\mathfrak{q}^n) - \ell(B'/B'\mathfrak{q}^n) \leq r\bar{P}_{\mathfrak{q}}(n) \leq \ell(B/B\mathfrak{q}^n) + r\bar{P}_{\mathfrak{q}'}(n),$$

where $\mathfrak{q}'$ is the ideal $(\mathfrak{q}+Aa)/Aa$ in the ring $A' = A/Aa$: in fact the length of $A'/\mathfrak{q}'^n = A/(aA+\mathfrak{q}^n)$ is the same whether we regard this ring as an $A$-module or as an $A'$-module. Since $a$ is a regular element in $A$, the polynomial $\bar{P}_{\mathfrak{q}'}(n)$ is of degree $d-1$ (Corollary 2 to Theorem 20, § 9). On the other hand $\bar{P}_{\mathfrak{q}}(n)$ and $\ell(B/B\mathfrak{q}^n) = \sum_i [B/\mathfrak{p}_i : A/\mathfrak{m}]\bar{P}_{\mathfrak{q}_i}(n)$ (for $n$ large) are polynomials of degree $d$, since $A$ is a local ring of dimension $d$ and since $B$ is a finite $A$-module. (Corollary 3 to Theorem 20, § 9.) It remains to study the term $\ell(B'/B'\mathfrak{q}^n)$ in (7). Since $B'$ is a finite $A$-module which is annihilated by $Aa$, it is also a finite $A'$-module ($A' = A/Aa$). The length of $B'/B'\mathfrak{q}^n$ is the same whether we regard this ring as an $A$-module or as an $A'$-module. Thus, since $A'$ is a local ring of dimension $d-1$, $\ell(B'/B'\mathfrak{q}^n)$ is, as above, a polynomial of degree $d-1$ for $n$ large.

This being so, inequalities (7) prove that the degree of $\ell(B/B\mathfrak{q}^n)$ is exactly $d$, and that *the polynomials* $\ell(B/B\mathfrak{q}^n)$ *and* $r\bar{P}_q(n)$ *have the same leading term.* In view of what was shown in the earlier part of the proof, the length $\ell(B/B\mathfrak{q}^n)$ of the $A$-module $B/B\mathfrak{q}^n$ is equal to the polynomial $\sum_i [B/\mathfrak{p}_i : A/\mathfrak{m}]\bar{P}_{\mathfrak{q}_i}(n)$ for $n$ large. This proves our assertion.

COROLLARY 1. *The hypothesis and notations being as in Theorem 24, suppose furthermore that all the local rings $B_{\mathfrak{p}_i}$ have the same dimension as $A$. Then*

$$(8) \qquad [B:A]e(\mathfrak{q}) = \sum_i [B/\mathfrak{p}_i : A/\mathfrak{m}]e(\mathfrak{q}_i).$$

In fact, all the polynomials $\bar{P}_{\mathfrak{q}_i}(n) = \bar{P}_{\mathfrak{q}B_{\mathfrak{p}_i}}(n)$ have then the same degree $d = \dim(A)$.

The hypothesis that all the local rings $B_{\mathfrak{p}_i}$ have the same dimension as $A$ is verified in the following cases:

(1) $B$ is a local ring.

(2) In most semi-local rings which occur in algebraic geometry.

(3) $A$ is an *integrally closed* local ring. In fact, since no element $\neq 0$ of $A$ is a zero divisor in $B$, we may apply the "going down" theorem proved in Vol. I, Ch. V, § 3, Theorem 6. Let $\mathfrak{v}_1 = \mathfrak{m} > \mathfrak{v}_2 > \cdots > \mathfrak{v}_{d+1}$ be a maximal chain of prime ideals in $A$. Given any maximal ideal $\mathfrak{P}$ of $B$, we have $\mathfrak{P} \cap A = \mathfrak{m}$, and the "going down" theorem provides us with a prime ideal $\mathfrak{V}_2$ of $B$ such that $\mathfrak{V}_2 < \mathfrak{P}$ and $\mathfrak{V}_2 \cap A = \mathfrak{v}_2$. By repeated applications, we get a chain $\mathfrak{P} > \mathfrak{V}_2 > \cdots > \mathfrak{V}_{d+1}$ of prime ideals of $B$ beginning with $\mathfrak{P}$. We thus have a chain of $d+1$ prime ideals in $B_{\mathfrak{p}}$, whence $\dim(B_{\mathfrak{p}}) \geq d$, and therefore $\dim(B_{\mathfrak{p}}) = d$ since a chain of $d+2$ prime ideals in $B_{\mathfrak{p}}$ would induce a chain of $d+2$ prime ideals in $A$ (cf. Corollary 3 to Theorem 20, § 9).

(4) *A is a local domain of dimension* 1 *and also B is a domain.* In that case it is clear that all the rings $B_{\mathfrak{p}_i}$ are 1-dimensional. Let us furthermore assume that $B$ is integrally closed. In this case each ring $B_{\mathfrak{p}_i}$ is a discrete valuation ring of the quotient field of $B$, since each of these rings is noetherian, integrally closed and has only one proper prime ideal (see VI, § 10, Theorem 16, Corollary 1). If $v_i$ is the valuation defined by $B_{\mathfrak{p}_i}$, then $v_i$ is non-negative on $A$ and has center $\mathfrak{m}$ (VI, § 5), and the $v_i$ give *all* the valuations of the quotient field of $B$ which are non-negative on $A$ and have center $\mathfrak{m}$, for any such valuation must be non-negative on $B$, and its center must be one of the $\mathfrak{p}_i$. The integer $[B:A]$ is in this case the relative degree of the quotient field of $B$ over the quotient field of $A$, and $B/\mathfrak{p}_i$ is the residue field of $v_i$. Since in $B_{\mathfrak{p}_i}$ every ideal is a power of the maximal ideal $\mathfrak{p}_i B_{\mathfrak{p}_i}$, it follows at once that $e(\mathfrak{q}_i) = v_i(\mathfrak{q}_i) = v_i(\mathfrak{q})$, where we denote by $v_i(\mathfrak{q})$ the minimum of the integers $v_i(\omega)$ as $\omega$ ranges over $\mathfrak{q}$. In the special case $\mathfrak{q} = \mathfrak{m}$, if we set $e(\mathfrak{q}_i) = v_i(\mathfrak{m}) = e_i$, $[B/\mathfrak{p}_i : A/\mathfrak{m}] = n_i$, $[B:A] = n$, formula (8) takes the form

$$ne(\mathfrak{m}) = \sum_i e_i n_i, \tag{8'}$$

and in this form it is an analogue of a formula derived for the extension of a valuation (VI, § 11, formula (13)). In fact, the two formulas overlap when $A$ is a discrete valuation ring. We note finally the special case in which $B$ is the integral closure of $A$ in its quotient field. In this case we have

$$e(\mathfrak{m}) = \sum e_i n_i, \quad (e_i = v_i(\mathfrak{m})), \tag{8''}$$

*always provided B is a finite module over A.*

Corollary 2. *Let A be a complete local ring,* $\mathfrak{m}$ *its maximal ideal, K a subfield of A over which* $A/\mathfrak{m}$ *is finite,* $\{x_1, \cdots, x_d\}$ *a system of parameters of A,* $\mathfrak{q}$ *the ideal generated by* $x_1, \cdots, x_d$. *Then A is a finite module over* $K[[x_1, \cdots, x_d]]$. *If no element of* $K[[x_1, \cdots, x_d]]$ *is a zero-divisor in A, then we have*

$$[A:K[[x_1, \cdots, x_d]]] = e(\mathfrak{q}) \cdot [(A/\mathfrak{m}):K]. \tag{9}$$

That $A$ is a finite module over $K[[x_1, \cdots, x_d]]$ has been seen in § 9 (Remark, p. 293). On the other hand, $K[[x_1, \cdots, x_d]]$ is a power series ring in $d$ variables (Corollary 2 to Theorem 21, § 9), whence the ideal $\mathfrak{X}$ generated by $x_1, x_2, \cdots, x_d$ in this ring has multiplicity one, since $P_{\mathfrak{X}}(n) = \binom{n+d-1}{d}$. Thus, since $\mathfrak{q} = A\mathfrak{X}$, our formula follows immediately from Corollary 1.

REMARK. Corollary 2 to Theorem 7 (§3) shows that if $A/\mathfrak{q}$, considered as a vector space over $K$, is generated by the residue classes mod $\mathfrak{q}$ of certain elements $y_1, y_2, \cdots, y_q$ of $A$, then $A$, regarded as a $K[[x_1, x_2, \cdots, x_d]]$-module, is generated by these elements $y_1, y_2, \cdots, y_q$. We may take $q=[(A/\mathfrak{q}):K]$, whence $[A:K[[x_1, \cdots, x_d]]] \leqq [(A/\mathfrak{q}):K]$, and formula (9) shows that $e(\mathfrak{q})[A/\mathfrak{m}:K] \leqq [(A/\mathfrak{q}):K]$. This shows again that the multiplicity of the ideal $\mathfrak{q}$ is at most equal to its length.

**§ 11. Regular local rings.** Let $A$ be a local ring of dimension $d$, $\mathfrak{m}$ its maximal ideal. We say that $A$ is a *regular local ring* if $\mathfrak{m}$ may be generated by $d$ elements. Then any system of $d$ elements of $A$ which generates $\mathfrak{m}$ is obviously a system of parameters and is called a *regular system of parameters* of $A$.

EXAMPLE. A power series ring $K[[X_1, \cdots, X_n]]$ in $n$ variables over a field $K$ is a regular local ring of dimension $n$.

If $A$ is a regular local ring, then its completion $\hat{A}$ is regular, since the maximal ideal of $\hat{A}$ is generated by $\mathfrak{m}$ and since $\hat{A}$ has the same dimension as $A$. Conversely if $\hat{A}$ is regular, and if $\{\xi_1, \cdots, \xi_d\}$ is a regular system of parameters of $\hat{A}$, we can find $d$ elements $x_1, \cdots, x_d$ of $\mathfrak{m}$ such that $x_i \equiv \xi_i \pmod{\hat{A}\mathfrak{m}^2}$. Then $x_1, \cdots, x_d$ generate $\mathfrak{m}\hat{A}$ by Theorem 7, § 3. Therefore, by Theorem 9, (a′), § 4, we have $\sum_i Ax_i = A \cap \left(\sum_i \hat{A}x_i\right) = A \cap \hat{A}\mathfrak{m} = \mathfrak{m}$, whence $A$ is regular.

THEOREM 25. *Let $A$ be a local ring of dimension $d$; $\mathfrak{m}$ its maximal ideal. Then the following statements are equivalent:*

(a) *$A$ is a regular local ring.*

(b) *The associated graded ring $G_{\mathfrak{m}}(A) = \sum_{n=0}^{\infty} \mathfrak{m}^n/\mathfrak{m}^{n+1}$ is a polynomial ring in $d$ variables over the field $A/\mathfrak{m}$.*

(c) *$\mathfrak{m}/\mathfrak{m}^2$ is a vector space of dimension $d$ over $A/\mathfrak{m}$.*

PROOF. We give a cyclic proof: (a) ⇒ (b) ⇒ (c) ⇒ (a). That (a) implies (b) follows directly from Theorem 21, § 9, if we take for $\{x_1, x_2, \cdots, x_d\}$ a regular system of parameters. Another proof can be obtained by using multiplicities. Namely, the ideal $\mathfrak{m}$ is generated by a system of parameters. Since the length of $\mathfrak{m}$ is equal to 1, its multiplicity must also be 1 (§ 10, Theorem 23), whence $G_{\mathfrak{m}}(A)$ is a polynomial ring in $d$ variables over $A/\mathfrak{m}$ (§ 10, Theorem 23). Therefore (a) implies (b).

The fact that (b) implies (c) is evident. Finally, if $\mathfrak{m}/\mathfrak{m}^2$ has dimension $d$ over $A/\mathfrak{m}$, let $x_1, \cdots, x_d$ be elements of $\mathfrak{m}$ whose residue classes mod $\mathfrak{m}^2$ form a basis of $\mathfrak{m}/\mathfrak{m}^2$ over $A/\mathfrak{m}$. Then, if $\mathfrak{q}$ denotes the ideal $\sum_i Ax_i$, we have $\mathfrak{m} = \mathfrak{q} + \mathfrak{m}^2 = \mathfrak{q} + \mathfrak{m}(\mathfrak{q} + \mathfrak{m}^2) = \mathfrak{q} + \mathfrak{m}^3$, whence $\mathfrak{m} = \mathfrak{q} + \mathfrak{m}^n$ for every $n$, by induction. Since $\mathfrak{q}$ is closed, this implies $\mathfrak{m} = \bigcap_{n=0}^{\infty} (\mathfrak{q} + \mathfrak{m}^n) = \mathfrak{q}$, and $\mathfrak{m}$ is generated by the $d$ elements $x_1, \cdots, x_d$. Therefore (c) implies (a).

COROLLARY 1. *A regular local ring $A$ is an integrally closed integral domain. The passage from elements of $A$ to their initial forms preserves products. The order function $v$ relative to $\mathfrak{m}$* (see § 1; *$v(x)$ is defined by $x \in \mathfrak{m}^{v(x)}$, $x \notin \mathfrak{m}^{v(x)+1}$*) *is a valuation of the quotient field of $A$. The completion $\hat{A}$ of $A$ is an integrally closed domain.*

All these assertions, except the last, follow from (b) and from Theorems 1 and 3 (§ 1). The last assertion follows from the fact that $\hat{A}$ is also a regular local ring.

COROLLARY 2. *Let $A$ be a regular local ring of dimension $d$, $\mathfrak{m}$ its maximal ideal. In order that a system $\{x_1, \cdots, x_d\}$ of elements of $\mathfrak{m}$ be a regular system of parameters of $A$, it is necessary and sufficient that the residue classes of the $x_i$* mod $\mathfrak{m}^2$ *generate $\mathfrak{m}/\mathfrak{m}^2$ over $A/\mathfrak{m}$, or equivalently, that these residue classes be linearly independent over $A/\mathfrak{m}$.*

Since the dimension of $\mathfrak{m}/\mathfrak{m}^2$ over $A/\mathfrak{m}$ is $d$, the two conditions about the $\mathfrak{m}^2$-residues of $x_1, \cdots, x_d$ are equivalent. The necessity of our condition is obvious. For the sufficiency we notice that if the condition is satisfied, then we have $\mathfrak{m} = \sum_{i=1}^{d} Ax_i + \mathfrak{m}^2$, whence $\mathfrak{m} = \sum_{i=1}^{d} Ax_i$ as at the end of the proof of Theorem 25.

REMARK. We noticed, in the proof of Theorem 25, that, if $A$ is a regular local ring, then the multiplicity of its maximal ideal $\mathfrak{m}$ is equal to 1. For any other open ideal $\mathfrak{q}$ of $A$, *we have* $e(\mathfrak{q}) > 1$. In fact, $(\mathfrak{q} + \mathfrak{m}^2)/\mathfrak{m}^2$ is then a proper subspace of $\mathfrak{m}/\mathfrak{m}^2$ (otherwise $\mathfrak{m} = \mathfrak{q} + \mathfrak{m}^2$, whence $\mathfrak{m} = \mathfrak{q}$ as above). Taking a suitable basis of the vector space $\mathfrak{m}/\mathfrak{m}^2$, we see that there exists a regular system of parameters $\{x_1, x_2, \cdots, x_d\}$ such that $\mathfrak{q} + \mathfrak{m}^2 \subset Ax_2 + \cdots + Ax_d + \mathfrak{m}^2$. As this latter ideal is $\mathfrak{q}' = Ax_1^2 + Ax_2 + \cdots + Ax_d$, and since $\mathfrak{q} \subset \mathfrak{q}'$ implies $e(\mathfrak{q}) \geq e(\mathfrak{q}')$, we have to prove that $e(\mathfrak{q}') > 1$. But this follows from the fact that $G_{\mathfrak{q}'}(A)$ is a polynomial ring in $d$ variables over $A/\mathfrak{q}'$, whence $e(\mathfrak{q}') = \ell(A/\mathfrak{q}') = 2$, by Theorem 23, § 10.

It may be proved (by using the structure theorems for complete local rings) that if a local ring $A$ of the type encountered in Algebraic Geometry admits an ideal $\mathfrak{q}$ of multiplicity 1 (i.e., if $e(\mathfrak{q}) = 1$), then $A$ is a regular local ring (and, necessarily, $\mathfrak{q} = \mathfrak{m}$).

We shall give later a partial proof of the fact that every quotient ring $A_\mathfrak{p}$ ($\mathfrak{p}$: prime ideal) of a regular local ring $A$ is a regular local ring. We now describe those *factor rings* $A/\mathfrak{b}$ of a regular local ring which are regular.

THEOREM 26. *Let $A$ be a regular local ring and $\mathfrak{b}$ an ideal in $A$. For $A/\mathfrak{b}$ to be regular, it is necessary and sufficient that $\mathfrak{b}$ be generated by a subset of a regular system of parameters of $A$* (i.e., *by a system of elements of* $\mathfrak{m}$ *which are linearly independent* mod $\mathfrak{m}^2$).

PROOF. The equivalence of the notions "subset of a regular system of parameters," "system of elements of $\mathfrak{m}$ which are linearly independent mod $\mathfrak{m}^2$" follows immediately from Corollary 2 to Theorem 25. Suppose now that $\{x_1, \cdots, x_d\}$ is a regular system of parameters in $A$, and that $\mathfrak{b}$ is generated by $x_1, \cdots, x_j$. By a formula proved in § 9 (p. 292) we have dim $(A/\mathfrak{b}) = d-j$. On the other hand the maximal ideal $\mathfrak{m}/\mathfrak{b}$ of $A/\mathfrak{b}$ is generated by $d-j$ elements, namely the $\mathfrak{b}$-residues of $x_{j+1}, \cdots, x_d$. Hence $A/\mathfrak{b}$ is a regular local ring.

Conversely, assume that $A/\mathfrak{b}$ is a regular local ring, say of dimension $\delta$. We consider the canonical mapping $\varphi$ of $\mathfrak{m}/\mathfrak{m}^2$ onto $\bar{\mathfrak{m}}/\bar{\mathfrak{m}}^2 = \mathfrak{m}/(\mathfrak{m}^2 + \mathfrak{b})$, where $\bar{\mathfrak{m}} = \mathfrak{m}/\mathfrak{b}$. Both are vector spaces over $A/\mathfrak{m}$ $(=(A/\mathfrak{b})/\bar{\mathfrak{m}})$, of dimension $d$ and $\delta$ respectively, and it is obvious that $\varphi$ is $(A/\mathfrak{m})$-linear. Therefore the kernel of $\varphi$ has dimension $d-\delta$, whence $\mathfrak{b}$ contains $d-\delta$ elements, say $x_1, x_2, \cdots, x_{d-\delta}$, whose $\mathfrak{m}^2$-residues are linearly independent over $A/\mathfrak{m}$. By Theorem 25, Corollary 2, these $d-\delta$ elements form a subset of a regular system of parameters. By the preceding half of the proof, the ideal $\mathfrak{b}' = \sum_{i=1}^{d-\delta} Ax_i$, has the property that $A/\mathfrak{b}'$ is a regular local ring of dimension $\delta$. Now the ring $A/\mathfrak{b}$ is a homomorphic image of $A/\mathfrak{b}'$, since $\mathfrak{b}' \subset \mathfrak{b}$, and has the same dimension $\delta$ as $A/\mathfrak{b}'$. Since $A/\mathfrak{b}'$ is an integral domain (Theorem 25, Corollary 1), it follows from Theorem 20, Corollary 2 (§ 9) that the canonical homomorphism of $A/\mathfrak{b}'$ onto $A/\mathfrak{b}$ is an isomorphism. Hence $\mathfrak{b} = \mathfrak{b}'$, and this completes the proof of the theorem.

REMARKS:

(1) In the last part of the proof it is not necessary to fall back on Corollary 2 of Theorem 20 (§ 9). It is sufficient to observe that $G(A/\mathfrak{b}')$ is a polynomial ring in $\delta$ independent variables over $A/\mathfrak{m}$ and that, were $\mathfrak{b}'$ a proper subset of $\mathfrak{b}$, the ring $G(A/\mathfrak{b})$ $(=G((A/\mathfrak{b}')/(\mathfrak{b}/\mathfrak{b}')))$ would be a proper homomorphic image of the polynomial ring $G(A/\mathfrak{b}')$, and this contradicts the fact that $G(A/\mathfrak{b})$ is itself a polynomial ring in $\delta$ independent variables over $A/\mathfrak{m}$.

(2) The proof that $\mathfrak{b}' = \mathfrak{b}$ is based essentially on two facts: (1) $\mathfrak{b}' \subset \mathfrak{b}$;

(2) $\mathfrak{b}$ and $\mathfrak{b}'$ have the same leading ideal in $G(A)$. In this connection, it is proper to call attention to a general lemma on Zariski rings which covers the case under consideration.

LEMMA. *Let $A$ be a Zariski ring, $\mathfrak{m}$ an ideal defining the topology of $A$, $\mathfrak{a}$ and $\mathfrak{a}'$ two ideals in $A$ such that $\mathfrak{a}' \subset \mathfrak{a}$. If $\mathfrak{a}$ and $\mathfrak{a}'$ have the same leading ideal in $G(A) = \sum_{n=0}^{\infty} \mathfrak{m}^n/\mathfrak{m}^{n+1}$, then $\mathfrak{a} = \mathfrak{a}'$.*

In fact the associated graded module $G(E)$ of the $A$-module $E = \mathfrak{a}/\mathfrak{a}'$ is reduced to (0) since $G(E)$ is isomorphic to $G(\mathfrak{a})/G(\mathfrak{a}')$. This implies that $E/\mathfrak{m}E = (0)$, whence $E = \mathfrak{m}E$. Thus Theorem 9, (f) (§ 4) shows that $E = (0)$, whence $\mathfrak{a} = \mathfrak{a}'$.

**§ 12. Structure of complete local rings and applications.** In this section we restrict ourselves to *equicharacteristic* local rings, i.e., to local rings $A$ which have the same characteristic (zero or a prime number $p$) as their residue field $A/\mathfrak{m}$. Most of the theorems we are going to prove admit analogues in the unequal characteristic case, i.e., the case in which $A$ is a ring of characteristic 0 or $p^n$ ($p$: prime number, $n > 1$) and $A/\mathfrak{m}$ a field of characteristic $p$.† It is easily seen (cf. proof of Corollary 2 to Theorem 17, § 7) that a local ring is equicharacteristic if and only if it contains a field.

We recall (cf. p. 281, § 7) the notion of a *field of representatives* (or representative field) for a local ring $A$ with maximal ideal $\mathfrak{m}$: it is a subfield $L$ of $A$ which is mapped *onto* $A/\mathfrak{m}$ by the canonical mapping $\varphi$ of $A$ onto $A/\mathfrak{m}$. Then, since $L$ is a field, the restriction of $\varphi$ to $L$ is an isomorphism of $L$ onto $A/\mathfrak{m}$.

THEOREM 27 (I. S. COHEN). *An equicharacteristic complete local ring $A$ admits a field of representatives.*

PROOF.‡ In the case in which $A$ and $A/\mathfrak{m}$ have characteristic 0 the theorem has already been proved as a consequence of Hensel's lemma (Corollary 2 to Theorem 17, § 7). We may therefore restrict ourselves to the case in which $A$ and $A/\mathfrak{m}$ have characteristic $p \neq 0$.

We first prove our assertion under the assumption $\mathfrak{m}^2 = (0)$. The proof in this case will make no use of the noetherian character of $A$ nor of the completeness of $A$, *and will in fact be valid for any ring $A$ in which $\mathfrak{m}$ is the only maximal ideal*, provided $A$ and $A/\mathfrak{m}$ have the same char-

† For these extensions we refer the reader to the paper of I. S. Cohen "On the structure and ideal theory of complete local rings," *Trans. Amer. Math. Soc.*, 59, 54–106, (1946), or to P. Samuel "Algèbre Locale," *Mem. Sci. Math.* No. 123, Paris, 1953.

‡ The method of proof given in the text is due to A. Geddes.

acteristic $p \neq 0$ and provided that $\mathfrak{m}^p = (0)$. Let $A^p$ be the set of all elements $x^p$, where $x$ ranges over $A$. The set $A^p$ is obviously a subring of $A$. Furthermore, if $x^p \neq 0$, then $x \notin \mathfrak{m}$, $x$ admits an inverse $y \in A$, and $x^p$ admits $y^p$ as an inverse in $A^p$. Therefore $A^p$ is a *subfield* of $A$. Among the subfields of $A$ which contain $A^p$, Zorn's lemma provides us with a maximal subfield $L$. Let $\varphi$ be the canonical homomorphism of $A$ onto $A/\mathfrak{m}$; we prove that $\varphi(L) = A/\mathfrak{m}$. In fact, assuming the contrary, take an element $\alpha \in A/\mathfrak{m}$, $\alpha \notin \varphi(L)$. Since $\alpha^p \in \varphi(A^p) \subset \varphi(L)$, the minimal polynomial of $\alpha$ over $\varphi(L)$ is $X^p - \alpha^p$. We take a representative $a$ of $\alpha$ in $A$ ($\varphi(a) = \alpha$). Then the polynomial $X^p - a^p$ is irreducible over $L$, since otherwise we would have $(a - a^1)^p = 0$ for some $a^1$ in $L$, and $\alpha$ would belong to $\varphi(L)$. Thus $L[a]$ is a subfield of $A$, in contradiction with the maximal character of $L$.

We now come back to the general case. Since $p \geq 2$, the maximal ideal $\bar{\mathfrak{m}} = \mathfrak{m}/\mathfrak{m}^2$ of the local ring $A/\mathfrak{m}^2$ satisfies the condition $\bar{\mathfrak{m}}^p = (0)$ and hence $A/\mathfrak{m}^2$ admits a representative field $K_2$. We now construct, by *induction* on $n$, a *representative field* $K_n$ of $A/\mathfrak{m}^n$ such that, if we denote by $\psi_n$ the canonical homomorphism of $A/\mathfrak{m}^{n+1}$ onto $A/\mathfrak{m}^n$, $\psi_n$ *induces an isomorphism* of $K_{n+1}$ *onto* $K_n$. Suppose that $K_n$ has already been constructed. The inverse image $\psi_n^{-1}(K_n)$ is a subring $R$ of $A/\mathfrak{m}^{n+1}$ which contains the kernel $\mathfrak{p} = \mathfrak{m}^n/\mathfrak{m}^{n+1}$ of $\psi_n$. Any element $\xi$ of $R$ which is not in $\mathfrak{p}$ has as $\psi_n$-image an element $\xi' \neq 0$ of $K_n$. Since $K_n$ is a field, $\xi'$ is a unit in $A/\mathfrak{m}^n$, and therefore $\xi' \notin \mathfrak{m}/\mathfrak{m}^n$. Hence $\xi \notin \mathfrak{m}/\mathfrak{m}^{n+1}$ (since the maximal ideal $\mathfrak{m}/\mathfrak{m}^{n+1}$ in $A/\mathfrak{m}^{n+1}$ is the full inverse image of the maximal ideal $\mathfrak{m}/\mathfrak{m}^n$ of $A/\mathfrak{m}^n$), and $\xi$ is a unit in $A/\mathfrak{m}^{n+1}$. Let $\xi\eta = 1$, $\eta \in A/\mathfrak{m}^{n+1}$. Then $\eta' = \psi_n(\eta) \in K_n$, and therefore $\eta \in R$, since $R$ is the full inverse image of $K_n$. Thus $\xi$ is invertible in $R$, and we have therefore proved that $\mathfrak{p}$ is the only maximal ideal of $R$. Since we obviously have $\mathfrak{p}^2 = (0)$ (as $\mathfrak{p} = \mathfrak{m}^n/\mathfrak{m}^{n+1}$ and $\mathfrak{m}^{2n} \subset \mathfrak{m}^{n+1}$), the first part of the proof shows the existence of a representative field $K_{n+1}$ of $R$. Since the canonical homomorphism $A/\mathfrak{m}^{n+1} \to A/\mathfrak{m}$ is the product of $\psi_n$ by the canonical homomorphism $A/\mathfrak{m}^n \to A/\mathfrak{m}$ and since $K_n$ is a representative field of $A/\mathfrak{m}^n$, it follows that $\psi_n(K_{n+1}) = K_n$ and that $K_{n+1}$ is a representative field of $A/\mathfrak{m}^{n+1}$.

We now conclude the proof by using the fact that, since $A$ is *complete*, it is the *projective limit* of the residue class rings $A/\mathfrak{m}^n$. In fact, given any sequence $\{\eta_n\}$ of elements $\eta_n \in A/\mathfrak{m}^n$ such that $\eta_n = \psi_n(\eta_{n+1})$ for all $n$, there exists one and only one element $y$ of $A$ admitting $\eta_n$ as $\mathfrak{m}^n$-residue for all $n$. To see this, we take, for every $n$, an element $y_n$ of $A$ admitting $\eta_n$ as $\mathfrak{m}^n$-residue. Since $\eta_n = \psi_n(\eta_{n+1})$, we have $y_n \equiv y_{n+1} \pmod{\mathfrak{m}^n}$, whence the sequence $\{y_n\}$ is a Cauchy sequence. If $y$ is the limit of this sequence then $y-y_n \in \mathfrak{m}^n$ and hence $y$ admits $\eta_n$ as $\mathfrak{m}^n$-residue for every $n$. The uniqueness of $y$ follows easily from the fact that $\bigcap_{n=0}^{\infty} \mathfrak{m}^n=(0)$. Now, for every element $\eta$ of $K_1(=A/\mathfrak{m})$ we consider the elements $\eta_2=\psi_1^{-1}(\eta_1) \in K_2, \cdots, \eta_{n+1}=\psi_n^{-1}(\eta_n) \in K_{n+1}, \cdots$, and we denote by $u(\eta)$ the above constructed element $y$ of $A$. It is readily verified that $u(\eta+\eta')=u(\eta)+u(\eta')$, and that $u(\eta\eta')=u(\eta)u(\eta')$ (conservation of sums and products by passage to the limit), whence $u(K_1)$ is a subring of $A$. Furthermore, for every $\eta \neq 0$ in $K_1$, there exists an element $\eta'$ in $K_1$ such that $\eta\eta'=1$, whence $u(\eta')$ is the inverse of $u(\eta)$ (note that from the uniqueness of the element $y$, established above, follows that $u(1)$ is the element 1 of $A$). Therefore $u(K_1)$ is a subfield of $A$. Since its image $\varphi(u(K_1))$ in $A/\mathfrak{m}=K_1$ is obviously $A/\mathfrak{m}$ itself, we have found a representative field of $A$.

A somewhat shorter proof of Cohen's Theorem, due to M. Narita, may be given; it uses properties of *p-bases* in fields of characteristic $p$, (see Vol. I, Ch. II, § 17, pp. 129–131). We again restrict ourselves to the case in which $A$ and $A/\mathfrak{m}$ have characteristic $p \neq 0$. Let $\{x_\alpha\}$ be a family of elements of $A$ such that their $\mathfrak{m}$-residues $\bar{x}_\alpha$ form a $p$-basis of $A/\mathfrak{m}$. For every integer $k$, we consider the subring $R_k=A^{p^k}[x]$ of $R$.

We first notice that $R_k \cap \mathfrak{m} \subset \mathfrak{m}^{p^k}$. In fact, since $x^{p^k} \in A^{p^k}$, every element of $R_k$ may be written in the form $\sum_s a_s m_s(x)$ where $a_s \in A^{p^k}$ and where the $m_s(x)$ are monomials in the $x_\alpha$ with exponents $\leq p^k-1$. If $\sum_s a_s m_s(x) \in \mathfrak{m}$, we have, by taking $\mathfrak{m}$-residues, $\sum_s \bar{a}_s m_s(\bar{x})=0$. Since the monomials $m_s(\bar{x})$ are linearly independent over $(A/\mathfrak{m})^{p^k}$ (this is a property of $p$-bases), this imples $\bar{a}_s=0$, i.e., $a_s \in \mathfrak{m}$. Since $a_s \in A^{p^k}$, we may write $a_s=b_s^{p^k}$, whence $b_s \in \mathfrak{m}$ since $\mathfrak{m}$ is a prime ideal. Therefore $a_s \in \mathfrak{m}^{p^k}$, and the inclusion $R_k \cap \mathfrak{m} \subset \mathfrak{m}^{p^k}$ is proved.

Now let $y$ be any element of $A$. We are going to construct a Cauchy sequence $\{y_k\}$ such that $y_k \in R_k$ for every $k$ and that $y \equiv y_k \pmod{\mathfrak{m}}$. We take $y_0=y$, and we suppose that $y_k$ is already constructed. We write $y_k=\sum_s a_s^{p^k} m_s(x)$, where $a_s \in A$ and where the $m_s(x)$ are monomials in the $x_\alpha$ with exponents $\leq p^k-1$. Since $A/\mathfrak{m}=(A/\mathfrak{m})^p[\bar{x}]$, we can write $a_s \equiv \sum_t b_{st}^p m'_t(x) \pmod{\mathfrak{m}}$, where $b_{st} \in A$ and where the $m'_t(x)$ are monomials with exponents $\leq p-1$. Setting $y_{k+1}=\sum_{s,t} b_{st}^{p^{k+1}}(m'_t(x))^{p^k} m_s(x)$, we have $y_{k+1} \in R_{k+1}$ and $y_{k+1}-y_k=\sum_s (a_s-\sum_t b_{st}^p m'_t(x))^{p^k} m_s(x) \in \mathfrak{m}^{p^k}$, whence $\{y_k\}$ is the Cauchy sequence we were looking for. Since $A$ is *complete*, the Cauchy sequence $\{y_k\}$ admits a limit $y' \in A$, and we obviously have $y \equiv y'$ (mod $\mathfrak{m}$). Furthermore, since the subrings $R_k$ form a decreasing sequence, we have $y_j \in R_k$ for every $j \geq k$, whence $y'$ belongs to the closure $\bar{R}_k$ of $R_k$. Therefore $y'$ belongs to $R = \bigcap_{k=0}^{\infty} \bar{R}_k$, which is a subring of $A$.

The relation $R_k \cap \mathfrak{m} \subset \mathfrak{m}^{p^k}$ implies $\bar{R}_k \cap \mathfrak{m} \subset \mathfrak{m}^{p^k}$ since the ideal $\mathfrak{m}^{p^k}$ is closed. Therefore we have

$$R \cap \mathfrak{m} = \left(\bigcap_{k=0}^{\infty} \bar{R}_k\right) \cap \mathfrak{m} = \bigcap_{k=0}^{\infty} (\bar{R}_k \cap \mathfrak{m}) \subset \bigcap_{k=0}^{\infty} \mathfrak{m}^{p^k} = (0),$$

whence the restriction to $R$ of the natural homomorphism $\varphi$ of $A$ onto $A/\mathfrak{m}$ is one to one. On the other hand, since we have seen that every element $y$ of $A$ is congruent mod $\mathfrak{m}$ to an element $y'$ of $R$, $\varphi$ maps $R$ *onto* $A/\mathfrak{m}$. Therefore $R$ is a field of representatives of $A$, and Cohen's Theorem is proved.

COROLLARY. *An equicharacteristic complete regular local ring $A$ is isomorphic to a formal power series ring over a field.*

Let $\mathfrak{m}$ be the maximal ideal of $A$, $K$ a representative field of $A$, $\{x_1, \cdots, x_d\}$ a regular system of parameters. Then the subring $B = K[[x_1, \cdots, x_d]]$ of $A$ is a power series ring in $d$ variables (Corollary 2 to Theorem 21, § 9). Let $\mathfrak{X}$ be its maximal ideal. Since $\{x_1, \cdots, x_d\}$ generate $\mathfrak{m}$ in $A$, we have $\mathfrak{m} = A\mathfrak{X}$. Since $B/\mathfrak{X} = B/(\mathfrak{m} \cap B)$ is identical to $A/\mathfrak{m}$, and since $B$ is complete, Theorem 7 (§ 3) shows that $A = B$ (identify in that theorem the ring $A$ with the present ring $B$, and both modules $E$ and $F$ with the present ring $A$).

We prove now an algebraic result whose geometric counterpart is the fact that, if a subvariety $W$ of a variety $V$ carries a point $P$ which is simple on $V$, then $W$ is simple on $V$.

THEOREM 28. *If $A$ is an equicharacteristic complete regular local ring and if $\mathfrak{p}$ is a prime ideal in $A$, then $A_{\mathfrak{p}}$ is a regular local ring.*†

PROOF. We set $d = \dim(A)$, $d - r = \dim(A/\mathfrak{p})$. The theory of chains of prime ideals in the power series ring $A$ (VII, § 10, Theorem 34, Corollary 1) shows that the dimension of $A_{\mathfrak{p}}$ is $r$. We thus have to prove that $\mathfrak{p}A_{\mathfrak{p}}$ may be generated by $r$ elements. We first prove that

† This theorem has been proved by I. S. Cohen also in the unequal characteristic case, under the assumption that $p \notin \mathfrak{m}^2$, where $p$ is the characteristic of the field $A/\mathfrak{m}$ (the so-called "unramified case"). The theorem has also been proved for non-complete regular local rings $A$ in special cases. Thus, Zariski has proved the theorem in the case in which $A$ is a "geometric" local ring, and Nagata has proved the theorem in the more general case in which the prime ideal $\mathfrak{p}$ is "analytically unramified." Recently Serre has proved the theorem quite generally for arbitrary regular local rings, by using cohomological methods.

there exists a regular system of parameters $\{x_1, \cdots, x_d\}$ in $A$ such that the ideal $(\mathfrak{p}, x_{r+1}, \cdots, x_d)$ is primary for the maximal ideal $\mathfrak{m}$. This is a particular case of the following sharper result:

LEMMA 1. *Let $A$ be a regular local ring of dimension $d$, $\mathfrak{m}$ its maximal ideal, $\mathfrak{b}$ an ideal in $A$, $q$ the dimension of $A/\mathfrak{b}$, and $\{\mathfrak{v}_j\}$ $(1 \leq j \leq h)$ a finite family of non-maximal prime ideals in $A$. Then there exists a regular system of parameters $\{u_1, \cdots, u_d\}$ of $A$ such that the ideal $(\mathfrak{b}, u_1, \cdots, u_q)$ is primary for $\mathfrak{m}$, and that $u_1 \notin \mathfrak{v}_j$ for every $j$.*

PROOF OF THE LEMMA. We proceed by induction on $q$, the case $q=0$ being trivial. Let $q \neq 0$. We denote by $\mathfrak{v}_i$, $i=h+1, h+2, \cdots, k$, the isolated prime ideals of $\mathfrak{b}$; these prime ideals are distinct from $\mathfrak{m}$. From the family $\{\mathfrak{v}_i, \mathfrak{v}_j\}$ of prime ideals we extract those which are not contained in any other ideal of the family. Let $\{\mathfrak{p}_s\}$ be this reduced family. For every $s$, we have $\mathfrak{m}^2 \not\subset \mathfrak{p}_s$, whence we can find an element $c_s$ such that $c_s \in \mathfrak{m}^2$, $c_s \in \mathfrak{p}_t$ for every $t \neq s$ and $c_s \notin \mathfrak{p}_s$. Now we fix an element $x$ of $\mathfrak{m}$ such that $x \notin \mathfrak{m}^2$, and we denote by $I$ the set of those indices $s$ for which $x \in \mathfrak{p}_s$. We set $u_1 = x + \sum_{s \in I} c_s$. For every $s$ there is one and only one of the terms of this sum which is outside of $\mathfrak{p}_s$, whence $u_1 \notin \mathfrak{p}_s$. Furthermore, since $x \notin \mathfrak{m}^2$ and $c_s \in \mathfrak{m}^2$, we have $u_1 \notin \mathfrak{m}^2$. We may thus begin the required regular system of parameters with $u_1$ (Corollary 2 to Theorem 25, § 11).

We now use the inductive hypothesis. Since $u_1$ is not in any isolated prime ideal of $\mathfrak{b}$, the local ring $A/(\mathfrak{b}+Au_1)$ has dimension $q-1$ (§ 9, Theorem 20 (b)). This local ring is a residue class ring $A'/\mathfrak{b}'$ of the ring $A'=A/Au_1$, where $\mathfrak{b}'=(\mathfrak{b}+Au_1)/Au_1$, and this latter ring $A'$ is a regular local ring (§ 11, Theorem 26). Applying the induction hypothesis to $A'$ and $\mathfrak{b}'$, we find a regular system of parameters $\{u'_2, \cdots, u'_d\}$ of $A'$ such that the ideal $(\mathfrak{b}', u'_2, \cdots, u'_q)$ is primary for the maximal ideal of $A'$. If we take for $u_i$ $(i=2, \cdots, d)$ a representative of $u'_i$ in $A$, then the system $\{u_1, \cdots, u_d\}$ satisfied the conditions of the lemma.

CONTINUATION OF THE PROOF OF THEOREM 28. Let $K$ be a representative field of $A$. We apply the above lemma to the case $\mathfrak{b}=\mathfrak{p}$, $q=d-r$. We change the notations of the lemma as follows: the elements $u_1, \cdots, u_{d-r}, u_{d-r+1}, \cdots, u_d$ will now be denoted by $u_{r+1}, \cdots, u_d, u_1, \cdots, u_r$. We set $B=K[[u_{r+1}, \cdots, u_d]]$. In $B/(\mathfrak{p} \cap B)$, the residue classes $\bar{u}_{r+1}, \cdots, \bar{u}_d$ generate the ideal of non units, and, in $A/\mathfrak{p}$, these elements generate an ideal which is primary for the ideal of non units. Since $B/(\mathfrak{p} \cap B)$ is contained in $A/\mathfrak{p}$ and is complete, $A/\mathfrak{p}$ is a finite module over $B/(\mathfrak{p} \cap B)$ (Remark, p. 293, § 9). Thus, by Corollary 3 to Theorem

20 (§ 9), the rings $A/\mathfrak{p}$ and $B/(\mathfrak{p} \cap B)$ have the same dimension. As the dimension of the former is $d-r$, $B/(\mathfrak{p} \cap B)$ has also dimension $d-r$. Now, $B$ itself has dimension $d-r$, since it is a power series ring in $d-r$ variables over a field. Since $\mathfrak{p} \cap B$ is prime, this implies, by Theorem 20 (§ 9), that $\mathfrak{p} \cap B=(0)$.

We have already seen that $A/\mathfrak{p}$ is a finite module over $B/(\mathfrak{p} \cap B)=B$. More precisely, $A/\mathfrak{p}$ is generated by any system of elements $\{y'_j\}$ whose residue classes modulo $\sum_{i=r+1}^{d} (A/\mathfrak{p})\bar{u}_i$ generate $(A/\mathfrak{p})/\sum_{i=r+1}^{d} (A/\mathfrak{p})/\bar{u}_i$ over $K$ (Theorem 7, § 3). An equivalent form of this condition on the elements $y'_j$ is the following: the $y'_j$ are the $\mathfrak{p}$-residues of elements $y_j$ of $A$ whose residue classes modulo $(\mathfrak{p}, u_{r+1}, \cdots, u_d)$ generate $A/(\mathfrak{p}, u_{r+1} \cdots, u_d)$ over $K$. We may thus take for elements $y_j$ the element 1 and a finite number of suitable monomials in $u_1, \cdots, u_r$. Therefore we have $A/\mathfrak{p}=B[u'_1, \cdots, u'_r]$ ($u'_i$: $\mathfrak{p}$-residue of $u_i$), whence $A=\mathfrak{p}+B[u_1, \cdots, u_r]$.

The elements $u_1, \cdots, u_r$ are not necessarily in $\mathfrak{p}$. However, we shall now construct $r$ suitable elements $a_1, \cdots, a_r$ *of* $\mathfrak{p}$ which will belong to the polynomial ring $B[u_1, \cdots, u_r]$. The $\mathfrak{p}$-residue $u'_i$ of $u_i$ has been seen to be integral over $B$. Let $p_i(X)=X^{n(i)}+b_{n(i)-1,i}X^{n(i)-1}+\cdots+b_{1,i}X+b_{0,i}$ ($b_{ji} \in B$) yield an equation of integral dependence for $u'_i$ over $B$. We set $a_i=p_i(u_i)$. Then relation $p_i(u'_i)=0$ shows that $a_i \in \mathfrak{p}$. Furthermore, since the $p_i(X)$ are monic polynomials, the elements $u_i$ are integral over $B[a_1, \cdots, a_r]$ and therefore also over $K[[a_1, \cdots, a_r, u_{r+1}, \cdots, u_r]]$. Therefore $A$ is a finite module over $K[[a_1, \cdots, a_r, u_{r+1}, \cdots, u_d]]$, generated by a finite number of monomials $m_\alpha$ in $u_1, \cdots, u_r$: $A=\sum_\alpha K[[a_1, \cdots, a_r, u_{r+1}, \cdots, u_d]]m_\alpha$. In a representation $z=\sum_\alpha \varphi_\alpha(a_1, \cdots, a_r, u_{r+1}, \cdots, u_d)m_\alpha$ of an element $z$ of $A$, we may single out the terms of $\varphi_\alpha$ which are independent of $a_1, \cdots, a_r$; these terms are in $B=K[[u_{r+1}, \cdots, u_d]]$; and the other terms are in the ideal $\mathfrak{a}=\sum_\alpha^{r} Aa_i$. We therefore have

$$A = \mathfrak{a}+B[u_1, \cdots, u_r].$$

We are now in good position for studying the quotient ring $A_\mathfrak{p}$. Since $\mathfrak{p} \cap B=(0)$, $A_\mathfrak{p}$ contains the quotient field $L$ of $B$, whence it also contains the polynomial ring $S=L[u_1, \cdots, u_r]$. The above constructed elements $a_i$ are in $S \cap \mathfrak{p}$, and $S$ is integral over $L[a_1, \cdots, a_r]$, since $u_i$ is integral over $B[a_i]$. Therefore the ideal $\sum_{i=1}^{r} Sa_i$ is of dimension 0 in $S$. Since $S(\mathfrak{p} \cap S)$ contains this ideal, it follows *a fortiori* that also the ideal $S(\mathfrak{p} \cap S)$ has dimension zero. Then, since $S(\mathfrak{p} \cap S)$ is thus a maximal ideal in the polynomial ring $L[u_1, \cdots, u_r]$ in $r$ variables, it may be generated by exactly $r$ elements (VII, § 7, Theorem 24). Our proof will thus be complete if we show that $\mathfrak{p}A_\mathfrak{p}$ is generated by $S(\mathfrak{p} \cap S)$.

Now this is immediate. The relations $\mathfrak{a} \subset \mathfrak{p}$ and $A = \mathfrak{a} + B[u_1, \cdots, u_r]$ show that $\mathfrak{p} = \mathfrak{a} + (\mathfrak{p} \cap B[u_1, \cdots, u_r])$, whence $\mathfrak{p}A_\mathfrak{p} = \mathfrak{a}A_\mathfrak{p} + (\mathfrak{p} \cap B[u_1, \cdots, u_r])A_\mathfrak{p}$. Since $\mathfrak{a}$ is generated by the elements $a_i$ which lie in $\mathfrak{p} \cap S$, and since $\mathfrak{p} \cap B[u_1, \cdots, u_r]$ is obviously also in $\mathfrak{p} \cap S$, we have $\mathfrak{p}A_\mathfrak{p} = (\mathfrak{p} \cap S)A_\mathfrak{p}$, and this proves Theorem 28.

REMARK. For every prime ideal $\mathfrak{p}$ in the power series $A = K[[x_1, \cdots, x_d]]$, $A_\mathfrak{p}$ is a regular local ring. The corresponding statement for a *polynomial ring* $A = K[x_1, \cdots, x_d]$ and a prime ideal $\mathfrak{p}$ of $A$ is easier to prove. If $q$ is the dimension of $\mathfrak{p}$, we extract from $\{x_1, \cdots, x_d\}$ a maximal system of elements which are algebraically independent mod $\mathfrak{p}$, say $\{x_1, \cdots, x_q\}$. Then $A_\mathfrak{p}$ contains the field $L = K(x_1, \cdots, x_q)$ and therefore also the polynomial ring $S = L[x_{q+1}, \cdots, x_d]$. The ideal $\mathfrak{p}A_\mathfrak{p}$ is generated in $A_\mathfrak{p}$ by the ideal $S \cap \mathfrak{p}A_\mathfrak{p}$, since this latter ideal contains $\mathfrak{p}$. Now $S \cap \mathfrak{p}A_\mathfrak{p}$ is a prime ideal of dimension 0 in $S$, since the $\mathfrak{p}$-residue of $x_j$ is algebraic over $L$ for $j = q+1, \cdots, d$. Therefore this ideal, and hence also the ideal $\mathfrak{p}A_\mathfrak{p}$, is generated by $d-q$ elements (VII, § 7, Theorem 24). Since $d-q$ is the dimension of $A_\mathfrak{p}$, our assertion is proved.

It may be noticed that the proof of Theorem 28 is essentially based upon the same idea which runs through the above short proof for polynomial rings.

The following theorem, due to I. S. Cohen, is a generalization of the theorem of Macaulay for polynomial ring (VII, § 8, Theorem 26):

THEOREM 29. *Let $A$ be an equicharacteristic regular local ring of dimension $d$, and $\mathfrak{a} = (a_1, \cdots, a_r)$ an ideal in $A$ such that $\dim (A/\mathfrak{a}) = d - r$. Then $\mathfrak{a}$ is an unmixed ideal* (i.e., *all the associated prime ideals $\mathfrak{p}_i$ of $\mathfrak{a}$ are of dimension $d-r$; in particular, $\mathfrak{a}$ has no imbedded components*).

PROOF. We proceed by induction on $r$, the case $r = 0$ being trivial. We first achieve a reduction to the case of a *complete* ring. Let $\hat{A}$ be the completion of $A$. Then $\hat{A}\mathfrak{a}$ has dimension $d-r$ (i.e., the local ring $\hat{A}/\hat{A}\mathfrak{a}$ has dimension $d-r$) and is generated by $r$ elements. Suppose our theorem is proved for the ring $\hat{A}$ (which is a complete equicharacteristic regular local ring). Then, every associated prime ideal $\mathfrak{p}^\star_i$ of $\hat{A}\mathfrak{a}$ has dimension $d-r$. Now Theorem 12 (§ 4) shows that, for every associated prime ideal $\mathfrak{p}$ of $\mathfrak{a}$, there exists a $\mathfrak{p}^\star_i$ such that $\mathfrak{p}^\star_i \cap A = \mathfrak{p}$. We thus have $\dim (A/\mathfrak{p}) = \dim (\hat{A}/\hat{A}\mathfrak{p}) \geq \dim(A/\mathfrak{p}^\star_i)$ since $\hat{A}\mathfrak{p} \subset \mathfrak{p}^\star_i$. Hence $\dim (A/\mathfrak{p}) \geqq d-r$, and therefore $\dim (A/\mathfrak{p}) = d-r$

since $\mathfrak{p} \supset \mathfrak{a}$. We may thus assume that $A$ is a complete ring. We suppose that one of the associated prime ideals $\mathfrak{p}$ of $\mathfrak{a}$ has dimension $< d-r$, and from this we shall derive a contradiction. In the local ring $A_\mathfrak{p}$, the ideal $\mathfrak{p}A_\mathfrak{p}$ admits the maximal ideal $\mathfrak{p}A_\mathfrak{p}$ as an associated prime ideal. By Theorem 28 $A_\mathfrak{p}$ is a regular local ring. The dimension $n$ of $A_\mathfrak{p}$ is $> r$, since we have assumed that $\mathfrak{p}$ has dimension $< d-r$, and $\mathfrak{a}A_\mathfrak{p}$ is an ideal generated by $r$ elements. Furthermore, since every isolated prime ideal of $\mathfrak{a}$ has dimension $d-r$ (Vol. I, Ch. IV, § 14, Theorem 30), every isolated prime ideal of $\mathfrak{a}A_\mathfrak{p}$ has dimension $n-r$. Thus, with a change in notations, we have to prove that the following situation is impossible: we have a regular local ring $A$ of dimension $n$, an ideal $\mathfrak{a}=(a_1, \cdots, a_r)$ in $A$, of dimension $n-r$, generated by $r$ elements, $r<n$, and the maximal ideal $\mathfrak{m}$ of $A$ is an associated prime ideal of $\mathfrak{a}$.

Since a principal ideal in a local ring of dimension $q$ has either dimension $q-1$ or dimension $q$, the dimensions of the ideals $(a_1, \cdots, a_r)$, $(a_1, \cdots, a_{r-1}), \cdots, (a_1)$, $(0)$ form a sequence of integers such that the difference of two consecutive terms is 0 or 1. Since the dimension of $\mathfrak{a}$ is $n-r$, and since the dimension of $(0)$ is $n$, this implies that all these differences are equal to 1. In particular, the ideal $\mathfrak{b}=(a_1, \cdots, a_{r-1})$ has dimension $n-r+1$, whence $\mathfrak{b}$ is unmixed by our induction hypothesis. By Lemma 1, there exists therefore a regular system of parameters $\{u_1, \cdots, u_n\}$ of $A$ such that $u_1$ does not belong to any associated prime ideals of $\mathfrak{b}$, nor to any isolated prime ideal of $\mathfrak{a}$.

We now express the fact that $\mathfrak{m}$ is an associated prime ideal of $\mathfrak{a}$: we have $\mathfrak{a}:\mathfrak{m} > \mathfrak{a}$ (Vol. I, Ch. IV, § 6, Theorem 11), or equivalently, there exists an element $c \notin \mathfrak{a}$ such that $c\mathfrak{m} \subset \mathfrak{a}$. Then $cu_1 \in \mathfrak{a} = \mathfrak{b} + Aa_r$, whence we can write $cu_1 - da_r \in \mathfrak{b}$, where $d$ is a suitable element of $A$, or again $da_r \in \mathfrak{b} + Au_1$. Suppose that we have shown that $a_r$ does not belong to any associated prime ideal of $\mathfrak{b} + Au_1$. Then the relation $da_r \in \mathfrak{b} + Au_1$ implies $d \in \mathfrak{b} + Au_1$, and hence $d - eu_1 \in \mathfrak{b}$, with suitable $e$ in $A$. Using the relation $cu_1 - da_r \in \mathfrak{b}$, we find that $(c - ea_r)u_1 \in \mathfrak{b}$, whence $c - ea_r \in \mathfrak{b}$ since $u_1$ has been chosen outside of all the associated prime ideals of $\mathfrak{b}$. We therefore have $c \in \mathfrak{b} + Aa_r = \mathfrak{a}$, in contradiction with the hypothesis $c \notin \mathfrak{a}$.

Thus it remains to be proved that $a_r$ does not belong to any associated prime ideal $\mathfrak{p}$ of $\mathfrak{b} + Au_1$. By the induction hypothesis, applied to the ideal $(\mathfrak{b} + Au_1)/Au_1$ in the regular local ring $A/Au_1$, such an ideal $\mathfrak{p}$ has dimension $n-r$. If $\mathfrak{p}$ were to contain $a_r$, it would contain $\mathfrak{a} = \mathfrak{b} + Aa_r$, and therefore $\mathfrak{p}$ would be an isolated prime ideal of $\mathfrak{a}$, in contradiction with the fact that $u_1$ has been chosen outside of all the isolated prime ideals of $\mathfrak{o}$. This completes the proof of the theorem.

THEOREM 30. *An equicharacteristic regular local ring $A$ is a unique factorization domain.*

PROOF. The completion $\hat{A}$ of $A$ is a power series ring over a field (Corollary to Theorem 27), whence $\hat{A}$ is a unique factorization domain (Chapter VII, § 1, Theorem 6). It is therefore sufficient to prove the following lemma:

LEMMA 2. *If the completion $\hat{A}$ of a local domain $A$ is a unique factorization domain, then $A$ itself is a unique factorization domain.*†

PROOF. We have to prove that *every minimal prime ideal* $\mathfrak{p}(\neq(0))$ *of $A$ is principal.*‡ For this it is sufficient to prove that $\hat{A}\mathfrak{p}$ is principal, for, if we have $\hat{A}\mathfrak{p}=\hat{A}a'$ $(a'\in\hat{A})$ and if $\{b_1,\cdots,b_n\}$ denotes a basis of $\mathfrak{p}$, we have $b_i=a'b'_i$ $(b'_i\in\hat{A})$ and $a'=\sum_i c'_ib_i$ $(c'_i\in\hat{A})$; thus $a'=\left(\sum_i c'_ib'_i\right)a'$, whence $1=\sum_i c'_ib'_i$ since $\hat{A}$ is a domain; then, since $\hat{A}$ is a local ring, at least one of the terms $c'_ib'_i$, say $c'_1b'_1$, is invertible, whence $b'_1$ is invertible; since $b_1=a'b'_1$, we have $\hat{A}\mathfrak{p}=\hat{A}a'=\hat{A}b_1$. Hence we have $\mathfrak{p}=\hat{A}\mathfrak{p}\cap A=\hat{A}b_1\cap A=Ab_1$ (§ 2, Theorem 5, Corollary 2).

Since $\hat{A}$ is a UFD, it is sufficient to prove that all the associated prime ideals of $\hat{A}\mathfrak{p}$ are minimal (i.e., have height 1). For such an associated prime ideal $\mathfrak{v}^\star$, we have $\mathfrak{v}^\star\cap A=\mathfrak{p}$ (§ 4, Corollary 1 to Theorem 12). Thus, if we denote by $S$ the complement of $\mathfrak{p}$ in $A$, $\mathfrak{v}^\star\hat{A}_S$ is an associated prime ideal of $\mathfrak{p}\hat{A}_S$ (Vol. I, Ch. IV, § 10, Theorem 17), and we are reduced to proving that all the associated prime ideals of $\mathfrak{p}\hat{A}_S$ have height 1.

Now we notice that $A$ is integrally closed. In fact, if we denote by $K$ the quotient field of $A$ (considered as a subfield of the quotient field of $\hat{A}$) we have $A=\hat{A}\cap K$: for, if the element $x/y$ of $K$ $(x, y\in A)$ belongs to $\hat{A}$, we have $x\in\hat{A}y\cap A=Ay$ (§ 2, Corollary 2 to Theorem 5), whence $x/y$ belongs to $A$. Since $\hat{A}$ is a UFD, it is integrally closed in its quotient field. It follows that also $A$ is integrally closed in its quotient field $K$.

Therefore, since $\mathfrak{p}$ is a prime ideal of height 1, the quotient ring $A_S=A_\mathfrak{p}$ is an integrally closed local ring of dimension 1, i.e., a discrete

† This lemma and its proof have been communicated to us by M. Nagata.

‡ That this condition is satisfied if $A$ is a UFD has been pointed out in Vol. I, Ch. IV, §14, p. 238. Conversely, assume that this condition is satisfied and let $b$ be an irreducible element of $A$. The ideal $Ab$ is contained in some *minimal* prime ideal $\mathfrak{p}$ (Vol. I, Ch. IV, § 14, Theorem 29). We have $\mathfrak{p}=Ac$, whence $b$ is a multiple of $c$, and since $b$ is irreducible we must have $Ab=Ac$, i.e., $Ab$ is prime. Therefore $A$ is a UFD by Vol. I, Ch. I, §14, Theorem 4. (Note that UF.1 is satisfied since $A$ is noetherian.)

valuation ring (Vol. I, Ch. V, § 6, corollary to Theorem 14). Hence $\mathfrak{p}A_S = \mathfrak{p}A_{\mathfrak{p}}$ is a principal ideal, and, since $\mathfrak{p}\hat{A}_S = \mathfrak{p}A_S\hat{A}_S$, $\mathfrak{p}\hat{A}_S$ is also a principal ideal. Thus, since $\hat{A}_S$ is integrally closed, all the associated prime ideals of $\mathfrak{p}\hat{A}_S$ have height 1 (Vol. I, Ch. V, § 6, Theorem 14). This proves lemma 2 and Theorem 30.

REMARK. Lemma 2 reduces the problem of unique factorization in *arbitrary* (i.e., not necessarily equicharacteristic) regular local rings to the case of complete regular local rings. The unique factorization property holds also in an arbitrary complete regular local ring $A$ of *dimension* 1 or 2: it is obvious in dimension 1 since $A$ is then a discrete valuation ring; in dimension 2 one uses an analogue of Hensel's lemma for homogeneous polynomials in two variables.† It may also be proved that, if the unique factorization property holds for regular local rings of dimension *three*, then it holds in any regular local ring. (This has been proved by O. Zariski in unpublished notes, in 1947; subsequently this has been proved by M. Nagata ["A general theory of algebraic geometry over Dedekind domains, II" (§ 5, Proposition 11), *Amer. J. of Mathematics*, v. 80, 1958].) Using these facts and methods of cohomological algebra, M. Auslander and D. A. Buchsbaum have recently proved the unique factorization theorem in any regular local ring in their paper, "Unique factorization in regular local rings", PNAS, v. 45 (1959), pp. 733–734. We present their proof in Appendix 7, reducing the cohomological prerequisites to the knowledge of the properties of chains of syzygies given in VII, § 13.

**§ 13. Analytical irreducibility and analytical normality of normal varieties.** In this section we intend to study the completions of the local rings which occur in algebraic geometry. Such local rings are the local rings $\mathfrak{o}(W; V)$, where $W$ is an irreducible subvariety of an irreducible variety $V$ (Ch. VI, § 14, p. 93). In other words, the local rings which will be considered in this section are quotient rings $k[x_1, \cdots, x_n]_{\mathfrak{p}}$ of finite integral domains with respect to prime ideals $\mathfrak{p}$.

The results we are going to prove hold in a larger class of local rings,‡ but not for the class of all local rings.|| Actually they are consequences of the following hypothesis, which is of algebraic nature:

† See W. Krull, "Zur Theorie der kommutativen Integritätsbereiche," *J. für d. reine u. angew. Math.*, v. 192 (1953), or unpublished notes of O. Zariski.

‡ See P. Samuel, *Algèbre Locale*, Ch. V, Paris (Gauthier Villars), 1953.

|| See M. Nagata: "An example of a normal local ring which is analytically reducible" (*Mem. Coll. Sci. Univ. Kyoto*, Ser. A, 31 (1958), 83–85) and "An example of a normal local ring which is analytically ramified" (*Nagoya Math. J.*, 9 (1955), 111–113).

(D) *The local ring $A$ is a domain, and there exists an element $d \neq 0$ in $A$ such that, if $\hat{A}$ denotes the completion of $A$ and $(\hat{A})'$ the integral closure of $\hat{A}$ in its total quotient ring, then $d(\hat{A})' \subset \hat{A}$.*

In the first part of this section we shall derive some consequences of hypothesis (D). In the second part we shall show that the local rings of algebraic geometry satisfy hypothesis (D), thus proving that the consequences of (D) hold true for these local rings.

We shall say that a local ring $A$ is *analytically unramified* if its completion $\hat{A}$ has no nilpotent elements (other than 0).

LEMMA 1. *If a local domain $A$ satisfies condition* (D), *then it is analytically unramified.*

PROOF. Let $a$ be a nilpotent element of $\hat{A}$. For every element $x \neq 0$ of $A$, we have that $x$ is not a zero-divisor in $\hat{A}$ (Corollary 6 to Theorem 11, § 4), whence $a/x$ is an element of the total quotient ring of $\hat{A}$. Since we have $(a/x)^q = 0$ for some exponent $q$, $a/x$ is integral over $\hat{A}$. Using condition (D), we see that $da/x \in \hat{A}$, $d$ being an element $\neq 0$ of $A$ independent of $x$. Therefore the element $da$ belongs to all the principal ideals $\hat{A}x$ ($x \in A$, $x \neq 0$). If $A$ is not a field, we fix an element $y \neq 0$ in the maximal ideal $\mathfrak{m}$ of $A$, and we apply the above result to the principal ideals $\hat{A}y^n$, $n = 1, 2, \cdots$. We then have $da \in \hat{A}\mathfrak{m}^n$ for every $n$, whence $da = 0$, and therefore $a = 0$ since $d$ is not a zero-divisor in $\hat{A}$ (Corollary 6 to Theorem 11, § 4). This proves that $\hat{A}$ has no nilpotent elements ($\neq 0$) if $A$ is not a field. If $A$ is a field, then $\hat{A} = A$ and our assertion is trivial.

LEMMA 2. *Let $A$ be an integrally closed local domain such that $A$ and all its residue class rings $A/\mathfrak{p}$ ($\mathfrak{p}$: prime ideal) satisfy condition* (D). *Then $\hat{A}$ is an integrally closed domain.*

[The statement that $\hat{A}$ is a domain is often expressed by saying that $A$ is *analytically irreducible*, and the statement that $\hat{A}$ is integrally closed is expressed by saying that $A$ is *analytically normal.*]

PROOF. (1) We first prove *that $\hat{A}$ is integrally closed in its total quotient ring*, i.e., that $(\hat{A})' = \hat{A}$. With the same notations as in condition (D), let $d$ be an element $\neq 0$ of $A$ such that $d(\hat{A})' \subset \hat{A}$. We may assume that $d$ is not a unit in $A$, for, otherwise, our assertion is evident. Let $z$ be any element of $(\hat{A})'$; the element $dz$ belongs to $\hat{A}$. If we prove that $dz$ belongs to the ideal $\hat{A}d$, we will have $dz = dz'$ with $z' \in \hat{A}$, whence $z \in \hat{A}$ since $d$ is not a zero divisor (in $\hat{A}$, and therefore also in $(\hat{A})'$).

We thus have to prove that, for every $z$ in $(\hat{A})'$, we have $dz \in \hat{A}d$. This will be achieved if we prove that, *for every associated prime ideal $\bar{\mathfrak{p}}_j$ of $\hat{A}d$, the quotient ring $\hat{A}_{\bar{\mathfrak{p}}_j}$ is a discrete valuation ring.* In fact, assume this has been proved. Then it will follow that $\bar{\mathfrak{p}}_j$ is a prime ideal of height 1 of $\hat{A}$, and that the ideals which are primary for $\bar{\mathfrak{p}}_j$ are its symbolic powers. Hence we will have

$$\hat{A}d = \bigcap_j \bar{\mathfrak{p}}_j^{(s(j))}. \tag{1}$$

We denote by $v_j$ the normalized valuation of $\hat{A}_{\bar{\mathfrak{p}}_j}$ and by $w_j$ the function on $\hat{A}$ defined by $w_j(x) = v_j(\varphi_j(x))$, $\varphi_j$ denoting the canonical homomorphism of $\hat{A}$ into $\hat{A}_{\bar{\mathfrak{p}}_j}$. The function $w_j$ takes the value $+\infty$ on the kernel of $\varphi_j$, and satisfies the same relations as a valuation does:

$$w_j(xy) = w_j(x)+w_j(y), \quad w_j(x+y) \geq \min(w_j(x)\ w_j(y)). \tag{2}$$

Furthermore the symbolic power $\bar{\mathfrak{p}}_j^{(s)}$ is the set of all elements $x$ of $\hat{A}$ such that $w_j(x) \geq s$; it follows that $s(j) = w_j(d)$. This being so, we come back to the element $z$ of $(\hat{A})'$, and write an equation of integral dependence for $z$ over $\hat{A}$:

$$z^n + a_{n-1}z^{n-1} + \cdots + a_1 z + a_0 = 0 \quad (a_i \in \hat{A}). \tag{3}$$

The element $y = dz$ belongs to $\hat{A}$ and we have

$$-y^n = a_{n-1}dy^{n-1} + \cdots + a_1 d^{n-1} y + d^n a_0. \tag{4}$$

If we set $w_j(y) = \alpha$ and $w_j(d) = \beta$, (2) shows that

$$n\alpha \geq \min_{0 \leq i \leq n-1}(i\alpha + (n-i)\beta) = n\beta + \min_{0 \leq i \leq n-1}(i(\alpha-\beta)).$$

If $\alpha < \beta$, then $\min_{0 \leq i \leq n-1}(i(\alpha-\beta)) = (n-1)(\alpha-\beta)$, and we get the inequality $n\alpha \geqq n\beta + (n-1)(\alpha-\beta)$, i.e., $\alpha \geq \beta$, in contradiction with $\alpha < \beta$. We therefore have $\alpha \geq \beta$, i.e., $w_j(y) \geq w_j(d)$ for every $j$. Hence $y$ belongs to $\bar{\mathfrak{p}}_j^{(w_j(d))} = \bar{\mathfrak{p}}_j^{(s(j))}$ for every $j$, and therefore to $\hat{A}d$, by formula (1).

It remains to be proved that, for every associated prime ideal $\bar{\mathfrak{p}}$ of $\hat{A}d$, *$\hat{A}_{\bar{\mathfrak{p}}}$ is a discrete valuation ring.* Since $A$ is an integrally closed domain, we have $Ad = \bigcap_i \mathfrak{p}_i^{(n(i))}$, where the $\mathfrak{p}_i$ are prime ideals of $A$, of height 1 (Vol. I, Ch. V, § 6, Theorem 14). Therefore we have (§ 4, Theorem 11, Corollary 2)

$$\hat{A}d = \bigcap_i \hat{A}\mathfrak{p}_i^{(n(i))}. \tag{5}$$

We consider any one of the ideals $\mathfrak{p}_i$, and we call it $\mathfrak{p}$. Condition (D) and Lemma 1 applied to $A/\mathfrak{p}$ show that $\hat{A}/\hat{A}\mathfrak{p}$ has no nilpotent elements, i.e., that $\hat{A}\mathfrak{p}$ is a finite irredundant intersection of prime ideals $\bar{\mathfrak{p}}_j$. We consider one of them, say $\bar{\mathfrak{p}}_1 = \bar{\mathfrak{p}}$, and study $\hat{A}_{\bar{\mathfrak{p}}}$. Taking $x' \in \bigcap_{j \geq 2} \bar{\mathfrak{p}}_j$, $x' \notin \bar{\mathfrak{p}}$, we have $x'\bar{\mathfrak{p}} \subset \hat{A}\mathfrak{p}$. Let $a$ be an element of $\mathfrak{p}$ which is not in $\mathfrak{p}^{(2)}$;

since $A_{\mathfrak{p}}$ is a discrete valuation ring, $\mathfrak{p}$ is an isolated primary component of $Aa$, and there exists an element $x''$ of $A$ such that $x'' \notin \mathfrak{p}$ and $x''\mathfrak{p} \subset Aa$. Since $x''$ is not a zero divisor mod $\hat{A}\mathfrak{p}$ (Corollary 6 to Theorem 11, § 4), we have $x'' \notin \bar{\mathfrak{p}}$,† whence the element $x = x'x''$ does not belong to $\bar{\mathfrak{p}}$. On the other hand we have $x\bar{\mathfrak{p}} = x''x'\bar{\mathfrak{p}} \subset x''\hat{A}\mathfrak{p} \subset \hat{A}a$. Denoting by $\varphi$ the canonical homomorphism of $\hat{A}$ into $\hat{A}_{\bar{\mathfrak{p}}}$, we deduce that $\varphi(x)\bar{\mathfrak{p}}\hat{A}_{\bar{\mathfrak{p}}} \subset \varphi(a)\hat{A}_{\bar{\mathfrak{p}}}$. Since $x \notin \bar{\mathfrak{p}}$, $\varphi(x)$ is a unit in $\hat{A}_{\bar{\mathfrak{p}}}$, and, since $a \in \mathfrak{p} \subset \bar{\mathfrak{p}}$, the last relation shows that the maximal ideal $\bar{\mathfrak{p}}\hat{A}_{\bar{\mathfrak{p}}}$ is the *principal* ideal generated by $\varphi(a)$. Therefore $\hat{A}_{\bar{\mathfrak{p}}}$, and similarly every $\hat{A}_{\bar{\mathfrak{p}}_j}$, is a *discrete valuation ring.*

*Now we prove that we not only have* $\hat{A}\mathfrak{p} = \bigcap \bar{\mathfrak{p}}_j$, *but also*

$$\hat{A}\mathfrak{p}^{(n)} = \bigcap_j \bar{\mathfrak{p}}_j^{(n)}. \tag{6}$$

This we prove by induction on $n$. The proof of the inclusion $\mathfrak{p}^{(n)} \subset \bar{\mathfrak{p}}_j^{(n)}$ is straightforward (we recall that $\mathfrak{p} = A \cap \bar{\mathfrak{p}}_j$). Conversely, consider any element $y$ of $\bigcap_j \bar{\mathfrak{p}}_j^{(n)}$. We have $y \in \hat{A}\mathfrak{p}$ (since $n \geq 1$), whence (using the same elements $a$, $x''$ as above) $x''y \in \hat{A}a$; we write $x''y = ay_1$ ($y_1 \in \hat{A}$). Let $\varphi_j$ be the canonical homomorphism of $\hat{A}$ into $\hat{A}_{\bar{\mathfrak{p}}_j}$, $v_j$ the normalized valuation of $\hat{A}_{\bar{\mathfrak{p}}_j}$, and $w_j$ the mapping $\varphi_j v_j$. We have $w_j(x'') = 0$ (since $x'' \notin \bar{\mathfrak{p}}_j$), $w_j(y) \geq n$, $w_j(a) = 1$ (since $\varphi_j(a)$ generates the maximal ideal of $\hat{A}_{\bar{\mathfrak{p}}_j}$), whence $w_j(y_1) \geq n-1$. Hence $y_1 \in \bigcap_j \bar{\mathfrak{p}}_j^{(n-1)}$, and therefore $y_1 \in \hat{A}\mathfrak{p}^{(n-1)}$ by the induction hypothesis. Since $a \in \mathfrak{p}$, we have $x''y = ay_1 \in \hat{A}\mathfrak{p}^{(n)}$, whence $y \in \hat{A}\mathfrak{p}^{(n)}$ since $x'' \notin \mathfrak{p}$. This proves (6).

Combining (5) and (6) (applied to each $\mathfrak{p}_i$) we see that $\hat{A}d$ is a finite intersection of symbolic powers $\bar{\mathfrak{p}}_k^{(n(k))}$, where the $\bar{\mathfrak{p}}_k$ are prime ideals of $\hat{A}$ such that $\hat{A}_{\bar{\mathfrak{p}}_k}$ is a discrete valuation ring. This proves the announced assertion.

(2) We now know that $\hat{A}$ is integrally closed in its total quotient ring $S$. By Lemma 1, $\hat{A}$ has no nilpotent elements, whence its zero ideal is an irredundant intersection of prime ideals $\bar{\mathfrak{q}}_i$. The elements of $\hat{A}$ which do not belong to $\bigcup_i \bar{\mathfrak{q}}_i$ are regular in $\hat{A}$, and are therefore units in $S$. Hence the zero ideal of $S$ is the intersection of the *maximal* ideals $S\bar{\mathfrak{q}}_i$. Therefore $S$ is isomorphic to the direct sum of the fields $S/S\bar{\mathfrak{q}}_i$ (Vol. I, Ch. III, § 13, Theorem 32). If the number of these fields were greater than 1, $S$ would contain an idempotent $e \neq 0, 1$. Since $e^2 - e = 0$, $e$ is integral over $\hat{A}$, whence it belongs to $\hat{A}$, in contradiction with the fact that a local ring cannot contain any idempotent $e$ distinct from 0

† In other words, we have $\mathfrak{p} = \bar{\mathfrak{p}} \cap A$.

and 1 (since $e$ and $1-e$ would then both be non-units). Hence $S$ is a field, and $\hat{A}$ has no zero divisors. This completes the proof of Lemma 2.

REMARK. We have only used the hypothesis (D) for $A$ and for the factor rings $A/\mathfrak{p}$, where $\mathfrak{p}$ is a prime ideal of $Ad$.

In the next lemma we use, for avoiding typographical complications, the notation $c(B)$ for the completion of a semi-local ring $B$, and the notation $R'$ for the integral closure of a ring $R$ in its total quotient ring.

LEMMA 3. *Let $A$ be a local domain satisfying condition* (D). *Then $A'$ is a semi-local ring. Furthermore, if, for every maximal ideal $\mathfrak{m}$ of $A'$, the local ring $A'_{\mathfrak{m}}$ and all its residue class rings $A'_{\mathfrak{m}}/\mathfrak{p}$ ($\mathfrak{p}$ prime) satisfy* (D), *then the ring $c(A')$ is canonically isomorphic to $c(A)'$.*

PROOF. Condition (D) applied to $A$, i.e., the existence of an element $d \neq 0$ in $A$ such that $dc(A)' \subset c(A)$, implies that $c(A)'$ is contained in a finite $c(A)$-module, whence it is itself a finite $c(A)$-module since $c(A)$ is noetherian. Therefore $c(A)'$ is a complete semi-local ring, and $c(A)$ is a topological subspace of $c(A)'$ (Theorem 15, § 6).

On the other hand, since no element $\neq 0$ of $A$ is a zero divisor in $c(A)$ (Corollary 6 to Theorem 11, § 4), the total quotient ring $S$ of $c(A)$ contains the quotient field $K$ of $A$. We have $K \cap c(A) = A$, since, if an element $a/b$ ($a, b \in A$) of $K$ belongs to $c(A)$, we have $a \in b \cdot c(A)$, whence $a \in b \cdot c(A) \cap A = Ab$, and $a/b \in A$. It follows that the relation $d \cdot c(A)' \subset c(A)$ ($d \in A$) implies $dA' \subset c(A) \cap K = A$. As above, this shows that $A'$ is a finite $A$-module, therefore a semi-local ring, and that $A$ is a topological subspace of $A'$. Therefore $c(A)$ may be identified with a subring of $c(A')$. By Theorem 16 (§ 6), an element of $c(A)$ which is not a zero divisor in $c(A)$ is not a zero divisor in $c(A')$. Hence the total quotient ring $T$ of $c(A')$ contains the total quotient ring $S$ of $c(A)$. Furthermore the relation $dA' \subset A$ gives, by passage to the limit, $d \cdot c(A') \subset c(A)$, thus proving that the elements of $c(A')$ admit $d$ as a common denominator. Therefore $c(A')$ is a subring of $S$, showing that $T = S$. The relation $d \cdot c(A') \subset c(A)$ proves also that $c(A')$ is a finite $c(A)$-module, whence that $c(A')$ is integral over $c(A)$. Therefore $c(A')$ is a *subring* of the integral closure $c(A)'$ of $c(A)$ is $S$.

For completing the proof it remains to be shown that $c(A') = c(A)'$, i.e., that $c(A')$ is *integrally closed* in its total quotient ring $S$. We know (§ 7, Corollary 2 to Theorem 18) that $c(A')$ is a direct sum of complete local rings $B_i$. If we prove that all the rings $B_i$ are integrally closed domains, everything will be proved. For, denote by $K_i$ the quotient field of $B_i$. The direct sum $\sum K_i$ is then the total quotient ring of $c(A') = \sum B_i$. Let $x = (x_1, \cdots, x_n)$ be an element of $\sum K_i$ that is integral over $\sum B_i$. Writing component-wise an equation of integral dependence for $x$ over $\sum B_i$, we see that $x_i$ is integral over $B_i$ for every $i$, whence that $x_i \in B_i$ and that $x \in \sum B_i$.

We now prove that every $B_i$ is an integrally closed domain. We know (§ 7, remark to Corollary 2 to Theorem 18) that $B_i$ is isomorphic to the completion of $A'_{\mathfrak{m}_i}$, $\mathfrak{m}_i$ denoting one of the maximal ideals of $A'$. Since $A'$ is an integrally closed domain, so is $A'_{\mathfrak{m}_i}$. For completing the proof it suffices to notice that, by the hypotheses, Lemma 2 may be applied to $A'_{\mathfrak{m}_i}$.

In order to be able to apply Lemmas 1, 2, 3 to the local rings of algebraic geometry (which we call, for short, "*algebro-geometric local rings*"), it suffices to make sure that, given an algebro-geometric local ring $A$, then $A$ itself, all the rings $A/\mathfrak{p}$ ($\mathfrak{p}$: prime) and all the rings $A'_\mathfrak{m}/\mathfrak{p}'$ ($A'$ integral closure of $A$, $\mathfrak{m}$ maximal ideal in $A'$, $\mathfrak{p}'$ prime ideal in $A'_\mathfrak{m}$) satisfy condition (D). It is easily seen that all these local rings are algebro-geometric. In fact

(a) As to $A/\mathfrak{p}$, we write $A = B_\mathfrak{q}$, where $B$ is a finite integral domain $k[x_1, \cdots, x_n]$ and $\mathfrak{q}$ a prime ideal in $B$. Then $B_1 = B/(\mathfrak{p} \cap B)$ is a finite integral domain, $\mathfrak{p} \cap B$ is contained in $\mathfrak{q}$, and $A/\mathfrak{p}$ is isomorphic to $(B_1)_{\mathfrak{q}_1}$, where $\mathfrak{q}_1 = \mathfrak{q}/(\mathfrak{p} \cap B)$ (Vol. I, Ch. IV, Formula (1) at the end of § 10). Thus $A/\mathfrak{p}$ is algebro-geometric.

(b) As to $A'_\mathfrak{m}$, we still write $A = B_\mathfrak{q}$ and observe that the integral closure $B'$ of $B$ is a finite integral domain (Vol. I, Ch. V, § 4, Theorem 9). Denoting by $S$ the complement of $\mathfrak{q}$ *in* $B$, the intregral closure $A'$ of $A = B_\mathfrak{q} = B_S$ is $B'_S$ (Vol. I, Ch. V, § 3, Example 2). Thus, by the transitivity of quotient ring formation (Vol. I, Ch. IV, § 10), the ring $A'_\mathfrak{m} = (B'_S)_\mathfrak{m}$ is equal to $B'_{\mathfrak{m} \cap B'}$, whence it is algebro-geometric.

(c) As to $A'_\mathfrak{m}/\mathfrak{p}'$, we apply (a) and (b).

This being so, it is sufficient to prove the following:

LEMMA 4. *An algebro-geometric local ring* $A = k[x_1, x_2, \cdots, x_n]_\mathfrak{q}$ *such that* $k(x)$ *is separable over* $k$ *satisfies condition* (D).

PROOF. We first prove the following strong variation of the normalization theorem: *if the prime ideal* $\mathfrak{q}$ *is zero-dimensional then it contains a separating transcendence basis* $\{z_1, z_2, \cdots, z_r\}$ *of* $k(x)/k$ *such that* $k[x]$ *is integral over* $k[z]$. Passing to the homogeneous ring $\mathfrak{O} = k[y_0, y_1, \cdots, y_n]$ and to the one dimensional homogeneous prime ideal $\mathfrak{P} = {}^h\mathfrak{p}$ (p. 186), it will be sufficient to prove that $\mathfrak{P}$ contains $r$ homogeneous elements $\zeta_1, \zeta_2, \cdots, \zeta_r$ such that $\{y_0, \zeta\}$ is both a system of integrity of $\mathfrak{O}$ and a separating transcendence basis of $k(y)/k$ for we can then set $z_1 = \zeta_i/y_o{}^{m_i}$, $m_i$ = degree of $\zeta_1$). By Vol. I, Ch. II, §17, Theorems 41 and 43, a separating transcendence basis of $k(y)/k$ is the same thing as a $p$-basis of $k(x)$ (if the characteristic $p$ of $k$ is $\neq 0$). By the lemma on p. 198 it is therefore sufficient that for each $j = 1, 2, \cdots, r+1$ the elements $\zeta_1, \cdots, \zeta_{j-1}$ of $\mathfrak{P}$ have the following properties: the ideal $\mathfrak{O}\zeta_0 + \mathfrak{O}\zeta_1 + \cdots + \mathfrak{O}\zeta_{j-1}$ (where $\zeta_0 = y_0$) is of dimension $r+1-j$, $\mathfrak{P}$ contains none of the isolated prime ideals $\mathfrak{P}_i$ of this ideal, and $\zeta_0, \zeta_1, \cdots, \zeta_{j-1}$ are $p$-independent elements of $k(y)$. Assume $\zeta_0, \zeta_1, \cdots, \zeta_{j-1}$ have already been constructed (note, for $j = 0$, that $y_0 \notin \mathfrak{P}$ and is a transcendental over $k(x)$). Let $u$ and $v$ be homogeneous elements of $\mathfrak{O}$ such that $u \notin \bigcup \mathfrak{P}_i$, $u \in \mathfrak{P}$; $v \in \mathfrak{P}$, $v \in \bigcap \mathfrak{P}_i$ and $v \notin k(y_0, \zeta_1, \cdots, \zeta_{j-1})(y^p)$ (the existence of $v$ follows from the fact that the elements of every non-zero ideal in $\mathfrak{O}$ generate $k(y)$ over $k$). Let $p^a g$, $p^b h$, be the degree of $u$ and $v$ respectively (where $p \nmid a$, $p \nmid b$). If $a < b$ we set $\zeta_j = u^{hp^{b-a}} + v^g$; if $a \geqq b$ we set $\zeta_j = u^{hp} + y_0^{p^{a+1-b}} v^g$. It is then immediate that $\zeta_0, \zeta_1, \cdots, \zeta_j$ also satisfy the above conditions. The existence of a separating transcendence basis $\{z_1, z_2, \cdots, z_r\}$ which is also a system of integrity of $\mathfrak{o} = k[x]$ allows us first to reduce the proof of the lemma to the case in which $\mathfrak{q}$ is zero-dimensional, by adjoining to $k$ a maximal subset of $\{z_1, z_2, \cdots, z_r\}$ consisting of elements which are algebraically independent over $k$ mod $\mathfrak{q}$. Assuming now that $\mathfrak{q}$ is maximal we choose $\{z_1, z_2, \cdots, z_r\}$ as above, we set $\mathfrak{r} = k[z]$, $\mathfrak{z} = (z_1, z_2, \cdots, z_r) = \mathfrak{q} \cap \mathfrak{r}$. It is clear that $\{z_1, z_2, \cdots, z_r\}$ is a system of parameters in $\mathfrak{A} = \mathfrak{o}_\mathfrak{q}$.

The local ring $A$ is not, in general, a finite module over $B = \mathfrak{r}_\mathfrak{z}$. However, if we denote by $S$ the complement of $\mathfrak{z}$ in $\mathfrak{r}$, then the ring $I = \mathfrak{o}_S$ is a finite module over $B = \mathfrak{r}_S$, and therefore is a semi-local ring. Furthermore $A$ is a quotient ring of $I$ with respect to some maximal ideal. Then $\hat{B}$, which is a power series ring in $r$ variables over $k$, is a subring of $\hat{I}$. By what has been seen in § 7 (Remark, p. 283), $\hat{A}$ is a direct summand of $\hat{I}$. If we denote by $\varphi$ the projection of $\hat{I}$ onto $\hat{A}$, $\varphi$ maps $z_i$ (considered as an element of $\hat{B}$ and $\hat{I}$) on $z_i$ (considered as an element of $\hat{A}$); in order to avoid confusions, we denote this latter element by $\varphi(z_i)$. Since the elements $\varphi(z_1), \cdots, \varphi(z_r)$ of $\hat{A}$ are analytically independent over $k$ (Corollary 2 to Theorem 21, § 9), $\varphi$ maps *isomorphically* $\hat{B} = k[[z_1, \cdots, z_r]]$ onto the subring $k[[\varphi(z_1), \cdots, \varphi(z_r)]]$ of $\hat{A}$ (subring over which $\hat{A}$ is a finite module; see § 3). Furthermore, by Theorem 16 (b) (§ 6), and since $\hat{B}$ has no zero divisors, no element $\neq 0$ of $\hat{B}$ is a zero divisor in $\hat{I}$. From this it easily follows, by taking into account the fact that $\hat{A}$ is a direct summand of $\hat{I}$, that no element $\neq 0$ of $\varphi(\hat{B})$ is a zero divisor in $\hat{A}$.

From all this we deduce that the total quotient ring $Z$ of $\hat{A}$ contains the quotient field $L$ of $\varphi(\hat{B})$, and is a finite dimensional vector space over

$L$. The integral closure $(\hat{A})'$ of $\hat{A}$ is the integral closure of $\varphi(\hat{B})$ in $Z$, since $\hat{A}$ is integral over $\varphi(\hat{B})$. Furthermore $Z$ is a direct summand of the total quotient ring $T$ of $\hat{I}$. Theorem 16 (§ 6) shows that $T$ is a finite dimensional vector space over $L$, and that, if $\{a_1, \cdots, a_q\}$ is a basis of the quotient field $k(x)$ of $I$ over the quotient field $k(z)$ of $B$, then it is also a basis of $T$ over $L$. Lemma 4 will be proved if we prove the existence of an element $d \neq 0$ *of* $B$ such that $d(\hat{B})' \subset \hat{I}$ ($(\hat{B})'$ = integral closure of $\hat{B}$ in $T$): for, applying the projection $\varphi$ and noticing that any element of $Z$ which is integral over $\varphi(\hat{B})$ is the projection of an element of $T$ which is integral over $\hat{B}$, we get $\varphi(d) \cdot \varphi(\hat{B})' \subset \hat{A}$, i.e., $\varphi(d)(\hat{A})' \subset \hat{A}$, and we have $\varphi(d) \in A$, $\varphi(d) \neq 0$.

For proving the existence of $d$, we may assume that the basic elements $a_i$ belong to $I$. The trace mapping $T_{k(x)/k(z)}$ extends in a unique way to an $L$-linear mapping $\tau$ of $T$ into $L$. Since $k(x)$ is separable over $k(z)$, there exist elements $a'_1, \cdots, a'_q$ of $k(x)$ such that $\tau(a_i a'_j) = \delta_{ij}$ for all $i, j$ (Vol. I, Ch. V, § 11, proof of Theorem 30). Now, if $y$ is an element of $T$ that is integral over $\hat{B}$, we see readily that the elements $\tau(a_i y)$ are integral over $\hat{B}$, and hence belong to $\hat{B}$ since $\hat{B}$ is integrally closed. Since $y = \sum_i \tau(a_i y) a'_i$, we have $y \in \sum_i \hat{B} a'_i$, and therefore $(\hat{B})' \subset \sum_{i=1}^{q} \hat{B} a'_i$. Taking for $d$ a common denominator in $B$ such that $da'_i \in I$ for every $i$, we get $d(\hat{B})' \subset \hat{I}$, and Lemma 4 is proved.

We now restate, in geometric language, the results obtained by combining Lemmas, 1, 2, 3 and 4:

THEOREM 31 (Chevalley). *Let $V$ be an algebraic variety, $W$ a subvariety of $V$, both irreducible over a perfect field $k$. Then $V$ is analytically unramified at $W$, i.e., the completion of the local ring $\mathfrak{o}(W; V)$ has no nilpotent elements.*

In particular the extension of a prime ideal $\mathfrak{p}$ of $k[X_1, \cdots, X_n]$ to $k[[X_1, \cdots, X_n]]$ is an ideal that is equal to its radical.

THEOREM 32 (Zariski). *If, furthermore, $V$ is normal at $W$, i.e., if $\mathfrak{o}(W; V)$ is integrally closed, then $V$ is analytically irreducible and analytically normal at $W$, i.e., the completion of $\mathfrak{o}(W; V)$ is a domain and is integrally closed.*

THEOREM 33. *With the hypothesis of Theorem* 31, *the integral closure of $\mathfrak{o}(W; V)$ is a semi-local ring whose completion is canonically isomorphic to the integral closure (in its total quotient ring) of the completion of $\mathfrak{o}(W; V)$.*

# APPENDIX 1

## RELATIONS BETWEEN PRIME IDEALS IN A NOETHERIAN DOMAIN $\mathfrak{o}$ AND IN A SIMPLE RING EXTENSION $\mathfrak{o}[t]$ OF $\mathfrak{o}$

Let $\mathfrak{o}$ be a noetherian domain and let $\mathfrak{o}'$ be a domain containing $\mathfrak{o}$ and such that $\mathfrak{o}'=\mathfrak{o}[t]$, where $t$ is some element of $\mathfrak{o}'$. We wish to investigate the relations between prime ideals in $\mathfrak{o}$ and in $\mathfrak{o}'$. We first prove the following lemma:

LEMMA 1. *Let $t$ be algebraic over the quotient field of $\mathfrak{o}$, let $\mathfrak{p}'$ be a prime ideal in $\mathfrak{o}'=\mathfrak{o}[t]$ such that the prime ideal $\mathfrak{p}=\mathfrak{p}'\cap\mathfrak{o}$ has height* 1 (i.e., $\mathfrak{p}$ *is a minimal prime ideal in* $\mathfrak{o}$). *Then the $\mathfrak{p}'$-residue of $t$ is algebraic over $\mathfrak{o}/\mathfrak{p}$.*

PROOF. Upon passing to the rings of quotients $\mathfrak{o}_\mathfrak{p}$ and $\mathfrak{o}'_{\mathfrak{o}-\mathfrak{p}}$ we achieve a reduction to the case in which $\mathfrak{o}$ is a local domain having $\mathfrak{p}$ as its only proper prime ideal (since $\mathfrak{p}'\mathfrak{o}'_{\mathfrak{o}-\mathfrak{p}}$ is obviously a prime ideal in $\mathfrak{o}'_{\mathfrak{o}-\mathfrak{p}}$ whose contraction to $\mathfrak{o}_\mathfrak{p}$ is $\mathfrak{p}\mathfrak{o}_\mathfrak{p}$). We therefore assume that $\mathfrak{o}$ is a local domain and that the minimal prime ideal $\mathfrak{p}$ of $\mathfrak{o}$ is also the maximal ideal of $\mathfrak{o}$ ($\mathfrak{o}$ is then a 1-dimensional local domain).

Let $T$ be an indeterminate, let $\mathfrak{O}'=\mathfrak{o}[T]$ and let $\mathfrak{M}'$ be the kernel of the homomorphism $\varphi$ of $\mathfrak{O}'$ onto $\mathfrak{o}'$ which is uniquely determined by the following two conditions: (1) $\varphi$ is the identity on $\mathfrak{o}$; (2) $\varphi(T)=t$. Then $\varphi$ is a proper homomorphism, i.e., $\mathfrak{M}'\neq(0)$. If we set $\mathfrak{P}'=\varphi^{-1}(\mathfrak{p}')$, then it is immediately seen that $\mathfrak{P}'\cap\mathfrak{o}=\mathfrak{p}$ and that the $\mathfrak{P}'$-residue of $T$ can be identified with the $\mathfrak{p}'$-residue of $t$. So we have to show that the $\mathfrak{P}'$-residue of $T$ is algebraic over $\mathfrak{o}/\mathfrak{p}$.

Since $T$ is an indeterminate, we have $\mathfrak{O}'\mathfrak{p}\cap\mathfrak{o}=\mathfrak{p}$. Hence there is a homomorphism of $\mathfrak{O}'/\mathfrak{O}'\mathfrak{p}$ onto $\mathfrak{O}'/\mathfrak{P}'$ which sends the $\mathfrak{O}'\mathfrak{p}$-residue of $T$ into the $\mathfrak{P}'$-residue of $T$ and which reduces to the identity on $\mathfrak{o}/\mathfrak{p}$. Thus, for the proof of the lemma it will be sufficient to show that this homomorphism is proper, i.e., that $\mathfrak{O}'\mathfrak{p}<\mathfrak{P}'$.

Assume the contrary: $\mathfrak{O}'\mathfrak{p}=\mathfrak{P}'$. We fix an element $x\neq 0$ in $\mathfrak{p}$. In the local domain $\mathfrak{o}$ the principal ideal $\mathfrak{o}x$ is primary and hence contains a power of $\mathfrak{p}$. Therefore $\mathfrak{O}'x$ contains a power of $\mathfrak{P}'$, and therefore

$\mathfrak{P}'$ is contained in every prime ideal of $\mathfrak{O}'x$, in particular in every isolated prime ideal of $\mathfrak{O}'x$. By the principal ideal theorem (Vol. I, Ch. IV, Theorem 29) it follows then that $\mathfrak{P}'$ is a minimal prime ideal in $\mathfrak{O}'$, in contradiction with $\mathfrak{P}' > \mathfrak{M}' > 0$. Q.E.D.

Before proceeding with the proof of the next proposition we shall give another proof of the preceding lemma, which does not make use of the principal ideal theorem.

In the first place, we can achieve a reduction to the case in which $t$ belongs to the quotient field of $\mathfrak{o}$. In fact, there exists an element $a \neq 0$ in $\mathfrak{o}$ such that the element $\tau = at$ is integral over $\mathfrak{o}$. We set $\mathfrak{o}'' = \mathfrak{o}[\tau]$, $\mathfrak{p}'' = \mathfrak{p}' \cap \mathfrak{o}'$. It is clear that the $\mathfrak{p}''$-residue $\bar{\tau}$ of $\tau$ is algebraic over $\mathfrak{o}/\mathfrak{p}$ (use a relation of integral dependence of $\tau$ over $\mathfrak{o}$) and that $\mathfrak{o}''/\mathfrak{p}'' = \mathfrak{o}/\mathfrak{p}[\bar{\tau}]$. Therefore it is sufficient to prove that the $\mathfrak{p}'$-residue of $t$ is algebraic over $\mathfrak{o}''/\mathfrak{p}''$. Now, since $\mathfrak{p}$ is minimal in $\mathfrak{o}$, $\mathfrak{p}''$ is minimal in $\mathfrak{o}''$ (Vol. I, p. 259), and since $\mathfrak{p}' \cap \mathfrak{o}'' = \mathfrak{p}''$ and $t$ belongs to the quotient field of $\mathfrak{o}''$, the desired reduction is achieved.

Assume then that $t$ belongs to the quotient field of $\mathfrak{o}$. We may also maintain our previous reduction to the case in which $\mathfrak{p}$ is the maximal ideal of the one-dimensional local domain $\mathfrak{o}$. Let $t = y/x$, where $x, y \in \mathfrak{o}$, and let $\mathfrak{A}$ be the ideal generated by $x$ and $y$ in $\mathfrak{o}$. If $x$ is a unit in $\mathfrak{o}$ then $\mathfrak{o}' = \mathfrak{o}$, and the lemma is trivial in this case. If $x$ is not a unit and $y$ is a unit in $\mathfrak{o}$, then $y = xt \in \mathfrak{o}'\mathfrak{p} \subset \mathfrak{p}'$, i.e., $y \in \mathfrak{p}' \cap \mathfrak{o}$, in contradiction with $y \notin \mathfrak{p}$. Hence we may assume that both $x$ and $y$ are non-units in $\mathfrak{o}$. Then $\mathfrak{A}$ is primary for $\mathfrak{p}$ and thus we know that for large $n$ the length $\lambda(\mathfrak{A}^n)$ is a polynomial in $n$, of degree 1 (VIII, § 8). Consequently $\lambda(\mathfrak{A}^{n+1}) - \lambda(\mathfrak{A}^n) = q = \text{const.}$, for $n$ large.† Therefore $\lambda(\mathfrak{p}\mathfrak{A}^n) - \lambda(\mathfrak{A}^n) \leqq q$, for all $n \geqq n_0$, where $n_0$ is a suitable integer. This implies that if $n \geqq \max(n_0, q)$ then the $n+1$ basis elements $x^n, x^{n-1}y, \cdots, y^n$ of $\mathfrak{A}^n$ are linearly dependent mod $\mathfrak{p}\mathfrak{A}^n$ over the residue field $\mathfrak{o}/\mathfrak{p}$($\mathfrak{A}^n/\mathfrak{p}\mathfrak{A}^n$ is a vector space over $\mathfrak{o}/\mathfrak{p}$, of dimension $\leqq q$). We have thus a relation of the form

$$a_0x^n + a_1x^{n-1}y + \cdots + a_ny^n = 0,$$

where the $a_i$ are in $\mathfrak{o}$ and not all in $\mathfrak{p}$. Dividing through by $x^n$ we conclude that the $\mathfrak{p}'$-residue of $t$ is algebraic over $\mathfrak{o}/\mathfrak{p}$ (the elements

†Actually, all we shall need in what follows is that $\lambda(\mathfrak{A}^{n+1}) - \lambda(\mathfrak{A}^n)$ is bounded from above. A direct proof of this is immediate:

Fix an element $x \neq 0$ in $\mathfrak{A}$. Then $\mathfrak{o}x$ is primary for $\mathfrak{p}$, whence $\mathfrak{A}^n \subset \mathfrak{o}x$ for large $n$. Therefore $\mathfrak{A}^n = \mathfrak{B}x$ where $\mathfrak{B} = \mathfrak{A}^n : (x)$. We have $\lambda(\mathfrak{A}^n) = \lambda(\mathfrak{B}) + l(\mathfrak{B}/\mathfrak{B}x)$, where $l$ refers to lengths of $\mathfrak{o}$-modules. Since $\mathfrak{B}/\mathfrak{B}x$ and $\mathfrak{o}/\mathfrak{o}x$ are isomorphic $\mathfrak{o}$-modules, it follows that $\lambda(\mathfrak{A}^n) = \lambda(\mathfrak{B}) + \lambda(\mathfrak{o}x)$, and therefore (since $\mathfrak{B} \supset \mathfrak{A}^{n-1})\lambda(\mathfrak{A}^n) - \lambda(\mathfrak{A}^{n-1}) \leqq \lambda(\mathfrak{o}x)$, for all large $n$.

$a_1, a_2, \cdots, a_n$ cannot all belong to $\mathfrak{p}$, for if they do we would have $a_0 = -(a_1 t + \cdots + a_n t^n) \in \mathfrak{o}'\mathfrak{p} \cap \mathfrak{o}$, i.e., $a_0 \in \mathfrak{p}$, a contradiction).

We shall use the preceding lemma for proving the following:

PROPOSITION 1. *Let $\mathfrak{o}$ be a noetherian domain and let $\mathfrak{o}' = \mathfrak{o}[t]$ be a domain which contains $\mathfrak{o}$ and is a simple ring extension of $\mathfrak{o}$. Let $\mathfrak{p}'$ be a prime ideal in $\mathfrak{o}'$, different from $\mathfrak{o}'$; let $\mathfrak{p} = \mathfrak{p}' \cap \mathfrak{o}$ and let $\tau$ be the $\mathfrak{p}'$-residue of $t$.*

(A) *If $t$ is transcendental over $\mathfrak{o}$, then* (a) *$h(\mathfrak{p}') = 1 + h(\mathfrak{p})$ and $\tau$ is algebraic over $\mathfrak{o}/\mathfrak{p}$, if $\mathfrak{p}' \neq \mathfrak{o}'\mathfrak{p}$, and* (b) *$h(\mathfrak{p}') = h(\mathfrak{p})$ and $\tau$ is transcendental over $\mathfrak{o}/\mathfrak{p}$, if $\mathfrak{p}' = \mathfrak{o}'\mathfrak{p}$. Furthermore, if $\mathfrak{p}$ is any prime ideal in $\mathfrak{o}$ then $\mathfrak{o}'\mathfrak{p}$ is a prime ideal in $\mathfrak{o}'$, and we have $\mathfrak{o}'\mathfrak{p} \cap \mathfrak{o} = \mathfrak{p}$.*

(B) *If $t$ is algebraic over $\mathfrak{o}$, then $h(\mathfrak{p}') \leqq h(\mathfrak{p})$; and if, furthermore, $\tau$ is transcendental over $\mathfrak{o}/\mathfrak{p}$ then $h(\mathfrak{p}') < h(\mathfrak{p})$.*

PROOF. We first make a *remark* which will be useful in the proof of either part of the proposition. Let $\mathfrak{q}'$ be a prime ideal in $\mathfrak{o}'$ such that $\mathfrak{p}' > \mathfrak{q}'$ and assume that $\mathfrak{p}' \cap \mathfrak{o} = \mathfrak{q}' \cap \mathfrak{o} = \mathfrak{p}$. Let $\sigma$ be the $\mathfrak{q}'$-residue of $t$. We have $\mathfrak{o}'/\mathfrak{p}' = \mathfrak{o}/\mathfrak{p}[\tau]$ and $\mathfrak{o}'/\mathfrak{q}' = \mathfrak{o}/\mathfrak{p}[\sigma]$. Thus the natural homomorphism of $\mathfrak{o}'/\mathfrak{q}'$ onto $\mathfrak{o}'/\mathfrak{p}'$ sends $\sigma$ into $\tau$ and reduces to the identity on $\mathfrak{o}/\mathfrak{p}$. Since this homomorphism is not an isomorphism, *it follows that $\sigma$ is transcendental over $\mathfrak{o}/\mathfrak{p}$, while $\tau$ is algebraic over $\mathfrak{o}/\mathfrak{p}$.* From this it follows also that *there exists no prime ideal in $\mathfrak{o}'$ which is properly contained in $\mathfrak{p}'$ and properly contains $\mathfrak{q}'$.*

We now begin with the proof of part (A) of the proposition. Since $t$ is transcendental over $\mathfrak{o}$, it is seen immediately that if $\mathfrak{p}$ is any prime ideal in $\mathfrak{o}$ then $\mathfrak{o}'\mathfrak{p} \cap \mathfrak{o} = \mathfrak{p}$ and the $\mathfrak{o}'\mathfrak{p}$-residue $\tau$ of $t$ is transcendental over $\mathfrak{o}/\mathfrak{p}$. Hence $\mathfrak{o}'/\mathfrak{o}'\mathfrak{p}$ $(= \mathfrak{o}/\mathfrak{p}[\tau])$ is an integral domain, *and $\mathfrak{o}'\mathfrak{p}$ is thus a prime ideal in $\mathfrak{o}'$* (and contracts to $\mathfrak{p}$ in $\mathfrak{o}$). This proves the last assertion of part (A) of the proposition, and, in view of the preceding "*remark,*" it also establishes the fact that $\tau$ is transcendental over $\mathfrak{o}/\mathfrak{p}$ if and only if $\mathfrak{p}' = \mathfrak{o}'\mathfrak{p}$. It also follows that

$$h(\mathfrak{o}'\mathfrak{p}) \geqq h(\mathfrak{p}), \tag{1}$$

and that, consequently, if $\mathfrak{p}'$ is a prime ideal in $\mathfrak{o}'$ then

$$\text{“}\mathfrak{p}' \cap \mathfrak{o} = \mathfrak{p},\ \mathfrak{p}' \neq \mathfrak{o}'\mathfrak{p}\text{”} \Rightarrow \text{“}h(\mathfrak{p}') \geqq 1 + h(\mathfrak{p}).\text{”} \tag{2}$$

To complete the proof of part (A) of the proposition it remains to be shown that $h(\mathfrak{p}') = 1 + h(\mathfrak{p})$ or $h(\mathfrak{p}') = h(\mathfrak{p})$ according as $\mathfrak{p}' \neq \mathfrak{o}'\mathfrak{p}$ or $\mathfrak{p}' = \mathfrak{o}'\mathfrak{p}$.

Let

$$\mathfrak{p}' > \mathfrak{q}' > \cdots > (0) \tag{3}$$

be a strictly descending chain of prime ideals in $\mathfrak{o}'$ beginning with $\mathfrak{p}'$ and having maximal length, and let $\mathfrak{q}' \cap \mathfrak{o} = \mathfrak{q}$. Assertion (A) being trivial if $\mathfrak{p}' = (0)$, we use induction with respect to $h(\mathfrak{p}')$. We consider separately the two cases: (a) $\mathfrak{p}' \neq \mathfrak{o}'\mathfrak{p}$ or (b) $\mathfrak{p}' = \mathfrak{o}'\mathfrak{p}$.

CASE (a): $\mathfrak{p}' \neq \mathfrak{o}'\mathfrak{p}$.

If $\mathfrak{q}' = \mathfrak{o}'\mathfrak{p}$, then $\mathfrak{q} = \mathfrak{p}$, and $h(\mathfrak{q}') = h(\mathfrak{p})$ by the induction hypothesis. Hence $h(\mathfrak{p}') = 1 + h(\mathfrak{q}') = 1 + h(\mathfrak{p})$.

If $\mathfrak{q}' \neq \mathfrak{o}'\mathfrak{p}$, then $\mathfrak{p} > \mathfrak{q}$ (again by the preceding "remark"), and hence $h(\mathfrak{p}) > h(\mathfrak{q})$. We also have $\mathfrak{q}' \neq \mathfrak{o}'\mathfrak{q}$, for in the contrary case we would have the strictly descending chain $\mathfrak{p}' > \mathfrak{o}'\mathfrak{p} > \mathfrak{q}'$, contrary to the maximality of the chain $\mathfrak{p}' > \mathfrak{q}' > \cdots$. Hence, by our induction hypothesis, we have $h(\mathfrak{q}') = 1 + h(\mathfrak{q})$. Therefore $h(\mathfrak{p}') = 1 + h(\mathfrak{q}') = 2 + h(\mathfrak{q}) \leqq 1 + h(\mathfrak{p})$, whence $h(\mathfrak{p}') = 1 + h(\mathfrak{p})$, in view of (2).

CASE (b): $\mathfrak{p}' = \mathfrak{o}'\mathfrak{p}$.

If $\mathfrak{q}' = \mathfrak{o}'\mathfrak{q}$, then $\mathfrak{p} > \mathfrak{q}$ and $h(\mathfrak{p}') = 1 + h(\mathfrak{q}') = 1 + h(\mathfrak{q})$ (by the induction hypothesis). Hence $h(\mathfrak{p}') \leqq h(\mathfrak{p})$, and thus $h(\mathfrak{p}') = h(\mathfrak{p})$, in view of (1).

Now assume that $\mathfrak{q}' > \mathfrak{o}'\mathfrak{q}$. By the induction hypothesis, we have $h(\mathfrak{q}') = 1 + h(\mathfrak{q})$. Since $\mathfrak{p}' = \mathfrak{o}'\mathfrak{p}$, we have necessarily that $\mathfrak{p} > \mathfrak{q}$ and also that $\tau$ is transcendental over $\mathfrak{o}/\mathfrak{p}$. This property of $\tau$ can also be expressed as follows: if $\sigma$ is the $\mathfrak{q}'$-residue of $t$ (whence $\sigma$ is *algebraic* over $\mathfrak{o}/\mathfrak{q}$, since $\mathfrak{q}' > \mathfrak{o}'\mathfrak{q}$), then the $\mathfrak{p}'/\mathfrak{q}'$-residue of $\sigma$ is transcendental over the ring $(\mathfrak{o}/\mathfrak{q})/(\mathfrak{p}/\mathfrak{q})$. By Lemma 1, this implies that $\mathfrak{p}/\mathfrak{q}$ is not a *minimal* prime ideal in the ring $\mathfrak{o}/\mathfrak{q}$. In other words: $h(\mathfrak{p}) \geqq 2 + h(\mathfrak{q})$. Hence, $h(\mathfrak{p}') = 1 + h(\mathfrak{q}') = 2 + h(\mathfrak{q}) \leqq h(\mathfrak{p})$, and thus $h(\mathfrak{p}') = h(\mathfrak{p})$, in view of (1).

This completes the proof of Part (A) of the proposition. Note that we had to use Lemma 1 only in the case $\mathfrak{p}' = \mathfrak{o}'\mathfrak{p}$, $\mathfrak{q}' > \mathfrak{o}'\mathfrak{q}$.

We now deal with part (B) of the proposition. Let $T$ be transcendental over $\mathfrak{o}$ and let $\mathfrak{O}' = \mathfrak{o}[T]$. We have a homomorphism of $\mathfrak{O}'$ onto $\mathfrak{o}'$ which sends $T$ into $t$ and reduces to the identity on $\mathfrak{o}$. Let $\mathfrak{M}'$ be the kernel of this homomorphism. Since $\mathfrak{M}' \cap \mathfrak{o} = (0)$ and since the $\mathfrak{M}'$-residue $t$ of $T$ is algebraic over $\mathfrak{o}$, it follows from part (A) that $h(\mathfrak{M}') = 1$. Now, let $\mathfrak{P}'$ be the prime ideal in $\mathfrak{O}'$ such that $\mathfrak{P}'/\mathfrak{M}' = \mathfrak{p}'$. Then $\mathfrak{P}' \cap \mathfrak{o} = \mathfrak{p}$ and

$$(4) \qquad 1 + h(\mathfrak{p}') \leqq h(\mathfrak{P}').$$

Now, the $\mathfrak{p}'$-residue $\tau$ of $t$ is also that $\mathfrak{P}'$-residue of $T$. Hence by part (A) of the proposition, we have $h(\mathfrak{P}') = h(\mathfrak{p}) + 1$ if $\tau$ is algebraic over $\mathfrak{o}$, and $h(\mathfrak{P}') = h(\mathfrak{p})$ if $\tau$ is transcendental over $\mathfrak{o}$. Using (4), we find, in the first case: $h(\mathfrak{p}') \leqq h(\mathfrak{p})$, and in the second case: $h(\mathfrak{p}') \leqq h(\mathfrak{p}) - 1$. This completes the proof of the proposition.

In Proposition 1 it was *assumed* that there exists a prime ideal $\mathfrak{p}'$ in $\mathfrak{o}'$ such that $\mathfrak{p}' \cap \mathfrak{o}$ is the given prime ideal $\mathfrak{p}$ of $\mathfrak{o}$. We express that assumption by saying that $\mathfrak{p}$ *is not lost in* $\mathfrak{o}'$. There arises naturally the question of whether a given prime ideal $\mathfrak{p}$ in $\mathfrak{o}$ is or is not lost in $\mathfrak{o}'$. The answer is simple: $\mathfrak{p}$ *is not lost if and only if* $\mathfrak{o}'\mathfrak{p} \cap \mathfrak{o} = \mathfrak{p}$. The condition is obviously necessary for if $\mathfrak{p}' \cap \mathfrak{o} = \mathfrak{p}$ then $\mathfrak{p} \subset \mathfrak{o}'\mathfrak{p} \cap \mathfrak{o} \subset \mathfrak{p}' \cap \mathfrak{o} = \mathfrak{p}$, whence $\mathfrak{o}'\mathfrak{p} \cap \mathfrak{o} = \mathfrak{p}$. The converse has been established in the course of the proof of Theorem 3 of Vol. I, Ch. V, § 2 (by a reduction to the case in which $\mathfrak{o}$ is a local ring and $\mathfrak{p}$ is its maximal ideal).

Another necessary and sufficient condition that $\mathfrak{p}$ be *not* lost in $\mathfrak{o}' = \mathfrak{o}[t]$, a condition which is valid also if $\mathfrak{o}$ is not noetherian, is the following: *if* $\mathfrak{O}'$ *denotes the integral closure of* $\mathfrak{o}_\mathfrak{p}$ *in the quotient field* $K'$ *of* $\mathfrak{o}'$, *then* $1/t$ *does not belong to the intersection of the proper prime ideals of* $\mathfrak{O}'$. For the proof we make use of Theorem 8 in VI, § 5:

$$\mathfrak{O}' = \bigcap_{v \in \mathfrak{N}} R_v,$$

where $\mathfrak{N}$ denotes the set of all valuations of $K'$ which have center $\mathfrak{p}$ in $\mathfrak{o}$. Assume that there exists a maximal ideal $\mathfrak{M}'$ of $\mathfrak{O}'$ such that $1/t \notin \mathfrak{M}'$. Applying the cited Theorem 8 of VI, § 5, to the integrally closed local domain $\mathfrak{O}'_{\mathfrak{M}'}$, we see that there exists a valuation $v_0$ in $\mathfrak{N}$ such that $v_0(1/t) \leqq 0$. Hence $v_0(t) \geqq 0$ and $v_0$ is non-negative in $\mathfrak{o}[t]$, and thus, if $\mathfrak{p}'$ is the center of $v_0$ in $\mathfrak{o}[t]$ then $\mathfrak{p}' \cap \mathfrak{o} = \mathfrak{p}$. Conversely, assume that there exists a prime ideal $\mathfrak{p}'$ in $\mathfrak{o}[t]$ such that $\mathfrak{p}' \cap \mathfrak{o} = \mathfrak{p}$. We fix a valuation $v_0$ of $K'$ which has center $\mathfrak{p}'$ in $\mathfrak{o}[t]$. Then $v_0$ has center $\mathfrak{p}$ in $\mathfrak{o}$, i.e., $v_0 \in \mathfrak{N}$, and furthermore $v_0(t) \geqq 0$, whence $v_0(1/t) \leqq 0$. This shows that $1/t$ does not belong to the center $\mathfrak{M}'$ of $v_0$ in $\mathfrak{O}'$.

We add a few remarks in the special case in which $\mathfrak{o}_\mathfrak{p}$ is *integrally closed* and $t$ belongs to the quotient field of $\mathfrak{o}$. In that case, the second of the above conditions takes the following simple form: $\mathfrak{p}$ *is lost in* $\mathfrak{o}[t]$ *if and only if* $1/t$ *is a non-unit in* $\mathfrak{o}_\mathfrak{p}$. Thus $\mathfrak{p}$ is not lost in $\mathfrak{o}[t]$ in the following (and only in the following) two cases: (1) $t \in \mathfrak{o}_\mathfrak{p}$; (2) $t \notin \mathfrak{o}_\mathfrak{p}$, $1/t \notin \mathfrak{o}_\mathfrak{p}$. In case (1) we have $\mathfrak{o}'_{\mathfrak{o}-\mathfrak{p}} = \mathfrak{o}_\mathfrak{p}$ and this implies at once that there is only one prime ideal $\mathfrak{p}'$ in $\mathfrak{o}'$ such that $\mathfrak{p}' \cap \mathfrak{o} = \mathfrak{p}$ and that $\mathfrak{o}'_{\mathfrak{p}'} = \mathfrak{o}_\mathfrak{p}$. In case (2), Theorem 10, Corollary, of VI, § 5, yields a good deal of information. Since the prime ideals in $\mathfrak{o}'$ which contract to $\mathfrak{p}$ are in (1, 1) correspondence with the prime ideals in $\mathfrak{o}'_{\mathfrak{o}-\mathfrak{p}}$ which contract in $\mathfrak{o}_\mathfrak{p}$ to $\mathfrak{p}\mathfrak{o}_\mathfrak{p}$, we may assume, for simplicity, that $\mathfrak{o}$ is a local domain and that $\mathfrak{p}$ is its maximal ideal. Under this assumption we see that, in case (2), $\mathfrak{o}'\mathfrak{p}$ is one of the prime ideals in $\mathfrak{o}'$ which contracts to $\mathfrak{p}$ and that the $\mathfrak{o}'\mathfrak{p}$-residue $\tau$ of $t$ is transcendental over $\mathfrak{o}/\mathfrak{p}$. The other prime ideals in $\mathfrak{o}'$ which contract to $\mathfrak{p}$ are in (1, 1) correspondence with the maximal ideals of the polynomial ring $\mathfrak{o}/\mathfrak{p}[\tau]$.

For purposes of generalization of Proposition 1 we shall now restate Proposition 1. The notation being the same as in that proposition, we denote by $\dim_{\mathfrak{o}} \mathfrak{o}'$ the transcendence degree of the quotient field of $\mathfrak{o}'$ over the quotient field of $\mathfrak{o}$. A similar meaning is attached to the notation $\dim_{\mathfrak{o}/\mathfrak{p}} \mathfrak{o}'/\mathfrak{p}'$. Then Proposition 1 is expressed by the following inequality:

$$h(\mathfrak{p}') + \dim_{\mathfrak{o}/\mathfrak{p}} \mathfrak{o}'/\mathfrak{p}' \leqq h(\mathfrak{p}) + \dim_{\mathfrak{o}} \mathfrak{o}' \tag{5}$$

with equality if $\dim_{\mathfrak{o}} \mathfrak{o}' = 1$.

A straightforward induction on $n$ yields at once the following generalization of Proposition 1:

PROPOSITION 2. *Let* $\mathfrak{o}' = \mathfrak{o}[t_1, t_2, \cdots, t_n]$, *where* $\mathfrak{o}$ *and* $\mathfrak{o}'$ *are noetherian integral domains, and let* $\mathfrak{p}$, $\mathfrak{p}'$ *be prime ideals in* $\mathfrak{o}$ *and* $\mathfrak{o}'$ *such that* $\mathfrak{p}' \cap \mathfrak{o} = \mathfrak{p} (\mathfrak{p}' \neq R)$. *Then inequality* (5) *holds, and we certainly have equality in* (5) *if* $\dim_{\mathfrak{o}} \mathfrak{o}' = n$.

We shall say that *the dimension formula holds for a noetherian integral domain* $\mathfrak{o}$ if for any integral domain $\mathfrak{o}'$ which is finitely generated over $\mathfrak{o}$ and for any pair of prime ideals $\mathfrak{p}$, $\mathfrak{p}'$ in $\mathfrak{o}$ and $\mathfrak{o}'$ respectively ($\mathfrak{p}' \neq R$) such that $\mathfrak{p}' \cap \mathfrak{o} = \mathfrak{p}$, we have

$$h(\mathfrak{p}') + \dim_{\mathfrak{o}/\mathfrak{p}} \mathfrak{o}'/\mathfrak{p}' = h(\mathfrak{p}) + \dim_{\mathfrak{o}} \mathfrak{o}'. \tag{6}$$

We say that *the chain condition holds for prime ideals in a noetherian domain* $\mathfrak{o}$ if for any prime ideal $\mathfrak{p}$ in $\mathfrak{o}$, $\mathfrak{p} \neq R$, all maximal chains of prime ideals in $\mathfrak{o}_{\mathfrak{p}}$ (different from $\mathfrak{o}_{\mathfrak{p}}$) have the same length (therefore have length equal to the dimension of the local ring $\mathfrak{o}_{\mathfrak{p}}$). It is clear that in order to check whether $\mathfrak{o}$ satisfies the chain condition for prime ideals it is sufficient to check whether the above condition concerning $\mathfrak{o}_{\mathfrak{p}}$ is satisfied for all the *maximal* ideals $\mathfrak{p}$ in $\mathfrak{o}$.

PROPOSITION 3. *Let* $\mathfrak{o}$ *be a noetherian integral domain and let* $T_1, T_2, \cdots, T_n$ *be transcendentals which are algebraically independent over* $\mathfrak{o}$. *If for any* $n$ *the domain* $\mathfrak{o}[T_1, T_2, \cdots, T_n]$ *satisfies the chain condition for prime ideals then the dimension formula holds for* $\mathfrak{o}$.

PROOF. Let $\mathfrak{o}'$, $\mathfrak{p}$, $\mathfrak{p}'$ have the same meaning as in Proposition 2 and let $\mathfrak{O}' = \mathfrak{o}[T_1, T_2, \cdots, T_n]$. We have $\mathfrak{o}' = \mathfrak{O}'/\mathfrak{M}'$, where $\mathfrak{M}'$ is a prime ideal in $\mathfrak{O}'$ such that $\mathfrak{M}' \cap \mathfrak{o} = (0)$. By the second part of Proposition 2 we have

$$h(\mathfrak{M}') + \dim_{\mathfrak{o}} \mathfrak{o}' = n. \tag{7}$$

Let $\mathfrak{p}' = \mathfrak{P}'/\mathfrak{M}'$, where $\mathfrak{P}'$ is a prime ideal in $\mathfrak{O}'$ such that $\mathfrak{P}' \supset \mathfrak{M}'$.

Then clearly $\mathfrak{P}' \cap \mathfrak{o} = \mathfrak{p}$. Therefore, again by the last part of Proposition 2 and in view of $\mathfrak{O}'/\mathfrak{P}' = \mathfrak{o}'/\mathfrak{p}'$, we have

$$(8) \qquad h(\mathfrak{P}') + \dim_{\mathfrak{o}/\mathfrak{p}} \mathfrak{o}'/\mathfrak{p}' = h(\mathfrak{p}) + n.$$

Since the chain condition for prime ideals holds in $\mathfrak{O}$, we have $h(\mathfrak{P}') = h(\mathfrak{M}') + h(\mathfrak{p}')$, and from this (6) follows in view of (7) and (8). Q.E.D.

By the dimension theory of algebraic varieties we know that the chain condition for prime ideals holds in the coördinate ring of any affine variety, i.e., in any finite integral domain over a field $k$ (see VII, § 7, Theorem 20, Corollary 2). Hence, Proposition 3 implies that the dimension formula holds for any finite integral domain $\mathfrak{o}$ (and hence also for any local ring of $\mathfrak{o}$ with respect to a prime ideal $\mathfrak{p} < \mathfrak{o}$). However, this conclusion follows also directly from the dimension theory of algebraic varieties, without the intermediary of Proposition 3; it is sufficient to observe that for any prime ideal $\mathfrak{p}$ of a finite integral domain $\mathfrak{o}$ (over a field $k$; $\mathfrak{p} \neq \mathfrak{o}$) we have $h(\mathfrak{p}) = \dim_k \mathfrak{o} - \dim_k \mathfrak{o}/\mathfrak{p}$.

A more interesting feature of Proposition 3 is its application to the construction of examples of noetherian integral domains which do not satisfy the chain condition for prime ideals [compare with the remarks made in Vol. I, p. 242]. To construct such an example we have only to find a noetherian integral domain $\mathfrak{o}$ for which the dimension formula does not hold. We shall construct a local domain $\mathfrak{o}$ and a semi-local domain $\mathfrak{o}' = \mathfrak{o}[t]$ having the same quotient field as $\mathfrak{o}$ such that, with the same notations as in Proposition 1, part (B), we have $h(\mathfrak{p}') < h(\mathfrak{p})$ with $\tau$ algebraic over $\mathfrak{p}$, where $\mathfrak{p}$ is the maximal ideal of $\mathfrak{o}$.† Then, by Proposition 3, the domain $\mathfrak{o}[T]$, $T$-transcendental over $\mathfrak{o}$, does not satisfy the chain condition for prime ideals.

EXAMPLE. We first prove several simple lemmas.

LEMMA 2. *Let $\mathfrak{o}$ be an integral domain having only a finite number of maximal ideals $\mathfrak{m}_1, \mathfrak{m}_2, \cdots, \mathfrak{m}_q$. If each ring $\mathfrak{o}_{\mathfrak{m}_i}$ is local* (i.e., *noetherian*), *then $\mathfrak{o}$ is noetherian* (*hence semi-local*).

PROOF. Let $\mathfrak{A}$ be any ideal in $\mathfrak{o}$. We can find elements $a_1, a_2, \cdots, a_s$ in $\mathfrak{o}$ such that for each $i = 1, 2, \cdots, q$ these elements generate in $\mathfrak{o}_{\mathfrak{m}_i}$ the ideal $\mathfrak{o}_{\mathfrak{m}_i}\mathfrak{A}$. Let $b$ be an arbitrary element of $\mathfrak{A}$. Then for each $i$ we have $bc_i \in \sum_j \mathfrak{o} a_j$ for some $c_i$ in $\mathfrak{o}$ and not in $\mathfrak{m}_i$. Since the elements $c_1, c_2, \cdots, c_s$ generate the unit ideal in $\mathfrak{o}$ it follows that $b \in \sum_j \mathfrak{o} a_j$, showing that $\mathfrak{A}$ has a finite basis. Q.E.D.

LEMMA 3. *Let $\mathfrak{O}_1, \mathfrak{O}_2, \cdots, \mathfrak{O}_q$ be local domains contained in some field, let $\mathfrak{M}_i$ be the maximal ideal of $\mathfrak{O}_i$, let $\mathfrak{o} = \mathfrak{O}_1 \cap \mathfrak{O}_2 \cap \cdots \cap \mathfrak{O}_q$ and let $\mathfrak{m}_i = \mathfrak{M}_i \cap \mathfrak{o}$.*

† The example of such a pair of rings $\mathfrak{o}$ and $\mathfrak{o}'$ (given below) is due to M. Nagata (see reference in Vol. 1, p. 242, footnote). Our observation that $\mathfrak{o}[T]$ provides a counter-example to the chain condition for prime ideals is new.

*If* $\mathfrak{O}_i=\mathfrak{o}_{\mathfrak{m}_i}$, $i=1, 2, \cdots, q$, *then* $\mathfrak{o}$ *is noetherian and its maximal ideals are in the set* $\{\mathfrak{m}_1, \mathfrak{m}_2, \cdots, \mathfrak{m}_q\}$.

PROOF. If $x\in\mathfrak{o}$ is a non-unit in $\mathfrak{o}$ it must be a non-unit in at least one of the rings $\mathfrak{O}_i$. Thus $x$ belongs to one of the ideals $\mathfrak{m}_i$. This proves the lemma, in view of Lemma 2.

Let $R=k[x, y]$ be a polynomial ring in two independent variables $x$, $y$ over a field $k$. We fix a zero-dimensional discrete valuation $v$, of rank 1, which is non-negative on $R$, has $k$ as residue field, and is such that $v(x)=1$ (see VI, § 15, Example 2). Furthermore, we assume that the center of $v$ in $R$ is the maximal ideal $(x, y)$ of $R$. We set $\mathfrak{O}_1=R_v=$ valuation ring of $v$. We then consider the point $x=1$, $y=0$ in the $(x, y)$-plane and we denote by $\mathfrak{O}_2$ the local ring of that point, i.e., the ring of quotients of $k[x, y]$ with respect to the maximal ideal $(x-1, y)$. If $\mathfrak{M}_1=\mathfrak{O}_1x$ and $\mathfrak{M}_2=\mathfrak{O}_2\cdot(x-1, y)$ denote the maximal ideals of $\mathfrak{O}_1$ and $\mathfrak{O}_2$ respectively, we set $\mathfrak{o}'=\mathfrak{O}_1\cap\mathfrak{O}_2$, $\mathfrak{m}'_1=\mathfrak{M}_1\cap\mathfrak{o}'$, $\mathfrak{m}'_2=\mathfrak{M}_2\cap\mathfrak{o}'$.

Since $\mathfrak{o}'\supset k[x, y]$ it is clear that $\mathfrak{o}'_{\mathfrak{m}'_2}=\mathfrak{O}_2$. It is also obvious that $\mathfrak{o}'_{\mathfrak{m}'_1}\subset\mathfrak{O}_1$. We show that $\mathfrak{O}_1\subset\mathfrak{o}'_{\mathfrak{m}'_1}$ and that consequently $\mathfrak{O}_1=\mathfrak{o}'_{\mathfrak{m}'_1}$. If $\xi\in\mathfrak{O}_1$, we write $\xi=\dfrac{A}{B}$, where numerator and denominator are in $k[x, y]$. Let $v(B)=n\geqq 0$. Then $v(A)\geqq n$ and hence $A=x^n a$, $B=x^n b$, with $a$, $b$ in $\mathfrak{O}_1$. Since $x$ is a unit in $\mathfrak{O}_2$, we have also $a, b\in\mathfrak{O}_2$, whence $a, b\in\mathfrak{o}'$. Furthermore, since $v(b)=0$ it follows that $b\notin\mathfrak{m}'_1$. Hence $\xi=\dfrac{a}{b}\in\mathfrak{o}'_{\mathfrak{m}'_1}$, which proves our assertion that $\mathfrak{O}_1=\mathfrak{o}'_{\mathfrak{m}'_1}$.

Since $\mathfrak{O}_1$ and $\mathfrak{O}_2$ are local rings, the ring $\mathfrak{o}'$ is noetherian, by Lemma 3. Since neither one of the two ideals $\mathfrak{m}'_1$, $\mathfrak{m}'_2$ is contained in the other ($x\in\mathfrak{m}'_1$, $x\notin\mathfrak{m}'_2$; $x-1\notin\mathfrak{m}'_1$, $x-1\in\mathfrak{m}'_2$), it follows again by Lemma 3 that $\mathfrak{o}'$ is a semi-local ring, with $\mathfrak{m}'_1$ and $\mathfrak{m}'_2$ as its only (distinct) maximal ideals. Furthermore, we have $\mathfrak{o}'\supset k=\mathfrak{o}'/\mathfrak{m}'_1=\mathfrak{o}'/\mathfrak{m}'_2$.

We now set $\mathfrak{o}=k+(\mathfrak{m}'_1\cap\mathfrak{m}'_2)$. It is immediately seen that $\mathfrak{o}$ has only one maximal ideal $\mathfrak{m}$, namely $\mathfrak{m}=\mathfrak{m}'_1\cap\mathfrak{m}'_2$ (since every element of $\mathfrak{o}$ which is not in $\mathfrak{m}$ is a unit in $\mathfrak{o}$). *We assert that* $\mathfrak{o}'=\mathfrak{o}+kx$. For, let $\xi\in\mathfrak{o}'$ and let $\xi-c_1\in\mathfrak{m}'_1$, $\xi-c_2\in\mathfrak{m}'_2$, where $c_1, c_2\in k$. Then $\xi-c_1+(c_1-c_2)x\in\mathfrak{m}$ (since $x\in\mathfrak{m}'_1$ and $x-1\in\mathfrak{m}'_2$), and this proves the assertion. Thus $\mathfrak{o}'=\mathfrak{o}[x]$.

We now prove that $\mathfrak{o}$ is noetherian† (whence $\mathfrak{o}$ is a local domain). If $\mathfrak{A}$ is any ideal in $\mathfrak{o}$, different from $\mathfrak{o}$, the ideal $\mathfrak{A}\mathfrak{o}'$ in $\mathfrak{o}'$ is *contained in* $\mathfrak{o}$ since $\mathfrak{A}\mathfrak{o}'=\mathfrak{A}+\mathfrak{A}x$ and $\mathfrak{A}x\subset\mathfrak{m}x\subset\mathfrak{m}\subset\mathfrak{o}$. Hence $\mathfrak{A}\mathfrak{o}'$ *is an ideal in* $\mathfrak{o}$. As an ideal in $\mathfrak{o}'$ it has a finite basis $\{a_1, a_2, \cdots, a_h\}$ consisting of elements of $\mathfrak{A}$. Then $\mathfrak{A}\mathfrak{o}'=\sum\mathfrak{o}'a_i=\sum\mathfrak{o}a_i+\sum ka_ix\subset\mathfrak{A}+\sum ka_ix$. This shows that $\mathfrak{A}\mathfrak{o}'/\mathfrak{A}$, regarded as a vector space over $k$ (by viewing both $\mathfrak{A}\mathfrak{o}'$ and $\mathfrak{A}$ as vector spaces over $k$), is finite dimensional. It follows that any strictly ascending chain of $\mathfrak{o}$-ideals between $\mathfrak{A}$ and $\mathfrak{A}\mathfrak{o}'$ is necessarily finite. Now, let $\mathfrak{A}_1\subset\mathfrak{A}_2\subset\cdots$ be an ascending chain of ideals in $\mathfrak{o}$. Since $\mathfrak{o}'$ is noetherian we must have $\mathfrak{A}_n\mathfrak{o}'=\mathfrak{A}_{n+1}\mathfrak{o}'=\cdots$

† The fact that $\mathfrak{o}[x]$ is noetherian does not automatically imply that $\mathfrak{o}$ is noetherian. For instance, if $u$ and $v$ are indeterminates over a field $k$ then the ring $R=k[u, u^2v, u^3v^2, \cdots, u^nv^{n-1}, \cdots]$ is not noetherian, but the ring $R[v](=k[u, v])$ is noetherian.

for some $n$. Without loss of generality, we may assume that this is so already for $n=1$. Then

$$\mathfrak{A}_1 \subset \mathfrak{A}_2 \subset \cdots \subset \mathfrak{A}_1 \mathfrak{o}',$$

and this shows, by what we have just proved above, that $\mathfrak{A}_q = \mathfrak{A}_{q+1} = \cdots$ for some $q$. Hence $\mathfrak{o}$ is noetherian.

We have $h(\mathfrak{m}'_1)=1$, $h(\mathfrak{m}'_2)=2$, whence dim $(\mathfrak{o}')=2$. By the dimension theory of semi-local rings we have dim $(\mathfrak{o})=2$, since $\mathfrak{o}'$ is integral over $\mathfrak{o}$. Therefore $h(\mathfrak{m})=2$. On the other hand, $h(\mathfrak{m}'_1)=1<h(\mathfrak{m})$, and the $\mathfrak{m}'_1$-residue of $x$ is algebraic over $\mathfrak{o}/\mathfrak{m}$ ($=k$; in fact, that residue is equal to 0). Therefore, the dimension formula (5) does not hold for $\mathfrak{o}$, $\mathfrak{o}'$, with $\mathfrak{p}=\mathfrak{m}$ and $\mathfrak{p}'=\mathfrak{m}'_1$. Consequently, $\mathfrak{o}[T]$ does not satisfy the chain condition for prime ideals.

# APPENDIX 2

## VALUATIONS IN NOETHERIAN DOMAINS

In Chapter VI we have derived a number of results concerning the dimension, the rank and the rational rank of valuations in algebraic function fields (see, for instance, VI, § 10, Corollary of Lemma; VI, § 14, Theorem 31, Corollary and VI, § 15, Theorem 36). Our purpose in this appendix is to generalize these results to valuations of quotient fields of arbitrary noetherian domains (and also of fields which are of finite transcendence degree, or are finitely generated, over such fields).†

Let $R$ be a noetherian domain, $K$ the quotient field of $R$ and $v$ a valuation of $K$ which is non-negative on $R$ (VI, § 9, p. 38). Let $\mathfrak{p}$ be the center of $v$ in $R$. The following characters of $v$ may be considered:

(1) the *rank* of $v$ (*rank* $v$);

(2) the *rational rank* of $v$ (*r. rank* $v$);

(3) the *relative R-dimension* of $v$ ($dim_R\, v$): this is the transcendence degree of the residue field $\Delta_v$ of $v$ over the field of quotients $F$ of $R/\mathfrak{p}$.

Then we may also consider the *height* $h(\mathfrak{p})$ of $\mathfrak{p}$. If we denote by $\mathfrak{o}$ the local domain $R_{\mathfrak{p}}$, then $v$ is also non-negative on $\mathfrak{o}$, and the center of $v$ in $\mathfrak{o}$ is the maximal ideal $\mathfrak{m}$ of $\mathfrak{o}$. The relative dimension of $v$ is not affected if we replace $R$ by $\mathfrak{o}$, since $\mathfrak{o}/\mathfrak{m}$ is the field of quotients of $R/\mathfrak{p}$. The height of $\mathfrak{p}$ is now also the dimension of the local domain $\mathfrak{o}$. We shall deal directly with $\mathfrak{o}$ and discard the domain $R$ altogether. We set $r = \text{rank}\, v$, $\rho = \text{r. rank}\, v$, $d = \dim_{\mathfrak{o}} v$, $s = \dim(\mathfrak{o})$. To express our assumption that $v$ is non-negative on $\mathfrak{o}$ and that the maximal ideal $\mathfrak{m}$ of $\mathfrak{o}$ is the center of $v$, we shall say that $v$ *dominates* $\mathfrak{o}$.

PROPOSITION 1. *If $v$ is a valuation of the field of quotients $K$ of a local domain $\mathfrak{o}$ and if $v$ dominates $\mathfrak{o}$, then*

$$\text{rank}\, v + \dim_{\mathfrak{o}} v \leqq \dim(\mathfrak{o}) \quad (\text{or, } r + d \leqq s) \tag{1}$$

(*and hence $r$ and $d$ are finite*).

† This generalization is due to S. Abhyankar and is given in his paper "On the valuations centered in a local domain," *Amer. J. Math.*, 78 (1956), pp. 321–348. Our proofs differ from those given by Abhyankar. In particular, our proofs of Propositions 2 and 3 make no use of Cohen's structure theorems for complete local rings.

PROOF. We first prove that $r \leqq s$. Let $\mathfrak{P} > \mathfrak{P}_1 > \cdots > \mathfrak{P}_q$ be a strictly descending chain of prime ideals in the valuation ring $R_v$ of $v$, where $\mathfrak{P}$ is the maximal ideal of $R_v$ ($q$—an integer $\geqq 1$). For each $i = 1, 2, \cdots, q$, we fix an element $t_i$ such that $t_i \in \mathfrak{P}_{i-1}$, $t_i \notin \mathfrak{P}_i$ ($\mathfrak{P}_0 = \mathfrak{P}$), we consider the ring $\mathfrak{o}' = \mathfrak{o}[t_1, t_2, \cdots, t_q]$ and we set $\mathfrak{p}' = \mathfrak{P} \cap \mathfrak{o}'$, $\mathfrak{p}'_i = \mathfrak{P}_i \cap \mathfrak{o}'$ ($i = 1, 2, \cdots, q$). We have $\mathfrak{p}' > \mathfrak{p}'_1 > \cdots > \mathfrak{p}'_q$, whence $h(\mathfrak{p}') \geqq q$. On the other hand, we have $\mathfrak{p}' \cap \mathfrak{o} = \mathfrak{m}$, since $\mathfrak{P} \cap \mathfrak{o} = \mathfrak{m}$. By Proposition 2, Appendix 1, we have therefore the inequality $h(\mathfrak{p}') + \dim_{\mathfrak{o}/\mathfrak{p}} \mathfrak{o}'/\mathfrak{p}' \leqq s$, whence

$$q + \dim_{\mathfrak{o}/\mathfrak{p}} \mathfrak{o}'/\mathfrak{p}' \leqq s. \tag{2}$$

Hence $q \leqq s$, showing that rank $v \leqq s$.

We next show that $\dim_{\mathfrak{o}} v$ is finite. Let $x_1, x_2, \cdots, x_q$ be elements of $R_v$ whose $v$-residues are algebraically independent over the field $\mathfrak{o}/\mathfrak{m}$. We set now $\mathfrak{o}' = \mathfrak{o}[x_1, x_2, \cdots, x_q]$ and $\mathfrak{p}' = \mathfrak{P} \cap \mathfrak{o}'$. Then again $\mathfrak{p}' \cap \mathfrak{o} = \mathfrak{m}$, and this time we have $\dim_{\mathfrak{o}/\mathfrak{m}} \mathfrak{o}'/\mathfrak{p}' = q$. Hence, again by Proposition 2, Appendix 1, we have

$$h(\mathfrak{p}') + q \leqq s,$$

showing that $q$ is bounded, whence $\dim_{\mathfrak{o}} v$ is finite.

From $r \leqq s$ follows inequality (1) in the case $\dim_{\mathfrak{o}} v = 0$. Since we know now that $\dim_{\mathfrak{o}} v$ is finite, we may proceed by induction from $d-1$ to $d$, assuming that $d > 0$. We fix an element $x$ in $R_v$ whose $v$-residue is transcendental over $\mathfrak{o}/\mathfrak{m}$, we set $\mathfrak{o}' = \mathfrak{o}[x]$ and $\mathfrak{p}' = \mathfrak{P} \cap \mathfrak{o}'$. We have now, by Proposition 2, Appendix 1: $h(\mathfrak{p}') + 1 \leqq s$. On the other hand, the dimension of $v$ relative to $\mathfrak{o}'$ is $d-1$, and hence, by our induction hypothesis, we have $r + d - 1 \leqq h(\mathfrak{p}')$, and this yields the desired inequality (1).

We note that Proposition 1 remains true if $v$ is a valuation of an algebraic extension of $K$, for the rank, rational rank, and the dimension of $v$ are not affected by an algebraic extension of $K$. (See VI, § 11.)

The following result is stronger than Proposition 1 (since $\rho \geqq r$):

PROPOSITION 2. *With the same assumptions as in Proposition* 1, *we have*

$$\text{r. rank } v + \dim_{\mathfrak{o}} v \leqq \dim(\mathfrak{o}) \quad (\text{or, } \rho + d \leqq s).$$

PROOF. We consider separately various cases.

CASE 1. rank $v = 1$, $\dim v = 0$.

In this case the value group $\Gamma$ of $v$ consists of real numbers. Since every element of $\Gamma$ is of the form $v(a) - v(b)$, with $a$, $b$ in $\mathfrak{m}$, the set $v\{\mathfrak{m}\}$ $(= \{v(a) | a \in \mathfrak{m}\})$ has the same rational rank as $\Gamma$. The elements of $v\{\mathfrak{m}\}$ are positive real numbers. If $\alpha \in v\{\mathfrak{m}\}$ we denote by $\mathfrak{A}_\alpha$ the set of elements $x$ in $\mathfrak{o}$ such that $v(x) \geqq \alpha$. Then $\mathfrak{A}_\alpha$ is an ideal in $\mathfrak{o}$ (a valuation ideal; see Appendix 3). If $\mathfrak{B}$ is any ideal in $\mathfrak{o}$ then $\min(v(y), y \in \mathfrak{B})$ exists since $\mathfrak{B}$ has a finite basis. We denote this minimum by $v(\mathfrak{B})$. We may normalize $\Gamma$ so that $v(\mathfrak{m})=1$. For any $\alpha$ in $v\{\mathfrak{m}\}$ we can find an integer $n$ such that $n \geqq \alpha$. Then $\mathfrak{m}^n \subset \mathfrak{A}_\alpha$, showing that $\mathfrak{A}_\alpha$ is primary, with $\mathfrak{m}$ as associated prime ideal. The set of valuation ideals $\mathfrak{A}_\alpha$ in $\mathfrak{o}$ ($\alpha \in v\{\mathfrak{m}\}$) is naturally ordered by set theoretic inclusion: $\mathfrak{A}_\alpha > \mathfrak{A}_\beta$ if $\alpha < \beta$. The fact that each $\mathfrak{A}_\alpha$ is $\mathfrak{m}$-primary shows that each valuation ideal $\mathfrak{A}_\beta$ is preceded by only a finite number of valuation ideals $\mathfrak{A}_\alpha$. Hence the ordered subset $v\{\mathfrak{m}\}$ of $\Gamma$ (and also the ordered set of ideals $\mathfrak{A}_\alpha$) *is a simple infinite sequence*, say $\alpha_1 < \alpha_2 < \cdots < \alpha_i < \cdots$, where $\alpha_1 = 1$ and $\alpha_i \to +\infty$. This set $v\{\mathfrak{m}\}$ is closed under addition.

The length $\lambda(\mathfrak{A}_{\alpha_i})$ of $\mathfrak{A}_{\alpha_i}$ is clearly $\geqq i$. For any given positive integer $n$ let $i(n)$ be the subscript such that $\alpha_{i(n)} = n$. Then $\mathfrak{m}^n \subset \mathfrak{A}_{\alpha_{i(n)}}$, and therefore

$$\lambda(\mathfrak{m}^n) \geqq i(n). \tag{3}$$

Let now $q$ be a positive integer such that the rational rank of $\Gamma$ (and hence also of $v\{\mathfrak{m}\}$) is $\geqq q$. We can then find in $v\{\mathfrak{m}\}$ elements $\tau_1, \tau_2, \cdots, \tau_q$ ($\tau_1 = \alpha_1 = 1$) which are rationally independent. We assume that $\tau_1 < \tau_2 < \cdots < \tau_q$. Denote by $\sigma_n$ the number of ordered $q$-tuples $(j_1, j_2, \cdots, j_q)$ of non-negative integers $j_1, j_2, \cdots, j_q$ (*not all zero*) such that

$$j_1\tau_1 + j_2\tau_2 + \cdots + j_q\tau_q \leqq n (= \alpha_{i(n)}). \tag{4}$$

Since the elements $j_1\tau_1 + j_2\tau_2 + \cdots + j_q\tau_q$ are among the $\alpha_i$'s and are distinct, it is clear that $i(n) \geqq \sigma_n$. Hence, by (3):

$$\lambda(\mathfrak{m}^n) \geqq \sigma_n. \tag{5}$$

We now proceed to find an estimate for $\sigma_n$. Let $n'$ denote the integral part of $\frac{n}{q\tau_q}$. If $1 \leqq j_i \leqq n'$ for $i = 1, 2, \cdots, q$ then $j_i\tau_i \leqq \frac{n}{q}$ (since $\tau_i \leqq \tau_q$), whence $j_1\tau_1 + j_2\tau_2 + \cdots + j_q\tau_q \leqq n$. Therefore

$$n'^q \leqq \sigma_n. \tag{6}$$

Since $\frac{n}{q\tau_q} - 1 < n'$, it follows from (6) that there exists a polynomial $P_q$, *of degree* $q$, such that $P_q(n) \leqq \sigma_n$ for all $n$. On the other hand,

$\lambda(\mathfrak{m}^n)$ is a polynomial $P_{\mathfrak{m}}(n)$, *of degree* $s$, for $n$ large (VIII, § 8, Theorem 19). Hence by (5) and (6) we find

$$(7) \qquad P_q(n) \leqq P_{\mathfrak{m}}(n),$$

and hence $q \leqq s$. We therefore have $\rho \leqq s$ in case 1.

CASE 2. rank $v=1$, $\dim_{\mathfrak{o}} v > 0$. The proof in this case is by induction from $d-1$ to $d$ $(=\dim_{\mathfrak{o}} v)$ and is identical to the inductive argument given in the last part of the proof of Proposition 1.

CASE 3. rank $v>1$. We now use induction with respect to $r$ $(=\text{rank } v)$. Let $v=v_1 \circ \bar{v}$, where $v_1$ is a valuation of $K$, of rank 1, and $\bar{v}$ is a valuation of the residue field $\Delta_{v_1}$ of $v_1$, of rank $r-1$. We denote by $\Delta_v$ the residue field of $v$; this is also the residue field of $\bar{v}$. Let $\mathfrak{p}_1$ be the center of $v_1$ in $\mathfrak{o}$. Then, by the case $r=1$, we have

$$(8) \qquad \text{r.rank } v_1 + \dim_{\mathfrak{o}} v_1 \leqq h(\mathfrak{p}_1),$$

where, if we set $F_1 =$ field of quotients of $\mathfrak{o}/\mathfrak{p}_1$, then

$$(9) \qquad \dim_{\mathfrak{o}} v_1 = \text{tr.d. } \Delta_{v_1}/F_1.$$

The valuation $\bar{v}$ dominates the local domain $\bar{\mathfrak{o}} = \mathfrak{o}/\mathfrak{p}_1$. Let $\bar{v}_0$ be the restriction of $\bar{v}$ to $F_1$. Then rank $\bar{v}_0 \leqq$ rank $\bar{v} = r-1$, and hence by our induction hypothesis we have

$$(10) \qquad \text{r.rank } \bar{v}_0 + \dim_{\bar{\mathfrak{o}}} \bar{v}_0 \leqq h(\mathfrak{m}/\mathfrak{p}_1).$$

Adding (8) and (10) we find

$$(11) \qquad \text{r.rank } v - (\text{r.rank } \bar{v} - \text{r.rank } \bar{v}_0) + \dim_{\mathfrak{o}} v_1 + \dim_{\bar{\mathfrak{o}}} \bar{v}_0 \leqq h(\mathfrak{m}) = s.$$

We sh all prove in a moment that

$$(12) \qquad \text{r.rank } \bar{v} - \text{r.rank } \bar{v}_0 \leqq \text{tr.d. } \Delta_{v_1}/F_1 - \text{tr.d. } \Delta_{\bar{v}}/\Delta_{\bar{v}_0},$$

where $\Delta_{\bar{v}_0}$ is the residue field of $\bar{v}_0$. Note that, by Proposition 1, the transcendence degrees on the right-hand side of (12) are all finite. In fact, the right-hand side is equal to $\dim_{\mathfrak{o}} v_1 - \dim_{\bar{\mathfrak{o}}} \bar{v} + \dim_{\bar{\mathfrak{o}}} \bar{v}_0$. Note also that $\dim_{\bar{\mathfrak{o}}} \bar{v} = \dim_{\mathfrak{o}} v$. Hence from (11) and (12) we find

$$\text{r.rank } v + \dim_{\mathfrak{o}} v \leqq s,$$

which completes the proof of the proposition.

As to (12), this relation merely expresses the following general lemma:

LEMMA 1. *Let $K$ be a field, $K_0$ a subfield of $K$, $v$ a valuation of $K$ and $v_0$ the restriction of $v$ to $K_0$. If $\Delta$ and $\Delta_0$ are the residue fields of $v$ and $v_0$ respectively, and if* tr.d. $K/K_0$ *is finite, then*

$$(13) \qquad \text{r.rank } v - \text{r.rank } v_0 \leqq \text{tr.d. } K/K_0 - \text{tr.d. } \Delta/\Delta_0.$$

PROOF. Let tr.d. $K/K_0=g$ and tr.d. $\Delta/\Delta_0=h$, so that $h\leqq g$. Fix $h$ elements $x_1, x_2, \cdots, x_h$ in $K$ such that their $v$-residues $\bar{x}_i$ are algebraically independent over $\Delta_0$. Let $K'=K_0(x_1, x_2, \cdots, x_h)$ and let $v'$ be the restriction of $v$ to $K'$. From the fact that the $\bar{x}_i$ are algebraically independent over $\Delta_0$ follows at once that if $f(x_1, x_2, \cdots, x_h)$ is any non-zero polynomial in $x_1, x_2, \cdots, x_h$, with coefficients $a_i$ in $K_0$, then $v'(f(x))=\min\{v_0(a_i)\}$. Hence the value group $\Gamma'$ of $v'$ is the same as that of $v_0$. On the other hand, a simple argument similar to the one given in the proof of the lemma in VI, § 10 (p. 50) shows that if $\Gamma$ is the value group of $v$ then

$$\text{r.rank}\ \Gamma/\Gamma' \leqq \text{tr.d.}\ K/K' = g-h,$$

and this establishes the lemma.

Combining Proposition 2 with the above Lemma 1, we have the following

COROLLARY. *The assumptions being the same as in Proposition* 1, *except that we now assume that $v$ is a valuation of an extension field $K'$ of $K$ such that* tr.d. *$K'/K$ is finite, we have*

$$\text{(14)} \qquad \text{r.rank}\ v+\dim_{\mathfrak{o}} v \leqq s+\text{tr.d.}\ K'/K.$$

For the proof it is only necessary to apply first Lemma 1 to $v$ and the restriction $v_0$ of $v$ to $K$, and then Proposition 2 to $v_0$ and $\mathfrak{o}$.

In either Proposition 1 or Proposition 2 we *may* have the equality sign, i.e., either the rank of $v$ or the rational rank of $v$ may have its maximum value $\dim(\mathfrak{o})-\dim_{\mathfrak{o}} v$. Since r.rank $v\geqq$ rank $v$, it follows by Proposition 2 that if rank $v=\dim(\mathfrak{o})-\dim_{\mathfrak{o}} v$ then also r.rank $v=\dim(\mathfrak{o})-\dim_{\mathfrak{o}} v$. Therefore, information about valuations $v$ for which the rational rank takes its maximum value $\dim(\mathfrak{o})-\dim_{\mathfrak{o}} v$ will yield also information about valuations $v$ for which the rank takes that maximum value. The results proved below deal precisely with the case in which either r.rank or rank $v$ has its maximum value. First we give the following definition:

Let $\Gamma$ be an ordered (additive) abelian group, of finite rank $r$, and let $(0)=\Gamma_0<\Gamma_1<\cdots<\Gamma_{r-1}$ be the isolated subgroups of $\Gamma$. Then $\Gamma$ is said to be *an integral direct sum* if and only if each group $\Gamma_i/\Gamma_{i-1}$, $i=1, 2, \cdots, r$ $(\Gamma_r=\Gamma)$, is a finite direct sum of cyclic groups. Note that if $\Gamma$ is of rank 1, so that $\Gamma$ is therefore a subgroup of the additive group of real numbers, then $\Gamma$ is an integral direct sum if and only if it is a direct sum of cyclic subgroups, i.e., if and only if there exist real numbers $\tau_1, \tau_2, \cdots, \tau_\rho$ in $\Gamma$ which are rationally independent and such that every element of $\Gamma$ is a linear combination of the $\tau$'s, with integral coefficients (in that case the rational rank of $\Gamma$ is $\rho$). If $r>1$, we know that the rational rank of $\Gamma$ is the sum of the rational ranks of the $r$ groups $\Gamma_i/\Gamma_{i-1}$. It follows that if $\Gamma$ is an integral direct sum and if, furthermore, rank $\Gamma$=r.rank $\Gamma$, *then each of the groups* $\Gamma_i/\Gamma_{i-1}$ *is cyclic, hence is discrete, of rank* 1, *and consequently* $\Gamma$ *is discrete.* We shall make use of this remark in Proposition 3.

We first prove the following complement to Lemma 1.

LEMMA 2. *If in* (13) *of Lemma* 1 *the equality holds and if furthermore* $K$ *is finitely generated over* $K_0$, *then* $\Gamma/\Gamma_0$ *is an integral direct sum and* $\Delta$ *is finitely generated over* $\Delta_0$ (*here* $\Gamma_0$ *denotes the value group of* $v_0$).

PROOF. We use the notations of the proof of Lemma 1. In the transition from $v_0$ to $v'$ there is no change in the value group (whence $\Gamma'=\Gamma_0$), and the residue field $\Delta'$ of $v'$ is a purely transcendental extension of $\Delta_0$ (of transcendence degree $h$). Since the equality holds in (13), we have that r.rank $\Gamma/\Gamma_0=g-h$. If we set $\rho=g-h$ and fix $\rho$ elements $y_1, y_2, \cdots, y_\rho$ in $K$ such that $v(y_1), v(y_2), \cdots; v(y_\rho)$ are rationally independent mod $\Gamma_0$, then $y_1, y_2, \cdots, y_\rho$ are algebraically independent over $K'$ (VI, § 10, Lemma, p. 50) and it is immediately seen that the restriction of $v$ to the field $K^\star=K'(y_1, y_2, \cdots, y_\rho)$ is a valuation $v^\star$ having the following two properties: (a) if $\Gamma^\star$ is the value group of $v^\star$ then $\Gamma^\star/\Gamma_0$ is an integral direct sum; (b) the residue field of $v^\star$ coincides with the residue field $\Delta'$ of $v'$. Now, $K$ is a finite algebraic extension of $K^\star$. Hence, also $\Gamma/\Gamma_0$ is an integral direct sum (compare with proof of Theorem 36 in VI, § 15), and the residue field of $v$ is a finite algebraic extension of $\Delta'$. Q.E.D.

PROPOSITION 3. *Let* $\mathfrak{o}$ *be a local domain,* $K$ *its quotient field,* $K'$ *a finitely generated extension of* $K$ *and* $v$ *a valuation of* $K'$ *which dominates* $\mathfrak{o}$. *If* r.rank $v+\dim_{\mathfrak{o}} v=\dim(\mathfrak{o})+$tr.d. $K'/K$, *then the value group* $\Gamma$ *of* $v$ *is an integral direct sum, and the residue field* $\Delta_v$ *of* $v$ *is finitely generated over the field of quotients* $F$ *of* $\mathfrak{o}/\mathfrak{m}$.

PROOF. We first achieve a reduction to valuations of rank 1. Let $r=$rank $v>1$ and assume that the proposition is true for valuations of rank$<r$. Let $v=v_1\circ\bar{v}$. With the same notations as in Case 3 of the proof of Proposition 2 and setting $\mathfrak{o}_1=\mathfrak{o}_{\mathfrak{p}_1}$, we have, by the corollary of that proposition:

$$\text{(15)}\qquad \text{r.rank } v_1+\dim_{\mathfrak{o}_1} v_1 \leqq \dim(\mathfrak{o}_1)+\text{tr.d. } K'/K;$$

$$\text{(15')}\qquad \text{r.rank } \bar{v}+\dim_{\bar{\mathfrak{o}}} \bar{v} \leqq \dim(\bar{\mathfrak{o}})+\dim_{\mathfrak{o}_1} v_1.$$

Hence, by addition, and observing that $\dim_{\bar{\mathfrak{o}}} \bar{v}=\dim_{\mathfrak{o}} v$:

$$\text{r.rank } v+\dim_{\mathfrak{o}} v \leqq \dim(\mathfrak{o}_1)+\dim(\bar{\mathfrak{o}})+\text{tr.d. } K'/K.$$

Since dim $(\mathfrak{o}_1)+\dim(\bar{\mathfrak{o}})\leqq \dim(\mathfrak{o})$, it follows that in (15) and (15′) we must have the equality signs (and also—incidentally—$\dim(\mathfrak{o}_1)+\dim(\bar{\mathfrak{o}})=\dim(\mathfrak{o})$, i.e., $h(\mathfrak{p}_1)+h(\mathfrak{m}/\mathfrak{p}_1)=h(\mathfrak{m})$). Therefore, by our induction hypothesis, applied to $v_1$, we have: (a) the value group of $v_1$ must be an integral direct sum, and (b) the residue field $\Delta_{v_1}$ is a finitely generated extension of the quotient field of $\bar{\mathfrak{o}}$. Since $\bar{v}$ is a valuation of $\Delta_{v_1}$, it follows from (b) and our induction hypothesis, that also the value group of $\bar{v}$ is an integral direct sum and that the residue field of $\bar{v}$ is finitely generated over the field of quotients of $\bar{\mathfrak{o}}/\bar{\mathfrak{m}}$. Since $\bar{v}$ and $v$ have the same residue field and since $\mathfrak{o}/\mathfrak{m}=\bar{\mathfrak{o}}/\bar{\mathfrak{m}}$, the proposition follows for the given valuation $v$.

We assume now that rank $v=1$. Next we achieve easily a reduction to the case $K=K'$. For let $v_0$ be the restriction of $v$ to $K$. Using the assumption of our proposition and applying Lemma 1 and Proposition 2 to the valuations $v$ and $v_0$ respectively, we find that (a) r.rank $v-$ r.rank $v_0=$ tr.d. $K'/K-$ tr.d. $\Delta/\Delta_0$, and (b) r.rank $v_0=s-$ tr.d. $\Delta_0/F$. If we assume the truth of our proposition in the case $K'=K$, it then follows from (b) that the value group $\Gamma_0$ of $v_0$ is an integral direct sum and that $\Delta_0$ is finitely generated over $F$. From (a) it follows, in view of Lemma 2, that $\Gamma/\Gamma_0$ is an integral direct sum and that $\Delta$ is finitely generated over $\Delta_0$. This shows that the proposition holds for the given valuation $v$.

We can therefore assume that $K'=K$ and that rank $v=1$.

Our next preliminary step is a reduction to the case in which $\dim_{\mathfrak{o}} v=0$. For assume that $\dim_{\mathfrak{o}} v>1$. Choose an element $t$ in $K$ such that the $v$-residue of $t$ is transcendental over $F$ (where $F=$ quotient field of $\mathfrak{o}/\mathfrak{m}$) and set $R_1=\mathfrak{o}[t]$. If $\mathfrak{p}_1$ is the center of $v$ in $R_1$ and if we set $\mathfrak{o}_1=R_{1\mathfrak{p}_1}$, then $\mathfrak{o}_1$ dominates $\mathfrak{o}$ and $v$ dominates $\mathfrak{o}_1$. Since the $\mathfrak{p}_1$-residue of $t$ is transcendental over $F$, we know (Appendix 1, Proposition 1, part B) that $\dim(\mathfrak{o}_1)\leqq\dim(\mathfrak{o})-1$. Now, r.rank $v+\dim_{\mathfrak{o}} v=\dim(\mathfrak{o})$, and $\dim_{\mathfrak{o}} v=1+\dim_{\mathfrak{o}_1} v$. Hence r.rank $v+\dim_{\mathfrak{o}_1} v=\dim \mathfrak{o}-1\geqq\dim(\mathfrak{o}_1)$, and consequently r.rank $v+\dim_{\mathfrak{o}_1} v=\dim(\mathfrak{o}_1)$. If we assume that the proposition is true for $v$ and $\mathfrak{o}_1$ (note that $\dim_{\mathfrak{o}_1} v=\dim_{\mathfrak{o}} v-1$) we may conclude that the value group of $v$ is an integral direct sum and that $\Delta_v$ is finitely generated over $F_1$ (=quotient field of $\mathfrak{o}_1/\mathfrak{m}_1$). Since $F_1$ is a simple transcendental extension of $F$, the truth of the proposition is established for $v$ and $\mathfrak{o}$.

We may therefore assume that $K'=K$, rank $v=1$ and $\dim_{\mathfrak{o}} v=0$. The assumption of our proposition is now that r.rank $v=\dim(\mathfrak{o})$ $(=s)$. We shall use the notation of the proof of Case 1 of Proposition 2, and as $q$ we now take the integer $s$. To prove that $\Gamma$ is an integral direct sum we have only to show that the subgroup $\Gamma_0$ of $\Gamma$ which is generated by $\tau_1, \tau_2, \cdots, \tau_s$ has finite index (compare with the proof of Theorem 36 in VI, § 15). We shall assume the contrary and show that this leads to a contradiction.

Under the assumption that the index $\Gamma/\Gamma_0$ is infinite we can find an infinite sequence of elements $\gamma_1, \gamma_2, \cdots, \gamma_\nu, \cdots$ in $v\{\mathfrak{m}\}$ such that for each $\nu$ it is true that $\gamma_\nu$ does *not* belong to the group $\Gamma_{\nu-1}$ generated by $\tau_1, \tau_2, \cdots, \tau_s, \gamma_1, \gamma_2, \cdots, \gamma_{\nu-1}$ (note that the elements of $v\{\mathfrak{m}\}$ generate $\Gamma$). Let $q_\nu$ be the least positive integer such that $q_\nu\gamma_\nu \in \Gamma_{\nu-1}$ ($q_\nu > 1$; such an integer exists since every element of $\Gamma$ is rationally dependent on $\tau_1, \tau_2, \cdots, \tau_s$). Let $\delta_\nu = [q_1\gamma_1 + q_2\gamma_2 + \cdots + q_\nu\gamma_\nu]$ ([ ] means "integral part"). We consider the elements $\alpha$ of $\Gamma$ which are of the form $\alpha = j_1\tau_1 + j_2\tau_2 + \cdots + j_s\tau_s + j_{s+1}\gamma_1 + \cdots + j_{s+\nu}\gamma_\nu$, where the $j$'s are non-negative integers and $j_{s+1} < q_1, j_{s+2} < q_2, \cdots, j_{s+\nu} < q_\nu$. These elements belong to $v\{\mathfrak{m}\}$, and distinct sets of integers $(j_1, j_2, \cdots, j_{s+\nu})$ give rise to distinct elements of $v\{\mathfrak{m}\}$. If $j_1\tau_1 + \cdots + j_s\tau_s \leqq n - \delta_\nu$, then $\alpha \leqq n$ and hence the valuation ideal $\mathfrak{A}_\alpha$ is contained in $\mathfrak{A}_{i(n)}$, where $\alpha_{i(n)} = n$. It follows that for any $n$ we have $\lambda(\mathfrak{m}^n) \geqq q_1q_2 \cdots q_\nu\sigma_{n-\delta_\nu}$ [see (3) and the definition (4) of $\sigma_n$]. By (6) there exists a polynomial $P_s(n)$, of degree $s$, such that $\sigma_n \geqq P_s(n)$, for all $n$. We therefore have:

$$(16) \qquad q_1q_2 \cdots q_\nu P_s(n - \delta_\nu) \leqq \lambda(\mathfrak{m}^n).$$

For fixed $\nu$, the leading coefficient of $P_s(n - \delta_\nu)$ is the same as the leading coefficient $c$ of $P_s$. Since $\lambda(\mathfrak{m}^n)$ is itself a polynomial of degree $s$, for $n$ large, its leading coefficient $c'$ must therefore satisfy the inequality $q_1q_2 \cdots q_\nu c \leqq c'$. Since this is true for every $\nu$ and since $q_1q_2 \cdots q_\nu$ tends to infinity with $\nu$, we have a contradiction. Thus $\Gamma$ is an integral direct sum.

There remains to prove that the residue field $\Delta_v$ of $v$ is finitely generated over the quotient field $F$ of $\mathfrak{o}/\mathfrak{p}$, or equivalently (since we are dealing with case in which $\Delta_v$ is algebraic over $F$), that $[\Delta_v : F] < \infty$. Assuming the contrary, we shall show that for *each integer* $N$ there exists an integer $N_0$ (depending on $N$) such that

$$(17) \qquad \lambda(\mathfrak{m}^n) \geqq NP_s(n - N_0),$$

for all large $n$, and this again contradicts the fact that, for large $n$, $\lambda(\mathfrak{m}^n)$ itself is a polynomial of degree $s$ in $n$.

Since $[\Delta_v : F] = \infty$, by assumption, given any positive integer $N$ we can find, in $\Delta_v$, $N$ elements $\zeta_1, \zeta_2, \cdots, \zeta_N$ which are linearly independent over $F$. We fix elements $\omega_1, \omega_2, \cdots, \omega_N$ in $K$ whose $v$-residues are $\zeta_1, \zeta_2, \cdots, \zeta_N$ respectively. Suppose that the $\omega$'s are written as quotients of elements $\mathfrak{o}$, with common denominator $x_0 : \omega_i = x_i/x_0$, and let $v(x_0) = \alpha$. Using the notations of the proof of Case 1 of Proposition 2, let $\mathfrak{A}_\beta$ be the valuation ideal in $\mathfrak{o}$ which is the immediate successor of $\mathfrak{A}_\alpha$. We assert that

$$(18) \qquad \lambda(\mathfrak{A}_\beta) \geqq \lambda(\mathfrak{A}_\alpha) + N.$$

In fact, $\mathfrak{A}_\alpha/\mathfrak{A}_\beta$ is a vector space over $F$ (since $\mathfrak{m}\mathfrak{A}_\alpha \subset \mathfrak{A}_\beta$) and clearly $\lambda(\mathfrak{A}_\beta) - \lambda(\mathfrak{A}_\alpha) = \dim \mathfrak{A}_\alpha/\mathfrak{A}_\beta$. Now, we have $v(x_i) = \alpha$, $i = 1, 2, \cdots, N$, whence $x_i \in \mathfrak{A}_\alpha$. If we have a relation of the form $a_1x_1 + \cdots + a_Nx_N \in \mathfrak{A}_\beta$, with $a_i \in \mathfrak{o}$, then $v(a_1\omega_1 + \cdots + a_N\omega_N) > 0$, and hence, if $\bar{a}_i$ denotes the $\mathfrak{m}$-residue of $a_i$, then $\bar{a}_1\zeta_1 + \cdots + \bar{a}_N\zeta_N = 0$. Therefore $\bar{a}_1 = \cdots = \bar{a}_N = 0$, i.e., $a_1, a_2, \cdots, a_N \in \mathfrak{m}$, and this shows that the $\mathfrak{A}_\beta$-residues of $x_1, x_2, \cdots, x_N$ are linearly independent vectors of the space $\mathfrak{A}_\alpha/\mathfrak{A}_\beta$. This proves (18).

We now *fix* some element $\gamma$ in $v\{\mathfrak{m}\}$ such that the $\omega$'s admit a representation of the form $\omega_i = z_i/z_0$ with $v(z_0) = \gamma$ and all the $z$'s in $\mathfrak{o}$. We now note that this property of $\gamma$ is shared by any element of $v\{\mathfrak{m}\}$ which is of the form $\gamma + \alpha_\nu$, $\alpha_\nu \in v\{\mathfrak{m}\}$, for we have only to take an element $z$ in $\mathfrak{m}$ such that $v(z) = \alpha_\nu$ and write $\omega_i = z_iz/z_0z$. Since $\gamma + \alpha_\nu + 1 > \gamma + \alpha_\nu$, it follows from (18) (as applied to $\alpha = \gamma + \alpha_\nu$) that

$$\lambda(\mathfrak{A}_{\gamma+\alpha_\nu+1}) \geqq \lambda(\mathfrak{A}_{\gamma+\alpha_\nu}) + N, \quad \nu = 0, 1, \cdots; \alpha_0 = 0.$$

Therefore

$$(19) \qquad \lambda(\mathfrak{A}_{\gamma+\alpha_\nu}) \geqq N\nu, \text{ all } \nu \geqq 1.$$

Let $N_0 = [\gamma] + 1$ and let (in the notations of the proof of Case 1 of Proposition 2) $\sigma_{n-N_0}$ denote the number of non-negative solutions $(j_1, j_2, \cdots, j_s)$ of the inequality $j_1\tau_1 + j_2\tau_2 + \cdots + j_s\tau_s \leqq n - N_0$. For any such solution we have $j_1\tau_1 + j_2\tau_2 + \cdots + j_s\tau_s \leqq n - \gamma$, i.e., the element $\alpha = j_1\tau_1 + \cdots + j_s\tau_s$ of $v\{\mathfrak{m}\}$ is such that $\gamma + \alpha \leqq n$. Thus the number of $\alpha_\nu$'s in $v\{\mathfrak{m}\}$ such that $\gamma + \alpha_\nu \leqq n$ is at least equal to $\sigma_{n-N_0}$, and since for each such $\alpha_\nu$ we have $\mathfrak{m}^n \subset \mathfrak{A}_{\gamma+\alpha_\nu}$, it follows from (19) that

$$\lambda(\mathfrak{m}^n) \geqq N\sigma_{n-N_0}.$$

This establishes (17) and completes the proof of the proposition.

COROLLARY 1. *If the assumption* r.rank $v + \dim_\mathfrak{o} v = \dim(\mathfrak{o}) +$ tr.d. $K'/K$ *of Proposition* 3 *is replaced by the stronger assumption* rank $v + \dim_\mathfrak{o} v = \dim(\mathfrak{o}) +$ tr.d. $K'/K$ (*the other assumptions remaining the same*) *then $v$ is discrete and $\Delta_v$ is finitely generated over $F$.*

This follows from Proposition 3 and from the remark made just before the statement of Lemma 2.

A valuation $v$ of the quotient field $K$ of $\mathfrak{o}$ is said to be a *prime* $\mathfrak{o}$-*divisor of* $K$ if $v$ dominates $\mathfrak{o}$ and if $\dim_{\mathfrak{o}} v = \dim \mathfrak{o} - 1$.

COROLLARY 2. *A prime* $\mathfrak{o}$-*divisor* $v$ *of* $K$ *is a discrete, rank* 1 *valuation, and the residue field of* $v$ *is finitely generated over the field of quotients* $F$ *of* $\mathfrak{o}/\mathfrak{m}$.

Obvious.

# APPENDIX 3

## VALUATION IDEALS

Let $R$ be an integral domain and $K$ the quotient field of $R$. The valuations of $K$ which are non-negative on $R$ lead to a special class of ideals in $R$ which we shall call *valuation ideals*. Their definition is as follows:

DEFINITION. *An ideal $\mathfrak{A}$ in $R$ is a valuation ideal if it is the intersection of $R$ with an ideal of a valuation ring $R_v$ containing $R$; if $v$ is the corresponding valuation we say that $\mathfrak{A}$ is a valuation ideal associated with the valuation $v$, or briefly: that $\mathfrak{A}$ a $v$-ideal in $R$.*

Let $v$ be a valuation of an extension field $K'$ of $K$ and let $v_0$ be the restriction of $v$ to $K$. It is clear that if $\mathfrak{A}_v$ is an ideal in $R_v$ and $\mathfrak{A} = \mathfrak{A}_v \cap R$ is a $v$-ideal in $R$, then $\mathfrak{A} = \mathfrak{A}_{v_0} \cap R$, where $\mathfrak{A}_{v_0} = \mathfrak{A}_v \cap R_{v_0}$, and hence $\mathfrak{A}$ is also a $v_0$-ideal. Hence in studying valuation ideals in $R$ we may, without loss of generality, restrict ourselves to valuations $v$ of the quotient field $K$ of $R$.

If $v$ is a valuation, non-negative on $R$, and $\mathfrak{A}$ is an ideal in $R$, then the following statements are equivalent:

(a) *$\mathfrak{A}$ is a $v$-ideal.*

(b) *If $a, b \in R$, $a \in \mathfrak{A}$ and $v(b) \geqq v(a)$, then $b \in \mathfrak{A}$.*

(c) *The following relation is satisfied*

$$R_v\mathfrak{A} \cap R = \mathfrak{A}. \tag{1}$$

That (a) implies (b) is immediate, for if $\mathfrak{A} = \mathfrak{A}_v \cap R$, where $\mathfrak{A}_v$ is an ideal in $R_v$, then $b = \frac{b}{a} \cdot a \in R_v \cdot a \subset \mathfrak{A}_v$. Now, assume (b). Any element $b$ of $R_v\mathfrak{A}$ can be written in the form $b = a_1c_1 + a_2c_2 + \cdots + a_nc_n$, with $a_i \in \mathfrak{A}$ and $c_i \in R_v$. If $v(a_i) = \min\{v(a_1), v(a_2), \cdots, v(a_n)\}$ then $v(b) \geqq v(a_i)$, and thus if $b \in R$ then $b \in \mathfrak{A}$. This proves (c). That (c) implies (a) follows from the definition of $v$-ideals.

If $\mathfrak{A}$ is an arbitrary ideal in $R$ and $v$ is a valuation of $K$ which is non-negative on $R$, then the ideal $R_v\mathfrak{A} \cap R$ is, of course, a $v$-ideal in $R$, and is the smallest $v$-ideal in $R$ which contains the given ideal $\mathfrak{A}$; it can be characterized as being the set of all elements $b$ of $R$ such that $v(b) \geqq v(a)$ for some $a$ in $\mathfrak{A}$.

Since the set of ideals in $R_v$ is totally ordered by set-theoretic inclusion (VI, § 3, Theorem 3) it follows that for a given valuation $v$, non-negative on $R$, the set of $v$-ideals in $R$ is also totally ordered by set-theoretic inclusion. In the special case of a noetherian domain $R$ this set is even well-ordered in view of the "maximum condition" in $R$ (see Vol. I, p. 156). We shall derive later on in this section some results concerning the ordinal type of the set of $v$-ideals in a noetherian domain $R$ (for a given $v$).

We shall now discuss some examples.

EXAMPLE 1. *Any prime ideal $\mathfrak{p}$ in $R$ is a valuation ideal.* This is obvious if $\mathfrak{p} = (0)$. If $\mathfrak{p}$ is a proper prime ideal in $R$ then the statement follows from the existence of valuations $v$ which are centered at $\mathfrak{p}$, for if $v$ is any such valuation and if $\mathfrak{M}_v$ is the maximal ideal of $R_v$ then $\mathfrak{M}_v \cap R = \mathfrak{p}$. We see here incidentally that *a valuation ideal in $R$ may be associated with more than one valuation $v$.*

EXAMPLE 2. Let $R$ be a Dedekind domain. *Then every primary ideal $\mathfrak{q}$ is a valuation ideal, and conversely.* For if $\mathfrak{p} = \sqrt{\mathfrak{q}}$ (and leaving aside the trivial case $\mathfrak{q} = (0)$), then $\mathfrak{q} = \mathfrak{p}^n$, for some $n \geqq 1$. Thus, if $v_{\mathfrak{p}}$ is the $\mathfrak{p}$-adic valuation of $K$ which is defined by the prime ideal $\mathfrak{p}$ we have that $\mathfrak{q}$ is the set of all elements $x$ of $R$ such that $v_{\mathfrak{p}}(x) \geqq n$, showing that $\mathfrak{q}$ is a valuation ideal, by the above criterion (b). The converse is also obvious, since every valuation $v$ of $K$ which is non-negative on $R$ is either the trivial valuation or is a $\mathfrak{p}$-adic valuation $v_{\mathfrak{p}}$ defined by a prime ideal $\mathfrak{p}$ of $R$, and in the latter case the $v_{\mathfrak{p}}$-ideals in $R$ are the powers of $\mathfrak{p}$.

EXAMPLE 3. *In the general case not every primary ideal is a valuation ideal, and not every valuation ideal is primary.* For instance, let $R = k[X, Y]$ be a polynomial ring in two indeterminates, over a field $k$, and let $\mathfrak{A}$ be the ideal generated by $X^2$ and $Y^2$. Then $\mathfrak{A}$ is a primary ideal, with $(X, Y)$ as associated prime ideal. If $v$ is any valuation of $k(X, Y)$, non-negative on $R$, and if, say, $v(Y) \geqq v(X)$, then $v(XY) \geqq v(X^2)$, while $XY \notin \mathfrak{A}$. Thus $\mathfrak{A}$ is not a $v$-ideal. On the other hand, let $m$ and $n$ be positive integers and let $\mathfrak{A}$ be the ideal $X^n \cdot (X, Y^m)$. This ideal is not primary, and its associated prime ideals are $(X)$ and $(X, Y)$. We show that $\mathfrak{A}$ is a valuation ideal. The quotient ring of $k[X, Y]$ with respect to the prime ideal $(X)$ is a discrete valuation ring of rank 1. Let $v_1$ denote the corresponding valuation of $k(X, Y)$. Then $v_1$ is a one-dimensional valuation of $k(X, Y)$ (namely the prime divisor of $k(X, Y)/k$ whose center in the $(X, Y)$-plane is the line $X=0$ (see VI, § 14)). The residue field of $v_1$ is the field $k(Y)$ (or can be canonically identified with this field). Let $v_0$ be the valuation of $k(Y)$ which is non-negative in $k[Y]$ and has as center in $k[Y]$ the ideal $(Y)$ ($v_1$ is then the prime divisor of $k(Y)$ whose center on the line $X=0$ is the origin $Y=0$). Let $v=v_1 \circ v_0$ be the composite valuation of $k(X, Y)$ obtained by compounding $v_1$ with $v_0$. Then $v$ is a discrete, rank 2 valuation, and its value group can be identified with the set of all pairs of integers $(i, j)$, ordered lexicographically (see VI, § 10, Remark (A) concerning discrete ordered groups of finite rank). For a suitable identification we may assume that $v(X)=(1, 0)$ and $v(Y)=(0, 1)$. Then our ideal $X^n \cdot (X, Y^m)$ consists of all elements $f$ of $R$ such that $v(f) \geqq (n, m)$, and is therefore a $v$-ideal.

However, the following is true quite generally: *the radical* $\sqrt{\mathfrak{A}}$ *of a valuation ideal* $\mathfrak{A}$ *is prime, and if* $\mathfrak{A}$ *is associated with a given valuation* $v$ *then also the prime ideal* $\sqrt{\mathfrak{A}}$ *is a valuation ideal associated with* $v$. In fact, if $xy \in \sqrt{\mathfrak{A}}$, so that $(xy)^n \in \mathfrak{A}$ for some $n \geqq 1$, then assuming that, say, $v(y) \geqq v(x)$, we have $v(y^{2n}) \geqq v(x^n y^n)$, whence $y^{2n} \in \mathfrak{A}$, by criterion (b) of $v$-ideals. This shows that $\sqrt{\mathfrak{A}}$ is prime. Furthermore, if $a^n \in \mathfrak{A}$, and $b \in R$ is such that $v(b) \geqq v(a)$, then $v(b^n) \geqq v(a^n)$, whence also $b^n$ is in $\mathfrak{A}$, i.e., $b \in \sqrt{\mathfrak{A}}$, showing that $\sqrt{\mathfrak{A}}$ is a $v$-ideal. We include this result in the following lemma and we leave it to the reader to prove the other assertions of that lemma (the proofs being straightforward):

Lemma 1. *If* $\mathfrak{A}$, $\mathfrak{B}$ *are* $v$*-ideals in* $R$ *and* $\mathfrak{C}$ *is an arbitrary ideal in* $R$, *then* $\sqrt{\mathfrak{A}}$, $\mathfrak{A} \cap \mathfrak{B}$ and $\mathfrak{A}:\mathfrak{C}$ *are* $v$*-ideals.*

Since $\sqrt{\mathfrak{A}}$ is prime, it follows that if a $v$-ideal admits a primary (irredundant) representation then only one prime ideal of $\mathfrak{A}$ is isolated. We now prove the following proposition:

Proposition 1. *If a* $v$*-ideal* $\mathfrak{A}$ (*associated with a given valuation* $v$) *in* $R$ *admits an irredundant primary decomposition* $\mathfrak{A}=\mathfrak{q}_1 \cap \mathfrak{q}_2 \cap \cdots \cap \mathfrak{q}_h$, *then the prime ideals* $\mathfrak{p}_i=\sqrt{\mathfrak{q}_i}$ *form a descending chain (in a suitable order) and are themselves* $v$*-ideals (associated with the given valuation* $v$). *If, say,* $\mathfrak{p}_1 > \mathfrak{p}_2 > \cdots > \mathfrak{p}_h$ *then also the ideals* $\mathfrak{q}_i \cap \mathfrak{q}_{i+1} \cap \cdots \cap \mathfrak{q}_h$ ($i=1, 2, \cdots, h$; i.e., *the isolated components of* $\mathfrak{A}$) *are* $v$*-ideals.*

Proof. If $\mathfrak{p}$ is a prime ideal of $\mathfrak{A}$ there exists an element $c$ in $R$ such that $c \notin \mathfrak{A}$ and $\mathfrak{A}:(c)$ is primary for $\mathfrak{p}$ (Vol. I, Ch. IV, § 5, Theorem 6). By Lemma 1, $\mathfrak{A}:(c)$ is a $v$-ideal and hence, again by Lemma 1, $\mathfrak{p}$ is a $v$-ideal. Since the set of all $v$-ideals in $R$ (associated with the given valuation $v$) is totally ordered by set-theoretic inclusion, we must have $\mathfrak{p}_1 > \mathfrak{p}_2 > \cdots > \mathfrak{p}_h$ for a suitable labeling of the prime ideals $\mathfrak{p}_i$ of $\mathfrak{A}$.

The last assertion of the proposition follows from Lemma 1 and from the relation $\mathfrak{A}:(\mathfrak{q}_1 \cap \mathfrak{q}_2 \cap \cdots \cap \mathfrak{q}_{i-1}) = \mathfrak{q}_i \cap \mathfrak{q}_{i+1} \cap \cdots \cap \mathfrak{q}_h$.

We now fix once and for always a non-trivial valuation $v$ of $K$ which is non-negative on $R$ and we study the totally ordered set of valuation ideals in $R$ which are associated with $v$. We shall find it often convenient to replace $R$ by the quotient ring $R_\mathfrak{p}$, where $\mathfrak{p}$ is the center of $v$ in $R$. This will not affect essentially the valuation ideals of $R$, in view of the following lemma:

LEMMA 2. *If $\mathfrak{A}$ is any $v$-ideal in $R$ then the extension $\mathfrak{A}^e$ of $\mathfrak{A}$ in $R_\mathfrak{p}$ is a $v$-ideal in $R_\mathfrak{p}$, and we have $\mathfrak{A} = \mathfrak{A}^{ec}$. The correspondence $\mathfrak{A} \to \mathfrak{A}^e$ maps in* (1, 1) *fashion the set of $v$-ideals in $R$ onto the set of $v$-ideals in $R_\mathfrak{p}$. If $\mathfrak{A} = \mathfrak{q}_1 \cap \mathfrak{q}_2 \cap \cdots \cap \mathfrak{q}_h$ is an irredundant primary decomposition of $\mathfrak{A}$ then $\mathfrak{A}^e = \mathfrak{q}_1^e \cap \mathfrak{q}_2^e \cap \cdots \cap \mathfrak{q}_h^e$ is an irredundant decomposition of $\mathfrak{A}^e$.*

PROOF. If $x \in \mathfrak{A}^e$ we have $x = y/z$, where $y, z \in R$, $y \in \mathfrak{A}$, $z \notin \mathfrak{p}$. Then $v(z) = 0$ and $v(x) = v(y)$. If $x' = y'/z' \in R_\mathfrak{p}$, where $y', z' \in R$ and $z' \notin \mathfrak{p}$, and if $v(x') \geqq v(x)$, then $v(y') \geqq v(y)$ since $v(x') = v(y')$. Therefore $y' \in \mathfrak{A}$ and $x' \in \mathfrak{A}^e$. This shows that $\mathfrak{A}^e$ is a $v$-ideal in $R_\mathfrak{p}$. We have $\mathfrak{A} \subset \mathfrak{A}^{ec}$, and, on the other hand, we have just seen that if $x$ is any element of $\mathfrak{A}^e$ we have $v(x) = v(y)$ for some $y$ in $\mathfrak{A}$. Since $\mathfrak{A}$ is a $v$-ideal in $R$ this implies that $\mathfrak{A}^{ec} \subset \mathfrak{A}$, whence $\mathfrak{A}^{ec} = \mathfrak{A}$. If $\mathfrak{A}'$ is any $v$-ideal in $R_\mathfrak{p}$, $\mathfrak{A} = \mathfrak{A}'^c$ is a $v$-ideal in $R$, and for every $x$ in $\mathfrak{A}'$ there exists an element $y$ in $\mathfrak{A}$ such that $v(x) = v(y)$. This implies at once that the two $v$-ideals $\mathfrak{A}^e$ and $\mathfrak{A}'$ must coincide, thus $\mathfrak{A}'^{ce} = \mathfrak{A}'$. The last part of the Lemma follows from Vol. I, Ch. IV, § 10, Theorem 17 and from Proposition 1 by observing that the prime ideals $\mathfrak{p}_i$ of Proposition 1 are all contained in $\mathfrak{p}$.

LEMMA 3. *If $v$ has rank* 1 *and $\mathfrak{p}$ is the center of $v$ in $R$ then every $v$-ideal in $R$* (*other than* (0) *and* $R$) *is primary for $\mathfrak{p}$. If $R$ is noetherian then these ideals form a simple infinite descending chain having zero intersection.*

PROOF. Since every proper ideal in the valuation ring $R_v$ is primary, with $\mathfrak{M}_v$ as associated prime ideal, and since $\mathfrak{p} = \mathfrak{M}_v \cap R$, the first assertion of the lemma is obvious. From this it also follows that if $R$ is noetherian every proper $v$-ideal $\mathfrak{q}$ in $R$ is preceded by at most a finite number of $v$-ideals. Furthermore, $(R_v \mathfrak{q}\mathfrak{p}) \cap R$ is a $v$-ideal strictly contained in $\mathfrak{q}$, and this shows that the sequence $\{\mathfrak{q}_i\}$ of $v$-ideals (different from (0) and $R$) is infinite. The intersection of the $\mathfrak{q}_i$ must be the zero ideal (it is true, quite generally, for a valuation $v$ of any rank, that the intersection of all the $v$-ideals different from (0) is the zero ideal, because if $0 \neq x \in R$ and $\mathfrak{p}$ is the center of $v$ in $R$ then $v(x) < v(y)$ for every element $y$ of $R_v(x\mathfrak{p})$ and hence $x$ is not contained in the $v$-ideal

$R_v(x\mathfrak{p}) \cap R$). This also shows that the sequence $\{\mathfrak{q}_i\}$ of $v$-ideals (different from (0) and $R$) is infinite.

We now restrict ourselves to *noetherian* domains $R$ and we study the well-ordered set of valuation ideals of a valuation $v$ of rank $r > 1$, non-negative on $R$ ($r$ is necessarily finite; see Appendix 2). We denote by $\mathfrak{p}$ the center of $v$ in $R$. Let $v = v_1 \circ \bar{v}$, where $v_1$ is of rank $r-1$ and $\bar{v}$ a rank one valuation of the residue field of $v_1$. Since every ideal in $R_{v_1}$ is also an ideal in $R_v$ it follows that the *$v_1$-ideals in $R$ are also $v$-ideals.*

LEMMA 4. *If $\mathfrak{A}$ is a $v$-ideal (different from (0)) there exist two consecutive $v_1$-ideals $\mathfrak{B}_1$ and $\mathfrak{B}_2$ such that $\mathfrak{B}_1 \supset \mathfrak{A} > \mathfrak{B}_2$, and we have $\mathfrak{B}_1\mathfrak{p}^\rho \subset \mathfrak{A}$ for some integer $\rho \geqq 0$. The number of $v$-ideals between $\mathfrak{B}_1$ and $\mathfrak{A}$ is finite.*

PROOF. Since $R$ is noetherian, every ideal $\mathfrak{A}$ in $R$ is finitely generated and therefore $v$ admits, on $\mathfrak{A}$, a smallest value. As in Appendix 2, we denote this smallest value by $v(\mathfrak{A})$.

Let $\mathfrak{B}_2$ be the first (i.e., the largest) $v_1$-ideal which is a proper subideal of $\mathfrak{A}$. Let $\mathfrak{B}_1 = R_{v_1}\mathfrak{A} \cap R$. Then $\mathfrak{B}_1$ and $\mathfrak{B}_2$ are $v_1$-ideals in $R$ such that $\mathfrak{B}_1 \supset \mathfrak{A} > \mathfrak{B}_2$. From the definition of $\mathfrak{B}_1$ it follows that $\mathfrak{B}_1$ is the smallest $v_1$-ideal which contains $\mathfrak{A}$. Hence there are no $v_1$-ideals between $\mathfrak{B}_1$ and $\mathfrak{B}_2$.

The value group $\Delta$ of $\bar{v}$ is an isolated subgroup of the value group $\Gamma$ of $v$, and $\Gamma/\Delta$ is the value group of $v_1$ (VI, § 10, Theorem 17). By the definition of $\mathfrak{B}_1$ we have $v_1(\mathfrak{B}_1) = v_1(\mathfrak{A})$, and since $\mathfrak{A} \subset \mathfrak{B}_1$, we have also $v(\mathfrak{A}) \geqq v(\mathfrak{B}_1)$. Hence $0 \leqq v(\mathfrak{A}) - v(\mathfrak{B}_1) \in \Delta$. We now consider the two possible cases: (1) $v(\mathfrak{p}) \notin \Delta$ and (2) $v(\mathfrak{p}) \in \Delta$.

In case (1) we have $v_1(\mathfrak{p}) > 0$ (in this case, $\mathfrak{p}$ is also the center of $v_1$ in $R$, since—on the one hand—the center $\mathfrak{p}_1$ of $v_1$ is the greatest prime ideal in $R$ such that $v_1(\mathfrak{p}_1) > 0$, and—on the other hand—$\mathfrak{p}_1$ must be contained in $\mathfrak{p}$ because $v$ is composite with $v_1$). We have therefore $v_1(\mathfrak{B}_1\mathfrak{p}) > v_1(\mathfrak{B}_1)$ and hence $\mathfrak{B}_2 \supset \mathfrak{B}_1\mathfrak{p}$, showing that $\mathfrak{A} > \mathfrak{B}_1\mathfrak{p}$.

In case (2) we have $v_1(\mathfrak{p}) = 0$ (in this case, the center of $v_1$ in $R$ is proper subideal $\mathfrak{p}_1$ of $\mathfrak{p}$), and hence $0 < v(\mathfrak{p}) \in \Delta$. Since we have also $0 \leqq v(\mathfrak{A}) - v(\mathfrak{B}_1) \in \Delta$ and since $\Delta$ has rank 1, there exists an integer $\rho \geqq 0$ such $v(\mathfrak{p}^\rho) \geqq v(\mathfrak{A}) - v(\mathfrak{B}_1)$. For such an integer $\rho$ we have $v(\mathfrak{B}_1\mathfrak{p}^\rho) \geqq v(\mathfrak{A})$, showing that $\mathfrak{B}_1\mathfrak{p}^\rho \subset \mathfrak{A}$ (since $\mathfrak{A}$ is a $v$-ideal).

At this stage we replace $R$ by $R_\mathfrak{p}$ (see Lemma 2) and we therefore assume that $R$ is a local domain, with $\mathfrak{p}$ as maximal ideal.

If $\mathfrak{B}$ is any ideal in $R$ and $q$ is any integer $\geqq 1$, then the ideals between $\mathfrak{B}$ and $\mathfrak{B}\mathfrak{p}^q$ correspond in (1, 1) fashion to the $R/\mathfrak{p}^q$-submodules of $\mathfrak{B}/\mathfrak{B}\mathfrak{p}^q$, where $\mathfrak{B}/\mathfrak{B}\mathfrak{p}^q$ is to be considered as a module over $R/\mathfrak{p}^q$. Now $R/\mathfrak{p}^q$ is a ring satisfying both chain conditions (since $R$ is noetherian and $\mathfrak{p}$ is maximal in $R$; see Vol. I, Ch. IV, § 2, Theorem 2). Since $\mathfrak{B}/\mathfrak{B}\mathfrak{p}^q$ is a finite $(R/\mathfrak{p}^q)$ module, it has a composition series (Vol. I, Ch. III, § 10, Theorem 18). It follows that any descending chain of ideals between $\mathfrak{B}$ and $\mathfrak{B}\mathfrak{p}^q$ is finite. In particular, there can only be a finite number of $v$-ideals between $\mathfrak{B}_1$ and $\mathfrak{A}$, since $\mathfrak{A} \supset \mathfrak{B}_1\mathfrak{p}^\rho$. This completes the proof of Lemma 4.

COROLLARY. *If $\mathfrak{B}_1$ and $\mathfrak{B}_2$ are two consecutives $v_1$-ideals* ($\mathfrak{B}_1 > \mathfrak{B}_2$) *then the $v$-ideals $\mathfrak{A}$ such that $\mathfrak{B}_1 \supset \mathfrak{A} > \mathfrak{B}_2$ are finite in number if $\mathfrak{p}$ is also the center of $v_1$, and form a simple infinite sequence in the contrary case.*

For, if $\mathfrak{p}$ is the center of $v_1$ then we have seen that $\mathfrak{A} \supset \mathfrak{B}_1\mathfrak{p}$ if $\mathfrak{B}_1 \supset \mathfrak{A} > \mathfrak{B}_2$. If $\mathfrak{p}$ is not the center of $v_1$, then the assertion follows from the last part of the lemma and from the fact that the $v$-ideals $(R_v\mathfrak{B}_1\mathfrak{p}^q) \cap R$, $q=1, 2, \cdots$, are all distinct and lie between $\mathfrak{B}_1$ and $\mathfrak{B}_2$.

PROPOSITION 2. *Let $r$ be the rank of $v$, let $\mathfrak{M} > \mathfrak{M}_1 > \cdots > \mathfrak{M}_{r-1} > (0)$ be the prime ideals of $R_v$ and let $\mathfrak{p} = \mathfrak{p}_0 > \mathfrak{p}_1 > \cdots > \mathfrak{p}_{h-1} (> (0))$ be the distinct prime ideals in the set $\{\mathfrak{M} \cap R, \mathfrak{M}_1 \cap R, \cdots, \mathfrak{M}_{r-1} \cap R\}$. The ordinal type of the well ordered set of $v$-ideals in $R$, different from* (0), *is $\omega^h$ (where $\omega$ is the first infinite ordinal number). If $\mathfrak{A}$ is any $v$-ideal in $R$, different from* (0), *and if the ordinal type of the set of $v$-ideals preceding $\mathfrak{A}$ is of the form $m_0\omega^{h_0} + m_1\omega^{h_1} + \cdots + m_q\omega^{h_q}$,† where $h > h_0 > h_1 > \cdots > h_q \geqq 0$ and where $m_0, m_1, \cdots, m_q$ are positive integers, then $\mathfrak{p}_{h_0}, \mathfrak{p}_{h_1}, \cdots, \mathfrak{p}_{h_q}$ are the prime ideals of $\mathfrak{A}$* (here $\mathfrak{p}_0 = \mathfrak{p}$).

PROOF. The proposition is obvious if $r=1$ (see Lemma 3). We shall therefore use induction with respect to $r$.

Let $v = v_1 \circ \bar{v}$, where $v_1$ is of rank $r-1$ and $\bar{v}$ is of rank 1. Then $\mathfrak{M}_1, \mathfrak{M}_2, \cdots, \mathfrak{M}_{r-1}$ are the prime ideals of $R_{v_1}$, other than (0). If $\mathfrak{p}$ is also the center of $v_1$ in $R$ (i.e., if $\mathfrak{M} \cap R = \mathfrak{M}_1 \cap R$), then the set $\{\mathfrak{M}_1 \cap R, \mathfrak{M}_2 \cap R, \cdots, \mathfrak{M}_{r-1} \cap R\}$ also consists of $h$ elements, and thus, by our induction hypothesis, the set of $v_1$-ideals in $R$ is of ordinal type $\omega^h$. Since in this case there is only a finite number of $v$-ideals between any two consecutive $v_1$-ideals (see above Corollary), it follows also that the set of $v$-ideals has ordinal type $\omega^h$. If $\mathfrak{p}$ is not the center of $v_1$, then our induction hypothesis implies that the set of $v_1$-ideals in $R$ has ordinal type $\omega^{h-1}$, and it now follows from Lemma 4 and its Corollary that the set of $v$-ideals in $R$ has ordinal type $\omega^h$.

Now let $\mathfrak{A}$ be any $v$-ideal, different from (0), and let $\mathfrak{B}_1, \mathfrak{B}_2$ be two consecutive $v_1$-ideals such that $\mathfrak{B}_1 \supset \mathfrak{A} > \mathfrak{B}_2$ (Lemma 4). We shall consider separately the two cases: (a) $\mathfrak{B}_1 = \mathfrak{A}$; (b) $\mathfrak{B}_1 > \mathfrak{A}$.

† Concerning the notation $m_0\omega^{h_0} + m_1\omega^{h_1} + \cdots + m_q\omega^{h_q}$ see F. Hausdorff, *Grundzüge der Mengenlehre* (1914), p. 112.

We denote by $\mathfrak{p}'_0, \mathfrak{p}'_1, \cdots, \mathfrak{p}'_{g-1}$ the distinct prime ideals in the set $\{\mathfrak{M}_1 \cap R, \mathfrak{M}_2 \cap R, \cdots, \mathfrak{M}_{r-1} \cap R\}$ (we assume that we have $\mathfrak{p}'_0 > \mathfrak{p}'_1 > \cdots > \mathfrak{p}'_{g-1}$).

CASE (a): $\mathfrak{B}_1 = \mathfrak{A}$. If $\mathfrak{M} \cap R = \mathfrak{M}_1 \cap R$ then $g = h$ and $\mathfrak{p}'_i = \mathfrak{p}_i$. From Lemma 4 and its Corollary follow that in this case the set of $v_1$-ideals preceding $\mathfrak{A}$ has ordinal type of the form $m'_0\omega^{h_0} + m'_1\omega^{h_1} + \cdots + m'_q\omega^{h_q}$, where the $m'_\alpha$ are positive integers. Hence, by our induction hypothesis, the prime ideals of $\mathfrak{A}$ are $\mathfrak{p}'_{h_0}, \mathfrak{p}'_{h_1}, \cdots, \mathfrak{p}'_{h_q}$, i.e., $\mathfrak{p}_{h_0}, \mathfrak{p}_{h_1}, \cdots, \mathfrak{p}_{h_q}$. If $\mathfrak{M} \cap R \neq \mathfrak{M}_1 \cap R$, then the ordinal type of the set of $v_1$-ideals preceding $\mathfrak{A}$ must be equal to $m_0\omega^{h_0-1} + \cdots + m_q\omega^{h_q-1}$ (always by Lemma 4 and its Corollary; note that, by the Corollary, $\mathfrak{A}$ can have in this case no immediate predecessor in this set of $v$-ideals, whence $h_q > 0$). Hence $\mathfrak{p}'_{h_0-1}, \mathfrak{p}'_{h_1-1}, \cdots, \mathfrak{p}'_{h_q-1}$ are the prime ideals of $\mathfrak{A}$, and since we have obviously $\mathfrak{p}'_i = \mathfrak{p}_{i+1}$ in the present case, the proof of our proposition is complete in Case (a).

CASE (b): $\mathfrak{B}_1 > \mathfrak{A}$. In this case, $\mathfrak{A}$ has an immediate predecessor in the set of $v$-ideals (either $\mathfrak{B}_1$ or some $v$-ideal between $\mathfrak{B}_1$ and $\mathfrak{A}$). Hence $h_q = 0$. Since $\mathfrak{B}_1\mathfrak{p}^\rho \subset \mathfrak{A}$ for some positive $\rho$ and since $\mathfrak{B}_1 \not\subset \mathfrak{A}$, it follows that $\mathfrak{p}$ itself is one of the prime ideals of $\mathfrak{A}$, i.e., $\mathfrak{p}_{h_q}$ $(h_q = 0)$ is one of the prime ideals of $\mathfrak{A}$. Since all the prime ideals of $\mathfrak{B}_1$ are contained in $\mathfrak{p}$, the set of prime ideals of $\mathfrak{A}$ consists of $\mathfrak{p}$ and of the prime ideals of $\mathfrak{B}_1$. The set of $v$-ideals, preceding $\mathfrak{B}$, has ordinal type $m_0\omega^{h_0} + m_1\omega^{h_1} + \cdots + m_{q-1}\omega^{h_{q-1}} + m'_q$, where $m'_q \geqq 0$. By case (a), applied to the $v$-ideal $\mathfrak{B}_1$ (instead of to the ideal $\mathfrak{A}$ of the present case), we have that the prime ideals of $\mathfrak{B}_1$ are either $\mathfrak{p}_{h_0}, \mathfrak{p}_{h_1}, \mathfrak{p}_{h_{q-1}}, \cdots, \mathfrak{p}$ (if $m'_q > 0$) or $\mathfrak{p}_{h_0}, \mathfrak{p}_{h_1}, \cdots, \mathfrak{p}_{h_{q-1}}$ (if $m'_q = 0$). In either case, the desired conclusion concerning the prime ideals of $\mathfrak{A}$ follows. This completes the proof of the proposition.

# APPENDIX 4

## COMPLETE MODULES AND IDEALS

The subject matter of this appendix is of considerable importance for algebraic geometry. It deals with a general algebraic concept which, when specialized to the field of algebraic geometry, leads not only to the concept of a complete linear system on an algebraic variety, but also to the concept of a complete linear system with so-called "assigned base loci," as it gives a simple and workable formulation of the intuitive geometric notion of "base conditions."

Throughout this appendix we shall deal with a fixed integral domain $\mathfrak{o}$ and a fixed field $K$ containing $\mathfrak{o}$ ($K$ is not necessarily the quotient field of $\mathfrak{o}$). We shall deal with $\mathfrak{o}$-modules $M$ *contained* in $K$. Certain additional conditions will be imposed on $\mathfrak{o}$ and the modules $M$ as and when these conditions are needed. Thus, we may have to assume sometimes that $\mathfrak{o}$ is integrally closed, or that $\mathfrak{o}$ is noetherian, or that $M$ is a finite $\mathfrak{o}$-module.

The following special situations are of particular importance in geometric applications: (1) $\mathfrak{o}$ is a field $k$ (the ground field), $K$ is a field of algebraic functions over $k$, and $M$ is a finite $k$-module (contained in $K$); (2) $\mathfrak{o}$ is integrally closed and $M$ is an ideal in $\mathfrak{o}$.

1. We denote by $S$ the set of all non-trivial valuations of $K$ which are non-negative on $\mathfrak{o}$ ($S$ = the Riemann surface of $K$ relative to $\mathfrak{o}$; see VI, § 17). If $v \in S$ we denote by $R_v$ the valuation ring of $v$.

DEFINITION 1. *If $M$ is an $\mathfrak{o}$-module (contained in $K$)* THE COMPLETION OF $M$ *is the $\mathfrak{o}$-module*

$$M' = \bigcap_{v \in S} R_v M. \tag{1}$$

*The completion of $M$ will be denoted by $M'$. The module $M$ will be said to be complete if $M = M'$.*

COROLLARY. *If $\bar{\mathfrak{o}}$ denotes the integral closure of $\mathfrak{o}$ in $K$ and if we set $\bar{M} = \bar{\mathfrak{o}}M$, then $\bar{M}' = M'$, where $\bar{M}'$ is the completion of the $\bar{\mathfrak{o}}$-module $\bar{M}$.*

For $S$ is also the set of all valuations of $K$ which are non-negative on $\bar{\mathfrak{o}}$, and hence

$$\bar{M}' = \bigcap_{v \in S} R_v\bar{M} = \bigcap_{v \in S} R_v M = M'.$$

It follows that *the class of complete $\bar{\mathfrak{o}}$-modules coincides with the class of complete $\mathfrak{o}$-modules.* In the study of complete $\mathfrak{o}$-modules it would be therefore permissible, without loss of generality, to restrict the treatment to integrally closed domains $\mathfrak{o}$.

We list at once a number of formal properties of the operation of "completion" consisting in passing from $M$ to $M'$. To what extent these properties characterize axiomatically the operation of completion will be briefly discussed later on in this appendix. We denote by $\bar{\mathfrak{o}}$ the integral closure of $\mathfrak{o}$ in $K$. In the following proposition, $M$, $N$ and $L$ denote $\mathfrak{o}$-modules contained in $K$.

PROPOSITION 1. *The operation $M \to M'$ satisfies the following conditions:*

(a) $\mathfrak{o}' = \bar{\mathfrak{o}}$.
(b) $M' \supset M$.
(c) *If* $M \supset N$ *then* $M' \supset N'$.
(d) $(M')' = M'$.
(e) $(MN)' = (M'N')'$, *where by the product $MN$ of two $\mathfrak{o}$-modules $M$, $N$ we mean the $\mathfrak{o}$-module generated by the products $mn$ ($m \in M$, $n \in N$).*
(f) $(xM)' = xM'$ ($x \in K$).
(g) *If* $(MN)' \subset (ML)'$ *and if the $\mathfrak{o}$-module $M$ is either finite or is the completion of a finite $\mathfrak{o}$-module, then* $N' \subset L'$.

PROOF. Property (a) follows from VI, § 4, Theorem 6, while (b) and (c) are self-evident. From (b) and (c) follows $(M')' \supset M'$, but on the other hand, we have for any $v$ in $S$: $(M')' \subset R_v M' \subset R_v(R_v M) = R_v M$, whence $(M')' \subset M'$, and this proves (d). The inclusion $(MN)' \subset (M'N')'$ follows from (b) and (c). On the other hand, we have, for any $v \in S$: $M' \subset R_v M$, $N' \subset R_v N$, whence $M'N' \subset R_v MN$. Therefore $M'N' \subset (MN)'$, and thus, by (c) and (d), $(M'N')' \subset (MN)'$, which proves (e). Property (f) is self-evident.

For the proof of (g) we observe that it is sufficient to consider the case in which $M$ is a finite $\mathfrak{o}$-module, for if $M$ is the completion of a finite $\mathfrak{o}$-module $M_0$, then we have $(MN)' = (M'_0 N)' = (M'_0 N')' = (M_0 N)'$, and similarly $(ML)' = (M_0 L)'$, and thus $(M_0 N)' \subset (M_0 L)'$. Assume then that $M$ is a finite $\mathfrak{o}$-module. Now observe, that if $Z$ is any $\mathfrak{o}$-module and $v \in S$ then $Z' \subset R_v Z$, and hence $R_v Z' = R_v Z$. We have therefore $R_vMN = R_v(MN)' \subset R_v(ML)' = R_vML$. Since $M$ is a finite $\mathfrak{o}$-module, $R_vM$ is also a finite, therefore a *principal*, $R_v$-module ($R_v$ being a valuation ring). Therefore we can cancel $M$ in the inclusion $R_vMN \subset R_vML$. We thus have $R_vN \subset R_vL$ for all $v \in S$, and this establishes (g).

COROLLARY. *We have*

$$(\mathfrak{o}x)' = \bar{\mathfrak{o}}x, \quad x \in K; \tag{2}$$

$$\bar{\mathfrak{o}}M' = M'. \tag{2'}$$

Relation (2) follows from (f), by setting $M = \mathfrak{o}$, and from (a). We have $M' = (\mathfrak{o}M)' = (\bar{\mathfrak{o}}M')' \supset \bar{\mathfrak{o}}M' \supset M'$, and this establishes (2'). Relation (2') shows that $M'$ is always an $\bar{\mathfrak{o}}$-module. Of course, we know that already, in view of the Corollary of Definition 1, but we have derived this here again as a formal consequence of relations (a)–(f). Other formal consequences of these relations (more precisely: of the relations (b), (c) and (d)) are the following:

(h) $(\sum M_i)' = (\sum M'_i)'$

(i) $\bigcap_i M'_i = (\bigcap M'_i)'$,

where $\{M_i\}$ is any (finite or infinite) collection of $\mathfrak{o}$-modules. Note that relation (i) implies that *the intersection of any (finite or infinite) number of complete modules is complete.*

For any non-negative integer $q$ we denote by $M^q$ the $\mathfrak{o}$-module generated by the monomials $m_1m_2 \cdots m_q$, $m_i \in M$ (here $M^0$ stands for $\mathfrak{o}$).

DEFINITION 2. *An element $z$ of $K$ is said to be integrally dependent on the module $M$ if it satisfies an equation of the form*

$$z^q + a_1z^{q-1} + a_2z^{q-2} + \cdots + a_q = 0, \; a_i \in M^i. \tag{3}$$

It is not difficult to see that the above definition is equivalent to the following one: *$z$ is integrally dependent on $M$ if there exists a finite $\mathfrak{o}$-module $N$ (contained in $K$) such that*

$$zN \subset MN, \tag{4}$$

*where $MN$ denotes the $\mathfrak{o}$-module generated by the products $mn$, $m \in M$, $n \in N$.* For, if (4) holds then (3) follows by using a basis of $N$ and determinants (see the proof of the lemma in Vol. I, p. 255). On the other hand, if (3) holds then (4) is satisfied by taking for $N$ the module $M_0^{q-1} + M_0^{q-2}z + \cdots + M_0z^{q-2} + \mathfrak{o}z^{q-1}$, where $M_0$ is a finite submodule of $M$ such that $a_i \in M_0^i$ for $i = 1, 2, \cdots, q$.

From criterion (4) it follows immediately that the set of elements $z$ of $K$ which are integrally dependent on $M$ is itself an $\mathfrak{o}$-module. We may call that $\mathfrak{o}$-module *the integral closure of $M$ in $K$.*

THEOREM 1. *The completion $M'$ of $M$ in $K$ coincides with the integral closure of $M$ in $K$.*

PROOF. Let $z \in K$ be integrally dependent on $M$. Using (4) we can write

$$zn_i = \sum_{j=1}^{h} m_{ij}n_j, \quad i = 1, 2, \cdots, h, \tag{5}$$

where $N = \sum_{i=1}^{h} \mathfrak{o}n_i$ and $m_{ij} \in M$. Let $v$ be any valuation in $S$ (i.e., $v$ is non-negative on $\mathfrak{o}$) and let $m_{\alpha\beta}$ be one of the $h^2$ elements $m_{ij}$ for which $v(m_{\alpha\beta})$ is minimum. Dividing (5) by $m_{\alpha\beta}$ and observing that $m_{ij}/m_{\alpha\beta} \in R_v$ for all $i$ and $j$, we see that $z/m_{\alpha\beta}$ is integral over $R_v$, whence $z/m_{\alpha\beta} \in R_v$, $z \in R_v m_{\alpha\beta} \subset R_v M$. Since this holds for all $v$ in $S$ we deduce that $z \in M'$.

Conversely, assume that $z \in M'$. Let $L$ denote the set of all quotients $m/z$, $m \in M$, and let us consider the ring $\mathfrak{o}[L]$. For any $v$ which is non-negative on $\mathfrak{o}[L]$ (and hence also on $\mathfrak{o}$) there exists an element $m$ of $M$ such that $v(z) \geqq v(m)$ (since $z \in M'$). Hence there exists no valuation $v$ of $K$ which is non-negative on $\mathfrak{o}[L]$ and such that $\mathfrak{M}_v$ contains the ideal $\mathfrak{o}[L] \cdot L$. Therefore this ideal must be the unit ideal in $\mathfrak{o}[L]$ (VI, § 4, Theorem 4). Thus, there exists a finite set of elements of $L$, say $m_1/z, m_2/z, \cdots, m_h/z$ such that

$$1 = \sum_{i=1}^{h} \frac{m_i}{z} \cdot f^{(i)}\left(\frac{m_1}{z}, \frac{m_2}{z}, \cdots, \frac{m_h}{z}\right), \tag{6}$$

where the $f^{(i)}$ are polynomials with coefficients in $\mathfrak{o}$. We can write each of these polynomials in the form

$$f^{(i)}(m_1/z, m_2/z, \cdots, m_h/z) = F_{q-1}{}^{(i)}(m_1, m_2, \cdots, m_h, z)/z^{q-1},$$

where $q$ is a suitable integer, independent of $i$, and where the $F_{q-1}{}^{(i)}$ are homogeneous polynomials in $m_1, m_2, \cdots, m_h, z$, of degree $q-1$, with coefficients in $\mathfrak{o}$. Then, multiplying (6) by $z^q$ we find at once that $z$ satisfies an equation of the form (3). This completes the proof.

REMARK. Every element of $M^i$ is a finite sum of products of the form $m_1 m_2 \cdots m_i$, where the $m_j$ are elements of $M$. It follows therefore from Definition 2 and from Theorem 1 that the completion of $M$ is independent of the choice of the ring $\mathfrak{o}$. Thus, if $M$ happens to be also a module over another ring $\mathfrak{o}_1$ (for instance, if $\mathfrak{o}_1$ is a subring of $\mathfrak{o}$) then the completion of $M$ as an $\mathfrak{o}_1$-module is the same as the completion of $M$ as an $\mathfrak{o}$-module. If we take for $\mathfrak{o}_1$ the prime subring $k$ of $\mathfrak{o}$ and we treat $M$ as a $k$-module, then the valuation space $S$ which occurs in our Definition 1 of completion becomes the set *of all valuations of* $K$. We have therefore

$$\bigcap_{v \in S} R_v M = \bigcap_{\text{all } v} R_v M.$$

We give here a direct proof of this equality.

We have to show if $x \in \bigcap_{v \in S} R_v M$ then $v(x) \geqq v(M)$ for any valuation $v$ of $K$ (and we may assume here that $v \notin S$). Let $\Gamma$ be the value group of $v$, let $H$ be the set of all elements $\alpha$ of $\Gamma$ such that $\alpha = v(z)$ for some $z$ in $R_v\mathfrak{o}$. Let $\Delta$ be the set of those elements $\alpha$ of $H$ which have the property that also $-\alpha$ is in $H$. We note that $H$ is closed under addition (in $\Gamma$) and that if $\alpha \in H$ and $\beta \geqq \alpha$ then also $\beta \in H$. From this it follows easily that $\Delta$ is an isolated subgroup of $\Gamma$ (see VI, § 10; note that $\Delta$ is non-empty since $1 \in \mathfrak{o}$ and since therefore $0 \in \Delta$). The isolated subgroup $\Delta$ determines a valuation $v_1$ of $K$ with which $v$ is composite and whose value group is $\Gamma/\Delta$ ($v_1$ is the trivial valuation if $\Delta = \Gamma$). Now, if $a$ is any element of $\mathfrak{o}$ then $v(a)$ is either in $\Delta$ or is a *strictly positive* element of $\Gamma$. Therefore $v_1(a) \geqq 0$, i.e., we have $v_1 \in S$, and thus $x \in R_{v_1}M$. There exists then an element $m$ of $M$ such that $v_1(x) \geqq v_1(m)$, or—equivalently—$v(x/m) \in \Delta \cup \Gamma_+$. Thus $v(x/m) \geqq v(a)$ for some $a$ in $\mathfrak{o}$, and $v(x) \geqq v(am)$, $am \in M$.

2. We shall now present Theorem 1 under a different form, using properties of graded domains (VII, § 2). We adjoin to the field $K$ a transcendental $t$, we set $M^\star = Mt$ and we regard $M^\star$ as an $\mathfrak{o}$-module. Let $M^{\star\prime}$ be the completion of $M^\star$ in $K(t)$. Using either criterion (3) or (4) of integral dependence over $M^\star$ and applying Theorem 1 we find at once that an element $z'$ of $K(t)$ belongs to $M^{\star\prime}$ if and only if $z'/t \in M'$. Hence

$$M' = \frac{1}{t} \cdot M^{\star\prime}. \tag{7}$$

Hence the determination of $M'$ reduces to that of $M^{\star\prime}$. We consider the ring $R^\star = \sum_{q=0}^{\infty} M^{\star q}$ ($M^{\star 0} = \mathfrak{o}$). From the fact that $t$ is a transcendental over $K$ follows that $R^\star$ is a graded domain, $M^{\star q}$ being the set of homogeneous elements of $R^\star$, of degree $q$. Let $F$ be the field of quotients of $R^\star$ and let $F_0$ be the subfield of $F$ consisting of the homogeneous elements of $F$ which are of degree zero. We have $F = F_0(t)$, and it is clear that $F_0$ is the smallest subfield of $K$ which contains the ring $\mathfrak{o}$ and all the quotients $m/m'$, where $m, m' \in M$, $m' \neq 0$. In other words, $F_0$ is the set of all quotients $m_q/m'_q$, where $m_q, m'_q \in M^q$ and

$q=0, 1, 2, \cdots$. Let $F'_0$ be the algebraic closure of $F_0$ in $K$, and let $\bar{R}^{\star\prime}$ be the integral closure of $R^\star$ in $F'_0(t)$. Then, by the Corollary to Theorem 11 of VII, § 2, $\bar{R}^{\star\prime}$ is a graded domain: $\bar{R}^{\star\prime} = \sum_{q=0}^{\infty} \bar{R}_q^{\star\prime}$. From Theorem 1, and using criterion (3) of integral dependence over modules, we see at once (compare with the Remark at the end of VII, § 2) that

$$M^{\star\prime} = \bar{R}_1^{\star\prime}. \tag{8}$$

Relation (8) expresses in a different form the content of Theorem 1. At this stage it will be convenient to introduce certain notations and terminology. Given the $\mathfrak{o}$-module $M$, the field $F_0$ introduced above shall be denoted by $\mathfrak{o}(M)$. The module $M$ shall be said to be *homogeneous* if the sum $R = \sum_{q=0}^{\infty} M^q$ is direct (so that $R$ is therefore a graded ring). It is immediately seen that if $M$ is a homogeneous module, then every element $m$ of $M$, $m \neq 0$, is transcendental over $\mathfrak{o}(M)$, and that if this last condition is satisfied by some element $m$ of $M$, $m \neq 0$, then $M$ is homogeneous.

The above transition from $M$ to $M^\star$ is only necessary if $M$ is not homogeneous. If $M$ itself is homogeneous, then it is not necessary to introduce a new transcendental $t$, and we can deal directly with the graded ring $R$ (instead of with $R^\star$). Summarizing, we can now state the following result:

THEOREM 2. *If $M$ is a homogeneous $\mathfrak{o}$-module, if $F'$ denotes the field generated (in $K$) by $M$ and the algebraic closure $F'_0$ of $\mathfrak{o}(M)$ in $K$ and if $\bar{R}'$ is the integral closure,† in $F'$, of the graded ring $R = \sum_{q=0}^{\infty} M^q$, then the completion $M'$ of $M$ is the module $\bar{R}'_1$ of homogeneous elements of $\bar{R}'$, of degree* 1. *If $M$ is not homogeneous, then the adjunction of a transcendental $t$ to $K$ reduces the determination of $M'$ to the case of the homogeneous $\mathfrak{o}$-module $tM$.*

COROLLARY 1. *The completion $M'$ of $M$ is not affected if the field $K$ is replaced by any field between $F'$ and $K$ (where $F'$ is the field defined in Theorem 2; in particular, we may replace $K$ by $F'$).*

As a special case of complete $\mathfrak{o}$-modules we have the so-called *complete ideals in* $\mathfrak{o}$, where an ideal $\mathfrak{A}$ in $\mathfrak{o}$ is said to be complete if it is complete as an $\mathfrak{o}$-module.

† Note that if $y$ is any element of $M$, different from zero, then the quotient field of $R$ is given by $F_0(y)$, where $F_0 = \mathfrak{o}(M)$, and $F' = F'_0(y)$. Therefore, by the Corollary of Theorem 11 in Ch. VII, § 2, $\bar{R}'$ is a graded domain.

COROLLARY 2. *If $\mathfrak{A}$ is an ideal in $\mathfrak{o}$ then the completion $\mathfrak{A}'$ of $\mathfrak{A}$ is a complete ideal in the integral closure $\bar{\mathfrak{o}}$ of $\mathfrak{o}$ in $K$, and if $S'$ denotes the set of all valuations $v$ of the quotient field of $\bar{\mathfrak{o}}$ which are non-negative on $\mathfrak{o}$ then*

$$\mathfrak{A}' = \bigcap_{v \in S'} R_v\mathfrak{A}. \tag{9}$$

For, in the case of the present corollary, the field $F'$ of Corollary 1 is precisely the quotient field of $\bar{\mathfrak{o}}$, and this implies (9). Since $\mathfrak{A} \subset \mathfrak{o} \subset R_v$ for all $v \in S'$ and since $\bigcap_{v \in S'} R_v = \bar{\mathfrak{o}}$ it follows that $\mathfrak{A}' \subset \bar{\mathfrak{o}}$. Since $\mathfrak{A}'$ is also an $\bar{\mathfrak{o}}$-module and is the completion of the ideal $\bar{\mathfrak{o}}\mathfrak{A}$ (see Corollary of Definition 1), $\mathfrak{A}'$ is a complete ideal in $\bar{\mathfrak{o}}$.

3. We now study briefly the important case of *complete ideals in an integrally closed domain* $\mathfrak{o}$.

If $\mathfrak{o}$ is integrally closed in $K$ and $\mathfrak{A}$ is an ideal in $\mathfrak{o}$, then the completion $\mathfrak{A}'$ of $\mathfrak{A}$ is a complete ideal *in* $\mathfrak{o}$ (Corollary 2 of Theorem 2). We have therefore

$$\mathfrak{A}' = \bigcap_{v \in S} R_v\mathfrak{A} = \bigcap_{v \in S} (\mathfrak{o} \cap R_v\mathfrak{A}).$$

Since $\mathfrak{o} \cap R_v\mathfrak{A}$ is a valuation ideal in $\mathfrak{o}$ (Appendix 3), we see that *every complete ideal in $\mathfrak{o}$ is an intersection of valuation ideals.* On the other hand, if $\mathfrak{B}$ is a valuation ideal in $\mathfrak{o}$, associated with a valuation $v$ ($v \in S$), then $\mathfrak{B} = \mathfrak{o} \cap R_v\mathfrak{B}$ (Appendix 3, formula (1)) whence $\mathfrak{B}' = \mathfrak{o} \cap \mathfrak{B}' \subset \mathfrak{o} \cap R_v\mathfrak{B} = \mathfrak{B}$, i.e., $\mathfrak{B}' = \mathfrak{B}$. Thus, every valuation ideal in $\mathfrak{o}$ is a complete ideal, and so is every intersection (finite or infinite) of valuation ideals [see property ($i$) of the 'operation]. Consequently, the *class of complete ideals in $\mathfrak{o}$ coincides with the class of ideals which are intersections (finite or infinite) of valuation ideals.*

If $K_0$ is the quotient field of $\mathfrak{o}$ and $v_0$ is the restriction of $v$ to $K_0$ then $\mathfrak{o} \cap R_v\mathfrak{A} = \mathfrak{o} \cap R_{v_0}\mathfrak{A}$. Therefore we may replace $K$ by $K_0$, and we shall assume from now on that $K$ is the quotient field of $\mathfrak{o}$.

If $\mathfrak{A}$ is a complete ideal then

$$\mathfrak{A} = \bigcap_{v \in S} (\mathfrak{o} \cap R_v\mathfrak{A})$$

is a representation of $\mathfrak{A}$ as intersection (generally infinite) of valuation ideals, but there may be other such representations, and among these there may be even some which are finite intersections (take as $\mathfrak{A}$, for instance, a valuation ideal). In the case of a noetherian domain $\mathfrak{o}$ we have the following result:

THEOREM 3. *Let $\mathfrak{o}$ be a noetherian domain, $K$ a field containing $\mathfrak{o}$ and $\bar{\mathfrak{o}}$*

*the integral closure of* $\mathfrak{o}$ *in the quotient field of* $\mathfrak{o}$. *If an ideal* $\mathfrak{A}$ *in* $\bar{\mathfrak{o}}$ *is the completion (in* $K$*) of an ideal* $\mathfrak{B}$ *in* $\mathfrak{o}$ *(in particular, if* $\mathfrak{o}$ *itself is integrally closed and* $\mathfrak{A}$ *is a complete ideal in* $\mathfrak{o}$*), then* $\mathfrak{A}$ *is a finite intersection of valuation ideals of* $\bar{\mathfrak{o}}$ *associated with discrete valuations of rank* 1.

PROOF. We first establish a lemma on complete $\mathfrak{o}$-modules, where $\mathfrak{o}$ *is not necessarily noetherian.*

LEMMA. *Let* $K$ *be a field containing* $\mathfrak{o}$, *let* $M$ *be a finite* $\mathfrak{o}$*-module contained in* $K$ *and let* $\{x_i\}$ *be a finite* $\mathfrak{o}$*-basis of* $M$. *For each* $i$ *let* $\mathfrak{o}_i$ *denote the ring generated over* $\mathfrak{o}$ *by the quotients* $x_j/x_i$, $j \neq i$, *and let* $\bar{\mathfrak{o}}_i$ *be the integral closure of* $\mathfrak{o}_i$ *in* $K$. *If* $M'$ *is the completion of* $M$ *in* $K$ *then*

$$M' = \bigcap_i \bar{\mathfrak{o}}_i x_i.$$

PROOF OF THE LEMMA. If $y \in M'$ and $v$ is any valuation of $K$ which is non-negative on $\bar{\mathfrak{o}}_i$, then $v \in S$ and $v(x_j) \geqq v(x_i)$ for all $j$. Thus $R_v M = R_v x_i$, $v(y) \geqq v(x_i)$ for all such $v$, and hence $y \in \bar{\mathfrak{o}}_i x_i$.

Conversely, let $y$ be an element of the intersection of the $\bar{\mathfrak{o}}_i x_i$ and let $v \in S$. If $i$ is an index such that $v(x_j) \geqq v(x_i)$ for all $j$, then $v$ is non-negative on $\bar{\mathfrak{o}}_i$ and hence $v(y) \geqq v(x_i)$. Thus $y \in R_v x_i = R_v M$, and this shows that $y \in M'$. The lemma is proved.

The proof of the theorem is now immediate. We identify the ideal $\mathfrak{B}$ of the theorem with the module $M$ of the lemma. Since, by assumption, the completion $\mathfrak{A}$ of $\mathfrak{B}$ (in $K$) is contained in the integral closure $\bar{\mathfrak{o}}$ of $\mathfrak{o}$ in the quotient field of $\mathfrak{o}$, $\mathfrak{A}$ is also the completion of $\mathfrak{B}$ in this quotient field. We may therefore assume that $K$ is the quotient field of $\mathfrak{o}$. Each ring $\mathfrak{o}_i$ is noetherian. Now, it can be proved† that the integral closure of a noetherian domain is a Krull ring (VI, § 13). Hence each of the rings $\bar{\mathfrak{o}}_i$ is a Krull ring.‡ Since $x_i \in \mathfrak{o} \subset \bar{\mathfrak{o}}_i$, the principal $\bar{\mathfrak{o}}_i$-ideal $\bar{\mathfrak{o}}_i x_i$ is a finite intersection of valuation ideals in $\bar{\mathfrak{o}}_i$ belonging to *essential* (therefore discrete, rank 1) valuations (VI, § 13). Taking intersections with $\bar{\mathfrak{o}}$ we see that the theorem follows from the above lemma.

COROLLARY. *If, under the assumptions of Theorem* 3, *the ideal* $\mathfrak{A}$ *admits an irredundant primary representation (in particular, if* $\mathfrak{A}$ *is a complete ideal in a noetherian integrally closed domain), then* $\mathfrak{A}$ *also admits an irredundant primary representation in which every primary component is itself a complete ideal.*

Since each essential valuation of $\bar{\mathfrak{o}}_i$ is of rank 1, the corresponding

† See M. Nagata, "On the derived normal rings of noetherian integral domains," *Mem. Coll. Sci., Univ. Kyoto*, 29, Mathematics No. 3, 1955.

‡ The cited general result of Nagata is not needed if $\mathfrak{o}$ is a ring of quotients of a finite integral domain, for in that case we know (Vol. I, Ch. V, § 4, Theorem 9) that the rings $\bar{\mathfrak{o}}_i$ are noetherian.

valuation ideals in $\bar{\mathfrak{o}}$, of which $\mathfrak{A}$ is the intersection, are primary ideals. Those which are associated with the same prime ideal in $\mathfrak{o}$ yield a partial intersection which is a primary complete ideal.

One often deals with complete fractional $\mathfrak{o}$-ideals, i.e., with finite complete $\mathfrak{o}$-modules contained in the quotient field of $\mathfrak{o}$. It is clear that any such complete fractional $\mathfrak{o}$-ideal is of the form $\frac{1}{z}\cdot\mathfrak{A}$, where $\mathfrak{A}$ is a complete (integral) ideal in $\mathfrak{o}$ (use property (f) of Proposition 1).

4. We shall now discuss briefly the axiomatic aspects of the properties (a)–(g) (see Proposition 1) of the $'$operation. The operation of completion of $\mathfrak{o}$-modules $M$, in $K$, is not the only $'$operation on $\mathfrak{o}$-modules which satisfies properties (a)–(g) of Proposition 1. If we examine the proof of that proposition we see that we have not used the fact that the set $S$ consists of *all* the valuations $v$ of $K$ which are non-negative on $\mathfrak{o}$, but only the fact that the intersection of all the valuation rings $R_v$, $v \in S$, is the integral closure of $\mathfrak{o}$ in $K$. Therefore, if we choose any subset $S_1$ of $S$ with the property

$$\bigcap_{v \in S_1} R_v = \bar{\mathfrak{o}} \tag{10}$$

and define for any module $M$ in $K$ its completion $M'$ by

$$M' = \bigcap_{v \in S_1} R_v M, \tag{11}$$

we obtain another $'$operation which satisfies conditions (a)–(g). An important special case is the one in which $\mathfrak{o}$ is a noetherian integrally closed domain, $K$ the quotient field of $\mathfrak{o}$ and $S_1$ the set of all essential valuations of $\mathfrak{o}$ (i.e., the set of $\mathfrak{p}$-adic valuations $v_{\mathfrak{p}}$, where $\mathfrak{p}$ is any minimal prime ideal of $\mathfrak{o}$). In that case the "complete" ideals in $\mathfrak{o}$ are the ideals whose prime ideals are all minimal in $\mathfrak{o}$, and the "completion" of an ideal $\mathfrak{A}$ in $\mathfrak{o}$ is obtained by deleting from an irredundant primary decomposition of $\mathfrak{A}$ those components which belong to prime ideals which are not minimal in $\mathfrak{o}$.

It can be proved that any $'$operation is "equivalent" to a $'$operation defined by a suitable set of valuations $S_0$ satisfying (10), two $'$operations being "equivalent" if they coincide on the set of all *finite* $\mathfrak{o}$-modules in $K$. For the proof we refer the reader to the paper of W. Krull, entitled "Beiträge zur Arithmetik kommutativer Integritätsbereiche," Math. Zeitschrift, vol. 41 (1946).

We note that if we have two $'$operations, say $'^1$ and $'^2$, defined by sets $S_1$ and $S_2$ of valuations satisfying (10), and if $S_1 \subset S_2$ then $M'^1 \supset M'^2$ for any module $M$. Applying this inclusion to the module $M'^1$

instead of to $M$ we find $M'^1 \supset (M'^1)'^2$, and since the opposite inclusion is obvious, we have

$$(12) \qquad M'^1 = (M'^1)'^2, \quad \text{if } S_1 \subset S_2,$$

i.e., every module which is complete with respect to the $'^1$ operation is also complete with respect to the $'^2$ operation. (We may say in that case that the $'^2$ operation is "finer" than the $'^1$ operation.) In particular, if we take for $S_2$ the set $S$ of *all* valuations which are non-negative on $\mathfrak{o}$ we conclude that *in any 'operation the complete modules are integrally closed* (in $K$).

5. We shall now discuss briefly the application of the concept of complete modules to the theory of complete linear systems in algebraic geometry.

Let $V/k$ be a normal projective variety, of dimension $r$, let $\mathfrak{o}=k$, let $K=k(V)$ (we shall assume that $k$ is maximally algebraic in $k(V)$), and let us first study the 'operation defined by the set $S_0$ of all prime divisors of $K/k$ which are of the first kind with respect to $V/k$ (VI, § 14). Condition (10) is satisfied (with $S_1$ replaced by $S_0$; see end of VI, § 14). In this case, given a *finite* $k$-module $M$ in $K$, the completion $M'$ of $M$ in $K$ is obtained as follows:

For any prime divisor $v$ in $S_0$ we denote by $W_v$ the center of $v$ on $V$ ($W_v$ is an irreducible $(r-1)$-dimensional subvariety of $V/k$). We set $n_v = \min\{v(m), 0 \neq m \in M\}$ and

$$(13) \qquad Z(M) = D = -\sum_{v \in S_0} n_v W_v.$$

Since $M$ is a finite $k$-module, $n_v$ is finite for every $v$ in $S_0$, and only a finite number of $n_v$'s are different from zero. Thus, the above sum is finite, and $D$ is an element of the free group of divisors on $V/k$ (see VII, § 4[bis]), or—in algebro-geometric terminology—*$D$ is a divisorial cycle on $V/k$.* The completion $M'$ of $M$ is then the set of all elements $x$ of $K$ such that $(x)+D$ is an *effective* divisorial cycle (i.e., a divisorial cycle of the form $\sum h_W W$, $h_W \geqq 0$). Here $(x)$ denotes the divisor of $x$ (we include the zero in $M'$). The set of effective divisors $(x)+D$ $(0 \neq x \in M')$ is called a *complete linear system* on $V/k$, or, *the complete linear system determined by the cycle $D$* (and is often denoted by $|D|$). It consists of all effective divisorial cycles $D'$ on $V$ which are *linearly equivalent to $D$*, i.e., which are such that $D'-D$ is the divisor of an element $x$ of $K$ $(x \neq 0)$.

A basic result in algebraic geometry is the following: *the above complete module $M'$ is finite dimensional* (as a vector space over $k$). We have all the tools for a short proof of this result. First of all, we replace $M$ by the homogeneous module $N=Mt$, where $t$ is a transcendental over $k(V)$. Then $N'=M't$, and we have only to prove that $N'$ is finite dimensional. We now use the notations of Theorem 2, where $M$ is to be replaced by $N$, $\mathfrak{o}$ by $k$ and $K$ by $k(V)(t)$. Since $N$ is a finite $k$-module, the graded ring $R=\sum_{q=0}^{\infty} N^q$ is a finite integral domain over $k$. On the other hand, the field $F'$ is a finite algebraic extension of the quotient field of $R$ (since we are dealing with subfields of an algebraic function field $k(V)(t)$). Hence $\bar{R}'$ is a finite $R$-module (Vol. I, Ch. V, § 4, Theorem 9). On the other hand $\bar{R}'$ is also a graded ring (VII, § 2, Theorem 11). Hence $\bar{R}'$ has a finite $R$-basis consisting of *homogeneous elements* (of non-negative degree). A basis of $N$ over $k$, together with those basis elements of $\bar{R}'$ over $R$ which are homogeneous of degree 1, will therefore constitute a set of elements which span $\bar{R}'_1$ over $k$. Since $N'=\bar{R}'_1$ (Theorem 2), $N'$ is finite dimensional.

The mapping $x \to (x)+D$ $(0 \neq x \in M', (x)+D \in |D|)$ is such that two elements $x_1, x_2$ of $M'$ are mapped into the same cycle in $|D|$ if and only if $x_2/x_1 \in k$. This shows that the complete linear system $|D|$ (if it is not empty) has a natural structure of a projective space of dimension $s$, if $s+1$ is the dimension of $M'$. We say then *that* $|D|$ *has dimension* $s$.

If $D_1$ is any divisorial cycle which is linearly equivalent to $D$ (not necessarily an effective cycle) then it is clear that $|D|=|D_1|$ (in view of the transitivity of linear equivalence: if $D_1-D$ is the divisor of a function $y$ in $K$ and $D_2-D$ is the divisor of a function $z$, then $D_2-D_1$ is the divisor of $z/y$). If $D_1-D=(y)$, then the module $M'_1=\frac{1}{y}M'$ consists of all functions $x_1$ in $K$ such that $(x_1)+D_1$ is effective. This module $M'_1$ is therefore also complete and serves to define the same complete linear system $|D|$ $(=|D_1|)$ as the one defined by $M'$. Observe that if we denote by $M_1$ the module $\frac{1}{y}M$, then $Z(M_1)=D_1$ (see (13)) and $M'_1$ is the completion of $M_1$.

Conversely, suppose we are given a divisorial cycle $D$ on $V/k$ and assume that there exist *effective* cycles which are linearly equivalent to $D$ (the set of all such cycles will be denoted by $|D|$). Then the set $L$ of all elements $x$ in $K$ such that $(x)+D$ is effective (we include 0 in that set) is a $k$-module of dimension $\geqq 1$. *We assert that $L$ is finite-dimensional and complete.* To see this, write $D=\sum_{i=1}^{q} n_i W_i$, where the

$W_i$ are distinct irreducible $(r-1)$-dimensional subvarieties of $V/k$ and the $n_i$ are integers different from zero (if $D$ is the zero-cycle then $L=k$, and our assertion is trivial). Let $v_i$ denote the prime divisor of $K/k$ defined by $W_i$. For each $i$ we fix an element $y_i \neq 0$ in $K$ such that $v_i(y_i) \leqq -n_i$ and $v_j(y_i)=0$ if $j \neq i$ (see VI, § 10, Theorem 18), and we set $y=y_1y_2 \cdots y_q$ and $N=k+k \cdot y$. It is immediately seen that the cycle $Z(N)$ (see (13)) is such that $Z(N)-D$ is effective. That implies that whenever $(x)+D$ is effective, also $(x)+Z(N)$ is effective; in other words: $L$ is a subspace of the completion $N'$ of $N$. Since $N'$ is finite-dimensional, so is $L$.

Let $\Gamma=Z(L)$. It is clear that $D-Z(L)$ is effective. If $x \neq 0$ is such that $(x)+Z(L)$ is effective, then *a fortiori* $(x)+D$ is effective, and hence $x \in L$. *It follows that $L$ is a complete module*, as was asserted.

The complete linear system defined by $L$ is $|Z(L)|$, not necessarily $|D|$. However, if $D_1$ is any member of $|D|$ then $D_1=(x)+D$, where $x \in L$, whence $D_1=\Delta-Z(L)+D$, with $\Delta$ in $|Z(L)|$. Conversely, if $\Delta \in |Z(L)|$, then $(\Delta-Z(L)+D)-D=\Delta-Z(L)=(x)$, $x \in L$, whence $\Delta+(D-Z(L)) \in |D|$. This shows that $|D|$ *consists of the cycles of* $|Z(L)|$ *augmented by the fixed effective cycle* $D-Z(L)$.

In view of the (1, 1) correspondence $D_1 \rightarrow D_1+(D-Z(L))$ between cycles $D_1$ in $|D|$ and those in $|Z(L)|$, we have a natural projective structure in $|D|$. Any subspace of $|D|$ is called a *linear system on* $V$.

6. We shall now discuss an extension of the notion of a complete finite $k$-module in $k(V)$ and of the corresponding notion of a complete linear system on $V$.

If $S_1$ is any set of valuations of $k(V)$ *such that* $S_1$ *contains the set* $S_0$ of the preceding section and if $M$ is any finite $k$-module, we can consider the $S_1$-*completion* of $M$, i.e., the completion of $M$ with respect to the set $S_1$ (see (11)). We denote this completion by $M'_{S_1}$, and we say that $M$ is $S_1$-complete if $M'_{S_1}=M$. We reserve the notation $M'$ for the $S_0$-completion of $M$. It is clear that $M'_{S_1}$ is a subspace of $M'$. If $S$ is the set of *all* valuations of $k(V)$ then we know, by (12), that $M'_{S_1}=(M'_{S_1})'_S$ and that therefore $M'_{S_1}$ is also $S$-complete. Thus all our new complete modules are $S$-complete. We shall say that a finite $k$-module is complete *in the wide sense* if it is $S$-complete, *strictly complete* if it is $S_0$-complete.

We note that *given* $M$ *and* $S_1$ *there exists a finite set of valuations* $v_1, v_2, \cdots, v_q$ *such that if* $\tilde{S}$ *is the union of* $S_0$ *and* $\{v_1, v_2, \cdots, v_q\}$, *then* $M'_{S_1}=M'_{\tilde{S}}$. For, if $M'_{S_1}=M'$ then we can take for $\{v_1, \cdots, v_q\}$ the empty set. If $M'_{S_1}$ is a proper subset of $M'$ then there exists a valuation

$v_1$ in $S_1$ such that $M' \not\subset R_{v_1}M$. Then, if we denote by $\tilde{S}_1$ the set $\{S_0, v_1\}$ we will have $M' > M'_{\tilde{S}_1} \supset M'_{S_1}$. If $M'_{\tilde{S}_1} > M'_{S_1}$ we can find a valuation $v_2$ in $S_1$ such that $M'_{\tilde{S}_1} > M_{\tilde{S}_2} \supset M'_{S_1}$, where $\tilde{S}_2 = \{\tilde{S}_1, v_2\}$. Since $M'$ is finite-dimensional, this process must stop after a finite number of steps.

It follows from this observation that if we set $v_i(M) = \alpha_i$, where $\alpha_i$ is then an element of the value group of $v_i$, then *$M'_{S_1}$ consists of all elements $x$ of $M'$ satisfying the inequalities*

$$v_i(x) \geqq \alpha_i, \quad i = 1, 2, \cdots, q. \tag{14}$$

Conversely, let $v_1, v_2, \cdots, v_q$ be a finite set of valuations of $k(V)$, let $\alpha_1, \alpha_2, \cdots, \alpha_q$ be arbitrary elements of their respective value groups, let $M'$ be a strictly complete (finite) module and let $N$ be the set of all elements $x$ of $M'$ satisfying inequalities (14). *Then $N$ is complete in the wide sense and is, in fact, $S_1$-complete*, where $S_1 = \{S_0, v_1, v_2, \cdots, v_q\}$. For, if $y$ is any element of $\bigcap_{v \in S_1} R_v N$, then $y \in \bigcap_{v \in S_0} R_v N \subset \bigcap_{v \in S_0} R_v M' = M'$, i.e., $y \in M'$, and $v_i(y) \geqq v_i(N) \geqq \alpha_i$, whence $y \in N$.

We note that $N$ is also the set of elements of $N'$ satisfying (14), since $N' \subset M'$.

By an *elementary base condition* $(v, \alpha)$ (to be imposed on elements $x$ of $k(V)$) we mean an inequality of the type $v(x) \geqq \alpha$, where $v$ is a given valuation of $k(V)$ and $\alpha$ is a given element of the value group of $v$. The foregoing considerations can be then summarized as follows: *every complete (finite) k-module N, in the wide sense, is obtained from a strictly complete (finite) k-module* (in fact, from $N'$) *by imposing on the elements of the latter module a finite number of elementary base conditions, and every module thus obtained is complete in the wide sense.*

The choice of the finite set of elementary base conditions $(v, \alpha)$ is not uniquely determined by $N$. We shall show now that *any elementary base condition $(v, \alpha)$, imposed on a given finite k-module M, is equivalent with a suitable elementary base condition $(\tilde{v}, \nu)$ such that $\tilde{v}$ is a prime divisor of $k(V)/k$* (equivalent in the sense that both conditions determine the same submodule of $M$). It will follow from that, that any complete (finite) $k$-module, in the wide sense, can be obtained from a (strictly) complete $k$-module by imposing on the latter a finite number of elementary divisorial conditions. Naturally, the prime divisors in question will be, in general, of the second kind with respect to $V/k$.

To prove our assertion, we denote by $M_1$ the submodule of $M$ consisting of those elements $x$ of $M$ which satisfy the inequality $v(x) >$

$v(M)$. We next define in a similar way a submodule $M_2$ of $M_1$ ($M_2 = \{x \in M_1 | v(x) > v(M_1)\}$, and so we continue until we reach the submodule $N = M_h$ of $M$ consisting of those elements $x$ of $M$ which satisfy the given elementary base condition $v(x) \geqq \alpha$. We thus have $M > M_1 > M_2 > \cdots > M_h = N$. For each $i = 0, 1, \cdots, h-1$ ($M_0 = M$) we fix an element $z_i$ such that $z_i \in M_i$, $z_i \notin M_{i+1}$, we consider the finite $k$-module $L_i = \frac{1}{z_i} M_i$ consisting of the elements $x/z_i$, where $x$ is any element of $M_i$, and we denote by $R$ the finite integral domain $k[L_0, L_1, \cdots, L_{h-1}]$. It is clear that $v$ is non-negative on $R$ and that if $\mathfrak{p}$ denotes the center of $v$ in $R$ then a quotient $x/z_i$, with $x$ in $M$, belongs to $\mathfrak{p}$ if and only if $x \in M_{i+1}$. This being so, we fix a prime divisor $\tilde{v}$ of $k(V)$ whose center in $R$ is the prime ideal $\mathfrak{p}$ (see VI, § 14, Theorem 35) and we set $\tilde{v}(N) = \nu$. We show that if $x \in M$ and $\tilde{v}(x) \geqq \nu$, then $x \in N$, and this will establish the equivalence of the two base conditions $(v, \alpha)$, $(\tilde{v}, \nu)$ with regard to the module $M$. We shall show that the assumption that $x \in M_i$, $x \notin M_{i+1}$, $i < h$, leads to a contradiction. We have that $\frac{x}{z_i} \notin \mathfrak{p}$, whence $\tilde{v}(x) = \tilde{v}(z_i)$. On the other hand, since all quotients $\frac{u}{z_i}$, $u \in N$, belong to $\mathfrak{p}$, we have $\nu = \tilde{v}(N) > \tilde{v}(z_i)$. Hence $\tilde{v}(x) < \nu$, a contradiction.

A simple consequence of this result is the following:

*Every complete (finite) k-module M, in the wide sense, is a strictly complete k-module with reference to a suitable projective model $\tilde{V}/k$ of $k(V)/k$.* For the proof it is sufficient to construct a model $\tilde{V}/k$ of $k(V)$ such that:

(1) $\tilde{V}$ dominates $V$ (see VI, § 17);

(2) each of the prime divisors $v_1, v_2, \cdots, v_q$ which occur among the elementary divisorial base conditions of definition of $M$ is of the first kind with respect to $\tilde{V}/k$.

To construct a model $\tilde{V}$ satisfying these two conditions, we have only to construct first a model $V'_i$ of $k(V)/k$ such that the prime divisor $v_i$ is of the first kind with respect to $V'_i/k$ (VI, § 14, Theorem 31) and then take for $\tilde{V}$ the normalization of the join of $V/k$, $V'_1/k$, $V'_2/k$, $\cdots$, $V'_q/k$ (the join of two models has been defined in VI, § 17, and the extension to any finite member of models is obvious). If we set $S_1 = \{S_0, v_1, v_2, \cdots, v_q\}$ and denote by $\tilde{S}_0$ the set of prime divisors of $K/k$ which are of the first kind with respect to $\tilde{V}/k$ then we have $S_0 \subset S_1 \subset \tilde{S}_0$, and hence $M$, being $S_1$-complete, is also $\tilde{S}_0$-complete (see (12)), i.e., $M$ is strictly complete with reference to $\tilde{V}/k$.

We have discussed so far only the extension of the notion of complete $k$-modules. The corresponding extension of the notion of complete linear systems on $V/k$ is a straightforward matter of re-interpretation of the preceding discussion in terms of linear systems, taking into account that every $k$-module $M$ defines a divisorial cycle $D=Z(M)$ (see (13)) and a linear subsystem $L(M)$ of the complete linear system $|D|$ (defined by $M'$). If $M$ is complete (in the wide sense) we call $L(M)$ complete (in the wide sense). We thus can speak of "elementary base conditions" to be imposed on a linear system and we can then easily restate the preceding results in the terminology of linear systems.

We note that our definition of a complete linear system (in the wide sense) is invariant under birational transformations. For any such complete system is defined by a module $M$ which is $S$-complete, where $S$ is the set of all valuations of $k(V)/k$, and which therefore defines a complete linear system (in the wide sense) on any other projective model of $k(V)/k$.

# APPENDIX 5

## COMPLETE IDEALS IN REGULAR LOCAL RINGS OF DIMENSION 2

The theory of complete ideals in polynomial rings $k[x, y]$ of two variables presents some particularly striking features. This theory, developed by the senior author in 1938,† will be presented in this appendix in a much simpler form and in greater generality. The generalization consists in dealing with arbitrary regular local rings of dimension 2, rather than with that special class of such rings which is obtained by taking quotient rings of $k[x, y]$ with respect to maximal ideals in $k[x, y]$.

Very little is known about complete ideals in regular local rings of dimension greater than 2. It is almost certain that the theory developed in this appendix cannot be generalized to higher dimension without substantial modifications both of statements and proofs.

1. Let $\mathfrak{o}$ be a regular local ring *of dimension* 2. We shall denote by $\mathfrak{m}$ the maximal ideal of $\mathfrak{o}$ and by $k$ the residue field $\mathfrak{o}/\mathfrak{m}$ of $\mathfrak{o}$. By the unique factorization theorem in $\mathfrak{o}$ (Appendix 7, Lemma 2), every prime ideal in $\mathfrak{o}$, other than $\mathfrak{m}$, is principal, and every ideal $\mathfrak{A}$ in $\mathfrak{o}$ is of the form $x\mathfrak{B}$, where $x \in \mathfrak{o}$ and $\mathfrak{B}$ is an ideal which is primary for $\mathfrak{m}$ ($x$ being the g.c.d. of the elements of $\mathfrak{A}$, different from zero). If $\mathfrak{A}$ is complete, so is $\mathfrak{B}$ (since $\mathfrak{B} = \mathfrak{A} : \mathfrak{o}x$), and conversely. This fact indicates that in our proofs below we shall have to be concerned primarily with ideals in $\mathfrak{o}$ which are primary for $\mathfrak{m}$.

For any ideal $\mathfrak{A}$ in $\mathfrak{o}$, $\mathfrak{A} \neq (0)$, we denote by $r$, or $r(\mathfrak{A})$, the integer with the property: $\mathfrak{A} \subset \mathfrak{m}^r$, $\mathfrak{A} \not\subset \mathfrak{m}^{r+1}$, and we call *$r$ the order of $\mathfrak{A}$*. Clearly, $r \geqq 0$, and $r = 0$ if and only if $\mathfrak{A} = \mathfrak{o}$. In particular, the order $r$ of an element $x$ of $\mathfrak{o}$, $x \neq 0$, is the order of the principal ideal $\mathfrak{o}x$; thus, $x \in \mathfrak{m}^r$, $x \notin \mathfrak{m}^{r+1}$.

We know that the associated graded ring of $\mathfrak{o}$ (with respect to $\mathfrak{m}$) is a polynomial ring $k[z_1, z_2]$ in two variables, over $k$ (VIII, § 11, Theorem

† See O. Zariski, "Polynomial ideals defined by infinitely near base points," *Amer. J. Maths.*, 60 (1938), pp. 151–204.

25). If $0 \neq x \in \mathfrak{o}$ and $x$ is of order $r$, then $x$ has an initial form (VIII, p. 249), which we shall denote consistently by $\bar{x}$; here $\bar{x}$ is a homogeneous polynomial in $k[z_1, z_2]$, of degree $r$. The form $\bar{x}$ depends not only on $x$ but also on the choice of a regular system of parameters $t_1, t_2$ of $\mathfrak{o}$, the effect on $\bar{x}$ of a change of parameters being the same as that of a linear homogeneous (non-singular) transformation of the variables $z_1$ and $z_2$, with coefficients in $k$. We shall fix once and for always a regular system of parameters $t_1, t_2$. The fact that $x \in \mathfrak{m}^r$ and $x \notin \mathfrak{m}^{r+1}$ signifies that $x$ has an expression of the form $x = f(t_1, t_2)$, where $f$ is a homogeneous polynomial of degree $r$, with coefficients in $\mathfrak{o}$, *not all in* $\mathfrak{m}$. If we then denote by $\bar{f}$ the polynomial expression obtained from $f$ by reducing the coefficients of $f$ mod $\mathfrak{m}$, then $\bar{x} = \bar{f}(z_1, z_2)$.

If $\mathfrak{A}$ is an ideal in $\mathfrak{o}$, of order $r$, we denote by $L_i(\mathfrak{A})$ the set of initial forms of those elements of $\mathfrak{A}$ which are exactly of order $i$ (we include the zero in $L_i(\mathfrak{A})$). The homogeneous ideal in $k[z_1, z_2]$ which is generated by the union of all the sets $L_i(\mathfrak{A})$ is called the *initial ideal of* $\mathfrak{A}$ (compare with VIII, § 1). It is clear that $L_i(\mathfrak{A}) = \{0\}$ if $i < r$, $L_i(\mathfrak{A}) \neq \{0\}$ if $i \geqq r$, and that $L_i(\mathfrak{A})$ is a vector space over $k$.

We shall be particularly interested in the form space $L_r(\mathfrak{A})$. We shall call $L_r(\mathfrak{A})$ *the initial form module of* $\mathfrak{A}$. We denote by $c(\mathfrak{A})$ the greatest common divisor of the forms belonging to $L_r(\mathfrak{A})$ (and different from zero). We call $c(\mathfrak{A})$ the *characteristic form* of $\mathfrak{A}$. If $s$ is the degree of $c(\mathfrak{A})$ then $0 \leqq s \leqq r$.

The order function $r(x)$ defines a valuation $v_{\mathfrak{m}}$ of the quotient field of $\mathfrak{o}$ (VIII, § 11, Theorem 25, Corollary 1). We call this valuation $v_{\mathfrak{m}}$ the $\mathfrak{m}$-*adic prime divisor* of $\mathfrak{o}$; this is a discrete, rank 1, valuation, centered at $\mathfrak{m}$. Since $v_{\mathfrak{m}}(t_1) = v_{\mathfrak{m}}(t_2) = 1$, the $v_{\mathfrak{m}}$-residue of $t_2/t_1$, which we shall denote by $\tau$, is $\neq 0, \infty$. If $\alpha$ is any element of the residue field of $v_{\mathfrak{m}}$ ($\alpha \neq 0$), and if, say, $\alpha$ is the residue of $x/y$, where $x, y \in \mathfrak{o}$, then $x$ and $y$ must have the same order $r$. We can write then $x = f(t_1, t_2)$, $y = g(t_1, t_2)$, where $f$ and $g$ are forms of degree $r$, with coefficients in $\mathfrak{o}$, not all in $\mathfrak{m}$. Then $x/y = f(1, t_2/t_1)/g(1, t_2/t_1)$ and $\alpha = \bar{f}(1, \tau)/\bar{g}(1, \tau)$, where $\bar{f}$ and $\bar{g}$ denote the reduced polynomials, mod $\mathfrak{m}$. This shows that $k(\tau)$ *is the residue field of* $v_{\mathfrak{m}}$. It is immediately seen that $\tau$ is *transcendental over* $k$. In fact, if $\bar{F}(Z)$ is a non-zero polynomial with coefficient in $k$, of degree $r$, fix a polynomial $F(Z)$ of, degree $r$, with coefficients in $\mathfrak{o}$, which reduces to $\bar{F}$ mod $\mathfrak{m}$, write $F(z_2/z_1) = f(z_1, z_2)/z_1{}^r$, where $f$ is a form of degree $r$ (with coefficients in $\mathfrak{o}$), and consider the element $\xi = f(t_1, t_2)/t_1{}^r$. Since not all the coefficients of $f$ are in $\mathfrak{m}$, we have $v_{\mathfrak{m}}(f(t_1, t_2)) = r$, whence $v_{\mathfrak{m}}(\xi) = 0$. Therefore the $v_{\mathfrak{m}}$-residue of $\xi$ is different from zero. But this residue is obviously $\bar{F}(\tau)$. Hence $\bar{F}(\tau) \neq 0$, which proves that $\tau$ is transcendental over $k$. Thus, the residue field of $v_{\mathfrak{m}}$ *is a simple transcendental extension of* $k$ $(=\mathfrak{o}/\mathfrak{m})$, and this justifies our term $\mathfrak{m}$-*adic* prime *divisor* (see Appendix 2).

Let $v$ be any other valuation of the quotient field of $\mathfrak{o}$, centered at $\mathfrak{m}$ and *different from the* $\mathfrak{m}$-*adic prime divisor* $v_{\mathfrak{m}}$. We set $\gamma = v(\mathfrak{m})$, i.e., $\gamma = \min\{v(t_1), v(t_2)\}$. Let, say, $v(t_1) = \gamma$. Since $v$ is not the $\mathfrak{m}$-adic prime divisor $v_{\mathfrak{m}}$, there exists an element $x$ in $\mathfrak{o}$ $(x \neq 0)$ such that $v(x) > r\gamma$, where $r$ is the order of $x$. If we write, as above, $x = f(t_1, t_2)$, then we find that $v(f(1, t_2/t_1)) > 0$. Thus, if we denote by $\zeta$ the $v$-residue of $t_2/t_1$ then we find $\bar{f}(1, \zeta) = 0$, and hence $\zeta$ *is algebraic over* $k$. This conclusion holds for every valuation $v$ which is centered at $\mathfrak{m}$ and is different from the $\mathfrak{m}$-adic prime divisor $v_{\mathfrak{m}}$ of $\mathfrak{o}$.

DEFINITION 1. *Let* $\bar{g}(z_1, z_2)$ *be the irreducible form in* $k[z_1, z_2]$ *such that* $\bar{g}(1, \zeta) = 0$ (*the form* $\bar{g}$ *is determined only to within an arbitrary non-zero factor in* $k$). *Then* $\bar{g}(z_1, z_2)$ *is called the* DIRECTIONAL FORM *of the valuation* $v$.

The directional form of $v$ is, of course, of positive degree. We agree to regard 1 as the directional form of the $\mathfrak{m}$-adic prime divisor $v_{\mathfrak{m}}$.

LEMMA 1. *Let* $v$ *be a valuation centered at* $\mathfrak{m}$ *and different from the* $\mathfrak{m}$-*adic prime divisor, and let* $\bar{g}$ *be the directional form of* $v$. *If* $x$ *is an element of* $\mathfrak{o}$ $(x \neq 0)$, *of order* $r$, *then* $v(x) > r\gamma$ *if and only if the initial form* $\bar{x}$ *of* $x$ *is divisible by* $\bar{g}$ (in $k[z_1, z_2]$).

PROOF. In the preceding notations we have $v(x) > r\gamma$ if and only if $\bar{f}(1, \zeta) = 0$, hence if and only if $\bar{f}(z_1, z_2)$ is divisible by $\bar{g}(z_1, z_2)$ in $k[z_1, z_2]$. Since $\bar{x} = \bar{f}(z_1, z_2)$, the lemma is proved.

In the sequel we shall also speak of *directional forms* of an arbitrary ideal $\mathfrak{A}$ in $\mathfrak{o}$. We give namely the following definition:

DEFINITION 2. *If* $\mathfrak{A}$ *is an ideal in* $\mathfrak{o}$, *different from zero, every irreducible divisor* $\bar{g}(z_1, z_2)$ *of the characteristic form* $c(\mathfrak{A})$ *of* $\mathfrak{A}$ *is called a directional form of* $\mathfrak{A}$. (*We agree to regard* 1 *as a directional form of* $\mathfrak{A}$ *if* $c(\mathfrak{A}) = 1$.)

2. At this stage we shall introduce a construction which associates with each irreducible form $\bar{g}(z_1, z_2)$ in $k[z_1, z_2]$ another regular local ring $\mathfrak{o}'$, of dimension 2, which *dominates* $\mathfrak{o}$ (i.e., which is such that $\mathfrak{o}$ is a subring of $\mathfrak{o}'$, and $\mathfrak{m}' \cap \mathfrak{o} = \mathfrak{m}$, where $\mathfrak{m}'$ is the maximal ideal of $\mathfrak{o}'$) and has the same quotient field as $\mathfrak{o}$. In algebraic geometry this construction is the well-known "locally quadratic" transformation of an algebraic surface, with center at a given simple point $P$ of the surface.

Let $\bar{g} = \bar{g}(z_1, z_2)$ be an irreducible form in $k[z_1, z_2]$. We denote by $\mathfrak{o}'_{\bar{g}}$, or simply by $\mathfrak{o}'$ (whenever the form $\bar{g}$ is fixed throughout the argument), the set of all quotients $y/x$, where $x$ and $y$ are elements of $\mathfrak{o}$, such that: (1) order of $y \geqq$ order of $x$ (i.e., if $x \in \mathfrak{m}^n$ and $x \notin \mathfrak{m}^{n+1}$ then $y \in \mathfrak{m}^n$); (2) $\bar{g}$ does not divide the initial form $\bar{x}$ of $x$.

PROPOSITION 1. *Assuming that $\bar{g} \neq z_1$ we set $\tau' = t_2/t_1$ (where $(t_1, t_2)$ is a fixed pair of regular parameters of $\mathfrak{o}$) and $R' = \mathfrak{o}[\tau']$.*† *The set of elements $F(\tau')$ of $R'$ ($F$-polynomial over $\mathfrak{o}$) such that the reduced polynomial* (mod $\mathfrak{m}$) *$\bar{F}(z)$ is divisible by $\bar{g}(1, z)$ ($z$ being an indeterminate) is a maximal ideal in $R'$, and $\mathfrak{o}'$ is the ring of quotients of $R'$ with respect to this maximal ideal. Furthermore, $\mathfrak{o}'$ is a regular local ring, of dimension 2, which dominates $\mathfrak{o}$ (and is different from $\mathfrak{o}$). The residue field of $\mathfrak{o}'$ is $k(\alpha)$, where $\alpha$ is a root of the irreducible polynomial $\bar{g}(1, z)$. The $\mathfrak{m}$-adic prime divisor of $\mathfrak{o}$ is non-negative on $\mathfrak{o}'$ and its center in $\mathfrak{o}'$ is the principal ideal $\mathfrak{o}'t_1$.*

PROOF. Let $\bar{G}(z) = \bar{g}(1, z)$ and let $\alpha$ be a root of the irreducible polynomial $\bar{G}(z)$ in some extension field of $k$ (note that since $z_1 \neq \bar{g}(z_1, z_2)$, $\bar{G}(z)$ has *positive* degree). The transformation $\varphi$ of $R'$ onto the field $k(\alpha)$ which associates with each element $F(\tau')$ of $R'$ the element $\bar{F}(\alpha)$ of $k(\alpha)$ is *a mapping*. To see this it is only necessary to show that $\bar{F}(\alpha) = 0$ whenever $F(\tau') = 0$ (Vol. I, Ch. I, § 11, Lemma 2). Let $n$ be the degree of $F$ and write $F(z)$ in the form $f(1, z)$ where $f(z_1, z_2)$ is a form of degree $n$, with coefficients in $\mathfrak{o}$. The assumption $F(\tau') = 0$ implies that $f(t_1, t_2) = 0$. Hence, by the basic property of regular parameters, the coefficients of $f$, i.e., of $F$, are all in $\mathfrak{m}$. Hence $\bar{F}(z) = 0$, which proves our assertion. The mapping $\varphi$ is therefore a homomorphism of $R'$ onto the field $k(\alpha)$, and $\varphi$ is not an isomorphism since $\varphi = 0$ on $\mathfrak{m}$. The kernel of $\varphi$ is a maximal ideal $\mathfrak{p}'$ of $R'$. An element $F(\tau')$ of $R'$ belongs to $\mathfrak{p}'$ if and only if $\bar{F}(\alpha) = 0$, i.e., if and only if $\bar{F}(z)$ is divisible by $\bar{g}(1, z)$ in $k[z]$. This proves the first assertion of the proposition.

If $y/x \in \mathfrak{o}'$ and $n$ is the order of $x$, then we can write $x = f(t_1, t_2)$, $y = h(t_1, t_2)$, where $f$ and $h$ are forms of degree $n$, with coefficients in $\mathfrak{o}$, and $\bar{f}(z_1, z_2)$ is not divisible by $\bar{g}(z_1, z_2)$. Dividing both $x$ and $y$ by $t_1^n$ we find $y/x = H(\tau')/F(\tau')$, where $H(z) = h(1, z)$ and $F(z) = f(1, z)$. The fact that $\bar{f}(z_1, z_2)$ is not divisible by $\bar{g}(z_1, z_2)$ implies that $\bar{F}(z)$ is not divisible by $\bar{g}(1, z)$. Hence $F(\tau') \notin \mathfrak{p}'$, $y/x \in R'_{\mathfrak{p}'}$, and thus $\mathfrak{o}' \subset R'_{\mathfrak{p}'}$. Conversely, let $H(\tau')/F(\tau')$ be an arbitrary element of $R'_{\mathfrak{p}'}$, where therefore $\bar{F}(z)$ is not divisible by $\bar{g}(1, z)$. Let $s = \deg F(z)$ and let $f(z_1, z_2) = z_1^s F(z_2/z_1)$. Then $f(z_1, z_2)$ is a form of degree $s$, and $\bar{f}(z_1, z_2)$ is not divisible by $\bar{g}(z_1, z_2)$. If $n = \max(\deg H, \deg F)$, then $H(\tau')/F(\tau') = h_1(t_1, t_2)/f_1(t_1, t_2)$, where $h_1$ and $f_1$ are forms of degree

† If $\bar{g} = z_1$ the roles of $z_1$, $z_2$ as well as of $t_1$ and $t_2$ in this proposition have to be interchanged.

$n$ and $f_1(z_1, z_2) = z_1^{n-s}f(z_1, z_2)$. Since $\bar{g}(z_1, z_2) \neq z_1$, also $\bar{f}_1(z_1, z_2)$ is not divisible by $\bar{g}(z_1, z_2)$, and this shows that $H(\tau')/F(\tau') \in \mathfrak{o}'$. Thus $R'_{\mathfrak{p}'} \subset \mathfrak{o}'$, and consequently $R'_{\mathfrak{p}'} = \mathfrak{o}'$. Furthermore, if $\mathfrak{m}'$ is the maximal ideal of $\mathfrak{o}'$ then $\mathfrak{o}'/\mathfrak{m}' = R'/\mathfrak{p}' = k(\alpha)$. Thus $k(\alpha)$ is the residue field of $\mathfrak{o}'$.

If $x \in \mathfrak{m}$ we have $x = a_1t_1 + a_2t_2$, where $a_1, a_2 \in \mathfrak{o}$, or $x = t_1(a_1 + a_2\tau') = F(\tau')$, where $F(z) = a_1t_1 + a_2t_1z$. Then $\bar{F}(z) = 0$, showing that $x \in \mathfrak{m}'$. Thus $\mathfrak{m} \subset \mathfrak{m}'$, $\mathfrak{m}' \cap \mathfrak{o} = \mathfrak{m}$, and $\mathfrak{o}'$ *dominates* $\mathfrak{o}$. We observe that we have now shown incidentally that

$$(1) \qquad R'\mathfrak{m} = R't_1.$$

The element $t_2/t_1$ does not belong to $\mathfrak{o}$, as follows immediately from properties of regular parameters (or observe that the residue of $t_2/t_1$ in the $\mathfrak{m}$-adic prime divisor $v_{\mathfrak{m}}$ of $\mathfrak{o}$ is a transcendental over $k$, while the $v_{\mathfrak{m}}$-residue of each element of $\mathfrak{o}$ is in $k$). *Hence* $\mathfrak{o}$ *is a proper subring of* $\mathfrak{o}'$.

Let $q$ be the degree of $\bar{g}(z_1, z_2)$. We fix an element $u$ in $\mathfrak{o}$ such that $\bar{u} = \bar{g}(z_1, z_2)$ and we set

$$(2) \qquad t'_2 = u/t_1^q.$$

Then it is clear that $t'_2 \in \mathfrak{m}'$. We proceed to show that $t_1$ *and* $t'_2$ *form a basis of* $\mathfrak{m}'$.

Consider any element of $\mathfrak{m}'$; it can be written in the form $y/x$, where, if $x \in \mathfrak{m}^n$, $x \notin \mathfrak{m}^{n+1}$, then $\bar{x}$ is not divisible by $\bar{g}(z_1, z_2)$ and $y \in \mathfrak{m}^n$. It is clear that $x/t_1^n$ is an element of $R'$ which does not belong to $\mathfrak{p}'$ and hence is a unit in $\mathfrak{o}'$. Therefore $y/t_1^n$ must belong to $\mathfrak{p}'$. If $y \in \mathfrak{m}^{n+1}$, then, by (1), $y \in R't_1^{n+1}$ and $y/t_1^n \in R't_1$, and thus $y/x \in \mathfrak{o}'t_1$. Assume $y \notin \mathfrak{m}^{n+1}$. Then $\bar{y}$ must be divisible by $\bar{g}(z_1, z_2)$. We fix an element $\omega$ in $\mathfrak{o}$ such that $\bar{\omega}$ is of degree $n-q$ and such that $\bar{y} = \bar{\omega}\bar{g}(z_1, z_2)$. Then $y - \omega u \in \mathfrak{m}^{n+1}$, or $y - \omega u \in R' \cdot t_1^{n+1}$. Since $\omega/t_1^{n-q} \in R'$, we find that $y/t_1^n \in R't'_2 + R't_1$, showing that $y/x \in \mathfrak{o}'t_1 + \mathfrak{o}'t'_2$. This proves our assertion that

$$(3) \qquad \mathfrak{m}' = \mathfrak{o}'t_1 + \mathfrak{o}'t'_2.$$

We now consider the $\mathfrak{m}$-adic prime divisor $v_{\mathfrak{m}}$ of $\mathfrak{o}$. It is clear from the definition of $\mathfrak{o}'$ that $v_{\mathfrak{m}}$ is non-negative on $\mathfrak{o}'$. Since $v_{\mathfrak{m}}(t_1) = 1$, the center of $v_{\mathfrak{m}}$ in $\mathfrak{o}'$ contains the principal ideal $\mathfrak{o}'t_1$. Conversely, if $y/x$ belongs to the center of $v_{\mathfrak{m}}$ in $\mathfrak{o}'$ then in the preceding notations we must have $y \in \mathfrak{m}^{n+1}$, and we have already shown that this implies that $y/x \in \mathfrak{o}'t_1$. Hence *the principal ideal* $\mathfrak{o}'t_1$ *is the center of* $v_{\mathfrak{m}}$ *in* $\mathfrak{o}'$. Thus $\mathfrak{o}'t_1$ is a prime ideal in $\mathfrak{o}'$. It is different from $\mathfrak{m}'$ since $v_{\mathfrak{m}}(t'_2) = v_{\mathfrak{m}}(u) - q = 0$ and since therefore $t'_2 \notin \mathfrak{o}'t_1$. This shows that the local ring $\mathfrak{o}'$ is of dimension $\geqq 2$. Then (3) leads to the conclusion that $\mathfrak{o}'$ is a regular ring of dimension 2 and that $\{t_1, t'_2\}$ is a pair of regular parameters of $\mathfrak{o}'$. This completes the proof of the proposition.

The local ring $\mathfrak{o}'$ which we have just constructed will be referred to as the "*quadratic transform*" *of* $\mathfrak{o}$, *relative to the directional form* $\bar{g}$.

If $\mathfrak{A}$ is an ideal in $\mathfrak{o}$ and $r$ is the order of $\mathfrak{A}$ then $v_{\mathfrak{m}}(\mathfrak{A})=r$, whence $\mathfrak{o}'\mathfrak{A}\subset\mathfrak{o}'t_1^r$, $\mathfrak{o}'\mathfrak{A}\not\subset\mathfrak{o}'t_1^{r+1}$. Hence

$$\mathfrak{o}'\mathfrak{A} = t_1^r\mathfrak{A}', \tag{4}$$

where $\mathfrak{A}'$ is an ideal in $\mathfrak{o}'$. We call $\mathfrak{A}'$ *the transform of* $\mathfrak{A}$ *in* $\mathfrak{o}'$.

PROPOSITION 2. *Let* $\mathfrak{A}$ *be an ideal in* $\mathfrak{o}$, *of order* $r$, *and let* $\bar{g}^\sigma$ *be the highest power of* $\bar{g}$ *which divides the characteristic form* $c(\mathfrak{A})$ *of* $\mathfrak{A}$. *Then:*

(a) *The order* $r'$ *of the transform* $\mathfrak{A}'$ *of* $\mathfrak{A}$ *in* $\mathfrak{o}'$ *is not greater than* $\sigma$, *but is positive if* $\sigma$ *is positive. Thus, in particular,* $\mathfrak{A}'$ *is the unit ideal if and only if* $\bar{g}$ *does not divide* $c(\mathfrak{A})$.

(b) *If* $\mathfrak{A}$ *is primary for* $\mathfrak{m}$ *then* $\mathfrak{A}'$ *is either primary for* $\mathfrak{m}'$ *or is the unit ideal.*

PROOF. We fix an element $x$ in $\mathfrak{A}$ such that the initial form $\bar{x}$ is of degree $r$ and is exactly divisible by $\bar{g}^\sigma$. Then $x'=x/t_1^r\in\mathfrak{A}'$. Let $\bar{x}=\bar{g}^\sigma\bar{\psi}$, where $\bar{\psi}$ is of degree $r-\sigma q$ ($q$ being the degree of $\bar{g}$), and let $v$ be an element of $\mathfrak{o}$ such that $\bar{v}=\bar{\psi}$. Then $x-u^\sigma v\in\mathfrak{m}^{r+1}$, where $u$ is the previously chosen element of $\mathfrak{o}$ such that $\bar{u}=\bar{g}$. Dividing through by $t_1^r$ and setting $v'=v/t_1^{r-\sigma q}$ we find (using (1)) that $x'-t'^{\sigma}_2v'\in\mathfrak{o}'t_1$ (recall (2)). Now, since $\bar{v}$ is not divisible by $\bar{g}$, $v'$ is a unit in $\mathfrak{o}'$. Hence $x'\notin\mathfrak{m}'^{\sigma+1}$, showing that $r'\leqq\sigma$. If $\sigma$ is positive then the above argument shows that if $x$ is an arbitrary element of $\mathfrak{A}$ and $x'=x/t_1^r$ then $x'\in\mathfrak{m}'$ (and in particular, $x'\in\mathfrak{o}'t_1$ if $x\in\mathfrak{m}^{r+1}$). This shows that if $\sigma$ is positive then $\mathfrak{A}'\subset\mathfrak{m}'$, completing the proof of part (a) of the proposition.

Assume now that $\mathfrak{A}$ is primary for $\mathfrak{m}$. If $c(\mathfrak{A})$ is not divisible by $\bar{g}$ then we know already that $\mathfrak{A}'=(1)$. If $c(\mathfrak{A})$ is divisible by $\bar{g}$, and if, say, as in the preceding part of the proof, $x$ is an element of $\mathfrak{A}$ such that $\bar{x}$ is of degree $r$ and is exactly divisible by $\bar{g}^\sigma$, then we have in $\mathfrak{A}'$ the element $x'$ such that $x'-t'^{\sigma}_2v'\in\mathfrak{o}'t_1$, where $v'$ is a unit in $\mathfrak{o}'$. On the other hand, since $\mathfrak{A}$ is primary for $\mathfrak{m}$, some power of $t_1$ belongs to $\mathfrak{A}$, say $t_1^n\in\mathfrak{A}$, where we may assume $n>r$. Then $t_1^{n-r}\in\mathfrak{A}'$, i.e., some power of $t_1$ belongs to $\mathfrak{A}'$. Since $x'\in\mathfrak{A}'$, this implies at once that also some power of $t'_2$ belongs to $\mathfrak{A}'$, showing that $\mathfrak{A}'$ is primary for $\mathfrak{m}'$. This completes the proof.

We shall now study the class of ideals $\mathfrak{A}$ in $\mathfrak{o}$ which are contractions of ideals in $\mathfrak{o}'$ (any ideal $\mathfrak{A}$ in this class is then necessarily the contraction of its extended ideal $\mathfrak{o}'\mathfrak{A}$). We shall refer to the ideals of that class as *contracted ideals.* (Later on we shall use this term in a wider sense, since we shall replace $\mathfrak{o}'$ by a semi-local ring $\mathfrak{o}'_1 \cap \mathfrak{o}'_2 \cap \cdots \cap \mathfrak{o}'_m$, where the local rings $\mathfrak{o}'_1, \mathfrak{o}'_2, \cdots, \mathfrak{o}'_m$ are quadratic transforms of $\mathfrak{o}$, relative to distinct directional forms $\bar{g}_1, \bar{g}_2, \cdots, \bar{g}_m$.) We observe that every power of $\mathfrak{m}$ is a contracted ideal, since $\mathfrak{o}'\mathfrak{m}^n = \mathfrak{o}'t_1^n$ and clearly $\mathfrak{o}'t_1^n \cap \mathfrak{o} = \mathfrak{m}^n$ (every element of $\mathfrak{o}'t_1^n$ has value $\geqq n$ in the $\mathfrak{m}$-adic prime divisor of $\mathfrak{o}$). We also observe that as a consequence of Proposition 2, part (a) we have the following

COROLLARY. *If $\mathfrak{A}$ is a contracted ideal and $\bar{g}$ does not divide the characteristic form $c(\mathfrak{A})$ of $\mathfrak{A}$ then $\mathfrak{A}$ is a power of $\mathfrak{m}$.*

For we have then $\mathfrak{A}' = \mathfrak{o}'$ and hence $\mathfrak{A} = \mathfrak{o}'t_1^r \cap \mathfrak{o} = \mathfrak{m}^r$, where $r$ is the order of $\mathfrak{A}$.

PROPOSITION 3. *Let $\mathfrak{A}$ be an ideal in $\mathfrak{o}$, primary for $\mathfrak{m}$, let $r$ be the order of $\mathfrak{A}$, let $s$ be the degree of $c(\mathfrak{A})$ and let*

$$\mathfrak{B} = \mathfrak{A} : \mathfrak{m}^{r-s}. \tag{5}$$

*If $\mathfrak{A}$ is a contracted ideal then $c(\mathfrak{A})$ is a power of $\bar{g}$ (possibly, $c(\mathfrak{A}) = 1$), and we have*

$$\mathfrak{A} = \mathfrak{B}\mathfrak{m}^{r-s}. \tag{6}$$

*Furthermore, $\mathfrak{B}$ also is a contracted ideal, and we have $r(\mathfrak{B}) = s$, $c(\mathfrak{B}) = c(\mathfrak{A})$.*

PROOF. If $\bar{g}$ does not divide $c(\mathfrak{A})$ then, by the above corollary, $\mathfrak{A} = \mathfrak{m}^r$, $c(\mathfrak{A}) = 1$, and all the assertions of the lemma are trivial ($\mathfrak{B}$ is now the unit ideal). We shall therefore now assume that $\bar{g}$ divides $c(\mathfrak{A})$.

Let $\bar{\varphi} = \bar{g}^\lambda$ be the highest power of $\bar{g}$ which divides $c(\mathfrak{A})$, and let $\sigma$ be the degree of $\bar{\varphi}$. We fix in $\mathfrak{A}$ an element $x$ such that $\bar{x}$ is of degree $r$ and $\bar{x} = \bar{\varphi}\bar{\psi}$, with $\bar{\varphi}$ and $\bar{\psi}$ relatively prime; here $\bar{\psi}$ is a form $\bar{\psi}(z_1, z_2)$ of degree $r - \sigma$, with coefficients in $k$. We assert that for each integer $j \geqq 1$ there exist elements $x_j$ and $y_j$ in $\mathfrak{o}$ such that

$$x - x_j y_j \in \mathfrak{m}^{r+j}; \quad \bar{x}_j = \bar{\varphi}, \; \bar{y}_j = \bar{\psi}. \tag{7}$$

The assertion is trivial for $j = 1$ (take for $x_1$ and $y_1$ any two elements of $\mathfrak{o}$ such that $\bar{x}_1 = \bar{\varphi}$, $\bar{y}_1 = \bar{\psi}$). Assume that, for a given $j$, a pair of elements $x_j$ and $y_j$ satisfying (7) has already been found. Since $\bar{\varphi}$ and $\bar{\psi}$ are relatively prime, every form in $z_1, z_2$, of degree $\geqq r-1$, belongs to the homogeneous ideal $(\bar{\varphi}, \bar{\psi})$ in $k[z_1, z_2]$.† We apply this

† This is trivial if $\bar{\psi} = 1$. If the degree $r - \sigma$ of $\bar{\psi}$ is positive we observe that the space of forms of degree $r-1$ which can be written as linear combinations $A_{r-\sigma-1}\bar{\varphi} + B_{\sigma-1}\bar{\psi}$, where $A$ and $B$ are forms of degree $r - \sigma - 1$ and $\sigma - 1$ respectively, has precisely the desired dimension $r(= (r-\sigma) + \sigma)$ and thus consists of all the forms of degree $r-1$.

fact. If $x - x_j y_j \in \mathfrak{m}^{r+j+1}$ we set $x_{j+1} = x_j$, $y_{j+1} = y_j$. In the contrary case, we express the initial form $\overline{x - x_j y_j}$ (of degree $r+j$) as a linear combination $A_{r+j-\sigma}\bar{\varphi} + B_{j+\sigma}\bar{\psi}$, we fix elements $v$ and $w$ in $\mathfrak{o}$ such that $\bar{v} = B_{j+\sigma}$, $\bar{w} = A_{r+j-\sigma}$ and we set $x_{j+1} = x_j + v$, $y_{j+1} = y_j + w$. Then it is seen at once that $x - x_{j+1}y_{j+1} \in \mathfrak{m}^{r+j+1}$, $\bar{x}_{j+1} = \bar{\varphi}$, $y_{j+1} = \bar{\psi}$. This establishes (7) for all $j$.

We now define $\mathfrak{B}$ by $\mathfrak{B} = \mathfrak{A} : \mathfrak{m}^{r-\sigma}$, whence

$$\mathfrak{A} \supset \mathfrak{m}^{r-\sigma}\mathfrak{B}. \tag{8}$$

Since $\mathfrak{B}$ is primary for $\mathfrak{m}$ we have $\mathfrak{m}^{r+j} \subset \mathfrak{m}^{r-\sigma}\mathfrak{B}$ for large $j$. For such a large $j$ relation (7) yields the inclusion $x_j y_j \in \mathfrak{A}$. Hence $\dfrac{x_j}{t_1{}^\sigma} \cdot \dfrac{y_j}{t_1{}^{r-\sigma}} \in \mathfrak{A}'$, where $\mathfrak{o}'\mathfrak{A} = t^r\mathfrak{A}'$. Since $\bar{y}_j$ is not divisible by $\bar{g}$ and is exactly of degree $r - \sigma$, $y_j/t_1{}^{r-\sigma}$ is a unit in $\mathfrak{o}'$. Hence $x_j/t_1{}^\sigma \in \mathfrak{A}'$ and $\mathfrak{o}'x_j\mathfrak{m}^{r-\sigma} \subset t_1{}^r\mathfrak{A}' = \mathfrak{o}'\mathfrak{A}$. Since $\mathfrak{o}'\mathfrak{A} \cap \mathfrak{o} = \mathfrak{A}$, it follows that $x_j\mathfrak{m}^{r-\sigma} \subset \mathfrak{A}$, $x_j \in \mathfrak{B}$, $x_j y_j \in \mathfrak{m}^{r-\sigma}\mathfrak{B}$, i.e., $x$ belongs to $\mathfrak{m}^{r-\sigma}\mathfrak{B}$. This holds for every element $x$ of $\mathfrak{A}$ which does not belong to $\mathfrak{m}^{r+1}$ *and* is such that $\bar{x}$ is not divisible by $\bar{g}^{\lambda+1}$. Now, if $y$ is any element of $\mathfrak{A}$ which does not satisfy either one of these two conditions, then we see that both $x$ and $x + y$ belong to $\mathfrak{m}^{r-\sigma}\mathfrak{B}$, and hence $y \in \mathfrak{m}^{r-\sigma}\mathfrak{B}$. We have thus shown that $\mathfrak{A} \subset \mathfrak{m}^{r-\sigma}\mathfrak{B}$, and this, in conjunction with (8), yields the equality $\mathfrak{A} = \mathfrak{m}^{r-\sigma}\mathfrak{B}$. This equality implies that $c(\mathfrak{A})$ is at most of degree $\sigma$. Since $\bar{\varphi}(= \bar{g}^\lambda)$ divides $c(\mathfrak{A})$ and is of degree $\sigma$ we conclude that $c(\mathfrak{A}) = \bar{g}^\lambda$ and that $\sigma = s$. We thus have (5) and (6). By (6) we have at once that $r(\mathfrak{B}) = s$ and that consequently $c(\mathfrak{B}) = c(\mathfrak{A}) = \bar{g}^\lambda$. Thus everything is proved except the assertion that $\mathfrak{B}$ also is a contracted ideal. Now, we have $\mathfrak{o}'\mathfrak{B} = t^s\mathfrak{A}'$, where $\mathfrak{A}'$ is the transform of $\mathfrak{A}$ (i.e., $\mathfrak{o}'\mathfrak{A} = t^r\mathfrak{A}'$). If then $y \in \mathfrak{o}'\mathfrak{B} \cap \mathfrak{o}$ then $\mathfrak{o}'y\mathfrak{m}^{r-s} \subset t^r\mathfrak{A}' = \mathfrak{o}'\mathfrak{A}$, whence $y\mathfrak{m}^{r-s} \subset \mathfrak{A}$, $y \in \mathfrak{A} : \mathfrak{m}^{r-s} = \mathfrak{B}$. This completes the proof.

COROLLARY 1. *If $\sigma$ is an integer such that $r \geqq \sigma \geqq s$ and we set $\mathfrak{A} : \mathfrak{m}^{r-\sigma} = \mathfrak{C}$, then $\mathfrak{A} = \mathfrak{m}^{r-\sigma}\mathfrak{C}$, $\mathfrak{C} = \mathfrak{m}^{\sigma-s}\mathfrak{B}$, and $\mathfrak{C}$ also is a contracted ideal.*

We have $\mathfrak{A} = \mathfrak{m}^{r-s}\mathfrak{B} = \mathfrak{m}^{r-\sigma} \cdot \mathfrak{m}^{\sigma-s}\mathfrak{B}$, hence $\mathfrak{m}^{\sigma-s}\mathfrak{B} \subset \mathfrak{C}$, and thus $\mathfrak{A} \subset \mathfrak{m}^{r-\sigma}\mathfrak{C} \subset \mathfrak{A}$. Consequently $\mathfrak{A} = \mathfrak{m}^{r-\sigma}\mathfrak{C}$. The proof that $\mathfrak{C}$ is a contracted ideal is identical with the above proof that $\mathfrak{B}$ is a contracted ideal (with $s$ replaced by $\sigma$). Furthermore, we have $\mathfrak{C} : \mathfrak{m}^{\sigma-s} = (\mathfrak{A} : \mathfrak{m}^{r-\sigma}) : \mathfrak{m}^{\sigma-s} = \mathfrak{A} : \mathfrak{m}^{r-s} = \mathfrak{B}$. Since $r(\mathfrak{C}) = \sigma$, $c(\mathfrak{C}) = c(\mathfrak{A})$, the relation $\mathfrak{C} = \mathfrak{m}^{\sigma-s}\mathfrak{B}$ follows by applying Proposition 3 to the ideal $\mathfrak{C}$ instead of to $\mathfrak{A}$.

COROLLARY 2. *If $\mathfrak{A}$ is a contracted ideal, primary for $\mathfrak{m}$, and $q$ is any integer $\geqq 1$, then $\mathfrak{m}^q\mathfrak{A}$ also is a contracted ideal.*

Let $\mathfrak{D} = \mathfrak{o}'\mathfrak{m}^q\mathfrak{A} \cap \mathfrak{o}$. Using the notations of Proposition 3, we have $r(\mathfrak{D}) \leqq q + r$ (since $\mathfrak{D} \supset \mathfrak{m}^q\mathfrak{A}$), and $v_\mathfrak{m}(\mathfrak{D}) = q + r$. Hence $r(\mathfrak{D}) = q + r$.

The characteristic form $c(\mathfrak{D})$ of $\mathfrak{D}$ divides $c(\mathfrak{m}^q\mathfrak{A})$, i.e., $c(\mathfrak{A})$. If then $\sigma$ is the degree of $c(\mathfrak{D})$ we have $\sigma \leqq s$. Since $\mathfrak{D}$ is a contracted ideal, we have, by Proposition 3:

$$\mathfrak{D} = \mathfrak{m}^{r+q-\sigma}\mathfrak{C},$$

where $\mathfrak{C} = \mathfrak{D} : \mathfrak{m}^{r+q-\sigma}$ is again a contracted ideal. Since $\mathfrak{o}'\mathfrak{D} = \mathfrak{o}'\mathfrak{m}^q\mathfrak{A}$, it follows that $\mathfrak{o}'\mathfrak{m}^{r-\sigma}\mathfrak{C} = \mathfrak{o}'\mathfrak{A}$. By Corollary 1, applied to the ideal $\mathfrak{D}$ instead of to $\mathfrak{A}$, $\mathfrak{m}^{r-\sigma}\mathfrak{C}$ is a contracted ideal. Hence $\mathfrak{m}^{r-\sigma}\mathfrak{C} = \mathfrak{o}'\mathfrak{m}^{r-\sigma}\mathfrak{C} \cap \mathfrak{o} = \mathfrak{o}'\mathfrak{A} \cap \mathfrak{o} = \mathfrak{A}$. Therefore $\mathfrak{D} = \mathfrak{m}^q\mathfrak{A}$, which proves our assertion.

COROLLARY 3. *If $\mathfrak{A}$ is a contracted ideal in $\mathfrak{o}$, primary for $\mathfrak{m}$, then the initial form module of $\mathfrak{A}$ is the set of all forms of degree $r$ (= order of $\mathfrak{A}$) which are divisible by $c(\mathfrak{A})$.*

This is a direct consequence of (6).

3. We now undertake an extension of the preceding results to the semi-local case, as explained below.

Let $\bar{g}_1, \bar{g}_2, \cdots, \bar{g}_m$ be distinct irreducible forms in $k[z_1, z_2]$ (distinct in the sense that no two are associates) and let $\mathfrak{o}'_i$ be the quadratic transform of $\mathfrak{o}$, relative to the directional form $\bar{g}_i$. We set

$$\mathfrak{o}' = \mathfrak{o}'_1 \cap \mathfrak{o}'_2 \cap \cdots \cap \mathfrak{o}'_m. \tag{9}$$

It is not difficult to see that $\mathfrak{o}'$ is a semi-local ring having $m$ maximal ideals $\mathfrak{M}'_1, \mathfrak{M}'_2, \cdots, \mathfrak{M}'_m$, where $\mathfrak{M}'_i = \mathfrak{o}' \cap \mathfrak{m}'_i$, $\mathfrak{m}'_i$ being the maximal ideal of $\mathfrak{o}'_i$, and that $\mathfrak{o}'_i = \mathfrak{o}'_{\mathfrak{M}'_i}$. This is obvious if $\bar{g}_i \neq z_1$ for $i = 1, 2, \cdots, m$, because in that case each $\mathfrak{o}'_i$ is a ring of quotients of $R'$ $(= \mathfrak{o}[t_2/t_1])$ with respect to a maximal ideal $\mathfrak{p}'_i$; and similarly if $\bar{g}_i \neq z_2$ for $i = 1, 2, \cdots, m$. If both $z_1$ and $z_2$ are among the $m$ forms $\bar{g}_i$ and if $k$ is an infinite field, we can choose a linear form $c_1z_1 + c_2z_2$ $(c_1, c_2 \in k)$ which is different from all the $\bar{g}_i$, and we can reduce the situation to the case $\bar{g}_i \neq z_1$ $(i = 1, 2, \cdots, m)$ by choosing a new pair of regular parameters $\tau_1, \tau_2$ such that $\bar{\tau}_1 = c_1z_1 + c_2z_2$. The following procedure will work, however, also in the case of a finite field $k$. We choose an irreducible form $\bar{h}(z_1, z_2)$ in $k[z_1, z_2]$ which is different from all the $\bar{g}_i$, we fix an element $\xi$ in $\mathfrak{o}$ such that $\bar{\xi} = \bar{h}(z_1, z_2)$ and, denoting by $\lambda$ the degree of $\bar{h}$, we consider the ring

$$S' = \mathfrak{o}\left[\frac{t_1^\lambda}{\xi}, \frac{t_1^{\lambda-1}t_2}{\xi}, \cdots, \frac{t_2^\lambda}{\xi}\right].$$

The ring $S'$ consists of all quotients of the form $\eta/\xi^n$, where $n$ is an arbitrary integer and $\eta \in \mathfrak{m}^{n\lambda}$. Since $\bar{g}_i$ does not divide $\bar{\xi}$, $S'$ is a subring of $\mathfrak{o}'_i$. Let $\mathfrak{m}'_i \cap S' = \mathfrak{p}'_i$. Then $\mathfrak{p}'_i$ is a prime ideal in $S'$, and we have $S'_{\mathfrak{p}'_i} \subset \mathfrak{o}'_i$. On the other hand, let $y/x$ be any element of $\mathfrak{o}'_i$, where $x, y \in \mathfrak{o}$, $\bar{x}$ is not divisible by $\bar{g}_i$ and order $y \geqq$ order $x$. Upon replacing, if necessary, $x$ and $y$ by $x^\lambda$ and $x^{\lambda-1}y$ respectively, we may assume that the order of $x$ is a multiple of $\lambda$, say $n\lambda$. Then writing $y/x$ as a quotient of $y/\xi^n$ and $x/\xi^n$ we conclude at once that $y/x \in S'_{\mathfrak{p}'_i}$. Thus $\mathfrak{o}'_i = S'_{\mathfrak{p}'_i}$, $i = 1, 2, \cdots, m$, showing in the first place that each $\mathfrak{p}'_i$ is a maximal ideal of $S'$ and that—since $S'$ is noetherian—$\mathfrak{o}'$ is a semi-local ring. The relations $\mathfrak{o}'_i = \mathfrak{o}'_{\mathfrak{M}'_i}$ are now obvious. It now also follows that $\mathfrak{o}'$ is the set of all quotients $y/x$, where $x, y \in \mathfrak{o}$, $\bar{x}$ is not divisible by any of the $m$ forms $\bar{g}_i$, and order of $y \geqq$ order of $x$.

PROPOSITION 4. *Let $\mathfrak{A}$ be an ideal in $\mathfrak{o}$, primary for $\mathfrak{m}$, and let $r$ be the order of $\mathfrak{A}$. We assume that $\mathfrak{o}'\mathfrak{A} \cap \mathfrak{o} = \mathfrak{A}$, where $\mathfrak{o}'$ is the semi-local ring defined in* (9). *We set*

$$(10) \qquad \mathfrak{A}_i = \mathfrak{o}'_i\mathfrak{A} \cap \mathfrak{o}, \quad i = 1, 2, \cdots, m.$$

*Then*

$$(11) \qquad \mathfrak{A} = \mathfrak{A}_1 \cap \mathfrak{A}_2 \cap \cdots \cap \mathfrak{A}_m.$$

*The characteristic form $c(\mathfrak{A}_i)$ of $\mathfrak{A}_i$ is a power of $\bar{g}_i$:*

$$(12) \qquad c(\mathfrak{A}_i) = \bar{g}_i^{\lambda_i}, \quad (\lambda_i \geqq 0)$$

*and we have*

$$(13) \qquad c(\mathfrak{A}) = \prod_{i=1}^{m} \bar{g}_i^{\lambda_i}.$$

*If $c(\mathfrak{A}) = 1$ then $\mathfrak{A}$ is a power of $\mathfrak{m}$. If $c(\mathfrak{A}) \neq 1$ and if for a suitable labeling of the indices we have $\lambda_i \geqq 1$ for $i = 1, 2, \cdots, n$ $(1 \leqq n \leqq m)$ and $\lambda_i = 0$ for $i = n+1, \cdots, m$, then setting*

$$\mathfrak{O}' = \mathfrak{o}'_1 \cap \mathfrak{o}'_2 \cap \cdots \cap \mathfrak{o}'_n$$

*we have already*

$$\mathfrak{O}'\mathfrak{A} \cap \mathfrak{o} = \mathfrak{A}.$$

PROOF. We set $\mathfrak{C}' = \mathfrak{o}'\mathfrak{A}$, $\mathfrak{C}'_i = \mathfrak{o}'_i\mathfrak{A} = \mathfrak{o}'_i\mathfrak{C}'$. From the theory of quotient rings we know (Vol. I, Ch. IV, § 11, Theorem 19) that $\mathfrak{C}'_i \cap \mathfrak{o}'$ is obtained from $\mathfrak{C}'$ by considering an irredundant decomposition of $\mathfrak{C}'$ into primary components and deleting those components which are not contained in $\mathfrak{M}'_i$ (recall that $\mathfrak{o}'_i = \mathfrak{o}'_{\mathfrak{M}'_i}$). Since $\mathfrak{M}'_1, \mathfrak{M}'_2, \cdots, \mathfrak{M}'_m$ are all the maximal ideals of $\mathfrak{o}'$, it follows that

$$\bigcap_{i=1}^{m} (\mathfrak{C}'_i \cap \mathfrak{o}') = \mathfrak{C}',$$

or equivalently, since $\bigcap_{i=1}^{m} \mathfrak{C}'_i \subset \mathfrak{o}'$:

$$\bigcap_{i=1}^{m} \mathfrak{C}'_i = \mathfrak{C}',$$

i.e.,

$$\bigcap_{i=1}^{m} \mathfrak{o}'_i\mathfrak{A} = \mathfrak{o}'\mathfrak{A}. \tag{14}$$

Using (10) and the assumption that $\mathfrak{o}'\mathfrak{A} \cap \mathfrak{o} = \mathfrak{A}$, we find (11). Now each $\mathfrak{A}_i$ is, by definition, the contraction of an ideal in $\mathfrak{o}'_i$. Hence, by Proposition 3, we have that $c(\mathfrak{A}_i)$ is a power of $\bar{g}_i$, which proves (12).

Since $\mathfrak{A} \subset \mathfrak{A}_i$, we have $r(\mathfrak{A}_i) \leqq r$. On the other hand, since $\mathfrak{A}_i \subset \mathfrak{o}'_i\mathfrak{A}$ we have $v_{\mathfrak{m}}(\mathfrak{A}_i) \geqq v_{\mathfrak{m}}(\mathfrak{A}) = r$, where $v_{\mathfrak{m}}$ is the $\mathfrak{m}$-adic prime divisor of $\mathfrak{o}$. Hence $r(\mathfrak{A}_i) \geqq r$, and thus

$$r(\mathfrak{A}_i) = r = r(\mathfrak{A}). \tag{15}$$

Applying again Proposition 3 we find that if we set

$$\mathfrak{A}_i : \mathfrak{m}^{r-s_i} = \mathfrak{B}_i, \tag{15a}$$

where $s_i$ = degree of $c(\mathfrak{A}_i)$, then

$$\mathfrak{A}_i = \mathfrak{m}^{r-s_i}\mathfrak{B}_i. \tag{16}$$

We set $s = s_1 + s_2 + \cdots + s_m$, and we observe that since $r(\mathfrak{B}_i) = s_i$ we have $\mathfrak{m}^{r-s}\mathfrak{B}_1\mathfrak{B}_2 \cdots \mathfrak{B}_m \subset \mathfrak{B}_1\mathfrak{m}^{r-s_1} \subset \mathfrak{A}_1$, and similarly that $\mathfrak{m}^{r-s}\mathfrak{B}_1\mathfrak{B}_2 \cdots \mathfrak{B}_m \subset \mathfrak{A}_i$, $i = 1, 2, \cdots, m$. Hence

$$\mathfrak{m}^{r-s}\mathfrak{B}_1\mathfrak{B}_2 \cdots \mathfrak{B}_m \subset \mathfrak{A}. \tag{17}$$

Since $c(\mathfrak{B}_i) = c(\mathfrak{A}_i) = \bar{g}_i^{\lambda_i}$, the characteristic form of the ideal on the left-hand side of (17) is $\prod \bar{g}_i^{\lambda_i}$. Hence it follows from (17) that $c(\mathfrak{A})$ divides $\prod \bar{g}_i^{\lambda_i}$. On the other hand, since $\mathfrak{A} \subset \mathfrak{A}_i$, $\bar{g}_i^{\lambda_i}$ must divide $c(\mathfrak{A})$. (In this argument one must bear in mind that the ideals $\mathfrak{A}$, $\mathfrak{A}_i$ and $\mathfrak{m}^{r-s}\mathfrak{B}_1\mathfrak{B}_2 \cdots \mathfrak{B}_m$ all have the same order $r$.) This proves (13).

If $c(\mathfrak{A}) = 1$, then all $s_i$ are 0, $\mathfrak{A}_i = \mathfrak{m}^r$ (by (16)), and thus $\mathfrak{A} = \mathfrak{m}^r$.

If $c(\mathfrak{A}) \neq 1$ and $s_i = 0$ for $i = n+1, \cdots, m$, then $\mathfrak{A}_i = \mathfrak{m}^r$ and $\mathfrak{o}'_i\mathfrak{A} = \mathfrak{o}'_i\mathfrak{A}_i = \mathfrak{o}'_i\mathfrak{m}^r$ for $i = n+1, \cdots, m$. Hence, by (14)

$$\mathfrak{o}'\mathfrak{A} = \mathfrak{o}'_1\mathfrak{A} \cap \mathfrak{o}'_2\mathfrak{A} \cap \cdots \cap \mathfrak{o}'_n\mathfrak{A} \cap \mathfrak{o}'\mathfrak{m}^r.$$

Now, each ideal $\mathfrak{o}'_i\mathfrak{m}$ is the center, in $\mathfrak{o}'_i$, of the $\mathfrak{m}$-adic prime divisor of $\mathfrak{o}$. Hence $\mathfrak{o}'\mathfrak{m}^r$ is the symbolic power $\mathfrak{P}'^{(r)}$, where $\mathfrak{P}'$ is the center of $v_{\mathfrak{m}}$ in $\mathfrak{o}'$. It follows that every prime ideal of $\mathfrak{o}'\mathfrak{A}$ is contained in one of the maximal ideals $\mathfrak{M}'_1, \mathfrak{M}'_2, \cdots, \mathfrak{M}'_n$. Therefore $\mathfrak{o}'\mathfrak{A} = \mathfrak{O}'\mathfrak{A} \cap \mathfrak{o}'$, and thus $\mathfrak{A} = \mathfrak{o}'\mathfrak{A} \cap \mathfrak{o} = \mathfrak{O}'\mathfrak{A} \cap \mathfrak{o}' \cap \mathfrak{o} = \mathfrak{O}'\mathfrak{A} \cap \mathfrak{o}$.

This completes the proof.

We call an ideal $\mathfrak{A}$ in $\mathfrak{o}$ a *contracted ideal* if we have $\mathfrak{o}'\mathfrak{A} \cap \mathfrak{o} = \mathfrak{A}$ for some semi-local ring $\mathfrak{o}'$ which is an intersection of suitable quadratic transforms $\mathfrak{o}'_1, \mathfrak{o}'_2, \cdots, \mathfrak{o}'_m$ of $\mathfrak{o}$.

COROLLARY 1. *Let $\mathfrak{A}$ be an ideal in $\mathfrak{o}$, primary for $\mathfrak{m}$, let $\bar{g}_1, \bar{g}_2, \cdots, \bar{g}_m$ be the distinct irreducible factors of the characteristic form $c(\mathfrak{A})$ of $\mathfrak{A}$, let $\mathfrak{o}'_i$ be the quadratic transform of $\mathfrak{o}$, relative to the directional form $\bar{g}_i$, and let $\mathfrak{o}' = \mathfrak{o}'_1 \cap \mathfrak{o}'_2 \cap \cdots \cap \mathfrak{o}'_m$. If $\mathfrak{A}$ is a contracted ideal then already $\mathfrak{o}'\mathfrak{A} \cap \mathfrak{o} = \mathfrak{A}$ (if $c(\mathfrak{A}) = 1$ then $\mathfrak{A}$ is a power of $\mathfrak{m}$, and we have $\mathfrak{o}'\mathfrak{A} \cap \mathfrak{o} = \mathfrak{A}$ for every quadratic transform $\mathfrak{o}'$ of $\mathfrak{A}$).*

COROLLARY 2. *The assumption and notations being the same as in Corollary 1, the initial form module of $\mathfrak{A}$ is the set of all forms of degree $r$ ($r$ = order of $\mathfrak{A}$) which are divisible by $c(\mathfrak{A})$.*

This follows from (17).

THEOREM 1. (*Factorization theorem for contracted ideals.*) *Let $\mathfrak{A}$ be a contracted ideal in $\mathfrak{o}$, primary for $\mathfrak{m}$, let $r$ be the order of $\mathfrak{A}$, $s$ the degree of $c(\mathfrak{A})$, let $c(\mathfrak{A}) = \bar{g}_1^{\lambda_1}\bar{g}_2^{\lambda_2} \cdots \bar{g}_m^{\lambda_m}$, where the $\bar{g}_i$ are the distinct irreducible factors of $c(\mathfrak{A})$, and let $s_i$ be the degree of $\bar{g}_i^{\lambda_i}$. There exists one and only one factorization of $\mathfrak{A}$ of the form*

$$\mathfrak{A} = \mathfrak{m}^{r-s}\mathfrak{B}_1\mathfrak{B}_2 \cdots \mathfrak{B}_m, \tag{18}$$

*such that each $\mathfrak{B}_i$ is a contracted ideal whose characteristic form $c(\mathfrak{B}_i)$ is a power of $\bar{g}_i$. If we denote by $\mathfrak{o}'_i$ the quadratic transform of $\mathfrak{o}$ relative to the directional form $\bar{g}_i$ and set*

$$\mathfrak{o}'_i\mathfrak{A} \cap \mathfrak{o} = \mathfrak{A}_i, \tag{19}$$

*then*

$$\mathfrak{B}_i = \mathfrak{A}_i : \mathfrak{m}^{r-s_i}, \tag{20}$$

*and we have*

$$\mathfrak{A}_i = \mathfrak{m}^{r-s_i}\mathfrak{B}_i, \tag{21}$$

$$\mathfrak{A} = \mathfrak{A}_1 \cap \mathfrak{A}_2 \cap \cdots \cap \mathfrak{A}_m, \tag{22}$$

$$r(\mathfrak{A}_i) = r. \tag{23}$$

*Furthermore, we have $\mathfrak{B}_1\mathfrak{B}_2 \cdots \mathfrak{B}_m = \mathfrak{A} : \mathfrak{m}^{r-s}$.*

PROOF. We first prove the uniqueness of the factorization (18). Let (18) and

$$\mathfrak{A} = \mathfrak{m}^{r-s}\tilde{\mathfrak{B}}_1\tilde{\mathfrak{B}}_2 \cdots \tilde{\mathfrak{B}}_m$$

be two such factorizations. We have $c(\mathfrak{A}) = c(\mathfrak{B}_1)c(\mathfrak{B}_2) \cdots c(\mathfrak{B}_m)$, hence $c(\mathfrak{B}_i)=\bar{g}_i^{\lambda_i}$. Furthermore, $r(\mathfrak{A})=r-s+r(\mathfrak{B}_1)+r(\mathfrak{B}_2)+\cdots+r(\mathfrak{B}_m)$, and since $r(\mathfrak{B}_i)\geqq \deg c(\mathfrak{B}_i)=s_i$, it follows that $r(\mathfrak{B}_i)=s_i$. Similarly, $c(\tilde{\mathfrak{B}}_i)=\bar{g}_i^{\lambda_i}$, $r(\tilde{\mathfrak{B}}_i)=s_i$. Since $\mathfrak{B}_i$ is a contracted ideal and $c(\mathfrak{B}_i)$ is a power of $\bar{g}_i$, we have $\mathfrak{o}'_i\mathfrak{B}_i\cap\mathfrak{o}=\mathfrak{B}_i$, by above Corollary 1. Similarly $\mathfrak{o}'_i\tilde{\mathfrak{B}}_i\cap\mathfrak{o}=\tilde{\mathfrak{B}}_i$. Now, assuming, as we may, that $z_1\neq\bar{g}_1(z_1, z_2)$, we have $\mathfrak{o}'_1\mathfrak{B}_i=\mathfrak{o}'_1t_1^{s_i}$ for $i>1$ (Proposition 2, part (a)). Hence $\mathfrak{o}'_1\mathfrak{A}=t_1^{r-s_1}\mathfrak{o}'_1\mathfrak{B}_1$. Similarly $\mathfrak{o}'_1\mathfrak{A}=t_1^{r-s_1}\mathfrak{o}'_1\tilde{\mathfrak{B}}_1$. Hence $\mathfrak{o}'_1\mathfrak{B}_1=\mathfrak{o}'_1\tilde{\mathfrak{B}}_1$ and hence $\tilde{\mathfrak{B}}_1=\mathfrak{o}'_1\tilde{\mathfrak{B}}_1\cap\mathfrak{o}=\mathfrak{o}'_1\mathfrak{B}_1\cap\mathfrak{o}=\mathfrak{B}_1$. Similarly $\tilde{\mathfrak{B}}_i=\mathfrak{B}_i$ for all $i=1, 2, \cdots, m$.

To prove the existence of the factorization (18) we define the ideals $\mathfrak{A}_i$ by (19) and the ideals $\mathfrak{B}_i$ by (20). Then, by Propositions 3 and 4, and by (15), the relations (21), (22) and (23) are satisfied, the $\mathfrak{B}_i$ are contracted ideals, and we have $c(\mathfrak{A}_i)=g_i^{\lambda_i}$, by (12). Furthermore, we have by (17):

$$\mathfrak{m}^{r-s}\mathfrak{B}_1\mathfrak{B}_2\cdots\mathfrak{B}_m\subset\mathfrak{A}.$$

To prove the opposite inclusion, let $x$ be any element of $\mathfrak{A}$ such that the initial form $\bar{x}$ of $x$ is precisely of degree $r$. Let $\bar{x}=\psi_1\psi_2\cdots\psi_m\psi_{m+1}$, where $\psi_i$ is a power of $\bar{g}_i$ for $i=1, 2, \cdots, m$, and $\psi_{m+1}$ is not divisible by any of the $\bar{g}_i$. Let $\rho_i$ be the degree of $\psi_i$ $(i=1, 2, \cdots, m+1)$, so that $\rho_i\geqq s_i$ for $i=1, 2, \cdots, m$, and $\rho_1+\rho_2+\cdots+\rho_{m+1}=r$. We assert that for any integer $j\geqq 1$ there exist elements $x_{1j}, x_{2j}, \cdots, x_{m+1,j}$ in $\mathfrak{o}$ such that

$$(24)\qquad x-x_{1j}x_{2j}\cdots x_{m+1,j}\in\mathfrak{m}^{r+j};\quad \bar{x}_{ij}=\psi_i,\quad i=1, 2, \cdots, m+1.$$

The assertion is trivial for $j=1$ (take for $x_{i1}$ any element of $\mathfrak{o}$ such that $\bar{x}_{i1}=\psi_i$). Assume that for a given $j$ we have already determined $m+1$ element $x_{ij}$ satisfying (24). If $x-x_{1j}x_{2j}\cdots x_{m+1,j}\in\mathfrak{m}^{r+j+1}$ we set $x_{i,j+1}=x_{ij}$. In the contrary case, we consider the initial form of $x-x_{1j}x_{2j}\cdots x_{m+1,j}$. This is a form of degree $r+j$. Now, since any two of the forms $\psi_i$ are relatively prime, it is a straightforward matter to show that the homogeneous ideal in $k[z_1, z_2]$ generated by the $m+1$ forms $\varphi_i=(\psi_1\psi_2\cdots\psi_{m+1})/\psi_i$ contains all forms of degree $\geqq r-1$ $(=\rho_1+\rho_2+\cdots+\rho_{m+1}-1)$. [The proof is by induction with respect to $m$, the case $m=2$ having been settled in the course of the proof of Proposition 3; see footnote on p. 368.] We can therefore write

$$\overline{x-x_{1j}x_{2j}\cdots x_{m+1,j}}=\sum_{i=1}^{m+1}A_i\varphi_i,$$

where $A_i$ is a form in $z_1, z_2$, of degree $\rho_i+j$ (with coefficients in $k$). We fix in $\mathfrak{o}$ an element $u_i$, such that $\bar{u}_i=A_i$ and we set $x_{i,j+1}=x_{ij}+u_i$. The $m+1$ elements $x_{i,j+1}$ satisfy all our requirements.

Since the ideals $\mathfrak{B}_i$ $(i = 1, 2, \cdots, m)$ are primary for $\mathfrak{m}$, we can take in (24) $j$ so large as to have

$$(25) \qquad \mathfrak{m}^{r+j} \subset \mathfrak{m}^{r-s}\mathfrak{B}_1\mathfrak{B}_2 \cdots \mathfrak{B}_m.$$

For such a large value of $j$ we will have, by (17): $x_{1j}x_{2j} \cdots x_{mj}x_{m+1,j} \in \mathfrak{A}$. We may assume that $\bar{g}_1 \neq z_1$. Since $\bar{\psi}_2, \bar{\psi}_3, \cdots, \bar{\psi}_m, \bar{\psi}_{m+1}$ are not divisible by $\bar{g}_1$ and since the initial form $\bar{\psi}_2\bar{\psi}_3 \cdots \bar{\psi}_{m+1}$ of $x_{2j}x_{3j} \cdots$ $x_{m+1,j}$ is of degree $r - \rho_1$ it follows that $\dfrac{x_{2j}x_{3j} \cdots x_{m+1,j}}{t_1^{r-\rho_1}}$ is a unit in $\mathfrak{o}'_1$. Hence, if we set $\mathfrak{o}'_1\mathfrak{A} = t_1^r\mathfrak{A}'_1$, then $x_{1j}/t_1^{\rho_1} \in \mathfrak{A}'_1$, and thus $x_{1j}\mathfrak{m}^{r-\rho_1} \subset \mathfrak{o}'_1\mathfrak{A} \cap \mathfrak{o} = \mathfrak{A}_1$, and $x_{1j} \in \mathfrak{A}_1:\mathfrak{m}^{r-\rho_1}$. Since $\rho_1 \geq s_1$, we have, by Corollary 1 of Proposition 3: $\mathfrak{A}_1:\mathfrak{m}^{r-\rho_1} = \mathfrak{m}^{\rho_1-s_1}\mathfrak{B}_1$. Hence $x_{1j} \in \mathfrak{m}^{\rho_1-s_1}\mathfrak{B}_1$. Similarly, $x_{ij} \in \mathfrak{m}^{\rho_i-s_i}\mathfrak{B}_i$, $i = 1, 2, \cdots, m$, and we have also $x_{m+1,j} \in \mathfrak{m}^{\rho_{m+1}}$. Hence $x_{1j} \cdot x_{2j} \cdots x_{m+1,j} \in \mathfrak{m}^{r-s}\mathfrak{B}_1\mathfrak{B}_2, \cdots, \mathfrak{B}_m$, whence by (24) and (25) we and that

$$x \in \mathfrak{m}^{r-s}\mathfrak{B}_1\mathfrak{B}_2 \cdots \mathfrak{B}_m.$$

This inclusion holds for every element $x$ of $\mathfrak{A}$ such that $x \notin \mathfrak{m}^{r+1}$, but then, as in the proof of Proposition 3, we see at once that it holds for every element $x$ of $\mathfrak{A}$. Hence $\mathfrak{A} \subset \mathfrak{m}^{r-s}\mathfrak{B}_1\mathfrak{B}_2 \cdots \mathfrak{B}_m$, and this establishes (18).

Since, by (18), $\mathfrak{m}^{r-s}$ factors out from $\mathfrak{A}$, it is clear that if we set $\mathfrak{D} = \mathfrak{A}:\mathfrak{m}^{r-s}$ then $\mathfrak{A} = \mathfrak{m}^{r-s}\mathfrak{D}$. To complete the proof of the theorem we have only to show that $\mathfrak{D} = \mathfrak{B}_1\mathfrak{B}_2 \cdots \mathfrak{B}_m$. We observe first of all that from $\mathfrak{o}'\mathfrak{A} \cap \mathfrak{o} = \mathfrak{A}$ follows that $\mathfrak{o}'\mathfrak{D} \cap \mathfrak{o} = \mathfrak{D}$, i.e., that also $\mathfrak{D}$ is a contracted ideal. In fact, if $x \in \mathfrak{o}'\mathfrak{D} \cap \mathfrak{o}$ then $x = \sum x'_j y_j$, where $x'_j \in \mathfrak{o}'$ and $y_j \in \mathfrak{D}$. Hence $x\mathfrak{m}^{r-s} \subset \mathfrak{o}'\mathfrak{A} \cap \mathfrak{o} = \mathfrak{A}$, $x \in \mathfrak{D}$, as asserted. We can therefore apply Theorem 1 to the ideal $\mathfrak{D}$. We have to find, first of all, the ideals $\mathfrak{o}'_i\mathfrak{D} \cap \mathfrak{o}$. Let $\mathfrak{A}_i:\mathfrak{m}^{r-s} = \mathfrak{D}_i$. Since $\mathfrak{o}'_i\mathfrak{A}_i \cap \mathfrak{o} = \mathfrak{A}_i$ and $r = r(\mathfrak{A}_i)$, $s \geq s_i = \deg c(\mathfrak{A}_i)$, it follows from Corollary 1 to Proposition 3 that $\mathfrak{A}_i = \mathfrak{m}^{r-s}\mathfrak{D}_i$. Hence, assuming—as we may—that $z_1 \neq \bar{g}_i$, we have $\mathfrak{o}'_i\mathfrak{A}_i = t_1^{r-s}\mathfrak{o}'_i\mathfrak{D}_i$. On the other hand, we have $\mathfrak{o}'_i\mathfrak{A}_i = \mathfrak{o}'_i\mathfrak{A}$ (by the definition (19) of $\mathfrak{A}_i$), whence $\mathfrak{o}'_i\mathfrak{A}_i = \mathfrak{o}'_i\mathfrak{m}^{r-s}\mathfrak{D} = t_1^{r-s}\mathfrak{o}'_i\mathfrak{D}$. Hence $t_1^{r-s}\mathfrak{o}'_i\mathfrak{D}_i = t_1^{r-s}\mathfrak{o}'_i\mathfrak{D}$, $\mathfrak{o}'_i\mathfrak{D}_i = \mathfrak{o}'_i\mathfrak{D}$. But from $\mathfrak{o}'_i\mathfrak{A}_i \cap \mathfrak{o} = \mathfrak{A}_i$ follows—as was just shown above—that also $\mathfrak{o}'_i\mathfrak{D}_i \cap \mathfrak{o} = \mathfrak{D}_i$. Hence we conclude that $\mathfrak{o}'_i\mathfrak{D} \cap \mathfrak{o} = \mathfrak{D}_i$, $i = 1, 2, \cdots, m$. Since $\mathfrak{A} = \mathfrak{m}^{r-s}\mathfrak{D}$, we have $c(\mathfrak{D}) = c(\mathfrak{A})$, so that the integers $s_i$ of Theorem 1 are not affected by passing from $\mathfrak{A}$ to $\mathfrak{D}$. We have now $r(\mathfrak{D}) = s$. By Corollary 1 to Proposition 3 we have that $\mathfrak{D}_i:\mathfrak{m}^{s-s_i} = \mathfrak{A}_i:\mathfrak{m}^{r-s_i}$, i.e., $\mathfrak{D}_i:\mathfrak{m}^{s-s_i} = \mathfrak{B}_i$. Thus also the ideals $\mathfrak{B}_i$ are not affected. Thus the analog of (18) for $\mathfrak{D}$ (instead of $\mathfrak{A}$) is $\mathfrak{D} = \mathfrak{B}_1\mathfrak{B}_2 \cdots \mathfrak{B}_m$. This completes the proof of the theorem.

COROLLARY 1. *If* $\mathfrak{A}$ *is a contracted ideal and* $q$ *is any integer* $\geqq 1$ *then also* $\mathfrak{m}^q\mathfrak{A}$ *is a contracted ideal.*

It is sufficient to consider the case in which $\mathfrak{A}$ is primary for $\mathfrak{m}$ since every ideal in $\mathfrak{o}$ is of the form $x\mathfrak{A}$, $x \in \mathfrak{o}$, $\mathfrak{A}$ primary for $\mathfrak{m}$. We use the notations of Theorem 1, we set $\mathfrak{o}'=\mathfrak{o}'_1 \cap \mathfrak{o}'_2 \cap \cdots \cap \mathfrak{o}'_m$ and $\mathfrak{D}=\mathfrak{o}'\mathfrak{m}^q\mathfrak{A} \cap \mathfrak{o}$. It is clear that $r(\mathfrak{D})=q+r$. We have $\mathfrak{o}'_i\mathfrak{D}=\mathfrak{o}'_i\mathfrak{m}^q\mathfrak{A}=\mathfrak{o}'_i\mathfrak{m}^{q+r-s_i}\mathfrak{B}_i$, and since $\mathfrak{m}^{q+r-s_i}\mathfrak{B}_i$ is the contraction of its extended ideal in $\mathfrak{o}'_i$ (Proposition 3, Corollary 2), it follows that if we set

$$\mathfrak{D}_i = \mathfrak{o}'_i\mathfrak{D} \cap \mathfrak{o},$$

then

$$\mathfrak{D}_i = \mathfrak{m}^{q+r-s_i}\mathfrak{B}_i.$$

Thus $c(\mathfrak{D}_i)=c(\mathfrak{B}_i)=\bar{g}_i^{\lambda_i}$. If we set $\mathfrak{D}_i : \mathfrak{m}^{q+r-s_i}=\tilde{\mathfrak{B}}_i$, then $\mathfrak{D}_i=\mathfrak{m}^{q+r-s_i}\tilde{\mathfrak{B}}_i$, and therefore, by the uniqueness of factorization of contracted ideals, we have $\tilde{\mathfrak{B}}_i=\mathfrak{B}_i$. By Theorem 1, applied to $\mathfrak{D}$ (instead of to $\mathfrak{A}$), we have therefore $\mathfrak{D}=\mathfrak{m}^{q+r-s}\mathfrak{B}_1\mathfrak{B}_2 \cdots \mathfrak{B}_m=\mathfrak{m}^q\mathfrak{A}$.

COROLLARY 2. *Let* $\mathfrak{B}_1, \mathfrak{B}_2, \cdots, \mathfrak{B}_m$ *be contracted ideals whose characteristic forms* $c(\mathfrak{B}_1), c(\mathfrak{B}_2), \cdots, c(\mathfrak{B}_m)$ *are two by two relatively prime (any number of the* $c(\mathfrak{B}_i)$ *may be equal to* 1). *Then the product* $\mathfrak{B}_1\mathfrak{B}_2 \cdots \mathfrak{B}_m$ *is also a contracted ideal.*

If some $c(\mathfrak{B}_i)$ is 1, say $c(\mathfrak{B}_1)=1$, then $\mathfrak{B}_1$ is a power of $\mathfrak{m}$ (Proposition 4), and from Corollary 1 of Theorem 1 it follows that it is then sufficient to prove that $\mathfrak{B}_2\mathfrak{B}_3 \cdots \mathfrak{B}_m$ is a contracted ideal. *We may therefore assume that* $c(\mathfrak{B}_i)\neq 1$ for $i=1, 2, \cdots, m$. A further obvious reduction is permissible, whereby we may assume that *every* $\mathfrak{B}_i$ *is primary for* $\mathfrak{m}$. Finally we can carry out a third reduction to the case in which *each characteristic form* $c(\mathfrak{B}_i)$ *is a power of an irreducible form.* In fact, by Theorem 1, each contracted ideal which is primary for $\mathfrak{m}$ is a product of contracted ideals having only one directional form.

Let $s_i=\deg c(\mathfrak{B}_i)$. If the order $r_i$ of $\mathfrak{B}_i$ is greater than $s_i$ then $\mathfrak{m}^{r_i-s_i}$ factors out from $\mathfrak{B}_i$, i.e., we have $\mathfrak{B}_i=\mathfrak{m}^{r_i-s_i}\mathfrak{C}_i$, where $\mathfrak{C}_i$ is a contracted ideal (Proposition 3), and Corollary 1 of Theorem 1 allows us to replace in the proof $\mathfrak{B}_i$ by $\mathfrak{C}_i$. We may therefore assume that $r_i=s_i$, $i=1, 2, \cdots, m$. We set $s=s_1+s_2+ \cdots +s_m$, we denote by $\mathfrak{o}'_i$ the quadratic transform of $\mathfrak{o}$ relative to the directional form of $\mathfrak{B}_i$ and we set $\mathfrak{A}=\mathfrak{o}'\mathfrak{B}_1\mathfrak{B}_2 \cdots \mathfrak{B}_m \cap \mathfrak{o}$, where $\mathfrak{o}'=\mathfrak{o}'_1 \cap \mathfrak{o}'_2 \cap \cdots \cap \mathfrak{o}'_m$. We have then $\mathfrak{o}'\mathfrak{A}=\mathfrak{o}'\mathfrak{B}_1\mathfrak{B}_2 \cdots \mathfrak{B}_m$, and hence $\mathfrak{o}'_i\mathfrak{A}=\mathfrak{o}'_i\mathfrak{B}_1\mathfrak{B}_2 \cdots \mathfrak{B}_m=\mathfrak{o}'_i\mathfrak{m}^{s-s_i}\mathfrak{B}_i$. Consequently, if we set $\mathfrak{A}_i=\mathfrak{o}'_i\mathfrak{A} \cap \mathfrak{o}$ then $\mathfrak{A}_i=\mathfrak{m}^{s-s_i}\mathfrak{B}_i$ (since both $\mathfrak{A}_i$ and $\mathfrak{m}^{s-s_i}\mathfrak{B}_i$ are contractions of ideals in $\mathfrak{o}'_i$). From this, by unique factorization, we conclude that $\mathfrak{B}_i=\mathfrak{A}_i : \mathfrak{m}^{s-s_i}$, and from Theorem 1 we deduce that $\mathfrak{A}=\mathfrak{B}_1\mathfrak{B}_2 \cdots \mathfrak{B}_m$.

COROLLARY 3. *With the assumptions and notations as in Theorem* 1, *the decomposition* (22) *of* $\mathfrak{A}$ *is the only decomposition of* $\mathfrak{A}$ *into contracted ideals* $\mathfrak{A}_i$ *satisfying* (23) *and such that* $c(\mathfrak{A}_i)$ *is a power of* $\bar{g}_i$.

Let $\mathfrak{A}=\tilde{\mathfrak{A}}_1 \cap \tilde{\mathfrak{A}}_2 \cap \cdots \cap \tilde{\mathfrak{A}}_m$ be another such decomposition. Then $c(\tilde{\mathfrak{A}}_i)$ divides $c(\mathfrak{A})$, and thus the degree $\sigma_i$ of $c(\tilde{\mathfrak{A}}_i)$ is not greater than $s_i$. Since $\mathfrak{m}^{r-\sigma_i}$ factors out from $\tilde{\mathfrak{A}}_i$, we can write $\tilde{\mathfrak{A}}_i=\mathfrak{m}^{r-s_i}\tilde{\mathfrak{B}}_i$, where $\tilde{\mathfrak{B}}_i$ is again the contraction of an ideal in $\mathfrak{o}'_i$. We have $\mathfrak{m}^{r-s}\tilde{\mathfrak{B}}_1\tilde{\mathfrak{B}}_2\cdots\tilde{\mathfrak{B}}_m\subset \mathfrak{m}^{r-s_i}\tilde{\mathfrak{B}}_i=\tilde{\mathfrak{A}}_i$, hence

$$(26)\qquad \mathfrak{m}^{r-s}\tilde{\mathfrak{B}}_1\tilde{\mathfrak{B}}_2\cdots\tilde{\mathfrak{B}}_m\subset \mathfrak{m}^{r-s}\mathfrak{B}_1\mathfrak{B}_2\cdots\mathfrak{B}_m\ (=\mathfrak{A}).$$

On the other hand, $\mathfrak{m}^{r-s}\mathfrak{B}_1\mathfrak{B}_2\cdots\mathfrak{B}_m\subset\mathfrak{m}^{r-s_i}\tilde{\mathfrak{B}}_i$, and passing to the extended ideals in $\mathfrak{o}'_i$ we find that $\mathfrak{o}'_i\mathfrak{m}^{r-s_i}\mathfrak{B}_i\subset\mathfrak{o}'_i\mathfrak{m}^{r-s_i}\tilde{\mathfrak{B}}_i$. Since both $\mathfrak{B}_i$ and $\tilde{\mathfrak{B}}_i$ are contracted ideals it follows that $\mathfrak{B}_i\subset\tilde{\mathfrak{B}}_i$. Therefore, by (26), we have

$$\mathfrak{m}^{r-s}\tilde{\mathfrak{B}}_1\tilde{\mathfrak{B}}_2\cdots\tilde{\mathfrak{B}}_m = \mathfrak{m}^{r-s}\mathfrak{B}_1\mathfrak{B}_2\cdots\mathfrak{B}_m\ (=\mathfrak{A}).$$

By the unique factorization property of contracted ideals it follows now that $\tilde{\mathfrak{B}}_i=\mathfrak{B}_i$, whence $\tilde{\mathfrak{A}}_i=\mathfrak{A}_i$ $(i=1, 2, \cdots, m)$.

We conclude the theory of contracted ideals with the following result:

THEOREM 2. *Any product of contracted ideals in* $\mathfrak{o}$ *is a contracted ideal.*

PROOF. Let $\mathfrak{A}_1, \mathfrak{A}_2, \cdots, \mathfrak{A}_n$ be contracted ideals in $\mathfrak{o}$. It is sufficient to give the proof in the case in which the $\mathfrak{A}_i$ are primary for $\mathfrak{m}$. Using Theorem 1 we begin by factoring each $\mathfrak{A}_i$ into a product $\mathfrak{m}^{s_i}\mathfrak{A}_{i1}\mathfrak{A}_{i2}\cdots\mathfrak{A}_{im_i}$ of contracted ideals such that $c(\mathfrak{A}_{ij})$ is a power of an irreducible form in $k[z_1, z_2]$. Then, in the set of $m_1+m_2+\cdots+m_n$ ideals $\mathfrak{A}_{ij_i}$ we group together those ideals whose characteristic forms are powers of one and the same irreducible form in $k[z_1, z_2]$, we form the product of the ideals belonging to one and the same group and we denote the various partial products thus obtained by $\mathfrak{B}_1, \mathfrak{B}_2, \cdots, \mathfrak{B}_h$. By Corollaries 1 and 2 to Theorem 1 it is sufficient to prove that each $\mathfrak{B}_\alpha$ is a contracted ideal. We therefore may assume that the characteristic forms $c(\mathfrak{A}_i)$ of the given $n$ ideals $\mathfrak{A}_i$ are powers of one and the same irreducible form $\bar{g}$ in $k[z_1, z_2]$, $\bar{g}\neq 1$. The proof of the theorem will now be based on (and, in fact, will be an immediate consequence of) the following two lemmas:

LEMMA 2. *If* $\mathfrak{A}$ *is an ideal in* $\mathfrak{o}$ *and* $R'$ *is the ring* $\mathfrak{o}[t_2/t_1]$, *a necessary and sufficient condition that we have* $R'\mathfrak{A}\cap\mathfrak{o}=\mathfrak{A}$ *is that the following equality be satisfied:*

$$(27)\qquad \mathfrak{A}:\mathfrak{o}t_1 = \mathfrak{A}:\mathfrak{m}.$$

PROOF. Assume that $R'\mathfrak{A} \cap \mathfrak{o} = \mathfrak{A}$ and let $x$ be any element of $\mathfrak{A}:\mathfrak{o}t_1$. Then $xt_1 = a \in \mathfrak{A}$, $xt_2 = a \cdot t_2/t_1 \in R'\mathfrak{A} \cap \mathfrak{o} = \mathfrak{A}$, and so

$$x \in \mathfrak{A}:(\mathfrak{o}t_1 + \mathfrak{o}t_2) = \mathfrak{A}:\mathfrak{m},$$

which proves (27).

Conversely, assume that (27) holds true and let $x$ be any element of $R'\mathfrak{A} \cap \mathfrak{o}$. Then $x = \sum_{i=0}^{n} a_i \left(\frac{t_2}{t_1}\right)^i$, $a_i \in \mathfrak{A}$. We see that $a_n$ is divisible by $t_1$ in $\mathfrak{o}$, say $a_n = t_1 b_n$, and since $a_n \in \mathfrak{A}$ it follows from (27) that $t_2 b_n$ also belongs to $\mathfrak{A}$. If, then, we set $t_2 b_n = b_{n-1}$ we find

$$x = \sum_{i=0}^{n-2} a_i \left(\frac{t_2}{t_1}\right)^i + (a_{n-1} + b_{n-1}) \left(\frac{t_2}{t_1}\right)^{n-1}.$$

This is a new expression of $x$ as a polynomial in $t_2/t_1$, with coefficients which still belong to $\mathfrak{A}$, but the degree of the polynomial is now at most $n-1$. Continuing the reduction of degree we arrive at the desired conclusion.

LEMMA 3. *If two ideals* $\mathfrak{A}_1, \mathfrak{A}_2$ *are such that* $\mathfrak{A}_1:\mathfrak{o}t_1 = \mathfrak{A}_1:\mathfrak{m}$ *and* $\mathfrak{A}_2:\mathfrak{o}t_1 = \mathfrak{A}_2:\mathfrak{m}$, *then we have also* $\mathfrak{A}_1\mathfrak{A}_2:\mathfrak{o}t_1 = \mathfrak{A}_1\mathfrak{A}_2:\mathfrak{m}$.

PROOF. Since $\mathfrak{o}/\mathfrak{o}t_1$ is a regular ring of dimension 1, hence a principal ideal ring, there exists in $\mathfrak{A}_i$ an element $x_i$ such that $(\mathfrak{A}_i, t_1) = (x_i, t_1)$, $i = 1, 2$. We observe that our assumptions on $\mathfrak{A}_1$ and $\mathfrak{A}_2$ imply that $\mathfrak{A}_i \not\subset \mathfrak{o}t_1$ $(i = 1, 2)$. Hence neither $x_1$ nor $x_2$ is divisible by $t_1$. Now, let $\xi$ be any element of $\mathfrak{A}_1\mathfrak{A}_2:\mathfrak{o}t_1$. Then

$$\xi t_1 = \sum (\alpha_{1j}x_1 + \beta_{1j}t_1)(\alpha_{2j}x_2 + \beta_{2j}t_1),$$

where the $\alpha$'s and $\beta$'s are in $\mathfrak{o}$ and $\alpha_{ij}x_i + \beta_{ij}t_1 \in \mathfrak{A}_i$. Since $x_i \in \mathfrak{A}_i$, it follows that $\beta_{ij}t_1 \in \mathfrak{A}_i$, and hence $\beta_{ij}t_2 \in \mathfrak{A}_i$. Furthermore, $\sum \alpha_{1j}\alpha_{2j} \cdot x_1 x_2$ is divisible by $t_1$, and therefore $\sum \alpha_{1j}\alpha_{2j}$ is divisible by $t_1$. We then find easily that $\xi t_1 t_2$ has an expression of the form $t_1(\gamma x_1 x_2 + \delta)$, where $\gamma \in \mathfrak{o}$ and $\delta \in \mathfrak{A}_1\mathfrak{A}_2$. Therefore $\xi t_2 \in \mathfrak{A}_1\mathfrak{A}_2$, and this completes the proof.

We now apply these lemmas. Let $\mathfrak{o}'$ be the quadratic transform of $\mathfrak{o}$ relative to the directional form $\bar{g}$. Since $\mathfrak{o}'\mathfrak{A}_i \cap \mathfrak{o} = \mathfrak{A}_i$, we have *a fortiori*, $R'\mathfrak{A}_i \cap \mathfrak{o} = \mathfrak{A}_i$, $i = 1, 2, \cdots, m$ (we assume that $\bar{g} \neq z_1$ and that therefore $\mathfrak{o}' \supset R'$). Hence, $\mathfrak{A}_i:\mathfrak{o}t_1 = \mathfrak{A}_i:\mathfrak{o}\mathfrak{m}$ (Lemma 2), $\mathfrak{A}:\mathfrak{o}t_1 = \mathfrak{A}:\mathfrak{m}$ (Lemma 3) and thus $R'\mathfrak{A} \cap \mathfrak{o} = \mathfrak{A}$ (Lemma 2). Now, $c(\mathfrak{A})$ is a power of $\bar{g}$. This implies that $R'\mathfrak{A} = t_1{}^r\tilde{\mathfrak{A}}$, where $\tilde{\mathfrak{A}}$ is an ideal in $R'$ which is either the unit ideal or is primary for the maximal ideal $\mathfrak{p}'$ in $R'$ such that $\mathfrak{o}' = R'_{\mathfrak{p}'}$. Hence all the prime ideals of $R'\mathfrak{A}$ are contained in $\mathfrak{p}'$. Therefore $\mathfrak{o}'\mathfrak{A} \cap R' = R'\mathfrak{A}$, and thus $\mathfrak{o}'\mathfrak{A} \cap \mathfrak{o} = \mathfrak{A}$. This completes the proof of Theorem 2.

4. We now apply the preceding theory to complete ideals in $\mathfrak{o}$. The application is possible since it is not difficult to see that *every complete ideal* $\mathfrak{A}$ *in* $\mathfrak{o}$ *is in fact a contracted ideal.* To prove this it is sufficient to consider the case in which $\mathfrak{A}$ is primary for $\mathfrak{m}$. Let $\mathfrak{A} = \mathfrak{q}_1 \cap \mathfrak{q}_2 \cap \cdots \cap \mathfrak{q}_n$ be a representation of $\mathfrak{A}$ as a finite intersection of valuation ideals (Appendix 4, Corollary to Theorem 3), and let $v_i$ be a valuation with which $\mathfrak{q}_i$ is associated. Each $v_i$ has necessarily center $\mathfrak{m}$ in $\mathfrak{o}$. Let $\bar{g}_i$ be the directional form of $v_i$ (Definition 1). For each $i$ such that $\bar{g}_i \neq 1$ (i.e., such that $v_i$ is not the $\mathfrak{m}$-adic prime divisor $v_\mathfrak{m}$ of $\mathfrak{o}$) we consider the quadratic transform $\mathfrak{o}'_i$ of $\mathfrak{o}$ relative to $\bar{g}_i$ and we denote by $\mathfrak{o}'$ the intersection of all those rings $\mathfrak{o}'_i$. In view of the definition of $\mathfrak{o}'_i$ it follows at once from Lemma 1 that $v_i$ is non-negative on $\mathfrak{o}'_i$. Hence each of the $v_i$ is non-negative on $\mathfrak{o}'$ (and this includes the case in which $v_i = v_\mathfrak{m}$, for $v_\mathfrak{m}$ is non-negative on every quadratic transform of $\mathfrak{o}$). Let $\mathfrak{q}_i = \mathfrak{o} \cap \mathfrak{Q}_i$, where $\mathfrak{Q}_i$ is an ideal in the valuation ring of $v_i$ and let $\mathfrak{q}'_i = \mathfrak{Q}_i \cap \mathfrak{o}'$. Then $\mathfrak{A}$ is the contraction to $\mathfrak{o}$ of the ideal $\mathfrak{q}'_1 \cap \mathfrak{q}'_2 \cap \cdots \cap \mathfrak{q}'_n$ of $\mathfrak{o}'$, which proves the assertion.

Let $\mathfrak{A}$ be a complete ideal in $\mathfrak{o}$, primary for $\mathfrak{m}$, of order $r$. We can write then

$$\mathfrak{A} = \mathfrak{m}^r \cap \mathfrak{q}_1 \cap \mathfrak{q}_2 \cap \cdots \cap \mathfrak{q}_n, \tag{28}$$

where $\mathfrak{q}_i$ is a valuation ideal in $\mathfrak{o}$, associated with a valuation $v_i$ which is non-negative on $\mathfrak{o}$ and is centered at $\mathfrak{m}$, and where we now may assume that each $v_i$ is different from the $\mathfrak{m}$-adic prime divisor of $\mathfrak{o}$ (in view of the presence of the component $\mathfrak{m}^r$ in (28)). We say that a decomposition (28) of $\mathfrak{A}$ into valuation ideals (*one of which is* $\mathfrak{m}^r$, where $r$ = order of $\mathfrak{A}$) is *irredundant* if no $\mathfrak{q}_i$ is superfluous.

LEMMA 4. *Each prime divisor of the characteristic form* $c(\mathfrak{A})$ *of* $\mathfrak{A}$ *is a directional form of one of the* $v_i$. *If the decomposition* (28) *is irredundant then, conversely, the directional form of each* $v_i$ $(i = 1, 2, \cdots n)$ *is a prime divisor of* $c(\mathfrak{A})$.

PROOF. Let $\bar{g}_i$ be the directional form of $v_i$, let $\mathfrak{o}'_i$ be the quadratic transform of $\mathfrak{o}$ relative to $\bar{g}_i$ and let $\mathfrak{o}' = \mathfrak{o}'_1 \cap \mathfrak{o}'_2 \cap \cdots \cap \mathfrak{o}'_n$. We have just seen that $\mathfrak{A}$ is then the contraction of an ideal in $\mathfrak{o}'$. The first part of the lemma follows therefore from the expression (13) of $c(\mathfrak{A})$ given in Proposition 4. To prove the second part of the lemma, assume that one of the $\bar{g}_i$, say $\bar{g}_1$, is not a divisor of $c(\mathfrak{A})$. We shall show that $\mathfrak{q}_1$ is superfluous in (28). By assumption, there exists an element $x$ in $\mathfrak{A}$ such that the initial form $\bar{x}$ is of degree $r$ and is not divisible by $\bar{g}_1$. We have then $v_1(x) = v_1(\mathfrak{m}^r)$ (Lemma 1), and since $x \in \mathfrak{q}_1$ and $\mathfrak{q}_1$ is a $v_1$-ideal, it follows that $\mathfrak{m}^r \subset \mathfrak{q}_1$, showing that $\mathfrak{q}_1$ is superfluous.

Using this simple lemma we can now prove the following important complement to the factorization theorem (Theorem 1):

THEOREM 1′. *If the ideal* $\mathfrak{A}$ *of Theorem* 1 *is complete, then the factors* $\mathfrak{B}_i$ *in* (18) *and the ideals* $\mathfrak{A}_i$ *in* (19) *are also complete.*

PROOF. We consider an *irredundant* decomposition (28) of $\mathfrak{A}$ into valuation ideals. By Lemma 4, the set of directional forms of the valuations $v_1, v_2, \cdots, v_n$ coincides with the set $(\bar{g}_1, \bar{g}_2, \cdots, \bar{g}_m)$ of the irreducible factors of $c(\mathfrak{A})$. For each $i=1, 2, \cdots, m$, let $\tilde{\mathfrak{A}}_i$ be the intersection of $\mathfrak{m}^r$ and of those $\mathfrak{q}_j$ for which $v_j$ has directional form $\bar{g}_i$. Then

$$\mathfrak{A} = \tilde{\mathfrak{A}}_1 \cap \tilde{\mathfrak{A}}_2 \cap \cdots \cap \tilde{\mathfrak{A}}_m,$$

where each $\tilde{\mathfrak{A}}_i$ is a complete ideal, and again by Lemma 4, $c(\tilde{\mathfrak{A}}_i)$ is a power of $\bar{g}_i$. Furthermore, we have obviously $r(\tilde{\mathfrak{A}}_i)=r$. From the uniqueness of the decomposition (22) (Theorem 1, Corollary 3) it follows that $\tilde{\mathfrak{A}}_i=\mathfrak{A}_i$, and thus $\mathfrak{A}_i$ is a complete ideal. The completeness of $\mathfrak{B}_i$ now follows directly from the relation (20) in Theorem 1.

COROLLARY 1. *If* $\mathfrak{A}$ *is a complete ideal and* $q$ *is any integer* $\geqq 1$, *then also* $\mathfrak{m}^q\mathfrak{A}$ *is a complete ideal.*

We may assume that $\mathfrak{A}$ is primary for $\mathfrak{m}$. Let $\mathfrak{B}=\mathfrak{m}^q\mathfrak{A}$, let $\mathfrak{B}'$ be the completion of $\mathfrak{B}$ and let $r$ be the order of $\mathfrak{A}$. It is clear that the complete ideal $\mathfrak{B}' \cap \mathfrak{m}^{q+r}$ (which is primary for $\mathfrak{m}$) has order $q+r$ (since $\mathfrak{B}$ has order $q+r$ and $\mathfrak{B} \subset \mathfrak{B}' \cap \mathfrak{m}^{q+r}$). If, then, we denote by $\sigma$ the degree of the characteristic form $c(\mathfrak{B}' \cap \mathfrak{m}^{q+r})$ of $\mathfrak{B}' \cap \mathfrak{m}^{q+r}$ we have, by Theorem 1, $\mathfrak{B}' \cap \mathfrak{m}^{q+r} = \mathfrak{m}^{q+r-\sigma}\mathfrak{B}_1$ where $\mathfrak{B}_1=(\mathfrak{B}' \cap \mathfrak{m}^{q+r}):\mathfrak{m}^{q+r-\sigma}$. Now, $c(\mathfrak{B}' \cap \mathfrak{m}^{q+r})$ divides $c(\mathfrak{B})$ and $c(\mathfrak{B})=c(\mathfrak{A})$, while the degree of $c(\mathfrak{A})$ is $\leqq r$. Hence $\sigma \leqq r$. Let $\mathfrak{C}$ be the completion of $\mathfrak{m}^{r-\sigma}\mathfrak{B}_1$. Then from $\mathfrak{B}' \cap \mathfrak{m}^{q+r} = \mathfrak{m}^q \cdot \mathfrak{m}^{r-\sigma}\mathfrak{B}_1$ follows that $\mathfrak{B}' \cap \mathfrak{m}^{q+r} = \mathfrak{m}^q\mathfrak{C}$ (since $\mathfrak{B}' \cap \mathfrak{m}^{q+r}$ is complete). We have $\mathfrak{m}^q\mathfrak{C} \subset \mathfrak{B}' = (\mathfrak{m}^q\mathfrak{A})'$ and $\mathfrak{m}^q\mathfrak{A} = \mathfrak{B} \subset \mathfrak{B}' \cap \mathfrak{m}^{q+r} = \mathfrak{m}^q\mathfrak{C}$, i.e., $\mathfrak{m}^q\mathfrak{C} \subset (\mathfrak{m}^q\mathfrak{A})'$ and $\mathfrak{m}^q\mathfrak{A} \subset \mathfrak{m}^q\mathfrak{C}$. Applying property (g) of Proposition 1, Appendix 4, and observing that $\mathfrak{A}$ and $\mathfrak{C}$ are complete ideals, we conclude that $\mathfrak{A}=\mathfrak{C}$ and that consequently $\mathfrak{m}^q\mathfrak{A} = \mathfrak{B}' \cap \mathfrak{m}^{q+r}$, showing that $\mathfrak{m}^q\mathfrak{A}$ is a complete ideal.

COROLLARY 2. *If the ideals* $\mathfrak{B}_1, \mathfrak{B}_2, \cdots, \mathfrak{B}_m$ *of Corollary* 2 *to Theorem* 1 *are complete, then also the product* $\mathfrak{B}_1\mathfrak{B}_2 \cdots \mathfrak{B}_m$ *is complete.*

We refer to the proof of Corollary 2 to Theorem 1. All the preliminary reductions carried out in that proof are applicable also in the present case. In the last part of that proof (where we dealt with the case $r_i=s_i$, $i=1, 2, \cdots, m$) we found that if we set $\mathfrak{A}=\mathfrak{B}_1\mathfrak{B}_2 \cdots \mathfrak{B}_m$ then $\mathfrak{A}=\mathfrak{A}_1 \cap \mathfrak{A}_2 \cap \cdots \cap \mathfrak{A}_m$, where $\mathfrak{A}_i=\mathfrak{B}_i\mathfrak{m}^{s-s_i}$. Since $\mathfrak{B}_i$ is complete, $\mathfrak{A}_i$ is also complete by the preceding Corollary 1, and thus also $\mathfrak{A}$ is complete.

The further development of the theory of complete ideals in $\mathfrak{o}$ depends on the repeated application of successive quadratic transformations. If $\mathfrak{o}'$ is a quadratic transform of $\mathfrak{o}$ then we may consider any (of the infinitely many) quadratic transforms $\mathfrak{o}''$ of $\mathfrak{o}'$, and this procedure can be continued indefinitely, leading to infinite, strictly ascending sequences $\mathfrak{o} < \mathfrak{o}' < \mathfrak{o}'' < \cdots < \mathfrak{o}^{(i)} < \cdots$ of regular rings $\mathfrak{o}^{(i)}$ of dimension 2, each $\mathfrak{o}^{(i)}$ being a quadratic transform of its immediate predecessor $\mathfrak{o}^{(i-1)}(\mathfrak{o}^{(0)} = \mathfrak{o})$. For each ideal $\mathfrak{A}$ in $\mathfrak{o}$ we have defined its transform $\mathfrak{A}'$ in $\mathfrak{o}'$ [see (4)]. The property of $\mathfrak{A}$ of being a contracted ideal is not preserved under quadratic transformations, i.e., the ideal $\mathfrak{A}'$ in $\mathfrak{o}'$ is not necessarily a contracted ideal (in the sense of the definition given immediately after the proof of Proposition 4, with $\mathfrak{o}$ being replaced by $\mathfrak{o}'$; see p. 373). However, for complete ideals we have the possibility of using an inductive process, in view of the following property of these ideals:

PROPOSITION 5. *If $\mathfrak{A}$ is a complete ideal in $\mathfrak{o}$ and if $\mathfrak{o}'$ is a quadratic transform of $\mathfrak{o}$, then the transform $\mathfrak{A}'$ of $\mathfrak{A}$ in $\mathfrak{o}'$ is also a complete ideal.*

PROOF. Since $\mathfrak{o}'\mathfrak{A}$ differs from $\mathfrak{A}'$ only by a principal ideal factor, it is sufficient to prove that $\mathfrak{o}'\mathfrak{A}$ *is a complete ideal in* $\mathfrak{o}'$. We may assume that $\mathfrak{A}$ *is primary for* $\mathfrak{m}$, for any ideal $\mathfrak{B}$ in $\mathfrak{o}$ differs from such an ideal $\mathfrak{A}$ only by a principal ideal factor (unless $\mathfrak{B}$ itself is a principal ideal, in which case $\mathfrak{o}'\mathfrak{B}$ is also principal, hence complete, as $\mathfrak{o}'$ is integrally closed).

Let the quadratic transform $\mathfrak{o}'$ of $\mathfrak{o}$ be relative to the directional form $\bar{g}$. We may assume that $\bar{g} \neq z_1$. Let $r$ be the order of $\mathfrak{A}$. If $\bar{g}$ does not divide $c(\mathfrak{A})$, then $\mathfrak{o}'\mathfrak{A} = \mathfrak{o}'t_1{}^r$ (Proposition 2, part (a)), and thus $\mathfrak{o}'\mathfrak{A}$ is complete. Assume therefore that $\bar{g}$ divides $c(\mathfrak{A})$ and let $\bar{g}_1, \bar{g}_2, \cdots, \bar{g}_m$ be the irreducible factors of $c(\mathfrak{A})$, where we assume that $\bar{g}_1 = \bar{g}$. We apply the factorization $\mathfrak{A} = \mathfrak{m}^{r-s}\mathfrak{B}_1\mathfrak{B}_2 \cdots \mathfrak{B}_m$ given in Theorem 1. We have $\mathfrak{o}'\mathfrak{B}_i = \mathfrak{o}'t_1{}^{s_i}$ if $i > 1$, since $c(\mathfrak{B}_i)$ is a power of $\bar{g}_i$ and hence $\bar{g}$ does not divide $c(\mathfrak{B}_i)$ if $i > 1$. Therefore $\mathfrak{o}'\mathfrak{A} = t_1{}^{r-s_1}\mathfrak{o}'\mathfrak{B}_1$ and it is sufficient to prove that $\mathfrak{o}'\mathfrak{B}_1$ is a complete ideal in $\mathfrak{o}'$. *We therefore may assume that $c(\mathfrak{A})$ originally was a power of $\bar{g}$.* (Recall that, by Theorem 1′, the ideals $\mathfrak{B}_i$ are complete.)

By Lemma 4, the valuation ideals $\mathfrak{q}_i$ which occur in some irredundant decomposition (28) of $\mathfrak{A}$ into valuation ideals are associated with valuations $v_i$ having $\bar{g}$ as directional form. Therefore, in order to prove that an element $\xi$ of $\mathfrak{o}$ belongs to $\mathfrak{A}$ it is not necessary to prove that we have $v(\xi) \geqq v(\mathfrak{A})$ for all valuations $v$ which are non-negative on $\mathfrak{o}$; *it is sufficient to prove this only for those valuations $v$, non-negative on $\mathfrak{o}$, whose directional form is either $\bar{g}$ or* 1 (in the latter case, $v$ is the

$\mathfrak{m}$-adic prime divisor of $\mathfrak{o}$). In other words: *it is sufficient to prove that* $v(\xi) \geqq v(\mathfrak{A})$ *for all valuations* $v$ *which are non-negative on* $\mathfrak{o}'$. We shall make use of this observation.

Let $y/x$ be any element of $\mathfrak{o}'$ which belongs to the completion of the ideal $\mathfrak{o}'\mathfrak{A}$; here $x$ and $y$ are elements of $\mathfrak{o}$, $\bar{x}$ is not divisible by $\bar{g}$, and if $n$ is the order of $x$ then $y \in \mathfrak{m}^n$. If $v$ is any valuation which is non-negative on $\mathfrak{o}'$ then we must have $v(y/x) \geqq v(\mathfrak{o}'\mathfrak{A}) = v(\mathfrak{A})$, $v(y) \geqq v(x) + v(\mathfrak{A})$. Hence $v(y) \geqq v(\mathfrak{m}^n\mathfrak{A})$. Now $\mathfrak{m}^n\mathfrak{A}$ is a complete ideal (Theorem 1′, Corollary 1), and its characteristic form is $c(\mathfrak{A})$, hence a power of $\bar{g}$. Therefore, by the above observation, applied to $\mathfrak{m}^n\mathfrak{A}$, the validity of the inequality $v(y) \geqq v(\mathfrak{m}^n\mathfrak{A})$ for all $v$ which are non-negative on $\mathfrak{o}'$ implies that $y \in \mathfrak{m}^n\mathfrak{A}$. Hence $y \in t_1{}^n \cdot \mathfrak{o}'\mathfrak{A}$, and since $x = t_1{}^n \cdot x'$ where $x'$ is a unit in $\mathfrak{o}'$, it follows that $y/x \in \mathfrak{o}'\mathfrak{A}$. This completes the proof.

By Theorem 3 of Appendix 4, every complete ideal $\mathfrak{A}$ in $\mathfrak{o}$ has a decomposition into valuation ideals belonging *to discrete valuations of rank* 1. Let

$$\mathfrak{A} = \mathfrak{q}_1 \cap \mathfrak{q}_2 \cap \cdots \cap \mathfrak{q}_n \tag{29}$$

be such a decomposition of $\mathfrak{A}$ and let $v_1, v_2, \cdots, v_n$ be the corresponding valuations. Let $\mathfrak{q}'_i$ be the $v_i$-ideal determined by the condition $v_i(\mathfrak{q}'_i) = v_i(\mathfrak{A})$. Since $\mathfrak{A} \subset \mathfrak{q}_i$, we have $v_i(\mathfrak{A}) \geqq v_i(\mathfrak{q}_i)$, hence $\mathfrak{q}'_i \subset \mathfrak{q}_i$ and $\mathfrak{q}'_1 \cap \mathfrak{q}'_2 \cap \cdots \cap \mathfrak{q}'_n \subset \mathfrak{A}$. Since $\mathfrak{A} \subset \mathfrak{q}'_i$ for all $i$, it follows that $\mathfrak{A} = \mathfrak{q}'_1 \cap \mathfrak{q}'_2 \cap \cdots \cap \mathfrak{q}'_n$. Thus we may impose on the decomposition (29) the following further condition:

$$v_i(\mathfrak{A}) = v_i(\mathfrak{q}_i), \quad i = 1, 2, \cdots, n. \tag{30}$$

A decomposition (29) of $\mathfrak{A}$ into valuation ideals *belonging to discrete valuations* $v_1, v_2, \cdots, v_n$, of rank 1, shall be called a *standard decomposition* of $\mathfrak{A}$ if the relations (30) are satisfied.

Each standard decomposition (29) of $\mathfrak{A}$ determines the non-negative integer $\max\{v_1(\mathfrak{A}), v_2(\mathfrak{A}), \cdots, v_n(\mathfrak{A})\}$. We denote by $w(\mathfrak{A})$ the minimum value attained by this integer as the decomposition (29) ranges over the set of all standard decompositions of $\mathfrak{A}$. Then $w(\mathfrak{A})$ is a numerical character of $\mathfrak{A}$. It is a non-negative integer, and it is clear that $w(\mathfrak{A}) = 0$ if and only if $\mathfrak{A} = \mathfrak{o}$.

Let now $\mathfrak{o}'$ be a quadratic transform of $\mathfrak{o}$, relative to a directional form $\bar{g}$ (which we shall assume to be different from $z_1$), let $r =$ order of $\mathfrak{A}$ and let $\mathfrak{o}'\mathfrak{A} = t_1{}^r\mathfrak{A}'$, so that $\mathfrak{A}'$ is the transform of $\mathfrak{A}$ in $\mathfrak{o}$. We wish to prove that *if* $\mathfrak{A}$ *is primary for* $\mathfrak{m}$ *then*

$$w(\mathfrak{A}') < w(\mathfrak{A}). \tag{31}$$

We need the following simple lemma:

LEMMA 5. *Let* (29) *be a decomposition of a complete ideal* $\mathfrak{A}$ *into valuation ideals* $\mathfrak{q}_i$ *and let* $v_i$ *be a valuation* (*non-negative on* $\mathfrak{o}$) *such that* $\mathfrak{q}_i$ *is a* $v_i$*-ideal* (*we do not assume that the* $v_i$ *are discrete*). *Assume that* $\mathfrak{A}$ *is primary for* $\mathfrak{m}$ *and that the decomposition* (29) *satisfies conditions* (30). *Let* $h$ *be an arbitrary integer, let* $r$ *be the order of* $\mathfrak{A}$ *and let* $\tilde{\mathfrak{q}}_i$ *be the* $v_i$*-ideal determined by the condition:* $v_i(\tilde{\mathfrak{q}}_i)=v_i(\mathfrak{m}^h\mathfrak{A})$. Then

$$\mathfrak{m}^h\mathfrak{A} = \tilde{\mathfrak{q}}_1 \cap \tilde{\mathfrak{q}}_2 \cap \cdots \cap \tilde{\mathfrak{q}}_n \cap \mathfrak{m}^{r+h}. \tag{32}$$

PROOF. Denote by $\mathfrak{D}$ the ideal on the right-hand side of (32). Since $\mathfrak{m}^h\mathfrak{A}$ has order $r+h$ and since $\mathfrak{m}^h\mathfrak{A}\subset\mathfrak{D}\subset\mathfrak{m}^{r+h}$, also $\mathfrak{D}$ has order $r+h$. Therefore $c(\mathfrak{D})$ divides $c(\mathfrak{A})$, and the difference between the order $r+h$ of $\mathfrak{D}$ and the degree of $c(\mathfrak{D})$ is at least equal to $h$. Therefore, $\mathfrak{m}^h$ factors out from $\mathfrak{D}$ (Theorem 1). Let $\mathfrak{D}=\mathfrak{m}^h\mathfrak{E}$. Since $\mathfrak{m}^h\mathfrak{A}\subset\mathfrak{D}\subset\tilde{\mathfrak{q}}_i$ and $v_i(\mathfrak{m}^h\mathfrak{A})=v_i(\tilde{\mathfrak{q}}_i)$, it follows that $v_i(\mathfrak{D})=v_i(\tilde{\mathfrak{q}}_i)$. Thus $v_i(\mathfrak{m}^h)+v_i(\mathfrak{E})=v_i(\mathfrak{D})=v_i(\mathfrak{m}^h\mathfrak{A})=v_i(\mathfrak{m}^h)+v_i(\mathfrak{A})$, i.e., $v_i(\mathfrak{E})=v_i(\mathfrak{A})$, and $\mathfrak{E}\subset\mathfrak{q}_i$. Consequently $\mathfrak{E}\subset\mathfrak{A}$ and $\mathfrak{D}\subset\mathfrak{m}^h\mathfrak{A}$, showing that $\mathfrak{D}=\mathfrak{m}^h\mathfrak{A}$. Q.E.D.

We now proceed to the proof of the inequality (31). We fix a standard decomposition (29) of $\mathfrak{A}$ such that

$$w(\mathfrak{A}) = \max\{v_1(\mathfrak{A}), v_2(\mathfrak{A}), \cdots, v_n(\mathfrak{A})\}. \tag{33}$$

If $c(\mathfrak{A})$ is not divisible by $\bar{g}$ then $\mathfrak{A}'=\mathfrak{o}'$, whence $w(\mathfrak{A}')=0$ while $w(\mathfrak{A})>0$. We may assume therefore that $\bar{g}$ divides $c(\mathfrak{A})$. Then $\bar{g}$ is the directional form of at least one of the $n$ valuations $v_i$ (Lemma 4). Let $\bar{g}_1(=\bar{g}), \bar{g}_2, \cdots, \bar{g}_m$ be the (distinct) directional forms of $v_1, v_2, \cdots, v_n$ [$m\leqq n$; if one of the $v_i$ is the $\mathfrak{m}$-adic prime divisor of $\mathfrak{o}$ (so that $\mathfrak{q}_i$ is a power of $\mathfrak{m}$) we omit that particular $v_i$]. Let $r$ be the order of $\mathfrak{A}$, let $\tilde{\mathfrak{A}}_j$ be the partial intersection of those $\mathfrak{q}_i$ for which the corresponding $v_i$ has directional form $\bar{g}_j$ and let $\mathfrak{A}_j=\tilde{\mathfrak{A}}_j\cap\mathfrak{m}^r$. Then

$$\mathfrak{A} = \mathfrak{A}_1 \cap \mathfrak{A}_2 \cap \cdots \cap \mathfrak{A}_m,$$

each $\mathfrak{A}_j$ is a complete ideal, $c(\mathfrak{A}_j)$ is a power of $\bar{g}_j$, and $r(\mathfrak{A}_j)=r=r(\mathfrak{A})$. We know that a decomposition of $\mathfrak{A}$ with these properties is unique (Theorem 1, Corollary 3). Hence, by (19) (Theorem 1), we have $\mathfrak{A}_1=\mathfrak{o}'\mathfrak{A}\cap\mathfrak{o}$, whence $\mathfrak{o}'\mathfrak{A}_1=\mathfrak{o}'\mathfrak{A}$, *and thus the* $\mathfrak{o}'$*-transform* $\mathfrak{A}'$ *of* $\mathfrak{A}$ *coincides with the* $\mathfrak{o}'$*-transform of* $\mathfrak{A}_1$. Now, if, for a suitable labeling of the $v_1, v_2, \cdots, v_n$ we have that $v_1, v_2, \cdots, v_{n'}$ are the valuations whose directional form is $\bar{g}$, then

$$\mathfrak{A}_1 = \mathfrak{q}_1 \cap \mathfrak{q}_2 \cap \cdots \cap \mathfrak{q}_{n'} \cap \mathfrak{m}^r.$$

This is a standard decomposition of the complete ideal $\mathfrak{A}_1$. In fact, we have $v_i(\mathfrak{q}_i)\leqq v_i(\mathfrak{A}_1)\leqq v_i(\mathfrak{A})=v_i(\mathfrak{q}_i)$, for $i=1, 2, \cdots, n'$, whence

$v_i(\mathfrak{q}_i)=v_i(\mathfrak{A}_1)$. Furthermore, $\mathfrak{m}^r$ is a $v_\mathfrak{m}$-ideal, where $v_\mathfrak{m}$ is the $\mathfrak{m}$-adic prime divisor of $\mathfrak{o}$, and since $r(\mathfrak{A}_1)=r$ it follows that $v_\mathfrak{m}(\mathfrak{A}_1)=v_\mathfrak{m}(\mathfrak{m}^r)=r$. Now, $\max\{v_1(\mathfrak{A}_1), v_2(\mathfrak{A}_1), \cdots, v_{n'}(\mathfrak{A}_1), r\} \leqq \max\{v_1(\mathfrak{q}_1), v_2(\mathfrak{q}_2), \cdots, v_n(\mathfrak{q}_n)\}$ (since $v_i(\mathfrak{q}_i)=v_i(\mathfrak{A}) \geqq r$, for $i=1, 2, \cdots, n$) $=w(\mathfrak{A})$. Hence $w(\mathfrak{A}_1) \leqq w(\mathfrak{A})$, and it will be sufficient to prove that $w(\mathfrak{A}')<w(\mathfrak{A}_1)$. We may therefore assume that already our original ideal $\mathfrak{A}$ has the property that $c(\mathfrak{A})$ is a power of $\bar{g}$. If none of the $v_i$ ($i=1, 2, \cdots, n$) is the $\mathfrak{m}$-adic prime divisor $v_\mathfrak{m}$, we can add to the standard decomposition (29) of $\mathfrak{A}$ the $v_\mathfrak{m}$-ideal $\mathfrak{m}^r$, i.e., we may write

$$\mathfrak{A} = \mathfrak{q}_1 \cap \mathfrak{q}_2 \cap \cdots \cap \mathfrak{q}_n \cap \mathfrak{m}^r, \tag{34}$$

and this will still be a standard decomposition of $\mathfrak{A}$, since from $\mathfrak{A} \subset \mathfrak{m}^r$, $\mathfrak{A} \not\subset \mathfrak{m}^{r+1}$ follows $v_\mathfrak{m}(\mathfrak{A})=r=v_\mathfrak{m}(\mathfrak{m}^r)$. Relation (33) is not affected, since from $\mathfrak{A} \subset \mathfrak{m}^r$ follows $v_i(\mathfrak{A}) \geqq r$. We therefore use the decomposition (34) and we now assume that $v_1, v_2, \cdots, v_n$ are different from $v_\mathfrak{m}$. Since $c(\mathfrak{A})$ is a power of $\bar{g}$, any $\mathfrak{q}_i$ such that the directional form of $v_i$ is different from $\bar{g}$ is superfluous in (34) (Lemma 4), and the omission of that particular component $\mathfrak{q}_i$ will obviously not affect condition (33). We therefore assume that *$\bar{g}$ is the directional form of each of the n valuations $v_i$.*

This being so, each $v_i$ is non-negative on $\mathfrak{o}'$, and its center in $\mathfrak{o}'$ is the maximal ideal $\mathfrak{m}'$. Let $\mathfrak{q}'_i$ be the $v_i$-ideal in $\mathfrak{o}'$ such that $v_i(\mathfrak{q}'_i)= v_i(\mathfrak{A}')$. Since $\mathfrak{o}'\mathfrak{A}=t_1{}^r\mathfrak{A}'$ and $v_i(t_1{}^r)=v_i(\mathfrak{m}^r)$ (in view of our assumption that $\bar{g} \neq z_1$), we have

$$v_i(\mathfrak{q}'_i) = v_i(\mathfrak{A})-v_i(\mathfrak{m}^r) < v_i(\mathfrak{A}),$$

and hence

$$\max\{v_1(\mathfrak{q}'_1), v_2(\mathfrak{q}'_2), \cdots, v_n(\mathfrak{q}'_n)\} < w(\mathfrak{A}).$$

We shall now show that

$$\mathfrak{A}' = \mathfrak{q}'_1 \cap \mathfrak{q}'_2 \cap \cdots \cap \mathfrak{q}'_n, \tag{34'}$$

and this will establish inequality (31).

We have only to prove the inclusion $\mathfrak{q}'_1 \cap \mathfrak{q}'_2 \cap \cdots \cap \mathfrak{q}'_n \subset \mathfrak{A}'$. Let then $\xi'$ be any element of $\mathfrak{q}'_1 \cap \mathfrak{q}'_2 \cap \cdots \cap \mathfrak{q}'_n$ and let us write the element $t_1{}^r\xi'$ in the form $y/x$, where $\bar{x}$ is not divisible by $\bar{g}$, and $y \in \mathfrak{m}^h$ if $h$ is the order of $x$. We have to show that $\xi' \in \mathfrak{A}'$, or—what is the same thing—that

$$y/x \in \mathfrak{o}'\mathfrak{A}. \tag{35}$$

Since $\xi' \in \mathfrak{q}'_i$, we have $v_i(\xi') \geqq v_i(\mathfrak{A}')$, or $v_i(y/x) \geqq v_i(\mathfrak{o}'\mathfrak{A})=v_i(\mathfrak{A})$. Since $x \in \mathfrak{m}^h$, it follows that

$$v_i(y) \geqq v_i(\mathfrak{m}^h\mathfrak{A}), \quad i = 1, 2, \cdots, n. \tag{36}$$

For the $\mathfrak{m}$-adic prime divisor $v_{\mathfrak{m}}$ of $\mathfrak{o}$ we have $v_{\mathfrak{m}}(\xi')\geqq 0$, $v_{\mathfrak{m}}(y/x)\geqq r$, whence $v_{\mathfrak{m}}(y)\geqq r+h$, showing that $y\in \mathfrak{m}^{r+h}$. From this and (36) we conclude, by Lemma 5, that $y\in \mathfrak{m}^h\mathfrak{A}$, and this establishes (35) and completes the proof of the inequality (31).

Proposition 5 and inequality (31) complete our preparation of a basis for the inductive proofs of the theorems concerning complete ideals in $\mathfrak{o}$, given below.

THEOREM 2′. *Any product of complete ideals in* $\mathfrak{o}$ *is a complete ideal.*

PROOF. Let $\mathfrak{A}_1, \mathfrak{A}_2, \cdots, \mathfrak{A}_n$ be complete ideals in $\mathfrak{o}$. It is obviously sufficient to consider the case in which each $\mathfrak{A}_i$ is primary for $\mathfrak{m}$. Using the factorization theorem for complete ideals (Theorem 1′) and also Corollary 2 to Theorem 1′, we achieve at once a reduction to the case in which all the characteristic forms $c(\mathfrak{A}_i)$ $(i=1, 2, \cdots, n)$ are powers of one and the same irreducible form $\bar{g}$. In this case, let $\mathfrak{o}'$ be the quadratic transform of $\mathfrak{o}$ relative to the directional form $\bar{g}$ and let $\mathfrak{A}'_i$ be the $\mathfrak{o}'$-transform of $\mathfrak{A}_i$. We now use induction with respect to $\max\{w(\mathfrak{A}_1), w(\mathfrak{A}_2), \cdots, w(\mathfrak{A}_n)\}$, for the theorem is trivial when that maximum is zero. Since $w(\mathfrak{A}'_i)<w(\mathfrak{A}_i)$, our induction hypothesis implies that $\mathfrak{A}'_1\mathfrak{A}'_2\cdots\mathfrak{A}'_n$ is a complete ideal in $\mathfrak{o}'$. If we denote then by $\mathfrak{A}$ the product $\mathfrak{A}_1\mathfrak{A}_2\cdots\mathfrak{A}_n$ and by $r$ the order of $\mathfrak{A}$, we have $\mathfrak{o}'\mathfrak{A}= t_1{}^r\mathfrak{A}'_1\mathfrak{A}'_2\cdots\mathfrak{A}'_n$ (assuming—as we may—that $\bar{g}\neq z_1$), and hence also $\mathfrak{o}'\mathfrak{A}$ *is a complete ideal.* On the other hand, we have by Theorem 2 that $\mathfrak{A}=\mathfrak{o}'\mathfrak{A}\cap\mathfrak{o}$. Hence also $\mathfrak{A}$ is a complete ideal. Q.E.D.

The culminating point of our theory of complete ideals is a theorem of *unique factorization of complete ideals into simple* (complete) *ideals.* We shall say that *an ideal* $\mathscr{P}$ *is simple if it is not the unit ideal and has no non-trivial factorizations,* i.e., if from $\mathscr{P}=\mathfrak{A}\mathfrak{B}$, where $\mathfrak{A}$ and $\mathfrak{B}$ are ideals in $\mathfrak{o}$, follows necessarily that one of the ideals $\mathfrak{A}$, $\mathfrak{B}$ is the unit ideal. A principal ideal in $\mathfrak{o}$ (not the unit ideal) is simple if and only if it is prime, for it is easily seen that every ideal factor of a principal ideal in $\mathfrak{o}$ must be principal. In a noetherian ring, the ascending chain condition leads immediately to the conclusion that every ideal (different from the unit ideal) can be factored into simple ideals. For complete ideals we have also the following fact: *if a complete ideal* $\mathfrak{A}\neq(1)$ *admits at all non-trivial factorizations, it also admits a non-trivial factorization into complete ideals.* For, if $\mathfrak{A}=\mathfrak{A}_1\mathfrak{A}_2\cdots\mathfrak{A}_n$, then $\mathfrak{A}=\mathfrak{A}'\supset\mathfrak{A}'_1\mathfrak{A}'_2\cdots\mathfrak{A}'_n\supset\mathfrak{A}_1\mathfrak{A}_2\cdots\mathfrak{A}_n=\mathfrak{A}$, whence $\mathfrak{A}=\mathfrak{A}'_1\mathfrak{A}'_2\cdots\mathfrak{A}'_n$, where $'$ denotes the operation of completion (and where, therefore, if $\mathfrak{A}_i\neq(1)$ then also $\mathfrak{A}'_i\neq(1)$). It follows again by the ascending chain condition, that *every complete ideal* $(\neq 0)$ *can be factored into simple complete ideals.*

THEOREM 3. (*Unique factorization theorem for complete ideals*): *In a regular local ring* $\mathfrak{o}$, *of dimension* 2, *every complete ideal* $\mathfrak{A}$ ($\neq 0, \mathfrak{o}$) *has a* UNIQUE *factorization into simple complete ideals.*

PROOF. We shall use induction with respect to the numerical character $w(\mathfrak{A})$ introduced earlier in this appendix, for if $w(\mathfrak{A})=1$ then necessarily $\mathfrak{A}=\mathfrak{m}$,† and $\mathfrak{m}$ is obviously a simple ideal. The induction is based on a lemma which we shall state immediately after the following observation.

If $\mathscr{P}$ is a simple *contracted* ideal (in the sense defined at the end of the proof of Proposition 4), primary for $\mathfrak{m}$, then it follows from the factorization theorem for contracted ideals (Theorem 1) that $c(\mathscr{P})$ is a power of an irreducible form $\bar{g}$ in $k[z_1, z_2]$ (a *positive* power of $\bar{g}$, unless $\mathscr{P}=\mathfrak{m}$). Let $\mathfrak{o}'$ be the quadratic transform of $\mathfrak{o}$, relative to the directional form $\bar{g}$, and let $\mathscr{P}'$ be the $\mathfrak{o}$-transform of $\mathscr{P}$. We shall refer to $\mathscr{P}'$ as *the transform of* $\mathscr{P}$ (if $\mathscr{P}=\mathfrak{m}$, then $\mathscr{P}'=\mathfrak{o}'$).

LEMMA 6. *If* $\mathscr{P}$ *is a simple* CONTRACTED *ideal, different from* $\mathfrak{m}$ (*but not necessarily complete*), *then the transform* $\mathscr{P}'$ *of* $\mathscr{P}$ *is a simple ideal.*

PROOF. If $\mathscr{P}$ is a principal ideal, then the statement is trivial. We therefore assume that $\mathscr{P}$ is primary for $\mathfrak{m}$. Let $r$ be the order of $\mathscr{P}$, whence $\mathfrak{o}'\mathscr{P}=t_1{}^r\mathscr{P}'$ (we assume that $\bar{g}\neq z_1$). Let $\mathscr{P}'=\mathfrak{A}'\mathfrak{B}'$ be a factorization of $\mathscr{P}'$ in $\mathfrak{o}'$. We have to show that either $\mathfrak{A}'$ or $\mathfrak{B}'$ is the unit ideal. Let $a$ be the smallest (non-negative) integer with the property that $t_1{}^a\mathfrak{A}'$ is the extension of an ideal in $\mathfrak{o}$ (there exist integers with that property, since any ideal in $\mathfrak{o}'$ has a finite basis consisting of elements of the ring $\mathfrak{o}[t_2/t_1]$). Similarly, let $b$ be the similar integer, relative to the ideal $\mathfrak{B}'$. Since an extended ideal in $\mathfrak{o}'$ is also the extension of its contracted ideal, it follows that if we set

$$\mathfrak{A} = t_1{}^a\mathfrak{A}' \cap \mathfrak{o}, \quad \mathfrak{B} = t_1{}^b\mathfrak{B}' \cap \mathfrak{o},$$

then

$$\mathfrak{o}'\mathfrak{A} = t_1{}^a\mathfrak{A}', \quad \mathfrak{o}'\mathfrak{B} = t_1{}^b\mathfrak{B}'.$$

We have $\mathfrak{o}'\mathfrak{m}^r\mathfrak{A}\mathfrak{B}=t_1{}^{r+a+b}\mathfrak{A}'\mathfrak{B}'=t_1{}^{r+a+b}\mathscr{P}'=\mathfrak{o}'\mathfrak{m}^{a+b}\mathscr{P}$. Thus $\mathfrak{m}^r\mathfrak{A}\mathfrak{B}$ and $\mathfrak{m}^{a+b}\mathscr{P}$ have the same extension in $\mathfrak{o}'$. On the other hand, both these ideals are contractions of ideals in $\mathfrak{o}'$ (Theorem 2). Hence $\mathfrak{m}^r\mathfrak{A}\mathfrak{B}=\mathfrak{m}^{a+b}\mathscr{P}$. Now, $\mathfrak{m}$ does not factor out from either $\mathfrak{A}$ or $\mathfrak{B}$, for if say, $\mathfrak{A}=\mathfrak{m}\mathfrak{A}_1$ then $\mathfrak{o}'\mathfrak{A}_1=t_1{}^{a-1}\mathfrak{A}'$, in contradiction with the minimality of $a$. Hence the characteristic form $c(\mathfrak{A})$ of $\mathfrak{A}$ is a power of $\bar{g}$ and its degree is

† If $r$ = order of $\mathfrak{A}$ and $\mathfrak{o}'\mathfrak{A}=t_1{}^r\mathfrak{A}'$, where $\mathfrak{o}'$ is a quadratic transform of $\mathfrak{o}$, relative to a directional form $\bar{g}$ ($\bar{g}\neq z_1$), then $w(\mathfrak{A}')<w(\mathfrak{A})$ (see (31)); i.e., $w(\mathfrak{A}')=0$, $\mathfrak{A}'=\mathfrak{o}'$, and hence $\bar{g}$ does not divide $c(\mathfrak{A})$ (Proposition 2). This holds true for any irreducible form $\bar{g}$ in $k[z_1, z_2]$. Hence $c(\mathfrak{A})=1$ and $\mathfrak{A}$ is a power of $\mathfrak{m}$ (Proposition 4), and therefore $\mathfrak{A}=\mathfrak{m}$ since $w(\mathfrak{A})=1$.

equal to the order $a$ of $\mathfrak{A}$ (Proposition 3), and similarly for $\mathfrak{B}$. Hence the order of $\mathfrak{m}^r\mathfrak{A}\mathfrak{B}$ is $a+b+r$ and the degree of $c(\mathfrak{m}^r\mathfrak{A}\mathfrak{B})$ is $a+b$. The degree of $c(\mathfrak{m}^{a+b}\mathscr{P})$ is obviously $r$. Hence $r=a+b$, and from Theorem 1 it now follows that $\mathscr{P}=\mathfrak{A}\mathfrak{B}$. Hence either $\mathfrak{A}$ or $\mathfrak{B}$ must be the unit ideal, showing that either $\mathfrak{A}'$ or $\mathfrak{B}'$ is the unit ideal. This completes the proof of the lemma.

The proof of the theorem can now be rapidly completed.

Suppose that we have two factorizations of a complete ideal $\mathfrak{A}$ into simple complete ideals. Among the simple factors there may occur the maximal ideal $\mathfrak{m}$. We therefore put into evidence the power of $\mathfrak{m}$ which occurs in both factorizations:

$$\mathfrak{A} = \mathfrak{m}^h\mathscr{P}_1\mathscr{P}_2\cdots\mathscr{P}_n = \mathfrak{m}^{\tilde{h}}\tilde{\mathscr{P}}_1\tilde{\mathscr{P}}_2\cdots\tilde{\mathscr{P}}_{\tilde{n}},$$

where the $\mathscr{P}_i$ and $\tilde{\mathscr{P}}_j$ are simple complete ideals, all different from $\mathfrak{m}$. Let $r$ be the order of $\mathfrak{A}$ and $s$ the degree of the characteristic form $c(\mathfrak{A})$ of $\mathfrak{A}$. The latter is a product of the characteristic forms $c(\mathscr{P}_i)$, $i=1, 2, \cdots, n$, and since each $\mathscr{P}_i$ is simple, the degree of $c(\mathscr{P}_i)$ is equal to the order of $\mathscr{P}_i$ (otherwise, by Proposition 3, $\mathfrak{m}$ would factor out from $\mathscr{P}_i$). Hence $h=r-s$. Similarly $\tilde{h}=r-s$, and thus $h=\tilde{h}$. Each directional form $\bar{g}$ of $\mathfrak{A}$ (i.e., each irreducible factor of $c(\mathfrak{A})$) is the directional form of at least one of the $\mathscr{P}_t$ and also of at least one of the $\tilde{\mathscr{P}}_j$, and, furthermore, *the* directional form of each $\mathscr{P}_i$ and of each $\tilde{\mathscr{P}}_j$ is a directional form of $\mathfrak{A}$. Let $\bar{g}_1, \bar{g}_2, \cdots, \bar{g}_m$ be the distinct directional forms of $c(\mathfrak{A})$, let $\mathfrak{B}_\alpha$ (or $\tilde{\mathfrak{B}}_\alpha$) be the product of those $\mathscr{P}_i$ (or $\tilde{\mathscr{P}}_j$) whose directional form is $\bar{g}_\alpha$. Then $\mathfrak{A}=\mathfrak{m}^{r-s}\mathfrak{B}_1\mathfrak{B}_2\cdots\mathfrak{B}_m=\mathfrak{m}^{r-s}\tilde{\mathfrak{B}}_1\tilde{\mathfrak{B}}_2\cdots\tilde{\mathfrak{B}}_m$, and the characteristic form of $\mathfrak{B}_i$ (or $\tilde{\mathfrak{B}}_i$) is a power of $\bar{g}_i$. Hence, by Theorem 1, we must have $\mathfrak{B}_i=\tilde{\mathfrak{B}}_i$, $i=1, 2, \cdots, m$. This reduces the proof of the theorem to the case in which $h=0$ and all the ideals $\mathscr{P}_i$, $\tilde{\mathscr{P}}_j$ have the same directional form, say $\bar{g}$. In this case we introduce, as usual, our quadratic transform $\mathfrak{o}'$ of $\mathfrak{o}$, relative to the directional form $\bar{g}$, and we denote by $\mathfrak{A}'$, $\mathscr{P}'_i$, $\tilde{\mathscr{P}}'_j$ the $\mathfrak{o}'$-transforms of $\mathfrak{A}$, $\mathscr{P}_i$, $\tilde{\mathscr{P}}_j$ respectively. Then, clearly, passing to extensions in $\mathfrak{o}'$ and cancelling the common factor $t_1{}^r$, we find $\mathscr{P}'_1\mathscr{P}'_2\cdots\mathscr{P}'_n=\tilde{\mathscr{P}}'_1\tilde{\mathscr{P}}'_2\cdots\tilde{\mathscr{P}}'_{\tilde{n}}=\mathfrak{A}'$. Since the ideals $\mathscr{P}'_i$ and $\tilde{\mathscr{P}}'_j$ in $\mathfrak{o}'$ are simple (by the above lemma) and complete (by Proposition 5), and since $w(\mathfrak{A}')<w(\mathfrak{A})$, we have, by our induction hypothesis, that $n=\tilde{n}$ and that for a suitable labeling of the $\mathscr{P}_i$ and the $\tilde{\mathscr{P}}_j$ we must have $\mathscr{P}'_i=\tilde{\mathscr{P}}'_i$. If $a_i$ is the least integer such that $t_1{}^{a_i}\mathscr{P}'_i$ is an extended ideal, then $t_1{}^{a_i}\mathscr{P}'_i$ is the extension of both $\mathscr{P}_i$ and $\tilde{\mathscr{P}}_i$, and we have $\mathscr{P}_i=\tilde{\mathscr{P}}_i=t_1{}^{a_i}\mathscr{P}'_i\cap\mathfrak{o}$. This completes the proof of the theorem.

REMARK. It is not difficult to show that every contracted ideal

$\mathfrak{A}$ ($\neq \mathfrak{o}$) can be factored into simple *contracted* ideals. The proof is as follows:

Let $\mathfrak{A} = \mathfrak{A}_1\mathfrak{A}_2 \cdots \mathfrak{A}_n$ be a non-trivial factorization of $\mathfrak{A}$. Let $\{\bar{g}_1, \bar{g}_2, \cdots, \bar{g}_m\}$ be the set of directional forms of $\mathfrak{A}_1, \mathfrak{A}_2, \cdots, \mathfrak{A}_n$ and let $\mathfrak{o}' = \mathfrak{o}'_1 \cap \mathfrak{o}'_2 \cap \cdots \cap \mathfrak{o}'_m$, where $\mathfrak{o}'_i$ is the quadratic transform of $\mathfrak{o}$, relative to the directional form $\bar{g}_i$. Let $\mathfrak{o}'\mathfrak{A}_i \cap \mathfrak{o} = \mathfrak{B}_i$ and let $\mathfrak{C} = \mathfrak{B}_1\mathfrak{B}_2 \cdots \mathfrak{B}_n$. It is clear that $\{\bar{g}_1, \bar{g}_2, \cdots, \bar{g}_m\}$ is the set of directional forms of $\mathfrak{A}$. Therefore $\mathfrak{o}'\mathfrak{A} \cap \mathfrak{o} = \mathfrak{A}$ (Proposition 4, Corollary 1). By Theorem 2 (and Proposition 4, Corollary 1) we also have $\mathfrak{o}'\mathfrak{C} \cap \mathfrak{o} = \mathfrak{C}$ (since each directional form of $\mathfrak{B}_i$ is also a directional form of $\mathfrak{A}$ and since, therefore, the directional forms of $\mathfrak{C}$ are in the set $\{\bar{g}_1, \bar{g}_2, \cdots, \bar{g}_m\}$). Since $\mathfrak{o}'\mathfrak{A} = \mathfrak{o}'\mathfrak{C}$, it follows that $\mathfrak{A} = \mathfrak{C}$, whence $\mathfrak{A} = \mathfrak{B}_1\mathfrak{B}_2 \cdots \mathfrak{B}_n$, and this yields a factorization of $\mathfrak{A}$ such that $\mathfrak{B}_i \supset \mathfrak{A}_i$. If one of the $\mathfrak{A}_i$ is not a contracted ideal then $\mathfrak{B}_i > \mathfrak{A}_i$. If one of the $\mathfrak{B}_i$ is not a simple ideal (and is not the unit ideal) we factor it into simple ideals. Proceeding in this fashion and using the ascending chain condition in $\mathfrak{o}$, we arrive after a finite number of steps at a factorization of $\mathfrak{A}$ into simple contracted ideals.

However, the theorem of *unique* factorization of complete ideals into simple complete ideals does not generalize to contracted ideals, i.e., a contracted ideal does not necessarily have a *unique* factorization into simple contracted ideals. The reason for this is that Proposition 5 does not generalize to contracted ideals, i.e., the extension of an ideal in $\mathfrak{o}$ which is the contraction of an ideal in $\mathfrak{o}'$ is not necessarily a contracted ideal in the regular ring $\mathfrak{o}'$. For example, let $\mathfrak{o}'$ be the quadratic transform of $\mathfrak{o}$, relative to the directional form $z_2$, and let $\mathfrak{A}$ be the ideal $(t_2^2, \mathfrak{m}^4)$ in $\mathfrak{o}$. It is easily seen that $\mathfrak{A}$ is the contracted ideal of its extension $t_1^2(t'^2_2, t_1^2)$ in $\mathfrak{o}'$, where $t'_2 = t_2/t_1$ ($t_1$ and $t'_2$ are regular parameters in $\mathfrak{o}'$). However, the transform $\mathfrak{A}' = (t'^2_2, t_1^2)$ of $\mathfrak{A}$ is not a contracted ideal in $\mathfrak{o}'$ (we have $c(\mathfrak{A}') = 1$, but $\mathfrak{A}'$ is not a power of $\mathfrak{m}'$; see Corollary to Proposition 2). Note that as a consequence of this and in view of Proposition 5, $\mathfrak{A}$ cannot be a complete ideal; but this is also easily seen directly, because we have $t_2^2, t_1^4 \in \mathfrak{A}$, $t_1^2t_2 \notin \mathfrak{A}$, while $t_1^2t_2$ is integrally dependent on $\mathfrak{A}$ ($(t_1^2t_2)^2 = t_1^4t_2^2$). We note also that $\mathfrak{A}$ is a simple ideal and that $\mathfrak{m}\mathfrak{A} = \mathscr{P}^3$, where $\mathscr{P} = (t_2, \mathfrak{m}^2)$ is also a simple (complete ideal). Thus the *contracted ideal* $\mathfrak{m}\mathfrak{A}$ (see Corollary to Theorem 1) admits two distinct factorizations into simple contracted ideals.

5. We shall conclude this appendix with some miscellaneous properties of simple complete ideals.

(A) Let $\bar{g}$ be an irreducible form in $k[z_1, z_2]$, let $\mathfrak{o}'$ be the quadratic transform of $\mathfrak{o}$, relative to the directional form $\bar{g}$, and let $M_{\bar{g}}$ be the set of all simple complete ideals in $\mathfrak{o}$ having $\bar{g}$ as directional form. Then $\mathscr{P} \to \mathscr{P}'$, *where* $\mathscr{P} \in M_{\bar{g}}$ *and* $\mathscr{P}'$ *is the transform of* $\mathscr{P}$ (in $\mathfrak{o}'$), *is a* (1, 1) *mapping of* $M_{\bar{g}}$ *onto the set of all simple complete ideals in* $\mathfrak{o}'$.

For the proof, we shall assume that $z_1 \neq \bar{g}$. We note that if $\mathscr{P} \to \mathscr{P}'$ then $\mathscr{P} = t_1^a\mathscr{P}' \cap \mathfrak{o}$, where $a$ is the least integer such that $t_1^a\mathscr{P}'$ is the extension of an ideal in $\mathfrak{o}$. This shows that the mapping $\mathscr{P} \to \mathscr{P}'$ is univalent. Now, given any simple complete ideal $\mathscr{P}'$ in $\mathfrak{o}'$, define $a$ and $\mathscr{P}$ as above. Then $\mathscr{P}$ is a complete ideal in $\mathfrak{o}'$, and we have $\mathfrak{o}'\mathscr{P} = t_1^a\mathscr{P}'$. Clearly, $c(\mathscr{P})$ must be a power of $\bar{g}$ (Proposition 3), and $\mathscr{P}$ must be simple, for in the contrary case we would find at once that $\mathscr{P}'$ is not simple. This completes the proof.

(B) Starting with a given simple complete ideal $\mathscr{P}$, different from $\mathfrak{m}$, we consider its transform $\mathscr{P}'$ (this transform is an ideal in $\mathfrak{o}'$ introduced in (A), if $\mathscr{P} \in M_{\bar{g}}$). If $\mathscr{P}' \neq \mathfrak{m}'$ ($\mathfrak{m}'$ = maximal ideal in $\mathfrak{o}'$) we may repeat the procedure and consider the transform $\mathscr{P}''$ of $\mathscr{P}'$ in a suitable quadratic transform $\mathfrak{o}''$ of $\mathfrak{o}'$. Since $w(\mathscr{P}) > w(\mathscr{P}') > w(\mathscr{P}'') > \cdots$, this process is finite. We thus obtain a finite strictly ascending sequence of regular rings

$$\mathfrak{o} < \mathfrak{o}' < \mathfrak{o}'' < \cdots < \mathfrak{o}^{(h)},$$

each ring in the sequence being a quadratic transform of its immediate predecessor, and in each ring $\mathfrak{o}^{(i)}$ we have a simple complete ideal $\mathscr{P}^{(i)}$ such that $\mathscr{P}^{(i)}$ is the transform of $\mathscr{P}^{(i-1)}$ ($\mathscr{P}^{(0)} = \mathscr{P}$) and such that $\mathscr{P}^{(h)}$ is the maximal ideal $\mathfrak{m}^{(h)}$ of $\mathfrak{o}^{(h)}$. This sequence of rings $\mathfrak{o}^{(i)}$ and the integer $h$ are uniquely determined by $\mathscr{P}$; we say that $\mathscr{P}$ is a simple ideal *of rank h*. We denote by $v_{\mathscr{P}}$ the $\mathfrak{m}^{(h)}$-adic prime divisor of $\mathfrak{o}^{(h)}$. Then $v_{\mathscr{P}}$ is a prime divisor of the quotient field of $\mathfrak{o}$, and the center of $v_{\mathscr{P}}$ in $\mathfrak{o}$ is $\mathfrak{m}$ (in other words: $v_{\mathscr{P}}$ is of the second kind with respect to $\mathfrak{o}$; see VI, § 5, p. 19). It is clear that we have $v_{\mathscr{P}} = v_{\mathscr{P}'} = \cdots = v_{\mathscr{P}^{(h-1)}} = v_{\mathscr{P}^{(h)}}$, where $\mathscr{P}^{(h)} = \mathfrak{m}^{(h)}$.

(C) $\mathscr{P}$ *is a* $v_{\mathscr{P}}$*-ideal in* $\mathfrak{o}$.

Proof is by induction with respect to the integer $h$, since if $h = 0$ then $\mathscr{P} = \mathfrak{m}$ and $v_{\mathscr{P}}$ is the $\mathfrak{m}$-adic prime divisor of $\mathfrak{o}$. The integer $h$ is the number of successive quadratic transforms which are necessary to transform $\mathscr{P}$ into the maximal ideal of a suitable regular ring $\mathfrak{o}^{(h)}$. It is clear that $\mathscr{P}'$ is a simple ideal of rank $h-1$. Therefore, by our induction hypothesis, $\mathscr{P}'$ is a $v_{\mathscr{P}}$-ideal in $\mathfrak{o}'$. Now, let $\mathfrak{o}'\mathscr{P} = t_1^a\mathscr{P}'$, where we assume that $\bar{g} \neq z_1$, so that $a$ is the order of $\mathscr{P}$. Let $\mathfrak{q}$ be the $v_{\mathscr{P}}$-ideal in $\mathfrak{o}$ determined by the condition: $v_{\mathscr{P}}(\mathfrak{q}) = v_{\mathscr{P}}(\mathscr{P})$. Then $\mathfrak{q} \supset \mathscr{P}$, and we have to show that actually $\mathfrak{q} = \mathscr{P}$. We first show that

$\mathscr{P} = \mathfrak{m}^a \cap \mathfrak{q}$. We have only to prove that $\mathscr{P} \supset \mathfrak{m}^a \cap \mathfrak{q}$. Let $x$ be any element of $\mathfrak{m}^a \cap \mathfrak{q}$. Since $x \in \mathfrak{m}^a$, we have $x = t_1{}^a x'$, with $x'$ in $\mathfrak{o}'$. Since $x \in \mathfrak{q}$, we have $v_{\mathscr{P}}(t_1{}^a x') \geqq v_{\mathscr{P}}(\mathscr{P}) = v_{\mathscr{P}}(t_1{}^a \mathscr{P}')$, $v_{\mathscr{P}}(x') \geqq v_{\mathscr{P}}(\mathscr{P}')$, whence $x' \in \mathscr{P}'$ (since $\mathscr{P}'$ is a $v_{\mathscr{P}}$-ideal) and thus $x \in t_1{}^a \mathscr{P}' \cap \mathfrak{o} = \mathscr{P}$, which proves the equality $\mathscr{P} = \mathfrak{m}^a \cap \mathfrak{q}$. We now have to show that $\mathfrak{q} \subset \mathfrak{m}^a$. Suppose the contrary to be true, and let $b$ be the order of $\mathfrak{q}$ (whence $b < a$). The characteristic form $c(\mathfrak{q})$ of $\mathfrak{q}$ is then at most of degree $b$, and since $\mathfrak{m}^{a-b}\mathfrak{q} \subset \mathscr{P}$, also the degree of $c(\mathscr{P})$ is at most equal to $b$. On the other hand, the order $a$ of $\mathscr{P}$ is greater than $b$. Hence, by Proposition 3, $\mathfrak{m}$ factors out from $\mathscr{P}$, in contradiction with the fact that $\mathscr{P}$ is a simple ideal.

(D) The method of proof in (C) can be used to derive a general result which concerns arbitrary valuations centered at $\mathfrak{m}$ and which we shall want to use later on.

Let $v$ be a valuation centered at $\mathfrak{m}$, different from the $\mathfrak{m}$-adic prime divisor of $\mathfrak{o}$, let $\bar{g}$ be the directional form of $v$ (we assume that $\bar{g} \neq z_1$), and let $\mathfrak{o}' = \mathfrak{o}'_{\bar{g}}$ be the quadratic transform of $\mathfrak{o}$, relative to $\bar{g}$. In the well ordered set of $v$-ideals in $\mathfrak{o}$ we consider the initial infinite simple sequence $\{\mathfrak{q}_i\}$ where $i = 1, 2, \cdots, n, \cdots$; here $\mathfrak{q}_1 = \mathfrak{o}$ (if $v$ has rank 1, then this sequence is in fact the entire set of $v$-ideals in $\mathfrak{o}$). Since we have assumed that $v$ is different from the $\mathfrak{m}$-adic prime divisor of $\mathfrak{o}$, $v$ is also centered at the maximal ideal $\mathfrak{m}'$ of $\mathfrak{o}'$. We consider in $\mathfrak{o}'$ the initial infinite simple sequence $\{\mathfrak{q}'_j\}$ of $v$-ideals, $j = 1, 2, \cdots, n, \cdots$ $(\mathfrak{q}'_1 = \mathfrak{o}')$. Since the characteristic form of any $\mathfrak{q}_i$ is a power of $\bar{g}$ (Lemma 4), we can speak of the transform of $\mathfrak{q}_i$ in $\mathfrak{o}'$. We denote this transform by $\mathfrak{Q}'_i$. For any $\mathfrak{q}'_j$ there exists a smallest integer $a_j$ such that $t_1{}^{a_j}\mathfrak{q}'_j$ is the extension of an ideal in $\mathfrak{o}$. Then we set $\mathfrak{Q}_j = t_1{}^{a_j}\mathfrak{q}'_j \cap \mathfrak{o}$, so that $t_1{}^{a_j}\mathfrak{q}'_j$ is also the extension of $\mathfrak{Q}_j$. We call $\mathfrak{Q}_j$ the *inverse transform* of $\mathfrak{q}'_j$.

The result which we wish to prove describes the relationship between the two sequences $\{\mathfrak{q}_i\}$, $\{\mathfrak{q}'_j\}$ and is as follows:

(1) *The transform* $\mathfrak{Q}'_i$ *of any* $\mathfrak{q}_i$ *is a member of the sequence* $\{\mathfrak{q}'_j\}$.

(2) *The inverse transform* $\mathfrak{Q}_j$ *of any* $\mathfrak{q}'_j$ *is a member of the sequence* $\{\mathfrak{q}_i\}$ *(and hence any* $\mathfrak{q}'_j$ *is the transform of some* $\mathfrak{q}_i$*).*

(3) *Any* $\mathfrak{q}_i$ *is of the form* $\mathfrak{m}^h\mathfrak{Q}_j$, *where* $\mathfrak{Q}_j$ *is the inverse transform of some* $\mathfrak{q}'_j$.

(4) *If* $\mathfrak{Q}_\alpha > \mathfrak{Q}_\beta$ *then* $\mathfrak{q}'_\alpha > \mathfrak{q}'_\beta$.

Assertion (1) follows directly from relations (34) and (34′), as applied to the ideal $\mathfrak{A} = \mathfrak{q}_i$.

Let $\mathfrak{q}$ be the $v$-ideal in $\mathfrak{o}$ such that $v(\mathfrak{q}) = v(\mathfrak{Q}_j)$. Since $\mathfrak{o}'\mathfrak{Q}_j = t_1{}^{a_j}\mathfrak{q}'_j$, $\mathfrak{Q}_j$ has order $a_j$. It is clear that $\mathfrak{Q}_j \subset \mathfrak{m}^{a_j} \cap \mathfrak{q}$. By the same reasoning as the one used in (C) we find that, in the first place, we must have $\mathfrak{Q}_j = \mathfrak{m}^{a_j} \cap \mathfrak{q}$, and—next—that $\mathfrak{q} \subset \mathfrak{m}^{a_j}$ (always using the fact that the degree of the characteristic form $c(\mathfrak{Q}_j)$ of $\mathfrak{Q}_j$ is equal to the order $a_j$ of $\mathfrak{Q}_j$). Hence $\mathfrak{Q}_j = \mathfrak{q}$, and this proves assertion (2).

Let $\mathfrak{Q}'_i = \mathfrak{q}'_j$ be the transform of $\mathfrak{q}_i$ and let $\mathfrak{q} = \mathfrak{Q}_j$ be the inverse transform of $\mathfrak{q}'_j$. If $a$ and $b$ are the orders of $\mathfrak{q}$ and $\mathfrak{q}_i$ respectively, then $b \geqq a$. The two ideals $\mathfrak{q}_i$ and $\mathfrak{m}^{b-a}\mathfrak{q}$ have the same extension in $\mathfrak{o}'$, namely the ideal $t_1{}^b \mathfrak{q}'_j$. Since they are both contracted ideals, it follows that $\mathfrak{q}_i = \mathfrak{m}^{b-a}\mathfrak{q} = \mathfrak{m}^{b-a}\mathfrak{Q}_j$, and this proves assertion (3).

Let $a$ and $b$ be the orders of $\mathfrak{Q}_\alpha$ and $\mathfrak{Q}_\beta$ respectively ($a \leqq b$), whence $\mathfrak{o}'\mathfrak{Q}_\alpha = t_1{}^a \mathfrak{q}'_\alpha$, $\mathfrak{o}'\mathfrak{Q}_\beta = t_1{}^b \mathfrak{q}'_\beta$, and thus $\mathfrak{o}'\mathfrak{m}^{b-a}\mathfrak{Q}_\alpha = t_1{}^b \mathfrak{q}'_\alpha$. Assume that assertion (4) is false and that we have therefore $\mathfrak{q}'_\alpha \subset \mathfrak{q}'_\beta$. Then it follows that $\mathfrak{o}'\mathfrak{m}^{b-a}\mathfrak{Q}_\alpha \subset \mathfrak{o}'\mathfrak{Q}_\beta$, and therefore $\mathfrak{m}^{b-a}\mathfrak{Q}_\alpha \subset \mathfrak{Q}_\beta$, since both these ideals are contracted ideals. The equality $b = a$ is excluded since $\mathfrak{Q}_\alpha > \mathfrak{Q}_\beta$. Hence $b - a > 0$, i.e., $\mathfrak{m}$ factors out from $\mathfrak{Q}_\beta$, and this is in contradiction with the fact that $b$ is the least integer such that $t_1{}^b \mathfrak{q}'_\beta$ is an extended ideal. This establishes assertion (4).

(E) *The correspondence* $\mathscr{P} \to v_{\mathscr{P}}$ *is a* (1, 1) *mapping of the set of all simple complete ideals in* $\mathfrak{o}$ *onto the set of all prime divisors of the quotient field of* $\mathfrak{o}$ *which are of the second kind with respect to* $\mathfrak{o}$.

We first observe that if $\mathscr{P} \neq \mathfrak{m}$ then $v_{\mathscr{P}} \neq v_{\mathfrak{m}}$. In fact, assuming that the directional form $\bar{g}$ of $\mathscr{P}$ is different from $z_1$, we have $v_{\mathfrak{m}}(x) = r$ for every element $x$ of $\mathfrak{o}$ such that $x \in \mathfrak{m}^r$, $x \notin \mathfrak{m}^{r+1}$, while if the initial form $\bar{x}$ of $x$ is divisible by $\bar{g}$ then $v_{\mathscr{P}}(x) > r$ (Lemma 1). Now, if $\mathscr{P}_1$ and $\mathscr{P}_2$ are two arbitrary distinct simple complete ideals in $\mathfrak{o}$, then our assertion that $v_{\mathscr{P}_1} \neq v_{\mathscr{P}_2}$ is, in the first place, obvious if the directional forms of $\mathscr{P}_1$ and $\mathscr{P}_2$ (which are also the directional forms of the valuations $v_{\mathscr{P}_1}$ and $v_{\mathscr{P}_2}$) are distinct, and, in the second place, if $\mathscr{P}_1$ and $\mathscr{P}_2$ have the same directional form then the assertion follows immediately by induction with respect to the integer $s = \max\{\text{rank } \mathscr{P}_1, \text{rank } \mathscr{P}_2\}$, by passing to transforms $\mathscr{P}'_1$, $\mathscr{P}'_2$, since we have just proved the assertion $v_{\mathscr{P}_1} \neq v_{\mathscr{P}_2}$ in the case $s = 1$. We have thus shown that the mapping $\mathscr{P} \to v_{\mathscr{P}}$ is univalent.

To complete the proof we now have to show that given any prime divisor $v$ of $\mathfrak{o}$ (i.e., any valuation of the quotient field of $\mathfrak{o}$ such that $v$ has $\mathfrak{o}$-dimension 1) there exists a simple complete ideal $\mathscr{P}$ in $\mathfrak{o}$ such that $v = v_{\mathscr{P}}$. If $v$ is the $\mathfrak{m}$-adic prime divisor of $\mathfrak{o}$, there is nothing to prove: we have $\mathscr{P} = \mathfrak{m}$. In the contrary case, $v$ has a well defined directional form $\bar{g}$, and if $\mathfrak{o}'$ is the quadratic transform of $\mathfrak{o}$, relative to the directional form $\bar{g}$, then $v$ is still of the second kind with respect to $\mathfrak{o}'$, its center in $\mathfrak{o}'$ being the maximal ideal $\mathfrak{m}'$. If $v$ is the $\mathfrak{m}'$-adic divisor of $\mathfrak{o}'$, then we have $v = v_{\mathscr{P}}$, where $\mathscr{P}$ is the simple ideal (of rank 1) in $\mathfrak{o}$ whose transform is $\mathfrak{m}'$. In the contrary case, $v$ has a well defined directional form in $\mathfrak{o}'$, and if $\mathfrak{o}''$ denotes the corresponding quadratic transform of $\mathfrak{o}'$, then $v$ is of the second kind with respect to $\mathfrak{o}''$. If $v$ is the $\mathfrak{m}''$-adic prime divisor of $\mathfrak{o}''$, then we have $v = v_{\mathscr{P}}$, where $\mathscr{P}$ is the simple complete ideal (of rank 2) in $\mathfrak{o}$ whose (second) transform in $\mathfrak{o}''$ is $\mathfrak{m}'$. In the contrary case we go on to a suitable quadratic transform $\mathfrak{o}'''$ of $\mathfrak{o}''$. We have to show that after a finite number of steps we obtain a ring $\mathfrak{o}^{(h)}$ such that $v$ is the $\mathfrak{m}^{(h)}$-adic prime divisor of $\mathfrak{o}^{(h)}$, where $\mathfrak{m}^{(h)}$ is the maximal ideal of $\mathfrak{o}^{(h)}$. We shall show that the assumption that the above process does not terminate after a finite number of steps leads to a contradiction. Under such an assumption we will have an infinite strictly ascending chain of rings

$$\mathfrak{o} < \mathfrak{o}^{(1)} < \mathfrak{o}^{(2)} < \cdots < \mathfrak{o}^{(h)} < \cdots$$

with the following properties:

(1) Each ring $\mathfrak{o}^{(h)}$ is a quadratic transform of its immediate predecessor.

(2) $v$ is non-negative on any $\mathfrak{o}^{(h)}$, and its center in $\mathfrak{o}^{(h)}$ is the maximal ideal $\mathfrak{m}^{(h)}$ of $\mathfrak{o}^{(h)}$.

We now fix an element $\omega$ in the quotient field of $\mathfrak{o}$ such that the $v$-residue of $\omega$ is transcendental over the residue field $k$ $(=\mathfrak{o}/\mathfrak{m})$ of $\mathfrak{o}$. Since the residue field of $\mathfrak{o}^{(h)}$ is an algebraic extension of the residue field of $\mathfrak{o}^{(h-1)}$, it follows that the $v$-residue of $\omega$ is also transcendental over $\mathfrak{o}^{(h)}/\mathfrak{m}^{(h)}$. Now let us write $\omega$ in the form $\omega = y/x$, where $x, y \in \mathfrak{o}$. Both $x$ and $y$ necessarily belong to $\mathfrak{m}$. Assuming—as we may—that the directional form of $v$ is different from $z_1$, then we can write $x = t_1 x_1$, $y = t_1 y_1$, with $x_1, y_1$ in $\mathfrak{o}'$. Then $\omega = y_1/x_1$ and $v(x) > v(x_1)$. Since $v$ is also of the second kind with respect to $\mathfrak{o}'$, it follows again that both $x_1, y_1$ are in $\mathfrak{m}'$, and thus we find another representation of $\omega$, of the form $\omega = y_2/x_2$, with $x_2, y_2$ in $\mathfrak{o}''$ and $v(x_1) > v(x_2)$. Proceeding in this fashion we obtain an infinite, strictly decreasing sequence $v(x) > v(x_1) > v(x_2) > \cdots$ of *positive* integers, which is absurd.

(F) Let $\mathscr{P}$ be a simple complete ideal, of rank $h$, let $\mathfrak{o} < \mathfrak{o}' < \mathfrak{o}'' < \cdots < \mathfrak{o}^{(h)}$ be the sequence of successive quadratic transforms of $\mathfrak{o}$ which is determined by $\mathscr{P}$ (see (B)), and let $\mathscr{P}_i$ be the simple complete ideal in $\mathfrak{o}$ whose transform in $\mathfrak{o}^{(i)}$ is the maximal ideal of $\mathfrak{o}^{(i)}$ $(i = 1, 2, \cdots, h)$. Then:

(1) $\mathfrak{m} > \mathscr{P}_1 > \mathscr{P}_2 > \cdots > \mathscr{P}_h = \mathscr{P}$.

(2) *Each of the* $h+1$ *ideals* $\mathfrak{m}, \mathscr{P}_1, \mathscr{P}_2, \cdots, \mathscr{P}_h$ *is a* $v_{\mathscr{P}}$*-ideal in* $\mathfrak{o}$, *and every* $v_{\mathscr{P}}$*-ideal in* $\mathfrak{o}$ *is a power product of these* $h+1$ *simple ideals.*

If $h=0$ then assertion (1) is vacuous, while (2) is obvious, since $v_{\mathscr{P}}$ is in that case the $\mathfrak{m}$-adic prime divisor of $\mathfrak{o}$, and therefore every $v_{\mathscr{P}}$-ideal is in that case a power of $\mathfrak{m}$. We therefore proceed by induction with respect to $h$.

Let $\mathfrak{Q}'_i$ be the transform of $\mathscr{P}_i$ in $\mathfrak{o}'$ $(i=1, 2, \cdots, h)$. Then $\mathfrak{Q}'_i$ is a simple complete ideal $\mathscr{P}'_{i-1}$ in $\mathfrak{o}'$, of rank $i-1$, and $v_{\mathscr{P}}=v_{\mathscr{P}'}$, where $\mathscr{P}'=\mathfrak{Q}'_h=$ transform of $\mathscr{P}$. We have therefore, by our induction hypothesis:

$$\mathfrak{m}' = \mathfrak{Q}'_1 > \mathfrak{Q}'_2 > \cdots > \mathfrak{Q}'_h = \mathscr{P}'.$$

Since $\mathscr{P}_i$ is the inverse transform of $\mathfrak{Q}'_i$ (in the sense of (D)), it follows from statement (4) in (D) that

$$\mathscr{P}_1 > \mathscr{P}_2 > \cdots > \mathscr{P}_h = \mathscr{P}.$$

Since the strict inclusion $\mathfrak{m} > \mathscr{P}_1$ is obvious, assertion (1) is proved.

That each of the $h+1$ ideals $\mathfrak{m}, \mathscr{P}_1, \mathscr{P}_2, \cdots, \mathscr{P}_h$ is a $v_{\mathscr{P}}$-ideal follows from statement (2) of (D) and from our induction hypothesis. Now, if $\mathfrak{q}'$ is any $v_{\mathscr{P}}$-ideal in $\mathfrak{o}'$, then by our induction hypothesis, we have $\mathfrak{q}'=\mathfrak{Q}'^{n_1}_1\mathfrak{Q}'^{n_2}_2 \cdots \mathfrak{Q}'^{n_h}_h$. The inverse transform of $\mathfrak{q}'$ is clearly $\mathscr{P}^{n_1}_1\mathscr{P}^{n_2}_2 \cdots \mathscr{P}^{n_h}_h$, and assertion (2) now follows from statement (3) in (D).

This result characterized the simple complete ideal $\mathscr{P}$, of rank $h$, by means of the sequence $\{\mathfrak{q}_i\}$ of valuation ideals in $\mathfrak{o}$ which are associated with the corresponding prime divisor $v_{\mathscr{P}}$: that sequence contains precisely $h+1$ simple ideals, and $\mathscr{P}$ is the smallest of these simple ideals ($\mathscr{P}$ is the last simple ideal which occurs in the sequence $\{\mathfrak{q}_i\}$).

The arithmetic theory of complete ideals in $\mathfrak{o}$ which we have developed in this appendix has also geometric interpretations, since all the known results of the geometric theory of infinitely near points on an algebraic non-singular surface are included in this arithmetic theory. For these geometric interpretations we refer the reader to the original paper of O. Zariski (quoted in the beginning of this appendix).

# APPENDIX 6

## MACAULAY RINGS

Let A be a noetherian ring, $\mathfrak{a}$ an ideal in $A$ and $a$ an element of $A$. We say that *$a$ is prime to $\mathfrak{a}$* if $\mathfrak{a}:Aa=\mathfrak{a}$. This means that $a$ does not belong to any associated prime ideal of $\mathfrak{a}$ (Vol. I, Ch. IV, § 7, Corollary 2 to Theorem 11). We say that a sequence $\{a_1, \cdots, a_n\}$ of non-invertible elements of $A$ is a *prime sequence* if, for every $j$, $a_j$ is prime to the ideal $Aa_1 + \cdots + Aa_{j-1}$. (For $j=0$ we follow our usual convention that the empty set generates the zero ideal. Thus, a single element $a$ constitutes a prime sequence if and only if it is not a zero divisor.) It follows easily from Vol. I, Ch. IV, § 14, Theorems 30 and 31, that, if $\{a_1, \cdots, a_n\}$ is a prime sequence in $A$, and if $\mathfrak{p}$ is an associated prime ideal of $Aa_1 + \cdots + Aa_n$, then we have

$$h(\mathfrak{p}) \geq n, \tag{1}$$

equality holding if and only if $\mathfrak{p}$ is isolated.

We note that *$\{a_1, a_2, \cdots, a_q\}$ is a prime sequence if and only if for each $i=1, 2, \cdots, q$, $a_i$ is not a zero divisor in the ring $A/(Aa_1 + Aa_2 + \cdots + Aa_{i-1})$.*

We are going to devote several lemmas to the study of prime sequences in *local rings.*

LEMMA 1. *Let $A$ be a local ring, $\mathfrak{a}$ an ideal in $A$, $b$ a non-invertible element prime to $\mathfrak{a}$, and $\mathfrak{p}$ an associated prime ideal of $\mathfrak{a}$. Then there exists an associated prime ideal $\mathfrak{p}'$ of $\mathfrak{a}+Ab$ such that $\mathfrak{p}' \supset \mathfrak{p}$ (thus $\mathfrak{p}' > \mathfrak{p}$).*

This has been implicitly established in the proof of Theorem 44, VII, § 13, but we prove it again for the reader's convenience. Suppose the conclusion is not true. Then, for every associated prime ideal $\mathfrak{p}'_j$ of $\mathfrak{a}+Ab$, we have $\mathfrak{p} \not\subset \mathfrak{p}'_j$, and hence there exists an element $x$ of $\mathfrak{p}$ such that $x \notin \mathfrak{p}'_j$ for every $j$ (Vol. I, Ch. IV, § 6, Remark, p. 215). We thus have $(\mathfrak{a}+Ab):Ax=\mathfrak{a}+Ab$. Now, if $v$ is an element of $A$ such that $xv \in \mathfrak{a}$, we have $xv \in \mathfrak{a}+Ab$, whence $v \in \mathfrak{a}+Ab$; setting $v=a'+v'b$ ($a' \in \mathfrak{a}$, $v' \in A$), we have $xv'b=xv-xa' \in \mathfrak{a}$, whence $xv' \in \mathfrak{a}$ since $b$ is prime to $\mathfrak{a}$. In other words, we have $\mathfrak{a}:Ax \subset \mathfrak{a}+b(\mathfrak{a}:Ax) \subset \mathfrak{a}+\mathfrak{m}(\mathfrak{a}:Ax)$

(where $\mathfrak{m}$ denotes the maximal ideal of $A$), whence $\mathfrak{a}:Ax=\mathfrak{a}+\mathfrak{m}(\mathfrak{a}:Ax)$. From this we conclude that $\mathfrak{a}:Ax=\mathfrak{a}$ (VIII, § 4, Theorem 9, (f)), in contradiction with the fact that $x$ belongs to the associated prime ideal $\mathfrak{p}$ of $\mathfrak{a}$. Q.E.D.

LEMMA 2. *Let $A$ be a local ring, $\{a_1, \cdots, a_n\}$ a prime sequence in $A$, and $j\rightarrow i(j)$ a permutation of the indexing set $\{1, 2, \cdots, n\}$. Then $\{a_{i(1)}, a_{i(2)}, \cdots, a_{i(n)}\}$ is a prime sequence in $A$.*

By elementary properties of permutations, it is sufficient to prove that, for every $j$, $\{a_1, \cdots, a_{j-1}, a_{j+1}, a_j, a_{j+2}, \cdots, a_n\}$ is a prime sequence. The property that $a_i$ is prime to the ideal generated by the elements $a_k$ which precede it in this sequence is obviously true for $i=1, \cdots, j-1, j+2, \cdots, n$. It remains to be proved that this property is also true for $i=j+1$ and for $i=j$. We set $\mathfrak{a}=Aa_1+\cdots+Aa_{j-1}$. If $a_{j+1}$ were not prime to $\mathfrak{a}$, there would exist an associated prime ideal $\mathfrak{p}$ of $\mathfrak{a}$ such that $a_{j+1}\in\mathfrak{p}$; by Lemma 1, $\mathfrak{p}$ would then be contained in an associated prime ideal $\mathfrak{p}'$ of $\mathfrak{a}+Aa_j$, in contradiction with the fact that $a_{j+1}$ is prime to $\mathfrak{a}+Aa_j$; thus $a_{j+1}$ is prime to $\mathfrak{a}$. Now we prove that $a_j$ is prime to $\mathfrak{a}+Aa_{j+1}$. If an element $x$ of $A$ is such that $xa_j\in\mathfrak{a}+Aa_{j+1}$, we have $xa_j=ya_{j+1}+b$ $(y\in A, b\in\mathfrak{a})$, hence $ya_{j+1}\in\mathfrak{a}+Aa_j$ and $y\in\mathfrak{a}+Aa_j$, since $a_{j+1}$ is prime to $\mathfrak{a}+Aa_j$. Setting $y=b'+za_j$ $(b'\in\mathfrak{a}, z\in A)$, we have $xa_j=b'a_{j+1}+za_ja_{j+1}+b$, whence $(x-za_{j+1})a_j\in\mathfrak{a}$, and $x-za_{j+1}\in\mathfrak{a}$ since $a_j$ is prime to $\mathfrak{a}$. We therefore have $x\in\mathfrak{a}+Aa_{j+1}$, and this proves our assertion.

In the case of a *local* ring $A$, it has therefore a meaning to say that a finite *subset* $S$ of $A$ is a prime sequence in $A$, since, by Lemma 2, the property of $S$ being a prime sequence is independent of the order in which the elements of $S$ are considered.

LEMMA 3. *Let $A$ be a local ring, $\mathfrak{m}$ its maximal ideal, $\{a_1, \cdots, a_n\}$ and $\{a'_1, \cdots, a'_n\}$ two prime sequences in $A$ with the same number of elements, and $\mathfrak{a}$, $\mathfrak{a}'$ the ideals they generate in $A$. Then the $A$-modules $(\mathfrak{a}:\mathfrak{m})/\mathfrak{a}$ and $(\mathfrak{a}':\mathfrak{m})/\mathfrak{a}'$ are isomorphic.*

We proceed by induction on $n$. In the case $n=1$ the hypothesis means that $a=a_1$ and $a'=a'_1$ are not zero-divisors. Let $T$ be the total quotient ring of $A$ and let $f$ be the $A$-linear mapping $x\rightarrow a'a^{-1}x$ of $T$ into itself. If $x\in Aa:\mathfrak{m}$, then $xa'\in Aa$ (since, by definition, the elements of a prime sequence are non-invertible elements), whence $f(x)\in A$. On the other hand, since $x\mathfrak{m}\subset Aa$, we have $f(x)\mathfrak{m}=a'a^{-1}x\mathfrak{m}\subset Aa'$, whence $f$ maps $Aa:\mathfrak{m}$ into $Aa':\mathfrak{m}$. Similarly, the $A$-linear mapping $g$ defined by $g(x)=aa'^{-1}x$ $(x\in T)$ maps $Aa':\mathfrak{m}$ into $Aa:\mathfrak{m}$. Since $fg$ and $gf$ are the identity mappings, it follows that $f$ is an isomorphism of the $A$-module $Aa:\mathfrak{m}$ onto the $A$-module $Aa':\mathfrak{m}$.

Since we have obviously $f(Aa) = Aa'$, we deduce that $(Aa:\mathfrak{m})/Aa$ and $(Aa':\mathfrak{m})/Aa'$ are isomorphic.

In the general case, since $a_n$ is non-invertible, no associated prime ideal $\mathfrak{p}_i$ of $Aa_1 + \cdots + Aa_{n-1}$ is equal to $\mathfrak{m}$; similarly no associated prime ideal $\mathfrak{p}'_j$ of $Aa'_1 + \cdots + Aa'_{n-1}$ is equal to $\mathfrak{m}$. Let $b$ be an element of $\mathfrak{m}$ which does not belong to any $\mathfrak{p}_i$ nor to any $\mathfrak{p}'_j$ (Vol. I, Ch. IV, § 6, Remark, p. 215). Then $\{a_1, \cdots, a_{n-1}, b\}$ and $\{a'_1, \cdots, a'_{n-1}, b\}$ are prime sequences in $A$; let $\mathfrak{b}$ and $\mathfrak{b}'$ be the ideals they generate. Applying the case $n=1$ to the ring $A/(Aa_1 + \cdots + Aa_{n-1})$ and to the prime sequences constituted by the classes of $a_n$ and of $b$ respectively, we see that $(\mathfrak{a}:\mathfrak{m})/\mathfrak{a}$ and $(\mathfrak{b}:\mathfrak{m})/\mathfrak{b}$ are isomorphic as $(A/(Aa_1 + \cdots + Aa_{n-1}))$-modules, and therefore also as $A$-modules. Similarly the $A$-modules $(\mathfrak{a}':\mathfrak{m})/\mathfrak{a}'$ and $(\mathfrak{b}':\mathfrak{m})/\mathfrak{b}'$ are isomorphic. We apply now the induction hypothesis to the ring $A/Ab$. For $x \in A$, let us denote by $\bar{x}$ the $(Ab)$-residue of $x$. Since $\{b, a_1, \cdots, a_{n-1}\}$ is a prime sequence in $A$ (Lemma 2), $\{\bar{a}_1, \cdots, \bar{a}_{n-1}\}$ is a prime sequence in $A/Ab$; similarly for $\{\bar{a}'_1, \cdots, \bar{a}'_{n-1}\}$. Thus the induction hypothesis shows, as above, that the $A$-modules $(\mathfrak{b}:\mathfrak{m})/\mathfrak{b}$ and $(\mathfrak{b}':\mathfrak{m})/\mathfrak{b}'$ are isomorphic. This proves our assertion since the product of three isomorphisms is an isomorphism. Q.E.D.

Formula (1) shows that, in a local ring $A$, the number of elements of a prime sequence is bounded by dim $(A)$. Thus, in a local ring, there exist *maximal* prime sequences.

THEOREM 1. *Let A be a local ring. Any two maximal prime sequences in A have the same number of elements.*

PROOF. Let $\{a_1, \cdots, a_p\}$ and $\{a'_1, \cdots, a'_q\}$ be two maximal prime sequences in $A$. It is sufficient to show that the assumption "$p<q$" leads to a contradiction. In fact, if $p<q$, let us consider the ideals $\mathfrak{a}, \mathfrak{a}'$ generated by $a_1, \cdots, a_p$ and $a'_1, \cdots, a'_p$ respectively. Since $\mathfrak{a}':Aa'_{p+1} = \mathfrak{a}'$ by hypothesis, we have *a fortiori* $\mathfrak{a}':\mathfrak{m} = \mathfrak{a}'$ ($\mathfrak{m}$: maximal ideal of $A$), whence $(\mathfrak{a}':\mathfrak{m})/\mathfrak{a}' = (0)$. By Lemma 3 we have therefore $(\mathfrak{a}:\mathfrak{m})/\mathfrak{a} = (0)$, i.e., $\mathfrak{a}:\mathfrak{m} = \mathfrak{a}$. Thus $\mathfrak{m}$ is not an associated prime ideal of $\mathfrak{a}$, and there exists an element $b$ of $\mathfrak{m}$ which does not belong to any associated prime ideal of $\mathfrak{a}$ (Vol. I, Ch. IV, § 6, Remark, p. 215). Then $\{a_1, \cdots, a_p, b\}$ is a prime sequence, in contradiction with the maximality of $\{a_1, \cdots, a_p\}$.

DEFINITION 1. *Let A be a local ring. The common number of elements of the maximal prime sequences in A is called the homological codimension (or the grade) of A, and is denoted by* codh $(A)$. *If* codh $(A)$ $=$ dim $(A)$, *we say that A is a Macaulay ring (or a Cohen-Macaulay ring).*

We have seen that in any local ring $A$, we have the inequality

$$\text{codh}(A) \leq \dim(A). \tag{2}$$

It follows from the definition of codh $(A)$ that every prime sequence in $A$ may be included in a prime sequence with codh $(A)$ elements. To say that $A$ is a Macaulay ring is equivalent to saying that there exists a system of parameters of $A$ which is a prime sequence (by formula (1)).

EXAMPLES. (1) *Any regular local ring $A$ is a Macaulay ring.* In fact any regular system of parameters of $A$ is a prime sequence by VIII, Theorem 26, § 11.

(2) Any local domain $A$ of dimension 1 is a Macaulay ring. In fact any single element $\neq 0$ of the maximal ideal of $A$ constitutes a prime sequence. More generally, for a local ring $A$ of dimension 1 to be a Macaulay ring, it is necessary and sufficient that the maximal ideal $\mathfrak{m}$ of $A$ is not an associated prime ideal of (0).

(3) Any integrally closed domain $A$ of dimension 2 is a Macaulay ring. In fact, if we choose a non-invertible element $x \neq 0$ of $A$, all the associated prime ideals $\mathfrak{p}_i$ of $Ax$ have height 1 (Vol. I, Ch. V, § 6, Theorem 14), and are therefore distinct from the maximal ideal $\mathfrak{m}$. Hence there exists an element $y \in \mathfrak{m}$ such that $y \notin \mathfrak{p}_i$ for every $i$, and $\{x, y\}$ is a prime sequence.

Before giving the main property of Macaulay rings, we need a lemma:

LEMMA 4. *Let $A$ be a local ring, $d$ its dimension, $a_1, \cdots, a_j$ distinct elements of $A$. For* $\dim(A/(Aa_1 + \cdots + Aa_j))$ *to be equal to $d-j$, it is necessary and sufficient that $\{a_1, \cdots, a_j\}$ be a subset of a system of parameters of $A$.*

The sufficiency has been proved in VIII, § 9 (see p. 292). Conversely, if $A/(Aa_1 + \cdots + Aa_j)$ has dimension $d-j$, we consider $d-j$ elements $a_{j+1}, \cdots, a_d$ whose residue classes form a system of parameters of $A/(Aa_1 + \cdots + Aa_j)$. Then the ideal in $A$ generated by $a_1, \cdots, a_j, a_{j+1}, \cdots, a_d$ is primary for the maximal ideal of $A$, whence $\{a_1, \cdots, a_d\}$ is a system of parameters of $A$. Q.E.D.

THEOREM 2. *Let $A$ be a Macaulay ring, $d$ its dimension, $a_1, \cdots, a_j$ distinct elements of $A$ and $\mathfrak{a}$ the ideal generated by these $j$ elements. If* $\dim(A/\mathfrak{a}) = d-j$, *then $\{a_1, \cdots, a_j\}$ is a prime sequence, and for every associated prime ideal $\mathfrak{p}$ of $\mathfrak{a}$, we have $h(\mathfrak{p}) = j$ and* $\dim(A/\mathfrak{p}) = d-j$. *In particular, $\mathfrak{a}$ has no imbedded prime ideals, and is unmixed.*

PROOF. We proceed by induction on $j$. If $j=0$, the given set is empty and is a prime sequence. We then have $\mathfrak{a} = (0)$. We consider an associated prime ideal $\mathfrak{p}$ of (0). Let $\{b_1, \cdots, b_d\}$ be a prime sequence in $A$. By repeated applications of Lemma 1, we find a strictly increasing sequence of prime ideals $\mathfrak{p} < \mathfrak{p}_1 < \cdots < \mathfrak{p}_d$ such that $\mathfrak{p}_i$ is an associated prime ideal of $Ab_1 + \cdots + Ab_i$. This proves that we have $\dim(A/\mathfrak{p}) \geq d$, whence $\dim(A/\mathfrak{p}) = d$ since $\dim(A) = d$. On the other hand we have $h(\mathfrak{p}) = 0$ since, otherwise, we would get a chain of prime ideals in $A$ with $d+2$ distinct terms. This proves Theorem 2 in case $j=0$.

We now suppose that Theorem 2 is true for $j-1$. We set $\mathfrak{a}' = Aa_1 + \cdots + Aa_{j-1}$. Since $\dim(A/\mathfrak{a}) = d-j$, $\{a_1, \cdots, a_j\}$ is a subset of a system of parameters (Lemma 4). Hence $\dim(A/\mathfrak{a}') = d-j+1$ (Lemma 4). By the induction hypothesis, $\{a_1, \cdots, a_{j-1}\}$ is a prime sequence, and all the associated prime ideals of $\mathfrak{a}'$ have height $j-1$ and dimension $d-j+1$. If $a_j$ were contained in some associated prime ideal $\mathfrak{p}'$ of $\mathfrak{a}'$, we would have $\mathfrak{a} \subset \mathfrak{p}'$, whence $\dim(A/\mathfrak{a}) \geq \dim(A/\mathfrak{p}') = d-j+1$, in contradiction with the hypothesis. Therefore $\{a_1, \cdots, a_{j-1}, a_j\}$ is a prime sequence. This prime sequence is contained in some maximal prime sequence, say $\{a_1, \cdots, a_j, a_{j+1}, \cdots, a_d\}$, which has $d$ elements (and is therefore a system of parameters), since $A$ is a Macaulay ring. The $\mathfrak{a}$-residues of $a_{j+1}, \cdots, a_d$ form a prime sequence and a system of parameters in the ring $A/\mathfrak{a}$, which is therefore a Macaulay ring. Applying the case $j=0$ to the ring $A/\mathfrak{a}$, we see that we have $\dim(A/\mathfrak{p}) = d-j$ for every associated prime ideal $\mathfrak{p}$ of $\mathfrak{a}$. On the other hand, such an ideal $\mathfrak{p}$ contains some associated prime ideal $\mathfrak{p}'$ of $\mathfrak{a}'$, and we have $\mathfrak{p} \neq \mathfrak{p}'$ since $a_j \in \mathfrak{p}$ and $a_j \notin \mathfrak{p}'$ ($\{a_1, \cdots, a_j\}$ being a prime sequence); we therefore have $h(\mathfrak{p}) \geq h(\mathfrak{p}') + 1 = (j-1) + 1 = j$. Since the inequality $\dim(A/\mathfrak{p}) + h(\mathfrak{p}) \leq d$ holds for every prime ideal in a local ring $A$ of dimension $d$ (otherwise $A$ would admit a chain of prime ideals with $d+2$ terms), the relations $\dim(A/\mathfrak{p}) = d-j$ and $h(\mathfrak{p}) \geq j$ give $h(\mathfrak{p}) = j$. Q.E.D.

REMARK. Since a regular local ring is a Macaulay ring, Theorem 2 gives a new proof of Cohen-Macaulay's Theorem (VIII, § 12, Theorem 29), and generalizes it to a regular local ring of unequal characteristic. It may also be noticed that Macaulay's Theorem about polynomial rings (VII, § 8, Theorem 26) is an easy consequence of Theorem 2. In fact, let $k$ be a field, and $\mathfrak{A}$ be an ideal in $R = k[X_1, \cdots, X_d]$ of dimension $d-h$ and generated by $h$ elements $u_1, \cdots, u_h$. Let $\mathfrak{P}$ be an associated prime ideal of $\mathfrak{A}$, and $\mathfrak{M}$ a maximal ideal in $R$ containing $\mathfrak{P}$. The local ring $R_{\mathfrak{M}}$ has dimension $d$ and, since $\mathfrak{M}$ may be generated by $d$ elements (VII, § 7, Theorem 24), is regular. Since $\dim(R_{\mathfrak{M}}/\mathfrak{A}(R_{\mathfrak{M}})) = d-h$, Theorem 2 shows that $\dim(R_{\mathfrak{M}}/\mathfrak{P}R_{\mathfrak{M}}) = d-h$ ($\mathfrak{P}R_{\mathfrak{M}}$ being an associated prime ideal of $\mathfrak{A}R_{\mathfrak{M}}$; see Vol. I, Ch. IV, § 11, Theorem 19).

Since this relation holds for every maximal ideal $\mathfrak{M}$ containing $\mathfrak{P}$, the depth of $\mathfrak{P}$, whence also its dimension, is $d-h$.

Theorem 2 has many consequences.

COROLLARY 1. *Let $A$ be a local ring. The following properties are equivalent:*

(a) *$A$ is a Macaulay ring;*

(b) *There exists a system of parameters of $A$ which is a prime sequence;*

(c) *Every system of parameters of $A$ is a prime sequence.*

The equivalence of (a) and (b) has already been established. It is clear that (c) implies (b) since a local ring admits at least one system of parameters. Now, if $A$ is a Macaulay ring and if $\{a_1, \cdots, a_d\}$ is a system of parameters of $A$, we have $\dim (A/(Aa_1 + \cdots + Aa_d)) = 0 = d-d$, whence Theorem 2 shows that $\{a_1, \cdots, a_d\}$ is a prime sequence; thus (a) implies (c). Q.E.D.

COROLLARY 2. *Let $A$ be a Macaulay ring. For a finite subset $S$ of $A$ to be a prime sequence, it is necessary and sufficient that it be a subset of some system of parameters.*

In fact, if $S$ is a prime sequence, it is contained in a maximal prime sequence, i.e., in a system of parameters. The converse follows from Corollary 1 ((a) implies (c)), since any subset of a prime sequence is a prime sequence (Lemma 2).

COROLLARY 3. *Let $A$ be a Macaulay ring. For every prime ideal $\mathfrak{p}$ in $A$, we have $h(\mathfrak{p}) + \dim (A/\mathfrak{p}) = \dim (A)$.*

In fact, among the prime sequences which are contained in $\mathfrak{p}$, we consider a maximal one, say $\{a_1, \cdots, a_j\}$. Let $\{\mathfrak{p}'_i\}$ be the set of associated prime ideals of $\mathfrak{a} = Aa_1 + \cdots + Aa_j$. We have $\mathfrak{p} \subset \bigcup_i \mathfrak{p}'_i$, for in the contrary case $\mathfrak{p}$ would contain an element $b$ such that $\mathfrak{a} : Ab = \mathfrak{a}$, and then $\{a_1, \cdots, a_j, b\}$ would be a prime sequence, in contradiction with the maximality of $\{a_1, \cdots, a_j\}$. Therefore there exists an index $i$ such that $\mathfrak{p} \subset \mathfrak{p}'_i$ (Vol. I, Ch. IV, § 6, Remark, p. 215). On the other hand, since $\mathfrak{p}$ contains $\mathfrak{a}$, $\mathfrak{p}$ contains some isolated prime ideal $\mathfrak{p}'_k$ of $\mathfrak{a}$. Since $\mathfrak{a}$ is generated by a subset of a system of parameters (Corollary 1), we have $\dim (A/\mathfrak{a}) = \dim (A) - j$ (Lemma 4), and therefore, by Theorem 2, $\mathfrak{a}$ is unmixed. Hence $\mathfrak{p} = \mathfrak{p}'_i = \mathfrak{p}'_k$, and $\mathfrak{p}$ is an associated prime ideal of $\mathfrak{a}$. Thus our assertion follows from Theorem 2.

COROLLARY 4. *Let $A$ be a Macaulay ring. For every prime ideal $\mathfrak{p}$ in $A$, the local ring $A_\mathfrak{p}$ is a Macaulay ring.*

In fact, as in Corollary 3, we construct a prime sequence $\{a_1, \cdots, a_j\}$ such that $\mathfrak{p}$ is an associated prime ideal of $Aa_1 + \cdots + Aa_j$. Let $f$ be the canonical mapping of $A$ into $A_\mathfrak{p}$. Since $\mathfrak{p}$ is an isolated prime ideal of $Aa_1 + \cdots + Aa_j$ and since $h(\mathfrak{p}) = j$ (Theorem 2), $\{f(a_1), \cdots, f(a_j)\}$ is a system of parameters of $A_\mathfrak{p}$, and it remains to be proved that it is also a prime sequence. Now this is immediate, since for every $q \leq j$, we have the formula $(f(a_1), \cdots, f(a_{q-1})):(f(a_q)) = ((a_1, \cdots, a_{q-1}):(a_q))A_\mathfrak{p}$ (see Vol. I, Ch. IV, § 10) $= (a_1, \cdots, a_{q-1})A_\mathfrak{p} = (f(a_1), \cdots, f(a_{q-1}))$.

COROLLARY 5. *Let $A$ be a Macaulay ring. For every prime sequence $\{a_1, \cdots, a_j\}$ in $A$, the local ring $A/(Aa_1 + \cdots + Aa_j) = A'$ is a Macaulay ring.*

In fact, the given prime sequence is contained in a maximal prime sequence $\{a_1, \cdots, a_j, a_{j+1}, \cdots, a_d\}$, i.e., in a system of parameters ($d = \dim(A)$). We have $\dim(A') = d - j$ (Lemma 4), whence the residue classes of $a_{j+1}, \cdots, a_d$ in $A'$ form a system of parameters. Since they obviously form a prime sequence, Corollary 5 is proved.

REMARK. It follows from Corollary 5 that, if $W$ is an irreducible subvariety of a variety $V$, and if $V$ is a hypersurface in affine or projective space (more generally a complete intersection of hypersurfaces), then the local ring $\mathfrak{o}(W; V)$ is a Macaulay ring.

COROLLARY 6. *Let $A$ be a local ring, $\hat{A}$ its completion. For $A$ to be a Macaulay ring, it is necessary and sufficient that $\hat{A}$ be a Macaulay ring.*

Let $a_1, \cdots, a_i$ be elements of $A$. By Corollary 5 to VIII, § 4, Theorem 11, and since $\mathfrak{b}\hat{A} \cap A = \mathfrak{b}$ for every ideal $\mathfrak{b}$ in $A$ (VIII, § 2, Corollary 2 to Theorem 5), the relations $(\hat{A}a_1 + \cdots + \hat{A}a_{j-1}):\hat{A}a_j = \hat{A}a_1 + \cdots + \hat{A}a_{j-1}$ and $(Aa_1 + \cdots + Aa_{j-1}):Aa_j = Aa_1 + \cdots + Aa_{j-1}$ are equivalent. Thus for a subset $S$ of $A$ to be a prime sequence in $A$, it is necessary and sufficient that it be a prime sequence in $\hat{A}$. Now, if $A$ is a Macaulay ring, we take for $S$ a system of parameters of $A$ (which is therefore a prime sequence in $A$). Then $\hat{A}$ is a Macaulay ring since $S$ is a system of parameters of $\hat{A}$. Conversely, if $\hat{A}$ is a Macaulay ring, then any system of parameters $S$ of $A$ is a prime sequence in $\hat{A}$ since it is a system of parameters of $\hat{A}$ (Corollary 1); thus $S$ is also a prime sequence in $A$, and $A$ is a Macaulay ring. Q.E.D.

THEOREM 3. *Let $A$ be a local ring. The following properties are equivalent:*

(a) *$A$ is a Macaulay ring;*

(b) *There exists an ideal $\mathfrak{q}$ in $A$, generated by a system of parameters, such that $e(\mathfrak{q}) = l(A/\mathfrak{q})$.*

(b′) *For every ideal $\mathfrak{q}$ in $A$ generated by a system of parameters, we have $e(\mathfrak{q}) = l(A/\mathfrak{q})$.*

(c) *There exists an ideal $\mathfrak{q}$ in $A$ generated by a system of parameters* *such that the associated graded ring* $G_q(A)$ *is isomorphic to a polynomial ring in* dim $(A)$ *variables over* $A/\mathfrak{q}$.

(c′) *For every system of parameters* $x_1, \cdots, x_d$ *of* $A$, *the initial forms* $\bar{x}_i$ *of the elements* $x_i$ *in* $G_\mathfrak{q}(A)$ $(\mathfrak{q}=Ax_1+\cdots+Ax_d)$ *are algebraically independent over* $A/\mathfrak{q}$ (*whence* $G_\mathfrak{q}(A)$ *is isomorphic to a polynomial ring in* $d$ *variables over* $A/\mathfrak{q}$).

PROOF. The equivalence of (b) and (c) follows from VIII, § 10, Theorem 23. Similarly (b′) and (c′) are equivalent. It is obvious that (b′) implies (b). We are going to show that (a) implies (b′) and that (c) implies (a), and the proof will then be complete.

For proving that (a) implies (b′), we can, if $A/\mathfrak{m}$ is an infinite *field* ($\mathfrak{m}$: maximal ideal of $A$), use the discussion preceding Theorem 23 in VIII, § 10. In fact, in the course of that discussion we have constructed a suitable system of parameters $\{y_1, \cdots, y_d\}$ generating $\mathfrak{q}$, and we have shown that if that system satisfies the condition $(Ay_1+\cdots+Ay_{d-1})$: $Ay_d=Ay_1+\cdots+Ay_{d-1}$, then $e(\mathfrak{q})=l(A/\mathfrak{q})$. Now the above relation obviously holds since every system of parameters in a Macaulay ring is a prime sequence (Corollary 1 to Theorem 2). The process of adjoining an indeterminate to $A$ could then take care of the case of a finite residue field $A/\mathfrak{m}$. However, we prefer to give a direct proof of the fact that (a) implies (b′), since this proof uses two lemmas which are of interest in themselves.

LEMMA 5. *Let* $A$ *be a Macaulay ring, and* $\mathfrak{a}$ *an ideal in* $A$ *generated by a prime sequence. For every exponent* $n$, *the ideal* $\mathfrak{a}^n$ *is unmixed (and admits, therefore, the same associated prime ideals as* $\mathfrak{a}$; see Theorem 2).

We proceed by induction on $n$. The case $n=1$ is covered by Theorem 2. We suppose that our assertion is proved for $n$, and prove it for $n+1$. We have to show that if $c$ is prime to $\mathfrak{a}$ and if $x$ is an element of $A$ such that $cx\in\mathfrak{a}^{n+1}$, then $x$ belongs to $\mathfrak{a}^{n+1}$. Since $cx\in\mathfrak{a}^n$, the induction hypothesis shows that $x\in\mathfrak{a}^n$. Let $\{a_1, \cdots, a_j\}$ be a prime sequence generating $\mathfrak{a}$. By a suitable grouping of the monomials of degree $n$ in $a_1, \cdots, a_j$, we see that $x$ may be written in the form $x=x_1a_1+\cdots+x_qa_q$, where $q\leq j$ and $x_i\in(Aa_1+\cdots+Aa_i)^{n-1}$. We prove that $x\in\mathfrak{a}^{n+1}$ by induction on $q$. The case $q=0$ is trivial. For $q>0$, we write $x=x'+x_qa_q$ (where $x'=x_1a_1+\cdots+x_{q-1}a_{q-1}$), and we denote by $\mathfrak{b}$ the ideal generated by $a_1, \cdots, a_{q-1}$, $a_{q+1}, \cdots, a_j$; we have $\mathfrak{a}=\mathfrak{b}+Aa_q$. Since $\mathfrak{a}^{n+1}=\mathfrak{b}^{n+1}+\mathfrak{a}^na_q$, the relation $c(x'+x_qa_q)=cx\in\mathfrak{a}^{n+1}$ shows the existence of an element $y$ of $\mathfrak{a}^n$ such that $cx'+cx_qa_q-ya_q\in\mathfrak{b}^{n+1}$. Since $x'\in\mathfrak{b}^n$, this implies $(cx_q-y)a_q\in\mathfrak{b}^n$. Now, $a_q$ being prime to $\mathfrak{b}$ (Lemma 2), the induction hypothesis on $n$

shows that $cx_q - y \in \mathfrak{b}^n$, whence $cx_q \in \mathfrak{a}^n$ since $y \in \mathfrak{a}^n$. Again the induction hypothesis on $n$ shows that $x_q \in \mathfrak{a}^n$ ($c$ being prime to $\mathfrak{a}$). From $x = x' + x_q a_q$ we then deduce that $cx'$ belongs to $\mathfrak{a}^{n+1}$. Therefore $x' \in \mathfrak{a}^{n+1}$ by the induction hypothesis on $q$. Since $x = x' + a_q x_q$ and since $x_q \in \mathfrak{a}^n$, we have $x \in \mathfrak{a}^{n+1}$. Q.E.D.

LEMMA 6. *Let $A$ be a Macaulay ring and $\mathfrak{a}$ an ideal in $A$ generated by a prime sequence $\{a_1, \cdots, a_j\}$. We have $\mathfrak{a}^n : Aa_j = \mathfrak{a}^{n-1}$ for every $n$.*

Let $x$ be an element of $A$ such that $xa_j \in \mathfrak{a}^n$. We set $\mathfrak{b} = Aa_1 + \cdots + Aa_{j-1}$. Since $\mathfrak{a}^n = \mathfrak{b}^n + \mathfrak{a}^{n-1}a_j$, there exists an element $y$ of $\mathfrak{a}^{n-1}$ such that $(x-y)a_j \in \mathfrak{b}^n$. As $a_j$ is prime to $\mathfrak{b}$, it is prime to $\mathfrak{b}^n$ (Lemma 5), whence $x - y \in \mathfrak{b}^n$. Therefore $x \in \mathfrak{a}^{n-1}$, and we have proved the inclusion $\mathfrak{a}^n : Aa_j \subset \mathfrak{a}^{n-1}$. Since the opposite inclusion is obvious, Lemma 6 is proved.

CONTINUATION OF THE PROOF OF THEOREM 3. We are going to prove that (a) implies (b′). For this we proceed by induction on the dimension $d$ of $A$. The case $d = 0$ is trivial since we then have $\mathfrak{q} = (0)$, $e(\mathfrak{q}) = l(A) = l(A/\mathfrak{q})$. For $d > 0$, let $\{a_1, \cdots, a_d\}$ be a system of parameters generating $\mathfrak{q}$. We set $A' = A/Aa_d$, $\mathfrak{q}' = \mathfrak{q}/Aa_d$. Since $\{a_1, \cdots, a_d\}$ is a prime sequence, we have $\mathfrak{q}^n : Aa_d = \mathfrak{q}^{n-1}$ (Lemma 6), whence the formula $P_{\mathfrak{q}'}(n) = P_{\mathfrak{q}}(n) - \lambda(\mathfrak{q}^n : Aa_d)$ (Lemma 3, VIII, § 8) gives $P_{\mathfrak{q}'}(n) = P_{\mathfrak{q}}(n) - P_{\mathfrak{q}}(n-1)$ and therefore $e(\mathfrak{q}') = e(\mathfrak{q})$. Since $A'$ is a Macaulay ring (Corollary 5 to Theorem 2), the induction hypothesis gives $e(\mathfrak{q}') = l(A'/\mathfrak{q}')$. As $A'/\mathfrak{q}'$ is isomorphic to $A/\mathfrak{q}$, we have $e(\mathfrak{q}) = l(A/\mathfrak{q})$. Thus (a) implies (b′).

We finally prove that (c) implies (a). Suppose that $\mathfrak{q}$ is an ideal generated by a system of parameters such that $G_{\mathfrak{q}}(A)$ is generated over $A/\mathfrak{q}$ by $d$ ($= \dim(A)$) algebraically independent elements $\bar{a}_i$, and let $a_i$ be an element of $\mathfrak{q}$ admitting $\bar{a}_i$ as ($\mathfrak{q}^2$)-residue. It is sufficient to prove that $\{a_1, \cdots, a_d\}$ is a prime sequence (since $d = \dim(A)$). We set $\mathfrak{a} = Aa_1 + \cdots + Aa_{j-1}$ and prove that $\mathfrak{a} : Aa_j = \mathfrak{a}$. Let $y$ be an element of $A$ such that $ya_j \in \mathfrak{a}$; we set $ya_j = x_1a_1 + \cdots + x_{j-1}a_{j-1}$ ($x_i \in A$).

This is a relation of the type $\sum_{i=1}^{d} z_i a_i = 0$. Let us denote by $v$ the order function in $A$ (for $x \in A$, we have $x \in \mathfrak{q}^{v(x)}$ and $x \notin \mathfrak{q}^{v(x)+1}$; see VIII, § 1). Let $I$ be the set of indices $i$ for which $v(z_i)$ takes its minimum value, say $s$. We have $\sum_{i \in I} z_i a_i \in \mathfrak{q}^{s+1}$, whence, by passage to the initial forms, $\sum_{i \in I} \bar{z}_i \bar{a}_i = 0$. Choosing a fixed index $k$ in $I$, we see that, in the polynomial ring $G_{\mathfrak{q}}(A) = (A/\mathfrak{q})[\bar{a}_1, \cdots, \bar{a}_d]$, $\bar{z}_k \bar{a}_k$ is in the ideal generated by the indeterminates $\bar{a}_i$ ($i \in I$, $i \neq k$). Thus $\bar{z}_k$ is in this ideal, and there exist elements $b_i$ of $\mathfrak{q}^{s-1}$ ($i \in I$, $i \neq k$) such that $\bar{z}_k = \sum_{i \in I, i \neq k} \bar{b}_i \bar{a}_i$, i.e., such that $z_k - \sum_i b_i a_i$ is an element $z'_k$ of $\mathfrak{q}^{s+1}$. Setting $z'_i = z_i + b_i a_k$ for $i \in I$, $i \neq k$, and $z'_i = z_i$ for $i \notin I$, we get a relation $\sum_{i=1}^{d} z'_i a_i = 0$ in which $v(z'_i) \geq v(z_i)$ for every $i$ and $v(z'_k) > v(z_k)$.

Now, among the relations $ya_j = \sum_{i=1}^{j-1} x_i a_i$ ($x_i \in A$) we choose one which has the following two properties: (a) $\min_i (v(x_i))$ has the greatest possible value, say $s$; (b) the number of indices $i$ such that $v(x_i) = s$ is the smallest possible. Then we have $s = v(y)$. In fact $s > v(y)$ is obviously impossible. On the other hand, if $s < v(y)$, we transform, as above, the relation $ya_j - \sum_{i=1}^{j-1} x_i a_i = 0$: the coefficient $y$ of $a_j$ is then unchanged, whereas, either $s$ is increased, or the number of indices $i$ such that $v(x_i) = s$ is decreased. This is impossible. Thus $v(y) = \min_i (v(x_i))$. Transforming, as above, the relation $ya_j - \sum_{i=1}^{j-1} x_i a_i = 0$, this time with $ya_j$ playing the part of $z_k a_k$, we get a relation $y_1 a_j - \sum_{i=1}^{j-1} x'_i a_i = 0$ with $y_1 \in \mathfrak{q}^{v(y)+1}$ and $y - y_1 \in \mathfrak{a}$. Since $y_1 \in \mathfrak{a} : Aa_j$, we can apply the same process to $y_1$. By repeated applications we get an element $y_n$ of $\mathfrak{q}^{v(y)+n}$ such that $y - y_n \in \mathfrak{a}$. We thus have $y \in \mathfrak{a} + \mathfrak{q}^{v(y)+n}$ for every $n$, whence $y \in \mathfrak{a}$ since $\mathfrak{a}$ is closed. Consequently we have $\mathfrak{a} : Aa_j \subset \mathfrak{a}$. The opposite inclusion being obvious, we have $\mathfrak{a} : Aa_j = \mathfrak{a}$. Q.E.D.

REMARKS. (1) Let $R = k[X_1, \cdots, X_n]$ be a polynomial ring over a field $k$, and $\mathfrak{A}$ an ideal of the principal class of $R$. By passage to quotient rings $R_{\mathfrak{M}}$ ($\mathfrak{M}$: maximal ideals) and using Lemma 5, one proves, as in the Remark following Theorem 2, that $\mathfrak{A}^n$ is unmixed for every $n$.

(2) Let $A$ be a Macaulay ring. It is easily seen that the local ring $A[[X]]$ is a Macaulay ring.

# APPENDIX 7

## UNIQUE FACTORIZATION IN REGULAR LOCAL RINGS

In the present appendix we are going to prove that every regular local ring is a UFD. The method of proof, due to M. Auslander and D. Buchsbaum, uses the notion of cohomological dimension (VII, § 13).

LEMMA 1. *Let $A$ be a local domain. The following assertions are equivalent:*

(a) *$A$ is a* UFD.

(b) *Every irreducible element of $A$ generates a prime ideal.*

(c) *For any two elements $a$, $b$ of $A$, the ideal $Aa \cap Ab$ is principal.*

(d) *For any two elements $a, b \neq 0$ of $A$, the cohomological dimension $\delta(Aa+Ab)$ is $\leq 1$* (i.e., *considered as an $A$-module, $Aa+Ab$ is isomorphic with a factor module $F/F'$ with $F$ and $F'$ free*).

For the equivalence of (a) and (b) we first notice that (b) is nothing else but condition UF.3 of Vol. I, Ch. I, § 14; on the other hand every non-unit of $A$ is a finite product of irreducible factors since $A$ is noetherian (Vol. I, Ch. IV, § 1, Example 3), whence $A$ satisfies UF.1.

It is clear that (a) implies (c) since the ideal $Aa \cap Ab$ is obviously generated by the least common multiple of $a$ and $b$.

We now prove that (c) implies (b). Let $p$ be an irreducible element of $A$, $x$ and $y$ two elements of $A$ such that $xy \in Ap$ and $x \notin Ap$. We set $Ax \cap Ap = Am$. Since $m$ divides $xp$, $mx^{-1}$ (which is an element of $A$) is a divisor of $p$; it is not a unit since $m$ is a multiple of $p$ and $x$ is not. Since $p$ is irreducible it follows that $mx^{-1}$ and $p$, and therefore also $m$ and $xp$, are associates. Thus $Ax \cap Ap = Axp$. The hypothesis $xy \in Ap$ implies $xy \in Ax \cap Ap = Axp$, whence $xy$ is a multiple of $xp$ and therefore $y$ is a multiple of $p$.

Let us prove that (c) is equivalent to (d). Let $f$ be the $A$-linear mapping of (the free $A$-module) $A \times A$ onto $Aa+Ab$ defined by $f(x, y) = xa - yb$. Its kernel $F_0$ is the set of pairs $(x, y)$ such that $xa = yb$, and the mapping $(x, y) \rightarrow xa$ is obviously an isomorphism of $F_0$ onto the ideal $Aa \cap Ab$. If (c) holds, this ideal is principal, hence a free $A$-module, and therefore (d) is true. Conversely, if (d) is true, $Aa+Ab$ is isomorphic with a factor module $F/F'$ with $F$ and $F'$ free. Then the kernel $F_0$ of $f$ is equivalent to $F'$ in the sense of VII, § 13 (VII, § 13, Lemma 2) and is therefore a free module, since $A$ is a local ring (VII, § 13, Lemma 3). Since $Aa \cap Ab$ is isomorphic with $F_0$, it is a principal ideal, and (c) is true. Q.E.D.

LEMMA 2. *A regular local ring A of dimension* 1 *or* 2 *is a* UFD.

Let $a$ and $b$ be any two elements of $A$. Since $Aa+Ab$ is a submodule of a free module, we have $\delta(Aa+Ab) \leq \dim(A)-1$ by the theorem on syzygies (VII, § 13, Theorem 43). Hence $\delta(Aa+Ab) \leq 1$, and we use Lemma 1.

Notice that, if $\dim(A)=1$ (or 0), $A$ is a discrete valuation ring (or a field), and that the unique factorization properly is obvious in this case.

LEMMA 3. *A regular local ring A of dimension* 3 *is a* UFD.

Let $a$ and $b$ be any two elements of $A$. By the theorem on syzygies, we have $\delta(Aa+Ab) \leq 2$. In the proof of Lemma 1, we have seen that $Aa \cap Ab$ is a first module of syzygies of $Aa+Ab$, whence $\delta(Aa \cap Ab) \leq 1$. Since $x \rightarrow ax$ is an isomorphism of $Ab:Aa$ onto $Aa \cap Ab$, we also have $\delta(Ab:Aa) \leq 1$. From this we are going to deduce that $Ab:Aa$ is free, therefore a principal ideal, and this will complete the proof since $Aa \cap Ab$ will then be principal.

We set $\mathfrak{q}=Ab:Aa$, we denote by $\mathfrak{m}$ the maximal ideal of $A$, and we pick an element $b' \in \mathfrak{q}$, $b' \notin \mathfrak{m}\mathfrak{q}$. We have $b'a=a'b$ with $a' \in A$. Since the relations $xa'=yb'$ and $xa=yb$ are equivalent, so are $xa' \in Ab'$ and $xa \in Ab$, whence $Ab':Aa'=Ab:Aa=\mathfrak{q}$. We are going to prove that $\mathfrak{q}=Ab'$. For this it is sufficient to prove that $\mathfrak{q}=Ab'+\mathfrak{m}\mathfrak{q}$ (apply Theorem 9, Condition (f), of VIII, § 4, to the local ring $A/Ab'$ and to the ideal $\mathfrak{q}/Ab'$). In the contrary case, there exists an element $c$ of $\mathfrak{q}$ such that the classes of $c$ and $b'$ mod $\mathfrak{m}\mathfrak{q}$ are linearly independent over $A/\mathfrak{m}$. We consider a system of elements $(b', c, c_1, \cdots, c_n)$ of $\mathfrak{q}$ the $\mathfrak{m}\mathfrak{q}$-residues of which form a basis of $\mathfrak{q}/\mathfrak{m}\mathfrak{q}$ over $A/\mathfrak{m}$; these elements generate $\mathfrak{q}$ (loc. cit.). Consider $\mathfrak{q}$ as a factor module $F/F'$ of a free module $F$ with generators $(\beta, \gamma, \gamma_1, \cdots, \gamma_n)$ (these generators being mapped onto $(b', c, c_1, \cdots, c_n)$). The module of relations $F'$ is free, since $\delta(\mathfrak{q}) \leqq 1$. We have $F' \subset \mathfrak{m}F$ since the elements $b', c, c_1, \cdots, c_n$ are linearly independent mod $\mathfrak{m}\mathfrak{q}$.

Let us write $ca'=db'$ with $d \in A$. We have $a'\gamma-d\beta \in F'$ and evidently also $b'\gamma-c\beta \in F'$. We take a free basis $(\alpha_j)$ of $F'$ and write

$$a'\gamma-d\beta = \sum x_j\alpha_j, \tag{1}$$

$$b'\gamma-c\beta = \sum y_j\alpha_j. \tag{2}$$

Since $b'(a'\gamma - d\beta) = a'(b'\gamma - c\beta)$, we have $b'x_j = a'y_j$ for every $j$, whence $y_j \in \mathfrak{q}$. On the other hand, each $\alpha_j$ is a linear combination of the elements $\beta, \gamma, \gamma_1, \cdots, \gamma_n$ of the basis of $F$. Let $m_j$ be the coefficient of $\gamma$ in this representation of $\alpha_j$. We have $m_j \in \mathfrak{m}$ since $F' \subset \mathfrak{m}F$. Comparing the coefficients of $\gamma$ in both sides of (2), we get $b' = \sum y_j m_j$, whence $b' \in \mathfrak{m}\mathfrak{q}$. This contradicts our choice of $b'$ and proves the lemma.

THEOREM. *Every regular local ring $A$ is a* UFD.

PROOF. We proceed by induction on dim $(A)$. By lemmas 2 and 3 we may assume that dim $(A) \geq 4$. We consider two elements $a, b$ of $A$, set $\mathfrak{b} = Aa + Ab$ and prove that $\delta(\mathfrak{b}) \leq 1$ (Lemma 1). Let $\mathfrak{m}$ be the maximal ideal of $A$. The ideals $\mathfrak{b}, \mathfrak{b}:\mathfrak{m}, \cdots, \mathfrak{b}:\mathfrak{m}^n, \cdots$ form an increasing sequence, whence there exists an integer $n$ such that $\mathfrak{b}:\mathfrak{m}^n = \mathfrak{b}:\mathfrak{m}^{n+1} = \cdots$. Setting $\mathfrak{a} = \mathfrak{b}:\mathfrak{m}^n$, we have $\mathfrak{a}:\mathfrak{m} = \mathfrak{a}$, whence $\mathfrak{m}$ is not an associated prime ideal of $\mathfrak{a}$, and there exists an element $x$ of $\mathfrak{m}$, not in $\mathfrak{m}^2$, such that $\mathfrak{a}:Ax = \mathfrak{a}$.† Since $A/Ax$ is a regular local ring of dimension dim $(A) - 1$ (VIII, § 9, Theorem 20, Corollary 2 and VIII, § 11, Theorem 26), the induction hypothesis shows that the cohomological dimension $\delta_{A/Ax}((\mathfrak{b} + Ax)/Ax)$ of $(\mathfrak{b} + Ax)/Ax$, considered as an $(A/Ax)$-module, is $\leqq 1$. We set $S = \mathfrak{b} + Ax$, $\bar{S} = S/Ax$, $\bar{A} = A/Ax$. Since $\delta_{\bar{A}}(\bar{S}) \leqq 1$, we have an exact sequence

$$0 \to \bar{F}' \to \bar{F} \to \bar{S} \to 0,$$

where $\bar{F}'$ and $\bar{F}$ are free modules over $\bar{A}$. Considering $\bar{F}'$ and $\bar{F}$ as modules *over* $A$ we have $\delta_A(\bar{S}) \leqq 1 + \max(\delta_A(\bar{F}), \delta_A(\bar{F}'))$ (VII, § 13, formula (7)). Now, $\bar{F}$ may be written in the form $F/xF$, where $F$ is a free $A$-module; since also $xF$ is a free $A$-module we see that $\delta_A(\bar{F}) \leqq 1$; similarly $\delta_A(\bar{F}') \leqq 1$. We therefore have $\delta_A(S/Ax) = \delta_A(\bar{S}) \leqq 2$. Since $Ax$ is free, it follows from the formula $\delta_A(S) \leqq \max(\delta_A(S/Ax), \delta_A(Ax))$ (VII, § 13, formula (5)) that $\delta_A(S) = \delta_A(\mathfrak{b} + Ax) \leqq 2$. It follows then from formula (4) of VII, § 13, that $\delta(A/(\mathfrak{b} + Ax)) \leqq 3$.

From this and from VII, § 13, Theorem 44 it follows that, if $\mathfrak{p}$ is any associated prime ideal of $\mathfrak{b} + Ax$, we have $h(\mathfrak{p}) \leq 3$. Since dim $(A) \geq 4$,

† The existence of such an element $x$ can be proved as follows:

Let $\mathfrak{p}_1, \mathfrak{p}_2, \ldots, \mathfrak{p}_h$ be the prime ideals of $\mathfrak{a}$ and let $y$ be an element of $\mathfrak{m}$, not in $\mathfrak{m}^2$. Assume that $y \in \mathfrak{p}_1 \cap \mathfrak{p}_2 \cap \cdots \cap \mathfrak{p}_g$ $(0 \leqq g \leqq h)$, $y \notin \bigcup_{j=g+1}^{h} \mathfrak{p}_j$. Since $\mathfrak{m}^2 \cap \mathfrak{p}_{g+1} \cap \cdots \cap \mathfrak{p}_h \not\subset \mathfrak{p}_i$, $i = 1, 2, \cdots, g$, it follows that $\mathfrak{m}^2 \cap \mathfrak{p}_{g+1} \cap \cdots \cap \mathfrak{p}_h \not\subset \bigcup_{i=1}^{g} \mathfrak{p}_i$ (Vol. I, Ch. IV, §6, Remark, p. 215). Let $z$ be an element belonging to $\mathfrak{m}^2 \cap \mathfrak{p}_{g+1} \cap \cdots \cap \mathfrak{p}_h$ and not to $\bigcup_{i=1}^{g} \mathfrak{p}_i$. Then set $x = y + z$.

$\mathfrak{m}$ is not an associated prime ideal of $\mathfrak{b}+Ax$. In other words we have $(\mathfrak{b}+Ax):\mathfrak{m}=\mathfrak{b}+Ax$, whence $(\mathfrak{b}+Ax):\mathfrak{m}^n=\mathfrak{b}+Ax$. Now, since $\mathfrak{a}=\mathfrak{b}:\mathfrak{m}^n$, we have $\mathfrak{a}\subset(\mathfrak{b}+Ax):\mathfrak{m}^n=\mathfrak{b}+Ax$, and evidently $\mathfrak{b}\subset\mathfrak{a}$. For every $a\in\mathfrak{a}$, we may write $a=b+cx$ with $b\in\mathfrak{b}$ and $c\in A$; since $\mathfrak{b}\subset\mathfrak{a}$, we have $cx\in\mathfrak{a}$, whence $c\in\mathfrak{a}$ since $\mathfrak{a}:Ax=\mathfrak{a}$. In other words we have $\mathfrak{a}\subset\mathfrak{b}+\mathfrak{a}x$, whence $\mathfrak{a}=\mathfrak{b}+\mathfrak{m}\mathfrak{a}$ and therefore $\mathfrak{a}=\mathfrak{b}$.

Now, since $\mathfrak{b}:Ax=\mathfrak{b}$ and since we obviously may assume that the elements $a$, $b$ belong to $\mathfrak{m}$ (whence $\mathfrak{b}\subset\mathfrak{m}$), we have $\delta(\mathfrak{b}+Ax)=1+\delta(\mathfrak{b})$ (VII, § 13, Lemma 6), whence $\delta(\mathfrak{b})\leq 1$. Q.E.D.

# INDEX OF DEFINITIONS

The numbers opposite each entry refer to chapter, section and page respectively. Thus the entry "Composite valuation, VI, 10, 43" means that a definition of composite valuations may be found in Chapter VI, § 10, page 43. In the text, all newly defined terms are usually either introduced in a formal DEFINITION or italicized.

# Mathematics–Bestsellers

HANDBOOK OF MATHEMATICAL FUNCTIONS: with Formulas, Graphs, and Mathematical Tables, Edited by Milton Abramowitz and Irene A. Stegun. A classic resource for working with special functions, standard trig, and exponential logarithmic definitions and extensions, it features 29 sets of tables, some to as high as 20 places. 1046pp. 8 x 10 1/2. 0-486-61272-4

ABSTRACT AND CONCRETE CATEGORIES: The Joy of Cats, Jiri Adamek, Horst Herrlich, and George E. Strecker. This up-to-date introductory treatment employs category theory to explore the theory of structures. Its unique approach stresses concrete categories and presents a systematic view of factorization structures. Numerous examples. 1990 edition, updated 2004. 528pp. 6 1/8 x 9 1/4. 0-486-46934-4

MATHEMATICS: Its Content, Methods and Meaning, A. D. Aleksandrov, A. N. Kolmogorov, and M. A. Lavrent'ev. Major survey offers comprehensive, coherent discussions of analytic geometry, algebra, differential equations, calculus of variations, functions of a complex variable, prime numbers, linear and non-Euclidean geometry, topology, functional analysis, more. 1963 edition. 1120pp. 5 3/8 x 8 1/2. 0-486-40916-3

INTRODUCTION TO VECTORS AND TENSORS: Second Edition--Two Volumes Bound as One, Ray M. Bowen and C.-C. Wang. Convenient single-volume compilation of two texts offers both introduction and in-depth survey. Geared toward engineering and science students rather than mathematicians, it focuses on physics and engineering applications. 1976 edition. 560pp. 6 1/2 x 9 1/4. 0-486-46914-X

AN INTRODUCTION TO ORTHOGONAL POLYNOMIALS, Theodore S. Chihara. Concise introduction covers general elementary theory, including the representation theorem and distribution functions, continued fractions and chain sequences, the recurrence formula, special functions, and some specific systems. 1978 edition. 272pp. 5 3/8 x 8 1/2. 0-486-47929-3

ADVANCED MATHEMATICS FOR ENGINEERS AND SCIENTISTS, Paul DuChateau. This primary text and supplemental reference focuses on linear algebra, calculus, and ordinary differential equations. Additional topics include partial differential equations and approximation methods. Includes solved problems. 1992 edition. 400pp. 7 1/2 x 9 1/4. 0-486-47930-7

PARTIAL DIFFERENTIAL EQUATIONS FOR SCIENTISTS AND ENGINEERS, Stanley J. Farlow. Practical text shows how to formulate and solve partial differential equations. Coverage of diffusion-type problems, hyperbolic-type problems, elliptic-type problems, numerical and approximate methods. Solution guide available upon request. 1982 edition. 414pp. 6 1/8 x 9 1/4. 0-486-67620-X

VARIATIONAL PRINCIPLES AND FREE-BOUNDARY PROBLEMS, Avner Friedman. Advanced graduate-level text examines variational methods in partial differential equations and illustrates their applications to free-boundary problems. Features detailed statements of standard theory of elliptic and parabolic operators. 1982 edition. 720pp. 6 1/8 x 9 1/4. 0-486-47853-X

LINEAR ANALYSIS AND REPRESENTATION THEORY, Steven A. Gaal. Unified treatment covers topics from the theory of operators and operator algebras on Hilbert spaces; integration and representation theory for topological groups; and the theory of Lie algebras, Lie groups, and transform groups. 1973 edition. 704pp. 6 1/8 x 9 1/4. 0-486-47851-3